Genetics and Conservation

BIOLOGICAL CONSERVATION SERIES

Christine M. Schonewald-Cox, *Editor*
National Park Service
Department of the Interior
Washington, D.C.

---

1. C. Schonewald-Cox,   *Genetics and Conservation: A*
   S. Chambers, B. MacBryde,   *Reference for Managing Wild*
   W. Thomas (editors)   *Animal and Plant Populations*

Other volumes in preparation

BIOLOGICAL CONSERVATION

Christine M. Schonewald-Cox, *Editor*
Department of the Interior
Washington, D.C. USA

Editorial Board

J.A. Beardmore	University of Swansea, U.K.
S. Bratton	National Park Service, USA
S.M. Chambers	U.S. Fish and Wildlife Service, USA
F. diCastri	International Man and the Biosphere Secretariat, UNESCO, Paris, France
R.H. Evans	Ralston Purina Company, St. Louis, USA
O.H. Frankel	C.S.I.R.O., Canberra, Australia
J.F. Franklin	United States Department of Agriculture, Forest Service, USA
W.J. Hamilton, III	University of California, Davis, USA
S. Leopold	University of California, Berkeley, USA
D.R. McCullough	University of California, Berkeley, USA
R.T. Orr	California Academy of Sciences, USA
N.L. Ramanathan	Department of Environment, New Delhi, India
O.A. Ryder	Zoological Society of San Diego, USA
U.S. Seal	V.A. Medical Center, Minneapolis, USA
B. Wilcox	Stanford University, USA
Secretary to the Editoral Board:	J.W. Bayless, National Park Service, USA

Genetics and Conservation

A REFERENCE FOR MANAGING WILD ANIMAL AND PLANT POPULATIONS

Edited by

Christine M. Schonewald-Cox
National Park Service
Department of the Interior
Washington, D.C.

Steven M. Chambers
U.S. Fish and Wildlife Service
Department of the Interior
Washington, D.C.
and
George Mason University, Fairfax, Virginia

Bruce MacBryde
U.S. Fish and Wildlife Service
Department of the Interior
Washington, D.C.

W. Lawrence Thomas
U.S. Fish and Wildlife Service
Department of the Interior
Washington, D.C.

1983

THE BENJAMIN/CUMMINGS PUBLISHING COMPANY, INC.,
Advanced Book Program
Menlo Park, California

LONDON · AMSTERDAM · DON MILLS, ONTARIO · SYDNEY · TOKYO

Artist for cover and wildlife sketches: Sylvia Feder.

Cover: Green sea turtle, Griffon vulture, Common morning-glory, Okapi.
Beginning of Parts: I. Black-tailed prairie dog (*Cynomys ludovicianus*), Great Basin bristlecone pine (*Pinus longaeva*); II. Four-toed elephant-shrew (*Petrodromus tetradactylus*), Griffon vulture (*Gyps fulvus*); III. Fruit fly (*Drosophila melanogaster*), Common morning-glory (*Ipomoea purpurea*); IV. Green sea turtle (*Chelonia mydas*), Peregrine falcon (*Falco peregrinus*); V. Okapi (*Okapia johnstoni*), Western yarrow (*Achillea lanulosa*).

Library of Congress Cataloging in Publication Data
Main entry under title:

Genetics and conservation.

   (Biological conservation series ; 1)
   Bibliography: p.
   Includes index.
   1. Germplasm resources. 2. Population genetics.
I. Schonewald-Cox, Christine M. II. Series.
QH75.G45 1983   639.9   83-3922
ISBN 0-8053-7764-6

Copyright © 1983 by The Benjamin/Cummings Publishing Company, Inc.
Published simultaneously in Canada.

All rights reserved. No part of this publication may be reproduced, stored in a retrieval system, or transmitted, in any form or by any means, electronic, mechanical, photocopying, recording, or otherwise, without the prior written permission of the publisher, The Benjamin/Cummings Publishing Company, Inc., Advanced Book Program, Menlo Park, California 94025, U.S.A.

MANUFACTURED IN THE UNITED STATES OF AMERICA

ABCDEFGHIJ-MA-89876543

The Editors wish to dedicate this book to The International Program of Man and the Biosphere and its national efforts. We hope MAB will continue to foster synthesis and assist managers to protect biological resources wisely for long-term success.

CONTENTS

Foreword vii
Preface ix
Acknowledgments xiii

1. The Place of Management in Conservation.
   Otto H. Frankel 1
2. Genetic Principles for Managers. *Steven M. Chambers* 15

PART ONE   Isolation: Introduction 47

3. Isolation, Gene Flow, and Genetic Differentiation among Populations. *Fred W. Allendorf* 51
4. Isolation by Distance: Relationship to the Management of Genetic Resources. *Ronald K. Chesser* 66
5. The Differentiation of Populations over Short Distances. *Edwin H. Liu and Mary Jo W. Godt* 78
6. Sibling Species. *John W. Bickham* 96

PART TWO   Extinction: Introduction 107

7. What Do We Really Know about Extinction? *Michael E. Soulé* 111
8. Extinction, Survival, and Genetic Variation. *John A. Beardmore* 125
9. Genetics and the Extinction of Butterfly Populations. *Paul R. Ehrlich* 152
10. Extinction: Lessons from Zoos. *Katherine Ralls and Jonathan Ballou* 164

PART THREE   Founding and Bottlenecks: Introduction 185

11. The Genetics of the Founder Effect. *Hampton L. Carson* 189
12. Evolutionary Consequences of Inbreeding. *Robert K. Selander* 201

13. The Founding of Plant Populations.   *Michael T. Clegg and A. H. D. Brown*   216
14. Molecular Approaches to Studying Founder Effects.   *Jeffrey R. Powell*   229
15. The Elimination of Inbreeding Depression in a Captive Herd of Speke's Gazelle   *Alan R. Templeton and Bruce Read*   241

PART FOUR   Hybridization and Merging Populations: Introduction   263

16. Some Merging of Plant Populations.   *Jack R. Harlan*   267
17. Sea Turtles and the Problem of Hybridization.   *Archie F. Carr III and C. Kenneth Dodd, Jr.*   277
18. Hybridization and Gene Exchange among Birds in Relation to Conservation.   *Tom J. Cade*   288

PART FIVE   Natural Diversity and Taxonomy: Introduction   311

19. Preserving Natural Diversity.   *Gene Namkoong*   317
20. The Distribution of Genetic Variation within and among Natural Plant Populations.   *James L. Hamrick*   335
21. Systematics, Conservation and the Measurement of Genetic Diversity.   *Steven M. Chambers and Jonathan W. Bayless*   349
22. Interspecific Interactions and the Maintenance of Genetic Diversity.   *Douglas J. Futuyma*   364
23. The Relevance of Captive Populations to the Conservation of Biotic Diversity.   *Thomas J. Foose*   374
24. The Impact of Research on the Propagation of Endangered Species in Zoos.   *Kurt Benirschke*   402

Conclusion   403

25. Conclusions: Guidelines to Management: A Beginning Attempt.   *Christine M. Schonewald-Cox*   414

List of Contributors   447
Glossary   451
Appendices   469
   1. Some Symbols and Abbreviations   471
   2. Equations and Population Management (*W. Lawrence Thomas and J. Ballou*)   474
   3. Questions Posed by Managers (*C. M. Schonewald-Cox and Jonathan W. Bayless*)   485

4. The Distribution of Genetic Variation within and among
   Plant Populations   (*James L. Hamrick*)   500
5. Calculating Inbreeding Coefficients from Pedigrees
   (*J. Ballou*)   509
6. Monogamy and Polygyny Models
   (*Ronald K. Chesser*)   521

Author Index   571
Species Index   579
Subject Index   615

FOREWORD

Management of populations of wild animals and plants is as old as settled civilization itself—after all, domesticated species started their careers as their wild ancestors, and the genetic distance between them may not be very great. Basic lessons and useful practices were learned from domestication, the most basic being the avoidance of consanguineous matings, which was enshrined in social and religious precepts of ancient communities and has remained the focal point in the management of wild animal populations to this day. But settled civilization brought a parting of the ways: domesticates began to be managed—and gradually to be selected—for productivity, and wild species were there to be exploited, for their products such as meat, skins, or pelts, or for the enjoyment they gave, mostly to the privileged, in killing them. Ancient civilizations saw nature as created to serve man's uses as he chose, and man's stewardship for nature, at best, was "a minority tradition" in western civilization (Passmore, 1974). Neither the authorities of the church nor the poets and philosophers of the enlightenment deviated from the basic assumption of man's dominance over nature. Indeed, the drastic change of attitudes is a contemporary development, the result of the dramatic changes in the ecology of the world brought about by the destructive exploitation of the world's resources.

Once the conservation of threatened environments and resources became a recognized objective, it was natural that the emerging science of ecology stepped in to provide the theoretical infrastructure and the principles of management for the reserved or protected areas that began to be set aside a hundred years ago. Recently, population biology became an integral part of what is now termed conservation biology (Soulé and Wilcox, 1980). When it was postulated that genetic diversity may be a condition of long-term survival (Frankel, 1970; Frankel and Soulé, 1981), population and evolutionary genetics were seen as having a significant role in conservation biology.

The symposium from which this book emerged was planned to make managers of nature reserves, botanical and zoological gardens, and other forms of preservation aware of genetic principles and technologies of relevance in the management of biological resources. The aim of the symposium was to bring this new knowledge with the least possible delay to those best placed to apply it. This book will make it available to a wider circle.

The book addresses itself to population dynamics, which are liable to affect rates of survival or extinction in protected areas, and to the genetic principles and practices by which survival can be enhanced. The alternatives are clearly evident in the choice of topics: decline and extinction of species, in juxtaposition with the founding of new populations; isolation of populations in protected areas versus the merging of separated populations and taxa. A focal theme is the maintenance of genetic diversity as the genetic base for continuing evolution.

Readers concerned with reserve management may discern a bias toward animal problems. This is clearly not deliberate. There are a number of botanical chapters, but they deal largely with the basic issues of population genetics. Indeed, it is hard to see how it could be otherwise. There is little scope for the management of plant species with the exception of those with scientific or economic connotations, such as forestry species or wild relatives of domesticates. For this there are two reasons. The first is the great diversity of breeding systems—for many species totally unknown—as against the bisexual simplicity in the majority of likely animal target species. The second is the ecological difficulty of "managing" plant species without acute interference with the ecosystem.

Indeed, the genetic management of particular species—usually rare or endangered ones—as a means of securing their survival may affect the stability of the ecosystem of which they form a part, if the effect of management is to increase the population size and hence the demand on resources within the ecosystem. There are, of course, situations in which the species of concern is so far removed from natural conditions or of such focal concern that an impact on other biota is either absent or irrelevant. Captive preservation is the paradigm of such situations. But, to paraphrase John Donne, under natural conditions "no species is an island." Managerial support for one species is likely to be at the expense of others, and managers will have to attempt a balance sheet of the short- and long-term effects that can be expected.

The opportunities for genetic management are determined by a number of variables—primarily the population size, which in turn depends on area size and on body size. This is brought out clearly in an ingenious model presented in the final chapter, with demographic data for three size classes of mammals related to area size. Clearly in the smaller reserves there is scope, and need, for genetic management to avoid or reduce inbreeding. Only large reserves have the potential to generate and maintain genetic diversity for adaptation to environmental change. This does not mean that smaller reserves have no evolutionary role to play. Indeed, they have, though perhaps not for the long-term preservation of the larger vertebrates, which now occupy the prime attention of conservationists. Moreover, the future is unknown, and long-term models, plausible as they seem, should not discourage concerted conservation efforts on a less than ideal scale.

Whatever the opportunities for genetic management, genetic information should prove valuable to reserve managers in shedding light on processes in populations that neither facilitate nor require management, but are full of interest for the observant biologist. Besides, awareness of genetic principles and, even better, of genetic variation in populations facilitates intelligent intervention, or helps to avoid disturbances and harmful interference. Estimates of levels of genetic diversity have become possible and are widely obtained; and, as this book shows, they provide information of managerial relevance and scientific interest. Indeed, it is perhaps the heightening of scientific insight, as much as the application in management, that is the most significant contribution made by this book.

O.H. Frankel

PREFACE

It is commonly agreed that genetic diversity should be preserved. Considerable confusion exists, however, as to what exactly is meant by the term genetic diversity, within its several different contexts of use. As it *can* mean

- species diversity,
- allele diversity (including polymorphisms),
- allele frequency differences (between individuals within populations and between populations), or
- the combination of species diversity with allelic variations (so vital in providing for long-term phylogenetic evolution),

we found ourselves wondering how often policy makers and field managers of wildlife resources understand the wide-ranging implications of mandates that involve the protection and preservation of species for future generations. Unfortunately, our observations at both the policy-making and field-managing levels suggested that very few often understand.

For example, if we would ask a number of wildlife resources' policy makers and field managers what they perceive is required to preserve a resource, we could probably expect that many would want to return the species to a former point in time and preserve those qualities of the species present at that particular time. They may not realize that this requires a tightly managed, often farmlike system. On the other hand, what they probably would want, if made aware of the choice, would be to preserve a resource's ecological and natural evolutionary processes. This, in contrast to the former strategy, requires allowance for processes to take place, including speciation and natural extinction.

Our goal in *Genetics and Conservation* is to contribute to the preservation of the natural diversity of species with a view to preserve the evolutionary potential of species by exploring the relationship between today's advances in genetics and their potential contribution to the quality of wildlife (animal and plant) conservation.

Our immediate objectives are to help clarify the meaning of genetic diversity and, simultaneously, to elucidate to those who plan or implement programs the various implications that accompany their objectives in managing, protecting, preserving, or restoring wildlife. As the entire field of conservation is too broad for one book, we have focused here upon the application of genetics to conservation. This follows from our primary concerns, which center around the genetics and evolutionary consequences of certain — sometimes perceived as negative — naturally occurring as well as management-induced phenomena: isolation, extinction, population bottlenecks, founding of new populations, and merging of naturally disjunct populations.

Those working in the field of conservation are accustomed to the urgency that overshadows restoration projects and programs that hope to stave off extinctions. Unfortunately, new knowledge and tools useful to conservation are slow to reach the field scientist or manager. In contrast to short-range, economically profitable enterprises, the conservation of wildlife (animals and plants) does not now provide the economic incentives to ensure that the best and newest information is immediately made available for use and for further testing. It remains the responsibility of concerned individuals to voluntarily bridge the gap between scientific findings and application. For example, unusually successful attempts have been made in the management of fire and of mammal populations in zoos, and in the development of live storage of genetic material such as semen, tissues, and seeds. Although developments such as these are revolutionizing practices in conservation in a few places, their beneficial effects are just beginning to be recognized across the entire field of conservation. Therefore, this book has been written *primarily* to contribute to the translation and transfer of knowledge from the fields of population and evolutionary genetics to the field of conservation, including its management practices.

Isolation, already a major conservation problem, will surely become the state of nearly all remaining natural areas and populations requiring undisturbed habitats (particularly large and migratory species) unless there is some intervention. Isolation predisposes species or groups — particularly those of small populations — to extinction, the fate that the majority of large vertebrates and other ecologically inflexible species will face. Conservationists have a choice, however, whether many of these foreseen extinctions will occur. The choosing of the positive answer sounds simple and obvious, but it is riddled with complexities and, therefore, neither simple nor obvious. Nonetheless, conservationists must choose whether they will maintain certain habitats for the majority of species and allow declining species to reach critically low population sizes and disappear, or whether they will help species estimated, at least, as having a chance of temporary survival by use of specifically developed management programs of restoration. The latter choice will involve reducing the actual genetic isolation between populations and potentially founding new populations.

As populations are manipulated, particularly if they are manipulated in ignorance of genetic or taxonomic consequences, they inadvertently may be brought into contact with other populations to which they are sufficiently closely related to interbreed. The probability that adaptation will occur is low, but, paradoxically, such interbreeding may be beneficial occasionally. Such merging of disjunct populations is of major importance to a resources manager, and it is a fortunate one who can anticipate both the beneficial and possible deleterious responses of species to this contact. It cannot be overemphasized that one should gather some knowledge of species *before* attempting to manipulate either species or their habitats.

Another problem we address is communication, for frequently we hear criticism of wildlife management curricula and publications, charging that these do not meet the professional standards required in academic fields; on the other hand, we also hear that academic curricula and publications do not provide new knowledge or new discoveries in usable form for those in applied disciplines. Therefore, all the chapters included in this book have been reviewed by geneticists, applied scientists, and managers. They have been included because of their high academic quality *and* practicality. Additionally, the presentation of the material is in a format that can be read and appreciated by a variety of professional audiences and students. The most that is required is that the reader have a working acquaintance with basic biology, botany, zoology, and ecology. Because of all these facts, the book should serve well as an upper-division or graduate-level text for courses in conservation biology, wildlife and plant (including forestry) conservation, and resources management. For field managers and scientists, it should prove a valuable reference for wisely planning the specific, optimal directions that conservation programs can take and the techniques that it may be necessary to employ. Naturally, *Genetics and Conservation*, with its focus on genetics, should be used in conjunction with similar references on topics of conservation that omit such a focus. For example, there is little detail on the role of botanical gardens in conservation in this book. For more on that subject, see Brumback (1981), Simmons et al. (1976), and Synge and Townsend (1979).

The stimulus for producing this volume was the United States' Man and the Biosphere symposium and workshop, "Application of Genetics to the Management of Wild Plant and Animal Populations" held in Washington, D.C., August 9–13, 1982. Several of the speakers at the symposium were asked to contribute to this book, which we, the editors, regard as an attempt to advance the state of the art in conservation techniques.

Genetics and Conservation is the first book in the series *Advances in Conservation Biology* to be published by Addison-Wesley Publishing

Company. The purpose of this series is to accelerate the transfer of scientific knowledge and technology as they develop to the applied scientist, manager, and policy makers. For the series as well as this book, the term conservation is inclusive for a range of topics: reserve management and design, forestry, range management, wildlife management, fisheries biology, management of zoological park populations, and other similar fields connected with the long-term protection of the entire spectrum of biological diversity. (It also includes the individual treatments of habitat types or major biomes that contribute to developing management techniques specifically required for different climates and ecosystems.) We hope that this effort will promote a long-term and responsible coexistence of man and the biosphere.

Note: The editors' contributions were not made as part of their government positions and no statements by them necessarily reflect any Federal policy.

Christine M. Schonewald-Cox

ACKNOWLEDGMENTS

We wish to thank Michael H. Smith, Michael E. Soulé, Hampton L. Carson, and Gene Namkoong for assembling their parts of the symposium program that led to the choice of some chapters for the parts of this book. For their logistics, editorial, and graphics support we should like to thank especially Jonathan Bayless, Jonathan Ballou, Fara Leigh Smith, John Bridehoft, Donna Brown, Timmothy Halverson, and Jacqueline Clinton. The constant support and encouragement of Richard Briceland, John Dennis, William P. Gregg, and Albert Greene, Jr., of the National Park Service, and Donald King, Phyllis Ruben, and Jay Blowers of the State Department U.S. Man and the Biosphere Program have been extremely helpful to us in developing the symposium that gave rise to this book. We thank Ronald Nowak, C. Kenneth Dodd, Jr., Paul Opler, Katherine Ralls, Bill Marks, and David Steadman for reviewing various parts of the book. We should all like to thank Becky and Laura Thomas, Olga Herrera MacBryde, Brendon D. MacBryde, Kristine Lofthus Chambers, and Robert George Cox for their support under severe time constraints. Of course, this book would never have been written had it not been for the sponsorship of this project by the National Park Service of the U.S. Department of the Interior, the U.S. Man and the Biosphere Program of the U.S. Department of State, the Federal Marine Mammal Commission, the National Geographic Society, the New York Zoological Society Animal Research and Conservation Center, the International Man and the Biosphere Secretariat of the United Nations Educational, Scientific and Cultural Organization, the American Association of Zoological Parks and Aquariums, the Peregrine Fund, the International Council for Bird Preservation, the Society of American Foresters, the George Wright Society, the American Association for the Advancement of Science, the Ecological Society of America, and the World Wildlife Fund–U.S. In addition we should like to express our appreciation for the letters of endorsement we received from the following organizations: the Smithsonian Institution, the U.S. Fish and Wildlife Service of the U.S. Department of the Interior, the State of Alaska Department of Fish and Game, the State of Arkansas Game and Fish Commission, the State of Arkansas Forestry Commission, the State of California Department of Forestry, the State of California Department of Parks and Recreation, the State of Colorado Department of Natural Resources, the State of New Jersey Department of Environmental Protection, the State of North Carolina Wildlife Resources Commission, the State of Texas Parks and Wildlife Department, the American Association of Zoo Keepers, the Natural Areas Association, The Wildlife Management Institute, The Wildlife Society, Trout Unlimited, and the Sierra Club International Earthcare Center. Our thanks go to Mrs. Alicelia Franklin for her excellent work on the indices and Sylvia Feder for designing the cover of this book and the wildlife sketches.

CHAPTER 1

The Place of Management in Conservation

O. H. Frankel

Introduction: Management in the Context of Conservation

When I received an invitation to prepare a key chapter for a book on the genetic management of wild species, the question arose in my mind: To what extent is management of wild species supportive of conservation, and to what extent could it negate the purposes of conservation? Of course, management of wild species can have many objectives other than conservation. Yet the guidelines I received established conservation as the central theme in emphasizing the role of genetics in attempts "to enhance the survival of wild plants and animals." Indeed, the application of genetics suggests a long-term concern, unless the main objective is commercial or recreational exploitation, which clearly is not the case here.

Enhancement of survival does not necessarily mean protection in formal reserves. Indeed, rare or endangered species can and do survive without specific protection, although it is widely recognized that in times to come many species, especially in the tropics, may have little chance of survival except through conservation. Of course, conservation may take various forms, from the strict nature reserve to the wildlife park, fishing reserve, zoo, or botanical garden. There is a similarly broad spectrum for the time factor of conservation, from the inevitably short-term, fire-brigade-like salvage of a near-vanished species, to ecosystem and habitat conservation, notionally forever. And obviously there is a similarly broad spectrum for forms and intensities of management.

Whatever its objectives, *management* is *not* a notional concept but a program and a procedure, executed or at least programmed in our time, with an impact intended to take effect on a measurable and predictable time scale. Yet this time scale—in other words, the period over which the management, its impact, or both are to be operative—can be a notional concept, since it can extend from the shortest conceivable period—say, one season or one generation of manipulating a breeding system—to a long-term commitment—for example, to maintain heterozygosity at a set level. The measures may be similar, or indeed identical, but the objectives, and, as a rule, the biological system in which management operates, will be widely different.

Thus, three basic elements of conservation are emerging. The *first* is the objective or the principal target—a species, an association, a community, an ecosystem, or a group of ecosystems. In the context of this book the target can be defined as a "rare or endangered species." The *second* element is what I have called the "time scale of concern" (Frankel, 1974), which is the time dimension for which a program is expected or projected to remain operative. This may be as short as a generation—for example, the preservation of a specific tree; or it may be in perpetuity—that is, without a perceived end point. The *third* element is management. All forms of management, even its ostensible absence, may drastically affect numerical relations between species, even their survival or extinction. Obviously these three elements of conservation interact. In general, the broader the objectives, the longer the projected time scale will be, and vice versa. For the *long-term* preservation of a single species, be it the orangutan (*Pongo pygmaeus*), the tiger (*Panthera tigris*), or the coast redwood (*Sequoia sempervirens*), preservation in a designated "orangutan," "tiger," or "redwood" reserve is a figure of speech or a public relations exercise: long-term conservation is conceivable only within the confines of the ecosystem, which becomes the real target of conservation. This will be discussed further.

The theme of this chapter is the role and effect of population management in conservation. Although the emphasis is on conservation in perpetuity, the consideration of management, as we have already seen, focuses attention on the present. This, we need to remind ourselves, is inevitable, since any generation, indeed any government of the day, has the power to maintain or to destroy the heritage from the past. Destruction or damage may come from obvious sources such as fire or the bulldozer, the shotgun or the axe, or simply from neglect. Even management measures intended to benefit the cause of conservation can be fraught with danger, often in unexpected or unrecognized ways. Preventive fires, culling, selective logging, and indeed the genetic management of particular species may set up chain reactions with effects throughout the ecosystem.

Each of these three elements—the target, the time scale, and management—will make its own contribution in the short and in the long term in affecting the composition and the stability of an ecosystem, and each, and the three in compound, must be weighed in conservation planning.

Conservation in Perpetuity: Variation and Survival

Genetic manipulation of wild species has obvious affinities with genetic management of domesticated plants and animals. Indeed, the perception of a genetic component in conservation was generated by the concern for the preservation of what we have come to call the genetic resources of crop and livestock species. This term encompasses the genetic diversity that is actually or potentially available to the plant or animal breeder for recombination and selection in the never-ending task of adaptation to ever-changing physical, biological, and economic environments.

Obviously, natural populations need corresponding reservoirs for adaptation if they are to survive, although selection pressures are unlikely to be as powerful as in domesticates. Long-term conservation, in distinction from the static preservation of a species, ecotype or genotype, needs to be conceived as a dynamic process, a process of *continuing evolution* (Frankel, 1970; Frankel and Soulé, 1981). The crucial question is whether the conditions in nature reserves facilitate, restrict, or inhibit this essential process. In contrast to primeval wild populations, many populations of species in nature reserves are small, disconnected, and subject to inbreeding, genetic drift, and random fixation of alleles, resulting in a gradual weakening and genetic impoverishment, quite apart from the risk of loss through accidental death of individuals. In the absence of the genetic paraphernalia that are available to the plant breeder—the germ plasm collections, the inbred lines and genetic stocks, the "secondary" and "tertiary" gene pools representing related species or genera (Harlan and de Wet, 1971)—wild species must have available a pool of genetic diversity if they are to survive environmental pressures exceeding the limits of developmental plasticity. If this is not the case, extinction would appear inevitable. These issues seemed to have escaped examination by conservationists. A first, rather tentative discussion (Frankel, 1970) in due course led to a more thorough examination in which Michael Soulé played such an important part (Frankel and Soulé, 1981).

Naturally, the first approach was to examine the conditions under which adaptive evolution was, or was not, likely to proceed. It was concluded that the spatial conditions in protected areas will restrict population size, and hence the genetic diversity of space-demanding species, so that requirements for their survival—let alone genetic adaptation—might not be met (Soulé, 1980). This inverse relationship between survival and space will in the first instance affect the larger animals, foremost the larger mammals and birds, but others, such as large tropical trees, are also likely to be affected by area restrictions.

In principle, the conservationist is faced with three possible strategies: (1) to provide the space required for all-round survival and continuing adaptive evolution; (2) to accept extinction or removal (for example, to zoological or botanical gardens or to other reserves) of biota whose space requirements cannot be satisfied, with drastic consequences for other species and for the ecosystem as a whole; (3) to manage population size and population structure of selected species presumed to be threatened.

The first strategy, in practical terms of the world of today and even more so the world of tomorrow, is likely to be an abstraction rather than a reality, considering the size of even the largest reserves when set against the space requirement of large animals (Soulé, 1980). With rare exceptions there will be a drift toward strategy 2, which involves species extinction and its ecological consequences (Frankel and Soulé, 1981, Chapter 2). In this chapter we are primarily concerned with strategy 3, although the broader perspective serves as a useful background.

The Links with Domestication

It has already been remarked that genetic conservation of wild populations has affinities with genetic management of domesticates. The most tangible link is the wild species that are related to domesticated ones and constitute their perhaps most valuable genetic resources, at least in plants, although new genetic techniques open up unforeseen possibilities for interspecific and intergeneric gene transfers in animals. The available evidence—admittedly, restricted to a small number of species—shows that populations of wild plant species contain large amounts of genetic variation for protein markers, a good deal larger than are found in primitive cultivars (or land races) of related domesticated species (Brown, 1978). A wide range of wild biota are actual or potential genetic resources for crops and livestock, for horticulture and forestry, with benefits for the many industries based on them. Protected areas provide security, and in return species of economic significance may help to sustain the value of reserves in public and political consciousness. Little is known of the whereabouts and the state of protection of many wild relatives. Of special value therefore is a preliminary inventory of the wild relatives of crops, carried out by the International Union for the Conservation of Nature and Natural Resources (IUCN) on behalf of the International Board for Plant Genetic Resources (IBPGR) (Prescott-Allen and Prescott-Allen, 1981). The report lists 36 endangered, vulnerable, or rare plant taxa, of which only a minority are known in protected sites. Considering the preliminary nature of the survey, the actual number is likely to be a good deal larger. This survey needs to be extended as a matter of urgency.

There is an obvious link between agricultural and conservation genetics through the application of population genetic theory. Two population genetic concepts are of basic importance in conservation genetics. The first is fitness and its impairment by inbreeding; the second is genetic variance and its importance for genetic adaptation. Both are related to number—that is, the effective population size. Heterozygote advantage is familiar to breeders of livestock and of cross-fertilized plants—pigs (*Sus*), poultry, maize (*Zea mays*). But we now have hybrid wheat (*Triticum aestivum*), a normally self-fertilized plant; and in some self-fertilized plants a relatively high level of heterozygosity is maintained, presumably owing to its selective advantage. Of greater concern in conservation genetics is the reverse—inbreeding depression. On the basis of the experience of animal breeders that an increase in inbreeding coefficient as high as 1% per generation is acceptable, Franklin (1980) suggested that an effective population of 50 might be regarded as the minimal size, according to Wright's formula:

$$\Delta F = \frac{1}{2N_e}$$

which describes increase in the inbreeding coefficient (ΔF) per generation in a population with an effective size N_e. This number is, of course, based on random mating—a rare phenomenon in natural populations.

In distinction from (short-term) fitness, depletion of genetic diversity has long-term effects that, as we have already stressed, can lead to extinction. Franklin (1980) discussed changes in the genetic variance of quantitative characters, including the most important adaptive ones, as a consequence of genetic drift and mutation. The additive variance is lost at the same rate as heterozygosity—$1/2N_e$ per generation. In a population of 50, the genetic variance remaining after 100 generations is 36% of the original level. By comparison, the added variance derived from mutation is insignificant. With some backing from work on bristle numbers in *Drosophila* flies, but largely on grounds of common-sense reasoning, he proposes 500 randomly mating individuals as the minimum population size for maintaining genetic variation at a level to enable the species to adapt to changes in the environment. Should mating be restricted, as it is in many mammals, or should generations overlap, as they do in many animals and plants, population sizes would need to be increased, perhaps by an order of magnitude.

As was already stressed, even the largest nature reserves—and their number is small indeed on a world scale—will fail to accommodate effective population sizes required for the survival of the more space-demanding animal species. According to the United Nations List of National Parks and Equivalent Reserves, published by the IUCN in 1975, 93% are less than 5000 km². Of the 144 biosphere reserves registered in May, 1979, only 14 were larger than 5000 km², and 29 were between 1000 and 5000 km² (UNESCO, 1979); and the majority of the large reserves are in inhospitable environments. To remind ourselves of the proper perspective, the population density of the gray wolf (*Canis lupus*) is about one adult per 20 km². Seeing that only about one third of the population breed, that generations overlap, and that there are frequent and severe seasonal population bottlenecks, the minimum of $N_e = 50$ would turn into some number like 600, requiring 12,000 km² merely for short-term preservation, which is larger than Yellowstone National Park, Wyoming (Soulé, 1980).

Plants, being nonvagile and, many of them, selective of microenvironments, present greater problems in establishing conservation strategies, especially since pollination (hence gene flow) may depend on the action of physical or biological vectors. Numbers alone provide insufficient guidance in determining area requirements. However, the ground rules for minimum population size may be applicable as long as due allowance is made for restrictions in distribution and mating. It is therefore encouraging to note that, in a survey of lowland rain forest in northeast Peninsular Malaysia, Whitmore (1973) found that the frequency of the more common species (more than 25 individuals per 1 km²) of forest trees ranged between 300 and 25 per 1 km², so that there was a good chance of finding the minimum number of even the largest tropical tree species in a reasonably sized reserve in the region, even should the frequency of a species be as low as 1 tree per 1 km².

How can management attempt to deal with this crucial problem of numbers? As was already said, we cannot expect to find additional protected areas to satisfy the needs of all, or even of a significant number, of space-restricted species (strategy 1), except perhaps for those adapted to the more desolate parts of the earth. There are two possibilities for managerial intervention. The first is to modify the environment so as to increase the carrying capacity for the target species; the second, genetic intervention, is discussed later in this chapter. Management measures familiar in crop or animal husbandry, such as improvement of nutrition, water supply, shelter, control of competitors, predators, or parasites, could, to a limited extent, find a place, at least in emergency or bottleneck conditions.

How far can one go without compromising the ecosystem and abandoning altogether the most essential feature of nature conservation—self-regulation? Here agricultural experience provides a lesson and a warning.

In agricultural systems, impacts such as the application of a major or minor plant nutrient need to be large enough to have a marked or at least a statistically significant effect. Thus, effects and interactions are measurable or observable, and we can learn from their ramifications. In natural systems management measures are apt to be of a less drastic nature, but they spread widely to species other than the target species. The impact may be far-reaching and unpredictable. Natural ecosystems greatly exceed agro-ecosystems in the diversity of components and their interactions, including chain reactions that are likely to affect even biota that are ecologically remote from the targets of the impact.

Genetic Management: Induced Migration

The aim of genetic management is not so much to increase the population size directly as it is to reduce the rate of erosion of the genetic variance by the introduction of genotypes from other sources. These could be populations in other reserves, or captive or cultivated populations held in wildlife parks or zoological or botanical gardens. Induced gene flow may result in increased vigor and reproductive rate, hence in increased population size.

The first question that arises is, how many individuals need to be involved, and the second, what should be the genetic relationship between receptor and donor populations? For operational reasons the best solution would be a reciprocal exchange of minimal proportions. Apparently a very small proportion of two reasonably sized populations (such as 1000 individuals in each) exchanged between them once in every generation will establish virtual panmixis between them. Lewontin (1974, p. 213) showed that, for neutral genes, the difference (d) in allele frequencies between two populations, each of N individuals of which they exchange a proportion of m in each generation, and an average allelic frequency of $\bar{p}$, will be

$$d = 2 \sqrt{\frac{\bar{p}(1 - \bar{p})}{(1 + 4Nm)}}$$

For two populations with $N = 1000$, $m = 1\%$, and an allelic frequency of 0.5, this works out as

$$d = 2 \sqrt{\frac{0.5 \times 0.5}{41}} = 0.156$$

In a much smaller population—for example, of 50 individuals—such a small difference could also be obtained, but the proportion of the interchanges would need to be as high as 20% (see Chapter 3 for a more detailed discussion).

The question is whether panmixis is desirable; in other words, is it likely to mitigate the problem of eroding genetic variances? Panmixis would have two effects. It would double the effective population size through joining the two populations, and it would reduce local adaptation to the extent that the two environments differ. The first effect is an advantage, although one that diminishes with population size. The second could be a disadvantage—for example, if one of the populations is exposed to a disease or pest, with selection for resistance being slowed down by interchange with the unaffected sister population.

One may in fact question the realism of the model. Does panmixis based on random mating ever occur? Surely not in plants, with the possible exception of some wind-pollinated species. In many animal species the compensation for restrictions to random mating could lead to astronomical numbers. And is panmixis desirable, even to the extent that it occurs?

Frankel and Soulé (1981), sidestepping the—at least partly—irrelevant issue of panmixis, expect migration to have beneficial effects in increasing N_e and in increasing heterozygosity, depending on the genetic structure of the populations involved, especially if they have been exposed to a degree of inbreeding. There may be deleterious effects because of chromosomal incompatibility of the parents or phenotypic disabilities of the hybrid. There are other pitfalls that must be avoided, especially in animals, such as territorial, social, or behavioral problems, which need to be explored; indeed, some less tractable species could present difficulties! And, obviously, introductions must be able to survive and thrive in the host environment.

It is more difficult to specify positive characteristics for the host–migrant relationship. If we jettison panmixis and mutual assimilation as neither inevitable nor even desirable, genetic differentiation emerges as a more plausible criterion for the relationship between the partners, provided ecological compatability is safeguarded. Allozyme surveys of populations could provide evidence for partner selection. Changes of partners might be preferable to long-continuing relationships. But, as was already suggested, operational difficulties increase in proportion to operational diversity in space and in time.

In this rather overgeneralizing treatment, no mention has been made so far of an all-important factor in conservation genetics—the breeding system. Clegg and Brown (Chapter 13) and Hamrick (Chapter 20) discuss the importance of understanding mating systems of plants and their potential effects on the design of conservation or management programs (also see Appendix 3, Figure 2). In a contribution to a book on the biology of rare Australian plants and animals, James (1982) relates the complex and highly diverse breeding systems in a number of plant species that occur mainly in small, ecologically isolated populations with rare interpopulation gene flow. Individual populations maintain a level of hybridity by various devices—such as complex translocation heterozygotes—controlling recessive lethal alleles. Interpopulation gene flow destabilizes the system, occasionally with disastrous effects, yet serves to counteract the accumulated consequences of inbreeding in some of the species, and provides the potential for rapid and drastic genetic restructuring. James extends the warning that more harm than good might be done by deliberately combining populations of such species, and further, that, while many of the species with widespread distribution may have systems that are constantly coadapting throughout their range, these species may yet contain isolated demes with population characteristics similar to those he describes. He, Clegg and Brown, and Hamrick all emphasize that manipulation of plant populations may be destructive rather than conservative without a thorough prior study of their breeding system.

Are evolutionary processes likely to continue under conditions of nature conservation—as James' observations suggest—at a level exceeding minor phyletic change? I can see no reason why they should not, presumably with the exception of the larger mammals, birds, reptiles, and amphibians. Historical evidence derived from the relations between island size and evolutionary record may have neither general nor predictive significance. Here it is appropriate to recall the emphasis on local differentiation as a condition of steady evolutionary advance, depending on random genetic drift in relatively small populations (Wright, 1978, p. 523).

From what has been said, it appears that migration is likely to be successful in animal species with a large proportion of the population participating in breeding (Chapter 4 by Chesser). It is probably of least value, and fraught with potential risk, in many or perhaps most plant species, many of which are adapted to reproduction within small subpopulations with little external gene flow. In the larger mammals, birds, and perhaps reptiles and amphibians, as Frankel and Soulé (1981) suggest, a transfer of from 1 to 5 individuals per generation could be expected to have the required result.

But even within this restricted scope, let alone the enormous number of other rare or endangered species, there is inevitably a need for selection—that is, a choice of species to be singled out for preferential preservation. This issue was raised by McMichael (1982) in a paper at a recent symposium on the biology of rare or endangered species in Australia. McMichael points to the cost of essential preliminary research on the biology of target species—a minimum of $100,000 per species spread over several years—and to the shortage of personnel in Australia, let alone, one may add, in most of the developing countries. In the choice of target species, scientific criteria tend to receive a good deal of emphasis. For example, priority may be given to species within genera, families, etc., with few species; to "living fossils" without recently evolved relatives; to species that are distinct from their congeners; or to species with evidence of incipient speciation. However, McMichael questions the justification for giving a high priority to scientific as against other important criteria, or for permitting any particular scientific criterion to override the principle that *all* species are valuable on scientific grounds. McMichael suggests that "aesthetic appeal and the place the species occupies in our culture are important." While one may side more readily with the latter view rather than with the purely scientific criteria, both point to the subjective and temporal nature of this important management decision. If we assume that scientists can reach agreement on priorities in our time, is it likely that they will be the same even half a century from now? Will people continue to love the tiger, the giraffe, or the rhinoceros sufficiently to justify the cost of environmental and genetic management?

Genetic Management and the Ecosystem

In the course of the preceding section some doubts have arisen regarding the scope and practicability of managed migration between populations in protected areas. We have seen that its application is subject to restraints on biological as well as managerial and economic grounds. Now we must extend our discussion to the effects that genetic management of particular species may have on the stability of the community or ecosystem of which they form a part. This has recently been examined by Gilbert (1980) and by Frankel and Soulé (1981, Chapter 5), so here I shall restrict myself to a brief summary. If it is postulated that some species play a prominent role in the regulation of the ecosystem, such species should be given first consideration in the target choice for genetic support. Foremost among these are the large predators and herbivores, species of low density but high community impact through their chain effect on a large range of plants and animals. Perhaps even greater significance attaches to what Gilbert (1980) has called "mobile links"—bees, ants, hummingbirds, bats, etc.—which play a significant role in the life processes of plant species and indirectly in various food webs. Plants with equivalent mediating functions are called "keystone mutualists." Their role is to support mobile links. They are mainly large tropical trees with low density. The extinction of such dominant or key species would set up chain reactions throughout the ecosystem, causing its gradual impoverishment and eventual collapse.

If this reasoning is accepted, the case for genetic intervention is substantially strengthened. The choice of targets is placed on a rational basis that is likely to retain validity over a foreseeable time span, and their number is restricted to what should be a manageable scale. The identification of key species would require much research, yet some no doubt are known, and to salvage a few is better than to save none.

Here we must recall some of the observations made in the earlier sections of this chapter. First, not only attrition of populations but also their enrichment may have impacts on the ecosystem, through competition for resources on the one hand and increased pressure on resources on the other. Both could have widespread effects. Moreover, it may become necessary to supplement scarce resources from outside the ecosystem, which in turn could have significant repercussions. Support for tree species with a "keystone" role would be through enrichment of the population, either by planting or by protecting seedlings, both requiring removal of competitors and possibly additional aids such as water, nutrients, pesticides, and protection from herbivores. The effects could be of the kind envisaged as following from environmental impacts in the second section of this chapter.

We must also consider the difference in the time scale of the effects of genetic versus environmental intervention. Environmental impacts, as was previously suggested, may have profound effects on the composition and balance within an ecosystem. Yet, when expertise and moderation are applied, this is an unlikely outcome, and treatment effects can be programmed for a limited period. The reverse is likely after the introduction of genetic materials that differ substantially from the host population, or that upset its coadapted balance (see preceding section). The effects of migration of genotypes, as of genes, may be profound and long-lasting. They also may be harmful. Under natural conditions of gene flow such effects would be subject to natural selection, with other (unaffected) populations standing by as reserves — scarce or altogether lacking under conditions of conservation. The time lag of genetic change adds a further reason for caution in genetic intervention.

Laissez Faire or Intervention?

I can scarcely conclude this chapter without touching upon the philosophical issue that underlies this discussion. Mankind has taken it upon itself to restrict the habitats for a large and increasing number of the earth's biota. It has set aside a small proportion of the land mass as refuges for the remnants. Increasingly these more or less protected areas are becoming the main or the only habitats in which many of the world's nondomesticated flora and fauna survive, a process that is likely to be near-complete in the next century. The question is whether these refuges, or at least those that are large enough to be ecologically self-sustaining, should be left to their own devices to the greatest possible degree — that is, subjected to management to the extent that harmful factors or destructive agents are kept out or controlled; or, alternatively, whether the aim should be to maintain and encourage the highest level of diversity attainable with the help of ecological and genetic management. One can call the first philosophy "laissez faire," the second "interventionist."

Opponents of these philosophies would visualize scenarios showing the worst consequences of the views they oppose. Laissez faire, in the view of interventionists, would lead to ecosystems depleted of large vertebrates and of other animals and plants caught up in their web. In turn, supporters of laissez faire claim that intervention would result in wildlife parks, with animals and plants numbered, tagged, fed, and mated. The absurdity of such visions does not disguise the fact that they hold more than a grain of truth. I admit—as will have become apparent—to doubts as to the effectiveness and practicability of genetic management in the form of induced migration. Of course, there are other genetic measures such as rational culling on genetic principles, but their scope is even more limited. I further admit that I do not view the eventual demise of large vertebrates with unmitigated horror, though with great regret. Nor do I regard the "collapse" of ecosystems in consequence of this event as inevitable or even as likely. New Zealand has no large vertebrates, but it has a rich and wonderful flora and, even after depletion by feral predators, a great diversity of birds. It is not obvious that an ecosystem collapse followed upon the extinction of the only large vertebrates, the moas (Dinornithiformes). Nor does it seem unlikely that other ecosystems with vanishing dominant vertebrates may reestablish a new balance after a period of readjustment, and our descendants, no doubt having a better understanding of community biology, may assist this process by judicious migration.

While I tend to side with the laissez faire philosophy of conservation, I have no philosophical objection to man's "taking a hand" in evolution. Since we have destroyed so much, we seem to have acquired the right to create something in its stead. I see no reason why man should not play a positive and creative part in evolution to supplement or, at least for the larger vertebrates, to supplant organic evolution. We have, and we shall have even more in the future, the means to do so, and to do it far more rapidly and presumably with less trial and error than nature herself. Perhaps man will create a new diversity of his own imagining, which will lessen his will to cling to the precarious remnants from the past. I prefer such fanciful thoughts to the dire predictions that nature reserves are doomed in the face of the needs of ever-growing numbers of our species.

In the meantime our nature reserves are still with us, and so is our responsibility to safeguard them in *our* time, with a time scale of intent without end. This involves two commitments: to defend their safety, and to protect their security and integrity. The first demands incessant political vigilance and, if need be, battle; the second, effective protection, and effective but reticent management. Genetic management in my view can have only a limited role, restricted, it appears likely, to animal species of great human interest and ecological significance.

The Time Scale Relaxed

In the preceding sections attention has been focused on conditions of conservation in perpetuity, with population size, transformed into space, as the foremost requirement. This can be met, or approached, only by the largest nature reserves. Protected areas of a lesser size are valued principally for their scientific, cultural, educational, or recreational role. As conservation sites their spatial limitations will restrict the "life expectancy" of the reserve—that is, of the community for which it provides the habitat. We must, however, recognize that, while the concept of conservation in perpetuity responds to our sense of evolutionary responsibility (see Frankel, 1974), a shorter-term time scale is more clearly attuned to the historical and biological time scales of our own species. Moreover, the preservation of species to which we have an emotional or scientific attachment is an end in itself, though its failure in time may be predictable. With the time scale of concern relaxed, the concern for the stability of the ecosystem can also be relaxed, as can be the restraints on environmental and genetic management discussed earlier in this chapter. In consequence the main emphasis in management can be placed on the representation of particular target species.

This distinction is of importance in conservation planning, because the two systems, one based on "in perpetuity" and the other on "relaxed" time scales, can complement each other to good effect. The former places emphasis on the conservation of ecosystems, on evolutionary opportunities, and on the preservation of habitats in which these can be realized. Conservation with a "relaxed" time scale is able to give preferential treatment to the preservation of particular target species. It provides management opportunities that have no more than limited scope in major long-term reserves. In this sense there *can* be orangutan or tiger reserves. One may expect that with a growing understanding of community biology such reserves may serve as sources for a safe reintroduction into major reserves of species that had been endangered or lost there.

Summary

The theme of this opening chapter, and indeed of this book, is the role of genetic management in conservation. Three elements of genetic conservation are recognized: the target, the time scale of concern, and management. The target must be some rare or endangered species, the (notional) time scale should be "in perpetuity," and the management is the object of this inquiry.

In analogy with the genetic resources of crops and livestock, there must be a reservoir of genetic resources available to natural populations if they are to survive. There are further links with domestication. Wild relatives of domesticates are important genetic resources that should as far as possible be preserved in nature reserves. Population genetics is another important link. From it is derived the minimum population size of 50 to preserve fitness, and of 500 to maintain genetic variance for genetic adaptation.

The role of management is to help maintain population size and genetic variance. Management of the environment (nutrition, water supply, etc.) has effects that may extend beyond a specific target and spread through the ecosystem. This possibility has a lesson for genetic intervention, which also may have diverse chain effects, and they may be far more extensive in time. The most effective genetic intervention is induced migration. The aim is to increase genetic variance and eventually population size. The effect depends on the breeding system, genetic relationships between donor and recipient populations, population size, etc. Extensive research is essential. Cost and logistic problems restrict the number of candidates, and a choice of targets is inevitable. Criteria could be scientific interest, human attachment, or, more rationally, the role and importance of the species within an ecosystem. The choice is likely to fall on a few dominant animal species, where a migration of 1 to 5 per generation is expected to have the desired effect.

There are two philosophies of conservation—"laissez faire," with management restricted to excluding or controlling harmful or destructive agents, and "interventionist," aiming to maintain the highest level of diversity, with the help of ecological and genetic management. Scenarios painted by opponents of either, though exaggerated, point to inherent risks. The author regards the risks of laissez faire as acceptable and to a degree inevitable, since the scope for management—and especially for genetic management—is limited. Defense and protection of nature reserves is essential, and management should be effective, but reticent.

Large area is recognized as an essential condition of conservation in perpetuity. Reserves of a lesser size, having a limited "life expectancy," can be subject to management in which the main emphasis is shifted from the stability of the ecosystem to the preservation of particular species. Once the biology of communities is more fully understood than it is at present, such reserves could interact with major long-term protected areas in maintaining populations of species that have become endangered or extinct in major reserves, and that could be reintroduced to these former habitats.

CHAPTER 2

Genetic Principles for Managers

Steven M. Chambers

Heredity and Environment

Genetics is the study of the biological transmission of characteristics from one generation to the next. Not all characteristics of an individual result from the biological information transmitted from its parents or parent; that is, not all characteristics are hereditary. Some characteristics owe their expression in an individual largely to the influences of environment on the individual. Most characteristics are influenced by both heredity and environment, and these influences can be illustrated by examining the relationship between the genotype and the phenotype. The genotype is the total hereditary information of an individual. The actual appearance of that individual is its phenotype. The phenotype is the result of the expression of the genotype and, within limits set by the genotype, of environmental modifications. For example, two individuals that are identical twins (technically called monozygotic or single-egg twins) share identical genotypes, yet their phenotypes differ, as all of us who have known identical twins can attest. Their phenotypes differ because they have experienced slightly different environments from the time they occupied different space within the womb, through one twin's having been born before the other, to the influence of differing playmates and other influences as the twins mature and grow. The phenotype includes not only an individual's outward appearance but also other aspects such as its behavior and internal "workings" (physiology). The difference in behavioral phenotypes between two humans that are identical twins may be far more impressive than their physical differences.

Another example that illustrates the differing qualities of genotype and phenotype is skin color in humans. The darkness of human skin is to some degree dependent on the amount of exposure to the sun, especially in lighter-skinned individuals. An individual's skin color phenotype darkens ("tans") during a summer in the field, yet there are limits set by the genotype to the darkness attained and the tone to which the skin reverts when exposure to the sun decreases.

1. Heritability

Heritability is a measure of the relative contributions of heredity and environment to a character. Heritability estimates are made of individual characters in a particular population. Heritability can be generally defined as the proportion of the variation observed in the character that is controlled by genetic factors. More restrictive definitions are applied when one is dealing with more specific aspects of heritability (Ehrman and Parsons, 1976; Feldman and Lewontin, 1975).

Chromosomes and Cells

Genes are the hereditary units that make up the genotype. The chemical substance of the gene is deoxyribonucleic acid (DNA). The information carried by genes is encoded in the variable sequences of the four chemical bases that are major constituents of the long, helical DNA molecules (Figure 1). DNA and some specific types of proteins are the major constituents of the chromosomes of eukaryotic organisms (that is, those organisms with nuclei in their cells, which includes animals and plants and excludes viruses, bacteria, and the blue-green algae).

At least one complete set of chromosomes is found in the nucleus of most cells of a plant or animal. The number of chromosomes in a cell type is usually constant for individuals of a particular species and sex. Figure 2 is a photograph of preparations of the chromosomes found in a single cell taken from the bone marrow of each of three species of leaf-nosed bats

CHAPTER 2 Genetic Principles for Managers   17

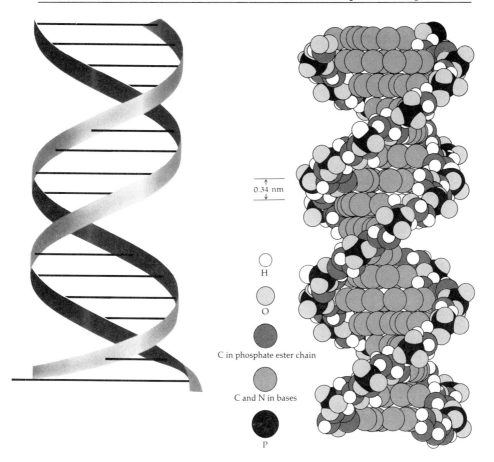

Figure 1. *The DNA double helix. The ribbons of the model on the left and the strings including dark atoms in the space-filling model on the right represent the sugar–phosphate "backbones" of the two strands. The bases are stacked in the center of the molecule between the two backbones. (nm = nanometer; 1 nm = 10^{-9} m). Figure and caption from Luria et al. (1981). Reprinted by permission from S. E. Luria, S. J. Gould, S. Singer's A View of Life.*

(Phyllostomadae). The chromosomes have been photographed and the images of the individual chromosomes have been cut apart and aligned for comparison. From this figure one can count the number of chromosomes in the cell, their relative sizes, and the position of the centromere (the narrow area on each chromosome that divides the chromosome into "arms"). This sort of picture of an individual's chromosomes is called a karyotype.

Karyotype 2A reveals that the cell has 24 chromosomes. (Each chromosome appears double because the chromosomal material has duplicated

Figure 2. *Karyotypes of three species of leaf-nosed bats.* (A) Vampyressa pusilla *from Colombia.* (B) Vampyressa melissa *from Peru.* (C) Vampyressa bidens *from Peru. After Gardner (1977).*

in preparation for cell division.) Chromosomes are usually examined in dividing cells because the chromosomes shorten and appear denser at that time. Each chromosome image is placed next to that of another chromosome that is very similar to it, and in fact these pairs normally carry genes controlling the same characters. These chromosome pairs are called homologous pairs. A sex chromosome, as discussed later in this chapter, will not necessarily have a homologue. The number of homologous chromosome pairs is referred to as n, the haploid chromosome number, which is 12 for the bat cell in Figure 2A ($n = 12$). The total number of individual chromosomes is $2n$, the diploid chromosome number, which is 24 ($2n = 24$) for this same cell. This cell therefore has two complete sets of chromosomes. Each complete haploid set of chromosomes is called a genome, and each genome contains a complete set of genes. This cell contains two genomes ($2n$) because the zygote, or fertilized egg, that was the ultimate ancestor of this cell (and all other cells in the bat's body) was formed at fertilization by the merging of a haploid egg ($1n$) and a haploid sperm ($1n$). Each chromosome in a homologous pair has therefore been derived from a different parent. The genotype of a cell therefore consists of two sets of chromosomes: a paternal haploid set, or genome, from the father, and a maternal haploid set, or genome, from the mother. Each homologous pair of chromosomes in turn consists of one paternal chromosome and one maternal chromosome. The term genome is also used in a more general sense to mean all the genetic material of an organism. Polyploidy (discussed later in this chapter; see also glossary) presents complications of the simple concept of chromosome sets presented here for diploid organisms.

Cellular Reproduction

Cells increase in number by dividing. Mitosis and meiosis are the two major types of cellular reproduction, or cell division, found in plants and animals.

1. Mitosis

After an egg is fertilized to form a zygote, the zygote divides in two to form two "daughter cells." Each of these cells contains the same number of chromosomes and amount of genetic material as the zygote. Each of these cells can then serve as a "mother cell" and divide in two to form two more daughter cells each. As this process continues, parts of the growing mass of cells form the increasingly complex structures of the embryo and, eventually, of the mature organism. At each division, the chromosomes are duplicated and sorted out to the daughter cells so that each of the daughter cells receives a diploid set identical to that which existed in the mother cell. This process of cell division that results in the formation of more cells without a change in chromosome number in those cells is called mitosis. Mitosis is responsible for growth and regeneration of an organism's tissues.

2. Meiosis

Meiosis is the second major type of cell division: it occurs only in some cells of the reproductive organs. During this process, diploid ($2n$) cells produce haploid (n) cells, which eventually form the gametes: sperm and eggs. The diploid number is restored at fertilization when two haploid gametes merge to form a diploid zygote. Unlike mitosis, each cell that begins meiosis gives rise to four haploid cells (not all of which will necessarily mature into gametes, depending upon the sex). To accomplish this, meiosis requires a sequence of two divisions, designated the first meiotic division and the second meiotic division (Figure 3). Another peculiarity of meiosis is that early in the first meiotic division each chromosome (each of which has already doubled to form two strands or chromatids, which remain attached) becomes aligned next to its homologue. Each double-stranded homologue then enters a different daughter cell as the cell divides.

This pairing of homologues is called synapsis and is not found in mitosis. The significance of pairing of homologues during synapsis will be discussed later. Each of the two cells resulting from the first meiotic division then undergoes a second meiotic division. During this division, each double chromosome divides to form two chromosomes. One of each of these is distributed to a different daughter cell. The meiotic process may be summarized as follows: a single diploid cell begins a sequence of two cell

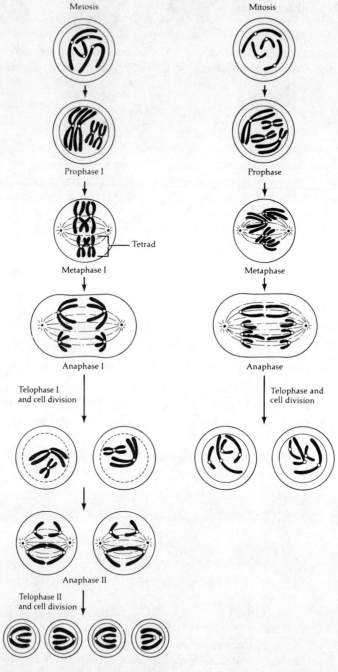

Figure 3. *A comparison of mitosis and meiosis. From Luria et al. (1981). Reprinted by permission from S. E. Luria, S. J. Gould, S. Singer's* A View of Life.

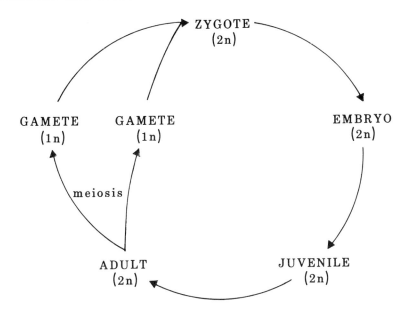

Figure 4. *A generalized life cycle for sexually reproducing plants and animals.*

divisions that results in the formation of four haploid cells. The gametes are derived from such haploid cells. Meiosis is illustrated and contrasted with mitosis in Figure 3.

An important consequence of meiosis is that each gamete contains a combination of chromosomal material from the paternal and maternal genomes of the individual that produced it. This genetic mixing occurs at synapsis, where maternal and paternal homologous chromosomes may exchange segments (recombination, described later in this chapter), and by the actual assortment of chromosomes into the gametes, where each chromosome has an equal chance of being derived from either the maternal or the paternal genome.

A general life cycle of sexually reproducing organisms is presented in Figure 4. The chromosome number of each stage is given. Extreme modifications of this basic plan, such as the free-living existence of a multicellular $1n$ stage (which produces the gametes), are found in ferns and some other modern representatives of early plants.

Mendelian Genetics

Two fundamental principles of genetics were published by Gregor Mendel in 1866. These principles, referred to as Mendel's principles or laws, are the *principle of segregation* and the *principle of independent assortment*. The inheritance of traits according to these principles is sometimes described as Mendelian. The principle of segregation holds that genes occur within individuals in pairs and that the pairs segregate (separate) completely during the formation of gametes. This is consistent with the observation that homologous chromosomes separate from one another and come to occupy different cells during the second division of meiosis. The principle of segregation holds that genes are assorted independently of one another in the formation of gametes and that these gametes combine randomly at fertilization.

Before describing some of the experiments on which Mendel based his principles, some additional definitions are needed. The term "gene" is a general term for heritable units. "Locus" and "allele" are more specific terms that refer to specific aspects of genes. Recall that genes occur in sequences along the chromosomes. Each gene usually has a particular location, called a locus or gene locus ("loci" is the plural of "locus"), on a particular chromosome. At other sites along the chromosome are loci of genes with different effects. It is accurate, therefore, to speak of a chromosome as consisting of a sequence of gene loci, or, simply stated, loci.

Genes at a locus that controls flower color were one of the subjects of Mendel's investigations of heredity in garden peas. Plants of a true-breeding red strain have a gene specifying red flower color at the same locus on each of one homologous pair of chromosomes. True-breeding with respect to a character means that individuals and their offspring that exhibit that character may be crossed for many generations with virtually all offspring displaying the same character (red flower color in this case). Since these plants have a gene specifying red at this locus on both homologues (homologous chromosomes), they are said to be *homozygous* for this gene. The genotype at this locus can be abbreviated as RR, each R representing an R gene on each homologue. The flower color phenotype is red. Genes with other effects on the flower color phenotype may be found at this same locus in other garden pea plants. A true-breeding white-flowered plant will have a gene at this locus on each homologue that specifies white flower color. An alternative genetic expression at a locus is called an *allele* (either syllable may be accented, but the final "e" is silent). The white-flowered plant is therefore homozygous for an allele specifying white, and its genotype at this locus can be abbreviated rr. Its flower color phenotype is white as opposed to the red of plants homozygous for the alternative R.

If a homozygous red-flowered pea plant is crossed with a homozygous white-flowered pea plant, the genotypes of this parental (P) generation are therefore RR and rr, respectively. Gametes are formed by meiosis within each parent. All gametes found in the red flowers will carry the allele for red (R), and all gametes formed in the white flowers will carry the allele for white (r). Since all the offspring begin as zygotes formed by the union of one gamete from each of two parents, if an RR and an rr plant are crossed,

Table 1. *Results of a Cross between Two Red-Flowered Pea Plants with the Genotype Rr*

Genotype	Phenotype	Proportion of Offspring	Genotypic Ratio	Phenotypic Ratio
RR	Red	1/4	1	3
Rr		1/2	2	
rr	White	1/4	1	1

the genotypes of the offspring (called the F_1, or first filial, generation to distinguish it from the parental generation) will all be *Rr*. Since these F_1 individuals carry two different alleles at this locus, they are termed *heterozygous*. Since the phenotype of these *Rr* individuals is red, the allele for red flower color is said to be *dominant* to the allele specifying white, which is then the *recessive* allele. When two of these F_1 individuals with *Rr* genotypes are crossed, their offspring (the F_2, or second filial generation) are found to be in a proportion of about three red individuals for every white individual. A summary of these results is given in Table 1. This ratio can be explained by examining the types of gametes formed by each of the F_1 individuals, which all have the same *Rr* genotype at this locus. As the chromosomes segregate at meiosis within one of these *Rr* individuals, one-half of the gametes formed will carry the dominant allele *R* and one-half will contain the recessive allele *r*. This occurs in both F_1 individuals involved in the cross. Plants in the F_2 generation may inherit either allele from either parent. The next step is to indicate the possible genotypes of the F_2 individuals with *Rr* parents: *RR*, *Rr*, and *rr*. The probability of obtaining each of the three genotypes from the *Rr* parents can now be calculated. First, the homozygous dominant genotype: the probability that an F_2 individual will inherit *R* from a particular *Rr* parent is 1/2. The probability that an individual has inherited the *R* allele from the other *Rr* parent is also 1/2 (since the other parent had the same heterozygous genotype). The probability that an individual would inherit *R* from both parents is the product of the separate probabilities of inheriting it from each parent, or $1/2 \times 1/2 = 1/4$.* This means that 1/4 or one in four F_2 offspring will be homozygous dominant (*RR*). Calculation of the proportion of the other homozygous genotype (*rr*) in the F_2 is carried out the same way as for *RR*. Half the gametes of each F_1 individual will carry *r* at this locus. The probability of an F_2 individual's inheriting an *r* from one parent is 1/2, and from the other parent it is also 1/2. The probability of an F_2 individual's being

---

*The same law of probability applies when a coin is flipped. The probability of flipping a head on any given flip is 1/2 (unless it is a very unusual coin). The probability of flipping two consecutive heads is $1/2 \times 1/2 = 1/4$. Briefly stated, the probability of two independent events (such as successive coin flips or segregation of gametes in two separate individuals) occurring together is the product of their separate probabilities.

formed by the union of two gametes carrying r and therefore being homozygous recessive (rr) is $1/2 \times 1/2 = 1/4$.

Calculation of the proportion of heterozygotes in the F_2 is complicated by the fact that F_2 heterozygotes may inherit either allele from either parent. If the two Rr parents in the F_1 generation are distinguished for the purposes of the present discussion as the first parent and the second parent, the probability that the F_2 individual was the result of the union of a gamete carrying the R allele from the first parent (with probability of 1/2) and a gamete carrying an r from the second parent can be calculated as $1/2 \times 1/2 = 1/4$. An F_2 individual can, however, obtain the same genotype by inheriting an r from the first parent (probability of 1/2) and an R from the second parent with a probability of $1/2 \times 1/2 = 1/4$. To obtain the total expected probability of finding heterozygotes at this locus among the F_2, the probabilities of each of the separate ways of attaining the genotypes must be added together: $1/4 + 1/4 = 1/2$. One in two or 1/2 of the F_2 generation is expected to have the heterozygous genotype. The genotypes, phenotypes, probabilities of occurrence, proportions of offspring expected, and phenotypic ratios of the F_2 individuals are presented in Table 1. These results can also be obtained by the use of a Punnet's square (Figure 5), which has all the possible types of gametes with respect to this locus from one parent arranged along one side of the square and all the possible gametes from the other parent along an adjacent side of the square. The genotypes of the offspring are then derived by combining a gamete from each parent. These genotypes are written within the smaller squares of the large square.

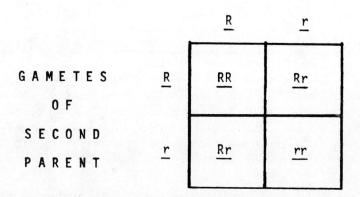

Figure 5. *Punnet's square, demonstrating the result of a cross between two individuals with the genotype Rr.*

1. Intermediate Inheritance and Codominance

Sometimes one allele does not exert clear-cut dominance over another allele. Heterozygotes in these cases are distinguishable from homozygous genotypes because both alleles contribute to the phenotype of the heterozygous individual. If red and white snapdragons (*Antirrhinum majus*) are subjected to the same crosses as in the garden pea example given above, the heterozygous individuals are pink rather than red, and therefore distinguishable from the *RR* individuals. The phenotypic ratio would be 1:2:1, the same as the genotypic ratio. This is termed intermediate inheritance, or incomplete dominance, since the phenotype of heterozygotes is intermediate between the corresponding homozygous phenotypes.

Codominance describes cases where both alleles at a heterozygous locus are equally and independently expressed in the phenotype. This is found in the human ABO blood groups where the alleles I^A and I^B, when homozygous, code for the blood group phenotypes A and B, respectively. An individual heterozygous for these alleles has the genotype I^AI^B and has the type AB phenotype. Alleles controlling the enzyme variation studied by electrophoretic methods (discussed in several chapters of this book) are generally codominant.

2. Multiple Alleles

More than two alleles can be found at a single locus, although any one individual would not normally carry more than two alleles (one per locus on each of two homologous chromosomes). A commonly used example is the ABO blood groups in humans, although the actual mode of inheritance for these groups is more complex than presented here. Various combinations of three alleles (I^A, I^B, and i) specify the blood group phenotypes A, B, AB, and O. Type O is the phenotype of the homozygous recessive ii; type AB is found in individuals that are I^AI^B; type A is specified by either I^AI^A or I^Ai; and type B is specified by I^BI^B or I^Bi. Note that, although I^A and I^B are condominant with respect to one another, both of these alleles are dominant to i.

3. Dihybrid Crosses: Inheritance at Two Loci Considered Simultaneously

The examples given so far have dealt with only one locus at a time. Mendelian explanations can be extended to cases where two or more loci are considered simultaneously, a situation called a dihybrid cross when two loci are involved. Two loci in the garden peas studied by Mendel will serve as an example.

Table 2. *Expected Phenotypes and Genotypes of Offspring of a Cross between Two Pea Plants with the Same YyTt Genotype*

Phenotype	Genotype	Genotype Probability	Phenotype Probability
Yellow, tall	YYTT	$(1/2 \times 1/2)(1/2 \times 1/2) = 1/16$	9/16
	YYTt	$(1/2 \times 1/2)[(1/2 \times 1/2) + (1/2 \times 1/2)] = 1/8$	
	YyTT	$[(1/2 \times 1/2) + (1/2 \times 1/2)](1/2 \times 1/2) = 1/8$	
	YyTt	$[(1/2 \times 1/2) + (1/2 \times 1/2)][(1/2 \times 1/2) + (1/2 \times 1/2)] = 1/4$	
Yellow, short	YYtt	$(1/2 \times 1/2)(1/2 \times 1/2) = 1/16$	3/16
	Yytt	$[(1/2 \times 1/2) + (1/2 \times 1/2)](1/2 \times 1/2) = 1/8$	
Green, tall	yyTT	$(1/2 \times 1/2)(1/2 \times 1/2) = 1/16$	3/16
	yyTt	$(1/2 \times 1/2)[(1/2 \times 1/2) + (1/2 \times 1/2)] = 1/8$	
Green, short	yytt	$(1/2 \times 1/2)(1/2 \times 1/2) = 1/16$	1/16

One locus controls seed color, with an allele specifying yellow color (*Y*) being dominant to one specifying green (*y*). The genotypes of plants with yellow may be *YY* or *Yy*, while plants with green seeds are *yy*. Another locus on a different chromosome controls height of the plant, with an allele specifying tallness (*T*) being dominant to an allele for shorter plant height (*t*). Tall plants have the genotype *TT* or *Tt*, and short plants are *tt*. A plant that has the genotype *YyTt* for these two loci will have yellow seeds and be tall. If two such plants are crossed, the frequencies of expected genotypes and phenotypes among the offspring can be calculated by first listing all possible genotypes among the offspring and then making calculations of their likelihood based on the probabilities of inheriting particular alleles at each locus (Table 2). These calculations are made by (1) separately computing the probability of the expected genotype for each genotype at each locus occurring in a single individual; (2) multiplying together the probability of each genotype at one locus with that of each genotype at the second locus to arrive at the probability of each two-locus genotype; and (3) adding together the probabilities of genotypes that specify the same phenotype to arrive at the probability of each phenotype.

As an example, the probability of obtaining a green-seeded, tall plant from a cross between two *YyTt* parents will now be calculated. A green-seeded plant has the genotype *yy*. A tall plant may have the genotype *TT* or *Tt*. The genotypes of the green-seeded tall plants that are the object of interest here may be represented as *yyT__* (since either allele may be in combination with *T* at the locus controlling height). Our objective is to calculate the proportion of offspring with this genotype that would be expected to result from a cross between two *YyTt* parents. All the possible phenotypes, their underlying genotypes, and their probabilities of occurrence are presented in Table 2. The two genotypes (*yyTT* and *yyTt*) that specify the desired phenotype have separate probabilities of 1/16 and 1/8. Added together, they represent a 3/16 probability of a given plant's being green-seeded and tall.

Sex Chromosomes and Sex Determination

The genetic loci discussed in the preceding examples are located on particular chromosomes called autosomes. Forty-four of the forty-six chromosomes in most human cells are autosomes. The other two chromosomes are called sex chromosomes because they determine the sex of an individual. Inheritance of genes located on sex chromosomes is predictable but does not follow the Mendelian expectations discussed in previous examples.

Humans and many other species have two types of sex chromosomes, designated X and Y. Both sex chromosomes in females are X's. Males have one X and one Y chromosome. All gametes produced by meiosis in a female will contain an X chromosome. All her offspring will inherit an X chromosome. Half of the sperm produced during meiosis by the father will contain a Y chromosome; the other half bear an X chromosome. There is an approximately even chance (probability of 1/2) that the X-bearing egg will be fertilized by a Y-bearing sperm. One-half of the offspring will be expected to be males (XY), and an equal (1/2) proportion females. It is the father's contribution that determines the offspring's sex with this type of sex determination.

1. Sex-Linked Inheritance

The gene loci on the sex chromosomes are referred to as sex-linked. This means that the probability of inheritance of a character specified by genes at such a locus depends on the individual's sex. The genes located on the X-chromosomes of mammals are among the best-studied sex-linked loci. Two generalizations can be made about sex-linked loci on the X-chromosomes. The first is that all alleles, even recessives, can be expressed in normal males. This is because a male has only one X-chromosome, which does not have a homologous X that could carry a dominant allele at the same locus that would mask a recessive. Entirely different loci are found on the Y-chromosome, which is not homologous to the X. The second observation is that females must be homozygous for a sex-linked recessive allele in order to exhibit the recessive phenotype. This explains why sex-linked recessive traits such as the "bleeder's disease" (hemophilia) and certain types of color blindness are far more common among males than females.

A normal male of a species with an XX-XY system of sex determination cannot be heterozygous for sex-linked traits because he has only one X-chromosome in addition to a Y. This is demonstrated by the tortoiseshell or calico color pattern in domestic cats. These patterns consist of orange and black markings on either a white (in the case of calico) or a black (in the tortoiseshell) background. Orange and black are specified by two codominant alleles at a locus on the X-chromosome. A cat must be heterozygous at this locus to exhibit the tortoiseshell or calico pattern. A male with normal sex chromosomes may carry either the orange or the black allele, but normally not both, since he has only one X-chromosome. Those rare male cats that exhibit both orange and black markings are likely to have an extra X-chromosome or some other chromosome abnormality that renders them sterile.

The XX-XY system of sex determination described in this section is found in those mammals that have been examined, in many insect species, and in some plant species that have male and female flowers on separate individuals (for plants see also Grant, 1975). A large number of different sex chromosome and sex determination systems are known among plants and animals. There are some sexual species for which identifiable sex chromosomes have not been discovered. In some nonmammalian vertebrates and insects, males have two sex chromosomes and females have only one. The males of these species therefore have a diploid chromosome number of one more than the female's, and it is the gamete of the mother, not the father, that determines the sex of the offspring. The mother also determines the sex of the offspring in birds with the ZZ-ZW sex chromosome system, where males are ZZ and females are ZW.

2. Sex-Limited Autosomal Traits Distinguished from Sex-Linked Traits

Sex-linked traits are controlled by genes that happen to be located on the sex chromosomes, but they do not necessarily influence sexual characteristics. Sex-linked traits should be distinguished from sex-limited autosomal traits such as the secondary sexual characteristics. Sex-limited autosomal traits are controlled by genes located on autosomes, but they are expressed in only one sex. The autosomal loci of genes that control milk production in dairy cows, for example, are also on the autosomes of dairy bulls.

Autosomal Linkage

Autosomal linkage between loci occurs when they are located on the same chromosome. The outcomes of crosses involving linked loci deviate from expectations based on the Mendelian principle of independent assortment. The following example concerns two loci with two alleles at each locus.

An individual with the genotype $AaBb$ is crossed with another that is $aabb$. According to independent assortment, the genotypes of the F_1 generation would be expected in the following proportions:

$1/4$ $AaBb$
$1/4$ $Aabb$
$1/4$ $aaBb$
$1/4$ $aabb$

If these loci are on the same chromosome, it is very unlikely that the expected proportions will be observed. The parents' genotype may be presented in terms of the structures of the homologous chromosomes as

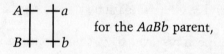

for the *AaBb* parent,

and

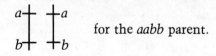

for the *aabb* parent.

Both the relative positions of the two loci along the chromosome and the alleles at each locus on each chromosome are indicated. Recall that whole chromosomes, not simply individual alleles, are passed on to the gametes during meiosis. On the basis of this information (incomplete information, as will be demonstrated below), half of the heterozygous parent's gametes would be expected to contain the chromosome carrying the alleles *A* and *B* and the other half of its gametes will carry *a* and *b*. All the gametes of the homozygous parent will contain a chromosome carrying *a* and *b*. Given this linking together of alleles, we would expect only two genotypes among the progeny: one half would be expected to be *AaBb*, and the other half would be expected to be *aabb*. This physical linking together of the two loci keeps the alleles at these loci from being independently assorted into gametes at meiosis, and therefore *Aabb* and *aaBb* offspring will not be found in the proportions predicted by the principle of independent assortment. The actual offspring from the above cross will, however, include *Aabb* and *aaBb* genotypes. This deviation is the result of the exchange of chromosomal material between homologous chromatids when they pair during synapsis at meiosis (Figure 6). This exchange results in the production by the heterozygous parent of gametes that contain the "crossovers," or crossover chromosomes *Ab* and *aB*. This process is called *crossing over*, and it must occur between the two loci in question to separate the linked alleles (*AB* and *ab* in this case) in order to form gametes that contain crossovers among their chromosomes.

The farther apart two loci are on the same chromosome, the more frequently crossing over will occur between them. The frequency of crossing over is reflected in the proportion of the progeny that are carrying crossovers. This property has been used to "map" chromosomes by providing a relative scale of the distance between loci.

Crossing over can therefore be very rare between loci that are very close to one another on a chromosome. Genes at such loci can be said to be tightly linked, and they may pose a problem to a plant or animal breeder who finds an undesirable allele at one locus that is closely linked to

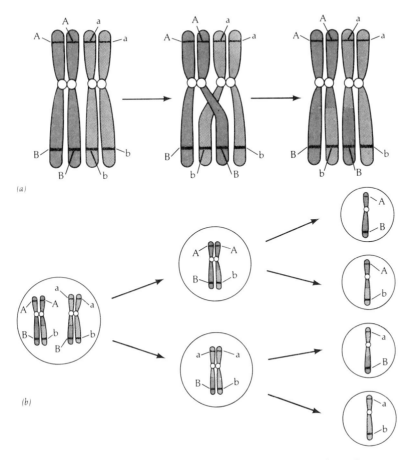

Figure 6. (A) Schematic diagram of crossing over. Two homologous doubled chromosomes, one bearing alleles A and B, the other alleles a and b, lie side by side during synapsis. Portions of an AB chromatid and an ab chromatid then exchange positions. The crossed-over chromatids break and are rejoined, producing one chromatid with the alleles A and b and one chromatid with the alleles a and B. In this way new combination of genes are created. (B) Each of these chromatids ends up in a different gamete at the end of second meiotic division. Figure and caption from Luria et al. (1981). Reprinted by permission from S. E. Luria, S. J. Gould, S. Singer's A View of Life.

another locus that has a highly desirable allele. The breeder must then breed his stock until a crossover individual is produced that represents the separation of the undesirable allele at one locus from the desirable allele at the other locus. This may take many breeding attempts or generations to accomplish.

Molecular Genetics

Genes were described early in this chapter as being made of the nucleic acid DNA, and the information of the genes was described as being encoded in the sequence of base compounds along the very long DNA molecules. The important links between DNA base sequences and phenotypic characters are proteins. Proteins not only form structural substances such as muscle and hair, they also form enzymes, which are essential biological catalysts of metabolic reactions taking place in all the cells of the body. Enzymes also participate in the regulation of these reactions, and even guide the replication of DNA. From these all-encompassing functions, it is easy to see how by controlling protein synthesis DNA controls the growth and maintenance of cells and organisms.

1. Protein Synthesis

The exact structural relationship between DNA and the chromosomes, which the DNA is a part of, is not completely known. Some major features of the process whereby DNA specifies proteins are known, and those features appear to be similar in a great variety of different plant and animal species. The following is a brief summary of the process as it is thought to function in plants and animals.

The first step in the process of protein synthesis in the cell is the transcription of the DNA base sequence into a complementary sequence of bases in a molecule of another nucleic acid, ribonucleic acid (RNA). This particular type of ribonucleic acid is referred to as messenger RNA (abbreviated mRNA). At this point the information from the DNA has been transcribed into a base sequence on mRNA.

The mRNA leaves the nucleus of the cell and enters the cytoplasm, the general substance of the cell outside the nucleus. There can be profound alterations in the length of the transcribed mRNA at this time. The mRNA contacts one or more cytoplasmic structures called ribosomes, and its sequence is translated into a sequence of amino acid residues (derived from amino acids). The completed chain of amino acid residues is the protein. Shorter chains of amino acid residues are called polypeptides, so named for the peptide type of molecular bond that links adjacent amino acid residues in polypeptides and proteins.

This greatly abbreviated description of protein synthesis mentions only some general features and omits numerous steps, all of which require the activity of various enzymes and other substances. The sequence of amino acid residues dictates the three-dimensional structure of the protein, which in turn determines the function of that protein. The human genome has been estimated to consist of about 100,000 gene loci (Dobzhansky, 1970). The possible number of different proteins is staggering and explains how an organism as complex as a mammal can be specified by "mere molecules."

2. The Genetic Code

The actual genetic "code" represented by the sequence of bases in DNA molecules has been worked out, and it appears to be identical for all organisms studied. There are four different bases in the base sequence of DNA, but the different arrangements in their order can code for an inconceivably large number of different sequences of the 20 amino acids that are common to living cells. The code is called a triplet code because a particular amino acid residue is specified by particular code "words" of three successive bases.

3. Enzymes, Isoenzymes, and Allozymes

At the molecular level a gene can be defined as a unit specifying a single protein. This remains a useful definition even though it has been subject to revisions and qualifications by molecular geneticists. Different proteins specified by correspondingly different alleles can be studied and formal studies made of their Mendelian patterns of inheritance. These different molecular forms of enzymes with the same general function are called isoenzymes or isozymes. When these forms are known to be coded by different alleles at the same locus, they are more precisely referred to as allozymes. At one time some evolutionary biologists believed that the study of these molecules would answer many long-standing evolutionary questions because proteins, being specified by DNA, were considered "closer to the genes." A great deal of this early optimism has now faded, as it has been realized that these molecules are but one aspect of the genotype. Those genes that code directly for proteins with structural or enzymatic functions are referred to as structural genes. Other genes, such as some regulatory genes, do not specify structural proteins or enzymes. One view is that these regulatory genes are responsible for most of what is recognized as evolutionary change (Wilson, 1976). Structural genes coding for allozymes are nevertheless an important aspect of the genotype, and these loci represent extremely useful genetic markers for studies of genetic divergence, evolutionary history, and population structure, which are of great interest to managers of wild populations as well as to geneticists, evolutionary biologists, and taxonomists.

Another theoretical argument concerns whether different allozymes and other alleles are maintained in natural populations by natural selection or whether they are selectively equivalent (of neutral value) or nearly so.

Again, the value of allozymes as genetic markers is not totally dependent on the ultimate correctness of either of these arguments.

Mutation

Up to this point in this chapter, it has been assumed that alleles are transmitted unchanged from one generation to the next, as indeed they usually are. An inherited change in a gene or chromosome is a *mutation*. Mutations are rare events, the average frequency of a mutant allele at a locus being estimated at 1 per 100,000 gametes (Dobzhansky, 1970; Nei, 1975). Eye color mutants in *Drosophila*, for example, arise at very low but predictable rates (*Drosophila* is a genus in the family Drosophilidae, which are known by the common name pomace flies, and includes the common fruit flies of genetics laboratories). Although most new mutations are considered harmful to individuals carrying them, mutation is the ultimate source of the great genetic diversity seen in natural populations and provides the "raw material" for evolution.

Although mutants (individuals carrying new mutations) have been observed by many generations of plant and animal breeders, the structural bases of many types of mutation have only recently been understood. Mutations can be classified as either point mutations or chromosomal mutations.

1. Chromosomal Mutations

Chromosomal mutations are changed in the observable structure or number of chromosomes. Most chromosomal mutations can be classified as belonging to one of six categories: polyploidy, aneuploidy, inversions, translocations, duplications, and deletions.

Polyploidy is the duplication of entire chromosome sets, often as a result of hybridization between chromosomally different individuals (Chapter 16). Polyploidy can also occur without hybridization. For example, a chrysanthemum with a chromosome number of $2n = 36$ can be derived by the doubling of a chromosome set of a chrysanthemum that has $2n = 18$. This can occur, for example, in a zygote if the chromosomes duplicate twice in the course of a single mitotic division. A polyploid individual can often be recognized by its having a $2n$ chromosome number that consists of multiples greater than two of the $1n$ chromosome number or numbers of the species from which it was derived. Polyploidy is common and extremely important in plant evolution, where it has contributed to the formation of about half of the flowering plant species and most ferns

(Grant, 1981). Polyploidy is very rare in most groups of animals, although cases have been reported, for example, in insects, earthworms, and fishes (White, 1973). Its possible long-term role in animal evolution remains very controversial (Heingardner, 1976).

Aneuploidy is the gain or loss of one or more chromosomes, but not in multiples of the haploid set. Aneuploidy can result from gametes that have either surplus or absent chromosomes. These gametes can form during meiosis when homologous chromosomes do not separate completely during cell division and one of the daughter cells receives an extra chromosome (or chromosomes) that normally would have entered another daughter cell. This failure of homologues to segregate is called *nondisjunction* because the homologues involved have not segregated—that is, become disjunct. Nondisjunction results in polyploidy when it involves the entire chromosome complement. A zygote formed by a gamete with one extra autosome and a normal gamete will have a chromosome number of $2n + 1$, and there will be three homologues of that particular chromosome. This is called trisomy, and individuals with this condition may experience abnormal development, as in the trisomy for chromosome 21 of humans that causes Down's syndrome. This genetic disease, commonly and misleadingly referred to as mongolism, causes symptoms similar to trisomies detected in other primates (Benirschke et al., 1980).

Nondisjunction can also result in the formation of aneuploid gametes that are deficient for one or more chromosomes and which will form a zygote with fewer than the normal diploid number of chromosomes.

The remaining classes of chromosomal mutations (summarized in Table 3) involve changes in the structure of individual chromosomes.

Inversions are the result of a segment of a chromosome becoming inverted, so that the order of loci is reversed within the inverted segment. For example, if loci *ABCDEFG* are in that order on a chromosome, an inversion involving the segment containing *C* through *E* would change the order to *ABEDCFG*. An inversion that involves the larger segment from *B* to *E* would result in an order of *AEDCBFG*. Individuals that are heterozygous for an inversion (one homologue has the "normal" sequence and the other

Table 3. *Examples of Chromosomal Mutations That Alter the Structure of a Portion of a Chromosome*

Sequence of loci (*A* through *G*) on a Chromosome Prior to Mutation	*ABCDEFG*
Inversion	*ABEDCFG*
Deletion	*ABCEFG*
Duplication	*ABCDDEFG*
Translocation (where XYZ are from a nonhomologous chromosome)	*ABCDEFGXYZ*

has the inverted sequence) may suffer a reduction in fertility due to disturbances of meiosis. Inversion polymorphisms are common in some *Drosophila* and some other groups of flies, although at least some of these flies are known to have mechanisms for avoiding this reduced fertility. Inversions have also been found in a variety of plants and animals, including primates. The commonness of inversions and their effects on fertility have not been established for natural populations of most plant and animal groups.

A *translocation* results when a portion of one chromosome becomes attached to a nonhomologous chromosome. Individuals derived from a gamete carrying a translocation may suffer reduced fertility because of disturbances of meiosis during synapsis caused by awkward patterns of pairing of homologous chromosome segments that are located on different chromosomes.

A chromosomal mutation that results in a chromosome with a segment that is repeated is a *duplication*. The presence of more than the normal number of active genes may cause disturbances in cell activities that result in the zygote's developing abnormally or not at all.

A *deletion* is the result of the loss of a segment of a chromosome. The effects of a particular deletion depend on the number and importance of the genes within the deleted segment. The effects may be considerably more severe in a diploid individual that is homozygous for a deficiency and therefore has some loci completely missing. A deletion and a duplication may be formed simultaneously during meiosis by an unequal crossover that leaves a segment from one homologous chromatid attached to the other; at the end of meiosis, one of the gametes formed will carry a deletion and another will carry a duplication.

2. Point Mutations

Point mutations normally do not cause observable changes in chromosome structure. Their effects are observed in the phenotype of the organism. The actual molecular basis of a point mutation has been discovered for the allele coding for human sickle cell hemoglobin (Hb^S), where the more common allele codes for "normal" hemoglobin (Hb^A). Hb^S causes the disease sickle cell anemia in individuals that are homozygous for this allele. An active hemoglobin molecule consists of two alpha polypeptide chains and two beta polypeptide chains. Sickle cell hemoglobin differs from normal hemoglobin in only 1 out of the 146 amino acid residues of the beta chain of a hemoglobin molecule. This difference can be accounted for by a change in a single base in the DNA of the gene coding for the beta chain that changes the three-letter code for one animo acid (glutamic acid for Hb^A) to the code for a different amino acid (valine for Hb^S).

Work on *Drosophila* and other organisms has revealed that many genes have a predictable rate of mutation and that these rates vary from locus to locus but average, as stated earlier, about 1 per 100,000 to 1 per 1,000,000 per locus per generation. Mutation rates can be increased by exposing individuals or cells to ultraviolet light, X-rays, or certain chemicals, although the great majority of all mutations are lethal or otherwise harmful.

Mutations usually have drastic and harmful effects on individuals carrying them; yet, in the long run of evolutionary time, mutation provides the variability that makes possible the adaptation and therefore the survival of individuals and populations.

Interactions between Loci

Most of the characters under genetic control are actually affected by genes at more than one locus. Many of the examples presented in this chapter and in genetics textbooks are relatively unusual in that they deal with completely heritable characters that are controlled by alleles at a single locus. These single-locus examples were necessary to demonstrate the basic laws of Mendelian inheritance. Once inheritance of alleles at a single locus is understood, then Mendelian explanations can be expanded to describe the interactions of genes at different loci and the effects of genes at a number of different loci on a single character.

Characters such as body size in humans and many other vertebrates and skin color in humans are controlled by genes at several loci, and the phenotype is the sum of the effects of genes at many different loci. This type of genetic control over a character is called quantitative or polygenic inheritance. The Mendelian basis of inheritance of alleles at any particular locus may not be well understood, but statistical methods enable geneticists and breeders to study these characters.

Many types of interaction between genes at different loci are not strictly additive. Epistasis is the influence of alleles at one locus on the phenotypic expression of alleles at another locus. For example, fruit color in garden squash is controlled by genes at two separate loci. A dominant allele at one locus results in white color regardless of alleles at the second locus. If an individual has only the recessive allele at the first locus, then one or two dominant alleles at the second locus will result in yellow color. Recessive alleles alone at both loci result in green color.

Alleles at a single locus often have effects on a number of different characters. Many examples of these effects, termed pleiotropic effects, have been described. For example, some alleles controlling coat coloration in mink (*Mustela vison*) have pleiotropic effects on fertility (Belyaev and Evsikov, 1967, cited in Dobzhansky, 1970). A recessive allele in the house mouse (*Mus musculus*) produces a white spot on the fur of the abdomen, a flexure of the tail due to fusion of some vertebrae, and anemia in homozygous mice (Silvers, 1979).

Deviations from Mendelian expectations at some loci are found when, for a variety of reasons, the genotype is not expressed in all cases (incomplete penetrance) or the intensity of genotypic expression (expressivity) varies.

Evolution by Natural Selection

The theory of natural selection was proposed by Charles Darwin and Alfred Russel Wallace in 1859 as a mechanism by which species change over generations in response to the pressures placed upon them by the environment. Although there has been vigorous debate among evolutionary biologists about the precise role and power of natural selection, few if any disagree that natural selection is the major creative force responsible for the adaptation of organisms to their environment and for the great diversity of the life on earth. Darwin and Wallace were unaware of the Mendelian laws of heredity when they proposed their theory, and it was not until the "new evolutionary synthesis" of the 1930s that genetics and natural selection were integrated into a unified evolutionary theory.

The Darwin–Wallace theory of natural selection can be simply stated as follows:

1. The numbers of individuals of a species tend to increase geometrically. If supplied with sufficient nutrients and protection, the total population size will increase at an ever-increasing rate. Geometric increase, however, is not maintained indefinitely because of high death rates.
2. All species are variable; that is, they are made up of variant individuals.
3. Those individuals with variations of characters that better enable them to survive and reproduce will leave proportionately more offspring than will other individuals and will pass on their heritable variations to their offspring. The result is an increase in the frequency of the advantageous characters in the next generation and the increased adaptation of the population to the environment.

The selective change in a character in one direction over a number of generations is called directional selection, and the selective factor or selection pressure is the result of environmental change. Selection is not always directional. Stabilizing selection is the elimination of extreme individuals and selection for an intermediate phenotype. Stabilizing selection ideally occurs in a constant environment.

Before we discuss the behavior of genes in populations, the term "evolution" needs to be defined. Evolution can be most simply defined as a change in the genetic composition of a population over time or as a change in the inherited component of a phenotypic change in a population over time. These two definitions are compatible and do not place restrictions on the time scale (short or long), the amount of change, or the direction of change necessary to consider a change as evolution, although evolution is usually considered to occur in a particular direction over a large number of generations.

1. Lamarkism and Teleology

Lamarkism and teleology are two scientifically obsolete concepts that were parts of early evolutionary theories. Teleology is the belief that "inner drives" within individuals dictate the direction of evolution toward some definite end. The processes of mutation and natural selection, however, can account for directional evolution, such as the increase in brain size during human evolution, without recourse to any inner drive.

Lamarkism is identified mainly with the belief that characteristics acquired during the lifetime of an ancestor can be inherited. If the sons of circumcised men were born without foreskins, this would be a case of Lamarkian inheritance. Jews have practiced circumcision for thousands of years, yet the effect of the operation is not passed on to the sons, who must undergo circumcision in each generation (Jerry Coyne, personal communication). Teleology and Lamarkism tend to creep into discussions of evolutionary change, and special care should be taken to frame questions in terms consistent with known genetic and evolutionary concepts.

2. Species, Populations, and Demes

The term "population" has been applied to a large number of differing groups of individuals within a species. Most strictly defined are the equivalent terms "deme" and "local population," both of which refer to breeding units of individuals that are essentially mating at random. If the genetic population structure of a species is such that definable local populations do not exist, then the term "population" applied to groupings within the species is more subjective. The term population is used in different senses by population geneticists for illustrative purposes, and their usage of the term may at times seem at odds with what a systematist, naturalist, or wildlife manager might consider a population. For example, Nei (1975) sometimes refers to an entire species as a "population" and various subunits of the species as "subpopulations," whereas a naturalist might consider the subpopulations as populations in their own right. The population, subpopulations, and colonies of Nei and other population geneticists are convenient subdivisions for describing particular cases and are normally defined in the context of a specific analysis.

The terms population and subpopulation are used by various authors of this book to refer to convenient theoretical or natural groups for analysis of variation within species. The terms "deme" and "local population" are used as defined in the preceding paragraph.

Population Genetics

1. Genotype and Allele Frequencies in Populations

As with Mendelian genetics, the behavior of genes in populations is best introduced by using examples concerning two alleles at a single locus. Many aspects of population genetics theory have yet to advance beyond simple cases involving one or two loci. The general case to be discussed here concerns a single locus in a hypothetical population. Two codominant alleles, A_1 and A_2, are found at this locus. Individuals in this population will have one of three possible genotypes at this locus: A_1A_1, A_1A_2, and A_2A_2, so that the sum of the frequencies of all three genotypes is equal to one. From these genotype frequencies, one can calculate the frequencies of the alleles A_1 and A_2 (designated p and q, respectively) in the population. The frequency of A_1 is calculated by taking the frequency of A_1A_1 genotypes and adding to it one-half the frequency of the heterozygotes, A_1A_2, since half the alleles at this locus in the heterozygotes are A_1. The frequency of A_2 can similarly be calculated as $q = 1/2$ the frequency of A_1A_2 + the frequency of A_2A_2, or alternatively by subtracting the frequency of the alternative allele from 1: $q = 1 - p$. At any locus where all the different genotypes can be identified and counted, genotype and allele frequencies for that locus can be calculated.

Evolutionary biologists and population geneticists are interested in the means by which genotype and allele frequencies change. In order to understand how various factors cause genetic change, the behavior of allele and genotype frequencies in an ideal population in the absence of these factors must first be examined. This behavior is described by the Hardy–Weinberg equilibrium.

2. The Hardy-Weinberg Equilibrium

The single-locus case introduced above will be used to demonstrate the predictions of the Hardy-Weinberg equilibrium. This expression is sometimes referred to as the Castle-Hardy-Weinberg equilibrium, but the former, more prevalent term will be used here to avoid confusion. The Hardy-Weinberg equilibrium applies to an ideal population of infinite size with non-overlapping generations (individuals may mate only with members of the same generation), and where mating among individuals in the population is completely at random. This ideal population is not subject to mutation, selection, or migration. These assumptions are never met in nature, yet genotype frequencies in natural populations often approach those predicted by the Hardy-Weinberg equilibrium.

If the frequencies of two alleles (A_1 and A_2) in the ideal population are p and q, the Hardy-Weinberg equilibrium describes the genotype frequencies of the next generation as $p + 2pq + q = 1$, where p^2 = frequency of A_1A_1, $2pq$ = frequency of A_1A_2, and q_2 = frequency of A_2A_2. The values of the *genotype* frequencies in the parental generation do not matter. The expression $p^2 + 2pq + q^2 = 1$ is the binomial expansion, or $(p + q)^2$, which has other applications in biology and other fields.

How this expression predicts genotype frequencies in the next generation may be seen by imagining the individuals in the first generation contributing gametes to a pool (the gene pool) that will form the zygotes of the next generation. Recall that the frequencies of A_1 and A_2 in this pool are p and q, respectively.

The probability of one gamete carrying A_1 uniting with another gamete carrying the same allele to form an A_1A_1 offspring is the product of their separate frequencies, or $p \times p = p^2$. The probability of a sperm bearing A_1 uniting with an egg carrying A_2 is similarly $p \times q = pq$. Since heterozygotes may also be formed by an egg carrying A_1 and a sperm carrying A_2 with a probability of $q \times p = qp$, the total frequency of heterozygotes will be $pq + qp = 2pq$. Calculation of the frequency of genotype A_2A_2 is $q \times q = q^2$, based on the same reasoning as the frequency of the A_1A_1 homozygote.

To summarize: The genotype frequencies in the second generation depend only on the parental allele frequencies, and not on the parental genotype frequencies. The genotype frequencies established in the offspring following this single generation of random mating are in the expected Hardy-Weinberg equilibrium frequencies and will remain the same throughout future generations as long as the underlying assumptions of the Hardy-Weinberg equilibrium are in force. Sample calculations are presented in Table 4.

The sum of the frequencies of the heterozygous genotypes is the *heterozygosity* of the population at that locus. Heterozygosity is usually

Table 4. *Sample Calculations of Allele and Genotype Frequencies after One Generation of Random Mating under the Assumptions of the Hardy-Weinberg Equilibrium*[1]

Generation 1
  Adults $p_1 = .7$ $q_1 = .3$
  Gametes $p_1 = .7$ $q_1 = .3$
Generation 2
  Zygotes   Genotype   Frequency
                   A_1A_1 $p_1^2 = .7^2 = .49$
                   A_1A_2 $2p_1q_1 = 2(.7)(.3) = .42$
                   A_2A_2 $q_1^2 = .3^2 = .09$

p_2 = frequency of A_1A_1 + 1/2 frequency of A_1A_2 = .49 + 1/2(.42) = .7
q_2 = 1/2 frequency of A_1A_2 + frequency of A_2A_2 = 1/2(.42) + .09 = .3

[1] The frequency of the allele A_1 is p_1 and p_2 in generations 1 and 2, respectively, and the frequency of A_2 is q_1 and q_2 in generations 1 and 2, respectively.

calculated from the expected Hardy-Weinberg genotype frequencies. This value of heterozygosity, more specifically referred to as expected heterozygosity, is an important measure of genetic variability. The expected heterozygosity over all loci is the mean of the heterozygosities of individual loci. When heterozygosity is calculated from observed frequencies of heterozygous genotypes, it is referred to as the observed heterozygosity.

Two general observations on the behavior of genes in populations in Hardy-Weinberg equilibrium are as follows:

1. The expected Hardy-Weinberg frequencies of genotypes at autosomal loci are attained after a single generation of random mating.
2. Allele frequencies stay the same for generation after generation. A less common allele will not become any more or less common or be driven out of the population simply because it is not the most common allele. Likewise, a recessive or dominant allele will remain at the same frequency in each generation and would not be expected to become less or more common.

If the assumptions of the Hardy-Weinberg equilibrium were always met, then there would be no changes in allele or genotype frequencies from one generation to the next. Natural populations, however, are subject to forces such as sampling error, mutation, selection, and migration in a finite population. Each of these mechanisms can change allele frequencies in populations. In simple cases, the theoretical extent to which these forces upset the Hardy-Weinberg equilibrium and change allele frequencies can be quantified and treated as simple mathematical elaborations of the Hardy-Weinberg equation.

A major difficulty in interpreting genotype frequencies in natural populations is that an infinite array of different combinations and intensities of these factors can explain any given set of genotype frequencies. Each of these major influences will be discussed in turn.

3. Causes of Deviations from Hardy–Weinberg Expectation

Sampling Error

Sampling error occurs because allele frequencies of the sample of gametes that form the zygotes of one generation may not be exactly representative of the allele frequencies of the parental generation that produced the gametes. If a very large number of samples of gametes are taken from this parental generation, the average allele frequencies of these samples will be close to those of the parental generation. The allele frequencies in any *single* gamete sample are likely to deviate from those of the parental generation. Any given sample will then have associated with it a certain probability of error, and the relative magnitude of this error becomes greater as the population becomes smaller. Allele frequencies may then change from one generation to the next, owing to this sampling error. This phenomenon is called random genetic drift. Some alleles will eventually be lost from the population, with a greater rate of loss being observed in small populations.

Mutation

Point mutations can change allele frequencies at a very slow rate. The majority of the new alleles that arise by mutation, even in the absence of selection, will be eliminated by sampling error. Mathematical models show that in larger populations mutation rates can alter the genetic composition of a population over long periods of time, if alleles have equivalent or nearly equivalent survival (selective) value (Kimura and Ohta, 1971).

Selection

Selection alters allele frequencies by lessening the likelihood that one or more genotypes will contribute to the next generation. The intensity of selection, or the degree to which particular genotypes make a lesser contribution to the next generation than predicted by the Hardy–Weinberg equilibrium, can be quantified as a selection coefficient. Selection coefficients indicate the relative deficiency of fitness of each genotype compared with others at the same locus. Relative fitness of a genotype is based on the proportion of alleles in the next generation that is contributed by individuals of that genotype relative to individuals of other genotypes.

Fitness can be broken down into two major components: viability and fecundity. Viability is the likelihood that an individual survives to a particular age. The most critical age is that at which the individual is capable of reproduction. The greater the chance that an individual will reach this age and persist through the reproductive period, the more offspring that individual is likely to leave. Fecundity, or reproductive output, is a second component of fitness that indicates the number of offspring produced during the life of an individual.

Migration

Population geneticists use the term migration in a different sense than it is used by ecologists. Most of the readers of this book probably think of migration as a two-way movement of an entire population of animals from one area to another, as in the seasonal migration of birds. Population geneticists use migration to describe any movement of successful breeding individuals or gametes from one population to another. Migrants therefore represent gene flow between populations. The term migration applies to all such movements, regardless of the number of migrants involved, periodicity, or the direction of their movement. Even plants are capable of migration, according to this definition, through the dispersal of seeds, pollen or spores. If two populations differ in allele frequencies, migration of individuals into one from the other will tend to cause allele frequencies in the recipient population to approach those of the donor population. This will occur even with low rates of migration. It has been estimated that a migration rate of less than one individual per generation is sufficient to keep the composition of neutral alleles in each of two populations similar (see Chapter 1 by Frankel and Chapter 3 by Allendorf).

4. Effective Population Number

The influences of sampling error (drift), selection, mutation, and migration depend on the number of individuals in the population being studied. An actual count of breeding individuals in a population yields the population number or population size. This number is not appropriate for population genetic models because such factors as unequal sex ratio, overlapping generations, generally nonrandom distribution of offspring, and nonrandom mating make the "effective" population number somewhat less than the census population number. This effective population size (abbreviated N_e) can be defined as the size of an ideal population in which genetic drift takes place at the same rate as in the actual population. The ratio of effective population size (N_e) to census population size (N) varies greatly for natural populations. The effective population number is almost always less than the census number, so that N_e/N is probably always less than 1 in natural populations and may be a small fraction in some cases (see Chapters 1, 4, 10, 15, and 23).

5. Inbreeding

One of the most applicable among the many definitions for inbreeding is the likelihood that two individuals share genes by descent. Inbreeding depression, as evidenced by loss of fitness and appearance of abnormalities in offspring resulting from breeding between close relatives, is something that most nonspecialists are aware of. Inbreeding allows the expression of deleterious alleles. Most deleterious alleles that persist in natural populations are recessive, since a dominant harmful allele would be exposed to severe selection and kept at a low frequency or removed from the population. Recessive harmful alleles may persist at low frequencies in populations because they are carried mainly in the heterozygous condition where their effects are masked by a dominant allele. This can be demonstrated by inserting sample allele frequencies into the formula for the Hardy-Weinberg equilibrium for two alleles at a single locus, with p being the frequency of the harmful recessive allele. Since $2pq$ (the frequency of heterozygotes) is very much greater than p^2 (the frequency of the homozygous recessive) for small values of p, most uncommon alleles are present in the heterozygous condition. Inbreeding increases the likelihood that two individuals heterozygous for the same harmful allele will mate and produce offspring that are homozygous for the harmful allele.

The effects of inbreeding depression may also result from the effects of decreasing heterozygosity for loci where heterozygotes have higher fitness (overdominance). If overdominant loci contribute greatly to overall fitness, then loss of heterozygosity in itself will cause a decline in fitness.

Evolutionary Genetics and Population Management

This chapter has presented a much-abbreviated discussion of the most basic genetic principles. The reader may note that there is often considerable difficulty in applying these simple concepts to natural populations and that authors of different chapters in this book vary somewhat in their views on the organization of genes within individuals and populations and on the forces that maintain or change the genetic composition of populations. These varying views reflect current controversies in genetics and evolutionary biology. Conservation biology, having evolutionary biology as one of its antecedents, has inherited many of these controversies. This book would not be a proper forum for settling these controversies, if that were even possible. Where these controversies arise, it is comforting to note that opposing viewpoints do not often result in different management recommendations. For example, conserving genetic variation is viewed as desirable by conservation geneticists who may differ among themselves on the relative importance of the various natural mechanisms that can maintain that variation. All agree that variation is needed to respond to unforeseen new selection pressures.

Although discussions of atypical gene loci and ideal populations may seem esoteric, their relevance will become more apparent with more experience in applying genetic principles to field problems. This crucial merging between theory and practice will be challenging and rewarding to field managers. The population geneticists and evolutionary biologists who join managers in addressing these problems will find unusually exciting research opportunities in some very basic and persistent problems in evolutionary genetics.

Acknowledgments

Linda Abbot and Tom Mullin reviewed this entire chapter and Jerry Coyne reviewed the later portions. I thank them for their helpful comments and suggestions, and Jonathan W. Bayless and Alfred Gardner for help with illustrations.

Isolation

PART ONE

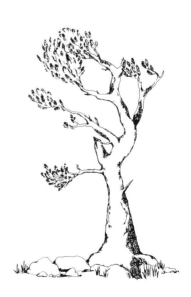

Introduction

Isolation plays a major role in promoting genetic divergence between populations and can ultimately lead to speciation. Over longer spans of geological time, isolation can result in the evolution of endemic higher taxa and faunas and floras on isolated land masses, such as the endemic radiation of land mammals during the "splendid isolation" of South America (Simpson, 1980). On the smaller scale, isolation can promote extinction as well as speciation among isolated populations. For every small isolated population that survives extreme isolation, many more probably become extinct. Genetic factors contributing to the extinction of these populations include within population inbreeding and genetic drift.

Park, reserve, and refuge populations are becoming increasingly isolated from other populations of the same species by the modification of intervening habitats. The demands made by increasing human populations are the cause of most if not all of this habitat modification. It is apparent that species that require a large area to maintain a sufficient population size or whose dispersal powers are poor will be especially impacted by this isolation.

The authors of this section of the book describe some of the mechanisms and consequences of isolation of populations. Allendorf describes the amounts and patterns of single-locus genetic divergence between populations that can be expected with given migration rates, population sizes, and selection patterns. Of special interest here is the amount of gene flow required to maintain the same alleles among semi-isolated populations.

Chesser then discusses strategies for maintaining genetic diversity in species or smaller groups of populations that are geographically subdivided so that natural gene flow can no longer occur and intervention by the manager is required to maintain overall genetic diversity and minimize inbreeding.

Liu and Godt describe examples that illustrate the special conditions under which genetic differentiation can take place over very short distances. Some of these examples also illustrate how the genetic variation in populations enables them to respond to and survive environmental changes, which often take place with extreme rapidity and intensity.

Speciation can take place without obvious divergence in visible characters. The resulting "sibling species" can cause problems to a manager who assumes that only a single species is present. Bickham discusses the divergence between sibling species and describes techniques for detecting their presence.

This section on isolation presents basic processes and problems that will be elaborated on in the other four sections of the book. Each of the other four topics—extinction, founding, merging and means of preserving diversity of populations—are either a direct or indirect consequence of isolation.

CHAPTER
3

Isolation, Gene Flow, and Genetic Differentiation Among Populations

Fred W. Allendorf

Introduction

The genetic resources of a species exist at two fundamental levels: genetic differences between individuals within a local population, and genetic differences between different local populations. Population genetics theory, developed principally by Wright (1943, 1951, 1969), has long emphasized the evolutionary importance of the genetic structure of species—that is, the distribution of genetic variation within and between populations. However, only in the last fifteen years, through the electrophoretic detection of genetic variation at many protein loci, has it been practical to describe the amounts and distribution of genetic variation in natural populations (Nei, 1975).

Efforts to preserve genetic resources must take into account the components of genetic variation, both within and between local populations. Management plans should consider the expected effects of human actions on the amount and distribution of genetic variation.

The pattern of genetic diversity in a species is largely determined by the effects of three fundamental evolutionary forces: genetic drift, migration (gene flow), and natural selection (Slatkin, 1980, 1981, 1982). We therefore need a good understanding of the effects of these three forces on genetic diversity in natural populations. The action of these forces on the distribution of genetic variation in natural populations is analyzed in this chapter. It also considers the implications of these results on the problem of maintaining genetic diversity in natural and managed populations of plants and animals. This is accomplished by using population genetics theory and a series of computer simulations.

The Model

The simulation program is an extension of that described by Allendorf and Phelps (1981). Consider a locus with two alleles in a diploid outbreeding species composed of 20 local random mating groups (demes or subpopulations), each with a population size of N. Each generation an individual has a probability m of breeding in a deme other than that of its birth. An emigrant is equally likely to migrate into any of the other demes; this is known as the "island model" of migration. Three different modes of natural selection are considered: (1) selective neutrality, (2) heterozygous advantage, and (3) opposing directional selection in different demes.

Changes in allele frequencies are simulated by using Monte Carlo simulations. Uniformly distributed (0–1) pseudorandom numbers are used to select each zygote. The first pseudorandom number determines whether the parent comes from the local deme or is an immigrant. The second number determines if the parent survives to reproductive age. The third number determines the allele transmitted by the parent. This process is then repeated for the other parent in order to form a zygote. The entire process is completed for N new zygotes for the next generation. The initial allele frequency in each deme is 0.5. The effects of mutation are assumed to be relatively small compared with those of the other three evolutionary forces and so are not included in this model.

1. Rationale for the Model

There are four primary variables that we must consider to understand the distribution of genetic diversity: population size, migration rates, and both the mode and intensity of natural selection.

Genetic drift is chance changes in allele frequencies as a result of random sampling among gametes from generation to generation (Hartl, 1980, p. 142). Because small samples are frequently not representative, genetic

CHAPTER 3 Isolation, Gene Flow, and Genetic Differentiation Among Populations   53

drift has a greater effect upon small populations. The effect of genetic drift is taken into account in this model by changes in the population size or number of individuals (N) in the 20 demes that constitute the species.

Migration is the exchange of reproductively successful individuals among demes (Hartl, 1980, p. 189). In this model, migration is equally likely to occur among all 20 demes at a rate m, as defined above. This pattern of migration (the island model) was chosen because its simplicity allows greater generality.

Natural selection is the differential success of genotypes in contributing offspring to the next generation (Hartl, 1980, p. 63). In this model natural selection occurs by differential survival probabilities. Three different patterns of natural selection are used: (1) selective neutrality, where all genotypes have equal probability of survival; (2) heterozygous advantage, where heterozygotes have a survival probability of 1 but the survival probability of both homozygotes is reduced by a selection coefficient with value s; and (3) differential directional selection, where one allele is favored by directional selection in 10 of the demes and the other allele is favored in the other 10 demes by the same intensity of selection in the opposite direction. With this mode of selection, the survival probability of the deleterious homozygous genotype is reduced by a value of t, and heterozygotes have a survival probability of $(1 - 0.5t)$ in all demes.

These three patterns of natural selection were chosen because they represent extremes in the effects of natural selection on divergence among demes. The pattern of heterozygous advantage will act to maintain a stable equilibrium at an allele frequency of 0.5 and therefore will restrict divergence among demes (Figure 1B) in comparison with the amount of divergence expected with selective neutrality (Figure 1A). Differential directional selection will have the opposite effect. That is, directional selection favoring different alleles in different demes will increase the amount of divergence among demes (Figure 1C) in comparison with that expected with selective neutrality. A comparison of the amount of divergence with these three patterns of selection for the same values of m and N will allow us to determine the potential effects of natural selection on genetic divergence among natural populations.

2. Simulations

The following four parameters were specified for each simulation: (1) the population size of (number of individuals in) each deme, N; (2) the migration rate, m; and (3) the mode and intensity of natural selection. The standard measure of divergence at individual loci, F_{ST} as defined by Wright (1943), was used to estimate the amount of allele frequency divergence among demes for each set of simulations:

$$F_{ST} = \frac{\sigma_q^2}{\bar{q}(1 - \bar{q})} \quad (1)$$

where $\bar{q}$ and σ_q^2 are the mean and variance, respectively, of allele frequencies among demes. Lower values of F_{ST} indicate less genetic divergence. Wright (1969) has shown that at equilibrium with the island model of migration and an infinite number of demes

$$F_{ST} = \frac{(1 - m)^2}{[2N - (2N - 1)(1 - m)^2]} \quad (2)$$

If m is small, this approaches the more familiar

$$F_{ST} = \frac{1}{(4Nm + 1)} \quad (3)$$

This is the ideal expected value for F_{ST} for a given population size and migration value with no selection (Table 1).

Nei and Chakravarti (1977) have shown that with a finite number of demes F_{ST} will eventually become zero because all demes will eventually

Table 1. *Simulation Results (Except Top Row) of Steady-State* F_{ST} *Values for 20 Demes and Selective Neutrality (s = 0) or Heterozygous Advantage in Which Both Homozygous Phenotypes Have a Reduction in Fitness of* s^1

s	mN						N
	0.5	1	2	5	10	25	
[Expected]	0.3333	0.2000	0.1111	0.0476	0.0244	0.0099	
0.00	0.3070	0.2043	0.1245	0.0418	0.0198	–	25
	0.3350	0.1826	0.1077	0.0484	0.0264	0.0120	50
	0.3216	0.1884	0.1061	0.0437	0.0251	0.0095	100
0.01	0.2826	0.1640	0.0666	0.0499	0.0220	–	25
	0.2431	0.1534	0.0824	0.0406	0.0232	0.0117	50
	0.1782	0.1236	0.0930	0.0383	0.0355	0.0109	100
0.05	0.1930	0.1259	0.0714	0.0441	0.0237	–	25
	0.1327	0.1072	0.0620	0.0341	0.0238	0.0092	50
	0.0827	0.0661	0.0432	0.0242	0.0185	0.0110	100
0.10	0.1217	0.1039	0.0533	0.0429	0.0216	–	25
	0.0938	0.0763	0.0503	0.0307	0.0207	0.0087	50
	0.0410	0.0290	0.0317	0.0217	0.0103	0.0070	100

[1] The three listed values for each intensity of s represent deme sizes (N) of 25, 50, and 100 individuals, reading downward. Each value is the mean of 20 repeats. Expected values were calculated using the formula $F_{ST} = 1/(4mN + 1)$.

become fixed for the same allele. I ignore this effect, however, assuming that in nature some migrants will be entering our "closed" system from the outside. In the finite deme model, F_{ST} will reach some steady-state decay distribution (Nei et al., 1977). The decay to $F_{ST} = 0$ will be extremely slow in these simulations because the total population size of all the demes together is 500 to 2000 individuals. The rate of approach to steady-state values depends upon both migration rate (m) and number of individuals in a deme (N) (Wright, 1951). Values of F_{ST} were estimated in these simulations after a sufficient number of generations to ensure that steady-state values had been reached; this is described in more detail in Allendorf and Phelps (1981).

Results and Discussion

1. Selective Neutrality

In the absence of natural selection, the amount of genetic divergence among demes is a function of the absolute number of migrants exchanged (mN) and not the proportion of exchange among demes (m). Thus, a given number of migrants will result in the same amount of allele frequency divergence regardless of the population size of the local demes (Table 1). For example, we expect to find the same amount of genetic divergence among demes of size 1000 with an m of 2.5% as among demes of size 50 with an m of 50%. The dependence of divergence on the number of migrants, rather than on the proportion of migrants, may at first seem counter-intuitive. The rate of divergence, however, results from the opposing forces of migration and genetic drift. The larger the demes are, the slower they diverge through drift; thus, proportionally fewer migrants are necessary to counteract the effects of drift. Small demes diverge rapidly through drift, and thus proportionally more migrants are required to counteract drift.

Low amounts of exchange (approximately one individual per generation) will maintain the presence of the same alleles in all demes (Kimura and Ohta, 1971, p. 122). Nevertheless, in contrast to what has been suggested by some authors (Spieth, 1974; Frankel and Soulé, 1981, p. 128), divergence in allele frequencies is often present even when there is a large amount of exchange among demes. For example, an exchange rate of 10 individuals per generation usually results in significant divergence among the 20 demes in the present model (Allendorf and Phelps, 1981). Thus, low amounts of exchange will maintain "qualitative" similarity among demes, but large amounts of exchange are needed to maintain "quantitative" similarity among them.

2. Heterozygous Advantage

As expected, this pattern of selection acts to maintain similar allele frequencies in different demes (Figure 1). That is, for a given amount of genetic exchange, there is less allelic divergence among demes in comparison with the neutral model. The amount of divergence is not simply a function of the absolute number of migrants with this model of selection. There is generally less divergence for a given number of migrants with increasing population size. This is because the effect of genetic drift is less in larger populations and natural selection is more effective in maintaining the same equilibrium frequency in all demes.

This effect can be seen by considering the two possible extremes. With a deterministic model, assuming N is infinite, any value of s greater than zero will maintain an allele frequency of 0.5 in every deme, even in the complete absence of genetic exchange among demes. In the other extreme, if s is 1.0 (homozygous lethality), then every deme will maintain an allele frequency of 0.5, irrespective of m and N. Thus, heterozygous advantage will significantly reduce the amount of genetic divergence among demes when either there is strong selection or there are large demes.

Heterozygous advantage is effective in reducing divergence among demes under a wide range of conditions (Table 1). For example, the amount of divergence with one migrant individual per generation and $N = 100$ with a 5% heterozygous advantage is reduced by approximately 67% in comparison with the ideal expected ($F_{ST} = 0.0661$ versus 0.2000). The effect on the distribution of allele frequencies of this intensity of selection can be seen in Figure 1.

Figures 2 and 3 display the interaction among the intensity of selection, migration rate, and deme size. Figure 2A shows the effect of variable mN with an s of 0.05 on all three deme sizes. There is a significant reduction in divergence for all three deme sizes for an mN of 2 or less. With an mN of 5, however, there is very little effect with an N of 25. And with an mN of 25, there is no detectable effect even with an N of 100.

Figure 3A shows the effects of variable selection intensity with an mN of 5 on all three deme sizes. With an N of 25, not even an s of 0.10 has an appreciable effect on divergence. With an N of 100, there is a reduction in divergence even with weak selection, $s = 0.01$.

These results can be generalized somewhat by comparing the relative magnitudes of s and m. In general, heterozygous advantage has no detectable effect on reducing divergence among demes when the quantity s/m is less than 1. The effectiveness of this pattern of selection increases as this quantity becomes increasingly greater than 1. For example, when mN is 1, m is equal to 0.04, 0.02, and 0.01, for deme sizes of 25, 50, and 100, respectively. Therefore, with deme sizes of 25 and an mN of 1, s must be approximately 0.05 or greater to be effective in promoting divergence in allele frequencies between demes.

CHAPTER 3 Isolation, Gene Flow, and Genetic Differentiation Among Populations   57

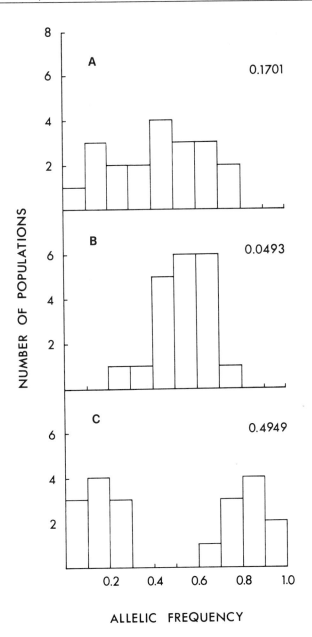

Figure 1. Simulation results showing distribution of allele frequencies in 20 demes with different modes of natural selection. In all cases N = 100 and m = 0.01 (mN = 1): (a) selective neutrality; (b) heterozygous advantage with s = 0.05; and (c) differential directional selection with t = 0.05. Numbers in the upper right corner of each graph represent the F_{ST} value for the particular simulation portrayed.

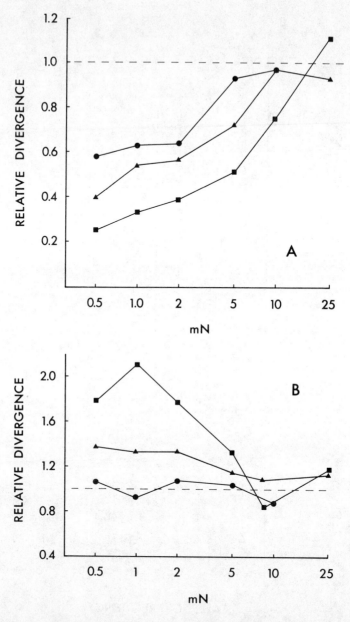

Figure 2. Simulation result showing the effects of variable mN on the relative amount of divergence with natural selection. The relative divergence is the observed F_{ST} divided by the F_{ST} expected with selective neutrality. Three different deme sizes (N) are shown: 25 (circles), 50 (triangles), and 100 (squares). (A) Heterozygous advantage, s = 0.05; (B) differential directional selection, t = 0.05.

CHAPTER 3 Isolation, Gene Flow, and Genetic Differentiation Among Populations

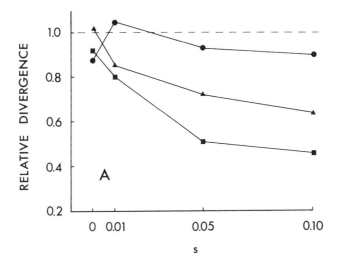

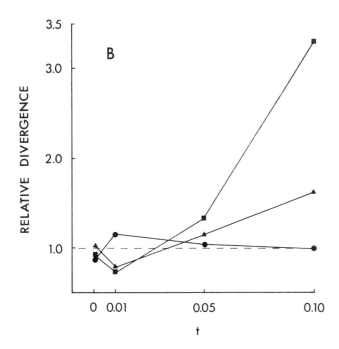

Figure 3. Simulation results showing the effects of variable intensity of selection on the relative amount of divergence with an mN of 5. (See Figure 2 for explanation of symbols.) (A) Heterozygous advantage; (B) differential directional selection.

Table 2. *Simulation Results (Except Top Row) of Steady-State F_{ST} values for 20 Demes with Differential Directional Selection*[1]

t	mN						N
	0.5	1	2	5	10	25	
[Expected]	0.3333	0.2000	0.1111	0.0476	0.0244	0.0099	
0.01	0.3343	0.1703	0.1070	0.0556	0.0220	–	25
	0.2979	0.1192	0.1000	0.0381	0.0256	0.0099	50
	0.2997	0.1850	0.1146	0.0354	0.0229	0.0105	100
0.05	0.3560	0.1857	0.1204	0.0497	0.0217	–	25
	0.4618	0.2679	0.1489	0.0550	0.0265	0.0113	50
	0.5950	0.4230	0.1982	0.0632	0.0207	0.0118	100
0.10	0.4700	0.2446	0.1632	0.0473	0.0289	–	25
	0.6242	0.3653	0.2611	0.0771	0.0356	0.0128	50
	0.8054	0.6575	0.4432	0.1589	0.0632	0.0193	100

[1] One homozygous genotype has a reduction in fitness of t in 10 demes; the other homozygous genotype has the same reduction in fitness in the other 10 demes. Heterozygotes have a reduction in fitness of one-half t in all demes. The three listed values for each intensity of t represent deme sizes (N) of 25, 50, and 100 individuals, reading downward. Each value is the mean of 20 repeats. Expected values are the same as in Table 1.

3. Differential Directional Selection

This pattern of selection acts to maintain allele frequency divergence among demes so that large differences in allele frequencies can be maintained even with extensive genetic exchange. This pattern of selection is more effective in promoting divergence with larger demes (Table 2). Figures 2B and 3B show the effectiveness of this mode of selection in increasing divergence under different conditions. For example, a selection intensity of $t = 0.05$ has no detectable effect with an N of 25 or with an mN of 10, irrespective of N (Figure 2B). As with heterozygous advantage, these results also can be generalized by using the quantity s/m. A detectable increase in divergence occurs only when this quantity is greater than 1.

Implications

1. Natural Populations

How can an analysis of the amount and pattern of divergence in allele frequencies among natural populations be used to estimate the amount of

CHAPTER 3 Isolation, Gene Flow, and Genetic Differentiation Among Populations

gene flow occurring between demes? The present model can be used to estimate gene flow only if its assumptions are realistic. One primary assumption is the island model pattern of gene flow. This model assumes that there are equal amounts of immigration and emigration for all demes and that there is no structure to the pattern of gene flow. That is, an emigrant from a particular deme is equally likely to migrate into any of the other demes.

The second primary assumption is the mode and intensity of natural selection. It has been notoriously difficult to estimate the effects of natural selection on genotypes at individual loci in natural populations (Lewontin, 1974, p. 236). In fact, there is still much debate whether the allelic variation of proteins detected with electrophoresis results from the action of natural selection or from selectively neutral mutations and genetic drift.

We have considered only three basic possible modes of natural selection. The simplest mode of selection is neutrality; this model is attractive for several reasons. *First*, it is falsifiable. The breeding structure (migration rates and local population sizes) is the same for all loci (Cavalli-Sforza, 1966). With selective neutrality, all loci are expected to show approximately the same pattern of divergence in allele frequencies. Statistical tests have been developed to determine if selective neutrality can account for the observed distributions of allelic frequencies at many loci in the same populations (Lewontin and Krakauer, 1973; see also Nei and Maruyama, 1975; and Robertson, 1975). Natural selection, however, is not falsifiable. There are an infinite number of models of natural selection that can be invoked to explain any set of data.

A *second* attractive feature of selective neutrality is its simplicity. In order to apply a model including selective differences among genotypes, it is necessary to make many assumptions about the mode and intensity of selection in each deme.

Thus, in analyzing divergence in allele frequencies in natural populations it is appropriate to consider selective neutrality as the null hypothesis. This hypothesis should be rejected only if there is evidence of significant differences in the distribution of allele frequencies at different loci or some other evidence for the action of natural selection.

I have applied this analysis to data of Ryman and Ståhl (1981) that describe divergence in allele frequencies at three loci among 5 local demes of brown trout (*Salmo trutta*) spawning in tributary streams of Lake Lulejaure, a large Swedish mountain lake (Table 3). There is significant allele frequency heterogeneity at all three loci. There is no geographical pattern to the distribution of allele frequencies; that is, geographically close demes are not more similar than distant demes. Thus, the island model of migration seems appropriate for these demes. The mean F_{ST} for the three loci is 0.069. From formula 3, it is estimated that 3.4 individuals migrate among demes each generation. Are the differences in divergence for the three loci compatible with selective neutrality? I have tested this assumption using

Table 3. *Allele Frequency Divergence among 5 Demes of Brown Trout Spawning in Tributaries of Lake Lulejaure, Sweden*[1,2]

Sample	Agp-2	Ldh-5	Mdh-2
Ruoktojokok	0.962	0.691	0.938
Såkasjokk (upper)	0.827	0.877	0.865
Såkasjokk (lower)	1.000	0.800	0.865
Tjegnaljokk	0.860	0.529	0.980
Appakisjokk	0.988	0.835	1.000
F_{ST}	0.074	0.082	0.052

[1] From Ryman and Ståhl (1981).
[2] There are two alleles at each locus; the frequency presented is that of the most common allele.

repeated simulations with an mN value of 3.4 among 10 demes, assuming an N of 50 (Figure 4). The patterns of divergence of all three loci are compatible with this breeding structure; 41% of the simulated F_{ST} values are less than the F_{ST} for *Mdh-2*, and 17% are greater than the F_{ST} for *Ldh-5*.

Thus, the observed patterns of divergence in allele frequencies are compatible with an average of 3.4 migrants per generation with selective

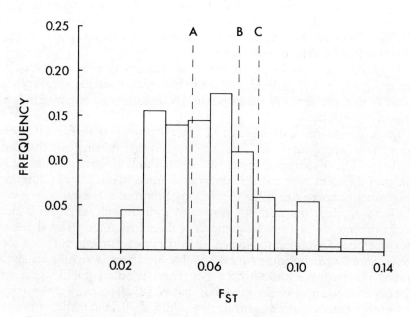

Figure 4. *Simulation results showing distribution of* F_{ST} *values in 200 repeats for* N = 50 *and* m = 0.068 *with 10 demes. The dotted lines show the* F_{ST} *values actually estimated by Ryman and Ståhl (1981) for three loci in brown trout from tributaries to Lake Lulejaure.*

neutrality under an island model of migration. This estimate of 3.4 migrants will be conservative if there is some tendency for increased exchange between adjacent demes (isolation by distance). In that case, a migrant is less effective in retarding divergence because it has a higher probability of reproducing in a deme that is similar to the one from which it emigrated (Chapter 4, and Chapter 19).

The importance of examining many polymorphic loci can be seen from this example. A single estimate of F_{ST} would be compatible with an extremely wide range of migration rates. As each additional locus and estimate of F_{ST} are added, the possible range of compatible migration rates is narrowed.

2. Managed Populations

These results also can be applied to design appropriate breeding structure for populations that are controlled by the action of humans—for example, populations in nature reserves, zoos, or botanical gardens. Franklin (1980) has argued that an effective population size of at least 500 is needed to preserve useful genetic variation within an outcrossing species. How should this population be subdivided? We should try to maximize both the adaptedness of individual demes to local conditions and the evolutionary potential of the species. To do this we should try to maintain a high amount of genetic variation both between individuals within each deme and between different demes.

We are likely to be limited by habitat and financial restraints as to the number and size of demes to be maintained. Therefore, the major aspect of the breeding structure that can be easily controlled is the amount of exchange among demes. There are two ways to approach this question: either we can base our choice on the particular genetic characteristics of individual species, or we can try to develop a general rule of thumb (Frankel and Soulé, 1981, p. 90). I am in complete agreement with Franklin (1980) that our theoretical and empirical understanding of the action of natural selection is currently insufficient to serve as a basis for application to individual species.

We should therefore identify the "ideal" amount of exchange among demes that can be generally applicable. An average exchange of one reproductively successful migrant individual per individual ($mN = 1$) among demes is preferred for at least two reasons. *First*, this amount of exchange is sufficient to avoid the loss of alleles in demes due to genetic drift in the absence of natural selection. Such losses occur with F_{ST} values of 0.33 or larger ($mN = 0.5$) (Kimura and Ohta, 1971, p. 122). *Second*, this amount of exchange is not sufficient to influence the frequencies of alleles in demes under natural selection. Thus, one migrant individual among demes per generation would ensure that all the genetic variation of a species is present in all local demes, but would allow genetic differentiation among demes in response to local selective pressures.

Summary

The interaction among genetic drift, gene flow, and natural selection in determining patterns of allele frequency divergence among a series of semi-isolated local populations (demes) is examined by using population genetics theory and a series of computer simulations. With selective neutrality, the amount of divergence among demes is a function of the number of migrants, and not the proportion of individuals exchanged. Therefore, some knowledge of effective local population sizes is needed to estimate the degree of reproductive isolation from allele frequency data. In addition, contrary to some previous claims, significant allele frequency divergence is expected to be present even when there is substantial exchange among populations. An average exchange of one migrant individual per generation will maintain the same alleles in all demes, but much larger amounts of exchange are required to maintain similar allele frequencies among demes.

Heterozygous advantage reduces the divergence among demes; that is, for a given amount of gene flow, there is less divergence in comparison with the absence of natural selection. The amount of divergence is not simply a function of the number of migrants with this mode of selection, since natural selection is more effective in larger demes because of reduced genetic drift. Heterozygous advantage significantly reduces divergence in allele frequencies among genes when there is strong selection or when there are large demes. In general, heterozygous advantage will be effective in reducing divergence when the selection coefficient is greater than the migration rate.

Differential directional selection acts to maintain divergence in allele frequencies among demes even when there is a large amount of gene flow. As with heterozygous advantage, this type of selection is effective when the selection coefficient is greater than the migration rate, although its effect on divergence is opposite that of heterozygous advantage. Thus, for a given number of migrant individuals, this mode of selection is more effective in promoting divergence with larger demes.

These results are considered with respect to the design of appropriate breeding structures for populations that are controlled by human actions. An average exchange rate of one reproductively successful migrant individual among demes per generation is desirable. This amount of exchange is sufficient to avoid the loss of alleles in demes due to genetic drift, but will allow the allele frequency in different demes to respond to local selective pressures.

Acknowledgments

I thank Nils Ryman, Gunnar Ståhl, and Steve Phelps for many helpful discussions on this topic and Kathy Knudsen for making the figures. This work was supported in part by a grant from the Great Lakes Fishery Commission.

CHAPTER 4

Isolation by Distance: Relationship to the Management of Genetic Resources

Ronald K. Chesser

Introduction

There is an ever-growing body of literature indicating the noncontinuous distribution of our biological resources. Genetic analyses of numerous species document the lack of continuous gene flow among populations separated by short geographic distances (Selander, 1970; Manlove et al., 1976; Ryman et al., 1980; Chesser, 1982; Chesser et al., 1982a, 1982b) as well as large ones (Smith et al., 1978b; Ryman et al., 1980; Chesser, 1981). Wright (1943) first introduced the concept of isolation by distance: populations within a species may diverge by way of random genetic drift due to the paucity of genetic exchange between populations imposed by geographic distance alone. Other factors that affect effective dispersal rates among populations may also act to reduce gene flow. Hence, barriers to continuous dispersal may be imposed by the following three factors: (1) geographic distance—distance alone may reduce the likelihood of movement of sufficient numbers of animals, seeds, or pollen among populations to prevent random events from affecting genetic divergence of populations (Wright, 1943; Mayr, 1970); (2) ecological distance—the presence of different habitat types between populations and/or physical barriers imposed by geographic formations (e.g., rivers, mountains) may inhibit gene flow (Mayr, 1970); and (3) behavioral distance—social structuring of populations may restrict successful entry of dispersing individuals into existing hierarchies even though movement among the units would otherwise be achieved easily (Selander, 1970; Daly, 1981; Chesser, 1982; Chesser et al., 1982a, 1982b).

The combination of these factors that limit gene flow among populations may have dramatic effects on the distribution of genetic resources. Isolated populations may accrue great genetic differences from others of the same species. Without recruitment of "new" genetic material polymorphism may be dramatically reduced (by loss of alleles) and inbreeding (breeding among related individuals) may increase. These conditions may be particularly pronounced in species whose distributions are limited to small management areas and/or zoos (Ralls et al., 1979; Chesser et al., 1980). Movement of individuals among these isolated areas may be accomplished only by sweepstakes dispersal (highly unlikely, but possible) or by active transportation by man.

Because inbreeding in normally outbreeding species is usually associated with a reduction in fitness and the loss of alleles, which prevents the reversal of inbreeding depression (Lerner, 1954; Falconer, 1960), management programs have become particularly concerned with the maintenance of genetic variation in captive and restricted biological resources (Lang, 1977; Flesness, 1977; Denniston, 1978; Foose, 1977; Chesser et al., 1980). Breeding schemes that involve substantially isolated and small populations (as in most zoos) must be devised to minimize the catastrophic effects of the loss of genetic variability and inbreeding depression. The purpose of this chapter is to present recommended breeding strategies for isolated populations of normally outbreeding species in which genetic exchange is limited and to demonstrate the outcomes of various breeding strategies on the maintenance of genetic variability and of viable individuals.

Maintenance of Genetic Variation

The primary objective of breeding programs that attempt to maintain genetic variability must be the preservation of existing alleles. There are two methods for maintaining these alleles: (1) maintaining a high frequency of heterozygous individuals (each individual carries two forms of the variant gene: See Foose, 1977; Flesness, 1977); (2) maintaining isolated subunits of the population that are homozygous for an allele and other subunits that are homozygous for alternative alleles (Chesser et al., 1980).

Maintaining a highly heterozygous population for long periods of time may not be possible without the influence of individuals being recruited into the population. In instances where populations are small, random loss of alleles by genetic drift may profoundly reduce both the number of alleles and heterozygosity. Populations with smaller effective population sizes (N_e; roughly equivalent to the number of breeding individuals in a population) will lose genetic variability faster than those with large population sizes.

This acceleration of loss of genetic polymorphisms with small breeding populations has caused some authors (e.g., Foose, 1977; Flesness, 1977; Denniston, 1978) to devise methods of increasing the effective population size within the isolated units. The predominant suggested method of increasing N_e has been to equalize the numbers of offspring in all matings (*op. cit.*). Equalizing family sizes reduces the variance in offspring numbers and thus increases the effective population size (Kimura and Crow, 1963). However, the increase of N_e may not be substantial enough to significantly curb the loss of genetic variation by genetic drift. Thus, the tendency of individual populations to lose their genetic variability may not be overcome by such a breeding program, and an exclusive focus on population size can have disastrous results for the management of genetic resources.

It is clear that small, isolated populations such as those common in zoological parks and management areas can become fixed for a single allele at a large proportion of their loci. Extreme isolation by distance, whether by geographic, ecological, or behavioral factors, will prevent the influx of new alleles, and the end result can be a highly inbred population with low viability and/or fecundity. However, species in natural environments usually comprise a number of populations that are relatively isolated from one another (Wright, 1978; Smith et al., 1978b). Although each of the populations approaches fixation for a single allele at each locus, each population may be becoming fixed for different alleles. Thus, when the entire array of populations is examined, all or most of the original alleles may be present; polymorphism is high over all populations, but heterozygosity is low within any given population.

Such a breeding program of maintaining high genetic variance between populations while genetic variance within populations is low was proposed by Chesser et al. (1980). They proposed that propagation of captive species in numerous isolated populations should be considered for the following reasons: (1) Genetic polymorphism is maintained by way of high homozygosity for alternative alleles in the numerous populations; (2) whereas heterozygosity is low owing to the isolated nature of the populations, high levels of heterozygosity can be regained by cross-mating among the individuals from different populations; (3) any loss of fitness due to inbreeding within the isolated populations can be compensated for by the outcrossing process; (4) outcrossing among inbred populations produces a temporary heterotic condition (greater viability for heterozygous individuals), which can be used to maintain genetic variability; (5) division of a captive group into numerous populations would lessen the likelihood of extinction by way of a local catastrophe or epidemic; and (6) this type of breeding program is economically feasible because most captive populations already consist of numerous isolated breeding units and the movement of animals among the isolated breeding units would not need to be continuous but rather could occur only when necessary to reverse inbreeding trends.

The breeding strategy proposed by Chesser et al. (1980) is probably very similar to that observed among natural populations of many species that exhibit isolation by geographic, ecological, and/or behavioral distance (Wright, 1978, 1980). In fact, exposure of the various populations to a heterogeneous environment may serve to preserve genetic variation (Levene, 1953; Deakin, 1966, 1968; Prout, 1968; Christiansen, 1974, 1975; Wright, 1931, 1980; Karlin and Campbell, 1980). However, as with any breeding scheme, the procedures proposed by Chesser et al. (1980) do have limitations. In their original presentation, they recognized that effective population sizes should be at least ten and that numerous populations, rather than few, would enhance genetic preservation. However, one problem not originally recognized is that exchange of individuals among a limited number of small populations would eventually result in a high degree of relatedness among *all* individuals in *all* populations. Thus, after many generations of cross-mating, the inbreeding coefficients could not be reversed easily and loss of alleles would again continue.

Although the breeding scheme of Chesser et al. (1980) would suffice for a large number of generations, additional measures should be taken to slow the rate of increase in relatedness among the individuals. The rate of increase in relatedness could be minimized by disallowing symmetrical exchange of individuals among the populations. The arrangement of populations into intermittently interbreeding units, or neighborhoods, within which exchange of individuals among populations is greater than that between neighborhoods, would considerably slow the approach to overall relatedness among individuals and better maintain genetic variation (Christiansen, 1974). Neighborhoods could be established with offspring from the original breeding colonies or by arbitrary divisions of the populations into groups, each such group or neighborhood containing approximately equal numbers of populations.

Exchange of individuals among the populations within the neighborhoods would follow the recommendations of Chesser et al. (1980), and the rate of exchange between neighborhoods probably should approach the reciprocal of the number of populations within neighborhoods multiplied by the exchange rate among populations within each neighborhood (unpublished data; manuscript in preparation). For example, if each neighborhood contained ten populations, then the exchange rate between neighborhoods should be approximately one-tenth the exchange rate among populations within each of these neighborhoods. As the population sizes grow, new neighborhoods should be established. Of course, after a long period of time neighborhood boundaries should be randomly reorganized among the total available populations to lessen the relatedness among the available mates. Limitations of population sizes and the enhancing effect of numerous populations are described by Chesser et al. (1980).

Dispersal and Inbreeding within Populations

The necessity of exchange of unrelated individuals among the "semi-isolated" populations (isolated for some period of time with subsequent influx of dispersing individuals) is to alleviate the inbreeding depression of fitness of individuals. The rate of exchange among the managed populations is dependent upon the rate of accumulation of inbreeding and thus is highly dependent upon the effective population size (Chesser et al., 1980). However, another important facet of the alleviation of inbreeding depression is the type of mating system each of these typically outbreeding species displays. To investigate the effects of different mating systems (polygyny and monogamy) and dispersal rates on inbreeding coefficients within semi-isolated populations, I calculated effective population sizes (N_e) with various dispersal distances (σ), using the neighborhood effective size equation (Hartl, 1981, p. 77). The neighborhood size depends upon the number of breeding individuals (δ) in each management unit (e.g., zoological park or game park) and the amount of dispersal between an animal's birthplace and the location where they produce their offspring (σ). Where the distance of dispersal from the birthplace follows a normal (bell-shaped) distribution, the quantity σ^2 will show that 39% of all animals reproduce within a radius of σ centered at their birthplace, 87% reproduce within a circle of radius of 2σ, and 99% reproduce within a circle of radius 3σ. The neighborhood effective size is calculated by $N_e = 4\pi\delta\sigma^2$ ($\pi = 3.14159$). The inbreeding coefficient or accumulation of inbreeding (F) is dependent upon N_e and is approximated by

$$F_t = \frac{1}{2N_e} + \left(1 - \frac{1}{2N_e}\right) F_{t-1}$$

(Crow and Kimura, 1970), where t represents the breeding period.

Models were generated with the following variables: (1) the number of animals in each management unit (values of 10 and 20 were used); (2) the exchange rate among the units as measured by σ^2 (values of 0.33 to 2.0 were used); (3) the type of mating (monogamous and polygynous); and (4) the number of females in each harem (for polygynous mating only; values of 2 and 5 were used).

Several conditions are assumed to be present within the populations, including (1) constant population size, (2) equal sex ratio, (3) equal probability of mating for each individual (for monogamy only), and (4) constant dispersal rate per generation. Although these assumptions may not be truly representative of biological populations, the results of the models provide meaningful information regarding the nature of inbreeding with different patterns of mating and dispersal.

Inbreeding within the populations increased even with relatively high rates of exchange of animals among the units (Figures 1 and 2). The accumulation of inbreeding depression was more rapid for the polygynous species than for those that were monogamous. Larger harem sizes cause a greater disparity in the number of breeding males and thus confer larger inbreeding rates (Figure 2). Greater exchange rates increase N_e by increasing the area from which animals can draw available mates. Exchange of animals among the units slows inbreeding accumulation but does not alleviate the probable loss of fitness indefinitely.

The above model demonstrates that inbreeding increases with subpopulations of a constant size (as in a single neighborhood). Hence a constant rate of exchange among the management units serves only to slow

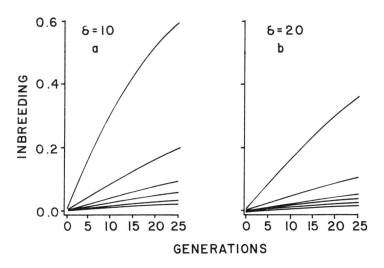

Figure 1. *The accumulation of inbreeding (F) for monogamous matings within populations with different amounts of movement between management areas over time. Graphs are presented for population sizes of 10 (a) and 20 (b). The dispersal distances (σ) represented by the lines from top to bottom are 0.33, 0.66, 0.99, 1.32, 1.65, and 2.0. Additional details of the model are given in the text.*

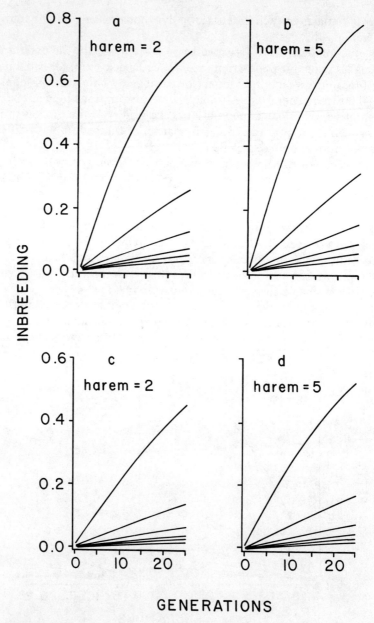

Figure 2. *The accumulation of inbreeding (F) for polygynous matings within populations with different amounts of movements between management areas over time. Harem sizes of 2 and 5 females were used for population sizes of 10 (a and b) and 20 (c and d). The lines from top to bottom represent dispersal distances (σ) of 0.33, 0.66, 0.99, 1.32, 1.65, and 2.0. Additional details of the model are given in the text.*

the rate of inbreeding. Additionally, it does not provide insight into the alleviation of inbreeding within the populations. However, the model's insufficiencies do illustrate the importance of maintaining separate reservoirs of nonrelated individuals (separate neighborhoods). A more effective approach would be to periodically promote the exchange of animals among previously isolated units.

Alleviation of Inbreeding Depression

To investigate the effects of different breeding types and dispersal on the alleviation of inbreeding coefficients within semi-isolated populations, I derived probabilities of occurrence of mating types and their associated inbreeding statistics (see Appendix 6). Assumptions of this model are the same as those of the previous model. The parameters varied were also identical, with the following exceptions: (1) population size was excluded, since this model uses the proportion of the population that is replaced by immigrants and the proportion of the different mating types (hence, population size becomes irrelevant); (2) immigrant males were always considered successful in polygynous matings.

The increase in the inbreeding coefficient (accumulation of inbreeding depression of fitness) for matings among *consanguineous* individuals is approximately

$$F_{xy} = F'_{xy} + (1 - F'_{xy}) F$$

where F'_{xy} is the coefficient of consanguinity between individuals x and y had there been no previous inbreeding. Therefore, F'_{xy} is equal to $0.25 + 0.75F$ between siblings. The above equation is virtually identical to that used to depict the increase in the proportions of heterozygosity for the offspring of consanguineous matings (Jacquard, 1974, p. 223), which is a primary aspect of the inbreeding coefficient (Wright, 1922a). Since the coefficient of consanguinity is identical to the inbreeding coefficient (Falconer, 1960, p. 89), the equation can be simplified to $F = 2F - F^2$ for consanguineous matings. Inbreeding coefficients of offspring produced by matings of native offspring with those produced by immigrant–native matings resulted in F values of $0.5F$; these animals were called relatives (half of the alleles fixed by common descent were diluted by the immigrant–native matings). Matings among relatives yielded F values of $0.25F$. The proportions of each type of mating were calculated, and the weighted mean inbreeding coefficient was calculated for each generation.

Results of the models involving monogamous matings (both sexes dis-

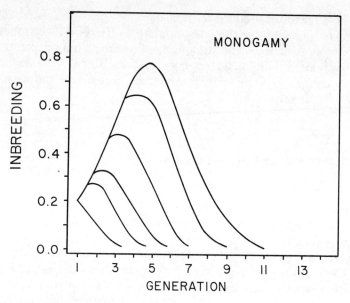

Figure 3. *Rate of alleviation of inbreeding with monogamous mating and either sex dispersing. The curves on the graph (from top to bottom) represent dispersal rates (σ^2) of 1%, 2%, 5%, 10%, 20%, and 50%.*

persing) are depicted in Figure 3. It is clear from these results that, when dispersal rates are low in relation to the population size, inbreeding may continue to increase rapidly owing to a high rate of consanguineous mating. Hence, in monogamous species, the influx of a relatively small percentage of individuals will not satisfactorily reduce the accumulation of inbreeding, and severe depression of fitness of many individuals may be inevitable. Inbreeding increases because of the slow diffusion of unrelated alleles into the population, while mating among kin only gradually decreases. For rapid alleviation of inbreeding depression in semi-isolated populations with monogamous breeding systems, the number of immigrants must be high relative to the population size. Populations that have a high degree of mating among relatives and that are separated by extreme isolation by distance may have little chance of overcoming the disastrous effects of fitness depression unless man-assisted exchange of individuals is enacted.

Polygynous mating systems are much more effective in alleviating inbreeding depression than is monogamy (Figure 4). The reversal of inbreeding trends was particularly rapid when the harem sizes were large and/or when dispersal rates were high. Immigrant, unrelated males controlling large harems produced many offspring and quickly reduced inbreeding values. Thus, in semi-isolated populations that have attained

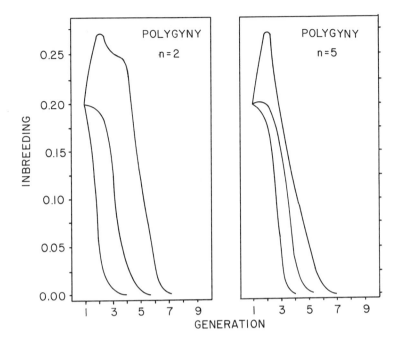

Figure 4. *Rate of alleviation of inbreeding with polygamous mating and male dispersal. Harem sizes (n) were 2 and 5. Curves on the graphs (from top to bottom) represent dispersal rates (σ^2) of 5%, 10%, and 20% for n = 2, and 2%, 10%, and 20% for n = 5.*

significantly large inbreeding coefficients, few dispersing males would be necessary to quickly counteract any fitness deficits. The maintenance of segregated neighborhoods to reserve a pool of unrelated individuals for subsequent cross-mating is therefore of primary importance to the long-term genetic variability and viability of populations.

There are distinct advantages and disadvantages to both monogamy and polygyny. Whereas inbreeding is alleviated slowly by the influx of monogamous individuals, effective population size is potentially at a maximum (if sex ratios are equal and with no variance in offspring numbers). Because of the rate of loss of alleles is proportional to the effective population size (Figures 1 and 2), with lower population sizes losing variability quickly, a system of monogamy may be superior to polygyny for the maintenance of genetic variation. The loss of genetic polymorphism in polygynous systems is largely due to the loss of "unique" alleles present in nonmating males. However, for species that regularly undergo cycles of inbreeding and outbreeding, a high degree of relatedness among individuals within the breeding units will minimize the loss of genetic material. Highly related individuals share alleles at a large proportion of their loci;

hence, breeding of any member within the group may transfer most of the same alleles carried by the "relatives" to future generations even if the relatives do not breed (inclusive fitness: see Bengtsson, 1978).

Recommendations

In considering the breeding schemes for maintenance of genetic polymorphisms proposed by Chesser et al. (1980) or the neighborhood segregation presented here, the type of breeding system and dispersal rates acquire primary importance. Management programs for monogamous species obviously must differ from those for polygynous species, and the periodicity and amount of exchange of animals must be considered separately for the different types. Equal exchange of animals among all management units is effective in slowing the accumulation of inbreeding, but eventual depression of fitness is likely. Segregation of the population into semi-isolated subunits (neighborhoods) where greater exchange of animals occurs within each neighborhood than between neighborhoods ensures the maintenance of unrelated individuals and the ability to alleviate inbreeding depression.

If inbreeding has accumulated in a population to a critical point, then measures should be taken to alleviate the inbreeding as quickly as possible. In most captive populations, isolation by geographic and/or ecological distance is sufficiently complete to prevent adequate natural dispersal for relieving the inertia of inbreeding accumulation. It is clear that man's assistance may be necessary to compensate for the incapabilities of many organisms to traverse these artificial distance barriers. Manipulated "matchmaking" efforts may also be necessary in some instances to assure participation of the majority of genetic entities (for example, removing continuously dominant males to allow subordinates to breed) in efforts to maintain continued representation of all polymorphisms.

The predominant foe of the maintenance of genetic variation is random genetic drift. Some studies advocate merely the slowing of the drift process by increasing effective population size. The goal of the proposals presented in this chapter is to make the drift process assist in the maintenance of genetic variation. This may be achieved by allowing the unavoidable decline of variation within each population to proceed while maximizing the stochastic accumulation of genetic variation among populations. We cannot, however, proceed solely on the basis of chance but must fully understand the odds associated with our maneuvers, for we are dealing with dwindling and unrenewable resources.

Summary

Variation for most natural populations is organized into rather discrete units as a result of either social or environmental structuring. This statement holds true even for large, highly vagile animals such as moose, white-tailed deer, and red deer. This type of isolation by distance is important to the maintenance of genetic variation as long as the isolation is not complete. The most effective model for the maintenance of genetic variation involving isolation by distance is that of interbreeding demes within a series of "neighborhoods." The common practice of confining populations to management areas is seen as the extreme case of isolation by distance. This practice may serve to enhance the maintenance of the species' genetic variation but can be detrimental to the fitness of the individual populations. The means of alleviating these detrimental effects on both monogamous and polygamous species are described and compared.

CHAPTER 5

The Differentiation of Populations over Short Distances

Edwin H. Liu
Mary Jo W. Godt

Introduction

The differentiation of populations on a microgeographic scale is now a well-recognized phenomenon (see Table 1). Adjacent populations that differ in their morphological, physiological, and genetic characteristics have been documented (e.g., Jain and Bradshaw, 1966; Snaydon and Davies, 1982; Silander, 1979). Although some of the most striking examples of microhabitat differentiation arise as a result of human activities, conditions in the natural environment can structure populations in a similar manner. Wise decisions concerning the management of natural populations require an understanding of the forces that shape and maintain the structure of these populations.

Many of the activities of our society affect the genetic structure and the dynamics of natural populations. For instance, the division of a population into smaller, semi-isolated units is a frequent result of man's construction activities. Roads, dams, fences, and buildings all can serve to isolate adjacent populations. These structures may restrict (and in some cases even eliminate) gene flow between adjacent populations. Oftentimes the consequence of such restricted gene flow is that the separate units diverge genetically. Generally speaking, however, this divergence is slow because selection pressures on opposite sides of these isolating physical barriers tend to be similar. Population divergence is accelerated, however, if population sizes are small.

There are many instances where the differentiation of adjacent populations occurs without any obvious barrier to gene flow (Ehrlich and Raven, 1969). Furthermore, this differentiation process is frequently rapid (e.g., Jowett, 1964; Snaydon and Davies, 1982) and can be observed within years rather than inferred over geologic time. Numerous examples illustrate the differentiation of populations on a microgeographic scale (Macnair, 1982; Smith et al., 1978b; Jones et al., 1977). Most of these involve plant populations, although animal examples are not unknown. Generally we see this phenomenon where the environment changes rapidly over small distances.

These changes may be due to natural heterogeneity of the environment, or they may be a consequence of man's activities. Abrupt transitions in levels of sunlight, temperature, moisture, wind, and salinity occur in natural environments. Extreme environmental changes also result from such man-caused conditions as the exposure of heavy metal-contaminated soils along mine sites, high-temperature inputs to aquatic systems, the introduction of toxins into the environment, and the indiscriminate release of polluting materials.

In this chapter we illustrate the special features of the biology of adjacent populations that have diverged genetically. In most of these examples, genetic differentiation has taken place in the absence of intrinsic barriers to gene flow. These examples elucidate the forces, and the relative magnitude of the forces, that cause genetic divergence on a microgeographic scale. Although these examples provide empirical support for many of the theoretical models of ecological genetics (such as Wright's shifting balance model and sympatric speciation) we shall not discuss these models. Rather, we present empirical evidence that illustrates the principles involved in microhabitat differentiation.

Patterns of Differentiation on Geographic and Microhabitat Scales

THEME: *Strong selective forces in the environment are capable of creating patterns of genetic differentiation on small local scales that are similar to patterns on a macrogeographic scale. These patterns can arise despite considerable gene flow among adjacent populations, in the form of dispersal of both gametes (pollen) and offspring.*

Table 1. *Observations of Adaptive Genetic Differentiation of Populations over Small Distances*

Cause of Divergence	Species	Distance Between Differentiating Populations	References
Water stress	*Avena barbata* (slender wild oat)	3–38 meters	Hamrick and Allard, 1972
Naturally occurring plant toxins	*Nuculaspis californica*[1] (a scale insect)	Adjacent	Edmunds and Alstad, 1978
Thermal stress	*Micropterus salmoides*[1] (largemouth bass)	100 meters	M.H. Smith et al., 1983
Wind exposure	*Agrostis stolonifera* (carpet bent grass)	100 meters	Aston and Bradshaw, 1966
Salt stress	*Atriplex nuttallii* (Nuttall saltbush)	Approx. 1 km	Goodman, 1973
Salt stress, perhaps stochastic factors also	*Spartina patens* (salt meadow cord grass)	Within 200 meters	Silander, 1979
Salinity, competition	*Salicornia europaea* agg. (marshfire glasswort)	Upper vs lower marsh	Jeffries et al., 1981
Fertilizer, liming regimes	*Anthoxanthum odoratum* (sweet vernal grass)	15½ meters	Davies and Snaydon, 1973, 1974; Snaydon and Davies, 1972
Copper, lead, zinc	*Agrostis tenuis* (colonial bent grass)	15½ meters	McNeilly and Bradshaw, 1967
Zinc	*Anthoxanthum odoratum*	10 meters	Antonovics and Bradshaw, 1970
Copper	*Chloridion cameronii*	10 meters	Drew and Reilly, 1972
	Trachypogon spicatus	10 meters	Drew and Reilly, 1972
	Becium homblei	3 meters	Drew and Reilly, 1972
	Rhynchosia monophylla (a rhynchosia)		Drew and Reilly, 1972

Table 1. (cont.)

Zinc	*Melilotus alba* (white sweet-clover)	Adjacent	Miller, 1982
Lead, zinc	*Agrostis stolonifera* (sheep fescue)	Adjacent	Gregory and Bradshaw, 1965
	Festuca ovina (sheep fescue)	Adjacent	Gregory and Bradshaw, 1965
	Festuca rubra (red fescue)	Adjacent	Gregory and Bradshaw, 1965
Zinc	*Agrostis canina* (velvet bent grass)	Adjacent	Gregory and Bradshaw, 1965
Copper and zinc	*Mimulus guttatus* (common monkeyflower)	Adjacent	Macnair, 1977
Arsenic	*Andropogon scoparius* (little bluestem)		Rocovich and West, 1975
Zinc	*Silene maritima* (sea silene)		Baker, 1978

[1]Animals; others are vascular plants, all grasses (Poaceae) except *Atriplex* and *Salicornia* (Chenopodiaceae, goosefoot family), *Becium* (Lamiaceae, mint family), *Melilotus* and *Rhynchosia* (Fabaceae, pea family), *Mimulus* (Scrophulariaceae, snapdragon family), and *Silene* (Caryophyllaceae, pink family).

Plants with large geographic ranges often develop locally adapted populations that are genetically distinct. These populations, called ecotypes, exhibit morphological and physiological characteristics that make the populations particularly suited for growth under the range of local climatic conditions they experience. The formation of these ecotypes requires the existence of genetic variation within the species, differing selection pressures in various portions of the species' range, and the isolation of these populations by distance (Chapter 4 by Chesser). Under conditions of limited gene flow, natural selection can mold adaptations of these populations to conditions of the local environment.

The differentiation of populations of slender wild oats (*Avena barbata*) in the climatic regions of California provides an example of responses to selective gradients that occur on both the macro- and microgeographic scales (Clegg and Allard, 1972; Hamrick and Allard, 1972). *Avena barbata* was introduced into California at the time of Spanish settlement. This species has spread over large geographic areas and appears to have differentiated into populations that are associated with particular climatic conditions. In this species, the association between the genetics of the populations and the environmental conditions is particularly clear-cut.

There is close correspondence between the genotypes of the plants, as marked by five loci, and habitat classification. Contrasting genotypes are characteristic of moist, cool geographic regions of California and dry, warm regions of that state. Transitional polymorphic genotypes can be found in regions with intermediate environmental conditions (Chapter 13 by Clegg and Brown).

These same genotypes have been found to segregate on a single hillside where the microhabitat conditions range from mesic to xeric. Dry portions of the hillside are populated by plants with the same genotype that is fixed in the warm, dry climatic regions of California, while adjacent wet sites contain those individuals that have genotypes characteristic of mesic climates. Since gene flow on the hillside is considerable, as measured by both pollen flow and seed migration, selective forces in the environment must be operating to maintain the differentiation of these populations.

The fine-scale relationships between genotype and environment suggest that the effective size of a microhabitat can be just a few meters. On the California hillside, microhabitats of the slender wild oat populations were estimated to be less than 3×5 meters in size. It is remarkable that populations can adjust genetically to these rapid environmental changes over small distances (Hamrick and Holden, 1979).

The slender wild oat genotypes associated with the different climatic regions of California have been shown, through reciprocal transplant experiments, to have highest fitness in climatic regions that are most similar to those from which they originated (Jain and Rai, 1980). The allozyme marker loci that were used to distinguish the different slender wild oat genotypes have no obvious functional relationship to survival or reproductive vigor. Yet they appear to be correlated with characters upon which selection can act.

Although the phenomenon of population differentiation on macro- and microgeographic scales has not been studied extensively, other studies show patterns of differentiation in small-scale transects that are similar to those observed over latitudinal distances. The California poppy, *Exchscholzia californica*, varies over climatic regions in a number of characters, including stamen number and the occurrence of annual or perennial growth forms. Examination of plants collected on small-scale transects of a few kilometers in a region with sharp environmental gradients reveals populations that have differentiated in ways that are similar to the divergence found over large geographic regions (Cook, 1962). Ecotypic differentiation along latitudinal and local gradients has also been reported in an arctic sedge, *Carex aquatilis* (Chapin and Chapin, 1981). These studies suggest an extremely intimate association between the genetic structure of these plants and the environment. Natural selection is the most likely means of maintaining the ecotypic differentiation of these associations.

Heavy Metal Stress in Plant Populations

THEME: *Genetic differentiation of plant populations can result from strong natural selection on heavy metal soils. Selection at the seedling stage overcomes the potential homogenizing effect of gene flow at the pollen and seed stages.*

The soils associated with mine tailings often contain toxic concentrations of heavy metals. Plants that grow on mine sites have evolved a tolerance for the toxic effects of heavy metals. These plants generally are preadapted genetic variants found within natural populations consisting mainly of nontolerant plants. The differentiation of tolerant populations from nontolerant populations can occur quickly, often within a few years. Some of the species that have evolved metal-tolerant populations are listed in Table 1.

Colonial bent grass (*Agrostis tenuis* [= *A. capillaris*]) has the ability to differentiate into tolerant populations when subjected to heavy metal stress. Adapted populations are specific in their tolerance for heavy metals. Populations of *A. tenuis* adapted to soils in copper mine sites are tolerant to high copper concentrations, but not to other metals such as lead or zinc. Likewise, other *A. tenuis* populations that are able to grow on lead mine sites do not exhibit tolerance to copper (Gregory and Bradshaw, 1965). Other plant species respond to heavy metal stress in a similar manner. Copper- and nickel-tolerant forms of *Agrostis stolonifera* (carpet bent grass) have been discovered (Jowett, 1958). The grasses *Festuca ovina, F. rubra, Anthoxanthum odoratum,* and a hybrid between *Agrostis tenuis* and *A. stolonifera* have developed zinc-tolerant forms on the same mine site (Gregory and Bradshaw, 1965). On copper-contaminated soils the legume *Rhynchosia monophylla,* as well as the grasses *Chloridion cameronii* and *Trachypogon spicatus,* has acquired copper tolerance (Drew and Reilly, 1972). In the United States, zinc-tolerant ecotypes of *Melilotus alba* (white sweet-clover) on mine sites in southern Wisconsin (Miller, 1982) and copper-tolerant populations of *Mimulus guttatus* (common monkeyflower) have been described (Allen and Sheppard, 1971). In every case, the tolerant populations were derived from adjacent natural populations that in general lacked tolerance to the metals.

The adaptive differentiation of populations subjected to heavy metal stress results in the modification of a wide range of characters, both physiological and morphological. The adapted variants have physiological modifications that confer heavy metal tolerance. The degree of this tolerance differs among mine site populations, as well as between mine site and control populations (McNeilly and Bradshaw, 1967). Morphological characters such as leaf length and plant height also distinguish the tolerant from the nontolerant populations (Jowett, 1964). Because many characters are involved in the adaptation of populations to heavy metal soils, it has been suggested that more than one genetic locus is involved in the evolution of such tolerance (Urquhart, 1971; Gartside and McNeilly, 1974).

Heavy metals are not the only chemicals that induce the differentiation of plant populations. Persistent morphological differences have been observed in grass plots given different fertilizer and liming treatments (Snaydon and Davies, 1972). *Anthoxanthum odoratum* (sweet vernal grass) in plots treated with different levels of mineral nutrients such as phosphate or calcium demonstrates adaptive differentiation in physiological properties (Davies and Snaydon, 1973, 1974).

Where direct observations of adaptive differentiation of plant populations have been made, it is apparent that evolutionary divergence is rapid and can be measured on a scale of years rather than millennia. Characters such as copper tolerance have been shown to develop quickly, in 10 to 15 years, in populations of *Agrostis stolonifera* (Wu et al., 1975). Genetic changes in populations of *Anthoxanthum odoratum* were observed 6 years after the initiation of liming treatments (Snaydon and Davies, 1982).

Undisturbed populations of the species that inhabit soils adjacent to mine sites appear to contain a low frequency of genotypes that are apparently preadapted to fitness under the strong selection of toxic levels of heavy metals. Tolerant seedlings can be obtained from undisturbed populations at the level of about one in a thousand (Wu et al., 1975; Walley et al., 1974). Species that are found adjacent to mine sites but that do not invade them generally do not have variants preadapted to these stresses (Gartside and McNeilly, 1974). Genetically distinct populations of tolerant plants are maintained on mine site soils despite considerable gene flow from nontolerant populations. While the metal-tolerant forms of the plants have large selective advantages on the contaminated areas, in uncontaminated areas they may have a lower fitness than the nontolerant forms (McNeilly, 1967).

Thus, strong selective forces in heavy-metal-contaminated soils can serve to isolate populations that are particularly adapted to those extreme environments. Under normal conditions, these preadapted genotypes are found in the population at low frequency. Their presence in the population allows the rapid exploitation of habitats that otherwise would be unavailable to the species.

Differentiation in Animal versus Plant Populations

THEME: *Adaptive genetic divergence is not as evident in adjacent animal populations as in plant populations.*

Relative vagility is the most significant distinguishing factor affecting the differentiation of animal and plant populations. Plants are generally sedentary, with relatively limited powers of movement, which are largely confined to dispersal of seed or pollen. Plants can escape temporary adverse conditions only by entering a dormant state. On the other hand, most animals can avoid adverse evironmental conditions simply by moving to another habitat (Wiens, 1976). Motile animals may experience a range of spatially varying selection pressures. These selection pressures may vary in intensity as well as direction. Such selective regimes are not likely to lead to strongly differentiated populations on a microgeographic scale.

Dispersal in plant populations has a tendency to be very localized (Levin, 1979). This localized gene flow encourages the formation of small demes adapted to the local environment. Gene flow is a less understood process in animal populations. Although it is generally accepted that many animals are capable of long-distance dispersal (Ehrlich and Raven, 1969), the extent to which such dispersal results in gene flow is not well known.

Plants tend to be more affected by the physical environment than animals (Bradshaw, 1972). Animals, in general, have more highly evolved homeostatic mechanisms than plants. These homeostatic mechanisms mitigate the effects of fluctuations in the physical environment.

Man is altering the selective forces on animal populations in a number of ways. Perhaps some of the most common and important alterations of the environment in reference to population differentiation are man-made barriers to animal movement. These include nearly all man-made structures such as roads, cities, dams, and reservoirs. Not only do these barriers have a large impact on gene flow, but in restricting animal movements they intensify the effects of adverse environmental conditions. Man is also altering the selective forces on animals by his use of widespread and pervasive toxins (for example, DDT and warfarin) and by the release of thermal effluents and pollutants. Animals are not adapted to these unnatural conditions, and frequently they are unable to escape these stresses. These man-made changes which restrict gene flow among local populations and intensify selection pressures on populations encourage genetic divergence among animal populations.

Differentiation in Insect Populations

THEME: *Microdifferentiation of populations can result from biological interactions between organisms as well as selective forces of the physical environment on organisms.*

Some of the most robust evidence of genetic differentiation and divergence of populations due to selective factors of biological origin in the natural environment is found in insect populations (e.g., Knerer and Atwood, 1973; Tauber and Tauber, 1977a; Chapter 22 by Futuyma). A particularly striking example is the relationship between black pineleaf scale insects (*Nuculaspis californica*) and their host trees, ponderosa pine (*Pinus ponderosa:* Edmunds and Alstad, 1978). The associations between pineleaf scale populations and their hosts suggest that some populations of these insects have evolved adaptations for specific trees. One of the characteristic features of scale infestations of ponderosa pine is the extreme variation found in the level of infestation among trees within the same area. Some trees may be heavily infested, while adjoining trees, even ones with intertwining branches, may remain insect-free for years. If insecticide is applied to control scale populations, reinfestation of trees follows the pattern of infestation shown prior to treatment. Trees that were heavily infested return to that state, while lightly infested trees become lightly infested again. Reciprocal tree grafts between infested and uninfested trees also tend to retain their characteristics. Scale insects on an infested tree do not colonize uninfested tree branches grafted to the infested tree. Likewise, scale insects infesting branches grafted to scale-free trees do not move to colonize these trees.

Ponderosa pines show tree-to-tree heterogeneity in their defensive compounds (Harborne, 1973). This heterogeneity (both in the kinds and amounts of compounds present) is genetically determined (Squillace, 1976). Natural selection acts to increase adaptation of scale populations to the specific array of chemical defenses in the host tree. Individual scale insects are adapted to living on particular host trees. Female scale insects are completely sessile, and most offspring remain on the tree on which they were born. This results in the differentiation of populations on a small scale. This differentiation is an adaptive strategy that apparently is the result of evolutionary interactions between the two species.

Gene flow between populations on different trees is restricted; it is confined to interpopulational mating by winged males and the dispersal of larvae to other trees by wind. The semi-isolation of these populations permits more specific adaptation of particular populations of scale insects to particular trees than would be the case in more randomly mating and dispersing populations.

The trees present a heterogeneous environment to scale insects. The defensive compounds of the various trees exert a selection pressure that is disruptive. Each tree, however, presents an environmental patch that poses relatively predictable and directional selection pressures on the scales through its defensive compounds. Scale populations are adapted to this situation in that they form relatively isolated populations between trees. Offspring of scales adapted to a particular tree may not be adapted to trees with different defenses. Maximum fitness is attained by parent scales whose descendants remain on the parent trees.

The microdifferentiation of populations in this case is the result of limited gene flow between populations, with strong directional selection in every environmental patch (tree) and overall disruptive selection, leading to different adaptive modes in each habitat patch.

Warfarin Resistance in Rats

THEME: *Genetic differentiation of animal populations can occur over short distances, even in populations of motile animals. Frequently this differentiation is due to the interaction of many factors, with limited gene flow often playing a dominant role.*

Genetic subdivision in animal populations is frequently found where the populations are organized into small, local breeding groups, or demes. Gene flow between local groups may be restricted because of social structure, as in prairie dogs (*Cynomys endovicianus*: Chesser, 1981; and Chapter 4 by Chesser, this book) and house mice (*Mus musculus*; Selander, 1970), or it may be limited because of differences in breeding habitat, as with host-specific insects. If local populations are relatively small and isolated, sampling accidents or stochastic (random) effects may overcome weak selection pressures (Wright, 1982). This results in populations that consist of many small, semi-isolated groups that diverge genetically owing to the random fixation of alleles.

If strong disruptive selection pressures are superimposed on such structured populations, genetic divergence is accentuated. This occurs not only when selection pressures are different for each local population, as is probably the case with host-specific insects (e.g., Edmunds and Alstad, 1978; Tavormina, 1982), but also when the same selection pressure is differentially applied to the various local populations (Bishop, 1981).

Man imposes strong selection pressures on populations by the use of toxins. Warfarin, a powerful anticoagulant chemical, has been used extensively in the United States and Great Britain as a rodenticide. Animals that are susceptible to warfarin poisoning die as a result of internal hemorrhaging due to breakdown of the blood-clotting process (Bishop and Hartley, 1976).

Five years after warfarin was introduced into Great Britain as a rodenticide, populations of warfarin-resistant brown rats (*Rattus norvegicus*) were discovered (Bishop, 1981). Resistance to warfarin poisoning in rats from Wales is conferred by a single, dominant allele. Associated with warfarin resistance is an increased requirement for vitamin K, a key factor in the blood-clotting process. Heterozygotes for the warfarin-resistant allele require two to three times the amount of vitamin K required by normal, susceptible animals, whereas warfarin-resistant homozygotes require nearly twenty times the normal amount (Bishop and Hartley, 1976).

Adjacent brown rat populations differing in their resistance to warfarin poisoning have been found. These populations are separated by distances as small as 5 meters (Bishop et al., 1977). Genetic differentiation of these populations of motile animals is due to a number of factors, including population size and structure, occurrence and constancy of the selective pressure of warfarin, other selective pressures of the environment, random events, and limited migration between populations.

Brown rat populations, even on a single farm, tend to be organized into a number of small, semi-isolated breeding groups. Gene flow (by successful migration and mating) between established populations is rare. Rat populations in some habitats periodically become extinct; these habitats are recolonized when favorable conditions occur. Hence, stochastic events are likely to play a large role in determining the genetic structure of these small, adjacent populations.

Superimposed on these stochastic events are strong selective forces, such as warfarin. When a brown rat population is poisoned by warfarin, all the susceptible homozygotes die. Heterozygotes and warfarin-resistant homozygotes survive. Warfarin-resistant homozygotes, however, face a strong selective disadvantage in some environments owing to their large requirement for vitamin K. Barn populations often subsist mainly on grain, a food deficient in vitamin K, while countryside populations have an abundant supply of vitamin K in leafy green vegetation (Partridge, 1980). Thus the relative selective disadvantage of the warfarin-resistant homozygote varies with habitat.

The selective pressure of warfarin also varies in the different populations. Rat populations in buildings are periodically poisoned, while countryside populations a few meters away are probably never subjected to the toxin (Bishop, 1981).

Strong directional selection through the continual use of warfarin should rapidly lead to the fixation of the warfarin-resistant allele in populations in which this allele is found. This directional selection is balanced, however, by several factors. First, there are strong selection pressures exerted against the resistant homozygotes because of their large dietary requirement for vitamin K. Second, in the absence of warfarin poisoning, heterozygotes are at a selective disadvantage to normal homozygotes, perhaps also due to their vitamin K requirement.

In summary: Known rat populations are subdivided over small distances into small, genetically distinct breeding units. Migration between these populations is low. Stochastic effects (such as a successful migration event) as well as strong selection interact to determine the genetic structure of these adjacent populations. The interaction between such deterministic and chance elements may be important in establishing the genetic structure of many populations (e.g., M.W. Smith et al., in press; Smith et al., 1978b).

Strong Selection in Largemouth Bass Populations

THEME: *Where there are large differences in selection pressures on either side of an isolating barrier, rapid genetic divergence can occur between populations.*

Adjacent populations separated by physical barriers will genetically diverge owing to drift. Where there are large differences in selection pressures on either side of the isolating barrier, selection also will promote divergence. Rapid genetic divergence can occur between populations under the influence of strong directional selection on either population (Kettlewell, 1973).

On the Savannah River Plant near Aiken, South Carolina, a large reservoir, Par Pond, serves as a cooling and recycling system for thermal effluents from a nuclear reactor. Two smaller ponds, B and C, feed into Par Pond and serve as initial cooling ponds. Pond C receives thermal effluents, which results in temperatures over 50°C in this pond. Pond B is currently a post-thermal pond. It received thermal effluents until 1964, but it is now at ambient temperatures.

Largemouth bass (*Micropterus salmoides*) populations in the reservoir system have been shown to diverge genetically under the influence of the strong selection pressure of the thermal effluents. Two forms of the malate-dehydrogenase-1 enzyme are found in all three largemouth bass populations. One of these enzymes has significantly greater thermal stability than the alternate form. The allele for the more thermally stable enzyme was shown to increase in frequency in the bass populations subjected to extreme thermal stress in Pond C. In post-thermal Pond B, frequencies of the allele coding for the more thermally stable enzyme returned to those frequencies characteristic of the natural populations of this region after the discharge of thermal effluents was discontinued. This return to "normal" took place within 20 years (M.H. Smith et al., 1983).

Strong, man-imposed directional selection operated to rapidly alter allele frequencies in the thermally affected ponds. This selection operated on pre-existing genetic variation within the populations and caused rapid divergence in the genetic characteristics of these adjacent, but isolated, largemouth bass populations.

Discussion

Genetic subdivision over short distances is a well-recognized feature of many plant and animal populations. Restrictions in migration or gene flow, often the consequence of specific behavioral or gamete dispersal mechanisms, result in patchiness of genetic structure within a contiguous population. For instance, genetic subdivision has been documented in populations of house mice within a single barn (Selander, 1970). This genetic heterogeneity was not attributable to selection, but rather was the result of limited movement and mating between mice from different areas of the barn.

In plant populations, specific mechanisms and modes of reproduction frequently lead to genetic patchiness. For example, populations of plants that self-fertilize and have limited seed dispersal mechanisms rapidly develop patchy genetic structure. Computer simulation models also suggest that patchiness can be affected in outcrossing plants by pollinators that fly between neighboring plants (Turner et al., 1982).

Stochastic genetic divergence of groups within a population is quite different from the adaptive differentiation of adjacent populations. Individuals within adaptively diverging populations often have specific physiological, morphological, or reproductive strategies that permit the successful exploitation of otherwise unsuitable habitats. In addition, adaptive genetic divergence between adjacent populations may confer greater average fitness to the populations by allowing each unit to respond to specific environmental conditions, and maximize its fitness under these conditions (Ford, 1956).

Over time, the pattern of genetic divergence in adaptively divergent populations is expected to remain constant. In contrast, although patches may always be found in randomly diverging populations, the orientation of these patches in space, and the specific genetic characteristics of each patch, are unpredictable.

The differences between genetic divergence over large distances and microgeographic divergence lie not in the evolutionary forces involved, but in the relative magnitude of these forces. Selection, migration, gene flow, mating patterns, and drift are factors that influence the extent and rapidity of genetic divergence in both cases. Limited gene flow over large geographic distances allows selection and drift to dominate evolutionary change in widely distributed populations. In contrast, gene flow, whether by gamete or offspring dispersal or both, may be considerable in the case of local divergence. Generally, strong selection operating at the postzygotic, or juvenile, stage counteracts this gene flow and, in so doing, creates genetically distinct populations. Thus, the gene pools of the adjacent populations may be effectively isolated, though gametes mix freely (Figure 1).

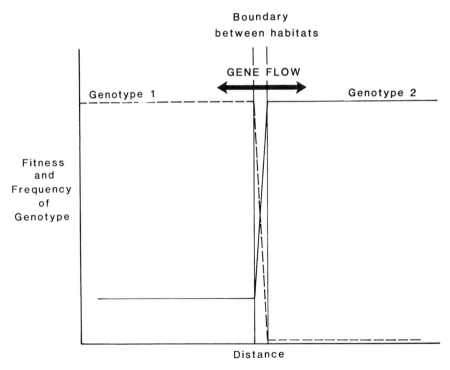

Figure 1. *Disruptive selection leads to genetic divergence of adjacent populations.*

Adaptive genetic differentiation over a small scale is not a ubiquitous phenomenon. Such differentiation is characterized by specific environmental features and special attributes of the species involved. In particular, heterogeneous (or patchy) environments produce the conditions necessary for microhabitat differentiation. Large environmental changes occur over short distances in these heterogeneous environments. Very often, these environmental changes are the consequence of human activity. For instance, heavy metal wastes resulting from mining activities, high-temperature inputs to aquatic systems from power-generating plants, and the release of toxins such as insecticides, pesticides, or herbicides for population control can be factors leading to the establishment of genetically distinct adjacent populations. Natural variation in salinity (Goodman, 1973), moisture (Hamrick and Allard, 1972), soils (Antonovics, 1971), or defensive chemical compounds in plants (Edmunds and Alstad, 1978) can also cause microhabitat differentiation of populations. Generally, the environmental changes in these habitats are so radical that few, if any, species can adapt by phenotypic plasticity.

Environmental variation, whether it is man-made or natural, can create strong selection pressures that change rapidly over short distances. This strong selection causes populations that have specific biological properties to undergo adaptive divergence on a microhabitat scale. Figure 2 illustrates how the interaction of selective forces with populations that differ in gene flow characteristics can affect the pattern of divergence in adjacent populations.

Not all species respond adaptively to disruptive selection pressures of their environment. For instance, the number of plant species that are able to colonize heavy-metal-contaminated sites is generally less than the

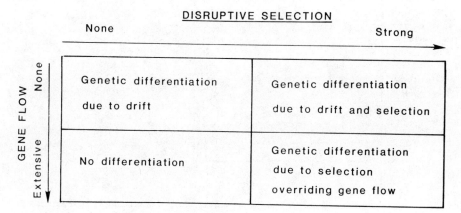

Figure 2. *The relative importance of selection and gene flow in the differentiation of adjacent small populations. Note that strong selection at the juvenile stage overrides extensive gene flow at the gamete stage.*

CHAPTER 5 The Differentiation of Populations over Short Distances

number of species found on adjacent, undisturbed sites. The ability of populations to differentiate on a microgeographic scale under the influence of strong selection is contingent upon the presence of genetic variants that either pre-exist in the population or are *de novo* mutations, which result in organisms that differ in fitness under the varying selection regimes. Of course, the probability of the occurrence of a *de novo* mutation depends on the mutation rate, the effective population size, and the time span. Those populations that are able to diverge rapidly either are composed of organisms with short generation times and have large population sizes, or are populations that contain preadapted genetic variants (Chapter 22).

In species that have small effective population sizes and long generation times, rapid evolution is unlikely to occur by means of *de novo* mutations. In these cases, when adaptation to a new selective regime occurs quickly, it is likely that the genetic variants were pre-existing in the population. The genetic variants may be due either to neutral mutations or to mutations that were deleterious under the current selection regime. These variants may have previously been selected for by other conditions, or they may be the results of recurrent mutations.

Man has the capacity to radically change environmental conditions over short periods of time. Furthermore, man-initiated environmental changes frequently result in selection pressures that are without precedent in the evolution of many species. Since man-caused environmental changes are seldom gradual over time, but are sudden relative to the life span of many organisms, those populations that are able to respond adaptively tend to be from (1) those species that are genetically flexible (that is, have high genetic variability pre-existing in the population) or (2) those with short generation times and large population sizes.

The construction of physical barriers across the range of a population frequently has the effect of isolating subpopulations on either side of the barrier. Each of these subpopulations will be smaller than the entire population and will tend to lose genetic variability owing to the effects of inbreeding and drift. In human activities, strong selective pressures often accompany the process of isolation. It is possible that local extinction could occur over small areas in demes that, because of the reduced genetic variability caused by population subdivision, lack of genetic resources to adapt to the new environmental conditions.

The process of genetic differentiation of populations over short distances merits consideration by managers of natural populations. Small-scale differentiation of populations is probably most common in habitats that have been altered by man. The two components of microhabitat differentiation, the application of strong selective pressures over short distances and the preservation of genetic variability, which allows populations to adapt, are both subject to manipulation by human activities on a local scale.

The adaptive differentiation of local populations is in general beneficial to the species because it allows the exploitation of otherwise unavailable habitats. Differentiation of populations may also serve to mitigate some of the adverse effects of human impacts on the environment. For example, the ability of grass species to adapt to heavy metal stress and cover spoil sites that would otherwise remain barren is certainly a positive aspect of microhabitat differentiation. Managers should understand that, as strong selective influences are introduced to the environment, the ability of populations to respond to these pressures may be compromised by the construction of physical barriers that partition genetic variability. Finally, it is important for managers of natural areas to recognize (1) that the species that can adapt to extreme radical environmental changes are relatively few, (2) that differentiation arises from specific interactions of populations with the strong selective forces of environments that are heterogeneous or patchy over short distances, and (3) that adaptively differentiated populations may represent unique genotypes worthy of preservation.

Summary

Heterogeneous environments, often with large environmental gradients over short distances, can exert strong evolutionary pressures on natural populations of plants and animals. Very often these environmental gradients are the consequence of human activity. When disruptive selection is imposed on populations that have particular characters such as limited mobility or mobility that is restricted by the construction of man-made barriers, adaptive differentiation of populations over short distances can result. If the populations contain mutants that are preadapted to the new selective regime, or if the effective population size is large and generation time is short, population differentiation can occur rapidly, often within a few years.

Genetic differentiation of adjacent populations may confer greater average fitness to the populations by allowing each unit to respond to specific environmental conditions and maximize its fitness under these conditions. Adaptive differentiation of populations can occur even though there is considerable gene flow between adjacent populations.

Adaptive differentiation is quite different from genetic subdivision of populations due to stochastic processes. In the latter case limited gene flow leads to genetic patchiness in the population, whereas strong disruptive selection overcomes the effects of gene flow in adaptively differentiated populations. In subdivided populations orientation of patches in space and the specific genetic characteristics of each patch are unpredictable. No overall increase in fitness can necessarily be inferred. On the other hand,

CHAPTER 5 The Differentiation of Populations over Short Distances

strong selective forces structure adaptively differentiated populations, and the dimensions of adapted patches remain constant over space and time.

The examples in this chapter illustrate the adaptive differentiation of populations on a microgeographic scale: (1) Moisture gradients are associated with particular genotypes, as marked by five loci, in wild oats. The same genotype–habitat associations have been found segregating over a small scale (a hillside) as over a large latitudinal range. (2) High concentrations of heavy metals on spoil sites have led to the evolution of metal-tolerant populations of grasses that are genetically distinct from adjacent nontolerant populations. (3) Populations of parasitic black pineleaf scale insects have been found to be genetically differentiated on a tree-to-tree basis. Populations of these insects are apparently adapted to the particular array of defensive compounds of the host tree, which vary on a tree-to-tree basis. (4) Adjacent rat populations have been observed that differ genetically in their resistance to warfarin, a rodenticide.

Managers of natural areas need to recognize (1) that the species that can adapt to extreme environmental changes are relatively few, (2) that adaptive genetic differentiation arises from specific interactions of populations with the strong selective forces of environments that are heterogeneous over short distances, and (3) that adaptively differentiated populations may represent unique genotypes worthy of preservation.

CHAPTER 6

Sibling Species

John W. Bickham

Introduction

The biological species concept can be stated as follows: "*Species are groups of interbreeding natural populations that are reproductively isolated from other such groups*" (Mayr, 1970; his italics). Most biologists accept the biological species concept at least in theory, although it has definite complications when applied to some plants (Grant, 1981), but it must be recognized that in practice species generally are distinguished by their morphological features (Sokal and Crovello, 1970). Populations of organisms that are reproductively isolated yet are not morphologically distinguishable are called sibling species or cryptic species. In this sense "cryptic" refers not to color or pattern that results in an organism's resembling its substrate or background, but to the fact that such a population may not be recognized as a species because of its morphological resemblance to another. Sibling species are fairly common, particularly in certain groups, and their existence can present problems to population managers. These problems can be circumvented by understanding the biology of the species involved. The purposes of this chapter are to discuss sibling species, including (1) how they can be recognized and the ways in which they differ, and (2) the kinds of problems that they present to population managers and how such problems can be circumvented. I present examples of sibling species of animals, although it should be noted that sibling species may be encountered in any group of animals or plants. These few illustrative examples should suffice to establish the general principles involved.

Are Sibling Species Less Genetically Differentiated Than Morphologically Different Species?

Genetic differences among organisms can be calculated from data developed by electrophoretic studies of proteins. A useful index to summarize this information is Nei's (1972) genetic distance, or D. This is "the accumulated number of codon substitutions per locus (genetic distance) since the time of divergence of two populations" (Avise, 1976). Therefore, the more two species (or populations) differ genetically, the greater the value of D. It is well known that different species generally show profound genetic differences. This is typically reflected in high D values when well-differentiated species are compared. For example, in the sunfish genus *Lepomis* the average D value for 10 species studied was 0.627 (Avise and Smith, 1974a, 1974b; Avise, 1976).

When sibling species are studied, the genetic distance is found on the average to be lower than that between morphologically distinct species. In the *Drosophila willistoni* fly group, morphologically distinct species have an average D value of 1.058, while for sibling species the average genetic distance is 0.750 (Ayala et al., 1974). Although sibling species may on the average show smaller genetic distances than morphologically distinguishable species, sibling species are each distinct and unique, and they forge separate evolutionary paths. For this reason, they deserve consideration in conservation and management decisions equal to that given to morphologically distinct species.

How Genetically Different Are Sibling Species?

It is best to answer this question by considering a couple of examples that represent opposite ends of a spectrum of possible differences. Because the critical factor in speciation is the attainment of reproductive isolation, not the accumulation of genetic differences, it is possible that some newly evolved species differ only by one or a few genes that result in reproductive isolation. How many gene differences does it take to provide for reproductive isolation? A relevant case has been presented in a series of papers on green lacewings (Chrysopidae) by Tauber and colleagues (Tauber and Tauber, 1977a, 1977b; Tauber et al., 1977).

1. The Green Lacewings *Chrysopa carnea* and *C. downesi*

Chrysopa carnea and *C. downesi* are sibling species that occur sympatrically (their ranges overlap) in the northeastern United States. A high degree of genetic similarity is implied by the fact that the two species are fully interfertile under laboratory conditions. Under natural conditions the species are reproductively isolated by differences in habitat and reproductive season. *Chrysopa carnea* is pale green and occurs in grassy areas and meadows during the reproductive season. It produces several generations each summer. At the end of summer it goes into a period of diapause during which the adults change to a reddish-brown color and move to their overwintering site—the senescent foliage of deciduous trees. *Chrysopa downesi* is dark green and occurs on coniferous trees throughout the year. It breeds only once a year, during the early spring.

Crossing experiments indicate that the difference in color between the two species is caused by fixation of alternative alleles at a single locus in each species (Tauber and Tauber, 1977b). Crossing experiments also show that fixation of alternative alleles at each of two unlinked loci in each species accounts for the differences in timing of reproduction (Tauber et al., 1977). Reproductive isolation between these sibling species thus can be accounted for by differences at as few as three loci (Tauber and Tauber, 1977b).

It should be emphasized, however, that other genetic differences are likely to occur rapidly after reproductive isolation is attained, and, in fact, these two species of green lacewings have other known genetic differences (Tauber et al., 1977).

2. The Fruit Flies *Drosophila melanogaster* and *D. simulans*

The sibling species *Drosophila melanogaster* and *D. simulans* are morphologically very similar, although they can be distinguished by a consistent difference in male external genitalia. The following information on the comparative evolutionary biology of *D. melanogaster* and *D. simulans* is taken from a review by Parsons (1975). These species occur together in the wild but seldom hybridize, and when they do the hybrids are sterile. The two species occupy some of the same habitats, and both are cosmopolitan in distribution. However, numerous subtle differences have been demonstrated in laboratory and field experiments. They differ, for example, in male courtship behavior, and in their responses to the effect of darkness on mating behavior. Chromosomes differ: there is a major inversion in chromosome III in *D. melanogaster* but not in *D. simulans*, and there are perhaps as many as 24 lesser rearrangements (cf. Chapter 2 for discussion of chromosomal mutations). Quinacrine staining has also shown consistent differences in both meiotic and mitotic chromosomes. The two species differ markedly in frequency of chromosome polymorphism, with *D. melanogaster* being highly polymorphic and *D. simulans* monomorphic. The two species also differ in tolerance to temperature fluctuations: *D. melanogaster* is more tolerant than *D. simulans*. Patterns of seasonal variation in population size differ, with *D. melanogaster* reaching peak population sizes in January and *D. simulans* in May in Melbourne, Australia. *Drosophila simulans* is less resistant to a variety of physiological stresses (such as desiccation) than is *D. melanogaster*. Additional quantitative differences were found between the species in nutritional requirements and in various behavioral patterns such as dispersal.

All differences found between the two are quantitative rather than qualitative, with the single exception of alcohol tolerance. It was shown in laboratory experiments and in natural populations that *D. melanogaster* larvae can utilize a medium containing a higher concentration of ethyl alcohol. *Drosophila melanogaster* and *D. simulans* should serve as a caveat to population managers. Although sibling species are morphologically very similar and may overlap in ecological characteristics, numerous genetic differences may be present.

How Are Sibling Species Identified?

In both examples of sibling species presented (*Chrysopa* and *Drosophila*), the species in each pair do possess slight differences in appearance. Most sibling species probably possess subtle morphological differences, but often these differences are inadequate for the routine identification of specimens. Discrete characters, sometimes discovered during electrophoretic or karyotype analyses, provide absolute differences between sibling species. These techniques are valuable not only for identifying sibling species, but also for evaluating the degree of reproductive isolation between populations.

Morphological, karyological, and biochemical analyses may be used on virtually any organism, but certain taxonomic groups possess special characters that are especially useful. Species-specific songs or calls may be useful in identifying sibling species of insects, birds, or frogs. Behavioral differences also are useful in animals with stereotyped visual displays; for example, the territorial and mating displays of lizards, and the light flashes of fireflies (*Photuris*: Barber, 1951) can sometimes be used to differentiate sibling species. Yet another method to identify sibling species is the study of parasites. A parasitic species may be specific for a host species, but not the host's sibling species. The following example on the sibling species of pocket gophers and their lice illustrates how these techniques can be used to study sibling species.

Sibling Species of Pocket Gophers of the *Geomys bursarius* Complex

Pocket gophers are fossorial animals that typically have very limited dispersal ability. They often are found in small, inbred populations in which genetic drift and inbreeding may be expected to play important roles in the determination of population genetic parameters and fixation of chromosomal rearrangements. This is supported by empirical data. Gophers often have low levels of heterozygosity, high F_{ST} values, and a great deal of karyotypic diversity.

The plains pocket gopher, until recently, has been considered to be a single, widely distributed variable species; Hall (1981) considered *Geomys bursarius* to comprise 21 subspecies and range from south Texas to Manitoba. Chromosomal studies have documented extensive karyotypic variation. Diploid numbers vary from $2n = 70$ to 74 and F.N. (fundamental number: the number of autosomal arms) varies from 68 to 98 (Baker et al., 1973; Kim, 1972; Hart, 1978; Honeycutt and Schmidly, 1979). Some of the chromosomal variability is due to polymorphisms within populations (Patton et al., 1980), but much of the variation exists as fixed differences among populations. Population genetic and cytogenetic studies have demonstrated that some of the karyotypically differentiated populations are actually reproductively isolated sibling species (Tucker and Schmidly, 1981; Bohlin and Zimmerman, 1982). However, not all karyotypically differentiated populations are reproductively isolated, and in one case a broad zone of hybridization occurs.

Populations of the *Geomys bursarius* complex have been intensively studied in Texas and Oklahoma. Three sibling species recently recognized to occur in this area are *G. attwateri, G. bursarius* in the stricter sense, and *G. breviceps*. Contact zones have been located between *G. breviceps* and each of the other two species (Tucker and Schmidly, 1981; Zimmerman and Gayden, 1981; Bohlin and Zimmerman, 1982). These contact zones are always narrow, with hybridization limited to a small area. A few interspecific F_1 hybrids are found (Tucker and Schmidly, 1981; Bohlin and Zimmerman, 1982), but backcross or F_2 progeny are unknown. Tucker and Schmidly (1981) and subsequent studies by R. C. Dowler (personal communication) have revealed pregnant F_1 hybrid females, the embryos of which possessed backcross karyotypes. However, these F_2 generation progeny from backcrosses must be inviable because they are never found as juveniles or adults. Gene flow does not occur across the contact zones, and, although the species are morphologically virtually indistinguishable, they are genically and chromosomally very different.

A contact zone occurs in Burleson County, Texas, between *G. attwateri* and *G. breviceps*. The *G. attwateri* karyotypes ($2n = 70$; F.N. = 72, 74) are superficially similar to the *G. breviceps* karyotype ($2n = 74$; F.N. = 72), but chromosome banding analyses reveal numerous chromosomal differences (R. C. Dowler, personal communication). Analyses of meiotic cells reveal numerous meiotic disturbances. The chromosomal differences appear to contribute to the observed hybrid breakdown. Genic differentiation has occurred subsequent to the speciation event. Electrophoretic analyses (R. C. Dowler, personal communication) reveal numerous fixed differences at enzymatic gene loci.

Historical factors are known to play an important role in pocket gopher population genetics (Patton and Yang, 1977; Smith and Patton, 1980). The genetic makeup of populations is often the result of founder events or population bottlenecks. Figure 1 shows the distribution of sandy and clay soils and the corresponding distribution of pocket gophers *G. attwateri* and *G. breviceps* in Milam County, Texas. Sandy soil (the preferred habitat of gophers) occurs as islands of suitable habitat, at least some of which have gophers. This distribution pattern illustrates two of the reasons why pocket gopher populations are highly differentiated; each isolated island of sand is likely to be colonized by a few founders, and resident populations are often small.

Although species of gophers are usually allopatric or parapatric, occasionally different species come into contact because of chance dispersal across a barrier. In Figure 1 it can be seen that populations of the two species are separated by belts of clay soil that occur along the Little River and its tributaries. South of the Little River the two species are separated by the Brazos River with *G. breviceps* each of the river and *G. attwateri* to the west. In Burleson County, just south of Milam County, *G. breviceps* is on the east side of the Brazos River at the site of an old meander scar in the river. Here a change in the course of the river has allowed contact between the two species; this is the only known site of hybridization between them.

Farther south, in Washington and other counties that border the west side of the Brazos River, other contact zones may exist. My students have collected both species a few miles apart within the same belt of sandy soil in Washington County. Other evidence of present or past contact in this area comes from the distribution of lice (Trichodectidae) that live on the gophers (Timm and Price, 1980). *Geomys breviceps* is infested with lice of the species *Geomydoecus ewingi*, while *Geomys attwateri* is infested mostly by *Geomydoecus subgeomydis*. However, the lice species' distributions do not exactly correspond to the gopher species' distributions. The

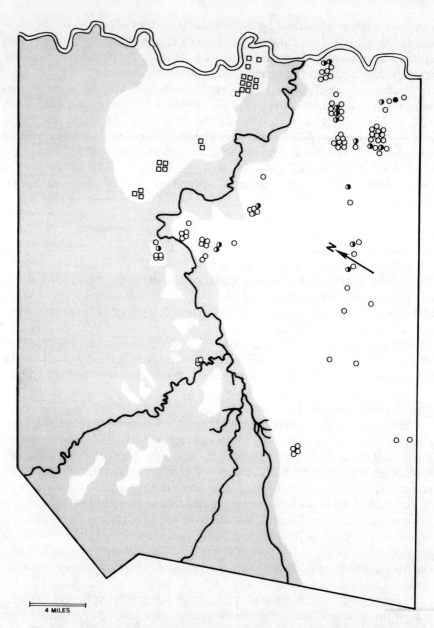

Figure 1. *Distribution of pocket gophers (*Geomys breviceps, *squares;* Geomys attwateri *race F, open circles; race G, solid circle, F × G hybrids, half solid circles) in Milam County, Texas. Clay soil is stippled; sandy soil is open. North of the Little River sandy soil is distributed as "islands," some of which are occupied by gophers. A clay belt prevents contact of the two species in this county, which is bounded on the east by the Brazos River.*

occurrence of *Geomydoecus subgeomydis* east of the Brazos River where it infests *Geomys breviceps* is likely the result of interspecific contact of gophers, either past or present.

The pocket gopher example illustrates the combined use of karyology, electrophoresis, meiotic analysis of hybrids, and parasite distributions for the recognition and evaluation of sibling species relationships. Further, it illustrates the need for detailed distributional data, especially in organisms that have highly specific habitat requirements.

Sibling Species and Population Management

The preceding examples serve to illustrate that sibling species are found among a diversity of animal groups. I now consider certain problems faced by population managers dealing with conservation issues.

For sound management practices to be developed it is necessary to understand the nature of interbreeding units among populations of a species or species complex. Because many species will become extinct in the wild during the next few decades (Frankel and Soulé, 1981), the only hope for their continued survival is propagation in zoos or botanical gardens. It is unfortunate that specimens obtained from the wild and used to found captive or cultivated populations often are not accompanied by reliable and accurate information on collecting locality and habitat. Undoubtedly, many captive populations are the descendants of wild-caught specimens that came from various geographic localities. It is easy to envision what happens when these belong to different sibling species. One can predict that, even if hybridization is possible (as between *Drosophila melanogaster* and *D. simulans*), a high degree of perinatal mortality, particularly in the backcross or F_2 and later generations, and sterility may be expected. This is apparently the case in interspecific crosses in the *Geomys bursarius* complex. Even under the most favorable circumstances where the species are fully interfertile, a hybrid population will result, the individuals of which may no longer be well adapted to the ecological niches of any of the parental species. The captive population would then be nothing more than a domesticated strain, unique in its genetic makeup and possibly unsuitable for future release into the wild.

The solution to the above problem is clear. Captive breeding programs should be designed such that the wild-caught founders come from known localities and habitats. Care must be taken when interbreeding organisms from different localities. Genetic and karyotypic data should be routinely obtained at least on the founders and, where practical, on subsequent generations. Populations with different karyotypes or fixed differences at many gene loci should be treated as different species and not routinely crossbred, even though they are considered conspecific, until it is determined that these populations can tolerate that degree of crossbreeding.

Another conservation-related aspect of the sibling species problem deals with the design of refuge or park systems. Such refuge systems are sometimes designed to preserve groups of species such as waterfowl or large grazing herbivores and their predators, or individual species such as the whooping crane (*Grus americana*), green turtle (*Chelonia mydas*), or Attwater's prairie-chicken (*Tympanuchus cupido attwateri*). Because basic genetic and karyological data are not available even for many United States species that are endangered, threatened, or otherwise in need of management, it is unknown how often refuge lines cut across unsuspected species boundaries. The resulting problem can be illustrated with a hypothetical example. If a refuge is designed to support 1000 individuals, it may be that there are only 700 and 300 individuals, respectively, of two sibling species. The refuge has, then, been set up in such a way as to contain fewer than the desired number of individuals of both species, with possible problems in their management as a result.

Another rather distressing aspect of the sibling species problem is a consequence of such species' often having very small natural distributions. As a result they may be highly vulnerable to habitat destruction and can be considered highly endangered. This problem can be illustrated with data on bats. The little yellow bat *Rhogeessa tumida* is generally considered to be a widely distributed, morphologically variable species (Hall, 1981). It ranges from Tamaulipas, Mexico, to Brazil. LaVal (1973) revised the species classification on the basis of multivariate analysis of morphometric data. He recognized the species as being extremely variable but could not divide the species into subspecies. Karyological investigations (Bickham and Baker, 1977; Honeycutt et al., 1980) and ongoing research have revealed extensive chromosomal variability. Populations with diploid numbers of $2n = 30, 32, 34, 42,$ and 52 are known. The $2n = 32, 42,$ and 52 populations are known from either a single locality or very small geographic region. Two chromosomal races probably represent sibling species because they occur together at one locality without interbreeding.

The *Rhogeessa* example serves to illustrate the need to know in a detailed manner the taxonomy and distribution of species and their populations. Some sibling species in this complex appear to be endangered through habitat destruction. The $2n = 42$ cytotype is known to occur in a strip of tropical forest along the Pacific coast of Chiapas, Mexico, and nowhere else. This forest has been mostly destroyed for agricultural purposes. The $2n = 32$ cytotype is known only from a tropical forest near Rama, Nicaragua. This forest also has been nearly entirely destroyed. Obviously, a refuge system designed to preserve the species *R. tumida*, as morphologically defined, likely would not preserve the genetic diversity that is in fact encompassed in this species complex.

Species and Population Management: Economically Important Species

Some population managers are concerned with economically important species, such as game animals and fisheries stocks. Many of the same problems discussed above under "conservation" also apply here. In addition, fish and game managers may be concerned with breeding programs designed to enhance desired characteristics of such species. Such examples as antler size in white-tailed deer (*Odocoileus virginianus*) and catchability and growth rate in largemouth bass (*Micropterus salmoides*) are well known. The problems may entail crossbreeding of individuals from different species or populations to arrive at the desired traits. Hybrids resulting from such crosses may not constitute a wild, self-sustaining population. The obvious implications of crossbreeding populations in species for which the genetics, cytogenetics, and systematics are not well known were established earlier in this chapter and do not need to be further elaborated (see also Chapter 24).

Conclusions

The ultimate message for the manager from this chapter is to understand the relationships between the populations that are to be managed. Basic genetic and cytogenetic data should be obtained on any species for which a management program is to be designed. It is obvious that basic systematic studies are very important and have significant practical applications. The problem of sibling species will continue as long as species are managed with an incomplete knowledge of the reproductive relationships between populations.

Summary

Improvements in the ability to identify sibling species have mostly been due to technical progress in fields such as biochemistry and cytogenetics and in the application of computers to morphometric studies. Methods useful for identifying sibling species differ for different groups of organisms.

Speciation is the process of populations attaining reproductive isolation. Sibling species attain a level of reproductive isolation in the absence of obvious morphological divergence, often with little genetic divergence. Rates, modes, and patterns of speciation vary among organisms according to the presence or absence of various biological characteristics. Organisms with low vagility or dispersal capability, for example, often develop reproductive isolation between contiguous populations, owing to the fixation of chromosomal or genetic changes in inbred demes. Organisms with different biological properties may speciate by way of gradual allopatric divergence or rapid divergence associated with founder events.

Knowledge of the particular biological properties of organisms that influence mode of speciation can be used by biologists interested in population management. This knowledge can allow for predictions as to what groups are most likely to have hard-to-detect sibling species as well as for the identification of techniques useful in studying them. Additionally, such knowledge can be useful in predicting how population isolation, which can occur with the establishment of refuges, will affect the course of speciation and genetic divergence.

Acknowledgments

I thank R. C. Dowler and R. G. Ruiz for assistance in preparing the figures. J. W. Sites, Jr., and R. C. Dowler read an earlier draft of the manuscript and offered helpful suggestions.

Extinction

PART TWO

Introduction

Myers (1979) and Ehrlich and Ehrlich (1981b) estimate that because of recent human pressures at least one species becomes extinct each day. All conservationists, wildlife managers, foresters, agriculturalists, and fisheries managers will have, if they have not already had, the occasion to observe extinctions among the resources under their protection and management. Extinction can be a natural process and one we might not regret if it occurred at a rate balanced by an equivalent rate of origin of new species. Patterns of habitat alteration and resource exploitation particularly are associated with accelerating extinction rates. This part describes extinction as an evolutionary process and explores the causes of extinction, the factors that can exacerbate extinction, and various guidelines and suggestions for mitigating the loss of fitness and probability of extinction in small populations.

Soulé's chapter begins this part on extinction with an outline and general discussion of the various causes of extinction. He concludes that virtually all recent extinctions are the result of human activities. He then discusses the life history characteristics and their interactions that predispose particular species to extinction.

Beardmore reviews the theoretical background and experimental work on the possible short-term and long-term adaptive values of genetic variation.

Some of the most intensely studied populations are the butterflies and associated host plants studied by Ehrlich and his colleagues. The life history characteristics of the butterflies described in Ehrlich's chapter suggest that, in the short run, ecological factors cause population extinction in these animals before populations reach a small enough size for genetic factors to become important. Ralls and Ballou discuss the evidence for inbreeding depression in small captive populations of mammals. Their conclusions apply to both captive and wild populations of normally outbreeding species that are forced by various circumstances to inbreed.

CHAPTER 7

What Do We Really Know About Extinction?

Michael E. Soulé

The Extinction Problem

Scores of birds, mammals, and flowering plants have suddenly become extinct, particularly on oceanic islands, during the last 100 years. It is as if some epidemic were raging among these species. Now the disease appears to be spreading to the continents. You already know the pathogen—it is man. During the last few centuries Europeans have conquered, colonized, and cannibalized the planet's islands. They brought guns, clubs, rats, cats, dogs, mongooses, goats, pigs, weedy plants, diseases and exotic birds to help finish the job. The results are well known (recent reviews include those by Frankel and Soulé, 1981; Myers, 1979; Ehrlich and Ehrlich, 1981b).

What concrete conclusions about extinction have we assembled from these island die-offs? There are really only two (Frankel and Soulé, 1981, Chapter 2). First, animals and plants, in adapting to islands that have few if any predators or herbivores, often change in ways that make them very susceptible to predation and herbivory. Birds tend to lose the ability to fly, and they also adopt ground nesting and foraging. Reptiles, birds, and mammals lose their fear and flight reactions in the absence of predators. Plants also lose their defenses (such as poisonous chemicals and spines) against herbivores (Carlquist, 1974).

The second consequence of isolation is an area effect. Predators on small or ecologically uniform islands can easily exterminate the few patches of a prey species. On continents, in contrast, it is very unlikely that all patches will be exterminated simultaneously — new patches are reestablished in the complex habitat mosaic before the source populations can all be expunged.

This completes our puny list of facts. It is sad to say that we can add little to these but speculation. We have witnessed and chronicled the extinction of dozens of species at the hands of man. Given a fresh world to start with, we could predict which island species would be most susceptible to anthropogenic extinction. But we are still unable to describe in detail the processes leading to *natural* extinction, especially on continents.

It is disappointing that we know so little about natural extinction. Since Western science has been taking notes, virtually all extinctions of large animals and of plants have been directly or indirectly attributable to man (Greenway, 1967; Frankel and Soulé, 1981, Chapter 2). As far as I know, no biologist has documented the extinction of a continental species of a plant or animal caused solely by nonhuman agencies such as competition, disease, or environmental perturbations in situations unaffected by man. Even if there are cases of the last remaining population of a continental species being wiped out by a natural catastrophe, there could not be adequate data on why the species was so restricted in the first place.

In other words, the extinction problem has little to do with the death rattle of the final actor. The curtain in the last act is but a punctuation mark — it is not interesting in itself. What biologists want to know about is the process of decline in range and numbers.

The "extinction problem" is analogous to the "speciation problem;" we do not have the tools to examine processes that occur with a time scale of thousands of years. Extinction, perhaps, is even more difficult to study than speciation. Speciation may occur relatively quickly under certain kinds of conditions (complete isolation with or without chromosomal rearrangements). The process of extinction, however, is probably a matter of the gradual loss of fitness. Why fitness decreases (if it does) is something that we know little or nothing about, although there is no shortage of ideas to account for it (e.g., Van Valen, 1973; Maynard Smith, 1976b; Rosenzweig, 1973).

The Taxoscope and the General/Local Extinction Continuum

The extinction problem would be easier to solve if we had the right equipment. Imagine being on a stationary satellite that was equipped with a special device sensitive to only one species and with sufficient resolution to detect every individual of the species. Such a hypothetical instrument, we might call it a "taxoscope," could produce a dynamic map of the entire distribution and abundance of our species, which we could record and play back like a time-lapse film. One thing would immediately be clear, at least for most species—the occupied range, when examined closely, is actually a mosaic of vacant and occupied habitat patches. If our taxoscope were able to record for a long enough period (geological in scale), we would see that extinction occurs when there is a higher rate of patch death than patch birth. Again, it is this difference in rates that is interesting (Pickett and Thompson, 1978), not the blinking out of the last individual in the last patch.

Our hypothetical taxoscope is capable of focusing on particular patches and can monitor the swelling and shrinking of such a patch or local population. We could then, given access to other appropriate tools, begin to determine the factors relevant to patch birth and death. (For a small extra charge, accessories to the taxoscope allow quantitative monitoring of microclimate, competition coefficients, disease, and other relevant variables.) For example, we might learn that the demise of a particular patch is attributable to a local drought (Ehrlich, Chapter 9) or to an increase in the abundance of a terrestrial herbivore (Willis, 1974; Karr, 1982). Armed with such information, a manager might be able to control the relevant variables so as to ensure the indefinite survival of the patch in question.

Paradoxically, however, information on patch size or survival may not be at all relevant to the causes of extinction in the species as a whole. Patch extinction and species extinction can be, and probably often are, completely distinct processes. For example, patch A of a butterfly species may be extinguished by a local drought; simultaneously, patch B may be going extinct because of overgrazing by a vertebrate. In the meantime patch C may be dying out because the butterfly's food plant is becoming less palatable (natural selection for herbivore resistance) while the parasite load on the butterfly is increasing. The point is that the three distinct patches are being extinguished for three different reasons.

While these particular patches are dying out, the species as a whole could be either expanding its range or experiencing a decline, and the reasons for the general expansion or decline could be completely independent of the factors accounting for the demise of patches A, B, and C. Of course, as the range of the species shrinks, the correlation between what is happening to a given local population and to the species as a whole will increase; the correlation coefficient becomes 1 when there is only a single population remaining.

Nevertheless, it is essential to recognize that species extinction and population extinction are not the same. The immediate cause for the extinction of a local population may often be stochastic ecological events leading to a demographic nose dive. At the same time, however, a much more widespread factor, such as the chronic presence of low levels of pesticides in the environment or a deterioration in the climate, may be causing a general decline in fitness and abundance throughout the entire range of the species. It is such regional trends that make the patches susceptible to local extinction by what appear to be purely local challenges and stresses.

For this reason it is quite conceivable that well-meaning managers could devote much of their time and resources to trying to rescue a local population of some species that is doomed, regardless of what they do. Managers must attempt to be cognizant of the complex issues related to general as well as local extinction.

In summary, contemporary biologists recognize levels of extinction, some less final than others. At one extreme, extinction means the reproductive failure or death of all individuals in a species. Of course the term "species" is an abstraction that attempts to freeze a collection of individuals that vary in space and time. Putting aside this perennial problem, we can agree, perhaps, on the operational use of the phrase "species extinction" as meaning the death of the last individual that is referred to by a certain scientific binomial. (Our lives are much too short ever to document the gradual phyletic transition of one taxonomic species into another, if such occurs.)

At the other end of the extinction continuum, extinction can refer to the demise of geographic or systematic units smaller than the entire species. These units may be so-called subspecies or varieties, or they may be local geographic groups, loosely referred to as populations. In this context, the term "population" always must be qualified by modifiers, as "local," "Pennsylvania," or "South Atlantic."

There are many kinds of groupings below the level of the species. Some may be distinct geographic entities that rarely if ever exchange genes with other such entities—for example, skipjack tuna (*Katsuwonus pelamis*) in the Eastern Pacific (G. D. Sharp, personal communication). Other entities referred to as populations may be ephemeral groups in relatively intimate genetic communication with other similar groups, such as the *Euphydryas editha* groups of butterflies on Jasper Ridge referred to by Ehrlich in Chapter 9.

Can Extinction Be Studied?

The simplest explanation for extinction is the inability to adapt to changing conditions. But this explanation is not only simple, it is trivial. Indeed it is true that colobus monkeys cannot survive if their forest habitats are clearcut. But all that this tells us is that monkeys cannot grow wings. Here we are dealing with the problem that the rate of environmental change (deforestation) in parts of Africa is several orders of magnitude higher than the rate at which genes are substituted in these populations of vertebrates. Even if the generation time of monkeys were minutes rather than years, it is still unlikely that monkeys could adapt to living only on a prairie. Even baboons (*Papio* and *Theropithecus*) sleep in trees, in caves, or on rocky ledges.

A more interesting question is the cause of rarity. This assumes, of course, that rarity always follows commonness in a temporal sequence. Such may not be the case. Some species might be rare from the time they become genetically independent of their forebearers and never become common relative to other species of a similar body size, taxonomic group, and trophic level. My guess is that many species belong to this category. If such a rare species also is confined to a small geographic area, it will not persist long because it will be expunged by a local catastrophe, or be driven to extinction by stochastic events such as genetic drift or demographic stochasticity. The fossil record will fail to record such provincial, ephemeral species.

Nevertheless, a significant fraction of species do, at some time, occupy a relatively large area. The question then becomes: what causes such a species to suffer a reduction in its range so that its existence is jeopardized by the merest random event? Such a question is subject to investigation. Changes in range are something that can be observed, even within the lifetime of a human. The problem is that most range changes will be explained by relatively short-term climatic variation. A more interesting case would be one for which a combination of physical and biological interactions such as competition, predation, and long-term climatic trends seems to explain the range changes.

Long-term studies of this sort may be the only way to obtain meaningful information on the process of extinction; by long-term, I mean for generations. After all, the great European cathedrals required the labors of generations — is our devotion to knowledge so fickle that research projects requiring 50 to 100 or more years are beyond consideration?

Factors in the Extinction of Local Populations

Table 1 gives many of the factors that might contribute to the extinction of a local population. Some of these factors have been discussed elsewhere (Frankel and Soulé, 1981, Chapter 2). Some of the conclusions reached were as follows:

1. *Island forms* are by far the most sensitive to environmental change of any kind.

2. *Isolation* caused by man or by behaviorally advanced ecological analogues is the forerunner of mass extinction.
3. *Predation* can annihilate local populations of prey, but the ability on the part of the prey to colonize empty patches of habitat decreases the probability that predation alone will cause extinction of the prey over a large area.
4. *Competition* can decrease the density and distributional range of interacting species, but, by itself, competition has not been observed to precipitate extinction, even on islands.
5. *Habitat destruction* is currently the major cause of extinction, especially in the tropics, in arid regions, and in regions subject to industrial impacts such as acid precipitation.
6. *Man* is without peer as an agent of extinction; in historical time no species of continental plant or animal has become extinct except by the direct or indirect hand of man.

Some of the factors listed in Table 1 do not become operative until one or more of the other factors have reduced the local population to a very small size. The thresholds of relevance of these "intrinsic" factors may be species-specific. Among these are (1) demographic stochasticity and (2) behavioral dysfunction. Also included in this category is (3) genetic deterioration—that is, loss of heterozygosity, inbreeding depression, and loss of genetic plasticity. It is because the genetic factor becomes operational only

Table 1. *Possible Factors Contributing to the Extinction of Local Populations*

1. Rarity (low density)
2. Rarity (small, infrequent patches)
3. Limited dispersal ability
4. Inbreeding
5. Loss of heterozygosity
6. Founder effects
7. Hybridization
8. Successional loss of habitat
9. Environmental variation
10. Long-term environmental trends
11. Catastrophe
12. Extinction or reduction of mutualist populations
13. Competition
14. Predation
15. Disease
16. Hunting and collecting
17. Habitat disturbance
18. Habitat destruction

at very low population sizes that we have referred to conservation genetics as the "genetics of scarcity." These factors require some discussion.

1. *Demographic stochasticity* is random fluctuation in population variables such as the distribution of age classes and the sex ratio. These variables, in turn, can cause random fluctuation in population size. Extinction can result from the random variation of these variables in a small population. Sex ratio extinction occurs when all the reproductive individuals in a given generation are the same sex (e.g., the dusky seaside sparrow, *Ammospiza maritima nigrescens*). A skewed sex ratio also increases the rate of inbreeding, thus depressing the fitness of the surviving individuals (Senner, 1980). Such an event is very improbable in any given generation unless numbers fall below 20 or so ($P < .000001$ for $N_e = 20$). Even for species that normally have a very skewed sex ratio, such as harem-breeding marine mammals, there is usually an abundance of sexually mature males waiting on the periphery should all the breeding males suddenly die.

The other principal cause of demographic extinction is random fluctuation in size. In general, the stochastic fluctuation of a population is proportional to $1/\sqrt{K}$, K being the carrying capacity of the habitat (May, 1973). Thus, a population of 100 will usually fluctuate between 90 and 110 (standard deviation of 10), or about 10%, while a population of 9 individuals will fluctuate about 33%, or between 6 and 12.

It is apparent that demographic stochasticity is unlikely to cause extinction for all but the smallest populations. Of course, if some extrinsic factor periodically decreases K, in local populations, the chances of extinction are increased. For example, for species whose population sizes are strongly influenced by weather or some other density-independent factor, as is the case for annual plants and many invertebrates, appropriate management measures must be taken. This matter is briefly discussed at the end of the chapter.

2. The *dysfunction of social behavior* is the second intrinsic factor that can precipitate a sudden and fatal decline in numbers. Species of animals that either forage in large groups or form large breeding congregations appear especially susceptible to catastrophic mortality. Halliday (1978), in his very readable account of the extinction of birds, makes a good case for the vulnerability of species that, like some whales and dolphins, come to the aid of dead or wounded conspecifics, or are dependent for breeding success on the formation of large breeding aggregations. Both forms of behavior expose the animals to several of the factors listed in Table 1, including epidemics, local catastrophic events, and, in recent time, extinction by gunfire and large-scale habitat destruction.

In the United States, the extinctions of the Carolina parakeet, the passenger pigeon, and the heath hen appear to have been facilitated by their extreme sociality. The Carolina parakeet (*Conuropsis carolinensis*) was doomed because it (a) foraged in large flocks that devastated fruit and nut crops; (b) nested in large tree cavities, which were apparently usurped by feral honey bees recently imported from Europe; and (c) went to the "aid" of a wounded member of the flock, thus leading to mass suicide in the presence of a determined hunter.

The passenger pigeon (*Ectopistes migratorius*) may be the best candidate for an extinction stemming from an inability to breed successfully below a certain numerical threshold. Halliday contends that, once the majority of these birds had been decimated by hunting, they were unable to form groups large enough to facilitate reproduction. It is interesting that this species laid only a single egg each year and, therefore, was a K-selected species. This low intrinsic rate of increase contrasts with the similarly endangered Attwater's greater prairie-chicken (*Tympanuchus cupido attwateri*) and extinct heath hen (*T. c. cupido*). Grouse, such as these, nest on the ground and lay clutches containing up to 17 eggs. Thus, even species with relatively high reproductive potential are vulnerable if their social behavior exposes them to wholesale slaughter.

In conclusion, what can be said about the minimum population sizes of such social animals, notwithstanding that they are a small minority of higher vertebrates? It is clear that the answer is species-specific. It depends on many factors that cannot be generalized. The appropriate criterion could be stated as follows: For highly social species, the minimum viable population could be larger by two or more orders of magnitude than are the minima estimated by demographic stochasticity or by reference to population genetics.

3. *Genetic deterioration* is the third intrinsic factor that can cause the extinction of a relatively small population. Genetic deterioration has two components. The *first* is inbreeding depression—the decrease in vigor, viability, and fecundity that is generally the result of inbreeding in a normally outcrossed species. Inbreeding depression occurs in all but a very small and unpredictable fraction of cases when the breeding system or the effective population size produces rates of inbreeding higher than 1% or 2% per generation (Chapter 10 by Ralls and Ballou; Franklin, 1980; Soulé, 1980). The *second* form of genetic deterioration is a decrease in the amount of genetic variability in populations. Any such decrease lowers the population's potential evolutionary adaptability. These topics have been covered in detail elsewhere (Flesness, 1977; Denniston, 1978; Franklin, 1980; Senner, 1980; Soulé, 1980; Frankel and Soulé, 1981; and other chapters in this volume).

Recent studies on tropical birds have pinpointed other factors that may precipitate local extinctions. For example, Terborgh and Winter (1980) showed that rarity was the best predictor of vulnerability to extinction. Using data from Willis (1980) on the bird species in remnant forest patches in the state of São Paulo, Brazil, they found that the species most prone to extinction were those that are constitutively rare or that specialize on resources from patchily distributed habitats. The sizes of Willis' study sites ranged in area from 21 to 1400 hectares. It is interesting to note, however, that some of the very rare species (initial abundance of 0 to 5 individuals) still persisted after several decades in the smaller forest patches.

Rarity and reliance on a patchily distributed resource, however, were not the only significant factors. Species of birds that rely on resources (fruit and nectar) that are vulnerable to climatic fluctuation also had high failure rates.

In contrast to Terborgh and Winter's relatively straightforward results, Karr (1982) reported a much more complex pattern in the vulnerability of bird species on Barro Colorado Island in Lake Gatún in Panama; the island was isolated from the mainland by an impoundment during the construction of the Panama Canal early in this century.

Karr does not list initial rarity as a particularly significant factor, although he does not exclude it, especially in the first years following isolation. One of the best predictors of disappearance is susceptibility to predation by ground and near-ground predators such as snakes and coatimundis (*Nasua nasua*), whose populations on the island have increased, probably as a consequence of the decrease in large predators and human hunters.

Another correlate of extinction found by Karr is the requirement for resources or conditions that fluctuate seasonally. Karr believes that most of the extinctions are related to stresses during the dry season. Especially adverse are the occasional droughts. Before the area was isolated by the rising waters of Lake Gatún, many species probably abandoned it during each dry season, or at least during especially severe droughts. They would return when conditions were more favorable. Against this is a background of decreasing rainfall in Panama in the last 70 years, adding to the stress of the dry seasons. Now, recolonization is very unlikely because even the narrowest of water barriers discourages the dispersal of most forest birds.

Predicting Vulnerability to Extinction

One-dimensional lists such as that given in Table 1 are of limited use because they contain no information on the interaction of the threatening factors. Table 2 is an attempt to provide a framework of suggested interactions. The factors have been divided into two groups: (1) those that involve some kind of environmental change; (2) those that are intrinsic for the species or that characterize a constant relationship of the species to its environment. In theory, the table could be used to analyze the vulnerability of any species. The more exacerbating factors that characterize a species and its habitat, the greater is its vulnerability to extinction.

What conclusions can we reach from such a qualitative analysis? As shown by the rankings, species with relatively poor dispersal mechanisms

Table 2. *Extinction Vulnerability Analysis: Interactions Between Threatening Factors*[1]

Environmental Changes	Exacerbating Qualities					
	Rarity (Clumped)	Rarity (Dispersed)	Big (Low r)	Highly Social	Insular	Nonvagile
Other species						
Competition					X	
Predation				X	X	X
Disease	X		X	X	X	X
Hunting/ Collecting	X	X	X	X	X	X
Loss of Mutualists						X
Habitat/resources						
Catastrophe	X			X	X	X
Succession	X					X
Environmental variation						X
Long-term trends					X	X
Disturbance	X	X	X	X	X	X
Sum of rankings	5	2	3	5	7	9

[1] An X indicates that the particular interaction is probable.

appear to be extraordinarily susceptible to extinction. Plant species that are dependent on animals for the dispersal of their seeds may be especially prone to local extinction when their animal mutualists are also vulnerable.

The next most susceptible category appears to be island-dwelling populations. The reasons are that (1) such species cannot shift their ranges in the face of changes in the biological or physical environment, (2) such populations are subject to evolutionary deterioration in antipredator and antiherbivore defenses, and (3) such populations are subject to area effects including catastrophes (Frankel and Soulé, 1981, Chapter 2).

The vulnerability analysis omits mention of two intrinsic factors, demographic stochasticity and genetic deterioration, discussed in the preceding section. The reason for the omission is that these factors apply universally, given small enough population sizes. To include them would require adding a third dimension to the table.

The Forked-Path Domino Theory of Ecological Disintegration

Another way to approach the extinction problem, especially in relationship to management, is to consider the consequences of a relatively sudden decrease in area. Indeed, most of the problems that concern conservationists involve such an event. (Exceptions include hobby collecting, poaching, and other forms of hunting.) The model depicted in Figure 1 suggests that a decrease in area triggers a causal chain or network that ultimately leads to the extinction of populations. The flowchart indicates the relationships of the factors.

Note that there are two major and independent consequences of a decrease in area. The first is a critical reduction in the population sizes of large and rare species, thus greatly increasing their probability of extinction, for reasons to be set forth below. In turn, these primary extinctions set in motion a chain of events that can lead to the extinction of many other species.

The second major consequence of a decrease in area is usually isolation. In other words, the fragmentation of a natural area creates barriers to the dispersal of organisms between the fragments. The seriousness of such an event depends on the dispersal abilities and migratory behaviors of the affected species. Some species will cross barriers of unfavorable habitat with little or no hesitation, depending, of course, on the width of the unfavorable terrain. Others are reluctant to venture into unfamiliar habitat. Many tropical birds and butterflies are prevented from crossing rivers or deforested terrain by intrinsic inhibitions to dispersal (Ehrlich and Raven, 1969; Diamond, 1975; Terborgh, 1975).

Isolation, therefore, is the trigger for an erosion of species diversity. Ecological processes that are normal and even necessary for the maintenance of species diversity on a regional scale can be fatal on a small scale. Catastrophes such as landslides and fires, for example, can maintain a healthy balance of successional stages in a large forest, but in an isolated patch they can extinguish entire populations.

In addition, early successional stages can, simply by chance, be absent in a small area. For example, habitats created by treefalls in tropical humid forests are essential for some of the life history stages of mobile mutualists such as pollinators (Gilbert, 1980). The smaller the area, the less likely it is that a treefall will occur within a given time interval. If the interval between treefalls becomes too great, the animals that are dependent on the unique plant resources of these habitats will disappear.

The most important point to be made from this analysis is that any significant decrease in area sets in motion a whole series of consequences that together lead to (1) the extinction of large animals, (2) the disappearance of certain kinds of habitats, (3) and changes in the abundance of many other species. These changes, in turn, cause another series of secondary effects, which further reduces species diversity, alters the preexisting pattern of ecological interactions, and (4) produces unpredictable changes in the abundance and distribution of species within the habitat island.

In summary, two kinds of event paths are initiated by a decrease in habitat size. This can be visualized as the knocking over of a domino at the

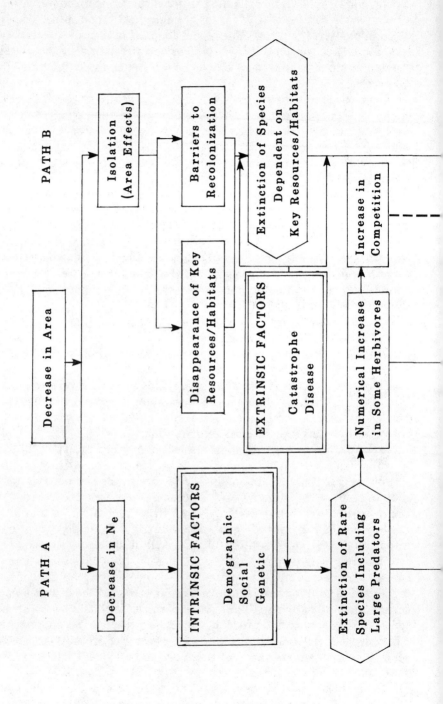

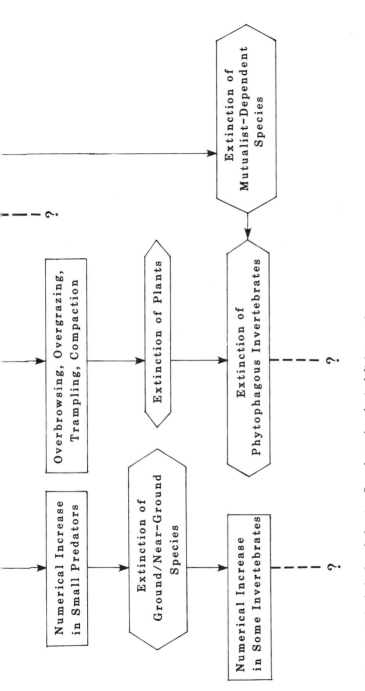

Figure 1. *The forked-path domino flowchart of ecological disintegration.*

head of a line of dominoes that divides into two lines. One of the paths eventually leads to the extinction of certain key species. In turn, this can initiate a chain of events leading to the loss of habitat and, indirectly, the extinction of other species. The second path (the effect of isolation and reduction in area, per se), is the absence or disappearance of certain habitats or resources, leading to the disappearance of species dependent on them. To carry the metaphor one step further: the rate at which dominoes fall depends on the steepness of the path. Analogously, the rate of ecological disintegration depends on the size of the habitat fragment.

Summary

Virtually all recent extinctions are anthropogenic. Except for island cases, we know very little about the process of natural extinction. The factors involved in the extinction of local populations, and their relative importance, may be entirely different from the syndrome of factors relevant to the extinction of a species. Three intrinsic factors may contribute to a relatively sudden extinction, especially in small populations; they are (1) demographic stochasticity, (2) social dysfunction, and (3) genetic deterioration. Isolation, per se, and associated area effects may precipitate causal cascades of extinctions. The major extinction factors, excepting demographic stochasticity and genetic deterioration, can be divided into two categories: (1) those associated with physical and biological changes in the environment, and (2) those that are fundamental characteristics of the species or that describe a constant interaction between the species and the environment. Such a classification can serve as the basis for a "species vulnerability analysis." Analysis of species vulnerability indicates that the populations that are most susceptible to extinction are (1) those with relatively poor dispersal mechanisms and (2) populations on natural or man-made islands. A decrease in habitat size (area) initiates two independent sequences of events: (1) the extinction of relatively rare species as a consequence of a decrease in population sizes; (2) isolation and the disappearance of essential resources or habitats and the subsequent extinction of species dependent on them. Ultimately, these two sequences of events interact to cause further disintegration of ecological interactions and further extinctions.

CHAPTER 8

Extinction, Survival, and Genetic Variation

John A. Beardmore

Introduction

The interest of geneticists in conservation is relatively recent and is not widespread, though Frankel, Soulé, and Wilcox have recently brought sharply into focus the critical role of genetical input to the formulation of conservation strategies (Frankel, 1981; Frankel and Soulé, 1981; Soulé and Wilcox, 1980).

While the origin of new species is a matter of considerable current interest, the subject of extinction is scarcely referred to in many excellent texts on evolutionary genetics. Even Lewontin (1974, p. 12), who says early in his influential book that "extinction is as much [an] aspect of evolution as is the phyletic evolution that is the subject of evolutionary genetics, strictly speaking" and "our present theories do not deal with [this] except on the most general and nonrigorous plane," does not return to the matter later. Omission of this sort seems likely to follow from a tacit belief that we in some way "understand" the genetical mechanisms of extinction better than those of speciation. It also may be that we are influenced by the thought that, as most of the extinction processes visible to us at present are caused directly or indirectly by man's activities, the relevance of studies of such processes to extinctions in an evolutionary perspective is slight. Both of these assumptions I believe to be unsound.

It is important to take a realistic attitude toward extinction. Paradoxical as it may appear, extinction is the norm for most forms of life, as was pointed out many years ago by Romer (1949). There will be some uncertainty as to whether this thesis necessarily holds for life forms such as bacteria and viruses, but it certainly does for more complex species. Of all the species that have ever existed, whether dinosaurs, ungulates, or trees, the majority have gone extinct. Thus the 1.5 million species very conservatively estimated to exist now represent less than 1% of the species ever present on this planet (Nei, 1975).

The reasons for this have their roots in the relative ability of species to adapt to altered circumstances. A rapid rise in temperature may pose insuperable problems of survival for a species that, though abundant, is localized in its distribution, whereas the same change does not bring about the extinction of a less abundant but more widely dispersed species. The precipitous decline in the numbers of the sun star (*Helianthus kubiniji*) in the Gulf of California has been linked to climatic changes in 1977–1978, which brought about a period of elevated sea temperature (Dungan et al., 1982). With a relatively narrow distribution this is, then, a species exposed to a risk of the type referred to above.

Similarly, epidemics of many kinds pose a constant threat to some mammalian species that may be limited to one or a few populations. Some examples are given by Frankel and Soulé (1981).

At a different level, parasitic species often will be totally dependent upon their hosts, and thus the evolutionary prospects of a parasite are firmly linked, although at secondhand, to the variety of biological, physical, and climatic factors that influence the host species (Chapter 22 by Futuyma).

The effects (usually unfavorable) of man's activities on natural populations are the immediate cause of the concern felt by most conservationists. However, we ought to remind ourselves that the pressure of natural forces on natural populations is often dramatic and can display both acute and chronic consequences. Thus the Mt. St. Helens volcanic eruption in 1980 in the state of Washington devastated, more or less immediately, a considerable area, with dramatic effects upon the local flora and fauna. Long-term effects of a less obvious nature are to be seen in the chemical changes in soils over much wider areas. For example, the content of mercury in soils exposed to volcanic fallout may be one or two orders of magnitude greater than that characteristic of soils not so exposed (Siegel and Siegel, 1982). (Note Chapter 5 by Liu and Godt for effects of some heavy metals.)

Some extinctions may take place as very generalized events involving a great many species occurring over relatively short periods of time. Thus, at the end of the Cretaceous period much extinction of both plant and animal species took place. This is likely to have resulted from a massive disturbance of some type, possibly a cometary event. In such circumstances there is little scope for selection to operate even in very large populations. However, this event is of particular interest because it has been linked to more wide-reaching consequences. Hsu et al. (1982) link the mass reduction in pelagic species to reduced ocean productivity, which led to increased atmospheric CO_2 and ultimately increased temperature (both aquatic and atmospheric). Some species were able to adapt to the changed thermal regime but others were not; Hsu et al. allude in particular to the dinosaurs as a group unable to tolerate the rapid increase in thermal stress.

What this example portrays is a principle that we always must bear in mind: isolating particular groups or species for consideration is a highly artificial procedure. A species is always part of a more or less complex ecosystem dependent in all sorts of ways upon other species. The removal or introduction of even a single species may have far-reaching consequences affecting numbers of other species in some cases having little direct connection with any target species (Frankel and Soulé, 1981).

It is also important that we bear in mind the immense resilience of many species to pressures caused by man. Thus the response of rabbit (Leporidae) populations to myxamatosis or the effects of Dutch elm disease (*Ceratostomella ulmi*) on the elm (*Ulmus*) in the United Kingdom, while initially dramatic, pose problems that are often ethical or esthetic rather than truly evolutionary (contrast Chapter 22 by Futuyma). A striking example of the ability of a natural population to adapt to, and indeed to exploit, a man-made environmental change is provided by the euglossine bee *Eufriesia purpurata*. In Brazil males of this species have been observed to collect DDT from the walls of sprayed houses, presumably for nutritional purposes. The bees display no ill effects from the DDT despite containing an average of 2000 μg per bee. For comparison the LD_{50} (50% lethal dose) for the honey bee (*Apis mellifera*) is estimated to be about 6 μg per bee (Roberts et al., 1982).

The foregoing discussion illustrates the types of pressures to which populations may be exposed and the variability of responses that may be observed. Central to the ability to adapt is the amount and nature of genetic variation available to a population; the purpose of this chapter is to assess the significance of genetic variation in general and heterozygosity in particular for survival and extinction. To do this I shall survey the nature of genetic variation, the levels characteristic of different situations, and the ways in which selection can operate upon it.

Genetic Variation

1. Organization of the Genetic Material

The genetic information encoded in DNA is distributed among a set of chromosomes that vary in number in a species-specific manner over about two orders of magnitude. Most complex organisms have a diploid number that falls into the band between 10 and 50. In eukaryotic species the haploid amount of DNA varies from about 3×10^7 nucleotide pairs in fungi through 5×10^9 nucleotide pairs in mammals to about 2×10^{12} nucleotide pairs in some vascular plants (Watson, 1970). The numbers of genes that could be coded for by such amounts of DNA can be calculated readily, but the numbers so obtained are vastly greater than seem likely to be required or used. The existence of noncoding segments (introns) within genes, as well as highly repeated sequences and bizarre elements such as transposons, together with the existence of nontranscribed regions between genes, collectively indicate that, in fact, much of the DNA in the genome does not have a coding function. Present thinking is that only a small fraction of the genome codes for polypeptide products. In the pomace fly *Drosophila melanogaster* this has been estimated to be about 3%, which by extension would indicate that there are about 100,000 functional structural loci in a typical mammal.

Genetic variation present in a population ultimately has its origin in mutation; the production of particular genomic constitutions results from the union of gametes, in the reproduction of which gene reshuffling typically occurs first by independent assortment of chromosome pairs and second by recombination within chromosome pairs (Chapter 2). The effect of the first of these is directly dependent upon the number of chromosome pairs, and the second upon the amount of crossing over that takes place. Both of these characters are themselves under genetic control, though, as chromosome number is constant in the vast majority of species, it is difficult to perceive that there are genetic determinants of it. In a few species, however, this can be shown. In the dog whelk (*Nucellus lapillus*) individuals may have a chromosome number between 26 and 36 depending on the extent to which fusion of arms (two acrocentric chromosomes fuse to form a single metacentric chromosome) takes place (Staiger, 1957; Bantock and Cockayne, 1975). There are also numerous cases of insects and plants with variable numbers of extra so-called B chromosomes (in addition to the normal chromosomal set), which appear to influence the fitness of their possessors (Berry, 1977). That recombination is under genetic control is demonstrated by selection experiments in which the amount of recombination in a particular segment of the *Drosophila melanogaster* genome could be increased from 15% to 22% and decreased to 8.5%. The genes controlling these changes are widely dispersed throughout the genome (Chinnici, 1971a, 1971b) and are likely therefore to exist in appreciable number.

The linear sequence of genes within chromosomes may be protected against recombination by means of chromosomal inversions, which, in effect, make crossing over ineffective largely through the production of inviable products from crossovers. Inversions can be very large as in *Drosophilia pseudoobscura* and may then be responsible for causing massive amounts of linkage disequilibrium. In the absence of such chromosome constraints, linkage disequilibrium over any appreciable genetic map distance is much more difficult to maintain, unless there are substantial viability differentials involved. The best-known examples of linkage disequilibrium are those involving the breeding system, as in the genes controlling the relative style and stamen lengths and related characters in heterostylous *Primula* (primrose) and the mimetic polymorphism of body color and form in *Papilio dardanus* (a swallowtail butterfly: Clarke and Sheppard, 1960). However, the HLA (human leukocyte antigen) system in man also has provided good evidence. The HLA system is very complex but in its simplest form comprises four loci, *D*, *A*, *C*, and *B*, with about 1.5% recombination between the two outer loci. Certain chromosome types— e.g., *A1 B8*, *C3 B40*, and *D3 B8*—are several times as frequent in Caucasian populations as would be expected under random distribution. Other more complex chromosome arrangements are also more frequent than expected— e.g., *A23 CW4 B12* (Svejgaard et al., 1979). The extent to which linkage disequilibrium not involving chromosomal rearrangements (such as inversions) exists in populations is unknown. It may be common over short genetic map distances (say about 1 cM), but in general the experimental evidence for its existence over larger distances is negative (Langley et al., 1974).

2. The Nature of Genetic Variation

It might be supposed that at any gene locus a best possible allele could be produced, which would be then "used" in such a way that all individuals were homozygous for this allele. Extended to all loci such reasoning would predict that the normal state of affairs would be a totally or nearly homozygous genome (Chapter 11 by Carson). In fact this is not so, and a typical plant or animal population contains an appreciable amount of genetic variation. We can divide this variation into four categories, though distinguishing between them is far from easy and in many cases impossible.

1. Variation arising as the result of recurrent mutational changes.
2. Variation arising from inflow of genes from other populations or species ("migration").
3. Variation arising from stochastic processes such as genetic drift.
4. Variation held in the population by some form of selection.

In this chapter I do not propose to devote much attention to category 1 except to point out that it is possible that the mutational input to some species is raised by some of man's activities, notably the poorly controlled use and disposal of substances that are mutagenic (Beardmore, 1980). Although the load of deleterious mutations is thereby increased, this is unlikely to prove a significant problem for the species concerned if it is reasonably abundant in number. In large populations of insects the presence of mutagenic insecticides (Fishbein, 1976) may indeed prove helpful in generating mutations conferring some degree of resistance to the insecticide. In cases of small population size, however, it would seem sensible to watch carefully for the presence of known or suspected mutagens in the habitat and to minimize exposure as far as possible.

Category 2 variation is of a type that is frequently open to man to influence or control. In a general way it can be argued that it is a good strategy for a species to be geographically divided with some gene flow between populations; other contributors to this volume discuss the effects in detail (Allendorf, Chapter 3; Foose, Chapter 23; Chesser, Chapter 4).

Categories of variations 3 and 4 have to be considered together. The central debate in evolutionary genetics today is the extent to which genetic variation in populations is maintained by deterministic or by stochastic processes. However, for the purposes of this book it is not, I submit, a matter of critical importance. What we mainly are concerned with here are questions relating to the amount of genetic variation, the factors that depress it, and the effects such depression has on the populations concerned. Clearly, however, the factor of population size is more important in this context. Broadly we can say that the smaller the effective population size, the more important stochastic events become (including the random loss of genes).

One other factor not considered in detail also ought to be mentioned here. This is the reproductive capacity, or, more accurately, the scale of zygote production. Species with large numbers of zygotes per breeding pair are much better able to generate, through recombination, rare genotypes that may be of considerable importance in survival under changing conditions (Chapter 19 by Namkoong).

3. The Scale of Genetic Variation

The advent of electrophoretic and refined immunological techniques over the last twenty years or so has led to a massive increase in knowledge of the gene pools of a great variety of life forms. The basic feature uncovered by these studies is that most species possess very large amounts of genetic variation; that is, they are highly polymorphic at many loci. The loci surveyed in such cases may not be random with respect to the genome (Lewontin, 1974), but we do not as yet know of any systematic bias inherent in the approaches made. While most of the genes examined are structural loci, the work of Powell (1979) shows that regulatory loci also display polymorphism.

Table 1. *Mean Levels of Polymorphism (P) and Heterozygosity (H) in a Range of Life Forms*[1]

Group	Species	P	H	Number of Individuals Sampled
Vascular plants	*Lycopodium lucidulum* (shining clubmoss)	0.10	0.060	241
	Phlox drummondii (Drummond's phlox)	0.19	0.040	100
Invertebrates	*Phoronopsis viridis* (phoronid "worm")	0.26	0.088	120
	Limulus polyphemus (horseshoe crab)	0.25	0.057	64
	Euphausia superba (krill)	0.14	0.058	127
	Drosophila pseudoobscura ("fruit" fly)	0.42	0.123	1265
Vertebrates	*Pleuronectes platessa* (European plaice, a flounder)	0.26	0.102	2270
	Mus musculus (house mouse)	0.26	0.085	99
	Mirounga leonina (southern elephant seal)	0.28	0.028	42
	Macaca fuscata (Japanese macaque)	0.10	0.014	1002

[1] Mainly from Nevo (1978).

Typically a species will have about 25% of its loci polymorphic and the average heterozygosity of an individual will be about 7%, though such figures conceal considerable variations, as the data for a range of life forms (Table 1) reveal. Some general conclusions are stated in the analysis carried out by Nevo (1978), drawing heavily on the work of others. These are as follows: Invertebrates have higher levels of heterozygosity on average than do vertebrates, plants being intermediate (Chapter 20 by Hamrick). Tropical species possess more genetic variation than do temperate or cosmopolitan species, and habitat generalists more than habitat specialists (Tables 2 and 3). Mainland populations also tend to show higher levels of genetic variation than do island populations of the same species, though the sample sizes are small and the differences not always statistically significant.

Finally, the values for polymorphism and heterozygosity generally are highly significantly correlated.

Table 2. Comparisons of Mean Values for Heterozygosity (H) in Different Groups of Organisms[1]

Lower genetic variation	H	Number of species	Higher genetic variation	H	Number of species
Vertebrates	0.049	135	Plants	0.071	
Cosmopolitan	0.048		Invertebrates	0.112	93
Temperate	0.066	168	Tropical	0.109	45
Habitat specialists	0.046	120	Habitat generalists	0.106	117

[1]Data from Nevo (1978).

Table 3. *Genetic Variation Expressed as Average Heterozygosity (H) for Different Groups of Habitat Specialists and Habitat Generalists*[1]

Habitat type	Plants		Invertebrates		Vertebrates		Overall	
	H	No.	H	No.	H	No.	H	No.
Specialists	0.04	2	0.06	33	0.04	82	0.05	117
Generalists	0.08	12	0.15	54	0.07	56	0.11	122

[1] No. = number of species. All four habitat comparisons are highly significant. From Nevo (1978).

Genetic Variation and Selection

1. Genetic Architecture

What Mather (1973) has termed the genetic architecture of phenotypic characters is extremely important in the consideration of how freely a character can vary. In general, the further removed the character is from the central components of Darwinian fitness like viability, mating behavior, and fertility, the more likely it is that the genetic variance affecting the character present in the population will have higher additive and lower dominance and epistatic components. This means that selection can act relatively quickly on simple morphological characters and only relatively slowly on a character like complex mating behavior. Even so there are significant differences between species. Great tits (*Parus major*) have considerable variation in clutch size, which allows some adjustment of the number of young birds produced to levels of food supply, while the griffon vulture (*Gyps fulvus*) lays only one egg per clutch. If a species such as this vulture is threatened with extinction because of egg losses, an upward move in clutch size would be highly adaptive but probably very difficult indeed to achieve through natural selection.

2. Types of Selection and Their Effects upon Genetic Variation

The most obvious selection type is directional, in which a character becomes progressively changed as a result of some advantage accruing to individuals in the population with a phenotype other than the mean phenotype of the population. Changes of this sort may be rapid or slow, linear or nonlinear, and involve few or many loci, depending upon the character and the species. Directional selection may be expected to reduce genetic variation, and if it is severe the diminution will be severe. In artificial breeding programs it is thus often better to reduce the selection differential below the maximum possible so as to include a greater diversity of phenotypes to go forward to the next generation.

There are, however, a number of other types of selection, and inevitably there have been many theoretical approaches relating genetic variation to selection regimes. These can be condensed into a variety of classifications of selection regimes. Thus, Wallace (1981) has made a useful operational distinction between what he terms "hard" and "soft" types of selection. By *hard* is meant that selection operates on the phenotype of the individual as an absolute criterion independent of all other factors (as, for example, with unconditional lethal genes), by *soft* that selection is an interaction of the phenotype of the individual with the kind, frequency, and distribution of phenotypes of the same and other species; that is, it is both frequency- and density-dependent. In this classification frequency-dependent density-independent and density-dependent frequency-independent are types of selection that are neither hard nor soft. However, the general logic of the division would lead me to include these under soft selection.

Accepting that the carrying capacity of a typical given habitat has a more or less stable boundary, it is clear that most selection in natural populations will be of a soft (or softish) type.

Other classifications of selection regimes distinguish between (1) heterozygote superiority (absolute or composite), frequency-dependent, or multiple-niche types of selection, or (2) between stabilizing and disruptive types. None of these classifications is completely satisfactory. Multiple-niche, frequency-dependent, and disruptive (or diversifying) types of selection all have in common that more than one optimum phenotype exists and therefore these types of selection promote the retention of genetic diversity within populations.

Stabilizing selection is a term used to describe the situation found when average types survive and reproduce better than more extreme types. A classical example is that of the house sparrows (*Passer domesticus*) observed by Bumpus (1896) to have succumbed to, or survived, a storm. Individuals of large and small sizes were overrepresented among the victims, and individuals of intermediate size were overrepresented among the survivors. Many other examples drawn from a wide variety of species are known, one interesting case being provided by Lack's work on the common swift (*Apus apus*). Clutch size is variable, and on general principles some of the variation must be genetic. The unweighted mean number of live progeny per brood over all clutch sizes in 4 years varied between 1.8 and 2.6 birds, and the proportions of clutches with 2, 3, and 4 eggs also varied (Table 4). However, over the whole period the mean progeny per brood for 2-egg clutches was 1.92, that for 4-egg clutches was 1.85, while for 3-egg clutches it was 2.6 (Berry, 1977). A further example is drawn from height in man (*Homo sapiens*), where Conterio and Cavalli-Sforza (1960) found that the variance of height in a large sample of Italian males was sig-

Table 4. *Mean Live Progeny per Female for Different Clutch Sizes and Different Years in the Common Swift* Apus apus[1]

	Clutch size			
Years	2	3	4	Yearly mean
1	1.9	2.8	2.0	2.23
2	2.0	3.0	2.8	2.60
3	1.9	2.3	1.2	1.80
4	1.9	2.3	1.4	1.85
Clutch size mean	1.92	2.6	1.85	

[1] From Berry (1977).

nificantly greater among the unmarried than among the married men. The consequences for differential reproduction and hence Darwinian fitness are clear.

What are the genetic consequences of stabilizing selection? Laboratory experiments using a selective pressure determined by the experimenter (e.g., Gibson and Bradley, 1974) suggest that genetic variation is decreased and that therefore overall long-term fitness also decreases. However, a distinction needs to be drawn between the results of artificial selection and the results of natural selection. In nature, stabilizing selection is thought to act on many characters, but the intensity will vary considerably from case to case. Little is known directly about the genetic effects of natural stabilizing selection, but one recent study bears on this point. In the guppy *Poecilia reticulata* very strong stabilizing selection operates on the caudal fin ray number. The intensity of this selection is such that under laboratory conditions individuals of the most extreme caudal fin ray phenotypes are only half as likely to survive to 1 year of age (= peak breeding period) as those of the optimum caudal fin ray number. It also appears that the central phenotype is markedly more heterozygous (about 50%) than the extreme types. Thus in this instance natural stabilizing selection leads to retention of heterozygosity (Beardmore and Shami, 1979).

3. What Is the Function of Genetic Variation?

Accepting that a fraction (at present not easily definable) of all genetic variation in a population derives from mutational, migrational, or stochastic causes, what function or functions do we ascribe to the rest? It is truly remarkable that despite the vast accretion of knowledge about the scale of genetic polymorphism in recent years so few hard data on its adaptive significance have been gathered. We are thus left with a small number of examples in which it has been possible to ascribe well-defined differences of fitness to different genotypes within a polymorphism and to show that

Table 5. *Cases of Dependence of Allele Frequencies in Polymorphisms upon Environmental Factors*

Species	Loci	Environmental variable	References
Balanus amphitrite (striped barnacle)	7 loci, especially *Est-3*	H₂O temperature 24°–35°C	Nevo et al., 1977
Catostomus clarkii (desert sucker)	*Est*	H₂O temperature	Koehn, 1969
Drosophila melanogaster (fruit fly)	*Amy*	Sucrose/starch in medium	de Jong et al., 1972
Homo sapiens (man)	*Hbβ*, several alleles *G6PDH*	Malarial parasite	Vogel and Motulsky, 1979

these are in turn dependent upon the nature of the environment; a sample is given in Table 5.

The notion that genetic variety is adaptive in that it enables better utilization of habitats diverse in time and space goes back at least to Darwin's statement (1859) that "the more diversified the descendants from any one species . . . by so much will they be better enabled to seize on many widely diversified places in the polity of nature, and so enabled to increase in numbers."

Necessarily much of the evidence for a viewpoint of this sort is drawn either from laboratory studies involving a small number of variables or from natural populations in which more elaborate surveys relating environmental variation to genetic variation may be possible.

One of the most convincing studies of an adaptive response of a polymorphism to ecological and climatic variation is Dobzhansky's elegant demonstration that the frequencies of the Arrowhead, Standard, and Chiricahua gene arrangements in *Drosophila pseudoobscura* follow a regular and cyclic pattern with the march of the seasons (Dobzhansky, 1943). It could thus be argued that such a polymorphic population is better adapted to its environment than a monomorphic population, although strictly speaking the changes are simply a reflection of the fitness values of the different genotypes altering differentially with altered ecological circumstances. However, it has been shown that polymorphic populations of *Drosophila* are able, in laboratory conditions, to utilize a defined level of resources more efficiently than monomorphic populations, and hence it is likely that in nature too larger population size and/or biomass can be maintained by polymorphic populations (Beardmore et al., 1960).

4. Comparisons of Populations with Differing Levels of Heterozygosity

For many years it has been known that outbreeding species, when inbred by mating of relatives, characteristically exhibit inbreeding depression, but the extent of this depression in such inbred organisms appears to be a function both of the extent of inbreeding and of the particular species concerned. Some examples of the relative advantages of outbred (heterozygous) genotypes in a range of species are given in Table 6, and other examples are provided in Chapter 10 by Ralls and Ballou.

The effects of inbreeding are progressively deleterious and are generally most severe in populations with no previous history of inbreeding. Frankel and Soulé (1981, p. 68) calculate that a change in the inbreeding coefficient of 10% (equal to roughly one generation of half-sib mating) could produce a loss of 25% in reproductive performance. This will obviously be a major consideration in species with low reproductive capacity, though some have a breeding system that normally involves inbreeding and, because many

Table 6. *Relative Advantages of Outbred to Inbred Individuals in Different Species*

Species	Character	Relative advantage outbred/inbred	References
Man (*Homo sapiens*)	Age at death	1.80	Roberts, 1980
Man (*Homo sapiens*)	Congenital cataract	1.36	Nakajima et al., 1980
Guinea pig (*Cavia porcella*)	Number young weaned	2.29	Wright, 1922c
Mouse (*Mus musculus*)	Litter size	1.69	Bowman and Falconer, 1960
Quail (*Coturnix japorica*)	Breeding performance	2.93	Abplanalp, 1974
Flax (*Linum usitatissimum*)	Seed yield	1.40	Carnahan, 1947

recessives have already been exposed, are not likely to experience such a large decrement of population fitness. However, Templeton and Read (Chapter 15) show that good management can considerably mitigate the deleterious effects of inbreeding depression.

Concern about the manifestation in small populations of part of the recessive genetic load carried is understandable. However, as Benirschke (1977) has pointed out, culling individuals only should be undertaken "with the greatest care and [with the] full knowledge that it will reduce genetic heterogeneity." A genetic equivalent of death is sterilization, and an example of this is to be found in the recent castration of a black rhinoceros (*Diceros bicornis minor*) in the Addo Park in South Africa because he lacked the right external ear — a deformity possibly caused by a sex-linked gene. The judgment to eliminate this animal from the rest of the gene pool was then based on evidence from which the balance of long-term advantage perhaps could not be inferred unequivocally (de Vos and Braack, 1980).

As already indicated, there is good evidence that under laboratory conditions some populations with greater degrees of heterozygosity are able to maintain larger population sizes or biomass than less heterozygous populations. Thus Ayala (1968) found interpopulational hybrid populations of *Drosophila* to be able to maintain a size about one-fifth larger than that for populations derived from a single locality. Very low levels of heterozygosity appear usually to depress the overall fitness of a population, though in natural populations this may be exceedingly difficult to see or to measure. However, there is evidence that in some species devices to avoid inbreeding, and therefore extreme homozygosity, have evolved (Hoogland, 1982). The work of Wallace (summarized in Wallace, 1981) suggests that in a highly homozygous population even randomly induced heterozygosity resulting from the use of a mutagen such as ionizing radiation may increase fitness.

Some investigators have attempted to correlate electrophoretically assayed heterozygosity with components of fitness such as growth rate (Singh and Zouros, 1978; Beardmore and Ward, 1977; Ferguson, personal communication). The data are not entirely consistent, and indeed, since usually only on the order of five loci are examined, it would be surprising if they were. However, the observation of Smith et al. (1975) that heterozygosity tends to increase in decreasing populations and decrease in expanding populations of small mammals does suggest that the advantage of heterozygosity may be expressed most readily under severe environmental conditions — a point of view discussed at length by Parsons (1973).

Recognition of the considerable reservoir of genetic diversity within most gene pools has led to many efforts to establish adaptive relationships between genetic variability per se (often measured simply as average heterozygosity) and components of the environments in which the populations concerned live. Some general correlations arising from the review by Nevo (1978) have already been mentioned, but these suffer from the fact

Table 7. *Habitat Type and Genetic Variation in Four European Species of Frogs and Toads*[1]

Species	Habitat	P	H	Number of populations studied
Pelobates syriacus (Syrian spadefoot)	Specialist	0.09	0.023	2
Rana ridibunda (a marsh frog)	Intermediate	0.38	0.073	7
Hyla arborea (common tree frog)	Intermediate	0.43	0.074	7
Bufo viridis (a European toad)	Generalist	0.47	0.141	7

[1] P = average level of polymorphism; H = average level of heterozygosity; 26–32 loci used. From Nevo (1976).

that they are drawn very broadly indeed. An example of a survey with somewhat greater resolving power is given in Table 7. Here we see within a group of four species of frogs and toads (*Anura*) a clear gradient of genetic diversity, measured either as average polymorphism (P) or average heterozygosity (H), and ecological diversity of habitat. It is also worth mentioning one large study of 44 decapod crustacean species in which positive correlations of genetic variation with trophic instability, trophic generality, and the number of species per genus were established. While the correlations are only of moderate size, the pattern for these three indices of environmental diversity (seen in Table 8) looks consistent (Nelson and Hedgecock, 1980).

A summary of some views about relationships between spatial and temporal variation of the environment and genetic variation is given in Table 9. It will be seen that considerable diversity of opinion exists among geneticists as to the factors promoting the retention of genetic variability, and a considerable debate rages in the literature (Hedrick et al., 1976).

Table 8. *Relationships (Measured by Spearman Rank Correlation Coefficients) between Heterozygosity (H) and Polymorphism (P) at Ten Loci, and Various Biological Parameters, in Forty-Four Species of Decapods*[1]

	Trophic instability	Trophic generalism	Species per genus
H	0.300*	0.228	0.318*
P_{95}	0.293*	0.302*	0.287*

*Significant at the 5% level.
[1] From Nelson and Hedgecock (1980).

Table 9. Views of the Effects of Environmental Variation upon the Genetic Variation in Gene Pools of Populations

Type of environmental variation	Postulated effect on gene pool	References	Comments
Multiple spatial niche (coarse grained)	Promotes polymorphism	Selander and Kaufman, 1973	"Ecotype" differentiation
Unpredictable + predictable short-term variation	Promotes heterozygosity → polymorphism	Soulé, 1976	
Unpredictable trophic variation	Promotes monomorphism	Ayala and Valentine, 1976	Data inconsistent
Predictable long-term variation	Promotes polymorphism → heterozygosity	Haldane and Jayakar, 1963	Not contentious; good laboratory data
Unpredictable long-term variation	Greater extinction where genetic variation lowest	Thoday, 1953	Few data available

In my view the best evidence on this matter comes from laboratory experiments, which, although only a pale abstraction of the complexity of nature, allow far better control. Table 10 summarizes the results of six experiments undertaken in different laboratories over a considerable period. The experimental procedures are different in each experiment, and not all the individual comparisons between populations in uniform and variable environments are statistically significant. What is impressive, however, is that all comparisons show greater genetic variability in the populations in the more variable environments. That fifteen comparisons would, by chance, produce such a result has a probability of only 3×10^{-5}. It seems safe, therefore, to conclude that regular spatial and temporal variation in the environment promotes the retention of genetic variability.

Many years ago Timofeeff-Ressovsky (1935) showed that strains of *Drosophila funebris* differed significantly in their ability to tolerate high and low temperatures in the laboratory. One can correlate estimates of the annual temperature variation in the locations yielding these strains and the mean viability in three temperature conditions used: it is positive and significant (Beardmore, 1966). It is interesting to compare these findings with those of Bryant (1974), who analyzed data from electrophoretic surveys of geographic populations of a range of species in relation to four measures of climatic variables. He found that about 70% of the variation in heterozygosity between populations could be accounted for by measures of temporal climatic variation. This is a surprisingly high figure, though Timofeeff-Ressovsky's results might have given a clue to the importance of seasonal variation in determining the gene pools of populations of *Drosophila* and similar organisms.

Somero and Soulé (1974) have argued, in a survey of teleost (Teleostei) fishes over a wide geographic range, that there is no support for a relation between genetic variation and environmental variation. However, this study embraced only 13 species drawn from 8 families, and only one environmental variable was considered. Studies of this kind in nature are best carried out on groups of related species in environments not too geographically dispersed. A recent survey of 9 goby (Gobiidae) fish species (Wallis, 1981) produced a reasonably convincing relationship between greater genetic variation (measured over 31 loci) and broader environmental heterogeneity (based on 6 factors).

Soulé (1976) has further developed his view in a provocative article in which he argues that the level of genetic variation is primarily determined by the factors of population size and the interval that has elapsed since the last restriction or bottleneck. That there is a relationship between population size and heterozygosity is undeniable, and this can be interpreted equally well on selectionist and nonselectionist grounds. The problem of trying to estimate how long populations have been in existence is clearly a large one, and tests of this hypothesis will not be easy. However, I should like to mention one set of observations that bear on this point. They con-

Table 10. *Effects on Genetic Variation of Relatively Variable Environments Expressed as Ratio of Genetic Variation of Variable Environment Population to Genetic Variation of Controls in Drosophila Laboratory Populations*[1]

References	Measure of genetic variation	Type of variation in environment				
		Short temporal	Medium temporal	Long temporal	Spatial	Spatial and temporal
Beardmore and Levine, 1963	V_A SP	1.60 (2)	—	—	—	—
Long, 1970	V_A SP	1.28 (3)	1.31 (3)	1.20 (3)	—	—
McDonald and Ayala, 1974	H (20 loci)	—	—	—	1.27 1v (4)	—
		—	—	—	1.38 2v (5)	—
		—	—	—	1.49 3v (2)	—
		—	—	—	1.27 4v (4)	—
Minawa and Birley, 1975	H (7 loci)	1.10 (4)	—	—	—	—
Powell and Wistrand, 1978	H (9 loci)	—	—	—	—	1.23 1v (15)
						1.36 2v (3)
						1.41 3v (6)
Mackay, 1981	V_A SP	—	3.34 (4)	3.83 (4)	2.11 (4)	—

Probability that by chance 15 comparisons would all have ratios $> 1 = 0.00031$

[1] Figures in parentheses = number of populations used, v = variable, V_A = additive genetic variance, SP = sternopleural bristle number, H = heterozygosity.

cern the brine shrimp *Artemia*, which we believe to have evolved in the Mediterranean region and to have spread outward, producing several bisexual and many parthenogenetic species. Estimates for heterozygosity based on 24 loci are given in Table 11. *Artemia urmiana* is thought to be the most recently derived species and is so far known from only one lake in Iran; *A. persimilis* is the next most recent and is known from only one locality in Argentina; *A. franciscana* is widespread in North America and the Caribbean; *A. salina* is the European species and presumed to be closest to the ancestral type. The values for average heterozygosity (H) run in the opposite direction to what might be expected from what is known of species age and abundance.

To what extent do higher levels of genetic variation offer greater chances of evolutionary survival? There are several pieces of evidence that bear on this. Long (1970) constructed a fitness index based upon a combination of productivity, competitive ability, and performance in two novel environments and examined this in relation to genetic variation in 12 populations of *Drosophila melanogaster*. Figure 1 shows the relationship, which is positive and significant. Beardmore (1970) assayed a large number of newly captured strains of *D. melanogaster* for productivity in four different environments and for genetic variation affecting sternopleural bristle number. The regression of mean productivity upon additive genetic variance was positive and significant at the 0.02 level. A negative correlation also could be demonstrated between mean productivity and between replicate variance in productivity. In other words, the best strains also were more uniform in their utilization of the environment to produce biomass.

A final question needs to be posed: Does rapid evolution or speciation tend to exhaust genetic variation in the species or base species? This appears to be a plausible hypothesis, but critical evidence to support it is notably lacking. Such evidence as is available suggests that it may not be tenable. Avise (1977) on the basis of work with two families of North American fishes demonstrated that the highly speciose Cyprinidae (carps

Table 11. *Estimates of Average Heterozygosity*, H *(Twenty-Four Loci), in Bisexual Species of* Artemia, Brine Shrimp[1]

Species	Distribution	Presumed age	H
A. salina	Europe	Closest to ancestral type	0.089
A. franciscana	North America and Caribbean	Third most recent	0.091
A. persimilis	Argentina	Second most recent	0.130
A. urmiana	Iran	Most recent	0.135

[1] From Beardmore and Abreu Grobois, in press.

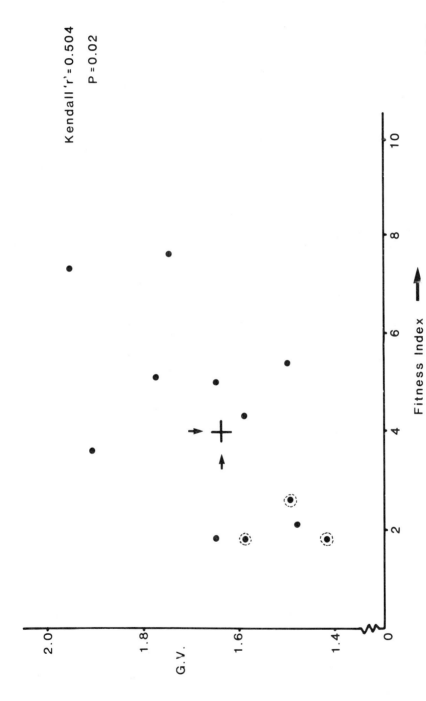

Figure 1. Correlation between genetic variation (G.V., as measured by additive genetic variance) and fitness in Drosophila fly populations. + marks the mean of all populations. Dotted circles indicate the constant-environment populations (see Table 10). Data from Long (1970).

145

and minnows) and the depauperate Centrarchidae (sunfishes and freshwater basses) have virtually identical heterozygosity levels. Other relevant data come from Skibinski and Ward (1982), who, in a large survey of 31 loci in vertebrates, concluded that there is a large and highly significant correlation between average heterozygosity and evolutionary rate of change. This survey considered only loci not species, but it would be reasonable to infer that this evidence also does not lend any weight to the view that more rapid rates of change, including speciation, tend to run down genetic variation. It is also relevant to note that an extremely slowly evolving species such as *Limulus polyphemus* (the horseshoe or king crab) has a level of genetic variation comparable with that found in many other faster-evolving species (Selander et al., 1970).

5. Heterozygosity, Buffering, and Phenotypic Variance between and within Individuals

It is a general feature of inbred lines of animals and plants that they display a considerable degree of phenotypic variation despite low levels of genetic variation (Lerner, 1954). Comparison of inbred and outbred stocks of outbreeding species shows that the phenotypic variance of inbreds often may be considerably greater than that of outbred material (Table 12). The general course of change in phenotypic variation with inbreeding is that, at the start, each generation of inbreeding leads to a reduction in phenotypic variance; that is, genetic variance and phenotypic variance proceed hand in hand. Once a certain level of inbreeding has been reached, however, the decrease in genetic variation caused by further inbreeding leads to an increase in phenotypic variance (Figure 2). While the precise pattern of this response will differ from species to species and from population to population, there seem to be good reasons for accepting that this pattern generally holds for outbreeding organisms.

Table 12. *Ratio of Inbred to Outbred Phenotypic Variances in Different Species*

Species	Character	Variance ratio	References
Gallus gallus (chicken)	Egg weight	1.465	Shultz, 1952
Mus musculus (house mouse)	Body weight	1.52	Falconer and King, 1953
Primula sinensis (Chinese primrose)	Stigma length	4.55	Mather, 1950
Zea mays (maize)	Plant height	1.46	Shank and Adam, 1960

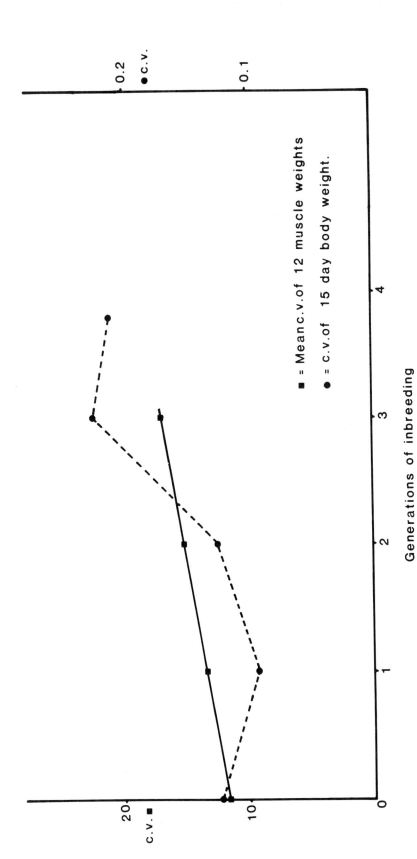

Figure 2. Mean coefficient of variation (C.V.) for (1) weight of muscles (mean of 12, scale at left), and (2) 15-day body weight (scale at right) during inbreeding of Japanese quail. From Benoit and Boesiger (1974), Boesiger (1974); see also Chai (1966).

Table 13. *Relationship of Heterozygosity at Electrophoretically Detectable Loci to Phenotypic Variance of Morphological Characters in Different Species*

Species	Number of characters	Number of loci	Genotypes in which the phenotype has greater variance	References
Uta stansburiana (side-blotched lizard)	8	20	Heterozygous	Soulé and Yang, 1973
Fundulus heteroclitus (mummichog, a killifish)	7	5	Homozygous	Mitton, 1978
Danaus plexippus (monarch butterfly)	2	6	Homozygous	Eanes, 1978
Poecilia reticulata (guppy)	1	4	Heterozygous	Beardmore and Shami, 1979
Zonotrichia capensis (rufous-collared sparrow)	11	4	Same	Handford, 1980
Pleuronectes platessa (European plaice, a flounder)	3	8	Same	McAndrew et al., 1982

The most generally accepted explanation of the lower susceptibility of relatively heterozygous genotypes to environmental variation is that heterozygotes have greater biochemical flexibility and hence are able to buffer the internal environment of the organism more effectively than homozygotes, though little direct critical evidence for this exists.

This greater buffering or homeostatic capacity also may be manifested at the intraindividual level. For example, inbred Japanese quail (*Coturnix japonica*) hens show greater variation of weight from egg to egg laid by the same bird than do outbred birds (Sittman et al., 1966). Soulé (1967, 1979) has shown that anole (*Anolis*) lizards with lower levels of heterozygosity display greater levels of morphological asymmetry (= less control of developmental stability) than do more heterozygous individuals. Comparable data derived from *Drosophila* flies tell the same story: the bilateral asymmetry of the number of bristles on a patch of the thorax appears to reflect the degree of developmental buffering, and laboratory populations as they adapt to new environments show a decrease over generations in asymmetry, representing an increase in developmental stability (Beardmore et al., 1960; Bradley, 1980).

The relation of electrophoretically assayed heterozygosity to phenotypic variation of outbred populations also has been examined in a range of animal species as shown in Table 13. The rationale for this approach is that it might *a priori* be expected that greater genetic variation would be reflected in greater phenotypic variation. In fact, the findings are extremely confusing and do not show any consistency of pattern, although as McAndrew et al. (1982) point out the differences may reflect the extent to which the characters studied are under stabilizing selection.

Genetic Variation and Extinction

What guidelines for conservation can be extracted from this account? Genetic variation is certainly necessary both for adaptation and for long-term survival; even inbreeding plants possess levels of genetic diversity that are often considerable (Chapter 13 by Clegg and Brown) and apparently adaptive (Hamrick and Allard, 1972).

On the other hand, apparently successful populations are known that possess little or no genetic variation when assayed electrophoretically. Such an example is the northern elephant seal (*Mirounga angustirostris*) of the western North American coast (Bonnell and Selander, 1974), which experienced a severe bottleneck of population size within the last century. At least three interpretations can be given to such observations:

1. Electrophoretic assay does not reveal variation at the loci more important for population survival.

2. Genetic diversity as revealed by electrophoresis is largely or wholly a reflection of population size and the time since the last severe restriction of size.
3. The niche of the northern elephant seal is not subject to significant competition from other species, and while the genotypes of individuals of the present day may be significantly distinct from the "best" possible genotypes, this does not matter, since all the selection is relative to other members of the same population.

Geographic subdivision of the population will tend to minimize both loss of genetic variance overall and disease risks. We may note here that assumptions about the effects of inbreeding on levels of genetic variation may not always be valid, since in any artificial inbreeding program natural selection also operates and fixation of genes often may be less than theory would indicate. Thus the loss of heterozygosity by inbreeding often may be considerably less than predicted by statistical considerations. Unfortunately, few systematic data on this are available, as their accumulation needs significant numbers of loci and lines to be studied. However, the classic work of Shultz and Briles (1953) with chickens (*Gallus gallus*) points the way. Eleven of twelve lines were still segregating at the B locus, and other loci showed lesser degrees of persistent polymorphism after inbreeding leading to inbreeding coefficient (F) values in the range 0.33 to 0.57. Nevertheless, the losses of vigor, fertility, and disease resistance associated with inbreeding depression should dictate a policy of maximum avoidance of inbreeding in outbreeding species. The main restriction on population size of some threatened species of mammals and birds often will be the area of suitable habitat available, and clearly political action is likely to be needed for improvement in most such cases.

For some species variety of habitat will be of considerable importance, for as we have seen this is expected to maintain appreciably higher levels of genetic variation and hence preserve a larger genetic capital, which can be drawn upon in the future and which also will have effects upon other components of the ecosystem.

Maintenance of the maximum possible population size is obviously of vital importance in minimizing risks of extinction; calculations of the minimum size needed have been discussed elsewhere (Frankel and Soulé, 1981; and Chapter 1 of this book). However, quite apart from expanding the breeding population, there are considerable virtues in endeavoring to expand reproductive capacity even if the net survival to the breeding stage is little greater than at the lower level of reproductive effort. This is because of the room for selection allowed. Quoting one or two cases from some simulation experiments which my colleague Dr. Roger Gilbert has performed, we may note that with a population size of 20 and the number of initial zygotes set at 50 the rundown of genetic variation is very much slower and the production of exceptional genotypes, on which selection may act, is much more common than with only 20 zygotes allowed. Thus, all other things being equal, it is worthwhile endeavoring to increase reproductive capacity even if the population size does not immediately benefit and the additional zygotes are mainly lost. Consider a case where the genetic variation we are interested in involves only ten loci each with two alleles. The variety of different genotypes that can then be generated numbers roughly 60,000, and it is clear therefore that, even in populations with abundant levels of genetic variation, a combination of modest population size and low reproductive capacity may not allow the exploitation of desirable genotypes simply because of the lack of room for selection to occur between zygote production and entry into the breeding pool.

Epilogue

The text for the day I fear is that the generalists—that is, the genetically diverse—will inherit the earth. Most of the important decisions diminishing the probability of extinction may be biologically based but will be political in nature. The genetical aspects of the management of wild populations ought to concentrate on the maintenance of genetic diversity and will benefit from attention to subdivision of populations, reproductive capacity, and, in particular, habitat diversity as increasing the chances of evolutionary survival.

Summary

The chapter starts with a brief discussion of the place of extinction in the natural order and of the ability of species to respond to pressures of various types. This is followed by a review of genetic variation including the basis, nature and scale of such variation.

The various ways in which natural and man-made selection can operate upon a population arc described together with the consequences of different types of selection particularly in terms of their effects upon the amount of genetic variation in populations.

The effects of inbreeding, habitat variety and reproductive capacity on genetic variation are discussed and some conclusions for effective management for survival are drawn.

CHAPTER 9

Genetics and the Extinction of Butterfly Populations

Paul R. Ehrlich

Introduction

Butterflies are both excellent indicators of the health of ecological systems (Pyle, 1976; Ehrlich and Ehrlich, 1981) and major tools used by biologists to investigate the properties of natural populations. Consequently, they are of central importance to all those interested in preserving and managing Earth's biotic resources. It is therefore appropriate to examine the state of knowledge of the genetics of extinction in this key group.

The relationship between the dynamics and the genetics of natural populations remains largely *terra incognita*. The dynamics of relatively few carefully defined demographic units (populations that change in size independently) have been followed for any length of time, and until recently it has been extremely difficult to study the genetics of most such populations because of the problem of determining gene frequencies at samples of loci.

A thorough understanding of the interactions between the size and the hereditary characteristics of populations nevertheless is crucially important to humanity. It would permit more effective manipulations of economically important organisms—for example, of insects that prey on pests of crops. And, more important, it would provide insight into the best tactics for ending the current epidemic of extinctions (Ehrlich and Ehrlich, 1981).

There is reason to hope, however, that knowledge of these dynamic genetic interactions, at least in certain groups of butterflies, will soon be much more extensive. The dynamics of some populations of Nearctic checkerspots (*Euphydryas* in the family Nymphalidae, the brush-footed butterflies, subfamily Nymphalinae) are now understood; changes in some demographic units have been traced for more than two decades (e.g., Ehrlich et al., 1975; Ehrlich and Murphy, 1981); and numerous population extinctions have been observed. Allozyme frequencies have been traced in a number of populations for 5 to 8 years, and these data are now being analyzed (Ehrlich et al., 1983).

Overall, then, the stage may be set for a substantial improvement in the understanding of how population size and gene frequencies interact, at least in butterflies. A breakthrough cannot come too soon, because of the increasingly precarious position of Earth's biota (Ehrlich and Ehrlich, 1981).

In this chapter, I attempt to summarize what is known about the extinction of butterfly populations, and to speculate about the genetics of extinction in these organisms. The speculations, I must emphasize, are just that — little enough is known about the genetics of *any* natural populations, and virtually nothing is known about the genetics of extinction.

Natural Extinctions

1. *Euphydryas*

Among the best documented extinctions of butterfly populations from natural causes are the two extinctions of the area G demographic unit of *Euphydryas editha bayensis* (San Francisco Bay region checkerspot) on Stanford University's Jasper Ridge Biological Reserve, California. From the time of the first census in 1960, this unit, which at its largest occupied an area of about 0.6 hectare, declined continuously from about 60 individuals until it became extinct in 1964–1965. It was naturally reestablished in 1966, reached a population size of several hundred in 1970, and then declined again to extinction in 1974–1975. During the same period, demographic units adjacent to G on either side (Jasper Ridge C and H) fluctuated in size but persisted (Ehrlich, 1965; Ehrlich et al., 1975).

All populations of *E. e. bayensis* (indeed, most *Euphydryas* populations in western North America) are extremely time-constrained in their development. The food plants of *E. e. bayensis* are vernal annuals that senesce rapidly in April–May. Butterflies lay their eggs on the plants, and before the plants dry up the larvae must complete the third instar and enter the resting state (diapause) in which the dry summer is passed. The larvae emerge from diapause after the fall–winter rains set in and the seeds of their food plants have germinated.

On the order of 99% of mortality in these populations occurs when pre-diapause larvae fail to reach the obligatory diapause instar before their food plants senesce (Singer, 1972). Year-to-year variation in that mortality, controlled largely by the complex phase relationship of the butterflies with the local climate (Singer and Ehrlich, 1979), is the cause of observed population trends. Adult populations tend to decline in springs following dry springs (because in dry springs the larval food plants senesce more rapidly, are sparser, and are of poorer quality because there has been little rainfall) and to increase following wet springs. In all years it is plant senescence that limits food availability—on Jasper Ridge, competition among larvae has not been a factor in more than two decades, and there is no sign of density-dependent population "regulation."

Within this general picture of the dynamics of *E. e. bayensis*, how does one explain the extinctions of the demographic unit in area G? It might be tempting to invoke a genetic explanation, since G is the smallest demographic unit. Could, perhaps, loss of heterozygosity as the population size shrank in the early 1960s (and as a result of a founder effect when it was reestablished) have made it more vulnerable to adverse conditions? There is little sign in our data that such an explanation is correct (Ehrlich et al., 1983). Populations of *E. editha* (Edith's checkerspot) show remarkable stability of both gene frequencies and heterozygosity through wide-ranging fluctuations of population size—including bottlenecking, as discussed below.

Instead, the explanation of the area G extinctions appears to lie in the relationship of *E. e. bayensis* with its adult nectar resources. Females lay masses of eggs every two days or so. In most years the time constraints are such that only larvae from the first one or two egg masses have any chance of reaching diapause. Females emerge with large numbers of mature eggs and large fat reserves, and the size of their first two egg masses is unaffected by the presence or absence of nectar resources. In wet years, however, larvae from third, fourth, and fifth egg masses may survive. *Those* egg masses will be larger if adequate nectar resources are available than they will be if sources are inadequate (Murphy et al., 1982).

It appears, then, that nectar resources play a key role in enhancing reproduction in favorable years. This permits demographic units to build to relatively large sizes, which provide buffering against extinction in these populations, as they have density-independent population regulation. In this context the probable cause of the area G extinctions becomes clear. Although larval resources in the area are adequate, nectar resources (unlike those in areas C and H) are both scarce and badly timed for the butterflies (Murphy et al., 1982). Therefore the G demographic unit has not been able to reach population sizes that would adequately buffer it against droughts, and it has twice gone extinct.

The dynamics of a number of California populations of *Euphydryas editha* and of its sympatric close relative, *E. chalcedona* (chalcedon checkerspot), have been monitored for a decade or more. The *E. editha* populations belong to an array of ecotypes, suites of populations adapted to diverse ecological conditions (Gilbert and Singer, 1973; White and Singer, 1974; Ehrlich and Murphy, 1981). Within ecotypes the populations tend to be more similar to one another in their ecological characteristics than to populations of other ecotypes, even when the latter are geographically closer.

This long-term monitoring has allowed our research group to take advantage of a "natural experiment" — the California drought of 1975-1977 — to study the response of *Euphydryas* demographic units to stress. Several populations of *E. editha* went extinct, and others were reduced in size by fivefold or more.

These responses differed among *E. editha* ecotypes. Within and between species, the responses appeared to be largely a function of the fine-tuning of the relationship of the populations to their larval food plants (Ehrlich, et al., 1980) and of habitat changes induced by human activities (such as grazing) that altered the vulnerability of the populations to extinction.

Information on allozyme frequencies is available for several populations that went through drought-induced bottlenecks in population size. Preliminary results show a surprising lack of influence of dynamic changes on allele frequencies and genetic variability. Polymorphic loci tend to maintain the same predominant alleles in roughly the same frequencies through dramatic changes in population size (Ehrlich et al., 1983). Furthermore, when there have been some indications of loss of genetic variability, it appears to have been rapidly restored.

There are three possible explanations, which are not mutually exclusive, for these observations:

1. The variation is "neutral," but population size at the bottleneck has been underestimated (some populations with effective population size, N_e, less than 20), so drift would not be expected to have had substantial effects on allele frequencies.
2. The variation is under selective control, and the polymorphisms are maintained by marginal overdominance (e.g., Gillespie, 1977), heterogeneity of the environment (Hedrick et al., 1976; Watt, 1977), or some other mechanism, and population size has been underestimated or the bottleneck has not existed long enough, as in explanation 1.

3. Genetic variability is lost, but there is sufficient gene flow from populations of the same ecotype (Ehrlich and Murphy, 1981) to restore the pre-bottleneck variability.

Preliminary data on the genetics of Great Basin populations of *E. editha* (Wilcox, Murphy, Ehrlich, and Brussard, in preparation) indicate that they have lower levels of heterozygosity than those in California (McKechnie et al., 1975) or Colorado (Ehrlich and White, 1980). These populations appear to be strongly isolated from one another and may not be subject to the large fluctuations in populations size that are found in some California ecotypes. This could indicate a mechanism-controlling level of variability that has little to do with the dynamic history of the populations.

One tentative conclusion may be drawn about the role of genetics in the extinction of *Euphydryas* populations. There is no sign of the sort of situation thought to be common in the extinction of populations of vertebrates or of *Drosophila* strains in laboratories—that is, small population size resulting in inbreeding depression, which in itself contributes to extinction (for overview see Frankel and Soulé, 1981). This conclusion is not grounded in evidence that *Euphydryas* are especially resistant to inbreeding depression. Rather it follows from the observation that *Euphydryas* populations *do not persist at the small sizes required for strong inbreeding to continue over several generations.* In short, when *Euphydryas* populations are reduced to 30 to 50 individuals, they appear either to go extinct promptly or to rebound to sizes an order of magnitude or more larger. Since the dynamics of most *Euphydryas* populations are largely density-independent, stochastic (random) extinction is the likely fate of any demographic unit that gets small enough to suffer a substantial decay of genetic variability. This is indicated not only by our frequent observations of population extinctions in nature, but by the difficulty of reestablishing extinct populations (Murphy, Ehrlich, and Wilcox, unpublished observations) and of transplanting species to previously unoccupied suitable habitats.

In the latter case (Holdren and Ehrlich, 1981, and unpublished observations), eggs and larvae of *Euphydryas gillettii* (Gillett's checkerspot) were transplanted south across the Wyoming Basin gap in the Rocky Mountains to sites in Colorado that had suitable larval and adult resources. Each transplant was made with the rough equivalent of the reproductive output of populations of 30 to 50 individuals. One transplant population fluctuated for three generations at an estimated size below 30 individuals, perhaps dropping below 10, and then increased in 1981 to more than 100 individuals. The other produced only one known adult in the first generation and then went extinct.

It may be, therefore, that for insect populations similar in their dynamics to *Euphydryas* conservation biologists need not concern themselves with the genetic effects of temporarily small population size. This is probably just as well, since the process of evaluating allozyme frequencies in a population of, say, 60 adults requires exterminating the population! What is of greater interest is sorting out the ecological factors that make populations of different ecotypes differentially susceptible to extinction (Murphy and Wilcox, in preparation).

2. *Glaucopsyche*

It also seems unlikely that genetic factors were directly involved in the extinction of a montane population of the silvery blue, *Glaucopsyche lygdamus* (family Lycaenidae, the gossamer-winged butterflies). A late season snowstorm destroyed the lupine inflorescences that are the basic resource for *Glaucopsyche* larvae, and a large population disappeared as the result of that single climatic event (Ehrlich et al., 1972).

That extinction illuminated a possible reason for the very early flowering of the perennial lupine plants (*Lupinus*; they normally set seed long before the end of the growing season). The risks of early flowering may be more than compensated by its impact on *Glaucopsyche*, which must oviposit on the buds. By sacrificing one season's reproduction, the plants gained a decade of virtual freedom from the attacks of their major seed predator (Ehrlich, 1982).

Nothing is known about the genetics of *Glaucopsyche* populations, but observations in connection with their food plant relationships (e.g., Breedlove and Ehrlich, 1972) make it seem unlikely that small populations would benefit greatly from the increased resources available to the average individual. In only the most unusual cases does it seem likely that density effects play crucial roles in the dynamics of populations of *Glaucopsyche* or of many other temperate zone lycaenines. My guess would be that, owing to their small size (wingspreads usually under 3 cm) and the frequent abundance of their food plants, they would more readily maintain substantial numbers than many *Euphydryas* demographic units. Inbreeding problems would be rather rare and extinctions a result of catastrophes rather than of declines in population size in response to "normal" variation in weather patterns.

Anthropogenic Extinctions

A number of species and many populations of butterflies have gone extinct as a result of human activities. One of the earliest recorded losses was that of a subspecies of the sthenele brown, *Cercyonis sthenele sthenele* (Nymphalidae, subfamily Satyrinae, the satyrs and wood nymphs). It disappeared under the spreading city of San Francisco in 1880 (Ehrlich and Ehrlich, 1981). *Glaucopsyche xerces* (Xerces blue) followed it in 1943, and five species of butterflies are now threatened or endangered in the San Francisco Bay area. Several populations of *Euphydryas editha bayensis* are known to have been paved over in that region, the most recent being one at Woodside that was largely destroyed in April 1980 while under study by our group. The sites of many other well-known butterfly populations are now under concrete in the Los Angeles basin and around other cities. In such cases, once again, one would not expect loss of genetic variability to play an important role. Inbred or outbred, organisms cannot persist without suitable habitat.

1. *Euphydryas editha*

The entire *E. editha bayensis* ecotype may be unusually susceptible to anthropogenic extinction, in spite of the existence of two demographic units on Stanford University's biological preserve. For at least several hundred years, its populations appear to have existed in isolated patches of grassland growing on serpentine soil. These patches are differentially affected by droughts and, presumably, other environmental stresses, and thus in any season the demographic units occupying them are differentially susceptible to extinction. Those units have probably gone extinct frequently and then have been reestablished in time by migrants from other colonies of these relatively sedentary insects. The ecotype thus has persisted as a shifting mosaic of fluctuating populations, not as static, permanent occupants of given sites.

As humanity has removed patch after patch from the mosaic of suitable habitat, the probability of successful recolonization has declined. The number of remaining patches is now so small that the entire ecotype is threatened with extinction. If the two remaining demographic units on Jasper Ridge go extinct, as they nearly did during the recent drought, recolonization from the other remaining major reservoir (Edgewood, 10 km away) would probably not occur before that colony went extinct as well (Murphy and Ehrlich, 1980).

The precariousness of the situation of *E. e. bayensis* was underscored in 1981-1982 when the Edgewood population, by two orders of magnitude the largest surviving demographic unit, was subjected to repeated Malathion sprayings as part of the program to attempt to control the Mediterranean fruit fly (*Ceratitis capitata*). The *Euphydryas* larvae were in diapause when the spraying occurred, but our data suggest that they nonetheless suffered considerable mortality, probably from "direct hits" on poorly sheltered individuals. The population declined from over 100,000 to considerably fewer than 10,000 adults. It is not at all clear that broadcast spraying is the appropriate response to medfly infestations (which will certainly recur), but the level of competence of the state and federal agencies with the responsibility for pest control provides little hope that a more sophisticated approach will be taken next time. The impact of the program on other butterfly populations in the treated areas is not known, but local pest outbreaks (Ehrlich group, unpublished observations) indicated substantial effects on nontarget organisms—as one would expect, the predators of pests were more severely affected than the herbivorous pests themselves.

The "mosaic" pattern of population regulation (Ehrlich and Birch, 1967; Ricklefs, 1973) exhibited by *E. editha bayensis* has also been described for populations of the checkered white, *Pieris protodice* (family Pieridae, the whites and sulphurs), in the Central Valley of California (Shapiro, 1978). The degree to which this sort of pattern applies to other butterflies, and the rates of population turnover that prevail, remain to be documented. But, as indicated above, I believe the pattern is a common one in populations of insects.

For the mosaic pattern to operate, individuals from one demographic unit must be genetically suited for life in the patch once occupied by another. Our group has used as a working hypothesis that within the *E. e. bayensis* ecotype this is the case—even though substantial differences in the detailed ecology between populations as close in distance (10 km) as those on Jasper Ridge and at Edgewood have been discovered. This assumption is soon to be tested with transplant experiments, but there is evidence from other butterflies that small genetic differences can prevent successful recolonization of empty habitat patches.

2. *Lycaena dispar*

The large copper, *Lycaena dispar* (Lycaenidae, Lycaeninae), became extinct in England in the middle of the last century when most of its marsh habitats were destroyed and collectors wiped out the few remaining colonies. In 1927, a colony was reestablished at Woodwalton Fen, using stock from Holland. With the help of constant management it survived until 1968, when a heavy flood wiped it out. In 1970, the butterfly was reintroduced again, but the scientist who has the most intimate knowledge of its biology believes that it cannot survive without constant husbandry (Duffey, 1977).

A major problem is the genetic differentiation of the Dutch stock from the now-extinct English stock. Subtle differences in the habitat requirements of the two strains make it unlikely that the reintroductions can persist unaided. To put it another way, if *L. dispar* in England once had a mosaic population dynamic pattern like that of *E. editha*, Dutch populations probably were not part of the mosaic. Instead they belonged to a different ecotype.

3. *Maculinea arion*

The problem of fine genetic adjustment to habitat requirements is beautifully exemplified by the story of the extinction of the large blue, *Maculinea arion* (Lycaenidae), in England (Thomas, 1980a, 1980b). Two of the basic habitat requirements of the species were its larval food plant, wild thyme (*Thymus praecox*), and one species of ant (*Myrmica sabuleti*) with which, like many lycaenids, it had a symbiotic relationship. The ants tended the larvae in return for droplets of sugary solution from larval honey glands. Last instar larvae were removed from the plants by the ants and taken into the ant nests where they lived as social parasites, being fed by the ants and eating ant brood.

The end for *M. arion* in England came when economic conditions changed and the grazing of sheep was discontinued in the areas where the blue still survived. Sheep kept areas in a condition similar to the downland localities where *M. arion* had originally thrived. The cessation of grazing permitted the thyme to grow luxuriantly, but made conditions unsatisfactory for *Myrmica sabuleti*. Under such circumstances that ant is rapidly replaced by *Myrmica scabrinodis*, an unsuitable host, whenever grazing is even slightly relaxed. Thus removal of sheep deprived *M. arion* of a necessary resource, and it went extinct.

4. *Papilio machaon*

Relatively little is known about the population biology of larger butterflies, in part perhaps because they tend to be more difficult to subject to mark–release–recapture experiments. An interesting exception is the English populations of the swallowtail *Papilio machaon* (family Papilionidae). The habitat of this species in Cambridgeshire was greatly reduced when once-extensive marshy peatland (fens) were drained. There is evidence that in the process a low-mobility phenotype evolved before the last remnant population went extinct at Wicken Fen in the early 1950s (Dempster et al., 1976).

Attempts to reestablish the species at Wicken failed repeatedly. Interestingly, investigations during the last attempt in the 1970s (Dempster and Hall, 1980) revealed no signs of inbreeding depression as the population got smaller. For instance, egg viabilities that were carefully monitored in the field showed no reduction as extinction approached.

Discussion

As stated at the beginning, butterflies are key indicator organisms for the health of ecosystems, systems that provide *Homo sapiens* with indispensible services without which civilization cannot persist (Ehrlich and Ehrlich, 1981). And butterflies are the only sizable group of invertebrates that are routinely studied and collected alive by large numbers of amateur naturalists. Only in large mammals, birds, a few groups of reptiles and fishes, and (in some regions) vascular plants are the decline and disappearance of populations and species equally likely to be noted.

In addition, far more is known about the host plant relationships of butterflies than about the diets of any comparable group of herbivores. Therefore not only are butterflies representative of the smaller animals in terrestrial ecosystems, they provide some index to the status of plant communities as well. It is probably fair to say that proximate causes of the vast majority of butterfly extinctions are changes in the populations of their larval food plants or adult resources.

Despite the great interest in butterflies, the monitoring of their populations is grossly inadequate in most overdeveloped countries (England represents an exception) and virtually nonexistent in less-developed nations—especially in those with rapidly disappearing stands of tropical moist forest. What is known, however, is not reassuring. There appears to be an accelerating global trend toward the loss of butterfly populations and species as a result of the expanding activities of *Homo sapiens* (e.g., Arnold, 1980, 1981, 1982; Bielewicz, 1967; Brown, 1970; Chew, 1981; Ehrlich et al., 1980; Emets, 1977; Kloppers, 1976; Lamas Mueller, 1974; Morton, 1982; Pyle, 1976; Pyle et al., 1981; Zukowski, 1959). This trend is caused primarily by habitat destruction, and, of course, it parallels that of a general despoliation of Earth's biota.

The steps necessary to arrest this general trend must be aimed at the conservation of entire ecosystems; they have been discussed in detail elsewhere (e.g., Myers, 1979; Ehrlich, 1980; Ehrlich and Ehrlich, 1981; Frankel and Soulé, 1981). Some of these steps could and should be initiated with no further knowledge of the biology of extinction in butterflies or other organisms. But further research should go on simultaneously, as it can provide insight into both the interpretation of observed extinctions and the tactics of conservation of butterflies and other organisms with similar population biologies.

We obviously need to learn much more about the relationship between genetics and dynamics in butterfly populations. Can inbreeding depression be ignored as a factor leading to extinction in many (or most) butterfly populations, as suggested above? Or is my hypothesis based on too restricted a sample of kinds of populations? Only further research will tell, especially on the dynamics and genetics of relatively vagile species (e.g., Brown and Ehrlich, 1980; Brussard and Ehrlich, 1970a, 1970b, 1970c).

We do not even know how frequently phenotypic changes accompany dynamic events. Ford and Ford (1930) in a classic paper associated phenotypic changes—especially changes in "variability"—with fluctuations in population size in *Euphydryas aurinia* (marsh fritillary). But modern attempts to do the same sort of thing have been few, and the results have been less than definitive (e.g., Ehrlich and Mason, 1966; Mason et al., 1968).

In fact, with the exception of the work by Dempster and his colleagues on *Papilio machaon*, I know of no studies successfully relating phenetic or genetic changes in butterfly populations to *any* human disturbance of the environment. There has been no equivalent of "industrial melanism," the darkening of moth species in areas subject to heavy pollution (Kettlewell, 1973), discovered in butterflies. One might, for example, expect model-mimic resemblances to become less precise where humanity has reduced populations of visual predators. And various forms of pollution, from pesticide drift to acid rains, might be placing selective pressures on populations. In spite of the existence of large butterfly collections made over long periods of time, people have not attempted to look carefully even at phenetic trends over decades. One reason for this undoubtedly is the non-random nature of the samples in most collections.

Thus, although more is known about the biology of extinction in the butterflies than in most groups of animals, the surface has just been scratched. Long-term monitoring of many more populations in diverse groups of butterflies should be started, so that information can be gathered on such things as rates of natural and anthropogenic extinctions and the dynamic and genetic events preceding them. If such research is not begun soon and pursued with skill and vigor, however, it seems likely that the phenomenon under study will itself terminate the opportunities for investigation.

Summary

Butterflies, because they are well-known biologically and the object of the attention of numerous amateur naturalists, serve as a crucial indicator of the health of ecosystems, and thus of humanity's life-support systems.

Natural extinctions of butterfly populations appear to be common occurrences, especially in response to stresses on their larval food plants. These extinctions, however, are normally followed by recolonization of habitat patches, and a mosaic pattern of population "regulation" prevails.

Anthropogenic extinctions of butterflies are becoming increasingly common as a result of widespread habitat destruction.

Little is known about the genetics of extinction in butterflies. What data there are indicate that inbreeding effects, as populations decline, are not important factors in the disappearance of populations. In contrast, fine genetic adjustments to habitat conditions often appear to make reestablishment of populations with stock from other ecotypes quite difficult.

Acknowledgments

My colleagues Dennis D. Murphy and Bruce A. Wilcox have collaborated in much of the work described here and have provided extensive and most helpful comments on the manuscript. Anne H. Ehrlich and Marcus W. Feldman (Stanford University) and Michael E. Soulé (Rocky Mountain Biological Laboratory) have been kind enough to read and criticize the manuscript also. Research by our group in the area of butterfly extinction has been supported by grants from the National Science Foundation, the most recent of which have been DEB 78 22413 and DAR 80 22413, and by a grant from the Koret Foundation of San Francisco.

CHAPTER 10

Extinction: Lessons from Zoos

Katherine Ralls
Jonathan Ballou

Introduction

The potential of zoological parks for conserving rare and endangered species and the necessity for developing captive, self-sustaining populations of these species has been widely recognized (Conway, 1967; Martin, 1975; Olney, 1977, 1980). However, as Pinder and Barkham (1978) pointed out, zoos had achieved only limited success by 1976. In an analysis of fifteen years of census data published in the *International Zoo Yearbook*, Pinder and Barkham concluded that zoos were maintaining self-sustaining populations for only 26 of 274 rare mammal species. According to a report compiled by Nancy Muckenhirn for the American Association of Zoological Parks and Aquariums (AAZPA) in 1980, American zoos collectively contain potential captive, self-sustaining populations of 96 species. Only one quarter of these species are listed in Appendix I of the Convention on International Trade in Endangered Species of Wild Fauna and Flora (CITES, U.S. Fish and Wildlife Service, 1981) and the International Union for the Conservation of Nature and Natural Resources Red Data Book (IUCN, 1966 *et seq.*). However, the future potential of zoos to maintain a variety of self-sustaining captive populations of rare and endangered species is greater than one would suppose from their historical performance, which resulted from breeding practices not based on sound genetic principles.

In the early part of the century there was little cooperation among zoos. Each was an independent, isolated institution, and relationships between them were largely competitive. Directors of major zoos strove to display the most diverse collection (Fisher, 1966). The result was that each zoo contained many species but, on the average, comparatively few individuals of each. Breeding animals in captivity was not a major goal of most zoos, as more stock could always be obtained from the wild. Some breeding did occur, but many breeding groups within individual zoos died out and new animals were then obtained. The National Zoological Park in Washington, D.C., for example, imported giraffes (*Giraffa camelopardalis*) four times between 1926 and 1961–1962. (S. Hamlet, personal communication).

The past few years have been a time of great excitement and change in the zoo world. Whereas only a few years ago (1977), for example, there was little evidence that inbreeding had deleterious effects in zoo animals, such effects have now been documented in a wide variety of species. Today the zoo community recognizes the need for careful genetic management of small populations and is making rapid progress in developing cooperative breeding plans for important taxa such as the Przewalski horse, *Equus caballus przewalski* (Ryder and Wedemeyer, 1982), the golden lion tamarin, *Leontopithecus rosalia rosalia* (Kleiman, 1982), and the Siberian tiger, *Panthera tigris altaica* (Foose and Seal, in press).

Our main purpose here is to summarize the evidence on the deleterious effects of inbreeding that helped to convince the zoo community of the importance of genetic management. We believe that this evidence is also of great importance to those concerned with the management of wild populations.

Effects of Inbreeding in Zoo Animals

Early warnings about the dangers of inbreeding in zoo animals, such as those by Bogart (1966), were unheeded. By the late 1970s many conservationists were concerned about the probable problems due to inbreeding and the loss of genetic variability in small populations both in captivity and in the wild (Denniston, 1978; Lovejoy, 1978; Seal, 1978). Most of these discussions as well as subsequent cautions (Franklin, 1980; Senner, 1980; Soulé, 1980) were primarily theoretical and were based on extrapolation from work on laboratory and domestic animals. There were few documented instances of deleterious effects of inbreeding in captive exotic animals, although Bouman (1977) and Flesness (1977) were able to show

Table 1a. Effects of Inbreeding on Juvenile Mortality in Ungulates

Species	Number of offspring		Percent juvenile mortality		Sign test*
	Noninbred	Inbred	Noninbred	Inbred	
Indian elephant† *(Elephas maximus)*	13	6	15.4	66.7	+
Zebra *(Equus burchelli)*	27	5	25.9	40.0	+
Pygmy hippopotamus† *(Choeropsis liberiensis)*	184	51	24.5	54.9	+
Muntjac *(Muntiacus reevesi)*	22	18	18.2	33.3	+
Eld's deer† *(Cervus eldi thamin)*	17	7	23.5	100.0	+
Père David's deer *(Elaphurus davidianus)*	17	22	11.8	13.6	+
Reindeer *(Rangifer tarandus)*	29	21	34.5	57.1	+
Giraffe *(Giraffa camelopardalis)*	14	5	21.4	60.0	+

Kudu (*Tragelaphus strepsiceros*)	14	11	28.6	27.3	–
Sitatunga† (*Tragelaphus spekei*)	16	59	6.3	47.5	+
Sable† (*Hippotragus niger*)	22	10	18.2	70.0	+
Scimitar-horned oryx† (*Oryx dammah*)	37	5	5.4	100.0	+
Wildebeest (*Connochaetes taurinus*)	7	41	14.3	29.3	+
Dik-dik (*Madoqua kirki*)	17	15	41.2	53.3	+
Dorcas gazelle† (*Gazella dorcas*)	50	42	28.0	59.5	+
Japanese serow† (*Capricornis crispus*)	73	62	28.8	56.5	+

*For the sign test, + indicates that juvenile mortality is higher in inbred than in noninbred young ($P = 0.0003$, one-tailed sign test).
†Difference between inbred and noninbred juvenile mortality significant at the 0.05 level by Fisher's exact test.

Table 1b. *Effects of Inbreeding on Juvenile Mortality in Primates*

Species	Number of offspring		Percent juvenile mortality		Sign test*
	Noninbred	Inbred	Noninbred	Inbred	
Ring-tailed lemur *(Lemur catta)*	70	16	18.6	31.3	+
Black lemur *(Lemur macaco)*	37	4	40.5	75.0	+
Brown lemur† *(Lemur fulvus)*	116	16	26.7	62.5	+
Greater galago *(Galago crassicaudatus crassicaudatus)*	146	44	30.8	38.6	+
Melanotic galago *(Galago crassicaudatus argentatus)*	19	33	26.3	36.4	+
Squirrel monkey *(Saimiri sciureus)*	441	9	27.9	33.3	+
Black spider monkey *(Ateles fusciceps robustus)*	11	12	18.2	58.3	+
Illiger's saddle-back tamarin† *(Saguinus fuscicollis illigeri)*	177	56	67.2	80.4	+

Saddle-back tamarin† *(Saguinus fuscicollis)*	381	25	32.3	60.0	+
Golden lion tamarin† *(Leontopithecus rosalia rosalia)*	369	145	50.1	63.4	+
Pig-tailed macaque *(Macaca nemestrina)*	2054	36	42.3	41.7	−
Crab-eating macaque *(Macaca fascicularis)*	206	31	31.1	32.3	+
Rhesus macaque *(Macaca mulatta)*	25	10	0.0	20.0	+
Celebes black ape *(Macaca nigra)*	75	11	30.7	54.5	+
Mandrill† *(Mandrillus sphinx)*	8	23	0.0	78.3	+
Chimpanzee *(Pan troglodytes)*	227	20	29.5	30.0	+

*For the sign test, − indicates juvenile mortality is higher in inbred than in noninbred young ($P = 0.0003$, one-tailed sign test).
†Difference between inbred and noninbred juvenile mortality significant at the 0.05 level by Fisher's exact test.

Table 1c. *Effects of Inbreeding on Juvenile Mortality in Small Mammals*

Species	Number of offspring		Percent juvenile mortality		Sign test*
	Noninbred	Inbred	Noninbred	Inbred	
Short-bare-tailed opossum† *(Monodelphis domestica)*	142	11	1.4	18.2	+
Elephant shrew *(Elephantulus rufescens)*	144	43	15.3	23.3	+
Climbing rat *(Tylomys nudicaudus)*	6	43	0.0	11.6	+
Wied's red-nosed mouse† *(Wiedomys pyrrhorhinos)*	18	5	0.0	100.0	+
Four-striped rat *(Rhabdomys pumilio)*	2	23	0.0	0.0	0
Salt-desert cavy† *(Dolichotis salinicola)*	11	6	0.0	66.7	+
Acouchi *(Myoprocta pratti)*	12	24	16.7	45.8	+
Zagoutis *(Plagiodontia aedium)*	4	8	0.0	25.0	+
Degu *(Octodontomys degus)*	66	51	10.6	23.5	+
Boris *(Octodontomys gliroides)*	11	43	18.2	34.9	+
Spiny rat *(Proechimus semispinosus)*	17	11	11.8	36.4	+
Punare *(Cercomys cunicularis)*	61	5	13.1	40.0	+

*For the sign test, + indicates that juvenile mortality is higher in inbred than in noninbred young ($P = 0.0005$, one-tailed sign test).
† Difference between inbred and noninbred juvenile mortality significant at the 0.05 level by Fisher's exact test.

that highly inbred Przewalski horse (*Equus caballus przewalski*) mares were less productive than less inbred animals.

Seal (1978) suggested that the rarity of such reports indicated only that few zoos had maintained and analyzed adequate records. Working with the unusually complete records of the National Zoological Park in Washington, D.C. (NZP), as well as with records from other institutions and studbooks, we soon provided evidence that his view was correct. Inbreeding leading to decreased breeding success and death of animals does exist in zoological parks.

To test for the presence or absence of an inbreeding problem we analyzed breeding records for 44 species: 16 ungulates (Ralls et al., 1979), 12 small mammals (Ralls and Ballou, 1982a), and 16 primates (Ralls and Ballou, 1982b). The species surveyed represented 7 orders, 21 families, and 36 genera. Because of the small sample sizes, we compared only two levels of inbreeding: *noninbred*, which included all young with unrelated parents, and *inbred*, which included all young with an inbreeding coefficient (F) greater than zero. Ideally, studies of the relationship between inbreeding and juvenile mortality should be based upon the total mortality before reaching reproductive age (Cavalli-Sforza and Bodmer, 1971), but zoo animals are often transferred to other institutions before this age. In order to circumvent this problem young ungulates and primates that survived six months or more were considered to have lived. We defined the died category to be all young surviving less than six months, including stillbirths and those born prematurely. This age criterion was inappropriate for the small mammal species because they become sexually mature much earlier. We used one half the age at sexual maturity as the cut-off point for these species.

The results were dramatic. Juvenile mortality of inbred young across taxa was higher than that of noninbred young in 41 of the 44 species (Table 1). This trend was statistically significant ($P < 0.001$) within each of the three species groups. Juvenile mortality of the inbred young was also significantly higher than for noninbred young within taxa for seven of the ungulate, five of the primate, and three of the small mammal species ($P < 0.05$, Fisher's exact tests).

We performed more detailed analyses on the ungulate data. The trend toward high mortality in the inbred young could be seen even at the level of the individual females. In 19 of 25 females belonging to 10 species, a larger percentage of young died when the female was mated to a related male than when she was mated to an unrelated male ($P < 0.01$, one-tailed sign test) (Ralls et al., 1979). An initial investigation of the species with the largest sample size, the Dorcas gazelle (*Gazella dorcas*), showed that the higher mortality rate of the inbred young could not be accounted for by birth order or birth season of the young, management changes, changes in population density, or differences between wild-caught and captive-born or inbred and noninbred mothers (Ralls et al., 1980). A later analysis of the data for all 12 ungulate species for which we had detailed management records gave similar results (Ballou and Ralls, 1982).

We also found that the very first stage of inbreeding usually produced an increase in juvenile mortality. In many of the ungulate species at NZP, much of the inbreeding was between wild-caught fathers and their daughters; young from such matings tended to have higher mortality rates than noninbred young (Table 2). Deleterious effects also occur at an early stage when wild mice (*Mus musculus*) are inbred (Lynch, 1977; Conner, 1975).

Senner (1980) developed a mathematical model to examine the effect of inbreeding depression on the probable survival time of a closed captive population. In this model, the probability of extinction during each generation is based on the probability of either no male or no female offspring's surviving to reproductive age. We combined our data on NZP's wildebeest (*Connochaetes taurinus*) herd from 1960 to 1978 with a simplified version of his model, considering only the effect of inbreeding on juvenile survival, to determine the probable effects of three different breeding plans on the future survival of this herd. We estimated a measure of inbreeding depression in viability as 0.69 from the average mortality rates of inbred and noninbred young and the effective population size (N_e) as 50% of the actual population size (N) from the usual adult sex ratio. We assumed that the carrying capacity of the zoo was 16 wildebeest and that the initial population was noninbred because NZP obtained several new wildebeest in 1979.

The three breeding plans were (1) obtaining a new, unrelated male each generation, which resulted in no inbreeding; (2) maximal avoidance of inbreeding ($N_e = 2N$) (Flesness, 1977); and (3) continuing the breeding practices used from 1960 to 1978 (which resulted in $N_e/N = 0.5$). With

Table 2. *Juvenile Mortality of Young Resulting from Matings between Wild-Caught Males and Their F_1 Daughters From Wild-Caught Mothers*

Species	Wild-caught male × daughter Number of births	Percent juvenile mortality		Sign test*
		Wild-caught male × daughter	Noninbred young (from Table 1)	
Indian elephant	2	50	15	+
Zebra	4	50	26	+
Pygmy hippopotamus	5	40	32	+
Giraffe	5	60	21	+
Sable	9	100	18	+
Scimitar-horned oryx	2	100	5	+
Wildebeest	4	25	14	+
Dorcas gazelle	16	44	28	+

*For the sign test, + indicates that juvenile mortality of young from father-daughter matings is higher than that of noninbred young ($P = 0.004$, one-tailed sign test).

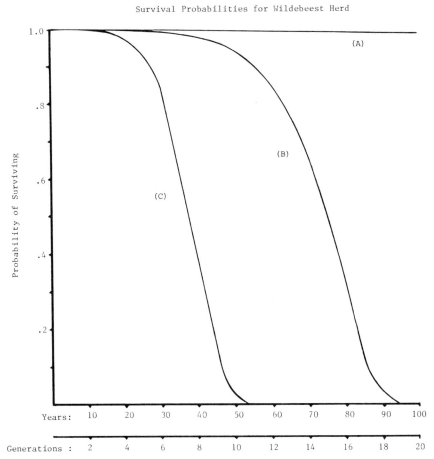

Figure 1. *The probability of the National Zoological Park's wildebeest herd surviving to 20 generations (100 years) under three different breeding plans: (A) acquiring a new unrelated male each generation (no inbreeding); (B) maximal avoidance of inbreeding; and (C) continuing the breeding practices used from 1960 to 1978.*

the first plan, the probability of the population's surviving for 20 generations is still above 99% (Figure 1, curve A), and it takes 150 generations for it to decrease to 95%. The population almost certainly becomes extinct by 20 generations if the second plan is followed (curve B) and by 12 if the third is used (curve C); however, our model probably underestimates the difference between these last two plans, as it does not include the effect of inbreeding on fecundity. Thus, careful breeding within a single zoo can increase considerably the probable life span of a small population, but the introduction of unrelated animals is needed to ensure long-term survival of a population.

Other Evidence on the Effects of Inbreeding

The deleterious effects of inbreeding were known long before the discovery of the mechanisms of Mendelian heredity. An example of pre-Mendelian experimentation is a study of inbreeding in rats (*Rattus*) begun in 1887 and carried on for some thirty generations (Ritzema-Bos, 1894). The population originated from a single pair (a wild male and an albino female), and both parent × offspring and full-sib matings were used. Although it is impossible to calculate the exact degree of inbreeding achieved because matings were apparently not carried out systematically, it is evident that the inbreeding resulted in a decrease in the rate of conception and average litter size, and an increase in juvenile mortality (Table 3).

Darwin (1868) summarized the experience of livestock breeders up to his time. He found a consensus that inbreeding usually leads to a reduction in constitutional vigor and fertility: "that any evil directly follows from the closest interbreeding has been denied by many persons; but rarely by any practical breeder; and never, as far as I know, by any one who has largely bred animals which propagate their kind quickly . . ."

Such early findings were confirmed by many later carefully designed studies on laboratory animals such as guinea pigs, mice, and rats, and on domestic animals such as sheep, pigs, and cattle. Inbreeding generally leads to a reduction in viability and fertility in these mammals (Falconer, 1981; Wright, 1977; Lasley, 1978). Characters expressed early in life, such as survival after birth and growth rate to weaning, are usually more severely affected by inbreeding than are characters that develop later in life and contribute little to fitness (Falconer, 1981; Lasley, 1978).

Table 3. *Results of an Early Experiment on Inbreeding in Rats*[1]

Year	Percentage of unsuccessful matings	Average litter size	Percentage mortality by 4 weeks of age
1887	0	7.50	3.9
1888	2.6	7.14	4.4
1889	5.6	7.71	5.0
1890	17.4	6.58	8.7
1891	50.0	4.58	36.4
1892	41.2	3.20	45.5

[1] From Ritzema-Bos (1894).

Many of these experiments were massive, involving large numbers of animals over many years, and the literature is extensive. Sewall Wright's classical experiments on inbreeding in guinea pigs (*Cavia porcellus*) resulted in the production of 29,310 inbred and 5105 control young from 1906 to 1924 (Wright, 1922b). In spite of the fact that only the 5 most vigorous inbred lines survived to the end of the experiment (30 others became extinct or declined so severely that breeding was discontinued), inbred young were consistently inferior to the controls in terms of number of young born, percentage of young born alive and raised to 33 days, and weight at 33 days. The literature on laboratory animals is summarized in Wright (1977).

The existence of successful inbred strains of laboratory mice does not imply that no deleterious effects are to be expected when inbreeding wild mammals (Frankel and Soulé, 1981; Ralls and Ballou, 1982a). The effects of inbreeding wild house mice, *Mus musculus*, are often severe (Conner, 1975; Lynch, 1977), and most attempts to develop inbred strains of laboratory mice are unsuccessful (Bowman and Falconer, 1960). Similar difficulties were encountered during the establishment of the inbred mouse strains used so widely for research purposes (Strong, 1978). The golden hamster, *Mesocricetus auratus*, is often cited as an example of an animal that was successfully inbred, because, until 1971, all captive animals had originated from three littermates captured in 1930 (Adler, 1948). However, if a population increases rapidly after passing through a single bottleneck, much of the genetic variation present in the founders may be retained (Nei et al., 1975). We calculated that the captive golden hamster could have retained about 60% of the genetic variation present in the wild population (Ralls and Ballou, 1982a). The existence of considerable genetic variation in hamsters descended from the original importation is also evidenced by the typical problems (reduced survivorship and fecundity) encountered when they are inbred (P. Newberne, personal communication). A second group of 12 golden hamsters was imported in 1971 (Murphy, 1971), and typical deleterious effects of inbreeding were encountered when these animals were inbred (Ralls and Ballou, 1982a).

The Père David's deer, *Elaphurus davidianus*, and the European bison, *Bison bonasus*, are also often cited by those who doubt the reality of inbreeding depression. As Frankel and Soulé (1981) point out, it is probable that many of the deleterious alleles in these two species were eliminated by very slow inbreeding and selection during a long period in which their populations were relatively small but still large enough to prevent the deleterious effects of rapid inbreeding.

The North Central regional dairy cattle breeding project (participating states: Iowa, Michigan, Minnesota, Missouri, Ohio, South Dakota, and Wisconsin), begun in 1947, is an example of the many extensive studies on inbreeding in livestock (Young et al., 1969). One of the original objectives of the project was to "investigate inbreeding (closed line

Table 4. *Average Effect of Three Levels of Inbreeding on Some Economically Important Traits in Dairy Cattle*[1]

Trait	Effect at level of inbreeding of		
	6.25%	12.5%	25.0%
Milk yield	− 300 lb	− 600 lb	− 1200 lb
Fat yield	− 9 lb	− 18 lb	− 36 lb
Fat	+ 0.02%	+ 0.04%	+ 0.12%
Birth weight	− 1.5 lb	− 3.0 lb	− 6.0 lb
Two-year weight	− 20 lb	− 40 lb	− 60 lb
Two-year height	− 0.4 cm	− 0.8 cm	− 2.4 cm
Two-year girth	− 1.2 cm	− 2.4 cm	− 4.8 cm
Mortality (% of outbred)	112%	125%	150%

[1] From Young et al. (1969).

matings) coupled with selection as a means of establishing improved strains of dairy cattle." The conclusion was that inbreeding usually increases juvenile mortality and depresses milk yield, fat yield, growth, and reproductive performance (Table 4). The three levels of inbreeding shown in Table 4 are those likely to occur in linebreeding (6.25%), half-sib or grandparent–offspring matings (12.5%), or full-sib or parent–offspring

Figure 2. *Effect of inbreeding and crossbreeding in sheep. Left, inbred; right, three-way cross of inbred lines. These two sheep had the same sire (from Wiener and Hayter, 1974). Courtesy of G. Wiener.*

Figure 3. Sam 951, an outstanding sire of the Charolais breed. He was mated with over 200 of his own daughters and no recessive defects were uncovered by these matings (from Lasley, 1978). Courtesy of Mrs. Charles Litton.

matings (25%). The average effects of inbreeding were approximately linear over this range of inbreeding, although Young et al. (1969) emphasize that effects in any particular herd were often much better or worse than these average effects. Inbreeding also resulted in delayed puberty, reduced conception rates, fewer multiple ovulations, and more loss of established pregnancies.

As a second example, the results of an inbreeding and crossbreeding experiment in domestic sheep (*Ovis aries*) are graphically portrayed in Figure 2. These two sheep had the same father, but the one on the left was highly inbred (inbreeding coefficient 0.59) while the one on the right was noninbred, the result of a three-way cross of inbred lines, its mother being a cross between two inbred lines. When the sheep were photographed at 10 months of age, the inbred lamb had attained only slightly more than 40% of the weight of its half-sib (Wiener and Hayter, 1974; Wiener and Wooliams, 1980). Lasley (1978) summarizes studies on livestock.

Inbreeding was used to fix desirable traits in modern breeds of livestock. However, reduced fertility was a major problem during the early periods of inbreeding, which led to the development of the modern livestock breeds (Wallace, 1923; Wright, 1923), and the inbred lines were repeatedly outcrossed to restore vigor and fertility. This option is often not available to those managing small captive populations (Chapter 15).

Inbreeding can be used to eliminate deleterious alleles, and impressive results have been obtained in livestock—for example, the bull shown in Figure 3. In order for deleterious alleles to be eliminated by selection before they become fixed in the population, however, the rate of inbreeding must be fairly low, not exceeding 2% to 3% per generation, in livestock (Frankel and Soulé, 1981, p. 73; and Templeton and Read, Chapter 15). Current texts on livestock breeding point out the generally deleterious effects of close inbreeding and recommend that it be used only under exceptional circumstances (Lasley, 1978). Many deleterious alleles have already been eliminated from modern livestock populations, so one would expect that they could be safely inbred at a faster rate than most undomesticated species. A maximum inbreeding rate of 1% has been recommended for undomesticated populations (Frankel and Soulé, 1981). However, when the goal is to produce an inbred stock and matings can be carefully planned, it may be possible to exceed this rate (Templeton and Read, Chapter 15).

Inbreeding depression has also been found in birds (Abplanalp, 1974; Kear and Berger, 1980; Greenwell et al., in press), fish (Mrakovcic and Haley, 1979), and insects and plants (Wright, 1977; Frankel and Soulé, 1981).

The results of our survey of the effects of inbreeding on juvenile mortality in zoo animals are thus very similar to those of the many studies on other animals. The deleterious effects of inbreeding in zoo animals were predictable, as the underlying genetic mechanisms are likely to be similar in all normally outcrossing and sexually reproducing species; in fact, these deleterious effects had been predicted (Bogart, 1966).

Inbreeding inevitably increases the proportion of homozygous individuals in a population. This increased homozygosity causes the typical decreased viability and fertility associated with inbreeding. Deleterious alleles tend to be recessive and are thus expressed only in the homozygote condition. Furthermore, the heterozygote may be superior to either homozygote at some loci. Inbreeding thus allows more deleterious recessives to be expressed and reduces the expression of heterozygote superiority (Falconer, 1981). In addition, inbreeding may disrupt polygenic characters, and thus act to decrease the fitness of populations as Carson points out (Chapter 11).

Inbreeding in the Wild

Sociobiological theory predicts that father–daughter mating should occur in polygynous species if the advantages of having a greater proportion of one's genes represented in one's offspring are greater than the disadvantages due to inbreeding (Bengtsson, 1978; Smith, 1979). The rapid rates of chromosomal evolution in several mammalian taxa also led to speculation that polygynous mammals often inbreed (Wilson et al., 1975; Bush et al., 1977).

However, the results of father–daughter matings in captivity indicate that the cost of inbreeding is high in many species (Table 4), and long-term field studies of individually known animals have documented the rarity of close inbreeding in a variety of polygynous species such as the spearnose bat, *Phyllostomus hastatus* (McCracken and Bradbury, 1977), Old World wild rabbit, *Oryctolagus cuniculus* (Daly, 1981), yellow-bellied marmot, *Marmota flaviventris* (Schwartz and Armitage, 1980), and black-tailed prairie dogs, *Cynomys ludovicianus* (Hoogland, 1982). The available evidence suggests that close inbreeding is also uncommon in many primates (Ralls and Ballou, 1982b) and ungulates (Ballou and Ralls, 1982).

Few data exist on the effects of inbreeding in the wild, but deleterious effects have been described in the great tit, *Parsus major* (Greenwood and Harvey, 1978) and olive baboon, *Papio anubis* (Packer, 1979).

Genetic Management of Captive Populations

The goal of most captive breeding should be to develop a captive, self-sustaining population that preserves as much of the genetic variability present in the wild species as possible rather than to develop an inbred line preserving only a small fraction of this variability (Franklin, 1980; Frankel and Soulé, 1981). An inbred line would preserve only one phenotype, and this phenotype might differ considerably from the most common phenotypes in the original wild population. Whether the ultimate goal is to maintain a vigorous captive population for exhibit or to produce animals for reintroduction into the wild, breeders of species in captivity should strive to maintain genetic diversity. Even successful inbred lines are often inferior to noninbred ones in terms of resistance to disease, reproductive success, and life span (Festing, 1976; Soulé, 1980). If genetically varied animals are reintroduced into the wild, it is more likely that some will survive. It is improbable that the alleles preserved in an inbred line will, by chance, be those that best suit individuals for survival in the wild.

Inbreeding zoo animals may sometimes be useful. For example, some zoos want to exhibit "white" tigers (actually off-white animals with gray-brown stripes) even though these individuals are known to have genetically-caused abnormalities of the visual pathways (Guillery and Kaas, 1973). A single, recessive autosomal locus is responsible for both the "white" color and the visual abnormality (Thornton, 1978). Development of a line homozygous for this recessive allele would enable efficient production of "white" tigers for exhibit purposes without using any of the limited space available in zoos for large felids to house heterozygous individuals, which are neither "white" nor particularly desirable members of a captive, self-sustaining tiger population. A breeding plan similar to that developed by Templeton and Read (Chapter 15) for Speke's gazelle, *Gazella spekei*, might be used to produce a line of white tigers. Their plan should also prove useful for captive populations of other species in which numbers are so small that inbreeding cannot be avoided.

Most existing captive populations were founded before the need for genetic management was widely recognized and have developed characteristics that are unsuited to the development of captive, self-sustaining populations. The most well-documented captive populations are those for which international studbooks are maintained. Each individual is permanently marked, and records of each individual's parentage and institutional location are maintained. A recent analysis of the Sumatran tiger (*Panthera tigris sumatrae*) studbook illustrates the general structure of many existing captive populations (Ballou and Seidensticker, in press).

The number of Sumatran tigers in captivity increased from approximately 25 in 1960 to 160 in 1980. Because this increase was mostly due to captive births, it might seem that a healthy, captive, self-sustaining population had been developed. However, a careful examination of the studbook revealed that there were three genetically distinct populations, founded in the United States, Jakarta, and Europe, with 2, 3, and 13 individuals, respectively. By 1980, these populations were spatially integrated. Tigers from different genetic populations coexisted within various countries and, in one case, within the same zoo. The three populations had different sizes, sex ratios, age structures, and survivorship and fecundity rates. Differential reproduction had resulted in unequal genetic contributions of the founding animals to the living population (Figure 4). Furthermore, the three populations had different levels of inbreeding, with mean inbreeding levels of 40% and 11% for the populations founded in the United States and Europe, respectively (Figure 5), and no inbreeding in the population founded in Jakarta. The genetic subdivision, unequal genetic contributions of founding individuals, and high levels of inbreeding found in these tigers are characteristic of many established captive populations.

Cooperative breeding programs to correct these characteristics now exist for some captive populations (Chapter 23). The main objectives of these programs are to maximize the retention of genetic diversity, equalize the genetic contributions of the founding individuals, and achieve a stable demographic structure. Genetic diversity can theoretically be maintained either by minimizing inbreeding throughout the population (Flesness, 1977) or by the alternation of mild inbreeding within zoos and outcrossing between zoos (Chesser et al., 1980).

When founding a new population, geneticists (Senner, 1980) also suggest choosing *at least* five founders that are not closely related, increasing the size of the population to the desired final size as rapidly as possible, and

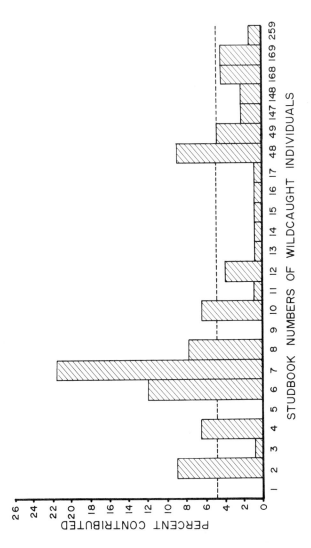

Figure 4. *Percentage of genetic material in the 1980 population of captive Sumatran tigers contributed by each of the founders. The dashed line indicates the percentage of genetic material which would have been contributed by each of the 21 reproducing wild-caught founders if their contributions were equal. Four of the wild-caught animals died before they bred and were not considered founders (from Ballou and Seidensticker, in press).*

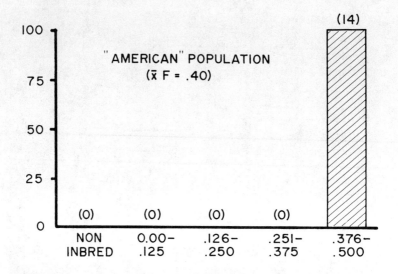
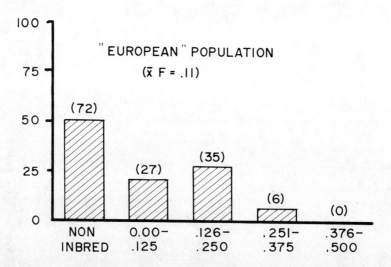

Figure 5. *Distribution of inbreeding coefficients in the Sumatran tiger populations founded in the United States (A) and Europe (B) (from Ballou and Seidensticker, in press).*

avoiding large fluctuations in population size. The final population size should be equivalent to an effective population of at least 50, which may mean maintaining many more than 50 individuals (Flesness, 1977; Senner, 1980; Frankel and Soulé, 1981).

Lessons from the Experience of Zoos

Although some of the techniques used to maintain genetic diversity in captive populations cannot be applied to wild populations, the difficulties experienced by zoos in maintaining small populations are directly relevant to the conservation of many species in the wild. As areas of suitable habitat decrease in size and become more isolated from each other, an ever-increasing number of species will exist in relatively small populations in which some degree of inbreeding will inevitably occur. These wild populations may be expected to exhibit a decrease in fitness similar to that seen in captive populations.

Genetic considerations are helpful in estimating whether a reserve is large enough to contain a viable population of a given species. Franklin (1980) suggests an effective population size of at least 500 for large mammals if the population is to be preserved indefinitely, and at least 50 for short-term maintenance. Natural populations with effective sizes of "less than 50 to 100 are in immediate danger and require immediate genetic management," such as the addition of unrelated animals (Frankel and Soulé, 1981). Soulé et al. (1979) have calculated that even the largest existing African reserves are too small and, if left alone, will probably lose the majority of the birds and large mammal species within a few hundred to a few thousand years.

Many wildlife managers are called upon to found new populations of game animals, and large numbers of animals have been captured and released in new areas. The success rate of these reintroductions could almost certainly be improved if they were planned in accordance with the principles of conservation genetics.

Conclusion

Many wildlife managers have not recognized the dangers of inbreeding. A recent textbook on wildlife management states that "many people believe that wildlife populations need new blood, or stocking. There is no evidence that extensive inbreeding occurs or that, where it does, it has deleterious effects on populations" (Giles, 1978). Whitehead (1980) feels that "adverse comment about the ill-effects of inbreeding is . . . often exaggerated," and Grieg (1979) believes that concern about inbreeding is based on "irrational fears" and that "this human prejudice is too easily transferred to the field of wildlife management."

Only a few years ago, there was almost no evidence that inbreeding in zoo animals had deleterious effects. Such effects have now been shown to be extremely common (Ralls et al., 1979, 1980; Ralls and Ballou, 1982a, 1982b; Ballou and Ralls, 1982). This demonstration would not have been possible without detailed records of individually known animals, which enabled comparisons between inbred and noninbred animals maintained under the same conditions. The NZP's Dorcas gazelle herd was generally regarded as a very successful breeding group. Ninety-three calves were born over an 18-year period, and so many young animals were sent to other zoos that most individuals of this species in the United States are related to the NZP herd. The American Association of Zoological Parks and Aquariums even presented NZP with a captive propagation certificate in 1979 "in recognition of the sustained captive breeding of Dorcas gazelles."

However, our analysis of the breeding records of this herd (Ralls et al., 1980) revealed evidence of two typical deleterious inbreeding effects: increased juvenile mortality in inbred calves (see Table 1), and delayed sexual maturity in the inbred females. Furthermore, all deaths from prematurity, inanition, and miscellaneous veterinary problems were of inbred calves.

It is highly probable that most wildlife populations in which substantial inbreeding supposedly occurs without ill effects are analogous to the NZP gazelle herd. Conclusive demonstration of the deleterious effects of inbreeding in such populations may be impossible owing to the lack of individual records. The wise manager, however, should consider the possibility of inbreeding depression in any small population of a normally outbreeding species that exhibits the classical symptoms of decreased fertility, increased juvenile mortality, and general lack of vigor. Inbreeding does not always have deleterious effects. Some species may have unusually small genetic loads of deleterious alleles, and even in forms that usually show inbreeding depression an occasional population may, by chance, be relatively free of harmful recessive alleles. However, the evidence that close inbreeding causes deleterious effects in most populations of a wide variety of species is overwhelming.

The zoo community has drawn upon the experience of animal breeders to develop sound genetic management plans for captive species. We hope those responsible for managing small populations in the wild will in turn build upon the knowledge gained from small populations in zoos.

Founding and Bottlenecks

PART THREE

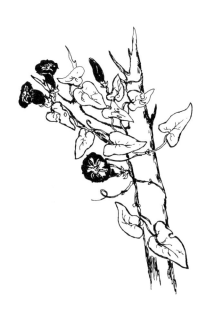

Introduction

Many natural and virtually all captive populations have experienced a bottleneck in population size. That is, the population has been at some time reduced to a relatively small number of individuals. These populations have in common with small founder populations the increased likelihood of change in genetic composition of the small population. Small populations may result from a decline in population size that takes place slowly over time or rapidly, as would result from some catastrophe; or may result from either a natural colonizing event or purposeful founding by human agents to establish or reestablish a species in a habitat.

This section on founding and bottlenecks was organized because much of conservation and management focuses upon small populations, many of which are the remnants of larger populations. In trying to avert subsequent reductions or in trying to encourage recovery in bottlenecked populations, we feel that managers can learn from the examples of founding events in nature and how species are adapted or can adapt to reduction in gene pool size.

The contributors to this section examine small animal and plant populations which have been reduced in size (either man-induced or naturally). Carson uses Hawaiian Drosophila *to illustrate how a founding event by a single pregnant female can be successful and lead to divergence and ultimately speciation. Selander examines how self-fertilization facilitates colonization of new habitats and also comments upon the diversity of mating systems existing in invertebrates. Clegg and Brown examine the factors which enable the success of colonizing in plants and develop guidelines to assist managers in attempts to found new plant populations. Powell describes molecular techniques that have been used to diagnose past population bottlenecks and proposes the application of mitochondrial DNA analysis for tracing maternal lineages and estimating founder population size. Finally, Templeton and Read describe their plan to purposefully inbreed a herd of Speke's gazelle. This unusual plan, based on research upon* Drosophila, *was applied to Speke's gazelle because very few individuals remain for this species, which may already be extinct in the wild. Templeton and Read's plan is an outstanding example of how basic research on founder events can be put to use in management.*

CHAPTER 11

The Genetics of the Founder Effect

Hampton L. Carson

Introduction

Over most of his existence, man has been at war with wild nature. Animals and plants that could not be put to use directly have been slaughtered, burned, and beaten back with ruthless abandon. Only very recently has a conservation ethic emerged as a powerful force. This trend is so new that those who find themselves charged with conserving the dynamic properties of populations of valuable species have almost no guiding principles, despite the fact that modern genetic biology of populations has been a mature science for over fifty years. Frankel and Soulé's *Conservation and Evolution* represents a new approach; this exciting pioneer work was published in 1981.

Conservation versus Improvement

The population manager of a valued natural species has goals that are quite different from those of the classical animal and plant breeder. Both manager and breeder want to maintain large, genetically variable populations. Whereas the goal of the breeder is to use this variability for the improvement of utilitarian characters of the species, the end of the conservationist manager is to provide for the preservation of an intrinsic biological state, considered valuable for its own sake. "Preservation," however, in the narrow sense, meaning the simple maintenance of individuals of a species in zoological or botanical gardens without regard to their population characteristics, is not enough. Frankel and Soulé (1981, p. 4) employ the term "conservation" to denote programs for the long-term retention of natural breeding communities that provide for the continuing operation of natural selection. I am sure everyone will agree that preservation in the narrow sense is useful only as a last resort for a seriously endangered species.

The accomplishments of the breeders of wheat, rice, maize, chickens, and cattle are among the great wonders of modern genetics. In this extraordinary age of the recombinant DNA molecule, however, the great accomplishments of biological engineering through selective breeding (which continues to increase the usefulness of many animals and plants) tend to be overlooked. In many of these domesticated species, the precise natural wild ancestors have virtually disappeared. The principal cause of this is that, by simple processes of recombination and selection of existing genetic variability, the breeder has so changed them genetically that the character of the progenitors has been lost. In a surprising number of cases, the progenitors themselves no longer exist even as a relict wild population.

When the population manager contemplates the situation the breeder has created for himself, he may feel that the principles of animal and plant breeding are not his concern. His goal, after all, is to conserve the natural characteristics of species, not to alter them. There is a dangerous fallacy here, which stems from a misunderstanding of the genetics of reproduction in a diploid, cross-fertilizing species. Reproduction in such species is not simply a device for transmitting sameness from one generation to the next. On the contrary, the geneticist has shown that the genetic processes that occur during reproduction provide the organism with a shifting and dynamic variability system. As egg or sperm are formed, not only is the chromosome number halved, but the process of meiotic recombination shuffles the alleles and deals out new unique combinations to the gametes. Further, each new organism results from the chance union of these gametes (egg and sperm) at fertilization. Each reproductive episode (generation) provides a new and extraordinary array of disparate genotypic combinations. The potential (number of unique phenotypes) is so vast that only a small finite number of the theoretically possible combinations are produced, and this is largely governed by chance. Many that do survive have low relative fitness. All this is an automatic consequence of reproduction under Mendelian heredity (Chapter 2).

Genetic variability is rife even in the small garden plot or the herd of the plant or animal breeder. In large measure, the breeder's success in his selective regimes has depended directly upon this variability. All the practices of the conservation manager must be based on recognition of the existence of this variability. He must learn to harness it in a manner that retains the inherent values that are perceived in these natural species.

The improver and the population manager, however, face a fundamental problem. In view of the ubiquitous generation of this field of genetic variability in his populations, how can the genetic basis of high fertility, fecundity, and somatic vigor be retained? The manager has many lessons to learn from the experience of the breeder in this regard; "improvements" or preservation efforts that sacrifice or abridge these qualities are flawed at their very base.

The Founder Effect

Both breeder and manager are in the business of guiding, by selection, the reproduction of the small populations that are under their control. The word "small" is crucial here, since in most significant instances the populations that are artificially worked by selection can be traced back to a very small number of founder individuals derived from the natural populations. The latter frequently are relatively very large populations, and when an artificial sample is drawn off and propagated, a novel genetic situation is created. Even though a managed population may currently consist of a very large number of individuals, its characteristics can be profoundly and permanently affected by the prior size of the population bottleneck, even though this bottleneck occurred a number of generations in the past.

The effects of such bottlenecks were explored in theoretical terms as early as 1921 by Sewall Wright; the resulting phenomena have been variously termed as *random drift, genetic drift,* or the *Sewall Wright effect.* The purpose of this part of the book is to deal with these population effects in the light of recent studies of natural, experimental, and theoretical population genetics. In the course of the following chapters, we hope to dispel some of the mistaken "conventional wisdoms" that have grown up around the concept of the effects of inbreeding and genetic drift on genetic variability in populations.

Natural and Artificial Founders

The potential for sudden genetic change engendered by the founder effect is especially powerful when the bottleneck is extreme. For example, the new population may have been established from a single gravid female under circumstances that prevent breeding contact with the ancestral population for a number of generations following the founder effect. This condition exists particularly clearly in the breeding of large, rare, and endangered animals in zoological gardens.

To many viewers, the establishment of such populations from one or very few founders is an example of a highly unnatural manipulation. What is perhaps not widely recognized is the fact that natural analogs of such artificial founder events exist in the case of chance natural colonization by simple founders on certain remote oceanic islands. The initial terrestrial colonizations by certain birds and insects of Norfolk Island, Tristan da Cunha, or Hawaii, for example, provide examples of cases wherein single founders for the origin of certain species or subspecies may be inferred. These are especially instructive "natural experiments" and require close attention.

Size of the Founding Group

Resistance to the idea that an entire biological species may arise from a single gravid female (a "fertilized Eve") is considerable, and most authors writing on the subject state the founder effect rather cautiously. Thus, mention is usually made of "a small number" of successful founders. I have taken a quite extreme view on this, arguing that the data on insular founding suggest that single "fertilized Eve" founders are indeed involved in a significant number of cases (Carson, 1971, 1982). The most striking data on this are from the picture-winged *Drosophila* flies of Hawaii. Each of the Hawaiian islands has been successively formed as the Pacific tectonic plate has moved over a fixed upwelling "hot spot" in the earth's mantle in the region of the central Pacific Ocean. Despite the fact that the islands are fairly close to one another, the great preponderance (about 95%) of the hundred or so species of picture-winged *Drosophila* on the six major islands are endemic to single islands or volcanoes. The implication is that both ability to arrive and success of colonization are low, resulting in the repeated proliferation of a number of full biological species arising from single female founder individuals, the ultimate bottleneck for a sexually reproducing cross-fertilizing species.

On the other hand, if migration is high and repeated colonizations are successful between two islands having similar environments, the expectation is that the gene pool from the older island would be essentially transported to the new locality without perceptible genetic change in population characteristics. This latter case has been observed, as indicated above, in only 3 out of 102 cases of picture-winged *Drosophila* (Carson and Yoon, 1982). For our present purposes, these observations are important in that they indicate that extreme attenuation of a population through a single founder is not necessarily a lethal practice for the manager and indeed has considerable precedent in natural populations. This remark should not be misconstrued. Successful founders in nature are probably very great rarities, and there is certainly a high risk of extinction for each individual potential founder.

The Diploid Cross-Fertilizing Reproductive Mode

In exploring to what extent genetic change might occur as a population is forced through a bottleneck, it is important to emphasize that our discussion centers on those species that are predominantly diploid and cross-fertilized. Such species characteristically carry large stores of genetic variability in a balanced state in their populations. Most of the rare, valued, and endangered animal and plant species that present practical problems for the manager are characterized by genetic systems of this sort. Genetic recombination, abetted by high cross-over frequencies and chromosome numbers, naturally generates zygotic diversity from the large field of genetic variability in the gene pool.

Not all species that colonize oceanic islands have genetic systems of this sort. In some beach-dwelling organisms, for example, the necessity for cross-fertilization has been removed by natural selection, which has favored development of systems that avoid fertilization, such as parthenogenesis and vegetative systems of reproduction. In these, the dynamism of the cross-fertilizing system is bypassed. A single colonist or propagule may encapsulate virtually the entire genetic endowment of the species; the new colony is essentially an invariant clonal reincarnation of the old.

There are, of course, various intermediate degrees of restriction or dampers on recombination. These are such things as polyploidy and patterns of partial inbreeding or self-fertilization, which occupy a sort of middle ground between the extreme examples considered above. These are dealt with in the chapters of Selander, Clegg and Brown, and Hamrick in this book. Our concern here centers on the diploid, cross-fertilizing population.

The Genetic Structure of Diploid Cross-Fertilizing Populations

Although considerable evidence existed in the early days of population genetics, it was not until the middle 1960s that the amazing magnitude of genetic variation in natural populations was fully realized. This, of course, was made possible by the development of ever-more-sensitive electrophoretic techniques for the detection of protein variability. Although the functional significance of the great amount of soluble protein polymorphism in populations is not yet entirely clear, there is no question that the data dramatically confirm the notion that most animal and plant populations carry large amounts of allelic heterozygosity.

Modern data on multiple factor inheritance, including the data from electrophoresis, have settled the old controversy between "classical" and "balance" views of population structure, as Dobzhansky once called them (1955). The "classical" hypothesis (which was synonymous with "wrong" in the Dobzhansky view!) held that a diploid species (giraffe, mouse, or man) had a genetic system in which the "norm" was a largely homozygous—that is, genetically fixed—state at each locus. Any variability seen was considered to be transitory.

The "balance" view has been stated by Dobzhansky in the following manner: "the adaptive norm is an array of genotypes heterozygous for . . . numerous gene alleles, allele complexes and chromosomal structures." Most of the animals and plants that present practical problems are characterized by such balanced systems. The system operates in the following manner. Genetic recombination naturally generates diverse genetic types from the large field of variability in the gene pool. In order to meet environmental challenges, natural selection in many such organisms tends to develop a system based on the higher fitness of heterozygotes. These are maintained under regimes of selection that exploit the advantages of heterozygosity for many alleles simultaneously. In these, the large amount of genetic variability is continually being recombined as balanced hybrid vigor is maximized. Homozygous segregants are less fit. Thus, the data of modern genetics do not favor the view that selection tends, classical style, to reduce heterozygosity and to render the organism homozygous and thus genetically invariable at most loci.

Putting the selective premium on the generation by recombination of superior heterozygotes has a side effect of great importance to the population manager. Thus, every population of this sort normally produces genetically less fit individuals on both sides of the maximum fitness mean. Many individuals are formed that carry doses of detrimental alleles that have been concealed in the more heterozygous individuals. Were the chance founder of a new population to be drawn from one of these individuals, which represent a sort of normal biological wastage produced by the system, the mean fitness in the new population may be permanently reduced and never have the variability necessary for the restoration of the highly fit heterozygotes characteristic of the ancestral population.

Genetic Loads

As Muller pointed out in a landmark paper (1950), the deleterious genetic conditions described above constitute a burden, which he described as a *genetic load*. Genetic loads in human populations and indeed in populations of various captive mammals are often fairly easy to perceive. Both breeder and pediatrician are well aware that poor genetic endowment of an individual may result from various lethal, semilethal, or subvital conditions produced by alleles or allele combinations. As is brought out in Chapter 10 by Ralls and Ballou and Chapter 15 by Templeton and Read, inbreeding (breeding of close relatives) accentuates the appearance of such deleterious conditions, but they can and do occur frequently in outbreeding systems as well. Extensive work on *Drosophila* populations has shown that such lethal alleles or allele combinations are characteristic of many wild species, some of which have vast populations numbering in the billions of individuals, such as *D. willistoni* of the American tropics. When we look for it, we can show that the biological success of such a species extracts a price: the lethals, semilethals, and subvitals manifested in a species represent a normal biological wastage occurring in all diploid species. In species where the individual receives great parental investment (in man, for example), the load appears as the high social and personal price we humans pay for our biological origin. Indeed, this load is normal but nonetheless substantial and tragic. *Drosophila* flies are subject to similar genetic loads and selective pressures, but the cost is more difficult to perceive.

Types and Typological Thinking

The emergence of the balance view discredits any view of the bisexual world that is based on a typological concept of species. The view that the phenotypes of all individuals of a species conform to that of a type specimen is utterly foreign to a system in which both genetic and environmental variance abounds. The human mind, nevertheless, unfortunately tends to gravitate toward typological thinking. The biological world in particular is so complex that it is mentally comforting for us to try to treat the populations we see as if the individuals they comprise were mere photocopies of one another. Any keen observer of mankind knows the powerful nature of innate individual differences. To transfer this notion to an individual maize plant, a mouse, or a *Drosophila* is difficult. To cope with the fact that there are no true biological replicates and that each individual is both genetically unique and has had environmentally unique experiences is a staggering assignment. Despite the designations of some taxonomists, certain idealist philosophers, or the racists, bisexual populations do not consist of individuals identical to a type specimen but consist only of arrays of genetic difference.

This concept is of great background importance for the animal and plant breeder. It is made more difficult for him by the fact that through selection he can achieve what superficially appears to be a certain sameness. Thus, cocker spaniels may appear superficially to conform to a "pure type" and may be held to this apparent identity by strong artificial selection. These man-imposed characteristics, however, like the flower colors and forms selected by the plant breeder, usually do not perturb the deep-set genetic variability system of the species. Most such changes are reversible when a less restricted gene pool is restored. The "balance" system appears to be retained by natural selection, which, perhaps paradoxically, pervades most systems of artificial selection.

Can Deleterious Genes Be Removed by Selective Breeding?

Deleterious alleles segregating in captive or cultivated populations are frequently both an annoyance and a threat to the success of both the breeder and the manager. Yet selective removal proves to be very difficult, or, if accomplished, it is accompanied by a decline in vigor of the purged stock. Alleles that behave in this way are likely to be somehow caught up in the system of balancing selection favoring heterozygotes. Thus, the deleterious allele may be hidden in a superior heterozygote. If this is indeed the case, then selection against the trait may remove from the population one of the essential ingredients of the high-fitness heterozygote. In such cases, if the maintenance of vigor is the main concern, as it frequently is in the case of the population manager, it may be more profitable to breed to maximize heterozygosity. This will require the acceptance of the segregating homozygotes as part of the price. With modern electrophoretic and cytogenetic methods, such individuals can be recognized at an early stage and culled from the breeding stock. This, short of a system of balanced lethals in which all homozygotes die as fertilized eggs, is the only known way of coping with the balanced genetic load. Maximizing of outbreeding is the usual breeding-plan prescription under these circumstances.

Does the Founder Effect Always Have a Deleterious Result?

In the generations following a founder event, some degree of inbreeding ensues. If the founder was a single fertilized female, the result in the next generation automatically requires brother-sister or mother-son mating, so that inbreeding is intense. Such a series of events has two correlated and important results, both of which are deleterious.

Enforced inbreeding can result in an increased probability of the appearance of deleterious homozygous allele combinations resulting from meiotic segregation of heterozygous gametes. Selection against such combinations is difficult if they owe their presence to the high fitness of individuals that carry these combinations in the heterozygous state. If the numbers of individuals are kept large during these early generations following the founder event, then the possibility increases that chance recombination will separate the deleterious allele from its linkage to beneficial effects of other heterozygous loci. Accordingly, the early generations following the founder event represent a crucial phase both for natural populations and for the greatly attenuated populations of the breeder and manager. There are extensive data on this subject from the *Drosophila* work. Once over this period, chances for successful survival of the managed population are greatly enhanced.

In introductory sections of this chapter, stress was laid on the fact that population attenuation (a bottleneck) and the founder effect are not to be looked upon simply as artificial constraints imposed only by man. The same phenomena operate in wholly natural situations and are most clearly observed on oceanic islands.

"Conventional wisdom" holds that simple, random genetic drift of allele frequencies, resulting in loss or fixation of alleles, is the major effect produced by a founder event. This might indeed be true under the classical view of population structure, but, under a balanced system, another effect may be superimposed and be much more important. Many key characteristics of such a species are likely to be based on complex polygenic balances. Polygenes, by definition, have small effects on a character individually, but because there are many of them the balance may be an elaborate one. The founder event and the inbreeding that necessarily follows may serve to destabilize these polygenic balances, leading natural selection to seek a new balance to replace the older perturbed one. As Templeton (1980) has emphasized, selection occurring in multilocus systems controlling integrated developmental, physiological, and behavioral traits is particularly sensitive to founder effects. He proposes that a shift may occur without a severe reduction in genetic variability (see Chapter 15 by Templeton and Read).

The question posed at the beginning of this section can be answered in the negative, especially for the conservation manager rather than the breeder. The single most important management practice that can be instituted is to maximize population sizes and allow natural selection to determine the best parents for each generation. In extremely attenuated populations this may be difficult, but the conservation manager must consciously avoid a selective breeding wherein he imposes his own criteria of what constitutes a successful reproductive event. This must be left to nature. As will be addressed below, this sometimes presents the manager with a severe dilemma. Unless he imposes some guidance, natural selection may remove the very characteristics he considers most valuable. As has been emphasized earlier, natural selection is a pervasive process operating in captive or cultivated, artificial populations as well as in natural ones.

Maintenance of Characteristics of Species and Subspecies / Varieties

Certain characteristics of species are valued above others. Although birds, for example, may have no esthetic sense, esthetics strongly guides the population manager. For example, the males of various bird species have acquired, through natural sexual selection, extraordinarily beautiful and complex courtship systems. In the birds of paradise (Paradisaeidae) and the bowerbirds (Ptilonorhynchidae), these attributes are among the greatest wonders of the biological world (Diamond, 1982). Plumages of intense and bizarre color combinations are woven together with modified feather clusters and shapes. Behavioral patterns keyed to the maximization of display effects are abetted by vocalization. The entire elaborate scheme takes years to develop in the individual bird and in many instances involves the defense by males of territorial arenas.

Recent genetic studies of certain Hawaiian *Drosophila* species have provided a means for understanding the genetic basis of such remarkable developments. Suffice it to say here that these secondary sexual characters are under genetic control; they are, furthermore, reflected in behavioral responses in the females. Most important for the current discussion is that such reproductive characters are apparently under balanced polygenic control. This means that, following a founder event, such characters are especially prone to undergo a shift in balance. The result is that, far from preserving all details intact, the population manager may find his populations altered, not by any act of artificial selection by him but by a naturally occurring shift under natural sexual selection in his attenuated populations.

Accordingly the manager may find that the animal or plant he wishes to preserve as a sample of esthetic interest has undergone a change. Indeed, such changes themselves represent a type of microevolutionary change of great theoretical interest. The genetic system is not a fixed and frozen entity but is dynamic and variable. The population manager may have to be content in preserving only a portion of the characteristics of the species he wishes to maintain. Indeed, as was shown long ago by Sumner (1932) with captive deer mice (*Peromyscus*) populations, subspecific characteristics, although mostly genetic in nature, may be difficult to keep in an intact state.

There is a tendency to designate certain subspecies as endangered, and management practices are instituted that target the preservation of particular characters. In view of what has been said about variability, the manager should be satisfied with a realistic goal. Resulting populations may not conform precisely with what is hoped for. What indeed may ensue is a true microevolutionary event. Each population manager should realize that, under his guidance, natural and artificial selection are producing a population that is genetically novel. The biological conserver, short of putting the DNA into liquid nitrogen, cannot hope to freeze the characteristics of any natural population, be it a deme, a subspecies, or a species.

Summary and Conclusions

The conservation ethic requires population managers to be devoted to maintaining valued natural characteristics of populations of species, not improving them. Yet management of any population must employ methods to minimize loss of fertility, fecundity, and vigor. Although population founding from a small inbred sample has some risks, it is not intrinsically bad; natural cases of population founding, apparently from single individuals, are numerous. Most valued animals and plants that represent problems for conservers are diploid and cross-fertilizing. By its very nature, this genetic system is inimical to the perpetuation of sameness. At each reproductive event an enormous field of genetic variability is produced. Most of the variability is held in sexual populations by a complex balancing selection based on the superiority of fitness of heterozygotes. At reproduction, this results in a genetic load that is a normal accompaniment of this kind of selection. The manager, by using large populations and methods of culling copied from breeders, can remove some but not all of the genetic load. Rather than set as his goal the preservation of "photocopies," he must be content to allow natural selection to operate in his populations. He must replace typological thinking with population thinking. The founder effect and inbreeding are not intrinsically bad, but the entire genetic system is such that the slavish maintenance of all subspecies characters probably cannot be realized. Rather, the manager is engaged in guiding an entirely new set of genetic events at the microevolutionary level. In this, his role of conserver has to be viewed realistically. He can only hope to retain a shifting variability system, not a fixed and invariable genetic entity.

CHAPTER 12

Evolutionary Consequences of Inbreeding

Robert K. Selander

Introduction

Inbreeding occurs as a consequence of the geographic division of a population into a number of subpopulations or through the choice of mates according to phenotype (assortative mating) or genetic relationship. Within geographic subdivisions, inbreeding may occur because of restriction of population number, even if mating is random, average heterozygosity being reduced as a consequence of change in allele frequencies from one generation to the next due to random genetic drift (the "inbreeding effect of a finite population;" Crow and Kimura, 1970; Chapter 4).

The most drastic alterations in the genetic structure of populations occur with changes in the breeding system, as, for example, from outcrossing to self-fertilization, which is very close inbreeding. Selfing may be regarded as random mating within a population of one; the inbreeding coefficient is one half each generation, and heterozygosity potentially is reduced by one half each generation.

The effect of inbreeding on the variance of characters is to decrease the variance within groups (whether geographic subdivisions or inbred strains in the same population), but to increase both the variance between groups and the total variance in the population that these groups constitute (see Chapter 3). However, the total variance will remain large only if many subdivisions or inbred lines (strains) persist. If, for example, there is a progressive loss of inbreeding lines through random extinction or as a result of competitive or other interactions among strains, the total variance in the population will decrease, eventually approaching zero when only one line survives.

From the standpoint of evolutionary theory, some major questions concerning breeding systems may be asked: What is the explanation for the circumstance that, among many species of higher organisms, there is a variety of mechanisms promoting outcrossing [e.g., self-incompatibility alleles in plants; incest avoidance behavior in prairie dogs (Foltz and Hoogland, 1981)], whereas other species reproduce by facultative, and sometimes obligate, self-fertilization? Are self-fertilization and outcrossing different reproductive "strategies" for optimizing the distribution of genetic variation? Can the evolution of breeding systems be explained in terms of Darwinian selection at the level of the individual, without recourse to group selection or teleological hypotheses? Under what sets of environmental conditions are the different breeding systems advantageous? What are the evolutionary consequences of the various genetic population structures generated by these very different breeding systems, with respect to, for example, rates of adaptational change and extinction?

These are difficult questions related to complex problems, particularly those concerning the "adaptive strategies" of breeding systems, that have occupied the attention of evolutionary biologists for a long time. At this point, there are no definitive answers or even very satisfactory hypotheses. Current lines of thought on the evolution and relative advantages of breeding systems are presented in recent books by Maynard Smith (1978), Bell (1982), and Shields (1982b). I cannot go into all these matters in detail in the space available, important as they are. Instead, I have limited my chapter to self-fertilization (selfing), and specifically to four topics concerning this mode of reproduction. *First*, I briefly outline some elementary population genetics theory relating to the evolution of selfing from outcrossing and the demographic circumstances in which this process might be expected to occur. *Second*, I illustrate the genetic consequences of selfing by reviewing recently obtained information on the breeding systems of terrestrial slugs. *Third*, I consider some aspects of the genetic interactions between selfing strains. And *finally*, I briefly discuss the long-range evolutionary prospects of organisms that have adopted the breeding system of selfing.

The Evolution of Selfing

How and under what conditions does selfing evolve in a population of outcrossers? Because there are several advantages to close inbreeding (some of which are shared with asexuality and some with outcrossing), the real problem is to explain why selfing or other forms of automixis are not universally practiced by organisms (see Table 1). The major advantage of self-fertilization over outcrossing is that the cost of meiosis is low or zero. Theoretically, this gives a mutation for selfing a large advantage over an allele for outcrossing, and it will lead to the evolution of selfing unless something else prevents it.

Maynard Smith (1977, 1978) developed a simple model of autoselection that considers the fate of a recessive mutation for selfing that occurs in low frequency in an outcrossing population of hermaphrodites. In descriptive terms, the model may be outlined as follows: The allele $+$ is dominant to another allele b at the same locus; genotypes $+/+$ and $+/b$ outcross, whereas b/b individuals fertilize their own eggs and also contribute sperm to $+/+$ and to $+/b$ individuals. Assume that each individual produces one egg (population fitness is constant): $+/+$ and $+/b$ individuals contribute one haploid genome to their own egg and one genome (via sperm) to the zygote produced by another $+/+$ or $+/b$ individual. But b/b individuals fertilize their own eggs, thereby contributing two haploid genomes, and also fertilize the egg of a $+/+$ or $+/b$ individual, for a total contribution of three haploid genomes. So at this point, b/b individuals enjoy a 3:2 advantage in fitness; and the b allele should quickly replace the $+$ allele, especially if the fitness of the selfed progeny is equal to that of the outcrossed progeny. But this generally will not be the case because of inbreeding depression in the selfed young. Still, the b allele will increase to fixation if the average fitness of selfed progeny is greater than one half that of outcrossed progeny. (If not, the selfing allele will be eliminated.) More-

Table 1. *The Relative Values of the Genetic Costs Hypothetically Associated with Different Breeding Systems*[1]

	Breeding system		
Potential costs	Outbreeding	Inbreeding	Asexuality[2]
Cost of meiosis	High	Low	None
Cost of males	High to none	Low to none	None
Recombinational load	High	Low	None
Segregational load	Low	High	None
Mutational load	Low	Low	High

[1] From Shields (1982a).
[2] Equivalent to ameiotic parthenogenesis = apomixis.

over, if outcrossing individuals have difficulty finding mates, this will have the effect of increasing the relative mean fitness of selfed offspring.

What is the mean fitness of selfed offspring relative to outcrossed offspring for an initially outcrossing species? There are no data for habitually outcrossing plants or animal hermaphrodites, but Maynard Smith (1978) believes that the usual picture of an inbred line in a species that outbreeds in nature is exemplified in his earlier work on sib-mating in *Drosophila subobscura* (Hollingsworth and Maynard Smith, 1955). (Note that selfing is a much more severe form of inbreeding than sib-mating.) Fitness fell to very low levels after a few generations, then rose a bit thereafter in the next 200 generations, but never reached a level equal to that of outbred flies. Prolonged inbreeding apparently does not eliminate inbreeding depression, which would be the case if it were entirely a consequence of recessive lethals. Rather, there must be fixation of alleles that lower fitness slightly in homozygous condition. There is evidence that the crossing of strains of organisms that normally self-fertilize or otherwise closely inbreed usually produces some heterosis (hybrid vigor) but relatively little compared with crosses between inbred strains of outcrossing species (Wright, 1977). The important point is that very good, successful genotypes can be produced by selfing or other forms of inbreeding, especially when large numbers of inbred strains are produced and the better ones are selected. There is nothing intrinsically weak or inferior about a homozygous genotype or a genome with little or no heterozygosity.

Following Maynard Smith's model, it would seem that the major obstacle to the evolution of selfing is the segregational load (Table 1), which in most outcrossing species must be reduced if the mean fitness of selfed progeny is to be high enough to cause an increase in the frequency of a selfing mutation. This can be achieved by mild inbreeding over extended periods of time in small populations, thus preparing the stage for an increase in the frequency of a selfing allele to fixation (or to an intermediate equilibrium).

The risk of an outcrosser's not being fertilized (while selfers are assured of fertilization) will be especially important in annual species and in those that are sparsely distributed. And if mates are not available, selfing is the only game to play, so that the value of the fitness of selfed progeny relative to unity for outcrossed progeny is inconsequential. [Charlesworth (1982) notes that Bell (1982) unduly minimizes the significance of the fact that asexuality will be most advantageous in environments where sexual individuals find it difficult to become fertilized. The same can be said for the evolution of selfing.]

Thus, we might expect to find an evolutionary correlation between selfing and (1) population density (or vagility of individuals) or (2) the average number of founders of colonies; and in general it is our impression that selfing forms of terrestrial slugs are not quite so abundant, even locally, as are some of the obligate outcrossers (see beyond). In plants, there are well-known associations between selfing and the annual habit and between outcrossing and the perennial habit (Stebbins, 1950; Grant, 1958; Baker, 1959; Harper, 1977; Chapter 13 by Clegg and Brown; see discussion in Shields, 1982a). Similar associations may exist in molluscs, but the details have not been worked out.

The type of population structure in which selfing might be favored because mates are not readily available (and for some other reasons; discussed later in this chapter) also may be one in which a reduction in inbreeding depression (from segregational and mutational loads) is likely to occur. This structure involves the fragmentation of a species into numerous subpopulations that are founded by small numbers of individuals. Either the subpopulations never become very large, or, if they do, extinction rates are relatively high, and there is frequent, repeated colonization. After a subpopulation is founded by one or a few individuals, its size may remain small, and hence the intensity of inbreeding remains high for many generations as a consequence of several factors, including (1) poor autocorrelation of environments between localities, (2) negative correlations of fitness effects of pleiotropic loci, and (3) low fertilization rates. Solbrig (1976) notes that the reduction in heterozygosity due to inbreeding in small colonies lessens the genetic diversity of the progeny of outcrossers; the latter he calls the "marginal benefit" from outcrossing. With increasing homozygosity, the mean fitness of outcrossed progeny may decrease in proportion to the degree that it depends on heterosis. There is increasing expression of the segregational load, causing an increase in outbreeding depression.

Breeding Systems in Terrestrial Slugs

Malacologists have long known from laboratory studies that many species of hermaphroditic pulmonate snails and slugs are capable of reproducing uniparentally (see Runham and Hunter, 1970; Duncan, 1975), presumably by self-fertilization, since parthenogenesis (apomixis) apparently is rare in molluscs. We have recently surveyed genetic diversity and breeding systems in species of two families of terrestrial slugs, the Arionidae and Limacidae. For many species, we have compared samples from North America (McCracken and Selander, 1980) and the British Isles (Foltz et al., 1982a, 1982b) in an effort to determine (1) if the introduced North American and native British populations have the same breeding system and similar amounts of genetic diversity, and (2) whether there are ecological and genetic correlates of colonization.

In each family, average heterozygosity over loci (H) varies widely across species (Tables 2 and 3), from zero (*Arion circumscriptus, A. silvati-*

Table 2. *Genetic Variation and Breeding System in Arionid Slugs*[1]

Species	Region	Populations	Individuals	P	H	Breeding system
Arion fasciatus	New York State	10	814	0	0	Self-fertilizing
A. circumscriptus	British Isles and New York State	9	358	0	0	Self-fertilizing
A. silvaticus	British Isles and New York State	7	77	.04	0	Self-fertilizing
A. intermedius	British Isles and New York State	15	217	.12	0	Self-fertilizing
A. distinctus	British Isles and New York State	11	205	.46	.180	Outcrossing
A. oweni	British Isles	2	13	.23	.044	Outcrossing
A. hortensis	British Isles and New York State	8	92	.14	.041	Outcrossing
		7	31	.42	.126	Outcrossing
A. subfuscus	British Isles	17	265	.27	.073	Mixed
	Eastern United States	6	105	.64	.197	Outcrossing
	Ireland	3	48	0	0	Self-fertilizing
	New York State	5	224	0	0	Self-fertilizing
A. lusitanicus	British Isles	3	35	.25	.082	Outcrossing
A. ater ater						
A. ater rufus	British Isles	33	631	.31	.059	Mixed

[1]Source: McCracken and Selander (1980); Foltz et al. (1982a); P = average polymorphism over loci, and H = average heterozygosity over loci.

Table 3. Genetic Variation and Breeding System in Limacid Slugs[1]

Species	Region	Populations	Individuals	P	H	Breeding system
Limax maximus	British Isles and Eastern United States	4 3	53 52	.08 .45	.027 .166	Outcrossing Outcrossing
L. marginatus	British Isles	5	55	.31	.034	Outcrossing
L. tenellus	British Isles	8	91	.14	.028	Outcrossing
L. pseudoflavus	British Isles	2	23	.14	.007	Outcrossing?
L. valentianus	Eastern United States	4	140	.33	.077	Outcrossing
Milax sowerbyi	British Isles	6	63	.44	.126	Outcrossing
M. budapestensis	British Isles	7	98	.39	.117	Outcrossing
M. gagates	British Isles	2	92	.08	.013	Outcrossing
Deroceras agreste	Scandanavia	11	106	.29	0	Self-fertilizing
D. reticulatum	British Isles and New York State	50	1051	.74	.191	Outcrossing
D. caruanae	British Isles	19	244	.23	.049	Outcrossing
D. laeve	Eastern United States	12	316	.26	.005	Mixed

[1] Source: McCracken and Selander (1980); Foltz et al. (1982b); L. Noble (unpublished data); P = average polymorphism over loci, and H = average heterozygosity over loci.

cus, A. intermedius, A. fasciatus, and *Deroceras agreste*) to 0.19 (*A. distinctus* and *D. reticulatum*). The species surveyed fall into three categories on the basis of the genetic structure of their populations. Thus, various species consist of (1) polymorphic and heterozygous populations that are locally panmictic; (2) one or more homozygous strains; and (3) both a heterozygous form and one or more homozygous strains, between which hybridization may or may not occur.

In species of the *first* category, the level of heterozygosity varies over an order of magnitude. For most of them, there is no evidence that the breeding system involves anything other than outcrossing, since genotypic proportions in local populations do not depart from Hardy–Weinberg equilibrium, and the observed heterozygosities are within the range characteristic of other outcrossing molluscs (Powell, 1975; Selander, 1976; Nevo, 1978). Three outcrossing species (*Limax maximus, Arion subfuscus* [form A], and *A. hortensis*) show much less heterozygosity in British than in North American populations, while *Deroceras reticulatum* and *A. distinctus* are variable in both regions. Increased heterozygosity is not usually expected in introduced populations, but it could arise through the multiple introduction and mixing of individuals from genetically differentiated source populations.

In the *second* category, five species (four arionids and one limacid) consist entirely of homozygous strains throughout their sampled ranges. Native and introduced populations of *Arion circumscriptus* and *A. silvaticus* consist of single strains, and *A. intermedius* is represented by three strains differing at single loci. *Arion fasciatus*, which was sampled only in North America, also is a single homozygous strain. In *Deroceras agreste*, almost 30% of the loci assayed are polymorphic (*P*), but heterozygosity is absent in local populations. Hence, this species is composed of a large number of homozygous strains differing, for the most part, at single loci.

Deroceras laeve may be included here, since local populations tend to be strongly polymorphic but only weakly heterozygous ($H = .005$ and inbreeding coefficient $F = .902$); thus, they usually consist of several strains between which some crossing occurs.

The *third* category includes *Arion ater* and *A. subfuscus*, both of which consist of both homozygous and heterozygous forms. Burnet (1972) determined that the monogenic form of *A. ater* is the subspecies *A. ater ater*, and that polymorphism in British populations results from hybridization between *A. a. ater* and *A. a. rufus*, which has been introduced there (Quick, 1960). In native populations of *A. ater rufus* in the Netherlands, genotypic proportions generally match Hardy–Weinberg expectations (Burnet, 1972), but polymorphic populations of *A. ater* in Britain and Ireland show marked deficiencies of heterozygotes, apparently arising from a mixed mode of reproduction in *A. a. ater*. In the laboratory, *A. a. ater* self-fertilizes preferentially, and even when crossbred offspring are produced, clutches invariably include some selfed offspring (Williamson, 1959).

Arion subfuscus consists of a heterozygous form native to England and a homozygous form native to Ireland, each of which has apparently been introduced throughout the British Isles. The two forms are genetically rather similar (I, which is Nei's measure of genetic identity, is equal to .816 for these two forms), and hybridization regularly occurs between them. Populations of heterozygous *A. subfuscus* in North America resemble polymorphic British *A. subfuscus* ($I = .909$), but a homozygous form in North America is very different from the homozygous form in Ireland ($I = .467$).

Is the absence or very low levels of heterozygosity in several species of slugs attributable to self-fertilization? We strongly favor this interpretation for a variety of reasons, including laboratory evidence of self-fertilizing capability in certain species, but other hypotheses may be suggested (Selander and Ochman, 1982). One possibility is that the homozygous forms are outcrossers that have recently experienced severe reductions in genetic variability as a consequence of founder effects and sampling drift in small populations. This is thought to be the case in several outcrossing animals for which extensive surveys of protein variation have failed to detect polymorphism (Bonnell and Selander, 1974; Schnell and Selander, 1981); but all these species have highly restricted geographic ranges and are known to have had histories of severe reductions in population size. In contrast, there is no convincing case of an absence of protein polymorphism in wide-ranging, outcrossing animal or plant species (see Chapter 20).

An hypothesis that depends on severe reduction in population size cannot reasonably account for our observations on slugs for several reasons. None of the homozygous forms is particularly restricted in distribution, either in its native European range or in North America (Chichester and Getz, 1973; Kerney and Cameron, 1979). Indeed, many of these forms are as widely distributed and abundant as are the highly polymorphic and heterozygous species. If some of the homozygous species are indeed outcrossers, it is difficult to imagine under what conditions their populations would have remained small enough for periods sufficiently long to become homozygous. Severe and sustained reduction in population size is required to achieve a substantial reduction in genetic variation in outcrossing populations (Nei et al., 1975).

In general, introduced species of limacid and arionid slugs are similar in both genetic structure and breeding system to native populations of the same species in the British Isles. The fact that homozygous forms of *Arion intermedius, A. subfuscus, A. circumscriptus,* and *A. silvaticus* occur in Britain demonstrates that the loss of variability antedates their colonization of North America. There does, however, appear to be a relationship between colonizing ability and breeding system. Among the native European species of arionids and limacids that have become firmly established and widespread in eastern North America, self-fertilizing forms are disproportionately represented. Of the 10 self-fertilizing species thus far identified in the British fauna, only two (*Deroceras agreste* and *Arion ater ater*) have failed to colonize North America. In contrast, 9 of 13 outcrossing species have not become established.

Self-compatibility is associated with colonizing ability in low-vagility species, particularly plants, because it dispenses with the requirement for finding a mate and assures reproduction (Stebbins, 1957; Baker, 1965; Ghiselin, 1969; Chapter 13 by Clegg and Brown). This is the simplest and most plausible explanation for the disproportionate represenation in the introduced North American slug fauna of species that are capable of self-fertilization.

Self-fertilizing molluscs have, in general, retained mating behavior and full hermaphroditic genital structure. Only in *Arion intermedius* is it likely that outcrossing is no longer possible, for mating behavior has never been observed, male genital structure is reduced, and a spermatophore has not been found (Davies, 1977). However, even in selfing species that have not completely abandoned outcrossing, most local populations nonetheless consist of a single homozygous strain, although the species as a whole may be composed of a number of distinctive genotypes (strains) variously distributed throughout its range. Where selfing strains co-occur, individuals of mixed genome are usually not evident, but whether this is because crossmating rarely occurs or because hybrid genotypes are inviable remains to be determined for individual cases.

Among slugs, we have found a number of species that appear to be outcrossers on the basis of genotypic frequencies in local populations matching Hardy–Weinberg expectations, but which apparently have very low values of polymorphism and heterozygosity. Is the low level of genetic variation in British populations of *Deroceras caruanae* (Table 3), for example, a consequence of a series of prolonged bottlenecks following initial founding from continental Europe? And is this a situation that prepares the way for the evolution of selfing?

Genetic Interactions between Selfing Strains

As long as selfing is not obligate, new gene combinations can be generated through outcrossing (hybridization) between strains. Thus, in a facultative selfing species, the number of distinctive lines may increase through both mutation within lines and recombination between them. Given this potential, it may be somewhat paradoxical that in animals at least, and perhaps also as a rule in selfing plants, regional populations consist of one or only a small number of lines. For example, the Mediterranean land snail *Rumina decollata*, which reproduces by facultative self-fertilization, is represented in southern France by only two homozygous strains that have different alleles at an estimated 50% (13 of 26) of their structural gene loci and also differ in body color (Selander and Hudson, 1976). At some localities, the two strains occur together without crossbreeding, but viable and fertile individuals of mixed genome (hybrids) are produced at other sites. Because continued outcrossing, even at low levels, is expected to reshuffle genetic variation, the persistence of only two strains in southern France has been interpreted as circumstantial evidence of strong coadaptation of genomes, with consequent inferiority of mixed genomes (Selander and Hudson, 1976). But there may be another, simpler explanation.

A field study of the genetic relationships of the light and dark strains of *Rumina* in two adjacent gardens in Montpellier, France, was made in 1973 by Selander and Hudson (1976). To measure temporal changes in genetic structure, we again studied this population in 1979, taking collections from 12 colony sites that were sampled in 1973 (Ochman and Selander, in preparation). Each snail was scored for five enzyme loci (*Est-6*, *Est-10*, *Est-11*, *Lap-1*, and *6-Pgd*) at which different alleles are fixed in the two strains. Individuals of mixed genome were those having combinations of alleles (over loci), either in homozygous or heterozygous state, of both the dark and light strains.

From 1973 to 1979, there was, overall, an increase in the proportion of the light strain, a roughly equivalent decrease in the proportion of the dark strain, and a small increase in the frequency of individuals of mixed genome (Table 4). This pattern is apparent both in the pooled sample and in the unweighted means for the paired samples from the same colonies. The proportion of the dark strain declined at all but two sites, and the frequency of individuals of mixed genome increased in all but four colonies.

The observed increase in frequency of the light strain is consistent with laboratory evidence that it has a higher reproductive potential than the dark strain or individuals of mixed genome, because its embryonic and posthatching development is more rapid, with the consequence that it begins reproducing at a much younger age (Selander, unpublished data). Ochman and Selander (in preparation) suggest that the resulting difference in fecundity between the strains was the primary factor responsible for the change in population structure that occurred in the Montpellier gardens between 1973 and 1979. Moreover, we propose that variation in fecundity can also account for the absence of extensive strain diversity in *Rumina* in the Montpellier region and in other parts of its range. This hypothesis eliminates the need to invoke coadaptation and hybrid inferiority (Selander and Hudson, 1976) to explain the presence of only two homozygous strains, despite hybridization, and it can be generalized to all interactions between self-compatible strains in which fecundity (r) and selfing rate (f)

Table 4. *Composition of Colonies of* Rumina *in 1973 and 1979*

Colony	Year	Number of individuals electrophoresed	Percentage		
			Light	Dark	Mixed
M-A	1973	62	74	18	8
	1979	81	67	16	17
M-D	1973	19	50	39	11
	1979	238	75	10	15
M-E	1973	71	75	11	14
	1979	205	85	4	11
M-F	1973	38	55	29	16
	1979	139	67	15	18
M-G	1973	30	3	97	0
	1979	21	81	19	0
M-H	1973	4	50	50	0
	1979	59	69	14	17
M-X	1973	10	50	50	0
	1979	33	85	6	9
A-A	1973	51	98	2	0
	1979	70	84	6	10
A-C	1973	23	87	4	9
	1979	46	98	0	2
A-I	1973	55	84	0	16
	1979	77	90	1	9
AqP	1973	52	63	33	4
	1979	169	89	4	7
Pooled[1]	1973	415	71.6	20.0	8.4
	1979	1137	80.2	7.8	12.0
Mean	1973		62.6	30.3	7.1
	1979		80.9	8.6	10.5

[1] $\chi^2_{(2)} = 57.64; P \ll .001$.

are variable. It specifies the conditions under which a given genotype (*a*), with associated values of *r* and *f*, will increase or decrease in a population consisting of one or more other genotypes.

In this model, we consider temporal variation in the relative abundance of strain *a*, whose mean relative fecundity r_a is greater than or equal to the relative fecundities of all other genotypes in the population; r_o is the mean relative fecundity of all other genotypes, including those of strain *b* and hybrids between the two strains; and f_a is the rate of selfing in strain *a*. In this version of the model, we assume that mating is nonassortative. In its simplest form, the model specifies that the proportion of strain *a* will increase only if

$$r_a - r_o > 1 - f_a$$

If strain *a* is an obligate selfer ($f_a = 1$), the situation is analogous to competition between pairs of reproductively isolated sexual species or parthenogenetic clones occupying the same niche; and the outcome of the interaction is determined by the fecundity of the genotype *a* relative to that of individuals not of that genotype. Therefore, strain *a* will increase owing to its higher fecundity. This is obvious; but what is not so apparent is that this outcome is also likely even if there is some outcrossing between the strains.

Consider the situation common to *Rumina* and to many species of plants in which self-compatibility has been established in a polymorphic population. Selfing is the predominant mode of reproduction, but individuals occasionally outcross. The simplest case is a population composed of two homozygous strains that are capable of interbreeding, but the model can incorporate any degree of genetic diversity. Thus, if strain *a* is only slightly more fecund than the mean of individuals of strain *b* and of mixed genome, the rate at which strain *a* selfs must be high if it is to increase in frequency in the population. Conversely, the difference in fecundity could be very large, but the frequency of strain *a* will decrease if the rate of outcrossing is sufficiently high. Note that even if the difference in fecundity is very small, a reduction in rate of outcrossing by strain *a* can further promote its increase in the population. We may also note that if the fecundity of strain *a* is lower than the average for individuals not of that strain, no degree of selfing will cause it to increase in the population. As a consequence of the interplay of these factors, we should expect self-compatible species to evolve arrays of highly adaptive genomes and to become particularly resistant to outcrossing.

In the process described by our model, the production of large numbers of hybrids leads to no change in the overall genetic diversity of the species. Thus, two strains meet locally and hybridize for a period, but eventually only one strain persists, and there remains no evidence of the occurrence of the other strain or of hybridization. Genotypic variation initially created by outcrossing between strains is gradually purged by repeated backcrossing of the hybrids to the predominant strain of higher fitness and/or lower outcrossing frequency. This hypothesis explains why only one or a few strains occur in any one region, when otherwise, through mutation and hybridization, one might expect *Rumina* to have generated large numbers of sympatric or microgeographically allopatric strains. According to our model, the better adapted homozygous lines persist despite the formation of heterozygous genotypes through hybridization, and the frequency of outcrossing between strains decreases as certain genotypes become better adapted. In this manner, self-fertilization becomes a mechanism for preserving particular adaptive combinations of genes (see Allard et al., 1968; Baker, 1972; Allard, 1975).

Evolutionary Fate of Inbreeders

What is the evolutionary fate of selfing populations or species? Insofar as molluscs are concerned, the answer seems to be an increased probability of extinction. Among slugs, the selfing species are members of genera most of which are outcrossers; there are no genera or families of selfers. *Arion fasciatus* may be exemplary—a single homozygous strain, and presumably thus vulnerable to environmental change (physical and biotic), and with no close relatives with which to crossbreed to produce new strains. On the other hand, *Rumina* is very successful, and apparently has been for a long time, since fossil shells like those of contemporary individuals have been discovered. Moreover, *Rumina* has no close relatives. But in this case great genetic diversity is represented in the form of dozens and perhaps hundreds of strains, so that the total genetic diversity is much greater than that in any comparable outcrossing species of mollusc. (But, we may ask, how many "species" of *Rumina* are there, and can all the strains cross with all others to produce viable progeny?) Somehow *Rumina* has managed to survive for a long time and to occupy a large geographic range in which it has generated a large array of genetically distinct strains. This has occurred despite the action of the process described by our model of the effect of backcrossing in reducing strain diversity. In any one region there are only one to three strains, but in the aggregate, over the whole Mediterranean region, there are numerous strains. This type of population structure must mean that dispersal is severely limited, so that geographic isolation has permitted the generation of complex strain diversity. Certainly it is significant that *Rumina* is among the more sedentary of all terrestrial molluscs, by reason of its occupying a region of arid or desert conditions (in which it is confined to very local "oases" and is active only intermittently following rains).

In conclusion, it is obvious that much further work on a variety of organisms will be required if we are to understand the complex of genetic and environmental factors responsible for the evolution of self-fertilization and other breeding systems. It has long been assumed that close inbreeding in natural populations is confined to plants, but the recent discovery that many pulmonate molluscs regularly self-fertilize has extended the range of organisms available for comparative analysis, thereby increasing the chance that significant advances in the theory of breeding systems can be made in the near future. Additionally, the study of closely inbred animals may contribute to other areas of population biology, including, for example, an understanding of the role of kinship in aggression and cannibalism (Jones, 1982).

Summary

Recent work on biochemical genetics of natural populations has demonstrated that certain groups of animals, notably the molluscs, have a variety of breeding systems fully equivalent to that occurring in plants. Close inbreeding in the form of obligate or facultative self-fertilization (or other automictic types of reproduction) is characteristic of populations of nearly one third of the sampled species of terrestrial slugs of the families Arionidae and Limacidae. Included among the selfers are some of the most widespread and ecologically versatile native European forms. As a consequence of close inbreeding, genetic variation in a selfing species is apportioned into monogenic strains; but the total genetic diversity in selfing species is reduced over that occurring in closely related outcrossers, since selfing species generally consist of only a few strains. This suggests that the long-term evolutionary potential of a species is reduced as a result of the evolution of selfing, although the short-term prospects for selfing forms may be superior to those of highly heterozygous outcrossers. As in plants, selfing in animals is associated with colonizing ability, selfing European species of slugs being disproportionately represented in the North American fauna (see Chapter 13).

A model of the evolutionary origin of selfing species from outcrossers is discussed. Important aspects of the model are a gradual reduction of the mutational load prior to the onset of selfing, and the gradual elimination of strains having relatively low fecundity and/or relatively low selfing frequency through repeated backcrossing with a strain of higher fecundity and/or higher frequency of selfing. Inevitably, close inbreeding becomes a mechanism for the development and preservation of highly adapted gene complexes and, hence, cannot be accounted for solely in terms of demographic models.

CHAPTER 13

The Founding of Plant Populations

Michael T. Clegg
A.H.D. Brown

Introduction

Plant species are usually composed of a network of populations that are dispersed in space and that are more or less genetically differentiated from one another. This network is a dynamic entity. The spatial mosaic of populations is in a continual state of flux, with local populations becoming extirpated and new populations appearing, in response to shifts in environmental conditions. Thus the founding of new populations is a constant feature of the ebb and flow of plant communities. Despite this fact, there is relatively little information on the genetic events that accompany the founding of plant populations. What information we do possess comes almost entirely from the colonization of new environments by weedy species.

While information on successful colonizing species promises some insight, it bears only indirectly on the problem of managing wild plant species. Moreover, Brown and Marshall (1981) concluded, in a recent review of genetic events associated with plant colonization, that colonizing success may be achieved through a variety of evolutionary pathways. There is no single genetic attribute that is an essential prerequisite for a successful founding event. Rather it is the integration of the major components of the genetic system that is essential, and there are a number of ways in which a successful integration can be achieved. It follows that the management of plant populations requires a knowledge of the major components and potential interactions of plant genetic systems.

In this chapter we first consider the principal components of plant genetic organization in the context of colonizing success. We then discuss the role of the mating system as the primary control on the expression of genetic variation. Finally, we outline some criteria for the genetic management of plant populations.

Genetic Variation

If the management of wild plant populations is to have a genetic dimension, then we must begin by asking how the founding of new populations affects the reservoir of genetic resources possessed by the species. Genetic variability is fundamental to the evolutionary process because the potential for an adaptive response to an environmental shift is strictly dependent on the genetic options open to the species. Thus our first concern in evaluating the founding of a plant population must be to consider the sample that is included in the newly founded population, from the reservoir of genetic variation possessed by the species.

There are several ways to quantify levels of genetic variation (Chapters 20 and 21). A crude measure is simply the average number of alleles per locus. A somewhat different although still crude measure is the fraction of loci that are polymorphic. More desirable measures should reflect the distribution of allelic frequencies over loci (e.g., gene or allelic diversity; Nei, 1975, p. 129). In a few cases of intercontinental colonization, comparisons of genetic variation between source populations and derivative (founder) populations are available. Table 1 gives comparisons for the grasses *Avena barbata* and *Bromus mollis*, which illustrate a marked reduction in genetic variation among the colonial populations relative to the source populations. Recent studies of *Emex spinosa* (Polygonaceae), which was introduced into Australia in the 1930s from the Mediterranean basin, also indicate that a loss of genetic variation is associated with colonization (Marshall and Weiss, 1982).

The loss of genetic variation associated with intercontinental colonization is to be expected because of (1) population bottlenecks associated with long-distance founding events, (2) the absence of repeated migration as a source of genetic enrichment, and (3) the possibility of novel selective pressures in the new environment. Indeed, we may expect to find rapid

Table 1. *Average Number of Alleles and Average Allelic Diversities per Locus over All Source and Colonial Populations*[1]

Species	Populations	n_a	H_T
Avena barbata (slender wild oat)	Source Mediterranean Colonial California, Region I California, Region II	 2.4 1.02 2.2	 0.435 0.001 0.180
Bromus mollis (soft brome)	Source England Colonial Australia	 2.06 1.88	 0.230 0.176

[1] n_a denotes average number of alleles per locus; H_T denotes average allelic diversity over all loci. The allelic diversity at a single locus is equal to $1 - \Sigma \bar{p}_i^2$, where $\bar{p}_i$ is the mean frequency of the ith allele in the combined subpopulations that comprise each source or colonial population. Data on *Avena* are from Clegg and Allard (1972); and data on *Bromus* are from Brown and Marshall (1981).

adaptive changes in recently founded populations. Studies of rose clover in California (Jain and Martins, 1979) and subterranean clover in Australia (Cocks and Phillips, 1979) are consistent with this expectation.

Rose clover (*Trifolium hirtum*) was introduced into the rangelands of California in the late 1940s. Subsequently, it was found to be colonizing roadside habitats. Jain and Martins (1979) have compared range and roadside populations for several demographic and genetic attributes. They found higher genetic diversity, lower survivorship, lower seed carryover during the winter, and higher fecundity among roadside populations. An increase in genetic diversity is, at first sight, surprising; however, the roles of mating structure and migration require quantification to explain this increase. A shift in the mating system toward increased outcrossing (discussed below) was uncovered in roadside populations, but the extent of migration has not been quantified. Subterranean clover (*Trifolium subterraneum*) was introduced into Australia near the end of the last century (Frankel, 1954). The species was subsequently domesticated in Australia and grown widely as a pasture legume. Cocks and Phillips (1979) have documented adaptive shifts in flowering time in extensive collections from South Australia. They have also found a large number of unique genotypes in this self-fertilizing species, presumably due to occasional outcrossing and subsequent recombination.

Unique patterns of genetic variation have been documented in other colonizing episodes. For instance, the slender wild oat (*Avena barbata*) exhibits a pattern of genetic variation in California not found in the Mediterranean source populations (Clegg and Allard, 1972; Kahler et al., 1980).

California populations can be broadly classified into two climatic zones. In the Central Valley region (region I) virtually all populations are monomorphic for the same seven-locus genotype, while in the intermontane valleys of the coastal zone (region II) extensive genetic polymorphism is found (Marshall and Jain, 1969; Clegg and Allard, 1972).

Regions of extensive monomorphism have been described in other colonizing plant species. For example, races of *Xanthium strumarium* (rough cocklebur) in Australia are almost entirely monomorphic over 13 allozyme loci (Moran and Marshall, 1978). This highly self-fertilizing species was introduced from the New World, but no information is presently available on genetic variability in the source populations. The common morning-glory (*Ipomoea purpurea*) is believed to have been introduced into the southeastern United States from the tropical highlands of Central America. Studies of weedy populations of this species in Georgia show low levels of allozyme variation, despite extensive flower color polymorphism (unpublished data). Thus we may conclude that high levels of genetic variation are not always a prerequisite for colonizing success.

Some plant genotypes exhibit a range of phenotypic responses to varying environments. This phenomenon is referred to as phenotypic plasticity when the phenotypic response allows survival and reproduction in the face of environmental change (Bradshaw, 1965). Theoretical calculations predict a negative correlation between phenotypic plasticity and genetic polymorphism (Levins, 1968). Comparative studies of related species appear to bear out the predicted relationship. Thus *Avena barbata* exhibits greater phenotypic plasticity than the more genetically variable species *A. fatua* (common wild oat: Marshall and Jain, 1968), and *Bromus rubens* (foxtail brome) is also more plastic, yet genetically less variable, than *B. mollis* (Jain, 1976).

This brief review of the role of genetic variability in colonizing success serves to highlight the multifarious and opportunistic nature of evolutionary processes. Nevertheless certain features are paramount. First, without genetic variation the adaptive response to environmental shifts is limited by the range of phenotypic plasticity. This range may prove inadequate in new or changing environments. Second, the founding of new plant populations is usually associated with novel selective regimes, and rapid genetic changes may occur, further eroding the reservoir of genetic variability. Third, the founding of new populations is frequently accompanied by a loss of variability through sampling and population bottlenecks. Mechanisms that enrich the pool of variation, such as migration (see Chapter 3 by Allendorf and Chapter 4 by Chesser) or hybridization (see Chapter 16 by Harlan) are likely to expand the adaptive range of the population.

Finally, the great majority of the colonizing plant species that have been the object of genetic study are weedy annuals that are predominately self-fertilizing or reproduce asexually. Indeed, among the world's 18 worst weed species, 9 reproduce by self-fertilization and 8 of the remaining 9 species reproduce asexually, in addition to outcrossing (Brown and Marshall, 1981). Apparently the mode of reproduction plays a crucial role in colonizing success and will therefore be a central factor in founding plant populations.

Reproductive Mode

The distribution of genetic variability in plants is dependent on the way genes are packaged in genotypes in the population. Individual genotypes are ephemeral, arising through reproduction and being eliminated through death in every generation. Thus the mode of reproduction, or mating system, determines the genotypic frequency distribution. Table 2 lists the genetic consequences of different mating systems.

Several features of plant mating systems are paramount. First, plants rely on a variety of agencies to effect pollination, including wind, gravity, and insect or other animal vectors. Because the role of the plant is relatively passive, it must depend on the reliability of these external agents for reproductive success. A plant transported to a new environment may undergo changes in the mating structure that can be deleterious. Second, plant mating systems are plastic and subject to evolutionary change (see Appendix 3, Figure 2). Finally, as emphasized in Table 2, genetic recombination is also controlled through the mating system. Thus the potential range of genotypic combinations is a function of the mating system.

A distinctive feature of plant reproduction is the great range of mating systems that occur in natural populations. On one extreme, some plant species (e.g., tobacco, *Nicotiana tabacum*) feature systems of negative assortative mating (genetically different individuals mate preferentially

Table 2. *Controls on the Distribution of Genetic Variation among Individuals*

Control		Consequences
Mating system	Outbreeding	High heterozygosity, high recombination
	Inbreeding	Low heterozygosity, low recombination
	Asexual	High heterozygosity, no recombination
Ploidy level	Diploid	High segregation, rapid response to selection
	Polyploid	Fixed heterozygosity through gene duplication, reduced segregation

with one another). In the case of tobacco, pollen cannot fertilize ovules of a maternal plant when the pollen and the maternal tissue share an allele at the self-sterility locus. This "mating rule" enforces heterozygosity at the self-sterility locus. A second consequence of the self-sterility system is that it selectively favors new mutants. Thus the mating rule can induce selective changes in the population. It should also be obvious that this mating rule precludes founding a new population from such a single plant. Clearly, the result will be extinction because there is no genetically different individual for the founder plant to mate with.

On the other extreme, many plant species feature systems of positive assortative mating (genetically similar plants mate preferentially). The most intense form of positive assortative mating is self-fertilization. The genetic consequences of self-fertilization were first studied by Mendel. Self-fertilization causes the loss of heterozygosity. Indeed, Mendel first showed that in any generation self-fertilization would halve the proportion of heterozygotes as compared with those in the previous generation. Clearly, when compounded over a few generations the result will be the elimination of heterozygotes from the population (Chapter 12 by Selander).

In point of fact, while a great many plant species do have mating systems that feature self-fertilization, in every case that has been studied some degree of outcrossing also occurs, albeit sometimes low. For example, the annual grass small fescue, *Festuca microstachys* (= *Vulpia microstachys*), which occurs in western North America, features less than 1% outcrossing in most years, but in some years the outcrossing rate may be as high as 6.7% (Adams and Allard, 1982).

While a great deal of variability in outcrossing rate exists within plant species (see Clegg, 1980, for a review), all plant species do have a characteristic mating structure. Any attempt at genetic management of wild plant species must include, as a primary consideration, information on the mating structure.

Inbreeding Depression

Having established the genetic consequences of plant mating systems, we now consider the related phenomenon of inbreeding depression. A classical observation of plant and animal breeding has been that a general loss of fitness accompanies forced inbreeding (Chapter 10 by Ralls and Ballou). The explanations for inbreeding depression are that (1) inbreeding causes an increase in homozygosity (as discussed above) and (2) increased homozygosity means an increased frequency of deleterious recessive homozygous genotypes and a decreased frequency of heterozygous combinations which may be overdominant.

Table 3. *Estimates of Inbreeding Depression from a Natural Population of* Gilia achilleifolia[1]

Generation	Net reproductive rate	Relative fitness
Outcrossed	19.87	1.0
Selfed	11.13	0.56
Second selfed	11.26	0.57

[1] From Schoen (1982b).

Experimental determinations of inbreeding depression in natural plant populations date back to Darwin. In a recent study Schoen (1982b) has measured inbreeding depression in the California annual wildflower *Gilia achilleifolia* (yarrow gilia). In these experiments, net reproductive rate was calculated as fecundity weighted by survivorship summed over age classes, for an outcrossed and two successively selfed generations. The results, shown in Table 3, reveal a marked loss in fitness upon one generation of selfing (due to reduced survival), but no additional loss after a second generation of selfing. The particular population studied has an outcrossing rate of about 96%, although *G. achilleifolia* can outcross at rates ranging from 15% to over 100% among different local populations (Schoen, 1982c). The results of this investigation are in general agreement with studies of selection in natural plant populations, which often show heterozygous advantage (e.g., Clegg and Allard, 1973; Marshall and Allard, 1970a). Finally, Brown (1979) has noted that inbreeding species frequently show an excess of heterozygotes, relative to expectations based on measured rates of outcrossing, while outbreeding species often show a deficit of heterozygotes compared to random mating expectations. Brown has termed this observation the "heterozygosity paradox" and has analyzed the potential causes of the phenomenon in detail. For present purposes, it is sufficient to note that a substantial loss in fitness accompanies large shifts in the mating system. Conversely, small shifts are tolerated, and, in fact, many plant species have a remarkably plastic mating structure.

Environmental Influences on Plant Mating Systems

Plant mating systems are subject to environmental influences and genetic factors. For example, many plants are pollinated by insects, and the absence of appropriate insect vectors may result in a drastic alteration of the breeding system. Moreover, there can be an interaction between genetic and environmental factors. For instance, populations of *Ipomoea*

Table 4. *Estimates of Pollination Behavior and Outcrossing Rate as a Function of Flower Color in a Natural Population of* Ipomoea purpurea[1]

Flower color	Pollinator visits		Outcrossing rate
	Observed	Expected	
White	54	95	0.35
Pink	432	438	0.53
Blue	691	643	0.62

[1] From Brown and Clegg (1982). $\chi^2_{(2)} = 21.62$.

purpurea in Georgia are characterized by a variety of flower color polymorphisms. Observations of pollinator behavior, summarized in Table 4, show that some colors are preferred by pollinators of this species.

Furthermore, the pattern of pollination visits is correlated with the outcrossing rate (Brown and Clegg, 1982). In fact, *I. purpurea* has a variety of genetically determined mechanisms that influence the mating system, including flower color polymorphism and variation in anther–stigma distance (Ennos, 1982). Further evidence that shifts in mating structure may accompany changes in habitat comes from Jain and Martins' (1979) study of rose clover (described above), where recently colonized roadside populations were found to have a higher outcrossing rate than range populations (5.1% versus 3.8%).

Influence of Mating System on Genetic Recombination

Beyond determining the genotypic frequency distribution, the mating system also exerts a direct control on the rate of genetic recombination in a population. Effective recombination occurs only in individuals heterozygous at many loci. Inbreeding reduces the frequency of heterozygotes, thereby reducing the effective rate of recombination. Allard (1975) has emphasized the role of inbreeding in facilitating the maintenance of multilocus associations in plant populations. In fact, multilocus associations may be common in highly inbred species. Thus Brown et al. (1977) report multilocus associations in wild barley, *Hordeum spontaneum*, similar to those observed by Clegg et al. (1972) in experimental populations of cultivated barley, *H. vulgare*. In contrast, specific multilocus genotype–environment correlations occur in California populations of *Avena barbata* (Allard et al., 1972; Hamrick and Allard, 1972), but are not observed in populations from Israel (Kahler et al., 1980). While more data are needed from other plant species that span the range of mating system options, it is clear that intense multilocus associations do occur in inbreeding species.

Ploidy Level

The second major control on the distribution of genetic variability in plants is ploidy level (Table 2). Many plant species are polyploid and consequently enjoy the advantages of extensive gene duplication. Stebbins (1971) estimates that between 30% and 35% of species of flowering plants are polyploid. In addition, polyploids are more common among successful colonizing species. In fact, all 18 of the world's worst weeds, listed by Brown and Marshall (1981), are polyploid.

There are many grounds for hypothesizing a wide adaptability for polyploid species. Among these grounds are (1) buffering against inbreeding depression due to fixed heterozygosity or due to reduced genetic segregation (see Chapter 16), and (2) wide environmental tolerance due to gene duplication and subsequent diversification. Perhaps because of these attributes, polyploids seem to be favored in habitats that experience large fluctuations in both climate and edaphic factors (Stebbins, 1971). In addition, polyploids are more likely to become weedy—that is, adapted to environments disturbed by human activity—than are related diploids. These very brief considerations make it apparent that ploidy level must be regarded as an important variable in founding plant populations.

Genetic Considerations in the Founding of Plant Populations

How can we use the facts on plant genetic organization developed in this chapter for the management of wild plant populations? To address this question we must recognize that genetic considerations cannot be divorced from the ecological setting of the plant population. Indeed, a plant population cannot be founded unless a suitable physical and biotic environment exists. In this important sense, we must regard ecological considerations as primary and genetic considerations as secondary. The primacy of ecological factors has been emphasized by Meagher et al. (1978) in one of the few genetic investigations of a rare and declining plant species (*Plantago cordata*, heart-leaved plantain).

Given that a suitable environmental setting exists, then the genetic correlates of founding a plant population, which are summarized in Table

Table 5. *Genetic and Demographic Correlates of Plant Colonization*

Situation	Correlate	Responses
New environment	Unique selective regime	Genetic change Phenotypic plasticity
Initial founding population of low density	Very restricted number of potential mates	Asexual reproduction Self-fertilization, inbreeding depression Shift in pollination mechanism, or agent
Initial population founded from very restricted numbers	Genetic drift and founder effect important	Random genetic change Loss of genetic variation Substantial genetic differentiation among populations
Initial population well below carrying capacity	Rapid increase in numbers	High survivorship High fecundity

5, must be examined. Some of these genetic correlates can be manipulated, while others must be regarded as fixed. To illustrate this point, consider, as a hypothetical example, a diploid annual that is outbreeding and adapted to a particular insect pollinator. Furthermore, suppose that this species has low fecundity, a restricted distribution, and a narrow environmental tolerance. Such characteristics contrast sharply with weedy species or successful colonizers and may be typical of endangered plant species. Clearly the fecundity and level of phenotypic plasticity are not subject to immediate manipulation. On the other hand, the level of genetic variability and the mating structure can be manipulated (within certain limits). Table 6 outlines the genetic options a manager might wish to consider. Clearly the *first* variable open to manipulation is the level of genetic variation.

While we have seen that many weedy species have undergone intercontinental colonization and large range expansions with limited levels of genetic variation, it does not follow that new populations should be initiated with low levels of genetic variability. This apparent contradiction illustrates the problem of generalizing from the current data base. Weedy species are atypical because they benefit from the environmental disruptions created by human activity and because their rapid range expansion is frequently facilitated by release from previous environmental constraints. As an illustration, the prickly-pear, *Opuntia stricta*, was among the cacti introduced into Australia during the nineteenth century. It became a

Table 6. *Genetic Management Options*

Variable	Options	Management technique
Level of genetic variation	Increase	Hybridization Wide founding sample Migration from source populations
	Maintain	Migration from source populations
	Explore new combinations	Small subdivided populations
Mating system	Maintain	Suitable pollinators available Potential mates available
Ploidy level	Increase	Artificial doubling of chromosomal complement

serious pest species, after an explosive population expansion. In 1925 an attempt to control the *Opuntia* spread was made by introducing some 2700 eggs of the phycitid moth *Cactoblastis cactorum*, the larvae of which feed on cactus in the plant's native South America. Within five years the density of *C. cactorum* had risen to 2.5×10^7 larvae per hectare in dense *Opuntia* populations (Murray, 1982). Very quickly the *C. cactorum* infection brought about effective control of the *Opuntia* population. Such a release from native parasites may be a common feature in intercontinental colonization events and could limit population growth in colonizers. In this instance, the initial spread of the cactus was facilitated by the absence of the moth. *Opuntia* populations in Australia have very little genetic variability and therefore may be especially sensitive to control efforts.

Indirect evidence pointing to the importance of genetic variability can be found in the biological control literature. For instance, Brown and Marshall (1981) have found a significant correlation between control success and mode of reproduction in the target plant species in 81 different control attempts. Specifically, they find that biological control agents are more than twice as effective against asexually reproducing species as compared with sexual species, presumably because of a greater tendency toward genetic homogeneity among the asexual species.

Finally, the evidence cited above for adaptive shifts following colonization events (such as in *Trifolium*) argues that genetic changes are a significant concomitant of founding events. For all these reasons it seems a prudent strategy to use genetically diverse materials if it is necessary to found a new plant population. It also may be a useful strategy to encourage genetic diversification by facilitating some genetic substructuring of the founding population.

A *second* feature of plant genetic organization that may be manipulated is the mating system. In most cases, the preferred strategy would be to maintain the mating system. This option may involve a prior study of pollination mechanisms and could involve the introduction of appropriate pollinators. Large increases in inbreeding should be avoided because of the loss of reproductive potential associated with inbreeding depression. (Increasing the level of ploidy probably would not be appropriate in managing wild plants, but it is a useful technique for the plant breeder involved with domestic plants.)

In an important sense the goals of the manager are either to maintain the status quo or to return the species to an earlier, more healthy state. To be most effective, management decisions should be based on genetic and ecological principles. The idea that conservation has a genetic and evolutionary dimension (Frankel and Soulé, 1981) is still fresh. The many ramifications of this important insight are only now being explored.

Summary

The level of organization of genetic variability in a population determines the adaptive potential of the population. Founding populations often encounter new environmental changes, or are sometimes released from previous environmental constraints. The response of a particular population to a founding event depends on both the adaptive potential and the demographic structure of the population, as well as on the environmental constraints imposed by the new setting. The management of wild plant populations requires an evaluation of these variables.

An evaluation of the adaptive potential of plant populations must begin with a consideration of the level and organization of genetic variability—that is, with the genetic structure of the population. The major determinants of genetic structure are the breeding system, the multilocus organization of alleles, and the ploidy level. Plant species exhibit a remarkable diversity of genetic structures. These range from the intense inbreeding often associated with weedy annual species, through to obligate outbreeding and high levels of genetic variability, associated with many long-lived woody species. Moreover, the reservoir of genetic variation latent in polyploid species provides a buffer against genetic erosion, associated with intense inbreeding or frequent founder events. By virtue of this plasticity of genetic organization certain plant species have been remarkably successful in rapidly colonizing new environments. Many "natural experiments" on the founding and subsequent spread of plant populations have occurred during the European colonization of North America and Australia. It is possible to exploit these experiments to gain some insight into the genetic events associated with the founding of plant populations.

Acknowledgments

We are indebted to Sir Otto Frankel for stimulating discussion and criticism. MTC gratefully acknowledges the support of a Guggenheim Foundation fellowship.

CHAPTER
14

Molecular Approaches to Studying Founder Effects

Jeffrey R. Powell

Introduction

Traditionally, genetics is studied by making a series of controlled crosses and then analyzing the offspring. Such an approach is impractical with the majority of species of plants and animals for a variety of reasons. Progress in molecular biology in the past twenty years or so has brought to light methodologies that make it possible to study the genetics of virtually any species. While one might think that molecular biology and population genetics are disparate disciplines, in fact the two fields have become closer in the last several years. A few years ago one of the leading figures in molecular biology, Francis Crick (1979), wrote: "A molecular biologist who wants to discuss the evolution of the eukaryotic genome will need not only to know a lot about the way DNA and its transcripts can behave but also something about modern ideas on population genetics." I submit that the converse is equally true: population geneticists who want to stay on top of their discipline need to know something about modern ideas in molecular biology. This is true for both theoretical and experimental population geneticists. In this chapter I shall discuss how molecular approaches to studying populations can be used. First, I discuss a technique that is quite familiar and has already proved its usefulness in population genetics—namely, electrophoresis of soluble proteins. Since the technique is so familiar, at times it has been viewed as a panacea for all population genetics problems. Thus my first task is to take a critical look at the technical and theoretical limitations of the technique. Second, I discuss two less-familiar methods of analyzing electrophoretic data. Finally I discuss more recently developed technology, which is now being assimilated by population genetics. This is the new DNA technology, which has gained so much publicity, if not notoriety, in recent years. While discussing the general issues of applying these technologies, I shall try as often as possible to relate results or potential results to the subject of this part of the book, founder events. However, the relevancy of my discussion of this subject may seem a bit remote at times.

Many of the chapters in this book discuss protein electrophoresis or isozyme or allozyme studies. There is an abundant, if not overabundant, literature on the subject, so I can hardly cover it here, nor is there any reason to attempt such a task. Rather, I shall try to make a few points that are less well known but are particularly relevant to conservation, hence to the subject of this book.

On The Limitations of Protein Electrophoresis

Protein electrophoresis is a popular technique for a variety of reasons. It is applicable to practically any organism. Variation at loci coding for proteins is fairly abundant, and it is only when there is variation that one can do genetics. It is reasonably inexpensive and technically straightforward. However, there is a fairly severe limitation on the ability of this technique to accurately measure amounts of variation, a point that is not widely appreciated, judging from the literature. If one estimates the fraction of a eukaryotic genome that is needed to code for all the proteins in an organism, it comes out to about 0.5%; this represents the fraction of the genome that actually codes for amino acid sequences and does not include such elements as introns and leader sequences. If one then estimates what fraction of nucleotide substitutions can be detected in this 0.5% of the genome, a figure on the order of 20% is obtained; this is for protein electrophoresis as usually performed. Thus, potentially, if every protein in an organism were subjected to electrophoresis, this technique would be able to detect 0.1% of nucleotide substitutions in the total genome (see Powell, 1975, for justification of these figures). Generally one does not actually study all ten to fifty thousand proteins in an organism, so in practice one studies an even smaller fraction of all possible variation. The point of this exercise is the following: allozyme studies are fine for determining that a population or species is genetically variable. However, and this is relevant to studying founder events, it cannot be used to prove that there is no genetic variation in a population or species. The technique is just incapable of detecting variation in the vast majority of the genome.

Theoretical Considerations

It is generally believed that when a population undergoes a bottleneck—for example, a founder event—the genetic variability should decrease. This is true, of course, but what is not so well appreciated is the actual quantitative decrease expected and what methods might be best suited for detecting the change in genetic variation. There is one study that is particularly relevant for electrophoretic surveys, in that it is concerned with neutral variation and a level of variation generally found in allozyme studies—that is, about 10% average heterozygosity. This is the work of Nei et al. (1975). I shall purposely avoid the issue of whether allozyme variation is strictly neutral; if it is subject to the forces of natural selection, the selection coefficients must be small enough that assuming neutrality will yield qualitatively relevant conclusions. I wish to emphasize four points stemming primarily from Nei et al. (1975).

1. With average heterozygosity taken as a measure of genetic variation, it is very difficult to detect a reduction in variation due to founder effects. This is true for two reasons. First, the decrease in heterozygosity is significant only if there is a very severe bottleneck. This can be understood intuitively by the following consideration. If a new population is begun by a single, randomly chosen hermaphroditic individual, his/her offspring would have 50% of the heterozygosity of the parental population. Two founders would produce 75% of the heterozygosity; five founders, 90%. The second problem concerns the statistical properties of the measure average heterozygosity, as usually calculated in electrophoretic surveys; it has a generally unappreciated large standard error associated with it, which is very dependent on the number of loci studied (Nei and Roychoudhury, 1974). Thus, unless there has been a quite severe bottleneck, say less than five founders, or the population remains small for a number of generations, and unless a rather large number of loci are studied, it is unlikely that calculating average heterozygosity from allozyme data would yield statistically significant evidence of a reduction in genetic variation.

2. Average heterozygosity will continue to decrease after the initial founder event for a considerable amount of time, on the order of 50 to 100 generations. This is so because, while the newly founded population is small, it will continue to lose heterozygosity owing to genetic drift faster than mutation restores it. Thus, contrary to intuition, the best time to detect a founder effect by the criterion of average heterozygosity is not immediately after the event but after several generations when the heterozygosity reaches the minimum. It is also interesting to note that this minimum value persists for a long time, thousands of generations, before mutation can significantly restore heterozygosity.

3. The decrease in heterozygosity is strongly dependent on the rate of

population growth (r). A newly founded population that expands rapidly (has a large r) will lose less variation than a slowly growing population. This is related to the point above; the longer the population remains small, the longer it will be subjected to strong genetic drift effects. This rate of population growth effect is very significant (see Figure 1 in Nei et al., 1975).

Taken together, these three points indicate that calculating average heterozygosity from an electrophoretic survey would have the best chance of detecting a statistically significant decrease in heterozygosity for a newly founded population if (a) the bottleneck was severe, (b) the number of loci studied was large, (c) the population was sampled several generations after the initial founder event when heterozygosity reaches a minimum, and (d) the population grew slowly. Studies in which these criteria are satisfied are probably rare.

4. There is an alternative measure of genetic variation, which is more sensitive to sampling errors (that is, founder effects) than is average heterozygosity: the average number of alleles per locus (also see Ewens, 1972). Unfortunately this measure also has a drawback: it is quite sensitive to sample size. One must sample a large number of individuals to obtain a reasonably accurate measure of this quantity, which is not possible in many cases.

We seem to have arrived at a rather negative view of the usefulness of electrophoretic surveys, at least with respect to using them to study founder events. I have intentionally sounded a negative note, as I do not believe that the limitations of allozyme surveys are as clearly understood as they deserve to be. Let me continue on a more positive note by showing how electrophoretic data may be useful in the management of wild populations.

A Couple of Analytic Procedures

One of the ironies of the application of molecular techniques to population genetics is that a tremendous amount of data can be generated fairly easily and rapidly. This brings up the problem of how to analyze table after table of allele frequencies. Many procedures have been used, some of which have become almost *de rigueur*. I should like to discuss two less widely used procedures that may have particular application in the management of wild populations.

The first procedure is an analytic one introduced by Lewontin (1972) that is relevant to notions of preserving genetic variation. The results of applying this procedure are somewhat surprising and may further explain why detecting founder effects is difficult. This analytic procedure is one that apportions genetic variation among different levels of organization. Lewontin posed the following questions: Of the total genetic diversity of the human species, how much is due to differences between individuals of the same population, how much is due to differences between populations of the same race, and how much is due to differences between races. The data he used were nine "biochemical markers," which included enzymes and blood groups. As a measure of diversity the Shannon information statistic was employed, a measure with several convenient properties and one that is used surprisingly sparingly in population genetics. The following conclusions were reached: 85% of the total diversity is due to differences between individuals of the same population, 8% is due to differences between populations of the same race, and only 6% is due to differences between races. If the gene loci used here are typical of the genome as a whole, then one can conclude that, if the human species were suddenly reduced to a single population, most of the genetic variation of the species would remain extant. A perusal of the allozyme literature indicates that most animal species have a pattern of genetic diversity not unlike that of humans. That is, most species have the greatest diversity within populations, and relatively small frequency differences between populations. Thus we may conclude that reducing a species to a single population (of a reasonable size) would still allow a considerable amount of a species' allelic diversity to survive.

It is important to stress that the above analysis concerns the apportionment of genetic diversity. It is only because there is so much variation within a population that interpopulation differences appear small. It is a completely different issue whether the degree of genetic differences between populations is sufficient to allow one to distinguish populations. Basically this latter issue is one of taxonomy, which is a matter of discrimination as opposed to apportionment. There is a rather powerful statistical technique that can be applied to allozyme data, which is often quite useful when the goal is to discriminate. This is a stepwise linear discriminant analysis (for details, see Jennrich, 1977). Basically what this procedure does is to maximize the ratio of between-group variance to within-group variance. Alleles with the greatest differences between groups are identified, and their frequencies are weighted to give the maximum discrimination. Once the maximum discrimination function is generated, the covariance with this first function is removed. New weighting coefficients are generated that maximize discrimination based on the remaining variance. The procedure can continue in such a manner to generate these canonical variables until all the variance is accounted for. There are commercially available programs that carry out this procedure—e.g., BMDP Biomedical Computer Programs, P-Series, University of California Press, 1977. As a demonstration of this technique, I shall present some of our electrophoretic data on the yellow fever mosquito, *Aedes aeqypti*. The species itself is of little consequence to what I am trying to demonstrate except that its pattern of allozymic variation is more or less like that of humans and most other species. Those familiar with allozyme studies will recognize that the data shown in Table 1 could be for almost any species. There is a lot of variation within populations and some differences in frequencies between populations, although nothing really striking. Certainly the differences between groups are not enough to allow one to easily distinguish populations. Subjecting these data to a stepwise linear discriminant analysis yielded the graph shown in Figure 1. In this figure are plotted the first two canonical variables; together they account for 92% of the dispersion in allele frequences, so going to the third or more canonical variable would add relatively little. If we had a sample of our mosquito and we did not know its origin, how accurately could it be assigned to one of these groups based solely on allozyme frequencies? We would be over 90% accurate. (See Powell et al., 1980, and Powell et al., 1982, for further details of the procedures and data summarized here.) The point of this demonstration is that, even if there are no truly diagnostic loci between groups, there

Table 1. *Mean Allele Frequencies at the Five Most Differentiated Loci in Seven Subsets of Populations of* Aedes aeqypti[1]

Gene and Allele	Subset						
	A	B	C	D	E	F	G
IDH-2							
100	0.911	0.526	0.932	0.431	0.622	0.660	0.822
116	0.083	0.474	0.067	0.570	0.370	0.340	0.118
PGM							
100	0.778	0.937	0.548	0.985	0.877	0.983	0.968
120	0.124	0.032	0.326	0.0	0.053	0.011	0.0
Others	0.097	0.031	0.125	0.015	0.070	0.006	0.033
MDH							
84	0.201	0.071	0.170	0.028	0.081	0.036	0.141
100	0.744	0.701	0.795	0.592	0.586	0.722	0.708
120	0.055	0.226	0.035	0.380	0.332	0.252	0.154
6PGD							
100	0.803	0.987	0.976	0.789	0.971	0.932	1.0
116	0.106	0.003	0.008	0.194	0.031	0.069	0.0
Others	0.091	0.0	0.016	0.008	0.010	0.0	0.0
HK4							
100	0.909	0.552	1.0	0.796	1.0	0.963	0.972
109	0.014	0.128	0.0	0.006	0.0	0.0	0.0
Null	0.045	0.320	0.0	0.198	0.0	0.037	0.028
Others	0.039	0.0	0.0	0.0	0.0	0.0	0.0

[1] From Powell et al. (1980).

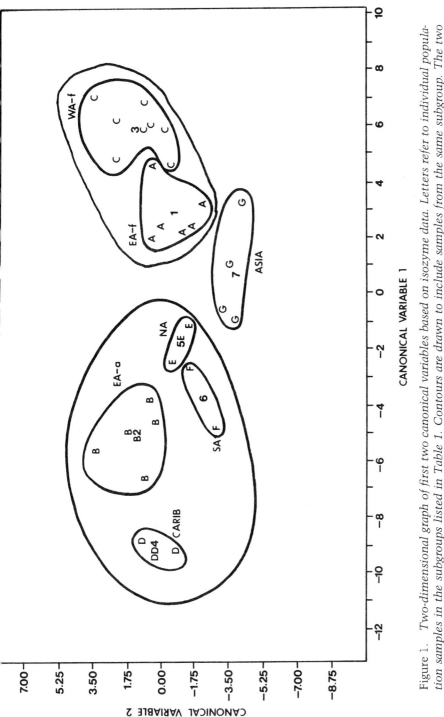

Figure 1. Two-dimensional graph of first two canonical variables based on isozyme data. Letters refer to individual population samples in the subgroups listed in Table 1. Contours are drawn to include samples from the same subgroup. The two largest contours include samples that fall into two described subspecies of Aedes aegypti. Asian samples (G's) do not fit easily into either subspecies. From Powell et al. (1980), which should be consulted for further details.

are powerful statistical procedures that may distinguish patterns that would otherwise remain unnoticed. It should also be noted here that this procedure will allow one to distinguish the origin of a population but not an individual.

The Potential of DNA Technology

So far I have mentioned technologies that have proved their utility in population genetics. I now want to turn to some newer molecular technology that I believe eventually will prove much more useful and informative than protein electrophoresis. Because this technology is so new, it is difficult to give many concrete examples of its use. Therefore I begin by making some general statements about the potentialities and give at least one example. I end by stating why I believe this technology will soon be readily available to practically anyone.

As is familiar to anyone who reads *The Wall Street Journal*, molecular biologists have made tremendous strides in recent years in isolating and manipulating specific pieces of DNA, manipulating them almost at will. For our purposes the most important breakthrough is the relative ease with which it is possible to study variation in DNA. A very powerful and relatively easy way is to use enzymes called restriction endonucleases. These remarkably useful enzymes recognize a set of four to six specific contiguous nucleotides and cleave the DNA at the recognition sites. The pieces of DNA generated can be separated according to size by electrophoresis in an agarose gel. If two DNAs differ in the number or position of the specific base sequence recognized by an enzyme, then digesting the DNA with the enzyme will produce products that differ in size. These differences will become evident when the pieces are separated on a gel.

As an example of the power of these tools, let us assume that we can isolate a specific piece of DNA about 10,000 base pairs in length from each of a large number of individuals. We then digest the DNA with ten different endonucleases, each of which recognizes a different set of bases. If on the average the enzymes recognize five bases and there are an average of five sites on the molecule, then each enzyme is assaying for variation at 25 nucleotide sites. So the ten endonucleases together detect variation at 250 sites. Thus if two DNAs differed at 1% of their bases, we would have a better than 90% probability of detecting the difference. These calculations are approximate, and there are many complicating factors. Nevertheless I present them here to indicate the level of sensitivity of the technique. The second important point is that by using restriction endonucleases one can study essentially any piece of DNA, unlike protein electrophoresis, which can only study a small fraction of the variation in the 0.5% of the genome that codes for amino acid sequences.

Thus it is clear that the 1% sensitivity of the restriction endonuclease method is a truly vast improvement in our ability to detect naturally occurring genetic variation. One can improve on this level of sensitivity to practically any degree either by increasing the number of enzymes used to cleave the DNA or by studying longer or more pieces of DNA. In studying variation in populations it is important to choose a methodology that detects differences at a level that is appropriate for the purposes intended. For instance, if one wanted to distinguish unambiguously the taxonomic status of an individual, DNA/DNA hybridization is generally good at about the family or genus level. Protein electrophoresis usually allows distinction to the species or subspecies level. The new DNA technology allows an almost arbitrary level of sensitivity right down to distinguishing individuals. One final general point. Unlike allozyme studies, it is possible to reconstruct phylogenies of different DNA patterns generated by restriction endonuclease analysis much as one does by using overlapping chromosome inversions. One can fairly confidently predict that one pattern arose from another in a relatively orderly manner. With protein electrophoresis it is impossible even to guess which alleles gave rise to which other alleles.

Mitochondrial DNA

One of the obvious problems with using restriction enzyme analysis is isolating a piece of DNA of convenient size. To work on the population level it must be relatively easy to repeat the isolation on a large number of individuals. There is one type of DNA that is surprisingly suitable for such analyses: the DNA found in mitochondria. This DNA is about 16,000 base pairs in most animal species and is circular. The available evidence indicates that this DNA is maternally inherited in the egg cytoplasm; sperm appear not to contribute mitochondrial DNA (mtDNA) to the zygote (Avise et al., 1979). Thus variation in mtDNA allows one to follow maternal lineages—a potentially very useful ability. What allows one to obtain large amounts of homogeneous mtDNA is that, within the level of detection of the experimental method, the mtDNA's in all mitochondria in an individual are identical. Lansman et al. (1981) present a particularly lucid discussion of the methodology as well as ways of interpreting data from restriction endonuclease analysis of mtDNA. Before proceeding to an example relevant to founder events, one further point about mtDNA needs to be made. The rate of evolution—that is, the rate of nucleotide substitution—in mtDNA appears to be very rapid compared with that in nuclear DNA. In primates, the best studied group for these comparisons, mtDNA appears to be evolving five to ten times as fast as nuclear DNA (Brown et al., 1979). Thus mtDNA should allow a finer distinction to be made between very closely related individuals than could be obtained with nuclear DNA.

As an example of the use of mtDNA in studying founder events, I shall briefly present some recent data we have collected on the fruit fly, *Drosophila pseudoobscura*. This species lives in oak–pine forests in the western half of North America, with populations extending from southern British Columbia to Northern Guatemala. Some 1500 km from the southern border of distribution is an isolated population in the vicinity of Bogota, Colombia. This population is partially reproductively isolated from North American populations (Prakash, 1972; Dobzhansky, 1974) and has been used as a paradigm for the early stages of speciation (e.g., Lewontin, 1974). An interesting question about this population is whether it represents a founder event and would then be a candidate for incipient speciation according to the theories of Carson (1971, 1975). Studies of the chromosomal constitution of the population reveal that it is variable, segregating for at least two third-chromosome inversions typical of southern Mexico and Guatemala. So, while there is some reduction in chromosome variation (as most populations of this species have several segregating inversions), there is still chromosomal variation. Originally, electrophoretic studies were thought to indicate a founder effect in Bogota (Prakash et al., 1969). However, more recent studies employing more thorough techniques have revealed considerable protein variation in this isolated population (Singh et al., 1976), which adds to the ambiguity of interpreting these data as indicating a founder event. I have studied seven strains of *D. pseudoobscura* from Bogota, collected at two different sites several years apart. The mtDNA of the seven fly strains were digested with eight different restriction endonucleases. All strains were identical. In North American populations of this species there is quite a bit of variation in mtDNA. For example, out of a sample of seven strains of one Mexican population, Sabinas Hidalgo near Monterrey, I detected four different maternal lineages. While I would like to study more Bogota material before reaching any strong conclusions, I do feel that the mtDNA studies give the best evidence to date that the Bogota *D. pseudoobscura* population was founded by a few individuals, perhaps a single maternal lineage.

While I have mentioned the possible value of mtDNA variation for studying founder effects, it is obvious that this technique could have wider applicability in the management of wild populations. Knowing the number of maternal lineages in a captive population may be of interest. If a population is founded by a known number of females, after several generations how many of the original maternal lineages can still be found? And in what frequencies? To date it is most practical to follow maternal lineages because of the convenience of mtDNA. In theory it is also possible to study paternal contributions. One need only find restriction endonuclease site differences in the nuclear genome of males, a variation that certainly exists. Then in theory one could determine the number of males that contribute to the subsequent gene pools.

Availability

This discussion of applying DNA technology to problems in the management of wild populations may seem rather esoteric and of little practical use. Most persons involved would not have the facilities or training to carry out such work. I predict that in the quite near future this will no longer be an obstacle. Because the kind of DNA technology I have been discussing has many practical applications, commercial enterprises have begun to be involved. For example, many of the problems in the genetics and management of wild populations are very similar to some medical problems (see Chapter 24). If a parent is heterozygous for a deleterious allele, can we perform amniocentesis to determine if the fetus received that allele and perhaps terminate the pregnancy? The whole question of proving paternity in humans is one of practical and legal import. The problem is that not every hospital will want to or be able to run its own DNA analysis laboratory. There will have to be either public centralized facilities or commercial firms that would offer the service for a fee. The fee should be very much less than the cost of maintaining one's own facility and would not require that users be familiar with all the largely irrelevant technical details.

Summary

Classical genetic techniques, a series of controlled crosses with large numbers of individuals scored, are impractical for most species of plants and animals. However, recent advances in molecular biology allow one to gather a considerable amount of genetic data on virtually any species, often with no harm to the individuals studied. Electrophoretic study of soluble proteins (primary gene products) is the most widespread technique. More recently developed techniques to study genes directly may prove even more useful. For example, studies on variation of mitochondrial DNA allow one to characterize and follow maternal lineages. Other techniques may allow one to determine paternity. Molecular data of these sorts may be practical in at least two respects. First, they may be used as markers of genetically differentiated populations. Second, in newly founded inbred populations, one could determine the actual numbers of males and females contributing to successive generations. The technology in this area is quickly evolving and becoming simplified. Furthermore, because such data are relevant to medical genetic problems, it is conceivable that in the near future diagnostic laboratories supplying the required data will be widespread.

CHAPTER
15

The Elimination of Inbreeding Depression in a Captive Herd of Speke's Gazelle

Alan R. Templeton
Bruce Read

Introduction

As more and more wildlife habitat is destroyed by increasing human activities, long-term captive breeding programs often become the only hope for the preservation of many species. Already certain species exist only in zoos with all natural populations extinct. Accordingly, there has been considerable interest in the management and genetics of captive zoo populations (Soulé, 1980; Frankel and Soulé, 1981; Chapter 10 by Ralls and Ballou; Chapter 23 by Foose). One of the primary genetic concerns is the role small population size can play in diminishing genetic variability and increasing inbreeding within captive populations. The increase in inbreeding is particularly relevant to the problem of population maintenance because it frequently induces an inbreeding depression—that is, a reduction in viability, birth weight, and fertility—that can directly endanger the survival of the entire population. That this is a real fear is evidenced by the experience of research scientists who deliberately try to create laboratory strains by inbreeding; although a few lines are successful, the vast majority become extinct. Moreover, inbreeding depression is a real phenomenon in many zoo populations (Bowman, 1977; Flesness, 1977; May, 1980; Ralls et al., 1979, 1980; Treus and Lobanov, 1971; Chapter 10 by Ralls and Ballou). Ignoring inbreeding depression is therefore an exercise in wishful thinking, and breeding programs for captive animals must be designed with the phenomenon in mind.

The predominant reaction to this recognition of the importance of inbreeding depression has been to design breeding programs that avoid or minimize inbreeding as much as possible. Unfortunately, even the minimal requirements for a breeding program with this goal are often far beyond what is practical to achieve for most zoos or wildlife management programs. To avoid inbreeding, small founder group sizes should be avoided (Senner, 1980). Second, changes in the inbreeding coefficient should be limited to no more than 1% per generation (Franklin, 1980), a figure that implies that the short-term, maintenance effective-population-size should be no fewer than 50 individuals (Frankel and Soulé, 1981; Franklin, 1980; Soulé, 1980). Effective sizes of populations are a function of the number of individuals actually contributing gametes to the next generation under highly idealized conditions. As a general rule of thumb, the effective population size is only one-third to one-fourth the actual size (Soulé, 1980), so the minimal requirement for short-term maintenance translates into populations of at least 150 to 200 animals. Finally, for long-term maintenance of the captive population, the effective population size must be a minimum of 500 (1500 to 2000 animals) (Soulé, 1980).

Unfortunately, even these minimal levels are impossible to meet in many cases. One such example is the Speke's gazelle (*Gazella spekei*). All captive animals of this rare species trace their ancestry to one male and three females imported to the United States in 1969 and 1972 (Read and Fruch, 1980). Because the sex ratio is unequal, the inbreeding effective size of this founder herd is limited by the relationship (Frankel and Soulé, 1981)

$$\frac{1}{N_e} = \frac{1}{4N_m} + \frac{1}{4N_f} \qquad (1)$$

where N_m is the number of males, N_f is the number of females, and N_e is the effective population size. Equation (1) yields an effective population size of three founders for the Speke's gazelle. However, even this number is an over-estimate because one of the founder females was bred much more than the other two, a factor that reduced the effective size closer to two. Because the natural habitat of this rare species is the border area between Ethiopia and Somalia—an area of prolonged guerrilla warfare—new individuals cannot be safely obtained from nature. Hence, there is no way to circumvent the impact of a very small number of founders in designing a breeding program for this species.

From this small founder population, the herd has grown to a total of 29 individuals (as of July, 1982) distributed over three zoos (22 at St. Louis,

Missouri, 4 at Brownsville, Texas, and 3 at San Antonio, Texas). This current size is a maximum for the captive population, so at no point in the herd's entire breeding history did it achieve the minimal short-term effective population size of 50 (150 to 200 animals), much less the minimal long-term effective population size of 500 (1500 to 2000 animals). Moreover, it does not seem feasible that these minimal effective population sizes will be achieved in the foreseeable future. Finally, as we shall document in this chapter, a very severe inbreeding depression occurred in this species. In light of the arguments given by Franklin (1980) and by Frankel and Soulé (1981), the prospect of breeding this species in captivity on a long-term basis would seem to be rather bleak. Unfortunately, the situation with respect to Speke's gazelle is not an uncommon one; for a great many species, zoos simply do not have the resources to achieve even the minimal recommendations needed to avoid inbreeding (see Chapter 23).

Fortunately, nature indicates that a second breeding strategy is feasible in which inbreeding is *not* avoided, but is actively used to eliminate the inbreeding depression. To understand this strategy, it is first important to note that inbreeding per se is not deleterious or maladaptive; that is, an inbreeding depression is *not* a universal companion of inbreeding in nature. For example, about a third of the vascular plants have extensive (although usually not exclusive) self-mating — the most extreme form of inbreeding possible in a sexually reproducing organism — yet they are evolutionarily quite successful (Chapter 13 and 20). Moreover, many animals also have systems of mating that promote inbreeding in nature (Chapter 12). In general, populations of plants and animals that regularly inbreed in nature display little or no inbreeding depression. The reason for this is that organisms adapt to their system of mating. Although we ordinarily think of evolutionary adaptation only in terms of an organism's adapting to the external environment, basic population genetic theory indicates that organisms also adapt to their internal genetic environment, which is determined in large part by their system of mating (Templeton, 1982). This theoretical expectation is confirmed by empirical studies that indicate that organisms can successfully adapt to even the most extreme forms of inbreeding (Templeton, 1979). Thus, species that are normally inbred do not display inbreeding depressions for the simple reason that they have adapted to the genetic consequences of inbreeding. With this in mind, we see that an inbreeding depression is similar to the maladaptive syndromes that appear in species when their external environment is rapidly altered. For example, if a warm-climate species is suddenly put in a cold-temperature climate, it would not be surprising to find a "depression" in viability, birth weights, or fertility. Similarly, when a species that normally outcrosses is forced into inbreeding, it is also likely to exhibit a depression in viability, etc. Given enough time, however, a species can adapt to an altered breeding environment just as it can adapt to an altered external environment. Thus, inbreeding per se is not necessarily a bad or harmful system of mating as is commonly believed; rather, inbreeding is simply an alternative to outcrossing that requires different adaptive responses.

Not only can an entire species adapt to inbreeding, but an inbred population extracted from an outcrossing species can also successfully adapt to inbreeding. For example, it is well known that most human populations are characterized by extensive outcrossing, but that when inbreeding does occur in such human populations, it often results in severe inbreeding depression (Morton et al., 1956). The severity of this inbreeding depression can be quantified by the number of lethal equivalents. Much, perhaps all, of an inbreeding depression is due to rare, deleterious, recessive alleles. Under outcrossing, these alleles are almost always found in heterozygous condition, so their deleterious recessive effects are not expressed. With inbreeding, however, these alleles can become homozygous and their deleterious consequences fully expressed. Some of these alleles are lethal; that is, homozygosity for that allele alone is sufficient to kill the individual. Many more, however, are deleterious but not lethal. Several of these deleterious alleles acting together, however, can be lethal. It is generally impossible to determine if an individual died from a lethal allele or several partially deleterious alleles. However, one can assume that all deaths resulting from deleterious or lethal combinations are due to independently acting lethal alleles and estimate this number. The resulting number, known as the "number of lethal equivalents," provides a convenient index as to the severity of the inbreeding depression. The number of lethal equivalents in human populations from France (a highly outcrossing population) is 3.0 to 5.0 lethal equivalents per individual (Morton et al., 1956). This means that on the average these outcrossed individuals bear what is effectively 3 to 5 alleles in heterozygous condition that if made homozygous would kill the individual. The mortality examined by Morton et al. was generally early in life, and if one looks over the entire human life span the number of lethal equivalents is 5 to 8 (Stine, 1977). These figures are typical for most human populations. However, the Tamils of India regularly have had a very large percentage of their marriages between relatives. These humans have no inbreeding depression whatsoever (Rao and Inbaraj, 1980). This example clearly demonstrates that humans can successfully adapt to high levels of inbreeding and as a consequence totally eliminate the phenomenon of an inbreeding depression.

Another example of an outcrossing species successfully adapting to inbreeding is provided by the captive populations of the European bison or wisent (*Bison bonasus*). This species almost went extinct early in the twentieth century and experienced a bottleneck with an effective size of no more than 12 (Slatis, 1960). Extensive inbreeding occurred, and the analysis of Slatis indicates the herd initially had about 6 lethal equivalents per individual (with respect to juvenile death). However, after a few generations of intense inbreeding, this inbreeding depression had vanished, and the number of lethal equivalents was no longer significantly different from 0.

The above examples clearly indicate that inbreeding depression is a temporary maladaptive syndrome that occurs during the transitional period from outcrossing to inbreeding. Consequently, our goal in designing a breeding program for the Speke's gazelle was not to avoid inbreeding — since that goal was clearly impossible — but rather to adapt them to inbreeding. One strategy is simply to allow the animals to inbreed and hope for the best (for example, the European bison), but recall that many inbred laboratory strains go extinct. Consequently, relying on chance alone is not a good strategy when one is dealing with a rare or endangered species. What we need to do, therefore, is to manipulate the breeding structure of the herd so as to maximize our chances of successfully adapting to inbreeding during this transitional phase. The appropriate population genetic theory already exists (Templeton, 1980), and fortunately it indicates that it is possible for a population to successfully and rapidly adapt to inbreeding even when the founder size and inbreeding effective population size are small and remain small. Hence, we are freed from the often impractical population sizes needed to avoid inbreeding. The essence of this theory states that the adaptation to inbreeding will be most successful and rapid when the genetic variability is maximized (within the constraints of small inbreeding effective size), both at the herd level and — very important — at the individual level. Moreover, under this breeding program, inbreeding is not avoided; rather, inbreeding is actively used as part of the selection procedure to eliminate the inbreeding depression.

We shall now document that a severe inbreeding depression did occur in the Speke's gazelle and show how the theory of Templeton (1980) can be translated into a very practical breeding program that results in a rapid elimination of the inbreeding depression.

The Breeding History of Speke's Gazelle through 1979

When the herd was founded in 1969 it consisted of one import male (M3) and two import females (F4 and F5). In 1972, an additional import female was added (F9). From 1969 through 1975, M3 was used as the sole breeding male. During this time, there were father–daughter matings and father–granddaughter matings. This resulted in a rapid increase in the inbreeding coefficient that was reflected in an increase in the infant mortality rate.

In 1976, M3 was removed as the sole breeder and allowed to breed only with the import females F4, F5, and F9. Two new males, M7 and M9, were now bred with generation captive born selected females. These two males were both the offspring of M3 with import or first females (F6 was the mother of M7, and F4 was the mother of M9). As a result, M7 and M9 are not inbred themselves. However, because all animals bred in captivity have M3 as an ancestor, all matings between these noninbred males and captive-bred females were necessarily between relatives, primarily half-sib and first-cousin matings. Consequently, the offspring of these matings were inbred. In subsequent years, other males were used as breeders, but in all cases matings between the noninbred but captive-bred animals resulted in inbred offspring.

This inbreeding had highly deleterious consequences. We first consider its impact upon survivorship or viability. Table 1 gives the number of animals surviving up to 30 days after birth and up to 1 year after birth (including abortions and stillbirths) as a function of the inbreeding coefficient of the animal. The inbreeding coefficient simply measures the intensity of the inbreeding in the offspring (or, alternatively, how closely related the parents are); it is determined by the standard pedigree calculations (Ralls et al., 1980). It is obvious from Table 1 that, as the inbreeding coefficient goes up, the viability goes down. For example, 22% of the noninbred offspring died before they were 30 days old, but 67% of the offspring with an inbreeding coefficient of 0.25 died before they were 30 days old. We can compare the severity of inbreeding depression encountered in the Speke's gazelle with that reported in other animals by applying a standard inbreeding regression analysis to this data set.

To quantify the severity of the inbreeding depression, we use the standard population genetic model that states

$$v_i = \exp(-A - BF_i) \tag{2}$$

where v_i is the viability or probability of surviving to a given age for an animal with inbreeding coefficient F_i; A measures the nongenetic or environmental causes of death [the quantity $1 - \exp(-A)$ represents the probability of dying by the index age from nongenetic causes]; and B is the number of lethal equivalents per haploid gamete (so $2B$ is the number of

Table 1. *Thirty-Day and One-Year Viability Data for the Offspring of Noninbred Parents*

	Thirty-day survival			One-year survival		
Inbreeding coefficient	Number surviving	Number dying	Total number of births	Number surviving	Number dying	Total number of births
0	22	6	28	20	8	28
1/16	2	0	2	2	0	2
1/8	8	4	12	3	9	12
3/16	1	0	1	1	0	1
1/4	7	14	21	5	16	21

lethal equivalents per individual). The observed viability is given by $v_i = x_i/n_i$, where x_i is the number of individuals with inbreeding level F_i that are still alive at the index age, and n_i is the total number of births (including stillbirths and abortions) with inbreeding coefficient F_i. Equation (2) can be put into the format of a standard linear regression by taking natural logarithms to yield

$$\ln(x_i/n_i) = -A - BF_i \qquad (3)$$

Thus, the probability of survival, x_i/n_i, should yield a straight line when plotted against level of inbreeding on a semilog graph. The only difficulty in applying equation (3) is that certain levels of inbreeding are represented by only one or two individuals in our sample. Hence, it is likely for an observed viability to be 0, which results in an undefined logarithm. We must therefore use a small sample size correction before performing the regression. The appropriate correction for a logarithm regression is

$$\ln \frac{x_i + 1}{n_i + 2} = -A - BF_i \qquad (4)$$

This smooths over the irregularities associated with viability values of 0, yet it minimally affects the estimated viability at those levels of inbreeding for which the sample size is large. Moreover, we shall perform a weighted least-squares regression using equation (4), so it is the inbreeding levels with large sample sizes that are the most important in determining the values of A and B.

The resulting regression for the 30-day viability data in Table 1 is shown in Figure 1. The regression explained 92% of the variance, indicating an excellent fit of the standard population genetic model of inbreeding depression. The estimates of A and B ($\pm$ standard deviations) stemming from the regression were $A = 0.22 \pm 0.08$ and $B = 3.09 \pm 0.53$. Since there are five levels of inbreeding in this regression and two parameters were estimated, we have 3 degrees of freedom to test the null hypothesis that $A = 0$ and $B = 0$. We tested these hypotheses using the standard t-tests associated with a weighted least-squares regression and obtained a t value of 2.65 for the null hypothesis that $A = 0$ (significant at the 5% level for a one-tailed t-test; note: a one-tailed test is used because only positive values of A and B have biological meaning) and a t value of 5.78 for the null hypothesis that $B = 0$ (which is significant at the 1% level). Hence, both nongenetic and genetic causes are a significant source of juvenile mortality, but the genetic factors are far more important as a source of mortality than the environmental factors for all inbreeding coefficients greater than 1/16. Moreover, the regression implies that the founders bore an average of $2B = 6.18$ lethal equivalents per individual that could kill an animal before it was 30 days old. This value is a bit higher than the corresponding

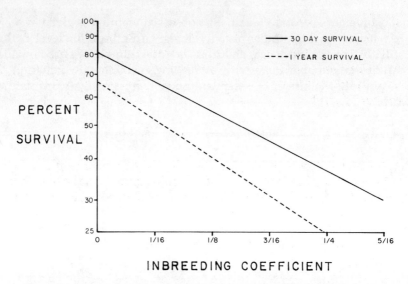

Figure 1. *The inbreeding depression in survivorship for 30 days and 1 year after birth in the offspring of noninbred parents. The figure shows a graph of the percentage of the animals surviving for 30 days (solid line) and 1 year (dashed line) after birth (including stillbirths and abortions) plotted against the inbreeding coefficient of the offspring. The lines are weighted least-squares regressions.*

value obtained from highly outcrossed human populations (Morton et al., 1956) and almost identical to that of the noninbred European bison (Slatis, 1960). Consequently, there can be no doubt that a severe inbreeding depression occurred with respect to juvenile mortality.

Figure 1 also shows the regression for the 1-year viability data. This regression explained 84% of the variance, which is still an excellent fit. The estimate of the environmental component of mortality has almost doubled to $A = 0.40 \pm 0.16$, and this value is significantly different from 0 at the 5% level (t value of 2.53 with 3 degrees of freedom). The value of B is now 4.01 ± 1.03 (significant at the 2% level with a t value of 3.90). Recall that the B value for juvenile mortality was 3.09. Therefore, although the nongenetic mortality increased by a large amount, the genetic mortality increased only slightly in going from the first 30 days to the entire first year. This indicates that the bulk of the genetic deaths occur early, whereas nongenetic deaths are more evenly distributed throughout the life span. This value of B yields an early adult lethal equivalent number of 8.02, a value very similar to the lifetime number of lethal equivalents in highly outcrossed humans (Stine, 1977).

The inbreeding depression was not limited to viability, but could also be seen in birth weights. Table 2 gives the average birth weights as a func-

CHAPTER 15 Elimination of Inbreeding Depression in Speke's Gazelle

Table 2. *Average Birth Weights for the Offspring of Noninbred Parents as a Function of the Offspring's Inbreeding Coefficient*[1]

Inbreeding coefficient	Average birth weight, kg	Number of animals
0	1.67	6
1/16	1.39	2
1/8	1.31	7
3/16	1.33	1
1/4	1.17	5

[1] Only live-born offspring are included in this sample.

tion of inbreeding level, and Figure 2 gives the weighted least-squares regression of the model:

$$w_i = \alpha + \beta F_i$$

where w is the weight (in kilograms) of a newborn animal with inbreeding coefficient F_i, α is an estimate of the birth weight of noninbred offspring, and β is a quantitative measure of the impact of inbreeding upon

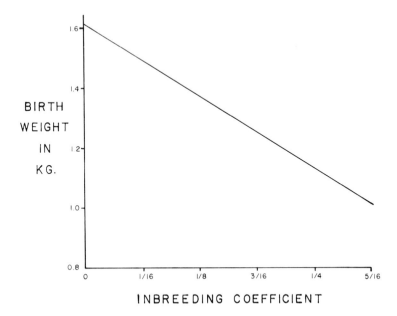

Figure 2. *The inbreeding depression in birth weight in the offspring of noninbred parents. The figure shows a graph of the birth weight of live-born animals plotted against the inbreeding coefficient of the animals obtained by a weighted least-squares regression.*

birth weight. Negative values of β imply a deleterious effect of inbreeding, and the larger the magnitude of β, the greater is the severity of the inbreeding depression. The regression shown in Figure 2 explains 89% of the variance and yields an α of 1.61 kg and a β of -1.93 (t value = -4.93, which is significant at the 2% level in a two-tailed t-test with 3 degrees of freedom). This represents a severe and highly statistically significant reduction in birth weight due to inbreeding.

A Breeding Design to Eliminate Inbreeding Depression

As the above section indicates, we faced an unfortunate but all too common situation with the Speke's gazelle. On the one hand, the small founder size meant that inbreeding could not be avoided. On the other hand, inbreeding induced a severe inbreeding depression that seriously endangered long-term maintenance. Thus, our only option was to eliminate the inbreeding depression. Hence, starting in 1980, our primary breeding goal became to put the theory of Templeton (1980) into practice so as to eliminate or reduce the inbreeding depression present in the Speke's gazelle (Templeton and Read, submitted).

The first recommendation stemming from this theory concerns the demographic management of the herd. During the transitional phase from outcrossing to inbreeding, many animals will inevitably be lost because of the inbreeding depression. To minimize the chances of extinction of the herd, it is therefore best to build up the size of the herd as much as possible *before* the full impact of the inbreeding depression takes effect. This is most easily done during the initial generations when breeding primarily involves the import animals and inbreeding can easily be avoided. Such an expanded population size not only protects against extinction, but it also maximizes the carryover of genetic variation into the captive herd from the founding individuals (Chapter 23; Templeton, 1980; Foose, 1980; Frankel and Soulé, 1981, Chapter 6). The rapidity of the adaptation to inbreeding depends upon how much genetic variation is present in the population, so the expansion of the population also decreases the number of generations during which the inbreeding depression is a serious problem. Moreover, it is also highly desirable to increase the population size during the phase of the inbreeding depression if at all possible, since this makes selection for inbred but healthy animals more efficient (Templeton, 1980).

During the first part of the management program, the population size was increased from the initial four founders to 19 animals (end of 1979). This expansion in size was accomplished not only by allowing the herd size at the St. Louis Zoological Park to increase, but also by establishing a

CHAPTER 15   Elimination of Inbreeding Depression in Speke's Gazelle   251

second herd at the Gladys Porter Zoo in Brownsville, Texas. During the second, post-1979 phase of the management program, we have been able to further increase the herd size from 19 animals to 29 (July 1982), 17 of which represent animals bred after the end of 1979. This second expansion was accomplished by growth of the herds at St. Louis and Brownsville and by the establishment of a third herd at the San Antonio Zoo. A fourth herd will soon be established in Dallas. This shows the importance of inter-zoo cooperation in designing such a selection program, since the resources available to any one zoo would not be sufficient to allow such a population expansion. Moreover, the cooperation between the participating zoos extended into the breeding program as well, with animals being transferred between zoos to meet our breeding goals. We now focus on the nature of these breeding goals.

The breeding program was initiated in February of 1980. Since the total pedigree of the herd was known, one easy way of monitoring the gene pool present in the starting herd was to calculate the percentage of the alleles found in any particular individual that trace back to one of the founders. These are then averaged over all individuals in the herd to give a description of the overall gene pool in terms of the percentages of the genetic contributions made by each of the four founding animals. Figure 3a

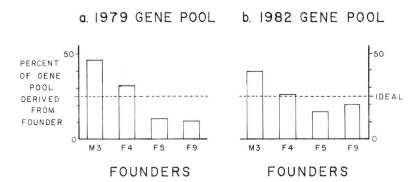

Figure 3. *The gene pool before and after the breeding program designed to eliminate the inbreeding depression. Because the total herd pedigree is known, it is possible to calculate how many of the genes found in the animals at any particular time can be traced to a particular founder. In this case, there are four founders indicated by the symbols M3 (the founding male), F4, F5, and F9 (the founding females). Part (a) shows the nature of the gene pool at the end of 1979 by indicating what percentage of the herd's shared genes can be traced to a particular founder. Part (b) shows the gene pool for the 17 animals alive in July 1982 that were produced by our breeding program. The goal is to equalize the genetic contribution of each founder, and this ideal goal is indicated by a dashed line in these graphs.*

shows the gene pool composition of the herd at the end of 1979, just before the breeding program began. As can be seen from that figure, the single founding male (M3) had the biggest genetic impact upon the herd's gene pool, having contributed close to 50% of all the alleles present in the individuals alive at the end of 1979. In addition, one female (F4) had a much larger genetic input into the gene pool than did the other two founding females, F5 and F9. The overall genetic variability in the herd's gene pool, given the constraint of four founding individuals, will be maximized when all four founding individuals contribute equally to the gene pool; that is, each founder should account for 25% of the alleles present in the herd. This is indicated by the dashed line marked "ideal" in Figure 3. As can be seen, the gene pool in 1979 was far from ideal. Basically, it was necessary to increase the genetic contributions of F5 and F9 and decrease those of M3 and F4. This was easily done by giving males and females bearing genes from F5 and F9 precedence in our breeding program. By preferentially using the males and females bearing the rarer genes, we were able to shift the gene pool rather rapidly, as illustrated in Figure 3b, which shows the gene pool of the 17 animals bred under our breeding program that were alive as of July 1982. As can be seen from that figure, we are still not at the ideal state, but we have made considerable progress in equalizing the ancestral genetic input and hence in maximizing the herd's genetic variability.

A second factor in choosing which animals should be parents and which should not was the inbreeding coefficient of the potential parent. Since we ultimately want to select for animals that do well even though inbred, animals that already are inbred but in good health are given precedence as parents. For example, males M20, M31, and MB2 all bear genes from the rare founders F5 and F9, and hence all are preferred by gene pool considerations alone. However, male M20 is not inbred, and therefore none of his alleles or allele combinations have been selected under the genetic state of inbreeding. However, M31 and MB2 are both inbred (with inbreeding coefficients of 1/16 and 1/8 respectively), and therefore we know that they have an allele combination that does well under at least a certain degree of inbreeding. Therefore, whenever possible we preferentially use inbred but healthy animals; for example, M31 and MB2 were used much more than M20. In all the crosses we designed, at least one parent was inbred, and by 1981 most crosses had both parents inbred.

The above two factors are criteria involved in choosing animals as parents, but now we must also choose with whom they should mate. Here the criteria of choice shift from the parental attributes to the offspring attributes. Our primary criterion is the level of individual genetic variability of the resulting offspring. This is easily measured by the percentages of founder contributions in the ancestry of a given individual. As with the overall gene pool, the ideal is to produce individuals with nearly equal ancestry from all four founding animals.

Equalizing founder ancestry in individual animals is important for three reasons. *First*, parents are chosen as mates that tend to have different founders in their ancestry. This results in what is technically known as disassortative mating with respect to predigree. Such a disassortative mating system can be very efficient in preserving large amounts of genetic variability in a small number of individuals (Averhoff and Richardson, 1976; Templeton, 1980). As noted earlier, such genetic variability is the necessary prerequisite in selecting inbred but healthy animals, and therefore it is important to preserve as much variation as possible during these critical transitional generations.

The *second* reason for the importance of this criterion is that disassortative mating by pedigree somewhat ameliorates the effects of inbreeding. By itself, this reduces the selective pressures imposed by inbreeding, and hence would slow down the rate of adaptation to inbreeding. However, many quantitative genetic experiments (Falconer, 1960) demonstrate that too intense a selective pressure makes inefficient use of genetic variability and can actually result in an inferior adaptation than a less intense selective regime. Moreover, when selective forces induced by inbreeding are extremely intense, transiliences (rapid changes in traits other than response to inbreeding) are more likely to occur (Templeton, 1980). Since the objective of this breeding program is to preserve the species as much as possible, transiliences are undesirable. Ideally, the only trait to undergo significant evolutionary change should be the response to inbreeding. As detailed in Templeton (1980), disassortative mating systems are very effective in making transiliences unlikely.

The *final* reason for this disassortative mating scheme also relates to the issue of genetic variation. Genetic variation exists not only in the form of alternative alleles for a particular kind of gene, but also as the number of combinations these genes can be arranged into during the process of gamete (egg or sperm) formation. Disassortative mating can greatly increase the opportunity for new gene combinations to arise, and thereby increase the overall level of genetic variation upon which selection can operate.

For all the above reasons, we chose as mates those individuals that would ideally tend to result in offspring bearing alleles from all four founding ancestors. In 1979, the animals in the herd were far from ideal. Only one male (M31) had genetic output from all four founders, and only six animals had genetic input from three ancestors. We were able to greatly alter this situation with our breeding program. For the animals bred under our program and alive as of July 1982, 71% had all four founders as ancestors versus 5% in 1979. Of the animals alive and bred since 1979, 100% have at least three founders in their ancestry, compared with only 37% before the breeding program began. The contrast between these two sets of figures clearly illustrates that we have been able to radically alter the individual genetic makeup of this herd in a very short period of time.

A secondary criterion in choosing mates is the inbreeding coefficient of the resulting offspring. This criterion was invoked to prevent the general avoidance of inbreeding. It was still possible to make a few matings from the 1979 animals that would result in noninbred offspring (the import female F4 was still alive), but it was obvious that such a strategy would not be feasible for very long. Since our goal was to produce healthy inbred animals, producing noninbred animals would only slow down our rate of progress. Hence, in contrast to most breeding programs for captive animals, we went out of our way to avoid matings resulting in noninbred offspring. However, we also tried to avoid extreme inbreeding coefficients, since too radical a change in inbreeding level often makes it difficult for effective selection to operate and may result in a transilience, as noted earlier.

In summary, then, our program consisted of five rules. The first was a demographic rule: increase population size before and during the selection procedure. The other four rules relate to the breeding program. Two of those rules involve the choice of parents: (1) choose animals to be parents that will maximize the genetic variability in the herds' gene pool, and (2) choose healthy, inbred animals as parents. The other two breeding rules involved the choice of how to pair the parental animals based on offspring attributes: (1) choose mates that result in offspring with maximal genetic variability in terms of founder ancestry, and (2) choose mating pairs that result in inbred offspring, although extreme inbreeding should be avoided. Note that all these breeding judgments can be made strictly from pedigree data; no special equipment or assays are required.

Results of the Breeding Program

The theory of Templeton (1980) indicates that adaptive response to inbreeding should be very rapid when inbreeding effective sizes are small, and the gazelle herd bore out this prediction. Table 3 presents the 30-day viability data for the selected animals. Since many of the selected animals are less than a year old, we cannot score 1-year viability except for a small number of animals. Hence, we shall confine our attention only to the 30-day viability data. As noted earlier, 30-day viability still picks up most of the genetic deaths. Figure 4 represents the logarithmic regression of these data. To see the progress of the breeding program, the previous regression from the noninbred parents (Figure 1) is drawn in Figure 4 for comparison. For the 30-day viability, the A component is virtually identical before and after selection (for the selected animals $A = 0.21 \pm 0.13$, which translates into an 81% chance of survival during the first 30 days versus the unselected value of 80% with $A = 0.22 \pm 0.08$). This shows that the intensity of nongenetic mortality is the same throughout the entire breeding history of the herd. In other words, if our breeding program did induce any

Table 3. *Thirty-Day Viability Data for the Offspring of Animals Selected as Part of the Breeding Program To Eliminate the Inbreeding Depression*

Inbreeding coefficient	Thirty-day survival		Total number of births
	Number surviving	Number dying	
1/32	1	1	2
3/32	1	0	1
7/64	3	0	3
1/8	14	4	18
5/32	1	1	2
3/16	1	1	2
13/64	1	0	1
7/32	1	0	1
1/4	3	3	6
9/32	3	1	4
5/16	3	3	6

herd management changes, these changes had no impact whatsoever on viability.

In contrast to the environmental component of mortality, our breeding program did have a large impact on the genetic causes of mortality. As is evident from Figure 4, inbreeding does not have nearly as large an effect on the selected animals as it did on the nonselected animals. In particular, the B value for 30-day viability has been reduced from 3.09 to 1.35 ($\pm$ 0.65). This B value of 1.35 is still significantly different from 0 at the 5% level, so the inbreeding depression has not been entirely eliminated. Nevertheless, it is significantly smaller at the 5% level than the original $B = 3.09$ (a one-tailed t-test with $t = 2.07$ and 12 degrees of freedom). This reduction in B values means that the number of lethal equivalents has been more than halved, from 6.18 to 2.70 per individual. Given the short time this breeding program has been in operation, we are extremely pleased by such a large and statistically significant reduction in the number of lethal equivalents.

Another major difference between the regressions on the selected versus the nonselected animals was that the semilog regression of the 30-day viability data for the nonselected animals explained 92% of the variance, but the regression for the selected animals explained only 32% of the variance. Part of this may be attributed to smaller sample sizes for the selected animals, but another reason is that the selected animals represent a very heterogeneous group with respect to the criterion we employed in the breeding program. For example, some animals had only one parent inbred, others both. Some animals had both parents with inbreeding coefficients of 1/4; others had both parents with inbreeding coefficients of 1/16. Some of the selected animals had all four founders in their ancestry; others

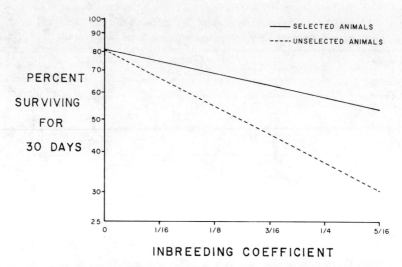

Figure 4. *The inbreeding depression in survivorship for 30 days after birth in the offspring of parents chosen as part of the program to eliminate inbreeding depression. In order to see the results of the breeding program, the inbreeding depression observed in the nonselected animals is redrawn from Figure 1 as a dashed line.*

only three. Hence, the inbreeding coefficient of the animal is not nearly so informative about the animal's breeding history as it was for the nonselected animals whose parents were all outcrossed. Ideally, we should like to examine the effects of all these potential sources of heterogeneity upon the success of our breeding program. Unfortunately, small sample sizes preclude such a finer dissection of the data at the present time. We shall, however, present one such partitioning of the data on the selected animals merely to indicate trends rather than to make strong statistical inferences.

As was mentioned earlier, the essence of our program is to use inbred animals as parents because genes passed through such inbred individuals will be more adapted toward the genetic state of inbreeding through the action of natural selection. This prediction may be tested by dividing the data on the selected animals into two sets: those with only one parent inbred versus those with both parents inbred. In the former case, the animals receive one set of genes *not* selected for inbreeding and one set that is selected (from the inbred parent). In the latter case, the animals receive both a paternal and a maternal set of genes that have been selected under inbreeding. Hence, we would expect the number of lethal equivalents to be higher in the animals with only one parent inbred.

To see if this prediction is true, we performed the semilog regression on 30-day viability data of selected animals with only one parent inbred. The resulting regression is shown in Figure 5, along with the regression for

CHAPTER 15 Elimination of Inbreeding Depression in Speke's Gazelle   257

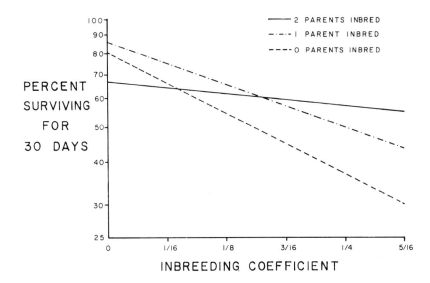

Figure 5. *The inbreeding depression in survivorship for 30 days after birth as a function of the number of inbred parents. The solid line is a weighted least-squares regression of the viability of offspring that had both parents inbred, the dashed line is for when only one parent is inbred, and the dot-dashed line is for when neither parent is inbred (redrawn from Figure 1).*

30-day viability of unselected animals from Figure 1 for comparison. The resulting B value is 1.74, which yields 3.48 lethal equivalents per animal. This value represents the average number of lethal equivalents of the inbred and noninbred parent. Since we had previously estimated the number of lethal equivalents in noninbred animals as 6.18, this implies the inbred parent had only 0.78 lethal equivalent in order to make the average come out to be 3.48. The number of lethal equivalents per inbred animal can be estimated directly by regression when both parents are inbred. This regression is also shown in Figure 5, and, as is obvious, there is almost no inbreeding depression whatsoever. The B value for this regression is 0.61, which implies 1.22 lethal equivalents per inbred animal. When coupled with the previous estimate of 0.78 for the number of lethal equivalents per inbred animal, these figures indicate that the inbred animals have only about 1 lethal equivalent per individual—a sixfold reduction from the lethal equivalent load found in their noninbred parents. This illustrates the rapidity of the adaptation to inbreeding. Because of sample sizes, it is meaningless to do a rigorous statistical analysis of these figures, but the effectiveness of using inbred parents to eliminate inbreeding depression is quite evident from Figure 5.

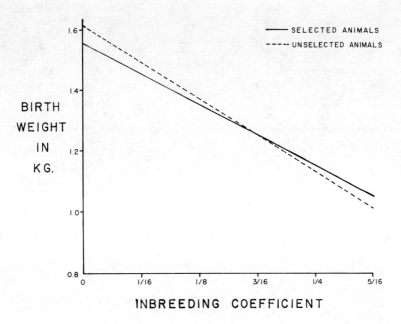

Figure 6. *The inbreeding depression in birth weight in the offspring of parents chosen as part of the program to eliminate inbreeding depression. In order to see the results of the breeding program, the inbreeding depression observed in the nonselected animals is redrawn from Figure 2 as a dashed line.*

The reduction in inbreeding depression occurred not only for viability, but also for birth weight. Table 4 gives the birth weight data for the selected animals, and Figure 6 plots the regression of birth weight versus inbreeding coefficient for the selected animals, with the regression from

Table 4. *Average Birth Weights for the Offspring of Selected Parents as a Function of the Offspring's Inbreeding Coefficient*[1]

Inbreeding coefficient	Average birth weight, kg	Number of animals
1/32	1.18	1
3/32	1.13	1
7/64	1.54	3
1/8	1.40	9
5/32	1.19	2
3/16	1.40	2
13/64	1.30	1
1/4	1.23	3
9/32	0.94	2

[1] Only live-born offspring are included in this sample.

Figure 2 redrawn for comparison. As before, the intercepts are nearly identical in the selected versus nonselected animals (1.54 kg versus 1.61 kg, respectively), but the slopes are different. In the nonselected animals the β value was -1.93, but in the selected animals the inbreeding depression has been reduced to a β value of -1.45 ± 0.79. As before, this indicates that an inbreeding depression still exists, but its severity has been reduced. However, this reduction is not statistically significant, so further data need to be gathered before we can definitely conclude that the birth weights were not as depressed by inbreeding in the selected animals as they were in the nonselected animals.

Discussion

The results given in this paper along with laboratory studies on fruit flies (Templeton, 1979) clearly demonstrate that animals can rapidly adapt to even extreme inbreeding. Moreover, this adaptation can occur in relatively small populations. Hence, the goal of eliminating inbreeding depression in captive zoo populations is a feasible and practical one. We therefore regard the breeding program outlined in this paper as an important addition to the breeding programs suggested by Frankel and Soulé (1981). However, note that we do not regard our breeding program as an alternative to theirs, only as an additional option. It is an addition rather than an alternative because the goals are so different: in ours, we are eliminating inbreeding depression while producing a highly inbred population; in the Frankel and Soulé program they are avoiding inbreeding and preserving the highly outcrossed nature of the founding animals. The principal advantage of our program is that it can be executed with a small number of founding and maintenance individuals and can be applied to species for which it is impractical to avoid inbreeding by periodically collecting new individuals from nature. These situations are very commonplace in zoos and represent situations in which the Frankel and Soulé program is not feasible. Our success with the Speke's gazelle demonstrates that small founder populations with no hope of additional wild collections should not be neglected; rather, these populations can be successfully maintained despite an initial inbreeding depression.

Our program, however, does have a major disadvantage. The ultimate reason behind many of these breeding programs is to save, at least in captivity, a rare and endangered species. In our breeding program we can save the species only by genetically altering it from its wild state. Undoubtedly, any captive breeding program will induce genetic alterations; the individuals will tend to be selected for "domestic" traits and perhaps adapt to other aspects of the captive environment as well. However, in our breeding program, we are deliberately adapting the individuals to a genetic environment that is very different from the genetic environment that exists in natural populations. Thus, it is not certain but it is possible that our breeding program might alter the individuals in such a way as to diminish their capabilities of surviving and reproducing in the wild. The Frankel and Soulé (1981) program might also do this, but drastic changes are less probable under their breeding program than under ours. If the option exists, their program is preferable to ours, particularly if reintroduction of the species back into the wild is eventually desired. However, these breeding options quite often do not exist, and our program offers a chance for the survival of the species (and even for ultimate reintroduction) that otherwise would not exist.

Summary

Inbreeding depressions have been commonly observed in many zoo populations, and the primary goal of many zoo breeding programs is to avoid inbreeding. However, in many cases, such a goal is impractical. For example, many individuals of Speke's gazelle (*Gazella spekei*) exist at several zoos, but all trace their ancestry to one male and three females acquired in the late 1960s. Moreover, collections from nature cannot be made because the natural habitat of this rare species is the border area between Ethiopia and Somalia—an area of prolonged guerrilla warfare. Because of the small founding population and the inability to collect new animals, high levels of inbreeding now exist in the herd. We show that an extremely severe inbreeding depression occurred for both viability and birth weight. Given this inbreeding depression and the impossibility of avoiding inbreeding, the prospects for long-term maintenance of this species may seem poor. However, population genetic theory and studies on natural and experimental populations indicate that it is possible to adapt to high levels of inbreeding and thereby eliminate inbreeding depressions. There are two major components to the design. First, we favor as parents those individuals who were already inbred but of proven viability. This actively selects for genes that do well under inbreeding. However, selection can operate only when there is genetic variability, so the second component of our design is to optimize conditions for selective response by maximizing genetic variability. This was accomplished at the herd level by equalizing the average genetic contributions of the four ancestral animals to the herd's gene pool and by allowing the herd size to expand during the selection procedure. Cooperation between several zoos aided the achievement of these goals by allowing a larger total herd size, by physically subdividing the herd (which protects against epidemics or other disasters destroying the entire herd), and by allowing breeding loans and exchanges to ensure maximal genetic variability in the herds kept at each zoo. Population genetic theory also indicates that the opportunity for selection to operate in small founder populations is greatly influenced by individual genetic variability as well as by gene pool variability. Individual genetic variability was increased by favoring crosses yielding animals that received nearly equal amounts of their genetic material from all four founding animals. The breeding program turned out to be extremely successful, with much, if not all, of the inbreeding depression eliminated in our selected animals. Consequently, severe inbreeding depressions do not necessarily imply that captive herds are doomed to extinction; rather, the inbreeding depression can be selectively eliminated, thereby making the long-term maintenance of the herd much more likely.

Acknowledgment

This work was supported in part by NIH Grant ROI GM 27021.

Hybridization and Merging Populations

PART FOUR

Introduction

Hybridization is the interbreeding of unlike individuals or populations either under human control or in nature. The type of interaction involved ranges from crossing between virtually identical individuals to crossing between individuals of different species. Crossing of virtually identical individuals from very similar habitats and latitudes presents no particular problem from a management point of view. Crossing of different species obviously is undesirable because even in those cases where "hybrid vigor" results in a strong F_1 generation, that vigor will generally be lost in succeeding generations, if reproduction in the F_1 individuals is even possible (recall the sterility of that vigorous interspecific hybrid, the mule).

Between these two extremes are the possibilities of crossing individuals separated by various degrees of genetic divergence. Judgements on the desirability of crossing individuals of different subspecies of animals, for example, is much more controversial. The argument against crossing individuals of different subspecies or genetically different populations is that adaptations to local conditions may be disrupted through the hybridization process. If both populations are healthy and of reasonable size, and as long as there are no pre- or postmating reproductive barriers, natural selection can be expected to result in adaptation of the hybrid to local conditions. The result is a viable population, but one different from the individual strains that produced it. This would be undesirable only if one of the parental populations is a taxon that is valued in itself and would be no longer discernible after hybridization. A remaining objection to hybridization concerns cases where populations have particularly intricate or sensitive behavioral or physiological mechanisms that are adaptations to a specific habitat or life history; disruption of these mechanisms by hybridization could be especially severe in an already small and declining population.

In this part of the book on population merging and hybridization, Harlan presents examples that illustrate the extreme evolutionary flexibility of genetic systems in plants. Merging of genetically different plant populations, often as a consequence of human-caused habitat modification, can result in the formation of highly adapted forms. Cycles of genetic differentiation and hybridization may be viewed as the natural mode of evolution in some plant groups and as important in the origin of new adaptive strains or species. Although the discussion in this chapter centers on crop plants, many of these plants are only partially domesticated and serve as analogs of completely wild populations. As some of our plant species become more restricted in range and require more management, the distinction between cultivated and wild populations will become even less definite.

Carr and Dodd discuss the possible impacts of hybridization on dwindling populations of sea turtles. Although many important aspects of sea turtle life history are unknown, these animals are known to have sensitive navigation systems that permit dispersal between widely spaced feeding areas and nesting beaches. Carr and Dodd describe the difficulty in making management decisions when the necessary data for defining a breeding or local population are unavailable.

In the following chapter, Cade reviews the range of different interactions between populations that can be grouped under the term "hybridization". This review also demonstrates that it can be difficult to distinguish whether a species or form has disappeared due to hybridization with another species or form, or it has been displaced by the other form by competitive interaction, without any exchange of genes between separate gene pools.

CHAPTER
16

Some Merging of Plant Populations

Jack R. Harlan

Introduction

Differentiation-hybridization cycles are considered by many biologists to be powerful systems for evolutionary advance. Sewall Wright (1931, 1932) pointed out some fifty years ago that one of the most efficient population structures from an evolutionary point of view is one in which the total population with capacity to interbreed is fragmented into subpopulations largely isolated one from the other but with occasional genetic interaction. Since then, population structures have been studied in various ways by many evolutionists. There seems to be a general consensus on the importance of hybridization as an evolutionary process in plants (Stebbins, 1959, 1977, p. 186; Ehrendorfer, 1959, 1964; Grant, 1981, p. 481; Dobzhansky et al., 1977), but some question as to its efficacy in animals (Mayr, 1970). It may be that the major difference between plants and animals in this respect has to do with the relative lengths of effective cycles—that is, the degree of genetic differentiation among subpopulations before hybridization. I have suggested that the length of the cycle is significant, self-regulating, and correlated to the degree of genetic buffering of the species in question (Harlan, 1966).

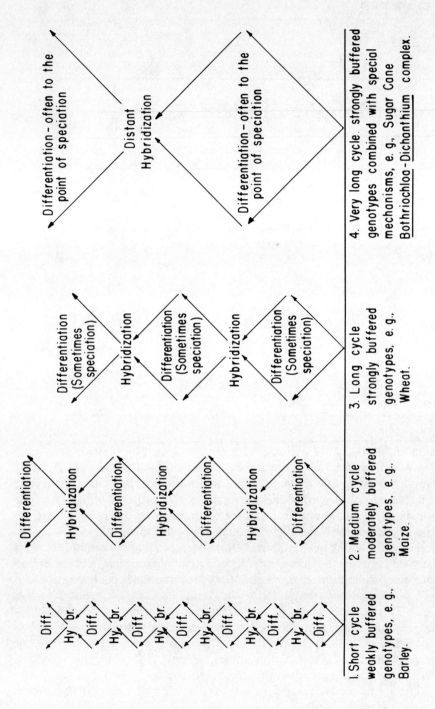

Figure 1. *Duration of differentiation–hybridization cycle in relation to degree of genetic buffering.*

Figure 1 is designed to illustrate these interactions. By genetic buffering I refer essentially to the amount of redundancy of genetic information. A diploid should be less well buffered than a tetraploid, and a tetraploid less well buffered than a high polyploid. In this scheme I assume that a self-fertilizing species is less well buffered than an outcrossing species (Chapter 8), although other factors may intervene to make this assumption an unsafe generalization. At any rate, experience supports the idea that differentiation in a group like the sugarcane subtribe Saccharinae (tribe Andropogoneae, family Gramineae) can proceed to the generic level and the cycle can still be closed by hybridization. Modern sugarcane (*Saccharum*) clones are all complex hybrid derivatives involving two or more genera and chromosome numbers in the 100 to 135 range. In both *Saccharum* and the largely Old World, warmer region bluestem grasses of the *Bothriochloa-Dichanthium* complex apomixis (asexual reproduction) of one kind or another provides an escape from sterility.

In effect, this chapter deals with the hybridization phase of differentiation–hybridization cycles. The consequences of bringing together well-differentiated populations that still have capability for genetic interaction can be rather spectacular and can be important economically as well as biologically. This chapter will review some of the work of myself and my colleagues with the subtribe Bothriochloinae (tribe Andropogoneae) and some crop–weed relationships from the point of view of merging plant populations.

Bothriochloa bladhii, a Genetic Aggressor, and Its Allies

Bothriochloa ischaemum is a widespread Eurasian grass adapted to mid-latitudes and upper elevations of lower latitudes. It is found from the Atlantic coasts of France and Iberia to the Pacific shores of China and the island of Taiwan. Over most of its range it is a tetraploid ($2n = 40$), but pentaploid ($2n = 50$) races are found in China. *Bothriochloa bladhii* (*B. intermedia*) is more tropical and ranges from southern Africa across East Africa and south Asia to Australia. It has limited cold tolerance and is confined to lower elevations. Over most of its range, this species is also tetraploid, but pentaploids, hexaploids, and octoploids do occur in restricted local regions. (The original *B. bladhii* may have arisen as an autotetraploid, by the doubling of the chromosome set from a diploid species resembling *B. longifolia*.)

Along the front ranges of the Himalaya–Karakoram mountain system, the two polyploid species have been brought into contact in fairly recent times. *Bothriochloa ischaemum* would naturally be confined to habitats above the deodar-hardwood forest, and *B. bladhii* to the plains below the forest. Neither grass is found in the forest itself. Deforestation and terracing of the mountainsides for farming have opened up a new habitat invaded from above by *B. ischaemum* and from below by *B. bladhii*. When this happened is not known with precision, but maize (*Zea*) is an important crop in the contact zone as it fills a niche between highland wheat (*Triticum*) crops and lowland rice (*Oryza*) crops. It is probable that the process of deforestation in this intermediate zone began with the introduction of maize in the late sixteenth century (Harlan, 1963a).

Hybridization and introgression along the suture or area of contact between the two species have resulted in an irruption of spectacular variability and the development of aggressive races adapted to the present state of disturbance. *Bothriochloa ischaemum* behaves as an obligate apomict. There is only a theoretical sexual potential within the species in that most ovules produce a sexual (n) embryo sac, but operationally these are swamped by one to several asexual ($2n$) sacs. One or more of these is precocious and preempts the sexual sac to form the $2n$ embryo directly (gametophytic apomixis). (From many thousands of emasculated spikelets we have succeeded only once in producing a hybrid with this species as a female, and this resulted from an unreduced female gamete.) The species is quite capable of functioning as a male, however. *Bothriochloa bladhii* has some actual sexual potential within the species. In fact, when facultatively sexual plants are intercrossed, about 1 plant of 16 in the offspring is fully sexual and self-incompatible. Such plants cross freely with everything in the neighborhood, and the introgression products are often remarkably vigorous. Apomixis also is dominant in these hybrid swarms, so that the F_1 plants set seed well without fertilization (agamospermy), even if they would have been sterile as sexual plants (Harlan et al., 1964).

Most of the clones and races of *B. bladhii* produce a complex array of volatile aromatic compounds that can be detected by odor at considerable distances on a warm day. While these compounds have been studied by gas chromatography and some have been characterized chemically, a quick, simple, and sensitive field technique is simply to chew pieces of leaf and taste for pungence. *Bothriochloa ischaemum* does not produce aromatics with pungent flavor. The presence and intensity of pungence are reliable clues to the movement of genetic information across the suture. Occasional plants with distinct pungence were detected at elevations as high as 1700 meters where most of the population was morphologically *B. ischaemum*. Plants without pungence were identified at elevations as low as 600 meters, where most of the population was morphologically *B. bladhii*. The morphological characters that distinguish the species were similarly blurred over the same range of elevation (Harlan, 1963a).

Careful examination of the plains race of *B. bladhii* involved in this introgression revealed that it was, itself, a product of hybridization between a more typical, tropical race of *B. bladhii* and a widespread lowland relative,

Dichanthium annulatum. This species ranges from Senegal and Morocco to Thailand and the Philippines and has a sibling species (*D. fecundum*) in Australia. It is mostly tetraploid, with hexaploid races in East Africa and Palestine and a diploid ($2n = 20$) race in India. The diploid race is completely sexual, and the other races have slight sexual potentials. The introgression product between *B. bladhii* and *D. annulatum* is common in the upper Pakistani plains, the Punjab, and the Gangetic plain of India. It is sufficiently stable and distinct that is has received the name *B. grahamii* (Harlan, 1982).

In northern Australia, *B. bladhii* introgresses with *Capillipedium parviflorum*, producing a series of intergrades, some of which have received the name *C. spicigerum* (scented top). In the same part of the world, *B. bladhii* also introgresses with *B. ewartiana* (desert blue grass), while in East Africa it interacts genetically with both *Capillipedium* and *B. insculpta*. All these intermediate products have been reconstructed experimentally. In all, *B. bladhii* appears to be a construct assembled from at least five additional species belonging to three genera; it served as a model for the compilospecies concept (Harlan and de Wet, 1963; de Wet and Harlan, 1966). The introgressing species of the three genera are genetically isolated from each other, except across the bridging *B. bladhii*.

The taxonomic consequences of the capacity for genetic absorption by this compilospecies are a dilemma difficult of resolution. Apparently, at one time there were three well-defined distinct genera: the world-wide *Bothriochloa* (20 species), and the Old World *Dichanthium* (15 species) and *Capillipedium* (10 species). But disturbance of the habitat and wide transport by human activities have brought *B. bladhii* into contact with members of each of the genera. Hybridization and introgression have followed. The integrity of the genera has broken down. They are in the process of being merged by genetic aggression.

There is an old, traditional dilemma that always troubles taxonomists: how different must taxa be in order to be ranked as separate species? In this case the dilemma is: to what extent must taxa merge in the field in order to be lumped taxonomically? For example, in different taxonomic works *Capillipedium* has been combined with *Bothriochloa*, and it has been kept distinct while *Bothriochloa* and *Dichanthium* have been combined; de Wet and Harlan (1970) suggested that the three taxa be united and recognized as sections of *Dichanthium*. This suggestion recently has been endorsed and followed up, so that the name *Dichanthium ischaemum* is used in *Flora Europaea* (1980), and Clayton (1977) has made the necessary formal name changes *D. bladhii* and *D. insculptum*. Merging taxa are characteristic of reticulate evolution so widely apparent throughout biological (or at least botanical) phylogeny. In this group we have an instructive example of extensive differentiation that can be closed by hybridization to produce a cycle.

The grasses of this particular group are rather weedy in nature; that is to say, they thrive under disturbance by man and his livestock. This is not all bad; they cover many millions of hectares, providing forage for livestock and protecting the soil from erosion. Other species of the same genera are not at all weedy and have extremely narrow distributions; several could be classed as endangered (Harlan, 1963b). What are the differences in genomes that allow one species to be a widespread, aggressive weed, and confine a related species to one or a few narrowly endemic sites? These differences are profoundly important to agriculture, forestry, range management, horticulture, and the suburban homeowner. I should like now to direct attention to a whole class of genetic interactions between differentiated populations that has a very practical economic effect on all of us.

Crops and Weeds

Most crop species have weed races. This is especially true of the annuals, although weedy forms of perennial crops are not rare—e.g., *Citrus* (oranges, lemons, etc.), *Psidium* (guava), *Punica* (pomegranate), *Mangifera* (mango), *Tamarindus* (tamarind), *Musa* (banana), *Prunus* (plum). The origins of weed races are sometimes obscure and in other cases quite clear. In the process of plant domestication, man has repeatedly taken wild plants and inserted them into an artificial habitat—the field, garden, or orchard. The wild plants responded by producing races adapted to the artificial conditions. These we call domesticated races. At the same time, weed races evolved with an adaptation similar to that of the domestic races but with reproductive habits resembling those of the wild ancestor.

The weed races of sorghum (*Sorghum bicolor*, subtribe Sorghinae, tribe Andropogoneae) are especially useful examples because their origins can be more clearly traced than the sources of weed races of some other crops. Wild *Sorghum bicolor* is a widespread grass in Africa, and displays great diversity in form and adaptation. One race is adapted to high rainfall and is more or less restricted to the forest zone. It is slightly weedy and tends to increase as forests are cut down or thinned by clearings and logging trails. It is common in vacant lots of towns and cities of the forest belt. Another race is adapted to the savanna zone and is one of the primary dominant climax species over vast reaches of East and southern Africa. In some areas it forms massive stands over thousands of hectares. As a wild plant of the climax vegetation it is not very weedy, but it may persist along roadsides or irrigation ditches when the grassland is plowed up for agricultural development (Harlan and de Wet, 1974).

If domesticated sorghum is grown in such an agricultural environment, however, weed races are produced inevitably and automatically.

These are products of hybridization between the remnant wild race and the cultivated race, whatever it may be. Such weeds are extraordinarily aggressive and cause serious crop losses in Sudan, Ethiopia, Kenya, Uganda, and southward. There is a tendency for the local weed to mimic the particular race of cultivated sorghum that it infests. If the cultivated race is a durra, which has a compact inflorescence, the weed race is likely to have a compact inflorescence; if the cultivated race is a bicolor, which has a loose, open inflorescence, the weed race will have a loose, open inflorescence. But the spikelet type itself of the weed races is typical of the weeds and differs sharply from the spikelets of cultivated sorghum (de Wet et al., 1976).

The mimicry in other characters is partly due to gene flow and partly due to selection pressures applied during weeding operations. Plants in vegetative condition that resemble the cultivated races very closely are missed during weeding and are allowed to mature seed and contaminate the soil. Plants that do not mimic cultivated forms so closely are pulled out and do not contribute to the next generation. It is not uncommon for fields to become so contaminated that rotation with other crops is forced on the farmer or the field must be abandoned altogether.

In the United States, another kind of weed sorghum or "shattercane" has evolved with a different genetic heritage. In the process of domestication, the shattering habit of wild sorghum was suppressed. A single recessive allele (sh) in homozygous condition, $shsh$, will suppress shattering. All cultivated sorghums carry a genotype of this sort, and seed is retained fairly well at maturity. The common American shattercane also has the $shsh$ genotype, however. It is derived from cultivated sorghum by the evolution of a secondary method of seed dispersal. Instead of forming an abscission layer that causes the inflorescences to shatter, as in wild sorghum, the inflorescence branch becomes very thin a few millimeters below the spikelet and the branch breaks at maturity. The abscission layer is effectively suppressed by the paired sh alleles, but the plant shatters anyway, and shattercanes have become very serious and troublesome pests in sorghum-growing areas of the United States (Harlan and de Wet, 1974).

But sorghum is remarkably versatile at producing noxious weeds. There is still another way. It is probable that sorghum was domesticated several different times in different regions. There are at least two or three genes (sh_1, sh_2, etc.), at different loci (they are nonallelic), any one of which will suppress shattering when homozygous. It is only necessary to cross cultivars with different shattering-suppression genes (e.g., $shsh\ Sh_2Sh_2 \times ShSh\ sh_2sh_2$) to recover the wild-type shattering habit (genotype $Shsh\ Sh_2sh_2$). There is more involved than shattering in the making of a successful weed, but shattering is an essential ingredient for seed dispersal. Shattercane spikelets have a characteristic morphology, and the seeds are much more dormant than those in cultivated sorghum. These characteristics seem to be rather tightly linked as a unit (Chapter 2), and one can produce weeds by simply crossing two particular cultivars. Since cultivated sorghum is some 5% to 10% outcrossing, such products can occur spontaneously.

Finally, there is a tetraploid ($2n = 40$) rhizomatous species, *Sorghum halepense*, called Johnson grass. It can be so aggressive and troublesome that sometimes farmers abandon the contest and graze it or cut it for hay even though prussic acid poisoning is always a hazard. Our analyses of the grass suggest that the U.S. Johnson grass, as such, originated right here in the United States. To be sure, the species *S. halepense* was introduced from the Mediterranean or Near East, but we have not been able to find anything in its native homeland as aggressive, vigorous, or pesky as U.S. Johnson grass. All the accessions we have grown from Mediterranean and near eastern countries have been relatively harmless and lack vigor and aggressiveness. Strains developed under winter rainfall regimes do not seem well suited to our climate. Collections from east of the Khyber Pass in monsoon climates are also quite different from the Johnson grass found in the United States. They are more vigorous than the winter rainfall materials, but are much taller and coarser in growth habit than our Johnson grass. We have put forth the theory that our kind of Johnson grass evolved in the United States by means of introgression of cultivated diploid sorghum into individuals of *S. halepense* introduced from the Mediterranean. The consequences of merging plant populations can be formidable indeed.

Sorghum may be somewhat more adept at producing weeds than are other cereals, but genetic interaction between wild, weed, and cultivated races is practically universal among seed crops, and it is very common among vegetatively propagated crops that also reproduce by seed. Weed rices (*Oryza*) have been particularly troublesome in Asia and Africa. Mimetic forms are common in which a particular weed race mimics the cultivar it infests so well that it goes unnoticed until it flowers and then it is too late to weed the crop. There are mimetic races of maize, wheat, barley (*Hordeum*), oats (*Avena*), rye (*Secale*), pearl millet (*Pennisetum glaucum*), finger millet (*Eleusine coracana*), and several other millets. The mimicry, as in sorghum, is due partly to gene flow and partly to the selection pressures of hand weeding.

Crop weeds are by no means confined to cereals. Weed sunflowers (*Helianthus*) are familiar to most people in the western and mid-western United States. Perhaps less familiar are the weedy races of radish (*Raphanus*), carrot (*Daucus*), mustard (*Brassica*), lettuce (*Lactuca*), beet (*Beta*), tomato (*Lycopersicon*), cotton (*Gossypium*), sesame (*Sesamum*), soybean (*Glycine*), cowpea (*Vigna*), okra (*Hibiscus*), gourd (*Lagenaria*), castor (*Ricinus*), and so on and on. Fields in the Andes are often overrun with weed potatoes (*Solanum*). The edible aroids (Araceae) are often weedy, and *Xanthosoma* (yautia), *Alocasia*, *Cyrtosperma* and *Colocasia* (taro) often escape into disturbed situations. The seed-producing bananas are often weedy in the tropics. Citrus, guava, papaya (*Carica*), mango, tamarind, and other fruits have escaped and become naturalized far from their native homelands. In most of these examples genetic interaction between weed and cultivated populations has been demonstrated, and in some cases wild populations have been involved as well.

The dynamics of these interactions are often very complex and have seldom been studied in adequate detail. Certainly, they have furnished us with some remarkably aggressive weeds that have expanded the geographic range of the species far beyond that of the wild progenitors. It is likely that the same genetic interactions have played major roles in improving our crop plants. Indeed, the intimate relationships between cultivated and weed races sometimes become so entangled that we cannot unravel them with present methodologies. Is crabgrass (*Digitaria sanguinalis*) so excellent a weed because it was once cultivated for food, or was it once cultivated as a millet because it is so excellent a weed (Portères, 1955)?

Summary

Differentiation-hybridization cycles are considered by many students of evolution to be powerful systems for speciation and evolutionary change. In effect, this chapter deals with the hybridization phase of differentiation-hybridization cycles. The dynamics of such systems are important in our discussion, and some examples of merging plant populations are described.

Bothriochloa ischaemum is a widespread Eurasian grass adapted to mid-latitudes and upper elevations of lower latitudes. It is found from the Atlantic coasts of France and Iberia to the Pacific shores of China and the island of Taiwan. Over most of its range it is a tetraploid ($2n = 40$), but pentaploid ($2n = 50$) races are found in China. *Bothriochloa bladhii* (syn. *B. intermedia*) is more tropical and ranges from southern Africa across East Africa and southern Asia to Australia. It has limited cold tolerance and is confined to lower elevations. Over most of its range, the species is tetraploid, but pentaploids, hexaploids, and octoploids do occur in restricted local regions.

Along the front ranges of the Himalaya-Karakoram mountain system, the two species have been brought into contact in fairly recent times. *Bothriochloa ischaemum* would naturally be confined to habitats above the deodar-hardwood forest, and *B. bladhii* to the plains below the forest. Neither grass is found in the forest itself. Deforestation and terracing of the mountainsides for farming have opened up a new habitat invaded from above by *B. ischaemum* and from below by *B. bladhii*. When this happened is not known with precision, but maize (*Zea mays*) is an important crop in the contact zone as it fills a niche between highland wheat crops and lowland rice crops. It is probable that the process of deforestation in this intermediate zone began with the introduction of maize in the late sixteenth century.

Hybridization and introgression along the suture has resulted in an irruption of spectacular variability and the development of aggressive races adapted to the present state of disturbance. Certain recognizable traits of *B. bladhii* can be detected at elevations as high as 1700 meters, where most of the population is morphologically *B. ischaemum*, and characteristics of the latter species can be detected at elevations as low as 600 meters, where the populations are morphologically *B. bladhii*. The parental species are both tetraploid in the region, but pentaploids and hexaploids appear in hybrid derivative populations.

Closer examination revealed that the race of *B. bladhii* involved was itself a derivative of introgression between a more tropical race of the species and the related *Dichanthium annulatum*. This hybrid product is probably older and is sufficiently stable that it has been given the name *B. grahmii*. In northern Australia, *B. bladhii* introgresses with *Capillipedium parviflorum*, producing a series of intergrades some of which have received the name *C. spicigerum*. In the same part of the world *B. bladhii* also introgresses with *B. ewartiana*, while in East Africa it interacts genetically with both *Capillipedium* and *B. insculpta*. All these intermediate products have been reconstructed experimentally. In all, *B. bladhii* appears to be a construct assembled from at least five additional species belonging to three genera; and it served as a model for the compilospecies concept (Harlan and de Wet, 1963).

Our analyses of cultivated sorghum and its near relatives have shown that profound modifications can be brought about by natural hybridization. The species *Sorghum bicolor* has a remarkable capacity to produce troublesome weeds, and the aggressive Johnson grass (*S. halepense*) of the United States appears to be a recent product of evolution derived from introgression of *S. bicolor* into an introduced Mediterranean race of *S. halepense*. In fact, genetic interaction between wild and cultivated races more often than not results in the production of weedy races of the species. It is so common as to be almost a general rule, if not a biological law.

CHAPTER
17

Sea Turtles and the Problem of Hybridization

Archie F. Carr III
C. Kenneth Dodd, Jr.

Introduction

There are seven, perhaps eight, species (and five genera) of sea turtles in the oceans of the world representing a lineage that goes back to the late Jurassic, 140 million years ago. With the exception of the Australian flatback, *Chelonia depressa*, which occurs only in northern Australia, all are faced with a variety of serious threats to their continued existence. The other species are generally circumglobal in distribution; they include the green turtle, *Chelonia mydas*, which occurs primarily in tropical waters although individuals may occur in temperate waters; the loggerhead, *Caretta caretta*, which regularly occurs in subtropical and temperate waters and occasionally in the tropics; Kemp's ridley, *Lepidochelys kempii*, the most endangered of all sea turtles, occurring only in the Gulf of Mexico and along the east coast of North America; the olive ridley, *Lepidochelys olivacea*, which occurs in tropical waters except in the northern Caribbean; hawksbill, *Eretmochelys imbricata*, a tropical species; and the leatherback, *Dermochelys coriacea*, the largest sea turtle, which nests in the tropics but regularly migrates to far northern and southern latitudes to feed. Some sea turtle biologists recognize an eighth species, the black turtle, *Chelonia agassizii*, which occurs off the west coast of North, Central, and South America, as distinct from *C. mydas* (A. Carr, Jr.; P.C.H. Pritchard, personal communications). Reviews of the biology, distribution, and threats to these species have been summarized in the many papers in Bjorndal (1982) and by Frazier (1980).

In 1979, biologists from around the world met in Washington, D.C., to review the status and biology of sea turtles and to make conservation and management recommendations to ensure their survival while allowing controlled exploitation for future generations. A conservation strategy plan was adopted (World Conference on Sea Turtle Conservation, 1979). The conference not only revealed a worldwide sense of obligation and need to initiate conservation activities, but also demonstrated that state-of-the-art turtle management is very elementary compared with bird and mammal game management. When asked how to reinforce a depleted or declining sea turtle population, most biologists would answer with a single recommendation: protect it.

Such a response is in most cases too vague. A manager concerned about the turtles within his area of authority would immediately respond: protect it how? Also, a simplistic response of "protect it" is really a passive response to the critical condition of many populations and may lead to a false sense of security once statutes for protection are in force.

In addition to statutory protection, a number of management options are available, including establishing egg hatcheries, moving nests to new locations on natural beaches (but not within a well-defined hatchery), massive relocation of eggs and/or hatchlings to new beaches in an attempt to supplement declining populations or reestablish extirpated ones, head-starting, captive breeding, predator control, farming and ranching, and many more. While it is beyond the scope of this chapter to discuss these options, the papers of Pritchard (1979, 1980), Ehrenfeld (1982), and Shabica (1982) provide thoughtful discussions and criticisms of them. In general, many of these options must be considered experimental. They are of limited use as conservation methods because there is no proof that they work. In some cases, it may turn out that they have been of negative value because they inadvertently diminished populations. For instance, for years hatcheries have been using Styrofoam boxes, which ensured a protected nesting environment but kept incubation temperatures perhaps lower than on beaches. With the discovery that temperature controls sex determination in sea turtles, it may be that males have been predominantly released for years in some operations (Mrosovsky and Yntema, 1980; Morreale et al., 1982). Experimental conservation procedures may also introduce the potential for hybridization between species or, more important, between demes. Hybridization, as we use it in this chapter, refers to the production of offspring between members of different populations brought together through human action, not that resulting from genetic exchange in natural populations due to immigration.

Management techniques such as head-starting and moving eggs or hatchlings to new beaches are thus still unproven, are in many cases cost-prohibitive, and require much experimental testing, some of which is underway. At present, it seems more biologically acceptable to emphasize measures designed to eliminate poaching, prevent predation by feral and domestic animals, eliminate commercial markets, ensure the protection of nesting, feeding, migratory, and developmental habitat, and promote the use of the turtle-excluder device developed by the U.S. National Marine Fisheries Service for shrimp trawls. These techniques are known to be effective and, in most cases, the most cost-effective methods for limited budgets.

The Problem with Sea Turtle Biology

One might ask why there are seemingly few proven management options available. One principle reason is that sea turtles are marine and, as such, are out of sight most of the time. We simply do not know much about their biological requirements and life histories. For instance, such important features as sex ratio, developmental habitats of hatchlings, age at first reproduction, growth rates, longevity, and recruitment into the adult population are virtually unknown for any species. Other features such as migratory routes and internesting movement patterns are poorly known. Turtle biologists spend much of their time piecing together life cycles from the results of mark-and-recapture programs. At best, as in the cases of the relatively well-studied loggerhead (*Caretta caretta*) and green turtle (*Chelonia mydas*), mark-and-recapture results are dependent upon the chance recovery of a few individuals in a tagged fraction of one sex of an adult population that happened to nest where researchers can get to them (Carr, 1980).

The hawksbill turtle (*Eretmochelys imbricata*) presents still another problem. Its nesting is diffuse; there is little tendency to aggregate (Carr et al., 1966). The biologist is thus denied the statistical advantage of substantial numbers of animals to tag on a single beach. Another species, Kemp's ridley (*Lepidochelys kempii*), is now almost extinct. Despite knowledge of where nesting occurs, there are so few Kemp's ridleys left that research on adults is badly frustrated. On a few stretches of shore, the giant leatherback turtle (*Dermochelys coriacea*) nests in large numbers, but leatherbacks are rarely recaptured away from the nesting beaches. Therefore, the mark-and-recapture procedure is of limited use (Carr, 1980).

Difficult research logistics and our lack of understanding of basic sea turtle life histories are thus obstacles to effective sea turtle management. Another obstacle is the inadequacy of our knowledge of their systematics (Smith et al., 1978). This weakness encumbers the fundamental decision on where or to which population to apply limited available resources in order to advance the most elementary phase of management—that is, protection. Without a clearer understanding of the geographical ranges of breeding populations within species, zoologists are hard-pressed to help wildlife managers organize specific protection priorities. The solution, then, is to attempt to protect every breeding aggregation.

In fact, our admittedly limited knowledge of the population genetics of sea turtles would indicate that this is the most justified strategy for conserving these species. Although karyological and chromosomal banding studies have not proved helpful in understanding population structure and limits (Bickham et al., 1980), electrophoresis may prove very valuable. In a preliminary study of 13 loci of green and loggerhead hatchlings, Smith et al. (1978) demonstrated that, while loggerheads had much less heterozygosity than greens, loci were polymorphic between populations such that populations were readily identifiable. Thus, they suggested that, for management purposes, populations be treated as functional units independent from one another. Hendrickson (1979) has also demonstrated genetic differences between species and individual populations of sea turtles in the amino acid composition of their keratins, although large amounts of variation within ratios and overlapping ranges between different categories at present preclude use in simple discrimination tests.

As spotty as management data may be, much research is being conducted around the world on many aspects of sea turtle biology. In time, management opportunities will be more diverse. However, as we continue to attempt to unravel the great mysteries surrounding sea turtle biology while attempting to secure the future of their populations and habitat, we must be careful to avoid management errors, such as inducing hybridization of wild stocks.

Our weakness in understanding sea turtle systematics is central to any consideration of hybridization. What are the genetic groupings and reproductive limits in sea turtles? Below the generic level the answer is in almost every case unclear. New sea turtle beaches are still being discovered, and there is evidence that the turtles at each nesting beach represent a more or less complete, separate breeding colony.

Studies of electrophoretic variation (Smith et al., 1978), keratin structure (Hendrickson, 1979), distribution of heavy metals in eggs (Stoneburner et al., 1980), and population movements of several species (for instance, Carr et al., 1978) all tend to confirm that each nesting colony is a separate deme, and the more colonies that are discovered, the more "kinds" of sea turtles we are compelled to recognize. If we interpret the amended U.S. Endangered Species Act of 1973 (16 U.S. Code 1531–1543) as a statute designed to preserve genetic diversity, demes would require protection and management, and we are obliged to prevent human actions that might jeopardize or weaken the genetic integrity of populations through hybridization, especially since we know so little of the effects of hybridization on fitness.

There are two principal management practices that might lead to hybridization of sea turtle populations. These can be expressed in general terms as follows (Bacon, 1975):

1. The stocking of turtle "farms" or ranches with turtles from disjunct breeding populations, followed by the deliberate or accidental release of these or their progeny.
2. Transfer of eggs or hatchlings from a productive beach to a second beach, with the aim of establishing or restoring a breeding colony.

In both cases, hybridization between populations within a species is thought to be more of a problem than interspecific hybridization, unless what is presently recognized as one species, such as *Chelonia mydas*, is actually a composite of sibling species. Interspecific hybrids of sea turtles are definitely known only from an *Eretmochelys* × *Chelonia* cross from a nest deposited in Surinam whose progeny were hatched and raised at Cayman Turtle Farm (Wood et al., *in press*).

Farming and Ranching

Sea turtle farming and ranching is attractive to commercial investors, since products such as shell, meat, and calipee are readily marketable and many wild populations are at least nominally protected from exploitation through national laws or international agreement. Farming (raising sea turtles in a closed cycle operation without relying on wild populations except to occasionally add heterozygosity to the breeding herd) and ranching (obtaining eggs from natural beaches and raising the hatchlings in controlled conditions for market) are represented by proponents as not just commercial enterprises, but as ones that contribute toward sea turtle conservation (Brongersma, 1980). According to this view, turtle farms and ranches can save wild populations by underselling illegal collectors and driving them out of business while at the same time producing a surplus of animals, some of which may be returned to the wild to repopulate depleted areas.

For a combination of economic and biological reasons, such arguments are considered questionable by the Sea Turtle Specialist Group of the International Union for the Conservation of Nature and Natural Resources

(IUCN) and many other biologists; the alleged conservation benefits do not bear up under close scrutiny (Ehrenfeld, 1974, 1980; Dodd, 1982) and have been deemed by United States courts as insufficient cause to allow the sale of farmed products in United States markets (Anon., 1979, 1982). Nevertheless, the appeal of farming and ranching turtles has found support in the Cayman Islands, Surinam, Reunion and Seychelles in the Indian Ocean, and islands in the Pacific (IUCN, 1971; Dodd, 1982).

We must point out that the potential for increasing hybridization of sea turtle demes has never been a major objection to sea turtle aquaculture. The main objection is that farming and ranching will in all likelihood increase the market for sea turtle products and thus encourage poaching, the exact opposite of what is claimed. Other objections are that it has not been demonstrated that farms can be made independent of wild stocks (Ehrenfeld, 1974), and that compensation for eggs removed from natural nesting grounds through the release of some hatchlings might be an empty gesture owing to mortality of the released animals and their inability to function or eventually breed due to lack of imprinting on their natal beach. Indeed, ranches make no pretense about relying on wild stocks.

Hybridization of farm turtles with turtles in surrounding waters is a distinct possibility from two sources: intentional and unintentional release. The prospect of unintentional release is aggravated by the fact that all turtle farms and ranches fall within the typhoon and hurricane belts of the world where accidental release could occur during these violent storms. Carr (1956) noted that the sea turtles in holding pens of soup-processing plants routinely escaped during hurricanes in the Caribbean. Deliberate release of farmed turtles into Caymanian waters has been allowed by Cayman Turtle Farms (Wood, 1982); since turtles at the Farm were obtained from various localities in the western Atlantic, the potential for hybridization with any remnant green turtle stocks in Cayman, or with migrants in the area, and the effects of such hybridization, is unknown.

Egg or Hatchling Transfer

Hybridization between populations of sea turtles that normally do not interbreed could result from another technique that has been used as a conservation device. This is the practice of transporting eggs or hatchlings from a productive nesting beach, such as at Tortuguero, Costa Rica, to a second beach where turtles once nested or where nesting is presently reduced. Based on the theory that hatchlings would somehow imprint to the new beach and return to nest when they reached sexual maturity, the hope is that the new beach would be recolonized. Such was the premise behind Operation Green Turtle in the 1960s, when thousands of eggs and hatchlings of green turtles were sent to beaches throughout the Caribbean (Carr, 1979).

Under certain circumstances, this procedure appears justifiable as an experimental tool, the best example being that of the United States and Mexican governments' attempts to establish a second nesting population of Kemp's ridley at Padre Island, Texas (Anon., 1978). Under other circumstances, such as Operation Green Turtle, however, such movements and their resulting success could lead to interbreeding between populations, perhaps in the long run to the detriment of the population in need of conservation. Admittedly, it is possible that there would be no detrimental impact from human-induced hybridization, but the potential would indicate that such schemes should be thoroughly evaluated for genetic consequences and other options explored.

Owing to the urgent need for sound turtle management procedures, it is perhaps unwise to disregard translocation of turtles as a recovery technique without a closer examination of the risk associated with it. Defining the risk as the probability of interbreeding by genetically distinct groups of turtles, what biological factors bear on that probability?

1. *Behavior of the wild-caught sea turtles released far from their point of capture.* Owing to the strong homing instinct of these animals and the navigational acuity they appear to show (Carr et al., 1978), it is possible that a released wild-caught turtle would leave the area of its release and return to its home waters, thus nullifying the risk of interbreeding with local animals. However, the possibility needs testing. Are the drive and ability to navigate affected by distance and by factors that the turtles are subjected to in captivity? What are the age-specific and sex-specific characteristics of navigation? Would, for example, the aggressive breeding behavior of male turtles override their homing drive if they were released among local females during the breeding season?

2. *Behavior of captive-bred turtles released to the wild.* The uncertainties surrounding this subject are manifold. There are indications that hatchling turtles may be imprinted by conditions at the native shore and that the imprints are crucial to later migratory success (Frick, 1976). In the absence of such cues, how will captive-bred turtles respond upon being released? Theory suggests one scenario. Owing to the absence of imprinted information, mortality in released captive-bred animals may be higher than that in released wild-caught turtles. If the captive-bred turtles are progeny of parents from separate demes, then heritable survival information may be lacking as well, and this would put the released animals at still more of a disadvantage. Thus, hybridization may not be a serious factor to be contended with, since these turtles may have a much decreased chance

of surviving and breeding successfully. However, captive-raised turtles at Cayman Turtle Farm appear to mate successfully, although at low rates of fertility, and it is unlikely that captive-bred turtles would not attempt to mate with wild turtles should the opportunity arise; hybridization with local populations remains a possibility with unknown consequences.

3. *Population ecology of sea turtles.* A reasonable understanding of sea turtle population ecology would greatly facilitate evaluation of the risk of hybridization. The subject is still in early stages of development, owing in large part to the overwhelming logistical problems the marine turtle student must face. In spite of the lack of much population ecology data, however, data from life history studies suggest that the effects of hybridization may be lessened on a population-wide scale. Specifically, sea turtles have low survivorship to maturity, perhaps 1% or less (Hirth and Schaffer, 1974), and maturation rates may be exceedingly slow. Thus, deleterious effects of hybridization, if they occur in a relatively large population, may have little influence owing to the normal small chance of survival of a particular hatchling coupled with any selective disadvantages caused by the hybridization.

4. *The behavior of hatchlings from transplanted eggs.* The above topics may be most germane to the situation in which turtles are released from farms or ranches. If done with great care, the hatching rate of transplanted eggs has been shown to be quite high. There is no known biological reason why the young animals produced by this method should not successfully recolonize the intended habitat, provided imprinting is a reality and provided enough eggs can be transplanted to overcome the high natural mortality rate of the hatchlings. Accordingly, there seems to be a high probability that transplants will eventually interbreed with whatever endemic turtles might occur in the transplanted habitat.

Despite our incomplete knowledge of sea turtle biology, it is obvious that there are certain circumstances under which hybridization would not attend translocation. One is the case in which the local population of the species is known to be extinct, such as the green turtle colony of Bermuda. Another is illustrated by the joint Mexico–United States translocation and head-starting program for the critically endangered Kemp's ridley (Anon., 1978). As late as 1947, 40,000 ridleys nested at one time on the Rancho Nuevo beach on the Gulf coast of Mexico. Today, only about 450 females come onshore in an entire season. The population crash is attributed to decades of relentless local killing and egg taking, combined with mortality in shrimp trawls in the Gulf of Mexico where shrimping has boomed since the 1940s. In the interagency Rancho Nuevo project an effort is being made to establish a colony on a beach on Padre Island, Texas. Because there is only one breeding population of the Kemp's ridley, even if the Texas transplants mature and breed with the Rancho Nuevo colony there is no genetic risk; they are obviously of the same genetic population.

Management and Hybridization Potential

Of equal importance to sound biological bases for carrying out or not carrying out particular wildlife management projects is a clear rationale for the particular management option decided upon. In choosing options, there are at least three reasons why programs that may increase the chances of hybridization of local wild populations of sea turtles should be avoided.

First, we know that natural selection through time produces individuals that are adapted for the habitat and local conditions in which they are found. In ways that may be very subtle, hybrid turtles might be at a biological disadvantage, since they possess traits from parents who came from other environmental conditions. In other words, the hybrids may have reduced fitness. Two examples might illustrate this.

In green turtles, one population that shows phenotypic and behavioral differences from most mainland colonies is the nesting assemblage at Ascension Island in the central equatorial Atlantic. A tag-and-recapture program there has revealed that the animals nesting on the tiny island come entirely from feeding grounds off the coast of Brazil, 1600 km away (Koch et al., 1968). The Ascension Island turtles are very large. Weights of over 225 kg are common in nesting adults, whereas females of 112.5 to 135 kg are more characteristic of other Atlantic colonies of greens. One possible reason for this difference is that the greater energy requirements for the 3200-km round-trip breeding migration explain the great bulk of the island turtle's population. Perhaps the advantage of this is that large size provides a favorable relation between drag and fuel storage space for the long migratory journey. If this characteristic is genetically determined, then hybridization of the Ascension turtles with any other green turtle deme would likely produce smaller progeny, and thus reduce capacity to use Ascension Island as a nesting site. Hybridization would tend to undo the work of natural selection, reducing the survival capacity of the progeny.

Another example involves hatchlings. When young turtles hatch, they characteristically move quickly down the beach and swim rapidly out to sea for a period of days in what is termed a swimming frenzy on their way to wherever baby turtles go during the "lost year." Presumably the amount of yolk to sustain the swimming and its duration are genetically determined. If populations of turtles are then adapted to the particular environmental conditions off their natal beaches such that they may reach appropriate developmental habitat, hybridization between demes with different environmental conditions offshore could produce hatchlings unable to reach appropriate habitat. Thus, hybrids might not swim long enough or have enough yolk to sustain them, or they might swim too far and eventually be caught in currents that take them far from the area.

A second objection to hybridization is an intellectual one. As we have pointed out, much remains to be learned about the biology of sea turtles. If widespread hybridization occurred, their systematic relationships and perhaps other aspects of their biology would be hopelessly confused. Similar systematic confusion has already resulted during studies of the biology of certain salamander and fish species due to introductions by fishermen who use them as bait.

A final category of objection involves ethical considerations. Ethical and moral arguments involve nonutilitarian reasons for saving species and prevention of excessive exploitation and habitat destruction. These arguments state that species, including demes in the context of the present discussion, have a right to exist and continue to evolve, and that mankind does not have the right to destroy them through hybridization or any other means.

Admittedly, the ethical arguments are tested when the threatened species is not an easily defined and differentiated one, but consists instead of a number of disjunct but biologically unique populations found throughout the world. They are tested again when the effect of the hybridization is not the immediate disappearance of the species or deme but an alteration of its genome. Even in this case, however, the original biologic entity is lost, and if humans are responsible for the loss, then, morally, humanity has erred and lost a wonderful treasure. Unfortunately, there is no space here for an exhaustive analysis of this ethic, and no doubt it is seen differently by different people. Excellent discussions on this topic can be found in Ehrenfeld (1976) and Ehrlich and Ehrlich (1981).

Conclusion

Because of the threats imposed by hybridization, translocation of eggs or hatchlings to effect sea turtle recovery can be recommended only *in extremis* where the species or population is locally extinct or so reduced as to make the risk of adverse effects of genetic contamination seem justifiable after careful weighing of all other options. Fortunately, very few turtle populations are so depleted as to invite translocations. Owing to a wide variety of arguments, including hybridization threats, farming or ranching cannot be justified as management options in sea turtle conservation. The most appropriate and biologically sound course for their conservation remains to close existing markets and prevent new markets from becoming established, to protect nesting, feeding, migratory, and developmental habitat, and to promote the use of trawls that will exclude turtles.

While interspecific hybridization appears rare in sea turtles, the localized breeding patterns, nest site fixity, and population movements of a number of species indicate that most populations are organized into discrete reproductive units or demes. Preliminary electrophoretic data tend to confirm this assumption. Therefore, wildlife managers are faced with a complicated task, which requires protection of each individual breeding population. A wide variety of management options are available, most of which are sound biologically and will not promote hybridization between adjacent demes.

Intraspecific hybridization is considered much more of a threat than interspecific hybridization to these endangered species. Management options should be thoroughly reviewed to determine their potential for promoting hybridization, and those programs in which it might occur should be eliminated or considered only as last-resort projects.

The two most common management practices that might promote intraspecific hybridization are the setting up of sea turtle farms or ranches, ostensibly to reduce commercial pressure on wild populations, and the translocation of eggs or hatchlings to beaches where populations are much depleted or extirpated, ostensibly to restore the viability of the populations. Hybridization with local stocks could result from the accidental or deliberate release of hatchlings from farms or ranches, as well as the interbreeding of translocated individuals with local turtles.

Threats from hybridization would tend to be minimized by the tendency of suddenly released captives to home or by the captive-raised individuals' inability to complete fertilization with wild turtles in a natural environment. Even if successful breeding and hatching occurred, the probable low survival rate of the hatchlings would further tend to minimize deleterious effects on the population as a whole, at least in large populations.

On the other hand, if carried out on a widespread scale, interspecific hybridization could disrupt local gene pools, producing offspring ill-adapted to survive in a particular developmental habitat, and thus further stress a declining population. Objections to schemes that might promote hybridization also involve scientific arguments that it may lessen capability to understand the biology and systematics of various populations, and ethical arguments concerning whether it is right or wrong to artificially disrupt a functioning, adapted gene pool. Our lack of understanding of sea turtle natural history and population genetics makes it difficult to predict what the results of hybridization will be.

It is recommended, however, that sea turtle farming or ranching not be viewed as a viable management option and that translocation, at least partly because of hybridization potential, be considered only after all other management options have been carefully considered.

Acknowledgments

We thank John L. Behler, Archie F. Carr, Jr., and Jack Frazier for their critical review of the manuscript.

CHAPTER 18

Hybridization and Gene Exchange among Birds in Relation to Conservation

Tom J. Cade

Introduction

The word "hybrid" like the word "exotic" often carries a bad connotation in biological conservation, because notions about "racial purity" and its converse, "mongrelization of the races," still persist and often influence political and administrative decisions about the management of wild species. Ideas about racial purity extend far back in the ethnic histories of particular groups of people, but they have their most immediate origins in pseudoscientific doctrines such as Nazism and in the practical breeder's goal of creating "purebred" stocks of domestic animals and plants, in which inbreeding and artificial selection for a preconceived phenotype are important components of the process.

This term "purebred" and its underlying assumptions about racial purity turn up in some rather strange places. Two examples taken from my recent experiences in dealing with the United States federal bureaucracy illustrate the kind of thinking still in vogue. The legal definition of "migratory bird" found in the federal regulations (50 CFR 10.12) promulgated under the Migratory Bird Treaty Act is " . . . any bird, whatever its origin and whether or not raised in captivity, which belongs to a species listed in §10.13, or which is a mutation or a hybrid of any such species . . . ." The official government reason published in the *Federal Register* for including the latter categories in the definition is as follows: "Because of the great difficulty in distinguishing mutations and hybrids from purebreds [sic], coverage of the former two is essential to adequate enforcement of the Act . . . ." Just the reverse reasoning has been taken in regard to the Endangered Species Act of 1973. Hybrids have been excluded from coverage under this act because "hybrids of some species might well interbreed with the few remaining purebreds [sic], so as to dilute or eliminate the original gene pool" (memorandum dated 2 August 1977 from Solicitor's Office, U.S. Department of the Interior, to Associate Director for Federal Assistance, U.S. Fish and Wildlife Service).

It is important to emphasize that hybridization and the exchange of genes between different populations (gene pools) are natural phenomena of wild species and often have important influences on speciation, adaptation, and other evolutionary processes. The notion of "racial purity" does not help us to understand either the genetic composition of particular populations or the biological consequences of hybridization and gene combinations between the gametes of individuals from different populations. In some instances, hybridization may lead to increased fitness or adaptation to new environments (Chapter 16 by Harlan). In others it can lead to loss of biological (phenotypic) diversity through the genetic assimilation of a small population by a large one, or by some other related influence; and particularly where small, relict, or insular populations are concerned, the principles and practices of biological conservation need to take the influences of hybridization into account. The extent to which man can intervene intelligently in processes associated with hybridization and gene exchange to the benefit of species diversity in natural communities remains problematical; but the question will no doubt be put to serious tests in the next two decades.

The Kinds of Hybridization among Birds

Following a period of genetic differentiation while geographically isolated, two avian populations that establish secondary contact can show a range of behavioral, ecological, and genetic interactions, from complete reproductive isolation and absence of ecological competition allowing for the development of extensive sympatry, in which case the two forms have achieved the status of full species, to complete reproductive and genetic compatibility with rapid exchange of genes between the two, in which case they are conspecific regardless of how phenotypically and genetically different they may be. Ernst Mayr (1963) defined hybridization as "the crossing of individuals belonging to two *unlike* natural populations that have secondarily come into contact," leaving open the question of how unlike they have to be, and he distinguished five intergrading kinds of hybridization within the range of possibilities that can occur when two populations come into secondary contact.

Table 1. *Classification of Hybridizing Forms in Birds*[1]

Forms involved	Pattern of distribution	Interactions
Subspecies or subspecies groups of polytypic species	Allopatric or parapatric	Primary intergradation or potential capability of so developing
Subspecies or subspecies groups of polytypic species	Parapatric	Secondary intergradation; random mating and hybridization among parental forms, F_1 and F_2 hybrids, etc., and backcrosses
Taxonomic border of species		
Semispecies (allospecies)	Parapatric	Secondary intergradation; complete assimilation of genes between two populations leading to hybrid swarm
Semispecies (allospecies)	Parapatric	Some secondary intergradation; but some assortative mating; F_1 hybrids common but backcrosses usually not; competition and reinforcement of isolating mechanisms
Taxonomic border of species		
Allospecies of a superspecies	Parapatric, often becoming sympatric	Rare, inconsequential, or no hybridization; little backcrossing; effective isolating mechanisms seldom breaking down; some competition in zone of overlap
Congeneric or related species but not allospecies	Parapatric or sympatric	Rare or no hybridization; no backcrosses; effective isolating mechanisms rarely breaking down; little or no competition

[1] Based in part on Mayr (1963) and Short (1969).

Short (1969) also has categorized degrees of hybridization in relation to kinds of species and intraspecific populations; Table 1 attempts to summarize these relationships. Hybridization between species with long-established sympatry (syntopy) is rare among birds, about one case in 60,000 individuals according to Mayr (1963), and without known ecological or evolutionary consequences because the hybrids are nearly always infertile or otherwise less fit than the parental forms. On the other hand, hybridization between parapatric populations is more frequent and can have evolutionary consequences by producing new gene combinations in the hybridized populations.

In their analysis of 516 nonmarine species and superspecies of North American birds, Mayr and Short (1970) found at least 52 (10%) of these taxa to be involved in hybridization under natural conditions. They identified 19 "hybrid zones" involving 16 taxa, in which random matings occur among hybrids and parental forms in all combinations between strongly differentiated geographic populations of the same species [there are many other such cases among less strongly differentiated populations in secondary contact—for example, the Rough-legged Hawks (*Buteo lagopus*) nesting in northwestern Alaska; Cade, 1955]; seven cases in which the extent of hybridization is limited or uncertain among such groups, and four cases in which polymorphic "color phases" are involved; five situations involving zones of overlap and extensive hybridization between parapatric "allospecies;" and twelve involving superspecies in which limited hybridization occurs between parapatric populations. Finally, limited hybridization occurs between sympatric species belonging to different superspecies or genera in six situations involving mainly galliforms (fowl-like birds, Galliformes) and hummingbirds (Trochilidae). Table 2 is a summary from Short (1972).

Anderson (1977) analyzed the same data on allospecies pairs in a slightly different way and identified eight examples involving extensive

Table 2. *Analysis of Avian Hybrid Situations in North America.*[1]

Situation	Number of			Extent of zone[2]	
	Superspecies	Species	Cases	Great	Small
Hybrid zones[3]	—	16	19	10	9
Zones of overlap and hybridization[4]	17	35	18	12	6
Hybrid swarms[5]	3	5	4	0	4

[1] From Short (1972).
[2] Great is 40 km or more in extent; small is less than 40 km.
[3] Zone where only hybrid phenotypes occur, or parental forms constitute less than 5% of total.
[4] Zone where parental forms overlap and constitute 5% or more of total population.
[5] Hybrid populations "out of genetic contact with parental forms."

hybridization and backcrossing, seven cases in which essentially only F_1 hybrids occur, and fourteen of parapatric species pairs in which little or no hybridization is known. The conclusion seems to be that hybridization could have played a role in the evolution of many or possibly most avian species; and Short (1972) has made the particularly relevant point that hybridizing species may have been favored by the fluctuating climatic conditions of the Pleistocene, owing to the role of hybridization in producing recombinant genotypes that can allow individuals to adapt rapidly to shifting environmental conditions.

In cases where partial reproductive isolation has developed between two populations, hybrid zones may remain stable for long periods of time without resulting in the movement of alleles from the gene pool of one population into that of the other beyond the region of immediate contact, owing to selection against the hybrids beyond the zone of contact, or inability to backcross with parental forms; while in other cases in which backcrossing is successful to some extent genes may be exchanged more widely through the two populations but without achieving panmixia, so that the two populations remain genetically distinct in some parts of their ranges. Also, the development of assortative mating between males and females of the parental forms in the hybrid zone can lead to sympatry and speciation, as seems to be taking place between Midwestern oriole (*Icterus*) populations (see Corbin et al., 1979, and compare with Sibley and Short, 1969). In still other cases in which no reproductive isolating mechanisms evolved during the period of geographic separation but measurable and sometimes marked phenetic differences did develop, the resulting "zone of secondary intergradation" will show extreme and unstable phenetic and genetic variability among the individuals, representing both parental forms and every kind of intermediate condition. Another situation involves the development of "hybrid swarms," in which the hybridized population comes to occupy an exclusive area where neither parental form occurs, including sometimes environments and types of habitat for which neither parental form shows suitable adaptations—thus resulting in an increase in biological diversity and the extension of a species into new range. Examples among birds include the hybrid swarms of the House Sparrow and Willow Sparrow (*Passer domesticus* and *P. hispaniolensis*) in Mediterranean regions (Meise, 1936; Mayr, 1963), the mixed populations of the Red-backed Shrike and Red-tailed Shrike (*Lanius collurio* and *L. phoenicuroides*) in Kazakhstan and surrounding regions (Panov, 1972), and some of the island populations of the Golden Whistler (*Pachycephala pectoralis*), a muscicapidae flycatcher of the Southwest Pacific (Galbraith, 1956; Short, 1969).

With this general background, we can survey some examples of avian hybridization in an effort to develop guidelines or principles about hybridization in relation to the conservation of biological diversity in birds.

CHAPTER 18 Hybridization and Gene Exchange among Birds   293

Dynamic Interactions between Populations in Recent Secondary Contact

Dynamic interactions between populations brought into recent secondary contact as a result of man-induced environmental changes or by the introduction of exotics are particularly instructive. Since they are also frequently the focus of political or managerial concern, they merit our special attention here.

1. Continental Populations

A. The Mallard Duck and Its Relatives. The Mallard (*Anas platyrhynchos*) is a highly successful species with considerable invasive capacity, and the species has expanded its range into many habitats that have been modified by human activities. Not only does the Mallard enjoy a widespread natural breeding distribution through much of the Holarctic, but it also has been introduced successfully outside its natural breeding range in places such as Australia and New Zealand, usually in a somewhat modified, "domesticated" form.

In addition, the Mallard has unusually broad reproductive and genetic compatibilities with other species of ducks and particularly with other mallard-like forms with which it has only recently come in contact owing to introduction or expansion of its range. In North America the Mallard has formed hybrids in the wild with all eight of the other resident species in the genus *Anas*, while in the Palearctic it is known to hybridize naturally with five of the other ten breeding species of *Anas* (Sibley, 1957). In captivity Mallards cross with many other species of ducks, including some in different genera. Introduced Mallards have hybridized extensively with the endemic (in the strict sense) Grey Ducks (*A. superciliosa*) of New Zealand and with the Australian Black Ducks (*A. superciliosa "rogersi"*) (Williams, 1970; Braithwaite and Miller, 1975), the former now considered to be a subspecies and the latter a population of the Spotbill Duck (*A. poecilorhyncha*), while the endemic Marianas Mallard (*Anas "oustaleti"*) comprises a hybrid swarm between Black Ducks dispersing from the south and Green-headed Mallards from the north (Yamashima, 1947; Braithwaite and Miller, 1975).

In North America the Mallard has made recent contact through extension of its range with the American Black Duck (*A. rubripes*) in the Northeast and on the Atlantic Coast, with the Mexican Duck (*A. diazi*) in the Southwest, and with the Mottled Duck (*A. fulvigula*) in the Southeast. There has been extensive hybridization and backcrossing between the Mallard and the Black Duck (Johnsgard, 1961, 1967) and between the Mallard and the Mexican Duck (Johnsgard, 1961; Finnley, 1977a, 1977b, 1978a,

1978b; Hubbard, 1977), but probably owing to the limited geographic contact so far (Nelson, 1980), only infrequent crossing with the Mottled Duck.

For a number of years waterfowl biologists have expressed concern about the possible assimilation of the gene pools of the Black Duck (*A. rubripes*) and the Mexican Duck (*A. diazi*) by the expanding Mallard (*A. platyrhynchos*) population. By assigning the Mallard populations (gene pools) of the Atlantic and Mississippi flyways an arbitrary value of 100, Johnsgard (1961) estimated the relative population sizes of the other North American forms as 17 for Black Ducks, 1.17 for Mottled Ducks (including the Florida Duck), and 0.52 for Mexican Ducks (including those north and south of the Rio Grande River). The small sizes of these populations relative to the Mallard indicate the ease with which they might theoretically be genetically assimilated through combination of their gametes with those of Mallards.

Indeed, Johnsgard (1967) and Heusmann (1974) have shown that hybridization between Mallards and Black Ducks has continued to increase in extent as well as shifting progressively eastward in recent years, owing primarily to a continuing spread of the Mallard population from the Midwest to the Atlantic Coast as farm ponds, city parks, and other man-created habitats favorable to the Mallard have been developed in localities where formerly only fostered habitat favorable to the Black Duck existed. Hunters shot more Mallards than Black Ducks in the Atlantic flyway for the first time in 1969; that year Mallards made up 18.5% of the total duck kill. The percentage of Mallards in the bag has increased steadily, so that by 1972 they made up 22.6%, while Black Ducks made up only 14.5%. This difference results in part from a decline in the number of Black Ducks but mainly from the great increase in Mallards. Consequently, Heusmann (1974) is rather pessimistic about the future of the Black Duck as a "distinct species," since he feels that some populations of Black Ducks are now separated genetically from Mallards only by a difference in habitat preference.

There is, however, good field evidence for a high degree of assortative mating where the two populations form mixed flocks in fall and winter during which pairing takes place, and it has not been determined yet whether the progeny of hybrids and backcrosses have the same survival and reproductive capabilities as the parental forms do; but the frequency of hybrids, while varying from locality to locality, up to a maximum reported 12.9% of all Mallards and Black Ducks banded at inland sites in Massachusetts (Johnsgard, 1961; Heusmann, 1974), is still well below that which would be expected if hybrids had the same fitness as the parental forms. Even in the insular situation of New Zealand, where gene exchange has been extensive between the introduced Mallard and the native Grey Duck population, a high rate of infertility (produced by a difference in reproductive timing of the two species) and a high rate of embryonic deaths restrict gene exchange between the two populations, as do differences in species recognition and courtship behavior (Williams and Roderick, 1973).

Hybridization between the Mallard and the Mexican Duck has produced some interesting biopolitical rationalizations in recent years (Finnley 1977a, 1977b, 1978a, 1978b). The United States population of Mexican Ducks was included on the list of endangered species in 1966 under the old Endangered Species Preservation Act because of the presumed threat of wetland drainage along the Rio Grande River and because of hybridization with the Mallard. At that time it was thought that 20% to 40% of the Mexican-like ducks in the United States were hybrids and that only 100 to 200 "pure" Mexican Ducks remained north of the border. By the time the U.S. Fish and Wildlife Service (FWS) finally decided to do something about the situation in 1977, the official estimate was 15,000 to 50,000 Mexican Ducks in Mexico and 1000 Mexican-like ducks in the United States, including hybrids and nonhybrids. That year the FWS closed the general waterfowl season to duck hunting in parts of 12 counties in Arizona, New Mexico, and Texas "to protect the endangered U.S. population of the Mexican Duck (*Anas diazi*)."

Needless to say, this action caused an uproar among the local duck hunters, and the concerned states immediately initiated actions to reverse the hunting closure. Hurried studies were carried out by the states and the FWS. The winter waterfowl surveys in January of 1978 yielded a minimum estimate of 22,470 Mexican and Mexican-like ducks south of the border (later increased to more than 50,000 by a survey done in May and June of 1978) and 1000 to 2000 Mexican-like ducks in the United States. Furthermore, farm ponds and other irrigation impoundments had more than compensated for loss of riparian wetlands, and the range of these ducks had actually expanded both in Mexico and in the United States since 1966. It was concluded that the duck populations had remained relatively stable for the past 10 to 15 years, existing under little hunting pressure. It was further concluded that the United States population of Mexican-like ducks consists entirely of intergrades between *A. diazi* and *A. platyrhynchos* and that hybridization was extensive as far south as the Mexican state of Durango (see Hubbard, 1977).

Conveniently, the Solicitor's Office of the Department of the Interior had rendered an opinion in 1977 that hybrids between an endangered and nonendangered form (species) are not covered under the provisions of the Endangered Species Act of 1973. At about the same time, the Committee on Classification and Nomenclature of the American Ornithologists' Union reached a decision to change the taxonomic status of *A. diazi* to that of a subspecies of *A. platyrhynchos*. In July of 1978 the Mexican Duck was officially removed from the United States federal endangered species list.

It is unclear exactly what role the issue of hybridization actually played in this decision, other than as a smoke screen. The U.S. Fish and Wildlife Service finally concluded that, while it recognizes the value of preserving populations of naturally interbreeding subspecies and species, to be listed for protection under the Act the *entire population* must be in jeopardy, not just one phenotype. Since the overall population of intermediate ducks in the United States and Mexico is stable and even expanding into Arizona and Texas, the requisite condition for endangered or threatened status does not exist. This correct conclusion could have been reached more directly and clearly simply by paying attention to the population size and population dynamics of the ducks in Texas, Arizona, and New Mexico without resorting to all the rhetoric about hybrids or maneuvering to change the taxonomic status of *A. diazi* from a species to a subspecies in order to make delisting easier. In retrospect, it is amusing to note that in 1966 one of the main reasons for placing the Mexican Duck on the endangered species list was because of the threat of hybridization with the Mallard, while one of the arguments for removing it from the list in 1978 was hybridization with the Mallard.

B. Blue-winged (*Vermivora pinus*) and Golden-winged (*V. chysoptera*) Warblers. Changes in breeding distribution, relative numbers in zones of contact, and hybridization have probably been studied more intensively in these two warblers than in any other pair of hybridizing species in North America (for good recent reviews see Gill, 1980, and Confer and Knapp, 1981). These two species are thought to have been allopatric at the time of European settlement in North America, the Blue-winged Warbler nesting mainly in the watersheds of the Ohio River and middle reaches of the Mississippi River in Missouri, southern Illinois, Indiana, Ohio, and in Kentucky and Tennessee west of the Appalachians, while the Golden-winged Warbler was distributed mainly in the southern Great Lakes regions eastward to the Atlantic coast. Both species have extended their breeding ranges northward and eastward, mainly in response to man-induced changes in habitats, so that they have been sympatric for many decades (Figure 1). The Golden-winged Warbler began its range expansion about 175 years ago, while the Blue-winged Warbler began a little more than 100 years ago, reaching Connecticut, for example, by 1870, but it has been accelerating its expansion in the last 50 years and is still moving northward, particularly in the Great Lakes region (Figure 2). Today the Golden-winged Warbler occupies exclusive breeding grounds only at the extreme northern limits of its range and at its highest nesting elevations in the Appalachian Mountains, and it has disappeared entirely as a breeder over much of its original southern nesting range, which is now occupied by Blue-winged Warblers.

The Golden-winged Warbler is a habitat specialist, nesting in early shrubby successional stages of old field regeneration, a relatively ephemeral stage of vegetation lasting between about 10 and 30 years after

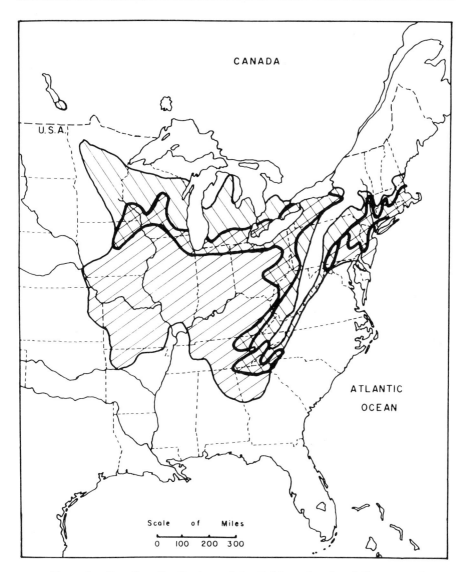

Figure 1. *Breeding distributions of the Golden-winged and Blue-winged Warblers* circa 1960. *The area of overlap is outlined by heavy black boundaries.* From Short (1963).

secondary succession has begun (Confer and Knapp, 1981), and much of the historical change in the distribution and numbers of this species can probably be attributed to the temporal and geographic pattern of field abandonment in the northeastern United States. The Blue-winged Warbler is more of a generalist, breeding not only in shrubby stages of old field succes-

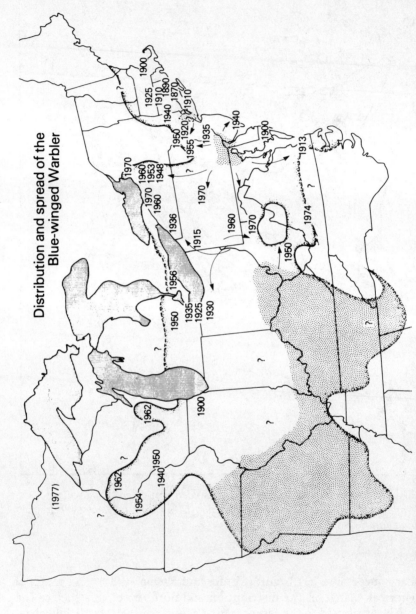

Figure 2. Breeding distribution of Blue-winged Warbler showing spread northward and eastward since late 1800s. The stippled areas indicate the approximate range in the mid-1800s. Note that populations in southeastern Pennsylvania and in the New York City region were formerly isolated from each other and from the main western population. From Gill (1980).

CHAPTER 18 Hybridization and Gene Exchange among Birds   299

sion, where it is often syntopic with the Golden-winged Warbler, but nesting in more wooded habitat as well, forest edges that are 60 to 70 years into secondary succession. Consequently, much more habitat is now available to the Blue-winged than is available to the Golden-winged Warbler.

Wherever these two species have come together, hybridization has been extensive, although Gill (1980) says that hybrids may obtain mates less frequently than the parental forms, and successful backcrossings are not frequent enough to consider the two populations members of the same species. The curious fact is that about 50 years after secondary contact has been established between the two species in any given locality only Blue-winged phenotypes remain, and the Golden-winged and hybrid phenotypes have disappeared (Figure 3).

It is not yet clear to what extent changes in habitat unfavorable to the Golden-winged populations, or interspecific competition, or genetic assimilation accounts for the disappearance of Golden-winged phenotypes. I feel that habitat changes and perhaps ecological competition are more important factors than gene exchange between the two species, particularly as there is no evidence for random mating or much backcrossing between the hybrids and parental forms. Gill (1980) has suggested, however, that

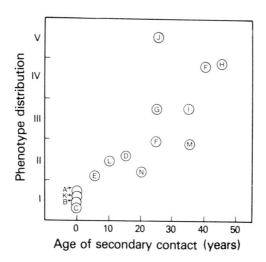

Figure 3. *Relation between phenotypes of hybridizing warbler populations and the number of years between establishment of Blue-wings at a given locality and the sampling of the population there. Each letter and circle represents a sample from a different locality. Phenotype distribution I includes mostly Golden-winged forms with a few Blue-winged and hybrid individuals, while phenotype distribution V includes mostly Blue-winged forms with some variability in wing-bar color and other hybrid characters. The other categories represent intermediate degrees of hybridization. From Gill (1980).*

following the initial period of hybridization and destruction of the Golden-winged genotypes, stabilizing selection favoring genotypes that produce "Blue-winged" individuals may eliminate hybrids with deviant phenotypes and thus maintain the norm of the Blue-winged populations. He has made the further intriguing speculation that incorporation of some (hidden?) Golden-winged genes into Blue-winged genotypes may broaden the ecological and physiological capabilities of "Blue-winged" Warblers, "thereby fueling their expansion into latitudes and altitudes previously occupied by Golden-wings, as described for insects (Lewontin and Birch, 1966)." If this is true, then some Blue-winged populations may actually be hybrid swarms.

Whatever the actual mechanisms favoring Blue-wings to Golden-wings prove to be, Gill (1980) and Confer and Knapp (1981) agree that the future of the Golden-winged Warbler is bleak. It could be gone in another 50 to 100 years, but possibly some Golden-wings may be able to persist in high-latitude and high-altitude refuges that Blue-wings cannot invade. Perhaps some kind of stabilized sympatry may yet evolve, although none is evident at this time.

Confer and Knapp (1981) feel that long-term preservation of the Golden-winged Warbler may entail both habitat manipulation and localized elimination of the species with a negative biological impact, as is now being done for the endangered Kirtland's Warbler (*Dendroica kirtlandii*). It should be emphasized, however, that, although species with large continental distributions may form hybrid zones or hybrid swarms with closely related congeners, they will usually have some part of their total range where their particular genotypes are best adapted to the environment and where hybrid genotypes are selected against. Our best hope for the continued existence of the Golden-winged Warbler lies in the probability that this generalization will apply to it.

2. Island Endemics

Genetic assimilation or other effects of hybridization are more likely to be a serious conservation problem for small, relict populations or island endemics, when they come in contact with a larger or reproductively more successful population with which individuals interbreed. Insular populations frequently develop marked phenetic (and genetic?) differences from close congeners without becoming reproductively isolated (e.g., *Pachycephala pectoralis*, the Golden Whistler: Galbraith, 1956). Two examples of this problem cited in the "ICBP Bird Red Data Book" (King, 1981) are the Seychelles Turtle Dove (*Streptopelia picturata rostrata*) and the Chatham Islands Yellow-crowned or Forbes Parakeet (*Cyanoramphus auriceps forbesi*). (See also Chapter 7 by Soulé regarding the vulnerability of island species.)

A. **Seychelles Turtle Dove.** This well-differentiated island form, which originally occurred on most of the Seychelles Islands, has been virtually, if not completely, eliminated as a distinct phenotype after a contact of a little more than 100 years with the introduced *Streptopelia picturata picturata* from Madagascar. Details of the introduction are sketchy, but Newton (1867) found *S. p. picturata* well established on Mahé, while *S. p. rostrata* occurred on other islands. No one knows how many of the Madagascar doves were introduced into the Seychelles, or whether there was a single introduction or several; but Penny (1974) supposes that the birds arrived aboard pirate ships, which often carried doves as part of their food stores, suggesting the possibility of multiple introductions over decades. In any event, by 1959 doves with endemic characters—vinaceous heads and short wings—were found on only two islands, while hybrids occurred on others, and birds phenotypically representative of *S. p. picturata* were on most but especially on Mahé, which must have been the location of the original introductions (Crook, 1960). By 1975 apparently no typical *S. p. rostrata* could be found on any of the islands, and many of the doves now have plumage characters close to the norm of *S. p. picturata*; however, *picturata*-like doves from the Seychelles have shorter wings than those from Madagascar (Gaymer et al., 1969), and it seems likely that some *S. p. rostrata* genes still are represented in the genotypes of these birds. The Seychelles population of turtle doves is probably approaching a stabilized hybrid swarm of *S. p. rostrata* and *S. p. picturata* genes, but phenotypically the birds look most like the Madagascar doves. Conservationists conclude, therefore, that a unique island endemic has been "swamped out of existence," but the truth in terms of the survival of alleles may be quite otherwise, not to mention the likelihood that the hybridized doves are better adapted to the current environment of the Seychelles than the original *S. p. rostrata* population would be.

B. **Chatham Islands Parakeets.** The Red-fronted Parakeet (*Cyanoramphus novaezelandiae*) and the Yellow-crowned Parakeet (*C. auriceps*) are broadly sympatric over most of New Zealand and also in the Chatham Islands (Figure 4), where more or less distinctive races of each species occur (*C. n. chathamensis* and *C. a. forbesi*) (Taylor, 1975). They show some differences in nesting habitat, the Yellow-crowned Parakeet being more of a forest-dwelling species nesting in tree cavities, while the Red-fronted is a more open country bird and nests in rocky crevices a well as in tree cavities at the forest edge. The two species are not known to hybridize extensively anywhere in New Zealand, nor formerly in the Chathams, but hybridization has become common in recent years on Mangere (112 hectares) and Little Mangere (16 hectares) Islands in the Chatham group (Taylor, 1975; Flack, 1976).

The Yellow-crowned originally occurred as a breeding bird in the Chathams only on these two small islands, while the Red-fronted occurred

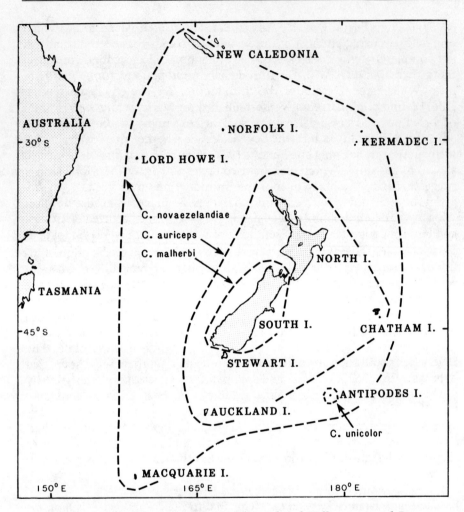

Figure 4. *Distribution of the Red-fronted and Yellow-crowned Parakeets and their relatives in the genus* Cyanoramphus, *in the New Zealand region. From Taylor (1975).*

more widely through the Chathams (Flemming, 1939). In 1937–1938, Flemming found Yellow-crowns only on Little Mangere, where he estimated the population to be 100 birds (probably too high); deforestation, overgrazing by sheep (*Ovis aries*), and predatory cats (*Felis catus*) rendered Mangere Island unsuitable for the Yellow-crowned. Later, the Red-fronted also disappeared from Mangere Island for a time; but following efforts at reforestation and the elimination of sheep and cats, both species of parakeets began to recolonize this island, Yellow-crowns appearing first around 1961. By 1968, however, Red-fronted Parakeets were more common than Yellow-crowned. In 1970 Taylor (1975) estimated that there were 60 parakeets on Mangere: 8% Yellow-crowned, 32% Red-fronted, and 60% hybrids. By 1973 there were 100 parakeets on Mangere: 6% Yellow-crowned, 47% Red-fronted, and 47% hybrids.

Why reproductive isolation has broken down between these two recolonizing populations is unclear. It may be related to the small number of colonists and to features of habitat that allow for a closer mingling of individuals of the two species than was formerly the case when the native vegetation was in place, or it might have to do with random genetic shifts that have accidentally produced reproductive compatibility between members of the two species.

In the 1930s Flemming (1939) had thought that the Yellow-crowned Parakeet would be secure on Little Mangere because this precipitous islet is "useless" for man's purposes, but Flack (1976) found otherwise. The 4 hectares of suitable forest habitat on top of the island had become progressively degraded by the development of an illegal helicopter landing pad for muttonbirders, by the actions of burrowing shearwaters (*Puffinus*), and by wind, and the 10 or so pairs of Yellow-crowns that Taylor (1975) had thought were sufficiently isolated from Red-fronted Parakeets to be safe from hybridization were in contact with breeding Red-fronts nesting on the rocky sides of the island. Flack (1976) collected several hybrids on Little Mangere between 1972 and 1976.

Environmental changes on Little Mangere Island are going against the habitat requirements of the Yellow-crowned Parakeets but are not disadvantageous to the more open-dwelling Red-fronted birds. At the same time, 100 hectares of grassland which has been relieved from sheep grazing on Mangere, has greatly increased the food supply for Red-fronts, and less than 100 meters of open water separate the two islands.

Without some kind of human intervention, it appeared that genetic assimilation of the Yellow-crowned population on these two islands was inevitable. Consequently, in 1976 the New Zealand Wildlife Service began a systematic program to shoot out all Red-fronted and hybrid parakeets on Mangere and Little Mangere in an attempt to promote population growth of "pure" Yellow-crowned Parakeets. More than 300 Red-fronted Parakeets and 15 hybrids were shot that year; approximately 12 pairs of Yellow-crowned Parakeets existed on the two islands then. Red-fronted and hybrid parakeets have been shot prior to each breeding season since, most of them being taken from mixed pairs. In 1982 there were approximately 40 Yellow-crowned and about half that number of Red-fronted Parakeets on Mangere, with very few hybrids (data from D.V. Merton, personal communication). Apparently the percentage of mixed matings on Mangere is now low, perhaps owing to the heavy, artificial selection against those Red-fronted and hybrid individuals that attempt to mate with Yellow-crowned individuals.

C. Collared and Pied Flycatchers in Gotland. Genetic assimilation is not the only way in which hybridization can influence small, insular populations. Recent studies on the extent of hybridization and breeding success of island populations of the Collared and Pied Flycatchers (*Ficedula hypoleuca* and *F. albicollis*), two European continental species that nest on Gotland in the Baltic Sea, show that hybridization also can have a "competitive influence" by reducing the effective reproductive output and rate of replacement in the species with the smaller population, given the condition that the hybrids are reproductively incompetent (Alerstam et al., 1978; Alatalo et al., 1982).

Conclusions — Hybridization and Management

Hybridization represents an extreme form of outbreeding. In the past, population geneticists have tended to believe that within species, particularly bisexual animal species, outbreeding is in the long run a more adaptive reproductive pattern than inbreeding; but there are obviously limits beyond which both outbreeding and inbreeding become maladaptive. Matings between closely related members of the same family (incest) in normally outbreeding species often may lead to inbreeding depression or other deleterious genetic consequences (Ralls et al., 1979; Ralls and Ballou, Chapter 10), and it is interesting to note the elaborate ploys and counterploys in social behavior by which some birds avoid close inbreeding (Koenig and Pitelka, 1979). On the other hand, matings between genetically very different individuals can disrupt favorable coadapted gene complexes (genotypes) because of recombinational load (Price and Waser, 1979) and can lead to a waste of gametes. A recent experimental study on preference for mates in domesticated Japanese Quail (*Coturnix japonica*) showed that individuals most often associated with first cousins in preference to siblings, third cousins or unrelated birds (Bateson, 1982).

Natural selection in sexually reproducing organisms may, therefore, usually favor matings between individuals of intermediate genetic similarity (Price and Waser, 1979), but the optimum genetic balance between too close inbreeding and overly distant outbreeding should vary among different species depending on their genetic constitutions and on the characteristics of their populations. Shields (1982b) has argued rather convincingly that inbreeding may be adaptively related to small effective population size, low fecundity, and well-developed philopatry (close ties to habitat), while outbreeding may be the better pattern for individuals with high fecundity and random or vagrant dispersal in large populations. Particularly in circumstances where close genetic tuning of the phenotype to local environmental conditions is critical to survival, resulting in so-called "ecotypes" or ecological races, inbreeding among members of small, "extended-family" demes may be advantageous, as long as recessive lethals or other deleterious alleles are not too frequent (Chapters 10, 11, 12, 15). Hybrid genotypes (recombinants), on the other hand, constitute a source of novel genetic diversity that can have adaptive potential for colonizing environments not occupied by either parental population or for increasing the fitness of individuals to reproduce in portions of the range occupied by one or both parental forms (Chapter 16 by Harlan). Viewed in this frame of reference, hybridization as a tool for restructuring populations or for creating new populations to be established in outdoor communities always should be considered carefully.

Hybridization can have both beneficial and deleterious consequences for the conservation of biological diversity. Depending upon whether crossbreeding should be prevented or promoted, the following kinds of procedures can be considered on a case-by-case basis.

1. Where a rare or endangered population comes in contact with a numerically larger or reproductively more vigorous population and hybridization results in measurable harm to the former, then it may be necessary to reduce or eliminate the less desired species or population in the zone of contact. Such control only would be practical in dealing with relatively small populations in circumscribed situations, such as the case of the Red-fronted and Yellow-crowned Parakeets on Mangere and Little Mangere Islands in the Chathams. It is doubtful whether control of Blue-winged Warblers could be effective enough to reduce hybridization with Golden-winged Warblers, as suggested by Confer and Knapp (1981), or whether Mallards could be effectively eliminated from the range of the Mexican Duck.

2. Alternatively, or in conjunction with the first procedure, it may be possible to manipulate features of the habitat in ways that favor the biological requirements of the vulnerable population at the expense of the other (for example, creating brushy successional stages of old fields to increase nesting habitat for the Golden-winged Warbler). A related possibility would be to create an ecological or geographic barrier between the two interbreeding populations. The easiest way to effect isolation between populations is to transplant the island endemic or relict species to another island or remote location where the other hybridizing population does not occur.

3. Populations that become genetically impoverished through drastic reduction in numbers and isolated from gene exchange with other conspecific populations may be reinvigorated and numerically increased by the transplantation of individuals with somewhat different genotypes from other populations of the same species (see Drury, 1974, and Chapter 3 by Allendorf and Chapter 4 by Chesser

for relevant discussion). Transplantations of the North American Wood Duck (*Aix sponsa*) over the last 40 years may indicate the potential of this procedure, although the genetic composition of the population involved can only be speculated upon. We tend to forget that 60 to 70 years ago the Wood Duck was a threatened, if not endangered, species. It responded dramatically to protection from hunting, to habitat management (such as supplying nest-boxes), and to captive propagation and release. In an early experiment, McCabe (1947) reestablished a breeding population in Wisconsin by releasing young ducklings hatched from parents in Illinois. Since then both captive-produced and wild-hatched ducklings have been translocated and established in many areas of North America remote from their geographic origin (Ripley, 1957; Lee and Nelson, 1966; Doty and Kruse, 1972; Capen et al., 1974), and today the "Woodie" is again one of the common dabbling ducks of North America.

The Dusky Seaside Sparrow (*Ammospiza maritima nigrescens*) is a population for which hybridization offers the *only* possibility for preserving its distinctive alleles. The entire gene pool now consists of the genotypes of only five captive males and their hybrid offspring with females of Scott's Seaside Sparrow (*A. m. peninsulae*; H.W. Kale II, and J.W. Hardy, personal communications). Further production of F_1 hybrids and backcrossing to the parental *A. m. nigrescens* males could result in a reconstituted population of "Dusky" Seaside Sparrows with essentially the same genotypes as the original wild birds, except that any special alleles that might have been associated with the distinctive female sex chromosome are now irretrievable (in birds the female is the heterogametic sex; Chapter 2 by Chambers). Unfortunately, political and legal decisions by the U.S. Fish and Wildlife Service have undermined financial and moral support for this prototypic experiment, and the five remaining males are required by governmental decree to live out their remaining years in celibacy because of the Department of the Interior Solicitor's opinion that hybrids are not condoned by the provisions of the Endangered Species Act. As Fran James (1980) noted, miscegenation is still frowned upon in some quarters, even among birds. (There is a current proposal that may yet rectify this situation.)

4. Habitats where a species has been extirpated and where there appears to be little chance for natural recolonization may be artificially restocked with nonindigenous individuals from other parts of the species' range. If the translocated individuals have some minimum ability to survive and to reproduce in their new environment, then natural selection can be expected to mold the genetic constitution of their genotypes over generations to produce a well-adapted

population. In instances where an ecotype similar to the original population is not available for restocking, individuals hybridized in captivity from geographically distant parents, so as to maximize genetic heterogeneity, may offer a better chance for establishing a new population than individuals drawn from one local or regional population (Drury, 1974), particularly if the environment also has been drastically modified from its original condition. This is the strategy that has been adopted for the recovery of the Peregrine Falcon (*Falco peregrinus*) in the eastern United States (Bollengier et al., 1979; Cade, 1980; Barclay and Cade, *in press*), initially against strong political and bureaucratic opposition based on the issue of introducing "exotics" (see U.S. Presidential Executive Order 11987, 1977) and on related questions about the legal and biological status of intraspecific hybrids [see Wade, 1978; also, American Ornithological Union (A.O.U.) Resolutions, 1979].

5. The natural existence of stabilized hybrid swarms in areas and habitats where the parental phenotypes do not occur suggests that the artificial creation and establishment of hybrids could, in some instances, lead to the formation of populations with new gene complexes that are better adapted to particular environments than either parental stock and so result in a net increase in biological diversity. It is particularly relevant to note the point that Anderson (1949) made about hybrid swarms of plants some time ago: "The production of hybrid swarms is limited to particular times and places at which man or nature may have hybridized the habitat . . . ." In other words, hybridized populations most often exist in habitats that have been drastically modified either by geological forces, catastrophes such as fire, or disturbances caused by man (see, also, Short, 1972; Chapter 16 by Harlan). Given the massive alterations of natural ecosystems currently under way around the world—particularly in the tropics where many allopatric and parapatric species of birds exist—we need to question the conventional wisdom that hybrids are always bad and should be eradicated in order to preserve "pure" races or species (that is, to prevent "contamination of native gene pools," in the words of an American Ornithological Union Resolution, 1979), because in many cases hybrids may prove to be better adapted to disturbed environments than their parental forms.

By attempting to eradicate hybrids or otherwise preventing the process of hybridization, we may be thwarting important evolutionary processes and unwittingly decreasing the potential for biological diversity in the future, especially in man-dominated environments. In any event, the occasional losses of biological diversity resulting from genetic assimilation or other influences of hybridization will be too small to weigh on the scale required to measure the losses resulting from habitat destruction and the outright elimination of whole communities, populations, and species of animals and plants.

Summary

Following a period of genetic differentiation while geographically isolated, two avian populations that establish secondary contact can show a range of behavioral, ecological, and genetic interactions. These interactions range from complete reproductive isolation and ecological compatibility allowing for the development of sympatry, to complete reproductive and genetic compatibility with rapid exchange of alleles between the two populations. Hybridization between sympatric species is rare among birds, but hybridization between allopatric populations in secondary contact is not. In many cases these "hybrid zones" and "hybrid swarms" remain stable for long periods of time without resulting in the movement of alleles from one population (gene pool) into the other beyond the immediate region of contact.

More dynamic situations involving recent secondary contacts brought about by man-induced environmental changes or by deliberate introduction of exotics are instructive. Interactions between the populations of Golden-winged and Blue-winged Warblers (*Vermivora chrysoptera* and *V. pinus*) in the northeastern United States and between the Mallard (*Anas platyrhynchos*) and its closely related, largely geographically non-overlapping counterparts, Black Duck (*A. p. rubripes*), Mottled Duck (*A. p. fulvigula*), and Mexican Duck (*A. p. diazi*), emphasize the difficulty of distinguishing the effects of hybridization and introgression, on the one hand, from the influences of competitive exclusion or direct responses to environmental changes, on the other, as far as temporal changes in the distribution and numerical strength of populations are concerned.

Genetic assimilation is likely to be a serious conservation problem only in the case of relict populations or island endemics. The latter frequently develop marked phenetic, and presumably genetic, differentiation from close congeners without becoming reproductively isolated. Two examples are the Seychelles Turtle Dove (*Streptopelia pictorata rostrata*), which was apparently "swamped out" by the introduced *S. p. pictorata* from Madagascar after a contact of little more than 100 years, and the Chatham Island Yellow-crowned Parakeet (*Cyanoramphus auriceps forbesi*), a small recolonizing population of which has been hybridizing with the well-established Red-fronted Parakeet (*Cyanoramphus novaezelandiae chathamensis*) on Mangere and Little Mangere Islands.

The Dusky Seaside Sparrow (*Ammospiza maritima nigrescens*) is a population for which hybridization represents the only possibility for preserving its distinctive alleles. The entire gene pool now consists of the genotypes of only five captive males and their hybrid offspring with females of *A. m. peninsulae*. Further production of F_1 hybrids and backcrossing them to the parental *A. m. nigrescens* males could result in a reconstituted population of "Dusky" Seaside Sparrows with essentially the same gene pool as the original wild population.

Hybridization resulting from human manipulations has both beneficial and deleterious implications for the conservation of genetic diversity, depending on the particular circumstances.

Natural Diversity and Taxonomy

PART FIVE

Introduction
by Gene Namkoong

The evolutionary dynamics of genetic variations obviously have a substantial impact on how populations and ecological systems can be best managed. The isolation of populations can lead to loss of genetic variations, declines in fitness, and higher rates of extinction of taxa and local populations. Higher extinction rates of local populations also can be directly related to recent environmental changes, some natural and some caused by human activity. The loss of populations and any consequent loss of genetic variation are assumed to reduce the species capacity to evolve in response to environmental crises and hence threaten species-level extinction. At the level of interpopulational genetic variation, the security of newly founded populations and the security of populations isolated from formerly connected gene pools are both threatened by the loss of alleles and of heterozygosity, and hence by reduced fitness. Perhaps a different problem exists for populations which drift genetically or which are selected for substantially different genotypic arrays than held in their original state. Then, the populations or the species may shift to different states than desired and may not be able to return to the original or to any other more desirable state. Similarly, a taxon may be swamped genetically by another and be so altered that, to all intents and purposes, it is lost.

The problems can be directly stated and specific recommendations for their solutions have been discussed in this book. There is a wide variety of techniques that can be used for any particular case. These range from simple programs of monitoring genotypic frequencies up to more heroic breeding efforts with relict populations to save species under threats of extinction. The choice of technique depends on the biological constraints of the particular species and the cost effectiveness of alternative management programs. The choice also might depend on the conservation or preservation objectives of the manager, though the "preservation of natural diversity" implies a genetic conservation objective and not the preservation of a given genotypic array.

To address the issue of preserving natural diversity in managerial terms requires that the problem is stated in terms of management objectives. If the management objective is to preserve natural diversity, then heterozygosity is but one indicator of the status of genetic variation. If the manager must work with one or a few isolated populations of an outbreeding species, then heterozygosity is often needed and serves as a useful measure of population quality. However, if the populations are being swamped by another species and the original taxon may not be recoverable, then heterozygosity caused by interspecific allelic combinations may be undesirable. Even local extinction may sometimes be viewed as an ordinary evolutionary process, while at other times it may be a harbinger of species extinction, depending on the dynamics of the species' evolution. The answer obviously depends on what else, if anything, is occurring with the species and what we mean when we say we wish to preserve natural or biological diversity.

For the manager, the appropriateness of controlling heterozygosity, migration rates, or isolation among populations depends not only on the nature of the problem but also on the concept of diversity that is the object of preservation. To form a concept of a desirable diversity, the nature of the organism must be known and the objective defined. In this part of the book, population genetic dynamics are described to better define management objectives, and some management concepts and decision rules are suggested. The first question to be addressed is what is meant by diversity. The complexity of the concept of diversity is brought out in the analysis of the structure of plant populations by Hamrick and found to vary substantially from species to species in particular ways. The degree of subvariation within species and its relationship to selective differences and reproductive biology is explored for a variety of species. Also predictive rules are derived for the distribution of genetic variation. The complexity of the concept of diversity is described in the chapter by Chambers and Bayless in terms of the elements which are measured and the measured characteristics used to analyze the structure of diversity. The relationship of these measures to species-level differences and the changes which can occur within taxa are described.

One of the driving selective factors in the evolution of species and their diversity is argued to be the variations in the biotic environment. In particular, the effects of coevolution on the interacting species are analyzed in the chapter by Futuyma and are considered to have a strong effect on the amount and distribution of genetic variations. He describes several forms of interaction and how patterns of variation are forced in host-pathogen (or predator-prey) systems.

In view of these results, it can then be argued that subpopulation structure and ecological complexity are not unusual features of species evolution and are often necessary features of the structure of natural diversity. In order to plan management interventions to preserve natural diversity, the structure of diversity would have to be considered as an objective of preservation and hence, preservation requires that multiple populations be managed. Various management techniques exist to preserve or enrich diversity. One way to enrich natural populations is the judicious use of captive or specially collected populations to supplement any present array. The use of such populations is discussed in the chapter by Foose in which he considers a captive population to be but one sample population, but one which happens to be more easily controlled than most. He discusses their use in controlled gene migration to enrich restricted populations.

The main management guideline to be drawn from these considerations is that genetic diversity within a single population requires large populations and high heterozygosity. With multiple populations, diversity can be created among different populations and, if multiple management units are used, diversity should exist in the populations' environments, biotas, and population sizes. In this sense, the earlier chapters of this book which called for a multiplicity of trials or of populations are echoed in these final chapters.

CHAPTER 19

Preserving Natural Diversity

Gene Namkoong

Introduction

Managers of biological resources who have the objective of preserving natural diversity must help to define the criteria of success by which they may be held accountable. This requires that agreement be reached on what is meant by diversity and on the measures by which a manager may say that diversity has been preserved. However, these measures are not easily defined because the frequencies of different phenotypes and genotypes of the constituent populations of a taxon change over time and hence the taxon is an evolving entity. Success is measured quite differently, depending on whether the underlying goal is to preserve the capacity to survive and evolve or to preserve a specific frequency distribution of phenotypes or genotypes. In the latter case, freezing the distributions that happen to exist at a specified time is required. However, simply freezing the mean of the distribution is very difficult to do and would not allow the taxon to evolve responses to various environmental challenges. The variance of the distribution may be of more interest, but then a determination must be made regarding the features of the distribution that are of importance to preserve.

In contrast to preserving a specific distribution of genotypic and phenotypic variations, a more reasonable goal may be to help ensure the survival and evolutionary capacity of taxa. In that case, the goal is to preserve the dynamical capacity of species to evolve and, hence, to preserve their genetic variability. I focus on the species level and below, to the exclusion of genera and higher taxonomic units, because it may be possible to reevolve subspecies variations from other populations of the species if necessary, but impossible to reevolve species from other species of the genus.

If consideration is given to diversity within species within which no crossability barriers exist, then a first approximation to an adequate criterion may be any of several diversity measures that provide weighted counts of allelic variations for multiple loci (Lewontin, 1974, Chapter 6). These measures differ in their specifics and usefulness (Gregorius, 1978); some provide a single measure of the amount of diversity, while others reflect a hierarchy of variations among infraspecific populational differences. Because it is often known or suspected that a hierarchy of population structures exists within species, the preservation of useful diversity may require preserving multiple populations or altering the architecture of the species to pack equivalent amounts of diversity into the new structure (see Chapter 21 by Chambers and Bayless). The manager must then decide what structure is desirable and how to manage it, and use the appropriate measure to evaluate the success of the program.

A further problem that affects the management program is that the biotic element of the environment is also evolving; in fact, when the genetic diversity of one species is affected, the genetic diversity of the other species components of at least the local ecosystem also may be substantially affected. If the manager can apply programs on only a finite number of localized ecosystems or management field units, and each unit includes several species, then the coevolution of multiple species components requires that a choice be made of which land areas to manage for a collective measure of diversity. Thus, the choice of populations and areas to manage with finite resources requires some evaluation of diversity for several species collectively.

In this chapter, I am assuming that the manager has an array of techniques available to preserve natural diversity and that some are more difficult and expensive to use than others. I assume that it is desirable for the manager to intervene as little as possible but still to ensure that the objective is met (cf. Chapter 1 by Frankel). The aim of this discussion is to examine how diversity is maintained or may be enhanced. I attempt to draw management implications from the previously highlighted phenomena of interpopulational structures and coevolutionary effects within ecosystems. A management strategy is then proposed for achieving a type of preservation of natural diversity.

Population Diversity

Within species, interpopulational differences are quite variable; they depend on the species' reproductive biology, environmental diversity, and specific traits or genes examined. Obviously, each species has evolved a survival mechanism, but the genetic variations may lie primarily within populations with little interpopulational differentiation, or primarily between isolated populations with less intrapopulational variability. The management implications of these species structures for preserving natural diversity need to be drawn.

If a species exists as a large population without genetic distinctions and without barriers to mating, then the presumptive evidence suggests that its future evolution would not be impeded if only a sufficiently large population were maintained. This assumes that microenvironmental differences do not selectively affect genotypes. Genetic diversity will not be dissipated any more rapidly than its "natural" rate would dictate. However, if the population size were to be curtailed severely, or if the environmental diversity actually is a selective agent for genotypic diversity and is masked by high migration rates, the potential exists for single population samples to lose the initial levels of diversity present in the species.

It may be quite common for a species to expand rapidly from normally small, isolated populations, into large, interbreeding epidemic populations with a transient homogeneity. This is true for some forest tree species and some insects that change behavior in the transition from endemic to epidemic populations (Namkoong et al., 1979). I suspect that this phenomenon may be true for vertebrates as well. In these cases, it would be a serious mistake to assume that a single subsample of the epidemic population would adequately preserve the diversity of the species, which more often would be found distributed among divergent, small populations. In fact the evolutionary dynamics of such species may require the alternating of endemic and epidemic populations, and hence a special non-steady-state management system may be required.

On the other hand, if a species exists with genetic divergence among populations, whether connected by migration or not, then the differences may be due to selective differences or to cryptic mating barriers, which allow alleles to diverge in frequency. While the divergence rarely can be ascribed solely to one or the other mechanism, it would be useful to know if the genetic variations are responses to selection pressures or are merely superficial artifacts of no evolutionary consequence. If they are superficial, and if all loci have similar evolutionarily trivial variations, or if any important genetic effects are shared commonly through all populations, then any single randomly chosen population could preserve significant diversity as well as any other. In such cases, the species' diversity is contained within the populations and the differences either are transient or are maintained by limitations on migration.

If divergence among populations exists at any of the loci due to selective effects of the environment, then the populational diversity reflects an important feature of the species' evolutionary capacities. It may be presumed that individual homeostasis, and indeed single populational genetic homeostasis, is insufficient for adaptation to the ecological diversity that a species experiences. Hence, management of diversity may require multiple populations. Very low migration rates among populations are sufficient to maintain alleles in the separate populations unless strongly selected against (Chapter 3 by Allendorf). Hence, a manager might easily compromise between a desire to maintain selectively important diversity among populations and the need to keep genetic variations within populations by introducing a few migrants among otherwise separate populations (Chapter 4 by Chesser).

It may sometimes be necessary or unavoidable to allow differences to go to extremes such that alleles, possibly different, are fixed in each population. At one extreme, nearly all loci may be forced to fixation while those that are critical for survival and have heterotic effects remain polymorphic. Adaptation to severe inbreeding may be required for survival and may even be enhanced as shown by Templeton and Read (Chapter 15). While this may not be feasible for many, the fixation of alternate alleles maximizes most measures of diversity and, as long as populations persist, the genetic polymorphism is protected at the species level. Thus, subdividing a species into populations that differ in allele frequencies and fixations can be an effective way for managers to protect genetic diversity. Since many species naturally exist in separate, environmentally distinct patches, we might examine evolution under these conditions.

The effect of subdividing a species into populations with different selection pressures was first studied by Levene (1953) for its consequences on the loss of alleles. The model he considered requires individual genotypes to migrate at random to one of several environmental patches, undergo selection in that locality in competition with other genotypes, and, if they survive, then rejoin the common species pool of survivors and random-mate to produce the next generation of migrants. The question he raised was whether a stable genetic polymorphism could be maintained even if the heterozygote was not superior to both homozygotes in any one environment. The answer he derived was that under certain conditions, such as when the harmonic mean fitness of the heterozygote over all environments exceeds that of the homozygotes, the polymorphism is protected. Thus, an average type of overdominance (marginal overdominance) is sufficient to ensure genetic diversity.

The selection–migration model used by Levene, however, is not very realistic, and many models have since been generated and analyzed for different migration models and selection patterns. For example, there is commonly a tendency for genotypes to stay within their locales of origin because of a limitation on migratory ability or by choice (Christiansen, 1974). General analyses by Karlin (1976) are quite comprehensive, including migration patterns of various forms such as migration restricted to neighboring populations and migration restricted to exchanges of small, peripheral "islands" with a larger central population. He also considered various forms of selection and different patterns of selective differences among populations. To oversimplify his wide-ranging conclusions, niche heterogeneity or environmental heterogeneity is a powerful force for maintaining genetic polymorphism (Chapter 5 by Liu and Godt). The location of the genetic diversity may lie within or between population units, depending on the specific migration patterns and types of gene action, but environmental heterogeneity is a strong protector of genetic diversity. Although it is not known that environmental heterogeneity and subdivided populations are necessary for genetic diversity, it may be concluded that managers can use diversity among populations to protect genetic diversity.

However, the manager must also consider that, in many species, the exchange of alleles by migration and even the effects of selection are not the same for the different sexes. In many plant species, pollen vectors widely disperse the male gametes while the seed zygotes fall close to the female parent. In some species, insect pollinators localize the mating to nearby males and females and the fertilized seed is then widely dispersed by wind or other vectors. In many animal species, similar sexual divergence occurs, affecting the population dynamics. Consider, for example, when mating is localized but zygotes disperse to multiple niches; viability selection within populations, each occupying a different niche, will cause allele frequency differences among them. This will then have the effect of increasing homozygosity over Hardy-Weinberg frequencies, (Wahlund effect) and selection effects will then be different. Alternatively, consider the situation where male gametes come from different populations, and, even if males and females are identically affected by selection, male and female allele frequencies will differ. A population that receives such an input of gametes will not be in Hardy-Weinberg equilibrium even if mating is random within this population, with respect to genotype.

In research conducted in my laboratory, Dr. H.R. Gregorius studied the particular case in which the female is sedentary and reproduces in her home patch, but the male is an indiscriminate, wide migrator. The primary objective was to examine the protectedness of allelic polymorphism in the sense that alleles reaching low frequency do not go to extinction in spite of an average selection for one allele in either or both sexes. In these analyses it was shown that this kind of differential sex migration has a strong effect on maintaining genetic diversity. Briefly, we showed that multiple populations can maintain genetic polymorphism if selection on the female strongly favors the heterozygote, in at least one population, or favors different homozygotes in different populations. This kind of protection does not require that the species as a whole have any kind of average overdominance, but only that the heterozygotes have sufficiently higher fitness than the more frequent homozygotes in a single population. If diversity is to be protected but only through the female line, regardless of any adverse selections that may occur in the males, then the heterozygote fitness must be twice the homozygote's, in at least one population. Thus, the manager of a single population can protect polymorphism and genetic diversity even if only one sex is controllable. If there is no net selection on the males and the allelic variations confer an equal average fitness when weighted over the whole species, then the female heterozygote fitness, naturally or artificially managed, need only be greater than that of the homozygote's. This holds even if the superiority exists or is forced in only one population, and hence it is a very strong factor in protecting polymorphism. Thus, even if a population is inundated with male gametes of one genotype, the polymorphism can be protected in the female population. On the other hand, if the protection of diversity were to rely on the control of the male gametes only, then the species-wide average fitnesses must favor the heterozygote.

Hence, Karlin's (1977) principle that polymorphism protectedness increases with increased environmental variance is reaffirmed. In fact, strong protection is afforded even when migration is global as long as certain kinds of sexual differences exist. As a management tool, in fact, sex differences may be more easily measured and so managed as to adjust migration rates among populations to ensure allelic polymorphism and hence genetic diversity.

The management guidelines can now be derived and fairly simply summarized for cases where populational ecological diversity is important. If only one unit can be managed, then ensure that heterozygotes are favored, and if only one sex is managed, then double the net fitness of the heterozygotes. If more than one unit can be managed, then diversify the selective pressures in the additional units.

Biotic Diversity

In addition to the physical factors of the environment, biotic factors often can be strong selective forces that affect fitness, differences in fitness, and variations among populations. To the extent that these factors are important, management plans must consider their effects and how to preserve diversity in a dynamically evolving system.

It is useful to review briefly the kinds of ecological models and their analyses before discussing the genetic variations that may exist and the

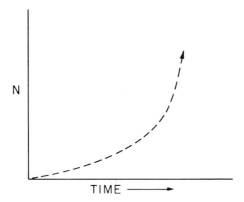

Figure 1. *Generalized population exponential growth curve based on equation (1).*

effect of ecological interactions on genetic diversity, and vice versa. It should be noted first that most genetic theory is built on the simplest ecological concepts, which rarely, if ever, are adequate. This conceptual model is one of exponential growth where at any particular time (t) the population size (N_t) determines the number of offspring in the next generation (N_{t+1}) by a constant rate of growth or decline (w). Thus,

$$N_{t+1} = wN_t$$

This growth relationship of change in population size (dN) with respect to change in time (dt) is conveniently put into an equation for the population change rate r, where $w = \log r$ (Figure 1):

$$\frac{dN}{dt} = rN \qquad (1)$$

Genetic variation may exist only in the rate constant, which is the only parameter of fitness, and evolution proceeds according to the relative sizes of r for the different genotypes. For this model to be adequate, death or removal of organisms must occur proportionately to their growth rate, r.

If this model is too ecologically naïve because the organisms also react to density effects in a different way than in proportion to r, then parameters reflecting these effects are required. A commonly used extension is (Figure 2)

$$\frac{dN}{dt} = \left(r - \frac{r}{K}N\right)N \qquad (2)$$

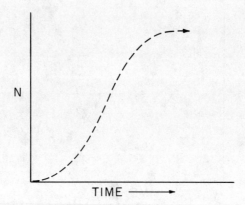

Figure 2. *Generalized population sigmoid growth curve based on equation (2).*

In this case, populations starting at low numbers but invading environmental niches with larger carrying capacities (K) initially grow at almost the exponential rates described by equation (1). When the population size gets larger, however, the population growth rate gets smaller and equilibrium exists when $N = K[(r - 1)/r]$. An interesting feature of this model is that, if r is very large, the population may overshoot its carrying capacity. The reaction to an ecological overload can then be a severe population crash, followed by an even larger overshoot (May and Oster, 1976). Populations in nature, however, do not seem to exhibit such pathological behavior (Mueller and Ayala, 1981) and therefore either regulate these parameters or possess mechanisms that induce self-damping behavior.

Among the mechanisms that may regulate population growth is the presence of pathogens or predators that benefit from large populations of a host or prey species. In general, the more contacts that are made between the antagonists, the greater the effect is felt on their population growth rates. As the pathogen or predator increases and attacks previously unattacked host or prey, the attacker population can grow larger — at least until the target population declines. An explicit expression of such an intuitively reasonable model is the Lotka–Volterra set of equations:

$$\frac{dN_a}{dt} = r_a N_a - \gamma_{ab} N_a N_b$$

$$\frac{dN_b}{dt} = - r_b N_b + \gamma_{ba} N_a N_b$$

(3)

where N_a is the host or prey population size, N_b is the pathogen or predator population size, and γ_{ab} is the effect that species b has on the survival and reproduction of species *a*. Population limitation is now dependent on the species' own self-regulation as much as on the population size of the alternate species. In both equations (2) and (3), genetic variations may exist in the r, K, and γ_{ab} parameters. Most reasonable biological concepts of these parameters indicate that there are physiological limitations to how widely reproductive, survival, and predation rates can vary. In particular, there are constraining relations among these activities, and genetic variation may simply reflect how alternate genotypes allocate energy among them.

A slightly more complicated but still linear model would include both self-regulation and interspecies effects. If we use general parameters, a type of competition model can be envisioned in which auto- and alloregulation may take on positive or negative signs. One mathematical model, commonly called the Gause competition set of equations, includes both damping effects:

$$\frac{dN_a}{dt} = \left(r_a - \frac{r_a}{K_a} N_a - \frac{r_a}{K_a} \alpha_{ab} N_b\right) N_a$$

$$\frac{dN_b}{dt} = \left(-r_b - \frac{r_b}{K_b} N_b - \frac{r_b}{K_a} \alpha_{ab} N_a\right) N_b$$

(4)

These equations could encompass cooperative effects among species if the signs were changed. Also, if more species interactions were important within a management unit, then each equation would have to be extended to include those effects, and equations for

$$\frac{dN_c}{dt}, \frac{dN_d}{dt}$$

would have to be added. Obviously, this gets more complicated, but the behavior of multiple-species systems can be analyzed. It is also obvious that these linear models of behavior do not perfectly reflect all the complexities of real organisms. There are self-regulatory effects that allow species to cooperate at low levels but then to compete during some parts of their lives and at certain densities. There may even exist a sequence of changes from cooperation to competition, to cooperation, etc. Nevertheless, within the ranges of population sizes that can be managed for these species, the simpler approximations may be adequate.

The main reason for introducing these ecological models is not to discuss ecology but to be able to discuss the relationship of genetic diversity to species survival and evolutionary capacity. If the simple ecology equation (1) applies, or any similar model applies with only a single parameter of genetic variation, then genetic diversity is maintainable only under certain conditions. Either the alternate alleles are favored in separate populations, or heterosis exists in one of the forms discussed in the previous section of this chapter.

However, if populations respond to density effects in a manner similar to that expressed by equation (2), then the results of management practices on genetic diversity are not as simply stated. In fact, heterosis at loci affecting r will not be sufficient to protect allelic polymorphisms in any populations. If there are two alleles (A and a) at a locus controlling K, then there are three possible genotypes (AA, Aa, and aa). To maintain this polymorphism this locus must exhibit overdominance so that $K_{AA} < K_{Aa} > K_{aa}$. The populations may in fact first change in gene frequency according to the dictates of the r parameters, then enter a phase where both r and K affect growth, but their eventual behavior is determined at large population sizes by K (Roughgarden, 1971). During this passage, there may even exist population sizes in which the heterozygote, with intermediate values of both r and K, can exhibit a transient state of fitness heterosis. However, since density continues to increase, the population size does not stay in the range that might cause us to observe heterosis. In these cases, management by limiting population sizes may actually protect polymorphisms by including a temporary but manageable heterosis. Thus, temporal sequences of site disturbances that actually reduce the population size of a species may have the effect of creating a more permanent heterosis. Managers of such species therefore have a means for allowing self-regulation to preserve genetic diversity without direct intervention at the genetic level. There is also a warning implicit in these results — that the observation of heterozygote superiority may not be a permanent phenomenon and hence that any observed heterosis cannot be relied upon to permanently maintain genetic diversity. At least, some continued eco-genetic monitoring is recommended.

If the species under management also responds to other species as modeled in equations (3) and (4), then the biological interactions become both cause and effect mechanisms to consider. Managing one species has effects that feed back through these other species. This may destabilize the target species as well as other species. It is at this level of ecological modeling complexity that the biotic diversity of the environment can be analyzed for its relationship to genetic diversity. It is also at this level that the analyses become very difficult, but the types of population and genetic behavior are more varied and offer more possibilities for regulating populations and maintaining diversity.

Consider the ecological models of species interactions in equations (3) and (4), or any of the more complex models, such as described by Goel et al. (1971). Even disregarding any genetic variations and limiting consideration to only two species, the numbers for two species that interact with each other can display stable behavior or oscillatory behavior, or even mixtures of these classes of behavior. For example, Lotka–Volterra equations (3), which describe predator–prey interactions, can be shown to induce an oscillatory behavior (Figure 3) with a certain value of γ. Where species a

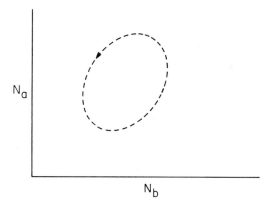

Figure 3. *Hypothetical interaction of two species (a, b) leading to in-phase oscillation of population size.*

and *b* are locked into a mutually regulated cycle such that, over time, they fluctuate in a manner that brings them slightly out of phase with each other (that is, the value of γ is different), the result is as shown in Figure 4.

However, when a self-regulating density-dependent effect is added and the model includes a positive or negative self-damping effect in addition to regulation by a competitor, the populations are not locked into a closed cycle but break into spirals of increasing or decreasing amplitude. In particular, in the Gause equations (4), the oscillatory behavior implied by closed cycles is not as much in evidence. Instead, depending on the sizes of r, K, and α_{ij} (effect of species i on species j), the populations may go toward

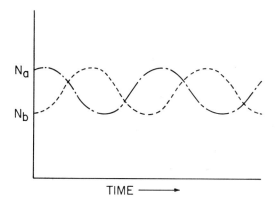

Figure 4. *Hypothetical interaction of two species (a, b) locked into a mutually regulated cycle in which oscillations in population size are slightly out of phase with each other.*

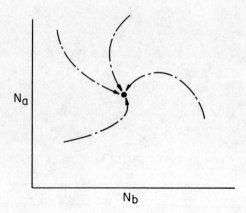

Figure 5. *Hypothetical interaction of two species (a, b) in which the sizes of the populations move toward a stable critical point.*

a stable critical point (Figure 5). Alternatively, they may be driven away from an unstable critical point and toward extinction of one or the other or both species (Figure 6). They may also be simultaneously attracted toward a critical point and repelled from it, a situation that is then called a saddle critical point (Figure 7).

The significant feature of these behaviors is that populations may be not only increasing or decreasing at any one time, but they may also be in the process of changing directions. The motions of the population changes in the different parts of the figure are what are important to the ultimate

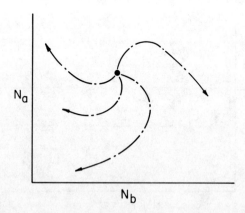

Figure 6. *Hypothetical interaction of two species (a, b) in which the sizes of the populations are deflected away from an unstable critical point and are driven toward extinction of population a or b or of both populations.*

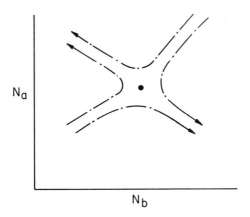

Figure 7. *Hypothetical interaction of two species (a, b) in which the sizes of the populations are simultaneously drawn toward a critical point and repelled from it, creating a saddle point.*

outcome. The manager's objective is satisfied either if a stable coexistence is created, or if each component is saved in at least one population, presumably in the management unit, under some degree of control.

It is possible that behavioral complexities such as switching from competitive to cooperative interactions occur with different relative population sizes of interacting species. The directions that populations take would then be very complicated but perhaps might reinforce the protection of species threatened by extinction.

The introduction of genetic variations to these systems adds interesting features. Simple predator–prey models can be shown to display the behavior of systems with a saddle critical point. Because of the possible increases and decreases in the two species, however, the gene frequencies can be shown to exhibit either a monotonic or a bidirectional behavior, depending on the initial population sizes and gene frequencies. That is, gene frequencies may also increase for a while as the populations achieve some balance, but then decrease during the local extinction of one of the species.

If we examine the kinds of gene effects that might vary within the species modeled in equation (4), which includes both an auto-and an alloregulator, we can surmise that the pure ecological model can be changed drastically. Not only could genetic polymorphisms stabilize otherwise unstable ecosystems, but they also might destabilize otherwise stable ecosystems. It is also true, however, that otherwise unstable genetic systems can be stabilized by ecological effects. Consider, for example, a heterozygous genotype with intermediate values for the r, K, and α parameters of equation (4). If it is intermediate in those values, then at low values of N_a and N_b the populations move in the direction dictated by the higher r. At high N_a and low N_b the growth of N_a is determined by K_a, and at low N_a and high N_b the growth of N_b is determined by K_b. In all these cases, heterozygote fitness is intermediate. However, there are combined values of N_a and N_b in which the heterozygote would have superior fitness, and, hence, at least a transient heterosis might exist with N_a and N_b in some balance. In contrast to the earlier case we examined with only self-regulation, it is now possible that N_a and N_b may be stabilized within the region that represents heterosis. Therefore, the stability of the genetic system may be dependent on the stability of the ecosystem, and vice versa.

While the details of these systems are very interesting to geneticists, the manager may not be as concerned with the details as much as with the fact that the systems are tied together and that one is not affected independently of the other. However, two important features emerge: (1) in complicated systems, multiple outcomes may result from independent trials; and (2) the present status of populations and genes is less important than their potential directions of evolution.

Management Strategies

In light of the intertwined effects that ecological and genetic variability have on species evolution, we might now wish to define our objective of preserving natural diversity in terms of preserving the dynamics of species evolution and its structures. The extent to which a species is subdivided and varies in its genetic content between component populations or maintains its diversity within a single, well-protected polymorphic system is part of the definition of the species' system. This structure is a feature of the natural diversity, which it might be necessary to preserve. Each species may have its variability structured in an unique way that reflects its mating pattern and reproductive biology, and the homeostatis contained within genetic loci, within individual genotypes, and within populations. To some extent, the genes themselves and what were once separate genes may have evolved into a single gene complex. Clearly also, individual genotypes have evolved a capacity to endure or escape environmental variations, and to the extent that they do, then genetic diversity is superfluous. But genetic variations do persist, and the demands of environments are often so wide that a subdivided species is more capable of survival than a single large species. The greater the environmental variation is relative to individual homeostasis, the greater is the advantage of structuring genetic variation within species.

At present there exists a heritage of species that reflect an array of forms in their present states, which may actually be transient states. They may be relict populations of some originally larger populations, or large epidemic populations of normally small endemics, or they may actually reflect a true steady-state environment to which they are well-adapted. Management of these species is required in some direct or indirect manner and, presumably, is conducted in an efficient way that uses the species' own dynamics to preserve their diversity and evolutionary capacity.

In this part of the book, the measures and meanings of diversity have been defined and the structures and dynamics of its maintenance have been described. It is now necessary to apply this understanding to management programs to preserve diversity with the goal of ensuring species survival and continued evolution. But management is required to stabilize species for a variable and uncertain future that may be even more variable than past environmental arrays have been. Increased human effects as well as natural events may well introduce greater variability and, by our restricting natural areas to less hospitable areas, may increase the extremes of adaptation required for species survival. If the goal is to ensure species survival and evolution, then the objective of preserving diversity may be too narrowly drawn. It might be more appropriate to seek to enhance diversity.

To execute any management strategy, of course, appropriate technologies must be chosen for the variety of problems faced. The techniques that are available to preserve or to enhance diversity have been reviewed in this book; they include intensive direct intervention, indirect reservation areas, and simple monitoring techniques.

The more intensive methods of direct intervention include features of both genetic and environmental choices of management. The environmental techniques include varying levels of selecting wilderness or natural areas up to constructing botanic gardens and zoological parks. The genetic techniques include founding new populations, introducing new genotypes into the breeding pool (Chapters 4 and 23), and the more heroic techniques of breeding for adaptation to particularly small population sizes. Because some degree of coordinated environmental and genetic management often is assumed when genes or populations are introduced, it is often assumed also that the initial gene frequency array will remain the same, and that these allele frequencies and population sizes are independent. Clearly, neither presumption is true, and if intensive management is to be ultimately useful, the initial conditions of allele frequencies and population sizes might require multiple starts, controlling individual populations and migration, and continued monitoring.

The less intensive methods of indirect intervention include primarily environmental management, which would affect genetic diversity and structure by the initial choice of areas and their management. There is no clear distinction between direct and indirect methods because one may incidentally introduce founding populations in carrying out a land or water management plan. Further, many direct interventions should be designed to provide initial populations that would then be self-regulating or be manageable through indirect environmental manipulations.

The initial choice in management design must be made between single large areas as opposed to multiple small areas. The debate on the point is quite lively (e.g., Simberloff and Abele, 1982), and at the species level the definitive answer has still not emerged. In terms of genetic diversity below the species level, there is clearly a need—for some species—to maintain a subdivided diversity, but not all species would benefit. In fact, for many species, an ideal structure is simply not known. Therefore, multiple populations should be chosen for the design, with effective population size considered as one of the variables along with physical and other biotic variables that should differ among management units. Because it is recognized that the target species' own population size as well as those of competitors influence genetic diversity, population and area sizes must be among the variables that are important in management design.

Within the large and small units, the manager may also choose to maintain a steady state or to design a sequence of temporal variations. In the multiple unit design, temporal variations may be sequenced out of phase among the units as would occur normally in harvesting or other population reduction schemes or in phasing disturbances such as controlled fires. If the manager is limited to a single small population, then the only choice is to maintain the natural state, which could be a steady state or successional.

In all of these programs, monitoring the genotypic arrays and tracking the changes in a large sample of genetic loci is essential. The averages of these statistics are important, but the structure of diversity is equally important, and all must be observed as parts of a semi-managed evolution. In large single populations, if the effective population size actually is large, then changes will occur slowly and alleles will not be easily lost, even if cryptic structures and inbreeding exists. When populations are small or if they locally respond rapidly to selection, then it is more critical to ensure that not all local units are losing the same alleles. In single small populations, of course, the tracking is critical and to maintain maximum diversity requires high levels of heterozygosity.

With this array of management methods, the problem is to develop priorities or decision rules for choosing methods when multiple species and areas require management. Since available management funds and efforts are always less than desirable, it might be appropriate first to use a triage rule: don't expend effort for those that can survive without our effort; don't expend effort on those that cannot survive no matter what is tried; and consider only those where management efforts can make a difference. Unfortunately, this still leaves too many species and areas to manage. Also, more sophisticated attitudes, knowledge, and techniques are consistently being developed to save species, so triage criteria for a particular situation can rapidly be outmoded. I suggest that the less intensive methods of management will have to be used for most species, and that for these species a structure of multiple populations is the safest strategy. Indirect management is far cheaper and easier to execute, but it does require research and monitoring. The many alternative designs cannot be elaborated on here, but population size ought to be included among the variables. If a species is limited to one population, then, clearly, maximum genetic diversity must be concentrated within it. If a species is limited to a few populations, then instead of preserving the levels of genetic diversity that happen to be found in the initial state, a better objective may be to increase diversity in those units over which management control exists. Thus, encouraging or simply allowing genetic drift and divergence to occur may be a useful approach in managing for genetic diversity.

There are, of course, many taxa in imminent danger of extinction. I suggest that in the same class as those situations that are clearly a crisis, where a species can be saved only with highly intensive efforts, are species with single populations that generally are not large enough to ensure their survival. Establishing second, third, or multiple populations of some species may be as important to their survival as increasing the population size of a species that is adapted to steady-state conditions. Thus, in choosing those few species on which intensive management will be practiced, consideration should be given to the maximization of diversity measured not only at the species level, but also at the intraspecies level.

The ultimate goal of intensive management of a species ought to be the evolution of the species to require less intensive management and to be buffered sufficiently to evolve through future ecological variations. This may require enhancing diversity on occasion, preserving diversity on another, and sometimes losing nonessential diversity. Errors will be made in losing useful diversity and in saving nonessential diversity. In the present infancy of genetic management, errors causing loss are more likely to be made.

Summary

We are beginning to define useful measures of diversity in eco-genetic terms, and we are finding that genetic diversity within species exists in varied patterns. Since we are also losing some of this diversity, management interventions to preserve natural diversity are required. In using natural areas or relict populations in zoological or botanical gardens, a problem exists in managing a stable system for all the allelic and species components.

In an example of a one-locus, two-allele system in two species, it is seen that an ecologically stable equilibrium can be maintained only if genetic variation is eliminated. However, this system is unstable in the sense that the introduction of genetic variants of either species can force the local extinction of one of the species. Since the system's behavior can change with environmentally caused selective changes, any single ecosystem would behave erratically. One way to achieve a meta-stability would be to manage multiple ecosystems, in a planned array of environments and populations of variable compositions and sizes. This type of management may be more manipulative than what is presently conceived as "laissez-faire" for "natural" ecosystems.

CHAPTER
20

The Distribution of Genetic Variation within and among Natural Plant Populations

James L. Hamrick

Introduction

Studies of genetic variation in natural populations usually involve two basic questions. The *first* is concerned with describing levels of genetic variation maintained within populations or species. To approach this question, population geneticists have turned to biochemical techniques, primarily electrophoresis, to obtain estimates of genetic variation in plant and animal populations. These estimates are usually quantified in terms of the number of loci polymorphic per population, the effective number of alleles per locus, or the mean number of loci heterozygous per individual. Recent reviews of the electrophoretic literature (Brown, 1979; Gottlieb, 1981; Hamrick et al., 1979; Nevo, 1978) have demonstrated that allozyme genetic variation differs among taxonomic groups (for example, vertebrates versus invertebrates: Nevo, 1978) and among species within taxonomic groups. These reviews have further demonstrated that allozyme variation within populations is associated with the life history traits of species—for example, geographic range, mating system, lifetime fecundity (Hamrick et al., 1979; Nevo, 1978).

The *second* question is of particular importance to the conservation of genetic resources, since it is concerned with the way in which genetic variation is partitioned within and among populations. In other words, does the majority of a species' genetic variation reside within populations, or does most of it occur among populations? To intelligently manage the genetic resources of naturally occurring species we must understand how genetic variation is distributed and what characteristics of the environment or of the species influence its distribution.

In plant species, characteristics that should influence the distribution of genetic variation include the effective population size, the geographic distribution of the species, the primary mode of reproduction, the mating system, the seed dispersal mechanism, and the community type in which the species most commonly occurs. Four of these factors (population size, reproductive mode, mating system, and seed dispersal) directly influence the effective size of a population (Wright, 1951). Species with small populations, vegetative reproduction, and limited pollen and seed dispersal would be expected to have relatively little variation within populations and relatively more variation among populations. Similarly, species with limited distributions or colonizing species might also be expected to have relatively high levels of interpopulation variation.

Unfortunately, there are few quantitative estimates of the effective population size of plant species (Beattie and Culver, 1979; Levin and Kerster, 1968). There are even fewer studies that have combined analyses of the genetic structure of populations with estimates of effective population size (Schaal and Levin, 1978). Thus, for the most part we are forced to base estimates of effective population sizes on what is known of the reproductive biology of the species.

In this chapter, the plant allozyme literature is summarized to obtain estimates of variation within and among populations. These estimates are used to determine whether relationships exist between the distribution of allozyme variation and certain life history traits of the species.

Procedures

Data from 122 studies representing some 91 species of seed plants in 47 genera and 19 families were used in this analysis (see Appendix 4). Each species was classified for its geographic range, mode of reproduction, mating system, seed dispersal mechanism, and the stage of succession in which it commonly occurs (Table 1).

Allele frequencies were used to calculate Nei's (1973) gene diversity

Table 1. *The Relationship between Five Life History Traits and the Distribution of Allozyme Variation within and among Natural Plant Populations*[1]

Characteristic	Number of studies	H_T	H_S	G_{ST}	A_P	P_A
Geographic range						
Endemic	10	.275	.208	.200	3.26	.639
Narrow	31	.261	.177	.275	2.96	.609
Regional	38	.238	.154	.312	3.23	.573
Widespread	43	.380	.293	.253	3.70	.702
Mode of reproduction						
Asexual	1	.172	.159	.080	3.29	.652
Sexual	108	.280	.194	.284	3.17	.620
Both	13	.325	.257	.209	3.30	.732
Mating system						
Selfed	35	.250	.141	.437	3.02	.559
Mixed—animal	26	.284	.181	.304	2.68	.644
Mixed—wind	3	.712	.560	.189	16.03	.425
Outcrossed—animal	23	.352	.238	.221	3.37	.633
Outcrossed—wind	35	.256	.248	.056	3.29	.737
Seed dispersal mechanism						
Large	22	.260	.181	.305	3.07	.635
Animal-attached	13	.262	.155	.425	3.77	.504
Small	37	.312	.196	.349	2.81	.846
Winged or plumose	38	.260	.238	.073	3.28	.706
Animal-ingested	12	.344	.210	.330	3.63	.526
Stage of succession						
Weedy and early	52	.295	.172	.394	3.37	.574
Middle	42	.262	.196	.236	3.15	.613
Late	28	.299	.275	.071	3.14	.786

[1] H_T = total allelic diversity; H_S = mean allelic diversity within populations; G_{ST} = the ratio of the allelic diversity among populations to the total allelic diversity; A_P = mean number of alleles per polymorphic locus; P_A = the proportion of the total number of alleles found within each population. See text for a further description of these statistics, and Appendix 4 for details on taxa covered.

statistics. For each polymorphic locus the total allelic diversity (H_T) for the species was calculated as follows:

$$H_T = 1 - \Sigma \bar{p}_i^2$$

where $\bar{p}_i$ is the mean frequency of the ith allele at a locus. The total allelic diversity (H_T) can be partitioned into the allelic diversity within populations (H_S) and the allelic diversity among populations (D_{ST}). Thus,

$$H_T = H_S + D_{ST}$$

In practice H_S is the weighted (by the population sample size) mean of the allelic diversities of each population. The proportion of allelic diversity due to the among-population component (G_{ST}) is the ratio D_{ST}/H_T. Mean values of H_T, H_S, and G_{ST} for each species were obtained by summing over all polymorphic loci.

Also of interest are the number of alleles maintained at each polymorphic locus (A_P). To obtain this value for each species, the total number of alleles observed were divided by the number of polymorphic loci. The proportion of alleles maintained within each population (P_A) was determined by summing the number of alleles present in each population and dividing by the total number of alleles observed within the species.

Weighted mean values of H_T, H_S, G_{ST}, A_P, and P_A were obtained for each category of the five life history traits. Weightings were the product of the number of populations and the number of loci.

Results and Discussion

Previous reviews (Brown, 1979; Gottlieb, 1981; Hamrick et al., 1979) of the plant allozyme literature demonstrated that plants maintain high levels of genetic variation within their populations. Approximately 37% of all loci are polymorphic within a population, and the proportion of loci heterozygous per individual averaged approximately 14%. In addition, Hamrick et al. (1979) demonstrated that plant species that have wide ranges, long generation times, wind pollination, outcrossed mating systems, and high lifetime fecundities and that occur in the later stages of succession tend to maintain higher levels of variation than do species with other combinations of these traits. The present study uses a similar data base to examine the influence of selected life history traits on the distribution of variation within and among populations.

1. Geographic Range

Widespread species, many of which are weeds, appear to maintain the highest levels of allozyme variation (H_T) at polymorphic loci (Table 1). Surprisingly, endemic species maintained the second highest levels of variation. Regionally distributed species appear to have somewhat less variation. This pattern is maintained for variation within populations (H_S), for the number of alleles per polymorphic locus (A_P), and for the proportion of total alleles observed within populations (P_A). Thus, the indication is that endemic and widespread species are more variable at polymorphic loci. The higher levels of heterozygosity for narrow and regionally distributed species reported by Hamrick et al. (1979) resulted from higher proportions of polymorphic loci per population, not from greater heterozygosity at polymorphic loci.

The G_{ST} values give a similar but opposite pattern; regional and narrowly distributed species have somewhat more genetic variation among their populations than widespread and endemic species. This may be due to the wider variety of environments that these first two groups of species face relative to endemic species. The somewhat lower genetic differentiation among populations of the widespread species may be explained by the presence of many weedy species within this group. Such plants might be expected to encounter relatively less site-to-site variation throughout their geographic ranges.

Direct comparisons of endemic and more widely ranging species belonging to the same genus are uncommon. For *Lupinus*, lupines (Babbell and Selander, 1974), and *Lycopersicon*, the tomato and its relatives (Rick and Fobes, 1975), the species with the more limited distributions have much larger G_{ST} values. This is not the case for *Hymenopappus* (Babbell and Selander, 1974), *Gaura* (Gottlieb and Pilz, 1976), and *Clarkia* (Gottlieb, 1973, 1974), since G_{ST} values are nearly identical for species with quite different ranges. Therefore, although the results indicate that more populations of narrow and regionally distributed species will be needed to maintain a given proportion of the species' variation, recommendations based on geographic range alone will not be dependable.

2. Mode of Reproduction

Most of the species that have been analyzed electrophoretically reproduce by sexual means. Only a single species was studied that reproduces completely asexually, and thirteen studies examined species that reproduce both sexually and asexually. As a result, comparisons among these groups may be of limited value. Nevertheless, sexually reproducing species appear to maintain somewhat less variation within species and populations but have more genetic differentiation among populations (Table 1). This may be explained in part by the observation that plants with both sexual and asexual reproduction are predominantly outcrossed and that none are self-pollinated (see below and Appendix 3, Figure 2).

In the only study that directly compares the distribution of genetic variation between asexual and sexual races, Usberti and Jain (1978) found that sexual populations of the grass *Panicum maximum* have higher within-population variation ($H_S = 0.381$) and lower between-population variation ($G_{ST} = 0.035$) than asexually reproducing populations ($H_S = 0.159$; $G_{ST} = 0.080$). Furthermore, 98.5% of the alleles within the species are found within sexual populations, whereas only 65.7% of the alleles are found within asexual populations.

3. Mating Systems

Plant mating systems have a marked effect on the level and distribution of genetic variation (Table 1). Animal-pollinated, outcrossed, and mixed-mated species have the highest levels of variation within species, whereas selfing species have the lowest H_T values. This pattern changes for variation within populations, since wind-pollinated outcrossing species have somewhat more within-population variation than animal-pollinated plants. Populations of selfing species are the least variable. The most striking influence of the mating system is seen, however, in the way that genetic variation is partitioned among populations. Selfing species have nearly 44% of their variation among populations, compared with only 6% for wind-pollinated outcrossing species. Predominantly outcrossed animal-pollinated species also have lower G_{ST} values than the mixed-mating animal-pollinated species. This relationship is mirrored in the proportion of alleles that are maintained within populations.

The results of this analysis are consistent with earlier reviews (Hamrick et al., 1979; Brown, 1979; Gottlieb, 1981) in indicating that the selfing mode of reproduction limits the movement of alleles from one population to another and as a consequence increases genetic differentiation among populations.

Direct comparisons of congeneric species that differ in their mating systems support the results in Table 1. In *Phlox* (Levin, 1975b, 1978), *Gilia* (Schoen, 1982a), and *Oenothera*, evening-primrose (Ellstrand and Levin 1980, 1982), selfing taxa have larger G_{ST} values than their more outcrossed congeners. Therefore, species with lower potential gene movement via pollen have more genetic differentiation among populations. This indicates that more populations of selfing species would need to be maintained to ensure the preservation of a predetermined level of genetic variation.

4. Seed Dispersal Mechanism

There are no striking differences among the five groups for H_T or H_S. Animal-ingested and small-seeded species have somewhat higher levels of variation at the species level, while winged-seeded and animal-ingested species maintain somewhat more variation within their populations. This result is generally consistent with the earlier results of Hamrick et al. (1979).

There are large differences among G_{ST} values. Species with winged or plumose seeds have little (7%) genetic differentiation, while animal-attached species have nearly 42% of their variation among populations. Large-seeded, small-seeded, and animal-ingested species also have rather high G_{ST} values. In addition, small-seeded plants and those with winged fruits maintain a significantly greater proportion of the total alleles within populations than do either animal-attached or animal-ingested species.

The occurrence of lower G_{ST} values and higher P_A values for winged- or plumose-seeded species is consistent with the expectation that these plants have relatively high rates of gene flow between populations. Furthermore, while the animal-attached mechanism of dispersal may be effective in the long-distance transport of seeds to new sites, it is doubtful that it is dependable enough to affect the distribution of genetic variation among populations. However, these results may be confounded by the association of certain mating systems with certain seed dispersal mechanisms. Most of the selfing plants included in this study have either animal-attached or small seeds, while the majority of the winged-seeded species are wind-pollinated conifers (Pinaceae). In this regard, it is interesting to observe that four species of the selfing but plumose-seeded genus *Tragopogon*, goat's-beard (Roose and Gottlieb, 1976), have G_{ST} values that average 0.254, much below the mean for selfed plants but much above that for other winged- or plumose-dispersed species.

5. Stage of Succession

The successional stage in which a species occurs appears to have an effect on the amount of genetic variation found within populations but not within species. Species characteristic of the later stages of succession have more variation within their populations (Table 1). This observation is consistent with the earlier findings of Hamrick et al. (1979) and indicates that higher levels of variation in these species are due to more equable allele frequencies. An illustrative example occurs in the conifers. Early successional species such as *Pinus contorta*, lodgepole pine (H_s = 0.160: Yeh and Layton, 1979) and *Pinus rigida*, pitch pine (H_s = 0.147: Guries and Ledig, 1982) have much lower within-population variation than species such as *Pinus ponderosa*, ponderosa pine (H_s = 0.284: Hamrick, unpublished), *Pinus longaeva*, Great Basin bristlecone pine (H_s = 0.465: Hiebert and Hamrick, 1983), or *Picea abies*, Norway spruce (H_s = 0.420: Bergmann, 1975b).

There is less population differentiation and a higher percentage of alleles are maintained in populations of species characteristic of the later successional stages. The result may be influenced by the high proportion of wind-pollinated outcrossing conifers represented in this group (21 of the 28 species). Nevertheless, the remaining species have a considerably lower mean G_{ST} value (0.135) than the other successional classes. It should be noted, however, that none of these eight species are self-pollinated.

6. Population Structure

Plant populations typically consist of individuals that are distributed in a clumped fashion. Furthermore, detailed studies of plant populations usually

Table 2. *Detailed Allelic Diversity Analyses of Four Plant Species with Contrasting Life Forms*[1]

Species	Life form	Population analyses[1]			Subpopulation analysis			Sources
		H_t	H_s	G_{st}	H'_s	H_{sp}	G_{sp}	
Avena barbata (slender wild oat)	Selfing annual	.289	.074	.736	.435	.280	.357	Clegg and Allard, 1972; Hamrick and Holden, 1979
Chrysanthemum leucanthemum (oxeye daisy)	Short-lived outcrossing perennial	.385	.367	.045	.470	.439	.050	Griswold, unpublished data
Liatris cylindracea (cylindric blazing-star)	Long-lived obligately outcrossing perennial	—	—	—	.328	.305	.069	Schaal, 1975
Pinus ponderosa (ponderosa pine)	Outcrossing tree	.289	.285	.015	.285	.272	.043	Hamrick, unpublished data
Pinus longaeva (Great Basin bristlecone pine)	Extremely long-lived outcrossing tree	.484	.465	.038	.478	.467	.022	Hiebert and Hamrick, 1983

[1]H_t = total allelic diversity; H_s = mean allelic diversity within populations; G_{st} = the proportion of the allelic diversity due to the among-population component; H'_s = mean allelic diversity within populations undergoing detailed subpopulation analysis; H_{sp} = mean allelic diversity among population subdivisions; G_{sp} = the proportion of the allelic diversity due to the among-subpopulations component.

demonstrate that alleles and genotypes are also distributed in a nonrandom fashion. Such genetic structure may be the result of microhabitat adaptation to physical environmental factors (e.g., *Avena barbata*, slender wild oat: Allard et al., 1972; Hamrick and Holden, 1979; *Anthoxanthum odoratum*, sweet vernal grass: Snaydon and Davies, 1972; Liu and Godt, Chapter 5), biotic factors (e.g., *Trifolium repens*, white clover: Turkington and Harper, 1979), vegetative reproduction (e.g., *Spartina patens*, salt meadow cord grass: Silander, 1979), or family structure (e.g., *Pinus ponderosa:* Linhart et al., 1981c; *Helianthus annuus*, common sunflower: Ellstrand et al., 1978).

The genetic structure of populations should also be affected by those factors that influence the distribution of genetic variation among populations. Brown (1979), for instance, has shown that plant mating systems affect the degree of microgeographic differentiation. Table 2 illustrates the degree of microgeographic variation found within populations of five plant species with widely divergent life forms. While the design of each study differs, it is clear that intrapopulation differentiation in *Avena barbata*, a selfing annual, is much greater than that in the wind-pollinated conifers or the outcrossing perennials. The obligately outcrossing long-lived perennial, *Liatris cylindracea*, cylindric blazing-star, is common to undisturbed prairies. Schaal's (1975) analysis of a single population indicates that while *L. cylindracea* has significant heterogeneity among population subdivisions most of its variation resides within the subdivision. *Chrysanthemum leucanthemum*, oxeye daisy, a species found in highly disturbed as well as undisturbed grasslands, provides a similar picture. The analysis of Griswold (unpublished) indicates that this predominately outcrossing animal-pollinated species has significant differentiation at both the geographic and the microgeographic levels but that much of the variation resides within population subdivisions. Thus, these more detailed studies are generally consistent with those of Table 1.

The existence of population subdivision, regardless of its source, makes it important for the resource manager to preserve as much of a natural population as is possible. The loss of some portion of a population may be as critical in a genetic sense as the loss of an entire population.

7. Allozyme versus Polygenic Traits

The use of electrophoresis to describe the distribution of genetic variation has a number of advantages: (a) genetic inheritance of electrophoretically detectable traits can be easily demonstrated; (b) most allozyme loci are codominant and allele frequencies can be calculated without the necessity of genetic crosses; (c) estimates of genetic variation can be compared directly between populations or species (Brown and Moran, 1981; Hamrick et al., 1981). The major disadvantage of electrophoretic traits is that the adaptive significance of allozyme variation remains obscure (Lewontin,

1974). We may ask, therefore, whether the patterns of allele frequencies and/or the distribution of allozyme variation bear any relationship to other genetically determined traits, especially those morphological and physiological traits that adapt individuals to their physical and biotic environments. If allozyme variation is not representative of other types of genetic variation, recommendations for genetic conservation based solely on allozyme data may not be trustworthy (Chapter 14 by Powell).

Unfortunately there are few studies that have examined the relationship between allozyme and polygenic variation. Those studies that are available produce mixed results. In *Avena barbata* and *A. fatua*, common wild oat, there is a close association between allozyme variation, single-gene morphological traits, and quantitatively inherited traits (Jain and Marshall, 1967; Marshall and Allard, 1970a). There is a similar correspondence between morphological and allozyme variation in *Liatris cylindracea* (Schaal and Levin, 1978). In other species, allozyme traits may not have interpretable patterns, while variation in morphometric traits is closely associated with environmental gradients. Forest trees consistently have high levels of genetically controlled morphometric variation associated with environmental variation (Libby et al., 1969). Yet, only 5% of the allozyme variation occurs among their populations. A specific example may be *Pinus ponderosa* on the Front Range of the Rocky Mountains in Colorado (Hamrick, unpublished; Mitton et al., 1980). Finally, Brown et al. (1978) have demonstrated that the distribution of variation within and among populations of *Hordeum spontaneum*, wild barley, is essentially reversed for allozyme and morphological traits.

To determine whether the distribution of quantitative variation is influenced by the life history traits considered above, variance components due to between-population genetic differences (σ_p^2) were related to estimates of the total genetic variation (V_G); see Table 3. Each species was grown under uniform garden or greenhouse conditions. The studies of *Abies concolor*, *Achillea lanulosa*, and *Mimulus guttatus* were based on collections from throughout the native range of the species. The *Avena barbata* populations were located within a radius of five miles in the Napa Valley of California. *Avena barbata*, slender wild oat, is a selfing annual species, while *Mimulus guttatus*, common monkeyflower, reproduces by vegetative and sexual means and is partially self-pollinated. *Achillea lanulosa*, western yarrow, is a widespread, predominantly outcrossing species, which also reproduces by rhizomes. *Abies concolor*, white fir, is a wind-pollinated tree.

For the most part, the quantitative traits show as much or more differentiation among populations than the allozyme traits show (Table 3). The mean G_{ST} value for conifer trees is approximately 0.05, whereas the ratio σ_p^2/V_G is between 0.20 and 0.35 for *Abies concolor*. The average ratio for the quantitative traits for *Achillea lanulosa* is nearly identical to the G_{ST} value for plants with similar mating systems. The ratio for *Mimulus gut-*

tatus is greater than the mean G_{ST} value for animal-pollinated species with mixed-mating systems. However, a more direct comparison can be made for this species, since McClure (1973) has studied its allozyme variation. The G_{ST} value for 11 populations located in a single watershed was 0.213, somewhat lower than the ratio for quantitative traits. This is not surprising, since the allozyme study was more restricted geographically than the morphometric study. The G_{ST} value from a rangewide study of *Avena barbata* in California was 0.736 (Clegg and Allard, 1972), which is considerably higher than that found for quantitative traits. However, an allozyme study restricted to a single environmental gradient in the Napa Valley produced a G_{ST} value (0.357) quite similar to the quantitative ratio (Hamrick and Holden, 1979).

The quantitative genetic ratios varied greatly from trait to trait. This indicates that some morphological characteristics vary greatly among populations, while others do not. There is some evidence that those traits with the largest differences are those that adapt the species to local environmental conditions (Hamrick, 1976).

We can conclude that there is some association between the distribution of allozyme and quantitative traits and that those factors that influence the distribution of allozyme variation also have some influence on the distribution of quantitative variation. However, other evolutionary factors such as selection prevent this association from being close.

Table 3. *Mean Ratios of the Between-Population Variance (σ_p^2) to the Total Genetic Variance (V_G) of Quantitative Traits for Four Plant Species Representing Different Life Forms*

Species	Number of traits	σ_p^2/V_G Mean	σ_p^2/V_G Range	Sources
Abies concolor (white fir)				
Elevational transect	13	.203	.000–.498	Hamrick, 1976
Rangewide	13	.346	.243–.661	Hamrick and Libby, 1972
Achillea lanulosa (western yarrow)				
Rangewide	24	.224	.025–.415	Hamrick, 1970
Mimulus guttatus (common monkeyflower)				
Rangewide	29	.468	.164–.965	Hamrick, 1970
Avena barbata (slender wild oat)				
Napa Valley	5	.385	.043–.666	Hamrick and Allard, 1975

Conclusions

The present results indicate that a species' mating system and its seed dispersal mechanism directly affect the distribution of genetic variation within and among its populations. Those plants with the greatest potential for gene movement, wind-pollinated species with winged or plumose seeds, have relatively little genetic differentiation among their populations, while selfing plants have much larger differences. In developing strategies to conserve genetic resources we must take the characteristics of individual species into account (Chapter 13 by Clegg and Brown).

Nei (1975) has shown that the number of populations needed to maintain a given level of H_T is a function of the rate of gene flow and the effective size of each population. This is consistent with the present results, since fewer populations of species with high rates of gene flow would be needed to preserve a given level of variation. Superficially it would appear that 99% of the variation found within outcrossing wind-pollinated species could be obtained by maintaining only 3 populations. For selfing species 99% of the total variation would be present within approximately 8 populations. To preserve 99% of all alleles detectable by standard electrophoresis, 4 populations of outcrossing species and 6 populations of selfing species would need to be preserved.

These conclusions are overly simplistic and subject to a number of qualifications. First, the typical electrophoretic study samples only a limited portion of a species' range. The mean number of populations sampled in the studies reviewed here was 12, and the mode was approximately 5. Thus, although the weightings used to calculate mean values of G_{ST} may help to rectify this problem, allelic diversity among populations may be seriously underestimated. Also standard electrophoretic techniques may not distinguish among alleles at a locus. Just how many cryptic alleles occur at loci is unknown, as is their distribution among populations. However, Shumaker et al. (1982) found the number and frequency of cryptic alleles to be low in three grass species. If this proves to be the general case for plants, the presence of cryptic alleles at polymorphic loci will have little impact on estimates of overall levels of genetic variation. Finally, one must also be careful in lumping species into categories. There is considerable variation within any one classification. Values for G_{ST} range from 0.001 to 0.258 for outcrossing wind-pollinated species and from 0.028 to 0.951 for selfing species.

A second serious problem is the inconsistent associations between quantitative and allozyme traits. One explanation for this result is the possibility that the two types of traits are sampling different sets of loci (Brown et al., 1978). Also the possibility exists that the substitution of one allele for another has different effects at the phenotypic level. In calculating allelic diversity, the substitution of one allele is equivalent to the substitution of any other, and there is no measure of the degree of genetic or phenotypic difference between the two alleles (Brown et al., 1978). It is those alleles that produce the greatest effect on the phenotype, which may be most essential to the long-term success of the species. All our evidence indicates that plants are highly differentiated morphologically and physiologically among geographic regions and locally heterogeneous environments. Any strategy for the preservation of genetic variation must take such heterogeneity into account by maximizing the conservation of locally adapted alleles (Brown and Moran, 1981).

A final concern is the preservation of alleles that are rare in present-day populations but may at some later time prove to be essential to the long-term survival of the species. This may be especially true of alleles that affect disease or pest resistance. An excellent example is the resistance of *Pinus lambertiana*, sugar pine, to white pine blister-rust. A rare dominant allele was discovered (Kinloch and Littlefield, 1977) that produces resistance to this lethal disease. Under many strategies of gene conservation this allele would have been lost.

In conclusion, it probably bears repeating that plant species are highly heterogeneous both within and among populations. Electrophoretic techniques, while extremely valuable as a means of evaluating genetic resources, probably are not adequate to describe all the genetic variation that occurs in plant species. To develop adequate strategies of gene conservation we need to combine electrophoretic, morphometric, and physiological studies that describe genetic variation within and among natural plant populations.

Summary

Nei's (1973) gene diversity statistics were applied to data from 122 electrophoretic studies to determine the distribution of allozyme variation within and among natural plant populations. The influence of five life history traits on the distribution of genetic variation was examined. Species that have regional distributions, sexual modes of reproduction, self-pollination, and animal-dispersed seeds and that are characteristic of the early stages of succession had more allelic diversity among their populations than did species with other combinations of these traits. In particular, outcrossed, wind-pollinated species of the later stages of succession with wind-dispersed seeds have the lowest amounts of allelic diversity among their populations.

Comparisons were also made between the distribution of quantitative genetic variation and allozyme data. It was concluded that generally greater differentiation occurs among populations for quantitative traits than for most allozyme loci.

The results of this review are discussed in light of the development of strategies for the conservation of genetic resources. It is suggested that the life history characteristics of a species must be well understood prior to the development of a management strategy. Furthermore, although electrophoretic studies provide the most efficient means of quantifying the distribution of genetic variation, they may not predict the variation patterns of other genetically determined traits. Allozyme studies done in conjunction with studies of the distribution of quantitative genetic variation should allow the development of successful management strategies.

Acknowledgments

I wish to thank K. J. Hamrick and J. R. Hamrick for assistance during the development of this chapter. G. B. Griswold helped with the data analyses and graciously shared her preliminary data. Discussions with R. Holt helped to formulate some of the conclusions.

CHAPTER
21

Systematics, Conservation, and the Measurement of Genetic Diversity

Steven M. Chambers
Jonathan W. Bayless

Introduction

The depictions of identifiable species of animals in prehistoric artifacts are evidence that a sensitive awareness of the diversity of life existed early in human history. Collections made by the early European explorers of the New World, Africa, and Asia revealed to European naturalists a diversity of organisms that was far greater than that of the relatively depauperate European flora and fauna. Linnaeus developed his system of naming and hierarchical classification of organisms as a means of organizing this enormous diversity. The fundamental features of this system remain the basis of biological classification to this day.

Organisms are classified according to their similarities or, more recently, their evolutionary or genetic relationships to one another. Overall morphological similarity or difference between taxa, as judged by a "competent" expert, has traditionally been the main criterion for placing organisms in different taxa. Several molecular and statistical procedures have recently been applied to systematic studies to provide more objective measures of the differences between, and therefore the diversity among, individuals and taxa.

In order to conserve natural diversity a manager must understand the distribution of diversity and the taxonomic status of individuals and populations. Conservation goals may not be met if a managed population has been misclassified or if the classification itself does not coincide with the distribution of genetic diversity (Laerm et al., 1982). The recognized classification may have been based on typological or other outdated concepts or on characters that are largely affected by age or environmental effects on the phenotype. The existence of sibling species in so many groups of plants and animals (Chapter 6 by Bickham) is another factor that should make managers extremely sensitive to the need for careful taxonomic evaluation of the populations under their care.

In this chapter methods of measuring diversity and differentiation will be briefly described and compared. The application of these methods to taxonomy will be discussed, with special attention being given to the problems of characterizing and conserving subspecies.

Measuring Elements of Diversity

The diversity of any group, whether it is a local population or a multispecies community, cannot be described unless the differences between individuals or species can be detected and measured. If little difference can be detected between individuals in a local population, then little diversity has been detected among those individuals and the population has low diversity. If differences between individuals are great, then there is great diversity among the individuals and the population can be described as containing significant diversity. The amount of diversity in the population can be quantified in terms of some measure of the average difference between individuals.

Biological diversity can be measured at any level from overall global diversity down to the diversity of genes within a single individual. Criteria for selecting communities that deserve special conservation efforts may range from the existence of a particular desirable species in the community to simply the total number of different species in the community. Ecologists generally do not consider the number of species in a community an adequate measure of diversity, so they often employ a measure of diversity, such as the Shannon information index, that takes into account not only the number of species in a community, but also their relative abundances (Odum, 1971, Chapter 6).

The object of conserving diversity is to maintain as many different kinds of organisms as possible at population sizes that allow them to continue to evolve. By "kinds" of organism, one usually means species. Conservation of species should whenever possible be accomplished by preserving entire ecosystems. If, however, a particular ecosystem is disturbed, there may be cause to be concerned about the ability of a particular species to survive without some sort of intervention. A species may also deserve particular attention for commercial or esthetic reasons or because, as a "keystone" species (Gilbert, 1980), many other species depend upon it.

This book is most directly concerned with diversity at the level of species and below. Since nearly all individuals of a sexually reproducing species are genetically different, it might seem that all such species are infinitely diverse. Individuals, however, can be grouped into populations of interbreeding individuals. Genetic diversity can be described in any grouping of individuals, based on the probability that component individuals or populations differ at particular loci or in some other defined amount of genetic material.

Several methods are available for measuring or estimating the genetic differences between, and therefore the diversity among, individuals and various groupings of individuals. Since the focus of this book is at and below the species level, this discussion will address the particular relevance of each method to species and subspecies. This is not an altogether arbitrary emphasis, since species may be viewed as genetically integrated groups, gene pools, or units of adaptation (Mayr, 1963), although in many cases it may be more appropriate to consider local populations as the units of adaptation (Ehrlich and Raven, 1969).

It will be seen that there is no particular method that unambiguously describes the "overall" genetic diversity of an individual, population, or species. This is at least partially because the genetic material itself is functionally heterogeneous. Each method described measures a somewhat different portion or expression of the genome. Even so, at least some concordance may be expected between data generated by the different methods. The method of choice may depend on the particular question being asked, the taxonomic distance between samples one may wish to compare, or, simply, the amount of money available.

1. Nucleic Acids

Evolutionary systematists have sought the means to measure the overall genetic similarity or differences between individuals and populations. The ultimate reductionist application of this approach is the comparison of chemical differences in the genetic material itself. This can be performed by determining the base sequences of the DNA that codes for individual gene loci. These procedures are still extremely costly in terms of both time and money and cannot be considered as a practical tool for population studies at this time.

Enzymes called restriction endonucleases can be used to construct a "map" of the sites at which particular enzymes cut a DNA molecule into fragments. Most systematic and evolutionary studies using these enzymes use mitochondrial DNA (mtDNA), which is experimentally accessible and of a conveniently small size (Lansman et al., 1981). Since these enzymes are specific for particular base sequences, some information (about 10% of the total) on the base sequence of the mtDNA is derived. Methods for quantifying within- and between-sample genetic diversity have been developed by Nei and Li (1979). Although these techniques are used to study the evolution and divergence of mitochondrial populations of the cells of eukaryotic species, they can lead to important inferences about genetic divergence between individuals and populations. The value of this technique in some of its specialized applications depends on the strict maternal inheritance of the mitochondria (Chapman et al., 1982), which needs to be verified in each species that the method is applied to. Studies of mtDNA hold special promise for studies of founder populations (Chapter 14 by Powell) and in defining nesting populations (Lansman et al., 1981).

Nucleic acid hybridization studies of DNA and/or RNA do not reveal particular base sequences, but give an overall measure of sequence similarity between different species. These methods can be used to compare populations divergent at higher taxonomic levels—from different genera, for example. The method, briefly, requires the preparation of RNA or single-strand DNA from the individuals to be compared. The strands are brought together and allowed to combine or "hybridize" *in vitro*. The more similar the base sequence of the two individuals, the greater will be their areas of complementary sequences, and the more stable will be these hybrid molecules. This relative stability, and therefore the similarity of the nucleic acids, can be measured by determining the resistance of the hybrid molecules to breakdown when exposed to increasing temperature or other quantifiable stress. Individuals with more nucleotide sequence differences will form less stable hybrid molecules. These techniques have been used to estimate within- and between-species diversity in DNA sequences in sea urchins (Britten et al., 1978; Grula et al., 1982). These studies also reveal that interpretation of the results of each analysis must take into account the type of DNA fraction being analyzed. As with DNA sequencing techniques, specialized laboratory facilities and procedures are required to carry out this work.

2. Protein Sequencing

The sequence of amino acids that form a specific protein is encoded in the sequence of bases in the DNA of the genes specifying that protein (Chapter 2 by Chambers). The genetic material can then be indirectly studied by determining the amino acid sequences of the primary gene products: polypeptides and proteins. Since each amino acid can be specified by more than one base sequence, amino acid residue sequencing is not as sensitive as DNA sequencing studies. Amino acid sequences of cytochrome c (Fitch and Margoliash, 1967) and globins (Goodman, 1976) have been used to infer phylogenetic relationships between a great variety of organisms. Differences between individuals or populations can be quantified as the probability that individuals differ in an amino acid residue at a particular site. These values can be averaged over all sites to give an average probability of difference per site.

3. Electrophoresis

Studies of protein differences by electrophoresis is at present the most convenient tool for determining genetic diversity within and between samples. The genetic basis of differences can usually be inferred, if proper precautions are taken, even in the absence of breeding data. It is relatively inexpensive and is not time-consuming in most of its applications. Although elaborate laboratory equipment is not required and a technician can be easily trained to carry out the laboratory work, the aid of a specialist should be enlisted in designing sampling procedures and in interpreting results. Although most methods of quantifying electrophoretically detected variation may be applied to other loci, multiple-locus data sets will normally be available only for electrophoretic data.

Within-population (or other category) diversity can be quantified as H, the proportion of heterozygous loci in an average individual in a population (average heterozygosity), or P, the proportion of loci that are polymorphic in a population. These quantities are related (Frankel and Soulé, 1981, p. 42), but each has statistical properties (discussed in Chapter 14 by Powell) that limits its use.

Nei's (1973) gene diversity analysis of allelic diversity in subdivided populations is based on expected heterozygosity at individual loci. Values representing diversity over all studied loci can be calculated by averaging the values for individual loci. The limitations of using these average values in describing overall diversity have been discussed by Gregorius (1978).

Uncommon or even rare alleles are important components of variability that may allow a species to respond evolutionary to unforeseen future evolutionary pressures (Chapter 22 by Futuyma), yet their contribution to heterozygosity and proportion of polymorphic loci may be small. Measurements that describe the total number of alleles are therefore of special interest to managers of genetic diversity. Clegg and Brown (Chapter 13) and Hamrick (Chapter 20) have presented such data in the form of the number of alleles per locus, and Hamrick (Chapter 20) has presented data on the proportion of alleles observed in a species that are found in a given population. Sampling for rare alleles is difficult because the number of alleles found will usually depend on sample size. Data with comparative value can still be gathered if the size of the sample relative to the size of the entire population can be estimated. More detailed discussions on the sampling of populations for the number of alleles at a locus are presented in Ewens (1972), Gregorius (1980), Templeton (1980), and Bryant et al. (1981).

Several methods of quantifying between-sample diversity based on allele frequencies are available, with those of Nei (1972) and Rogers (1972) being the most frequently used. Nei's normalized identity of genes (I) becomes unreliable as values approach zero (Nei and Roychoudhury, 1974), as is often found between samples belonging to different genera. The greatest value of electrophoretic methods is therefore usually in studies below the generic level, although this generalization depends on the generic concept employed, and useful information has been derived at higher levels in some cases.

A drawback of electrophoresis is that it underestimates the amount of variation that would be detected if amino acid or DNA sequences were determined for the same loci. Much of this "hidden" variation can be detected with the more elaborate sequential electrophoretic procedure (Coyne, 1976; Singh et al., 1976; Aquadro and Avise, 1982). The results of electrophoretic studies should be interpreted in terms of the differences detected rather than the similarities seen, since these similarities may vanish when the alleles are subjected to a more sensitive technique. For example, although samples from five breeding colonies of the northern elephant seal (*Mirounga angustirostris*) revealed no variation at the 24 loci sampled (Bonnel and Selander, 1974), it cannot be said that these populations contain no genetic variation, only that no variation has been detected. It is still a reasonable inference that this population has very low genetic variation compared with populations of other species that have been electrophoretically studied.

4. Immunology

Immunological methods provide a means of comparing taxa that have diverged beyond the resolution of electrophoresis and are otherwise not too distantly related. Although these methods may not differentiate between closely related species, numerous phylogenies of higher categories have been prepared based on immunological distances (Fitch, 1976). Since rabbits or other laboratory animals are required for producing the antibodies whose reactions are studied by these methods and purified protein samples are required for the more specialized immunological techniques of microcomplement fixation, immunological methods are relatively time-consuming and expensive.

5. Chromosomes

Analysis of chromosomes can yield useful information on genetic diversity at any taxonomic level. This information, however, rarely leads to clearcut taxonomic decisions, since chromosomal variation is known to occur both within and between species. Chromosomal polymorphism is very common in *Drosophila* and is now being found in a variety of other taxa (White, 1973). Some "good" species have identical karyotypes, such as the "homosequential" species of Hawaiian *Drosophila* studied by Carson (1970). Neither differences nor similarities per se can indicate whether two allopatric taxa belong to the same or different species. The behavior of different chromosomal markers can, however, give less ambiguous taxonomic information in areas of sympatry or in studies of meiosis in the offspring of individuals who are suspected of belonging to different species. The relationship between chromosomal changes and speciation remains controversial (Bush, 1975; White, 1978a; Futuyma and Mayer, 1980).

6. Relationships between Loci

Many of the techniques discussed so far allow the characterization of individual gene loci. What has not been discussed is the number of different genetic combinations of different alleles at different loci in a population. The relationships between loci are not always completely random; certain alleles at one locus may be more often associated with a particular allele at another locus, particularly if both loci are on the same chromosome. Lewontin (1974) has argued that interactions between loci deserve greater attention in evolutionary investigations. Gregorius (1978) has advocated the description of statistical relationships between loci by determining the hypothetical frequency distribution of gametic types. Although the statistical relationships between loci in electrophoretic surveys can easily be described, true chromosome "mapping" requires a far greater effort.

7. Morphometrics

Morphometrics includes the quantification of overall phenotypic similarities of populations by multivariate statistical analysis of measurements of many different characters. The results of these analyses are commonly presented in two–dimensional diagrams of points whose graphic distance from one another is a measure of overall phenotypic distance. The relative value of each measured character in assigning individuals to particular groups can also be estimated.

These methods have the advantage that they can be applied to the skeletal and other remains of dead organisms from museum collections or the field so that extant and extinct populations can be compared. Computer programs are commercially available for many of these techniques, so the limiting factors in their application are the time and expense of making measurements to produce the raw data. The interpretation of these analyses and the choice of characters used is still the subject of some debate.

Morphometric analysis measures the effects of a large number of gene loci. It is not usually known, however, how many loci are responsible for the measured divergence in morphology. Another problem is that these techniques also measure nongenetic variation. Characters influenced by nongenetic factors can sometimes, with careful testing, be identified and eliminated from the analysis.

8. Breeding Studies

Laboratory breeding studies can determine reproductive compatibilities between individuals in different populations (Moore, 1949; Oliver, 1972). These studies should not be overlooked as a means of providing additional estimates of evolutionary divergence and genetic diversity in organisms that may be conveniently reared under controlled conditions. The results must be carefully evaluated, especially in species with behavioral isolating mechanisms that may be disrupted under artificial conditions. Breeding studies can also determine the Mendelian inheritance of morphological and physiological traits. The ease of evaluating large numbers of loci by electrophoresis has led to some neglect of breeding studies, which were the only means of measuring genetic differences before the introduction of molecular techniques. Breeding studies should be, whenever practical, an adjunct to electrophoretic studies to evaluate the Mendelian interpretations of protein variation.

Comparison of Methods of Measuring Diversity

Each of the methods of estimating genetic diversity discussed above examines a different aspect of the genotype and/or phenotype. The results of different analyses should not be expected to be entirely congruent, since different criteria are appropriate for the evaluation of each method. Any systematist, geneticist, or manager should be pleased to have data from any or all of these approaches.

Managers require some criteria on which to base their selection of a technique, since they will rarely be able to afford them all. These criteria include the taxonomic level that the manager is interested in and, of course, the relative expense of the different methods. For example, electrophoretic methods are more appropriate at lower taxonomic levels. Information at higher levels may come from immunology or sequence data. The least expensive techniques are electrophoretic, morphometric, and chromosomal studies. Sequencing of DNA or proteins is most expensive and time consuming. The more elaborate sequential technique of electrophoresis (Coyne, 1976; Singh et al., 1976), which employs multiple conditions of gel concentration and pH to separate alleles, may actually come close to the resolving power of protein sequencing for detecting protein variants. Ramshaw et al. (1979) found that sequential electrophoresis could separate about 90% of the different sequence variants of hemoglobin that they studied.

Although these techniques measure different aspects of the genome, some concordance might be expected when two or more techniques are applied to the same populations. For example, Patton et al. (1975) found congruence between morphometric and electrophoretic data on introduced roof rats (*Rattus rattus*) in the Galápagos Archipelago. Even when the results of two more methods are not entirely concordant, careful interpretation of the data does not necessarily lead to different classifications (Gould and Woodruff, 1978; Seidel and Lucchino, 1981). Larson and Highton (1979) found congruence between electrophoretic, immunological, and DNA hybridization data on some *Plethodon* salamanders. King and Wilson (1975) found little difference between humans and chimpanzees based on DNA hybridization, electrophoresis, and immunology. They preceived these findings as being in contrast to the differences between these species in ecology and behavior.

A very general conclusion is that the patterns of relationship and diversity revealed by these techniques are congruent with one another within their appropriate taxonomic ranges and that they commonly agree with the results of careful biosystematic studies. Many of the exceptions consist of sibling species comparisons and/or cases of convergent or parallel evolution in superficial morphological characters.

Species

Mayr (1969) defines species as "groups of interbreeding natural populations that are reproductively isolated from other such groups." The different types of reproductive barriers between species have been reviewed by Mayr (1963); they include behavioral and developmental incompatibilities. Attempts have been made to quantify the amounts of genetic divergence required to establish reproductive isolation. The most detailed of these studies is that of Ayala et al. (1974) on the *Drosophila willistoni* group. These workers compared the average genetic identities and distances based on electrophoresis between populations at varying degrees of taxonomic divergence. These comparisons reveal that genetic distance increases with increasing taxonomic distance between samples. Semispecies pairs, which have partial reproductive isolation between them, have about an 80% similarity ($I = .80$) in their electrophoretically sampled loci. The observation of pairs of full species that have identities of greater than .9 (Johnson et al., 1977; Sene and Carson, 1977; Ryman et al., 1979) demonstrates that speciation can occur with very little divergence at the loci sampled by electrophoresis. This is not surprising, since the genes or possibly the chromosomal rearrangements responsible for determining reproductive isolation are qualitatively different from those coding for the mostly intracellular enzymes sampled by electrophoresis.

Mayr (1969 and elsewhere) has argued that the species is the only taxonomic category for which there is an objective definition. All other categories, including subspecies, need to be defined by more subjective criteria.

Subspecies

Varying concepts of what constitutes a subspecies have been employed in systematics. In plant systematics the term variety may replace the term subspecies, or occasionally both ranks may be recognized within a species (Stace, 1980). Some past practices, now discredited, include describing subspecies from single diseased or immature individuals and describing two or more subspecies within a single variable population. Some systematists have attempted to salvage the subspecies concept from these meaningless applications by applying a subspecies definition that requires that subspecies be either distinct allopatric populations or populations that intergrade very little according to a "90% rule" (90% of individuals of one population differ from 90% of the individuals in a population of another subspecies in the character used to discriminate; see Mayr, 1969). Many evolutionary biologists remain highly critical of the subspecies concept (Wilson and Brown, 1953; Futuyma, 1979) because the characters on which subspecies are based may not be concordant with other traits that are equally important. A thorough discussion of the validity of the subspecies concept accompanies Wiens (1981).

A strong argument against describing subspecies on the basis of minor character differences is that such characters may rapidly change, owing to environmental effects or rapid evolutionary change. Johnston and Selander (1964) noted that racial differences exist in house sparrows (*Passer domesticus*) in the United States. These differences apparently have rapidly evolved in the approximately 50 years since house sparrows were introduced into the United States. Sumner (1924) noted significant changes in measured characters of subspecies of *Peromyscus* after only four to seven generations in captivity. Inadvertent selection for survival under laboratory conditions can result in genetically based behavior changes (Bush et al., 1976). A transplanted population of the Devil's Hole pupfish (*Cyprinodon diabolis*) displayed significant increases in a number of measured characters (Williams, 1977) and brighter colors (Williams, personal communication). Rainbow smelt (*Osmerus mordax*) populations stocked into lakes displayed an increase in size and growth rates during early generations (Rupp and Redmond, 1966), but in subsequent generations reverted back to values near those of their ancestral populations (Copeman and McAllister, 1978). In each of these examples with fishes, the actual genetic component of these differences is not known. Each of these examples also suggests that changes in characters are likely to take place in managed populations.

Despite questions of the biological validity of the subspecies concept, conservation of subspecies is being actively pursued. This is largely because formally named subspecies are the only available indication of variation below the species level for the great majority of species. Managers whose objective is to conserve the genetic diversity within a species may often find that subspecies are the best available description of how that diversity is apportioned. In cases where the subspecific taxonomy has been based on careful study of geographic variation and reflects a number of concordant characters, subspecies may be a good indicator of genetic diversity.

The importance of carefully considering subspecies in conservation efforts can be illustrated by considering North American elk (*Cervus elaphus*), which are managed as subspecies. Some subspecies are considered to be near extinction, individuals are moved to colonize habitats where the local subspecies has been extirpated, and some individuals are considered exotic in the area where they have been introduced because they can be placed in a different subspecies from that of the previous elk population in that area. In these cases, the subspecies classification was based on studies of three type specimens and one thorough study of one subspecies, the tule elk (*C. e. nannodes*) (McCullough, 1969). The relationships of the subspecies to one another and the range of variation within any single subspecies had not been investigated. Although the subspecies classification is the best available description of subspecific diversity, that classification is based on little supporting data. A morphometric study of Pacific Coast elk populations (Schonewald-Cox et al., submitted) reveals that one of the subspecies, *C. e. roosevelti*, is actually a composite of morphologically different populations that are as distinct from one another as each is from *C. e. nannodes*.

The value of the subspecies concept in identifying important units of diversity for conservation can be evaluated by examining the uniqueness of subspecies as judged by electrophoretic data or some other measure of genetic diversity. Subspecies classifications generally correspond to genetic units defined by electrophoresis in *Speyeria* butterflies (Brittnacher et al., 1978), *Peromyscus pectoralis* (Zimmerman et al., 1978), northern oriole (Corbin et al., 1979), and bluegill (Avise and Smith, 1974a). Little congruence between electrophoretically similar populations and subspecies classification has been found in the lizard *Lacerta mellisellensis* (Gorman et al., 1975). Lack of congruence between karyotypes and subspecies classifications has been reported in the pocket gopher *Thomomys bottae alenus* (Patton, 1972), where the subspecies was found to be a composite of three chromosomal races, each of which was derived from a different ancestral stock.

The appropriateness of a subspecies classification for describing the allelic diversity within a particular species can be assessed by applying the gene diversity analysis (Nei, 1973, 1975). In its simplest application (described in Chapter 20 by Hamrick), this method partitions the total allelic diversity (H_T) in a species into the component found within populations (H_S), and that attributed to differences between the populations (D_{ST}) so that

$$H_T = H_S + D_{ST}$$

The proportion of the total diversity attributed to differences between populations is G_{ST}:

$$G_{ST} = D_{ST}/H$$

which is similar to F_{ST}, Wright's coefficient of gene differentiation (Nei, 1973, 1975). Values of G_{ST} have been found ranging from .07 (three races of man) to .7 (*Dipodomys ordii*, the Ord kangaroo rat). Most G_{ST} values are found toward the lower end of this range, which means that most diversity is found within individual populations. The analysis of the three races of man indicated that this classification accounts for very little (7%) of the variation in human populations. This means that this racial classification does not strongly correspond to biological groupings of humans. Some other classification may account for more diversity between groups, but it is likely that most allelic diversity will probably still be found within populations. Nei (1973) has shown that, if populations are arranged in hierarchy, the total allelic diversity can be partitioned into components that correspond to each level of the hierarchy. This relationship can be stated as

$$H_T = H_S + D_{CS} + D_{ST}$$

where H_T is, again, the total diversity, H_S is the within-population diversity,

Table 1. *Apportionment of Allelic Diversity of Samples of the* Goniobasis floridensis *Complex of Freshwater Snails*[1]

Between drainage systems	45.8%
Between localities within drainage systems	20.0%
Within localities	34.2%
Total	100.0%

[1] From Chambers (1980).

and D_{ST} and D_{CS} are the diversities among succeedingly lower levels of the hierarchy.

Chambers (1980) carried out a hierarchical allelic diversity analysis on 14 populations of a species complex of freshwater snails occupying five river drainage systems (Table 1). Most of the allelic diversity was found between drainages (46%), with the remainder between localities within drainages (20%) and within localities (34%). The large between-drainage and small within-drainage diversity suggests that the snails in each drainage system can be considered part of a different genetic and evolutionary unit. To conserve a large portion of the allelic diversity in this species complex, populations within each drainage must be protected.

In the case of two or more subspecies, each consisting of a number of populations, the total gene diversity (H_T) can be apportioned between individuals within populations (H_S), between populations within subspecies (D_{CS}), and between subspecies (D_{ST}).

In cases where diversity between subspecies (D_{ST}) is a large portion of the total diversity, conserving subspecies will be a good strategy for preserving the genetic diversity of the whole species. If the diversity between subspecies, or the subspecies level of classification, accounts for relatively little of the total diversity, then conservation of populations or some grouping of populations other than the recognized subspecies might be a better strategy.

An allelic diversity analysis of cutthroat trout (*Salmo clarki*), including the subspecies level of classification, is presented in Table 2. We calculated these values from the allele frequency data of Loudenslager and Gall (1980) on trout from inland drainages in the western United States. Those calculations exclude *Salmo clarki pleuriticus* because that subspecies was represented by only one sample. All samples were weighted equally

Table 2. *Apportionment of Allelic Diversity of Samples of Subspecies of Cutthroat trout (*Salmo clarki*) from Inland Drainages in the western United States*

Between subspecies	42%
Between localities within subspecies	25%
Within localities	33%
Total	100%

(a sampling procedure designed specifically for the present analysis would include a more even distribution of sample numbers among the subspecies and sample diversities would be weighted, probably on the basis of estimated population size). Since the subspecies level of classification accounts for a large portion (over 40%) of the observed allelic diversity, conservation of subspecies is a reasonable approach to conserving allelic diversity in these trout populations.

Allelic diversity analysis has broader uses in this context than the evaluation of subspecies classification. Subdivision of a species may be based on factors such as geographic or political boundaries. These analyses can be used to estimate the relative amount of the total allelic diversity that can be conserved by adopting different criteria for subdividing the population. The effectiveness of using governmental units as conservation districts can be evaluated. For example, would county-by-county conservation of a species provide the best means of maintaining allelic diversity, or is there another arrangement that might result in the conservation of the same or more allelic diversity at a lower cost? Diversity analysis can then be employed to suggest an apportionment of conservation effort that best coincides with the observed apportionment of allelic diversity. This strategy would maximize the amount of diversity conserved if limited financial resources or land use conflicts dictate that only a portion of the existing populations may be protected.

The type of analysis described above should not provide the sole criteria for managing the genetic diversity of a species or population. Other criteria for protecting a particular group might include its possession of rare or unique alleles that would have only a slight influence on an allelic diversity analysis. The proportion of alleles in a species possessed by a particular population, described by Hamrick in the preceding chapter, is another important criterion that may not be completely reflected in an allelic diversity analysis. Futuyma describes the value of rare alleles in Chapter 22. It must also be noted that the partitioning of allelic diversity as described above apportions *relative* amounts of allelic diversity, and the absolute amount (D_m in Nei, 1975) can be calculated and evaluated along with the apportionment.

A manager who is not satisfied with using electrophoresis as a sole measure of genetic diversity can employ another measure or, if resources permit, an additional measure or measures. Apportionments of diversity by different methods can be compared. For example, morphometric data can also be subjected to hierarchical diversity analysis. Individual characteristics can be subjected to a nested analysis of variance to determine the proportion of total variance that is accounted for by the subspecies level of classification. The characters subjected to such an analysis should be those that have the highest discriminating power (factor loading), as determined by a study of many characters.

Although in a given case subspecies may not correspond to distribution of major genetic diversity, their conservation need not be based on this sole criterion. The characteristics that subspecies are based on, which are often color or size, are of interest in themselves for their own biological, commercial, or esthetic value. One may not be able to defend the continued recognition of so many subspecies of milk snakes (*Lampropeltis triangulum*) or tiger salamanders (*Ambystoma tigrinum*) on biological grounds, but the attractiveness of these animals in an esthetic sense or their significance as biological phenomena is undeniable.

These analyses are suggested under the assumption that conserving maximum genetic diversity is the manager's stated aim. Ecological, behavioral, and other biological factors all interact with genetic factors so that management decision should be based on as much information as possible on all aspects of the organism's biology. Studies of the genetics and systematics of wild populations should be viewed as an essential component of a sound management program.

Summary

The Linnaean system of classification is the primary framework for organizing and describing the diversity of life. The evolutionary genetic relationships among individuals are the primary criteria for placing organisms within this classification in evolutionary systematics. There are several means of assessing these genetic relationships, that can be used to describe both within- and between-population diversity.

Individual species and subspecies are often the targets of management efforts. The appropriateness of subspecific classifications for describing overall genetic diversity has been strongly questioned. Techniques are now in use that can aid in assessing the proportion of observed variation that is accounted for by a subspecific level of classification. These techniques include electrophoretic, karyotypic, and morphometric analyses. In Pacific Coast tule elk the subspecific classification does not adequately describe the morphometric variation among populations. In some populations of cutthroat trout in the western United States, the subspecies classification accounts for a large proportion of the variation at isozyme loci. The appropriateness of using recognized subspecies for conserving genetic variation will vary from species to species.

Acknowledgments

We thank Christine Schonewald-Cox and Jack Williams for suggesting relevant examples during informative discussions on the subjects of this chapter.

CHAPTER
22

Interspecific Interactions and the Maintenance of Genetic Diversity

Douglas J. Futuyma

From the viewpoint of evolutionary genetics, extinction of populations can have two major causes. In an unchanging environment—that is, one in which the fitnesses of genotypes retain the same relative values from one generation to the next—a reduction in the sizes of the populations of an outbreeding species may result in inbreeding depression, and a loss of fitness to the extent that the populations may dwindle to extinction. This is not an inevitable consequence of inbreeding, for we know of populations such as the northern elephant seal (*Mirounga angustirostris*: Bonnell and Selander, 1974) that appear to have lost genetic variation in the course of a population bottleneck, yet still continue to flourish. Nevertheless, as Ralls and Ballou (Chapter 10) and Templeton and Read (Chapter 15) have indicated, and as laboratory geneticists have long been aware, inbreeding depression can threaten populations with extinction.

The other major cause of extinction, the one that presumably has been responsible for the extinction of more than 95% of all species in the course of earth history, is the failure of species to adapt to changes in the ecological environment that lower the fitness of the prevalent genotypes. Extinction in such cases follows if the populations do not have sufficient selectable genetic variation to replace prevalent, now inferior genotypes with new ones on a time scale commensurate with the rate of environmental change (Chapter 12 by Selander). One line of evidence supporting this classical explanation of extinction is the observation that almost all exclusively asexual forms of higher plants and animals, which are presumably not capable of evolving as rapidly as sexual species (but see Williams, 1975; Maynard Smith, 1978), are very closely related to sexually reproducing species, and thus, as Mayr (1963) put it, have "all the earmarks of recency." Except for abundant, rapidly reproducing microorganisms, in which mutation replenishes genetic variation at a high rate, few asexual taxa are diverse enough and distinct enough to give evidence of long persistence through geological time. Thus long-term persistence appears to require the genetic capacity for evolutionary responses to environmental change (Chapter 12 by Selander).

Much of the last decade's electrophoretic analysis of genetic variation seems to indicate that population size and structure strongly influence the level of heterozygosity, more or less as the theory of random genetic drift predicts. Whether or not most variation in enzymes is selectively neutral continues to be debatable, but in any case it may not be too relevant to the problem of extinction. For the persistence of a plant population in the face of, say, a change in soil moisture depends on whether or not the plant is genetically variable for those specific traits that confer adaptation to soil moisture and (perhaps just as important) to other ecological changes that accompany a change in soil moisture, such as changes in the community of competitors, pathogens, and insects that constitute much of the plant's biotic environment (Chapter 5 by Liu and Godt). Genetic variation in traits that adapt the species to such ecological variables may or may not be correlated with summary indices of genetic variation such as allozyme heterozygosity. There is little information on this point. Certainly if there is strong selection on such traits, a population may be monomorphic for alleles that confer local adaptation to some ecological variables, even if allozyme analysis reveals extensive variation.

Changes in the physical environment have surely played an important role in the extinction of species, but there is every reason to suppose that biotic changes have been even more important. Paleontologists can point to instances such as the extinction of South American mammals after the incursion of the North American biota, and instances of local extinction because of the recent introduction of competitors, predators, and parasites are numerous. DeBach (1966), for example, has described the competitive exclusion of one hymenopteran parasitoid introduced as a biological control agent of scale insects in California, by a superior competitor that was subsequently introduced. The impact of introduced predators is illustrated by cases such as the devastating effect of the sea lamprey (*Petromyzon marinus*) on fishes in the North American Great Lakes, and of an African cichlid fish that has extinguished several native fishes in Gatún Lake in Panama (Zaret and Paine, 1973; Christie, 1974). Successful instances of biological control are case histories of biotic extinction: for example, the chrysomelid beetles (*Chrysolina gemellata*) introduced to control the exotic Klamath weed or common St. John's-wort (*Hypericum perforatum*) in western North America have severely reduced the populations of their host. Perhaps the most striking and important instances of biotically caused extinction are those due to pathogens: the possible eventual extinction of the American chestnut (*Castanea dentata*) and of the American elm (*Ulmus americana*) because of introduced fungi are sad examples that are all too familiar.

Especially because of human transport, introductions of pathogens and insects will doubtless continue to pose an important threat of extinction; but if agro-ecosystems can be taken as models of natural systems at all, genetic changes in parasites and pathogens also may have an important impact. In response to the widespread planting of resistant strains of crops, fungi and insects can rapidly evolve new, virulent forms that overcome the resistance. Wheat (*Triticum aestivum*) breeding in the United States entails a continual race against the evolution of new, virulent strains of the Hessian fly [*Mayetiola* (= *Phytophaga*) *destructor*: Hatchett and Gallun, 1970; Sosa, 1981]; resistant strains of rice (*Oryza sativa*) in Asia are threatened by newly evolved strains of the brown planthopper (*Nilaparvata lugens*: Pathak, 1970); and there are many similar cases in the literature of plant pathology. For example, the wheat stem rust (*Puccinia graminis* var. *tritici*) evolved the ability to attack resistant wheats within a few decades after these strains were widely planted (van der Plank, 1963).

There is a widespread myth that parasites and pathogens evolve toward a benign relation with their hosts, the supposition being that it is disadvantageous to threaten the resource on which they depend. But this myth of the balance of nature depends implicitly, in many cases, on a group-selectionist argument, and the role of group selection in evolution is a strongly contested issue (Williams, 1966; Wade, 1978; Wilson, 1980). Where the individual fitness of a parasite is increased by prolonging the life of its host, so that the parasite can reproduce repeatedly, individual selection can indeed favor reduced virulence. But in many instances the fitness of the parasite is contingent on the sterility or death of its host. For example, some botflies (Diptera: Oestridae) and other parasites apparently prolong the life of their hosts by castrating them, so that the energy otherwise devoted to reproduction is diverted toward maintenance of the individual host and its parasites (Kuris, 1974; Baudoin, 1975). The transmission of the viruses that attack many lepidopteran caterpillars is accomplished by death of the caterpillar, from which body liquids containing viral particles then exude. In such instances, selection in the parasite favors increased virulence, even if the host population ultimately becomes extinct (May and Anderson, 1983; Roughgarden, 1983). Thus evolution may often work to destabilize communities (Roughgarden, 1979); indeed, Van Valen (1973) has postulated that most extinctions in geological time have been caused by the failure of species to adapt rapidly enough to genetic changes in other species.

There is abundant evidence from agricultural systems that genetically uniform populations are especially susceptible to outbreaks of newly evolved, virulent strains of pathogens (Barrett, 1981, 1983). For example, in 1970 as much as a 50% decrease in corn (*Zea mays*) yield occurred in many states in the United States because of the spread of a new race of the southern corn leaf blight fungus, *Helminthosporium maydis*, to which all the widespread cultivars of corn were susceptible. The genetic uniformity that led to the susceptibility of these cultivars was an incidental outcome of selective breeding for the possession of cytoplasm bearing a male-sterility factor. This factor had been used to improve the efficiency of artificial breeding programs in the development of high-yield cultivars. In many other such instances, epidemics of new strains of pathogens have been facilitated by the abundant planting of crops that are uniform for chromosomally inherited resistance factors.

Plant breeders are aware that wild species of plants harbor extensive genetic variation for resistance to pathogens and sometimes use these species as sources of resistant germ plasm. For example, wild relatives of the cultivated potato (*Solanum tuberosum*) in Central and South America are genetically variable for resistance to potato blight (*Phytophthora infestans*) and have been used for breeding resistant cultivars (van der Plank, 1968). The question, then, is how such genetic variation is maintained.

In the simplest models (e.g., Pimentel, 1968; Levin and Udovic, 1977), two alleles that confer resistance (*A*) or susceptibility (*a*) segregate in the population and maintain polymorphism if in the absence of the pathogen the susceptible genotype (*aa*) has highest fitness, as it may if resistance entails some physiological cost. These models may be extended to the "gene-for-gene" systems (Flor, 1956; Person, 1966) that have been described for many interactions between crop plants and their fungal pathogens. [A gene-for-gene system is also known for the interaction between an insect, the Hessian fly, and wheat (Hatchett and Gallun, 1970).] In such systems, each of several loci in the host confers resistance to a subset of pathogen genotypes, and each of the host's resistance alleles is matched by an allele in the pathogen that enables it to overcome the host's resistance. If each of the alleles (for resistance or for virulence) entails a physiological cost, complex polymorphisms may be maintained either as stable gene frequencies, stable cycles, or chaotic genetic oscillations (see May and Anderson, 1983, for references to this literature, and Chapter 19 by Namkoong). Clarke (1976) has emphasized that frequency-dependent selection will be important in maintaining such polymorphisms and has hypothesized that frequency-dependent selection for disease resistance could be a factor in the maintenance of enzyme polymorphism.

The critical point to emerge from these models is that if, in the absence of pathogens (or predators or parasites), the susceptible genotypes have highest fitness, then genetic variation for resistance will be lost if the natural enemies are not sufficiently abundant to maintain the selection. There is, however, little empirical evidence of the cost of resistance. The literature on the defenses of plants against herbivorous insects lays great stress on the defensive role of plant secondary compounds (e.g., Rosenthal and Janzen, 1979), some of which constitute a quite appreciable fraction of plant biomass, and may entail substantial energetic cost (Chew and Rodman, 1979). Cates (1975) provided evidence that a form of wild ginger (*Asarum canadense*) that is relatively unpalatable to slugs has a lower growth rate than the susceptible form, but much more information is required before the generality of the cost of resistance can be established.

Studies of genetic variation affecting the interaction of species in nature are only beginning. The plant pathology literature is almost exclusively concerned with cultivated plants, and little is known of the prevalence of polymorphic gene-for-gene systems (or other forms of complementary variation) in nature. Some information on genetic variation in plants and phytophagous arthropods is beginning to appear in the literature, but complementary variation in interacting species is seldom the focus. For example, there is some evidence that geographic variation in the frequency of cyanide-producing genotypes of *Lotus corniculatus* (bird's-foot trefoil) and *Trifolium repens* (white clover) is in part a response to herbivory (Jones, 1973). Sturgeon (1979) has provided circumstantial evidence that the frequency of different defense terpenoids in populations of ponderosa pine (*Pinus ponderosa*) has been affected by outbreaks of pine beetles. Edmunds and Alstad (1978) have shown that individual pine trees in a natural stand are occupied by specifically adapted demes of a scale insect, suggesting that variation in the host may maintain variation in the insect; but there is no evidence that herbivory is the selective factor accounting for variation in the pines (Chapter 5 by Liu and Godt). Many plant species are variable in characteristics that affect their susceptibility to insects; for example, genotypes of the rough cocklebur (*Xanthium strumarium*) differ in chemical, morphological, and phenological traits that are correlated with their infestation by two species of seed-eating insects (Hare and Futuyma, 1978), but there is no clear evidence that the herbivory maintains the variation.

Just as plants are genetically variable for resistance to insects and pathogens, the latter are genetically variable for traits enabling them to attack different plants. From the standpoint of management, this has two implications. The genetic variability of parasites may enable them to become so virulent as to threaten genetically homogeneous populations of hosts, as agricultural systems show. Conversely, the persistence of the parasites themselves may be in jeopardy if ecological changes alter the abundance of their hosts, and if the parasites are not genetically variable enough to adapt to alternate hosts that become more prevalent. If we define "parasite" broadly enough to include herbivorous insects such as butterflies, as Price (1980) does, then this becomes a serious concern even if the only foundation for a program of conservation is esthetic (Chapter 1 by Frankel and Chapter 9 by Ehrlich).

The very limited research that has been done on fungal parasites of wild hosts indicates that fungi are genetically variable in virulence, and that different species of hosts harbor genetically different fungi (e.g., Wahl et al., 1978; Eshed and Dinoor, 1981). Somewhat more work has been done on genetic variation in phytophagous arthropods. For example, artificial selection on laboratory populations of spider mites (*Tetranychus urticae*: Gould, 1979) and of bean weevils (*Acanthoscelides obtectus*: Wasserman and Futuyma, 1981) has been successful in developing improved adaptation to inferior hosts in a few generations. Insect pests of crop plants have become genetically altered both in their behavioral responses to widely cultivated hosts and in their phenological and physiological adaptations to these crops; in addition to the rice brown planthopper and the Hessian fly mentioned previously, populations of the codling moth (*Laspeyresia pomonella*), a pest of apple (*Malus domestica*), have become adapted to English walnut (*Juglans regia*) and plum (*Prunus*) in California (Phillips and Barnes, 1975) and the sulfur butterfly (*Colias philodice eriphyle*) has become adapted to alfalfa (*Medicago sativa*: Tabashnik, in press). In a few cases, populations of polyphagous insects appear to be genetically subdivided, with different genotypes occupying different species of hosts within a single breeding population (Tavormina, 1982; Via, 1982; Clarke et al., 1963). Such host-associated polymorphism appears to be less prevalent in sexually reproducing species than in those in which reproduction is largely or exclusively asexual. Most of the examples of "biotypes" associated with different cultivars or with different species of crops have been described for aphids and other parthenogenetic Homoptera (reviewed by Futuyma, 1983; Gould, 1983). In my laboratory, extensive study of the largely parthenogenetic fall cankerworm (*Alsophila pometaria*: Lepidoptera, Geometridae) has shown that populations of this moth consist of a variety of clonally reproducing genotypes, some of which are more abundant in stands of oak (*Quercus*) and others in stands of maple (*Acer*). The genotypes differ at least in phenological and behavioral attributes that appear to confer adaptation to one or the other host (Mitter et al., 1979; Futuyma et al., 1981).

Edmunds and Alstad (1978) demonstrated that scale insects in a single small locality are so differentiated as to be adapted to different individual pine trees of the same species: the insects performed better on their own tree than on others to which they were transferred (see Chapter 5 by Liu and Godt for a more detailed discussion). In the majority of described cases, however, the genetic variation in insects has been in adaptation to different species of host plants, rather than to different genotypes of a single host species. Whether or not genetic variation in hosts commonly maintains genetic variation in the insects (or other parasites) that attack them is not known. Moreover, it is not clear whether genotypes of insects that are adapted to different hosts commonly coexist within a single population, or if the genetic variation exists on a larger spatial scale. Population genetics models of the balance between selection and gene flow indicate that, unless balancing selection is quite strong, little polymorphism for host utilization may be expected within populations. Accordingly, most of the examples of genetic variation in the adaptation of insects to different hosts describe variation among local populations, each of which is adapted to the locally abundant host. Thus, for example, the genotypes of codling moth that are adapted to apple, English walnut, and plum have been found in local areas where each of these trees is most abundantly cultivated (Phillips and Barnes, 1975), and the same is true of *Colias* butterflies adapted to wild legumes versus alfalfa (Tabashnik, in press). Similarly, Rausher (1982) has found that spatially segregated populations of a checkerspot butterfly (*Euphydryas editha*) are adapted to locally abundant wild hosts (Chapter 9 by Ehrlich). These observations suggest that selection may tend to promote local monomorphism for genotypes best adapted to local conditions, and that a substantial fraction of the genetic variation exists among, rather than within, local populations. The existence of genetic variation in interacting species of predators and prey, insects and plants, or parasites and hosts implies that the interactions may evolve. It is far from certain, however, that such groups of species are locked into a continual, escalating "arms race." It seems possible (Schaffer and Rosenzweig, 1978; Maynard Smith, 1976a) that such systems of interacting species rapidly achieve a genetic equilibrium and that further coevolution proceeds only slowly, as novel mutations or other perturbations of the gene pool arise. If so, one threat to the genetic stability of such systems may be random drift, which may destabilize gene frequencies, so that, say, pathogens respond genetically to the progressive genetic impoverishment that their hosts may suffer through drift. The other major threat is the incursion of new species. An important question, then, is whether populations harbor genetic variation that can enable them to evolve rapid adaptation to new parasites (or predators) or hosts (or prey).

Although populations may fortuitously harbor genetic variations that could provide adaptation to new species, much as insects often have proved capable of evolving resistance to insecticides, it seems likely that the best preadaptation to an unfamiliar species may be adaptation to familiar ones. We must ask, then, whether genetic variations have correlated effects, providing adaptation both to naturally interacting species and to novel ones. There is extraordinarily little information on such genetic correlations in natural populations; there is need for studies such as Arnold's

(1981) model analysis of the western terrestrial garter snake (*Thamnophis elegans*) in which behavioral responses to different prey items such as slugs and leeches are genetically correlated. From the literature on plant–insect interactions (with which I am most familiar), there is reason to believe that such correlations may be common. First, the responses of insects to novel plants suggest that the insects respond to chemical factors that are shared among plant species, especially related ones. Thus specialized insects often attack novel plants that are related to their natural hosts. The Colorado potato beetle (*Leptinotarsa decemlineata*) moved readily onto the cultivated potato (*Solanum tuberosum*) from wild species of *Solanum*; *Caligo* butterflies, a major pest of banana in Central America, have as their natural hosts other members of the banana's order (Zingiberales). Rodman and Chew (1980) have described the case of a pierid butterfly (Lepidoptera: Pieridae) in western North America that lays eggs on introduced European crucifers (Cruciferae, the mustard family) even though the larvae fail to survive on these plants, apparently because the females respond to glucosinolates that are shared with a native crucifer host. Moreover, many of the secondary compounds of plants appear to be toxic to or repel a wide variety of insects other than those that are specifically adapted to overcome these defenses. Such observations only suggest the possibility of genetic correlations in resistance to various enemies. A more rigorous approach is that exemplified by Gould's (1979) study of correlated effects of selection. Spider mites (*Tetranychus urticae*) selected for adaptation to a cultivar of cucumber (*Cucumis sativus*, family Cucurbitaceae) that presumably owed its toxicity to cucurbitacin proved to have achieved a higher level of adaptation to unrelated solanaceous plants (Solanaceae) which do not contain cucurbitacin) than a control line of mites manifested.

The plant breeding literature suggests that factors that improve resistance to one insect pest may or may not affect resistance to others. For example, in cotton (*Gossypium*: Niles, 1980), increased hairiness of the leaf seems to reduce damage by leafhoppers (Homoptera: Cicadellidae) and perhaps the boll weevil (*Anthonomus grandis*), but the hairless condition appears to reduce damage by mirid bugs (Hemiptera: Miridae) and pink bollworm (*Pectinophora gossypiella*). The absence of extrafloral nectaries confers some protection against mirids and bollworms (*Pectinophora* and *Heliothis*), but not against spider mites. A genetic alteration of the bracts around the flower bud, known as frego bract, confers some resistance to the boll weevil, but increases susceptibility to mirid bugs. And a high level of gossypol, a sesquiterpene, reduces damage by leafhoppers, bollworms, and aphids (Homoptera;Aphidoidea), but apparently has no effect on spider mites and may actually increase damage by thrips (Thysanoptera) and whiteflies (Homoptera:Aleyrodidae). In many of these instances, it is not certain whether resistance is conferred by the trait under observation or by linked genes.

The existence of both positive and negative correlations in the effect of plant resistance factors on different species of insects has, I believe, great practical significance. On the one hand, the positive correlations provide a basis for the likelihood that factors that have protected plants against their natural enemies will provide preadaptation to novel ones. On the other hand, the negative correlations suggest the possibility that a full complement of insect species will impose conflicting selection pressures, and so maintain polymorphism for traits that confer resistance to some enemies, but susceptibility to others. Genetic variation for traits that can respond to selection by newly imported enemies may thereby be maintained. In this way, a diverse complement of natural enemies may be instrumental in enabling populations of their hosts to respond to changes in the biotic environment.

I see little reason to doubt that these ideas, which I have presented primarily in the context of interactions between plants and phytophagous insects, should apply equally to interactions among prey and pedators generally, or to interactions between animals or plants and pathogenic microorganisms. If my chapter is phrased in largely speculative terms, it is only because of the paucity of information on genetic aspects of multispecies associations, a topic that has yet to be explored in any systematic way.

The major practical implications of this all too speculative discussion are three. *First*, one of the major ecological threats to the persistence of managed populations, surely, is the introduction of species to which they are not adapted. Persistence may depend on the availability of genetic variation in traits that confer resistance to newly introduced species. *Second*, the traits that affect ecological interactions among species may often be under strong local selection, so that any particular local population can be less genetically variable for such traits than is an ensemble of populations occupying different habitats and thereby exposed to different local selective pressures. This implies that conservation of a desirable species should, ideally, include the preservation of areas that are large enough to contain subdivided populations in each of the ecological associations in which the species occurs. *Third*, the genetic variation that a species may mobilize in response to the incursion of new species of prey, predators, or parasites is likely to be actively maintained by the natural complement of species with which it interacts, and very probably will be eroded in the absence of the balancing selection that these species impose. Therefore, even if the esthetic ideals on which conservation programs are usually based do not include fungi, insects, parasites, and other unpopular organisms among the desirable species, these organisms may well be important to the long-term persistence of the species to which the conservation ethic is applied. The conservation of whole ecological communities, in as intact a state as possible, is therefore a desirable ideal, not only from the purely ecological consideration that species often depend on one another in subtle ways of which we are unaware, but from a longer-term, evolutionary consideration as well.

CHAPTER 22   Interspecific Interactions and the Maintenance of Genetic Diversity

The bark beetles (Coleoptera: Scolytidae), rusts (Uredinales), and gall aphids (Homoptera: Eriosomatidae) that a forest manager would gladly see extinguished may provide a kind of immunization against dangers unforeseen.

Summary

This chapter draws primarily on genetic studies of herbivorous insects and their host plants, but many of the conclusions probably apply also to interactions between hosts and their parasites and pathogens, and to other antagonistic interactions among species. Outbreaks of insects and pathogens have been documented to cause very high mortality, sometimes even extinction, of both agricultural and native plants. The impact is especially strong on genetically uniform populations. In some instances pest outbreaks are caused by genetic changes in pest populations. Maintenance of stable species associations may therefore depend on retention of genetic variation that provides adaptation to genetically changed species—that is, on coevolution. Genetic variation is promoted by interspecific interactions in several ways; instances are described in which natural herbivory by insects maintains variation in plants, and variation in insects is maintained by adaptation to different plant species. Because adaptation to one species may provide cross-adaptation to others, the members of species-rich communities are more likely to harbor genetic defenses against novel enemies than are species in impoverished communities. Because the level of genetic variation within a population depends on a balance between selection and gene flow, the genetic variance among populations in resistance to natural enemies may be greater than the genetic variance within populations. Therefore conservation of genetic diversity should emphasize maintenance of preserves that are large enough to contain subdivided populations.

CHAPTER 23

The Relevance of Captive Populations to the Conservation of Biotic Diversity

Thomas J. Foose

Introduction

With the accelerating destruction of the world's wildlife and wildlands, gene pools are becoming diminished and fragmented into gene puddles. As affirmed in the *World Conservation Strategy* of the International Union for the Conservation of Nature and Natural Resources or IUCN (1980), captive or cultivated populations and propagation can and must be integral parts of the global strategies to preserve the biotic and genetic diversity of the planet.

There are at least six main ways by which zoological parks can contribute to these strategies:

1. By serving as refugia for taxa destined for extinction in the wild.
2. By providing propagules for repopulation of natural habitats.
3. By reinforcing natural populations that are so small and fragmented that they are not genetically and demographically viable.
4. By maintaining repositories for germ plasm as an adjunct or an alternative to populations of living animals.

CHAPTER 23 Relevance of Captive Populations to Conservation of Biotic Diversity

5. By conducting research that will improve management for wild as well as for captive populations.
6. By educating the public to support conservation of wildlife in natural habitats.

The first four of these methods are direct, the last two indirect. Educating the public is the traditional role that has been recognized for zoos in wildlife conservation. But the crisis of our times requires that zoos perform more active functions. Obviously, serving as refugia for taxa destined for extinction in the wild is a last and less-than-optimal resort. But by performing functions (2), (3), and (5), zoological parks can contribute to programs for conservation of wildlife in natural habitats and diminish the need for function (1).

Indeed, we appear to be moving toward a world where the survival of many taxa will depend on the interactive management of both wild and captive populations. Moreover, as wild populations are reduced and fragmented, while captive collections and zoo facilities become larger, more naturalistic, and better coordinated, wild and captive populations are converging in terms of the kinds of management that must be employed for their survival.

Two major biological problems confront development of the captive programs: (1) selection of taxa, and (2) management of populations. Resolution of these problems will require a collective and strategic approach by zoological parks. In North America, zoological parks are attempting such a strategy through a program known as the Species Survival Plan (SSP) of the American Association of Zoological Parks and Aquariums (AAZPA). The SSP provides the best available case study to demonstrate how the major biological problems might be resolved.

Selection of Taxa

Selection of taxa must be considered in terms of both the number and the nature of the "entities" to be preserved. Chambers and Bayless (Chapter 21) have emphasized the difficulties of equating the biotic and genetic diversity that probably should be preserved with taxa as they are currently defined. Zoos are engaged in developing guidelines and assigning priorities concerning the taxonomic diversity and refinement sustainable in captive facilities and programs. In this endeavor, close consultation with population biologists and other wildlife managers is essential and again demonstrates the need for an integrated strategy to conserve wildlife.

Presently, taxa are being designated for captive programs by three very general criteria: (1) endangerment in the wild; (2) representation of diversity (taxonomic, zoogeographic, genetic); and (3) feasibility in captivity. Unfortunately, even with scientific and coordinated management, the capacity of zoos for populations large enough to be viable is very limited in relation to the great and growing number of taxa requiring sanctuary in captivity. Thus, selection of taxa becomes a process of assigning priorities for allocation of the space and resources available in zoos. For genetic reasons, captive populations should be as large as possible (Figure 1). But there are many taxa competing for the captive habitat. Consequently, it seems necessary to establish a "carrying capacity" for each taxon. This carrying capacity must be a compromise between maintaining large populations for genetic diversity and demographic stability and providing sanctuary for as many taxa as possible.

The problems and process of taxa selection for zoos can be exemplified by how the SSP is treating the family Rhinocerotidae, one of the most critically endangered groups of mammals on the planet. All five of the recognized species of rhinos are to some degree endangered (Table 1). It may already be too late for the two rarer species of Asian rhino: the Sumatran (*Dicerorhinus sumatrensis*) and the Javan (*Rhinoceros sondaicus*). Only the southern subspecies of the white rhino (*Ceratotherium simum simum*) is considered temporarily secure in the wild, although that situation could change drastically and rapidly.

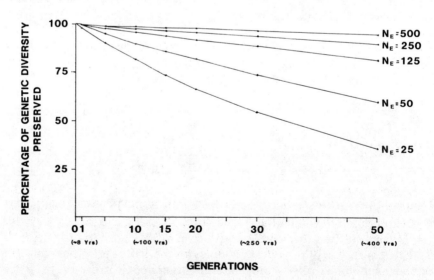

Figure 1. *Loss of genetic diversity (as measured by heterozygosity) due to random drift for various effective population sizes (N_e) for a total number of 250 animals, based on a rate of decline in heterozygosity of $1/2N_e \times 100\%$ per generation. The generation time of 8 years applies to Siberian tiger (*Panthera tigris altaica*). From Foose and Seal (1981).*

Table 1. Rhinos (Rhinocerotidae) in the Wild

Species	Estimated numbers	Distribution	Population trends
African			
Black *(Diceros bicornis)*	14,000–24,000	Many populations in sub-Saharan Africa	Declining precipitously
White *(Ceratotherium simum)*			
Northern (*C. s. cottoni*)	1,000	Two main populations	Decreasing rapidly
Southern (*C. s. simum*)	2,600–2,800	Several populations; more being established	Increasing
Asian			
Indian *(Rhinoceros unicornis)*	2,000	Several populations in India and Nepal	Increasing or stable temporarily
Javan *(R. sondaicus)*	< 57–66	One population	Increasing
Sumatran *(Dicerorhinus sumatrensis)*	200	Small and fragmented populations over a wide range in Southeast Asia	Decreasing

Table 2. Rhinos in Captivity[1]

	Indian rhino	Black rhino	White rhino Southern	White rhino Northern	All rhinos
North America (ISIS, 31/12/81)					
Current population	11/11 = 22	26/30 = 56	77/95 = 172	1/0 = 1	115/136 = 251
Institutions with species	8	24	48	1	62
	(9 owners)	(25 owners)			
Institutions with singletons	1	4	1	1	Not applicable
Institutions with pairs	3	9	28	0	Not applicable
Recent reproduction					
1977	0	3/1 = 4	3/1/1 = 5	0	6/2/1 = 9
1978	0/1 = 1	3/0 = 3	3/0/1 = 4	0	6/1/1 = 8
1979	0	1/1 = 2	5/6/1 = 12	0	6/7/1 = 14
1980	1/0 = 1	2/0/1 = 3	1/5 = 6	0	4/5/1 = 10
1981–82	0/1 = 1	3/1 = 4	2/2/4 = 8	0	5/4/4 = 13
World (Studbooks, 31/12/80)					
Current population	38/33 = 71	76/92 = 168	245/294 = 539	8/11 = 19	367/430 = 797
Institutions with species	32	67	118	7	Not calculated

[1] Ratios are number of males/number of females.

CHAPTER 23 Relevance of Captive Populations to Conservation of Biotic Diversity

Analysis of the capacity of existing facilities, as measured by the numbers North American zoos are currently maintaining, suggests that there is captive habitat for about 300 rhino (Table 2). How should this habitat be allocated? The initial proposal has been for the SSP to concentrate on three taxa representing the most endangered genera of rhinos: Indian (*Rhinoceros unicornis*), black (*Diceros bicornis*), and Sumatran (*Dicerorhinus sumatrensis*). A population of 100 has been proposed for each species.

A population of 100 is lower than the genetically effective population size (N_e) of 500 recommended for long-term survival by Franklin (1980). However, 100 seems justifiable for the three following reasons:

1. If intensively managed, an actual number of 100 can be enlarged toward a theoretical maximum of 200; that is, N_e can equal $2N$.
2. As discussed later in this chapter, loss of genetic diversity due to random drift is a function not only of effective population size (N_e) but also of generation time (Figure 1). The generation time indicated in Figure 1 applies to a Siberian tiger and is computed as 8 years; generation time of a rhino, depending on the species, would be 1.5 times to twice as long. Thus if the objective is to preserve specified levels of genetic diversity for prescribed periods of time (e.g., the next 200 years), animals like rhinos with relatively longer generation times perhaps can be viably preserved for longer time periods by somewhat smaller populations.
3. It is expected that other regions (e.g., Europe, Asia, South America) will develop captive programs and populations on a scale comparable to what is occurring in North America. Hence, the total captive population for each taxon may reach approximately 250 to 500 animals (actual number).

However, complicating this situation is the fact that the southern white rhino, which is not currently endangered in the wild, is the most populous form of rhino in captivity (largely because zoos have been serving as repositories for surplus removed from wild populations in order to stabilize them). Moreover, many zoos maintain the southern white rhino in simple pairs, a social situation in which they do not reproduce well. In contrast, both the Indian and black rhinos (and presumably Sumatran) do reproduce well in such situations. Thus it seems sensible to convert from the southern white rhino to these other rhinos in this kind of captive habitat. Consequently, there will be an attempt to relocate many of these southern white rhinos in zoological parks to new repositories (e.g., private ranches) that can accommodate relatively large herds of the subspecies. Such a program will achieve several objectives:

1. It will expand the capacity of captive facilities in North America for rhinos and thereby enable a program to be developed for all genera of the family.
2. It will place the southern white rhino in a situation more conducive to propagation.
3. It will create more habitat in zoos for species than are more immediately in need of close management and that can propagate well in such circumstances.

Although not critically endangered as a species in the wild, an SSP program for the southern white rhino seems justifiable (1) on the grounds of preserving biotic diversity as the taxon represents a unique genus; (2) since it seems inevitable that even this genus of rhino has a limited future in the wild (see below); and (3) because of the simple fact that it would, for a variety of reasons, be very difficult to destroy all these specimens in North American zoos.

But there is yet another problem; there are two subspecies of white rhino. While the southern subspecies (*Ceratotherium simum simum*) is considered temporarily "safe" by the IUCN, the northern (*C. s. cottoni*) is one of the most endangered rhinos. However, until recently the prevalent opinion has seemed to be that the two subspecies were not very different and so the southern subspecies would adequately preserve the uniqueness of the species and the genus. The situation has now changed somewhat. Benirschke (Chapter 24) mentions the recent research by Ryder and colleagues at the San Diego Zoo, which may reveal (though the sample size is very small) a more significant distinction between these two taxa. Should the SSP zoological parks attempt a massive effect with the northern white rhino? Such an endeavor would seem to require (1) appreciable expansion of facilities, (2) massive elimination of the southern white rhino, and/or (3) abandonment of one of the other species already designated. Problems of this nature will not be easy to resolve, but the white rhino dilemma does emphasize the need for determining as much as possible about systematic relationships to provide direction to conservation programs.

Some more perspective on the problem of inadequate facilities can be developed by considering a more extensive group such as the larger felids (Felidae, Table 3). The same conclusion emerges; there simply is not enough captive habitat even for all the endangered taxa. For example, there are going to have to be decisions on how many kinds of tigers zoos will maintain. Maintenance of populations large enough to be viable for all subspecies is impossible with the current capacity of zoos; maintenance of three tiger subspecies (e.g., Siberian, *Panthera tigris altaica*; Bengal, *P. t. tigris*; and Sumatran, *P. t. sumatrae*) will be difficult. Reduction to "temperate" and "tropical" tigers may be logical. Eventually, preserving just "the tiger" may be necessary. White tigers (of *P. t. tigris*) discussed by

Table 3. Capacity of ISIS Captive Facilities for Larger Felids (Felidae)[1]

Species	Extant subspecies	Subspecies in Red Data Book	ISIS institutions	ISIS population	Number of subspecies if population is:		
					100	250	500
Panthera leo (lion)	11	1	97	381	4	1	1
Panthera tigris (tiger)	8	8	110	450	4	2	1
Lions and tigers (subtotals)	19	9	120	831	8	3	2
Panthera onca (jaguar)	8	8	65	178	2	1	0
Panthera pardus (leopard)	15	15	72	246	2	1	0
Panthera uncia (snow leopard)	1	1	35	128	1	0	0
Felis concolor (puma)	29	2	69	173	2	1	0
"Intermediate" cats (subtotals)	53	26	N.C.	725	7	3	1
Neofelis nebulosa (clouded leopard)	4	4	20	63	1	0	0
Acinoynx jubatus (cheetah)	6	6	32	166	1	0	0
"Other large felids" (subtotals)	10	10	N.C.	229	2	1	0
Totals	82	45		1785	18	7	3

[1]N.C. = not calculated.

Table 4. Apparent Capacity of ISIS Facilities for Ungulates (Perissodactyla and Artiodactyla)

Family	Genera	Species	Subspecies	Taxa cited in Red Data Book				Specimens in ISIS 31-12-81	Taxa maintainable in population of:		
				G	S	SS			100	250	500
Cervidae (deer)	15	40	205	9	18	35		2139	21	9	4
Tragulidae (chevrotains)	2	4	60	0	0	0		31	0	0	0
Antilocapridae (pronghorns)	1	1	5	1	1	2		76	1	0	0
Bovidae	51	127	486	22	38	60		6827	69	28	14
Bovinae (cattle, etc.)	4	12	26	3	9	18		880	9	4	2
Caprinae (sheep, goats)	11	26	112	5	8	15		1656	17	7	3
"Antelope"	36	89	348	14	21	27		4291	43	17	9
Camelidae (camels, llamas)	3	6	7	2	2	2		891	9	4	2
Giraffidae (giraffes, okapis)	2	2	10	0	0	0		319	3	1	0
Suidae (pigs)	5	8	78	2	2	5		11	0	0	0
Tayassuidae (peccaries)	2	3	20	1	1	1		88	1	0	0
Equidae (horses, zebras, etc.)	1	8	21	1	5	9		1187	12	5	2
Tapiridae (tapirs)	1	4	7	1	3	3		147	1	0	0
Rhinocerotidae (rhinos)	4	5	9+	4	5	8		208	2	1	0
Hippopotamidae (hippos)	2	2	5	1	1	2		186	2	1	0
Total	89	210	914	44	76	127		12,110	121	48	24

Ralls and Ballou (Chapter 10) are a further serious complication because they also compete for the "tiger habitat" in zoological parks.

Examination of similar data for ungulates (Perissodactyla and Artiodactyla, Table 4) leads to the same conclusion: the current capacity of captive facilities is woefully inadequate. There may indeed not be space for such subspecies as magnificent as the giant sable antelope (*Hippotragus niger variani*).

Management of Populations

The kind of surveys presented in Tables 2-4, as well as the genetic and demographic analyses requisite for populational management, are possible only when adequate data are available. The International Species Inventory System or ISIS (Flesness et al., 1982) attempts to provide such data by compiling vital statistics on zoo animals. Moreover, for populations in the wild where identification of individuals is possible, ISIS also may be of service. There is already a proposal to enter the black rhino of Amboseli and perhaps other national parks in Kenya into the ISIS data bank.

1. Increasing Captive Habitat

Realizing the need for more captive habitat, zoo managers are attempting to expand zoo capacities. Their most relevant methods to the subject of this book are by increasing gene flow within captive populations through periodic redistribution of individuals and by regulating such factors as sex ratio and family sizes of reproducing animals. These activities are enlarging the effective population sizes of the species actually being maintained in zoological parks. Even with maximal management, however, there still will not be enough captive habitat. Thus, there is a need for more and larger zoos.

Territorial expansion is occurring both by zoological parks' enlarging their own facilities (e.g., the San Diego Wild Animal Park, the U.S. National Zoological Park's Conservation and Research Center at Front Royal, Virginia, and the St. Catherine's Island Survival Center of the New York Zoological Society), and also by their developing cooperative relationships with private establishments possessing vast amounts of land and a sincere commitment to conservation. The SSP is pursuing such a project on Grevy's zebra (*Equus grevyi*) with a ranch in Texas.

Some discussion of how this project is being developed may be instructive in terms of the genetic and demographic factors that are being considered in what is essentially the founding of a new population. Hence, this project may be valuable as a model for establishment of populations in natural or native habitats.

Attempts at establishing new populations, especially from captive stock, have often failed. There seems to be reason to suspect that the founder stock for such projects has been deficient in genetic diversity. Stock appears frequently to have been derived from highly inbred lineages. In such cases, inbreeding depression of survival and fertility could have caused the failure of the introductions. Success seems more likely if a genetically diverse set of founders is selected. Founders should be as unrelated and as noninbred as possible.

The number of founders is not nearly as important as their subsequent management (Senner, 1980; Foose and Foose, 1982). Usually, 5 to 10 pairs of prudently selected founders will be adequate.

For the Grevy's zebra population being established by the SSP in Texas, an analysis has been performed of the various genetic lineages or "bloodlines" existing in North American zoos. Different "bloodlines" are defined in terms of lineages from various wild-born founders that had been imported to the zoos from Africa over the years (Figure 2). It has been decided to assemble a foundation stock of 10 male and 10 female zebras for the Texas herd. By careful analysis and selection, it has been possible to obtain this stock from 15 completely distinct bloodlines.

Demographic factors also should be considered when establishing a new population. Perhaps the most basic guideline is based on a demographic parameter known as reproductive value, which is computable for any population where age- and sex-specific characteristics can be distinguished (Conley, 1978; Goodman, 1980). As with all demographic parameters and models, reproductive value is calculated from age- and sex-specific patterns of survival and fertility. Reproductive value is a measure of what the relative contributions will be from animals of various ages to the next generation, considering the probabilities of surviving to and then reproducing at various times of life.

Figure 3 depicts the reproductive value for the Asian wild horse (*Equus przewalski*), which may be considered a representative equid. Lenarz and Conley (1980) discuss the relevance of reproductive value to establishment of new populations. Generally, a population founded by animals from the age class with the highest reproductive value under prevailing conditions will grow most vigorously and hence attain the largest possible population after a particular period of time.

Thus, empirical evidence and theoretical stimulations suggest the characteristics of stock to found a new population, whether captive or wild. The founders should be as genetically diverse and have as high a reproductive value as possible. Five to ten pairs are normally adequate although in some cases fewer may suffice. The specification of pairs emphasizes the point that, even for highly polygynous species, approximate equality of sex ratio is advised to mitigate the depression of N_e that occurs

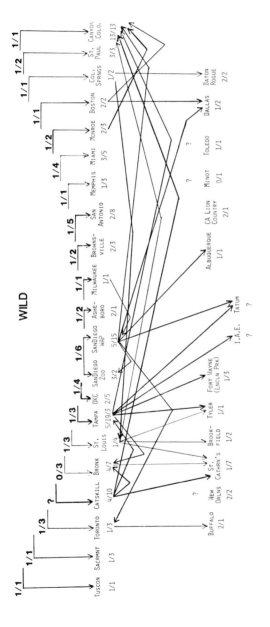

2. Bloodline analysis of Grevy's zebra (Equus grevyi) performed as a basis for selecting stock to found a new population on a Texas ranch as part of a cooperative program with North American zoos. Ratios are numbers of males/females.

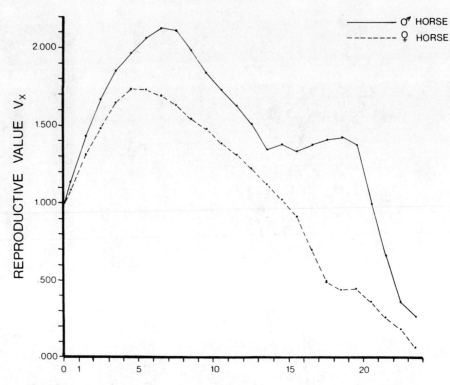

Figure 3. *Reproductive value by age class for the captive population of Asian wild horse (*Equus przewalski*). From Foose (1980, 1981).*

when there is a great disparity in the number of males and females reproducing (Franklin, 1980; Seal and Flesness, 1979). In wild populations, of course, there are limits to how much management of sex ratio is possible, but the manager can at least provide the potential for, if not actual implementation of, rotation of males in the producing groups.

Zoological parks also may increase their capacities through technological expansion. Benirschke (Chapter 24) describes many of the developments, possibilities, and problems in this area. In particular, the application of cryogenic techniques in conjunction with artificial insemination and embryo transplantation may augment the living populations by reducing the actual number of animals that must be maintained to achieve a certain effective population size (N_e).

Whatever taxa are selected, they can be viably propagated and preserved in a captive situation only if they are managed intensively as biological populations. Many endangered taxa reproduce well in captivity, as illustrated by Figure 4, which depicts the captive history of the snow leopard (*Panthera uncia*). However, even if many of the species cited seem self-

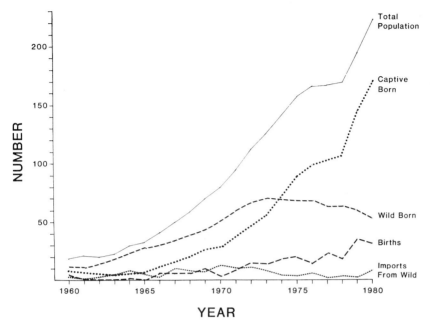

Figure 4. *History of snow leopard (*Panthera uncia*) in captivity.* From Foose (1983).

sufficient because they are reproducing well, the viability or vitality of these populations in an evolutionary sense may be problematic. For example, the Pere David's deer (*Elaphurus davidianus*) is considered one of the great successes of captive preservation, but the entire captive population probably descends from no more than 3 founders (Jones, 1982) and manifests almost no variability upon electrophoretic examination (Ryder et al., 1981).

2. Genetic Management

To contribute to the conservation of the evolutionary and ecological potential of taxa, captive populations must be analyzed and managed genetically and demographically. The primary objective of the genetic management of captive populations must be to preserve as much as possible of the heritable diversity that has evolved and exists in the wild gene pools.

An important development has been the recent emergence of general agreement among many North American "zoo geneticists" on a basic strategy for genetic management of captive populations (see Foose and

Foose, 1982). The basic components of this strategy are summarized below.

1. *Acquire an adequate number of founders.* Since no more diversity can be preserved in captive populations than has been obtained from the wild, as many founders as possible are desirable. Relatively few founders may be available, however, especially for rare taxa. Moreover, with such forms there must be care not to decimate the natural populations. If prudently selected, a few founders (5 to 10 pairs) can provide an astonishingly significant sample of the average diversity of the pertinent gene pool (Figure 5). A simple guideline is that founders should be unrelated, noninbred, and of course interfertile.

Complicating this simple prescription is the fact that founders may not always enter the population at the same time—that is, at its inception. For the captive populations, an animal entering from the wild at any time has been considered a "founder." This terminology has seemed justified, since gene flow has tended to be unidirectional from the wild to captivity. However, as such a gene flow continues, the distinction between later

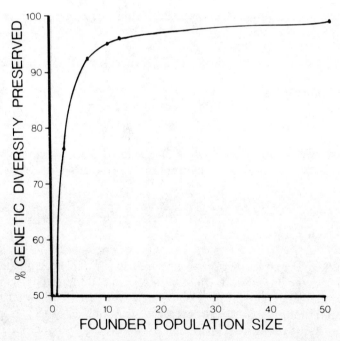

Figure 5. *Average percent of genetic diversity (as measured by heterozygosity) contained in founder populations of various sizes. It is assumed that founders are unrelated and noninbred. Diversity preserved is equal to $(1 - 1/2N_e) \times 100\%$. From Foose and Foose (1981).*

founders and migrants is obscured. In any case, a strategy also is emerging for optimal incorporation of the new "founder" or migrant into the managed population (K. Jones, personal communications).

2. *Expand the population as rapidly as possible from these founders to the carrying capacity determined*, with attention to other components of the strategy—for example, equalization of founder representation or bloodlines.

3. *Perhaps, subdivide the population.* The number and size of these subdivisions is not a point of general agreement and indeed may vary depending on the taxon being managed. For example, zoo populations of each continent may be considered the "neighborhood" as described by Chesser (Chapter 4) or the "population" discussed by Flesness (1977) or Foose and Foose (1982). Whatever the scheme of subdivision, there should be periodic, closely regulated exchange of genetic material between the subdivisions to minimize effects of inbreeding depression and perhaps genetic drift. Toward this objective, there recently has been a very significant exchange of Przewalski's or Asian wild horses between the North American and Soviety zoo populations.

4. Within the significant subdivisions recommended in 3:

A. *Maximize effective population size* (N_e). Figure 1 depicts the decline in genetic diversity due to genetic drift. This process has been discussed extensively elsewhere (Seal and Flesness, 1979; Franklin, 1980; Soulé, 1980; and other chapters in this book). Nonetheless, there are several points that should be repeated or further accentuated.

Loss of genetic diversity depends upon the size of a population. In general, the smaller the population, the faster genetic diversity is lost. However, the critical size is not merely the total number of individuals in the population, but rather the genetically effective size, N_e, which is a

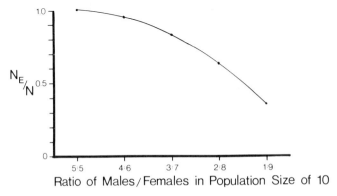

Figure 6. *Depression of effective population size (N_e) due to disparity in sex ratio (m/f) of reproducing animals. From Foose and Foose (1982).*

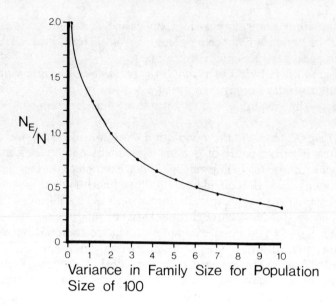

Figure 7. *Depression of effective population size (N_e) due to inequality of lifetime family sizes—that is, the total number of surviving and reproducing offspring each animal produces. From Foose and Foose (1982).*

function of the structure and the dynamics of the population. The same total number of individuals can produce very disparate values of N_e. Populations can be managed so that their effective population size can vary from a very small fraction to twice the actual number of individuals. Significant factors determining effective population size are the number, sex ratio, and family sizes (that is, total number of offspring produced by an individual in its lifetime) of the individuals that actually reproduce. The smaller the fraction of animals that actually reproduce and the more disparate their sex ratio and family sizes, the lower the effective population size will be (Figures 6 and 7).

For example, of the 558 southern white rhinos in captivity at the end of 1980, only 20 males and 65 females were reproducing. Using the simple formula (another form of equation (1) in Chapter 15) for effect of sex ratio on N_e,

$$N_e = \frac{4N_m N_f}{N_m + N_f}$$

where N_m is the actual number of breeding males, and N_f is the actual number of breeding females, the genetically effective size is only

$$N_e = \frac{(4)(20)(65)}{20 + 65} = 61$$

Maximizing effective population number for the carrying capacity of the captive population will therefore require that (1) as many animals as possible be recruited to the reproducing population, and (2) the sex ratio, and especially the family sizes, of these individuals be equalized as far as is feasible. With the intensive genetic management that is feasible in zoos, N_e can be increased to be approximately equal to or perhaps even greater than the total number of animals (N) in the populations. This possibility is in contrast to the situation in wild populations where N_e is usually significantly less than N.

The same possibilities for manipulating N apply to wild populations, although the management implied normally will be much more difficult to implement. Selective removal of dominant males or of already prolific females could regulate sex ratio and family size, although at the cost of interfering with natural selection. Consequently such intervention should

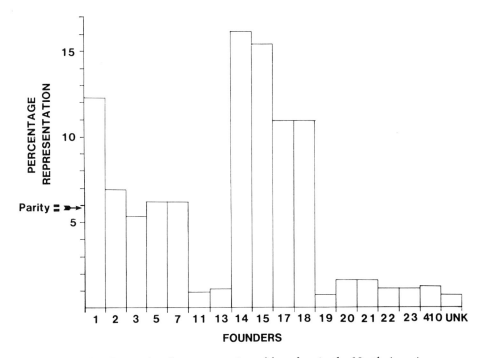

Figure 8. Proportional representation of founders in the North American population of Siberian tiger (Panthera tigris altaica) alive on 1 January 1981. From Foose and Seal (1981).

occur only after a thorough analysis of the objective and realities of the genetic management of the particular population.

Another important point is that genetic drift depends upon both the effective population size and the generation time. Effective populations of 250 to 500 individuals will preserve a high fraction of the original genetic diversity for 100+ generations, a period of time that will, for most large vertebrates, be centuries or even millennia.

B. *Equalize founder representation.* Preservation of genetic diversity will be maximized by equalizing the representation of founders through time. Unfortunately, founders or bloodlines are usually very unequally represented in captive populations (Figure 8, and Chapter 15 by Templeton and Read). However, zoos are becoming increasingly aware that equalizing founder representation is a more important criterion for managing captive populations than merely minimizing inbreeding coefficients. Rectifying disparity in founder representation has motivated the exchange of individuals of several very rare taxa between zoos during 1981–1982. These include the Siberian tiger, Asian wild horse, and okapi (*Okapia johnstoni*).

C. *Manage inbreeding coefficients.* In most cases, minimization of inbreeding coefficients is the best course (Flesness, 1978; Foose, 1981; Ryder et al., 1981). Templeton and Read (Chapter 15) describe a breeding program that includes inbreeding, which applies to cases where severe inbreeding is unavoidable.

3. Demographic Management

Demographic management is inextricably interrelated with genetic management and hence also is essential for captive populations. Captive populations whose reproductive husbandry has been reasonably mastered can possess explosive potentials for expansion (Figure 9). The sizes of these populations must be within the carrying capacities of zoological parks. In simplistic terms, diversity is the major objective of the genetic management of captive populations, and stability is the goal of the demographic regulation. Stability is particularly important for genetic reasons. If populations fluctuate significantly in number, the effective population size will be closer to the minimum than to the maximum and the rate of loss of alleles due to drift will increase (Franklin, 1980). Goodman (1980) and Foose (1980, 1981) discuss in detail the demographic regulation of closely managed captive and wild populations.

Figure 9. *Capacity for growth of captive population of Siberian tiger (*Panthera tigris altaica*) under various schedules of permitted reproduction. From Foose and Seal (1981).*

CHAPTER 23 Relevance of Captive Populations to Conservation of Biotic Diversity

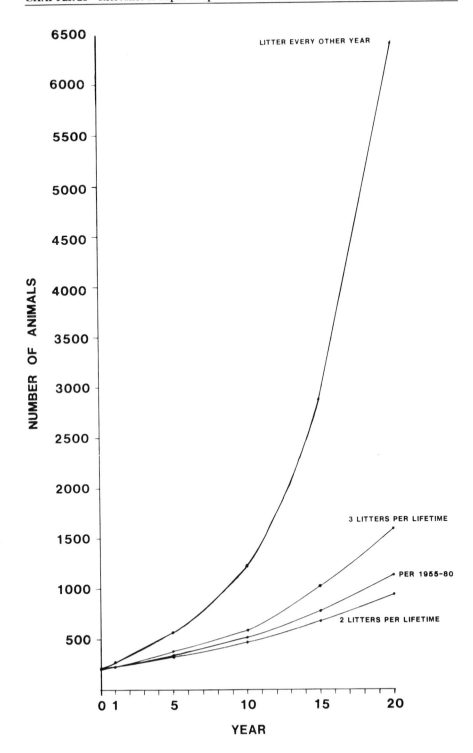

A general strategy also has emerged for the demographic management of captive populations (Foose 1980, 1981), as follows.

1. *Establish a carrying capacity that represents a compromise between maintaining a population large enough for genetic diversity and providing sanctuary for as many taxa as possible.* For many species of vertebrates, (Franklin, 1980) recommends a minimum effective population size of 500 for long-term preservation of evolutionary potential. With reasonable management, N_e can equal the actual number of animals maintained in a captive population. More intensive management can enlarge N_e even further, as discussed earlier.

2. *Expand population to carrying capacity as rapidly as genetic management permits.*

3. *Stabilize population at the carrying capacity.* Demographically, a population can be characterized by and hence managed through control of three basic parameters:

   A. The age and sex structure of the population at particular times — that is, how many animals of each sex are in each age class.
   B. The age and sex-specific survivorships — that is, how long animals live on the average, or equivalently, what the chances are that an animal will survive to a certain age.
   C. The age and sex-specific fertilities — that is, how well animals of various ages usually can be expected to reproduce.

Basically, there are two ways to stabilize a population demographically, as follows.

   A. *Prevention.* The fertilities can be modified by regulating reproduction through various kinds of birth control, thereby preventing more animals from entering the population.
   B. *Removal.* Survivorship can be modified by removal of individuals from certain age classes. It will be desirable to place the removed individuals in natural habitats, but in many cases euthanasia will be an inevitable method of regulation.

Whatever the management program, the animals specifically selected for removal or reproduction will best be identified by genetic criteria so as to maximize preservation of genetic diversity. Naturally, both genetic and demographic management must be applied within constraints imposed by the sociobiology of the species.

The age specificity of these managerial modifications emphasizes an important principle of demographic management. It is not enough, or even possible, merely to regulate numbers. Age distributions also must be managed. Indeed, critical to demographic management of captive popula-

CHAPTER 23 Relevance of Captive Populations to Conservation of Biotic Diversity   395

tions is stabilization of age structures. Constancy of number at a carrying capacity is impossible without stability of the age distributions, unless the population size is being maintained by imports of immigrants from elsewhere. This kind of demographic subsidization will not be possible or desirable for many or most endangered species. The stable age distribution may be as important a concept as inbreeding for management of captive populations.

A necessary, but not sufficient, trait of the stable age distribution is that each age class must contain an equal or greater number of animals than any older age class. This characteristic confers a recognizable configuration on a stable age distribution. Not every such pattern will represent a stable configuration for a population; the relative number in each class must represent the proportions specified by the stable age distribution defined by the suvivorship and fertilities. If an age distribution does not have this kind of configuration, it is not stable. Populations without stable age distributions will behave erratically, often oscillating drastically and detrimentally.

Hence, to equilibrate a captive population at the carrying capacity, management should attempt to produce a stable age distribution at approximately this desired size. In demographic terminology, a population stable in both total numbers and age distribution is designated "stationary."

Unfortunately, when confronted by proliferating populations, a comprehensible and common response by zoos in the past has been drastic curtailment of reproduction. The problem is that this kind of management can badly distort age distributions, placing more animals in older than in younger age classes, thus creating a demographically very unstable situation. A prime example is presented by the recent history of Siberian (or Amur) tigers in North America (Seal and Foose, in press). Over the period 1967–1977, the age distribution became more and more unstable (Figure 10). If this trend would continue for another 10 years, about 45% of the females and 19% of the males would be well past the demographic prime for this subspecies. The populations would be well on the way to demographic senescence and instability. Senescence of the age structure has probably been an appreciable cause of effective extinction in individual zoos and of instability in captive populations as a whole, and would be more conspicuous if zoos had not been able frequently to rejuvenate their populations by stock obtained from other institutions or from the wild.

If the patterns of survival and fertility were to continue at the level computed as average for the period from 1955 to 1980, the population could be stabilized at 250 by annually removing either 46% of the 0- to 1-year-olds, or 7% from every age class (Table 5). Other options where animals are permitted to reproduce multiple times during their lives require other levels of removal or harvest.

The computations suggest that the populations also could be stabilized entirely by regulating reproduction. However, the calculations may be

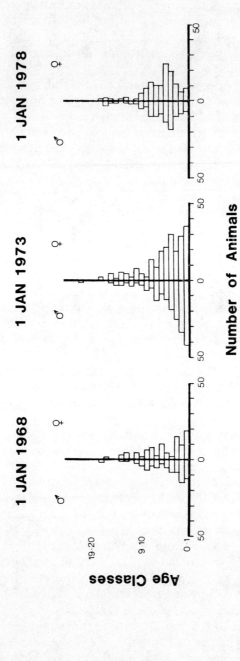

Figure 10. *Age distribution of Siberian tigers (Panthera tigris altaica) in North American zoos from 1969–1978.*

Table 5. Some Options To Attempt Demographic Stabilization of Siberian Tiger Population at Carrying Capacity of 250

	Option 1	Option 2	Option 3	Option 4	Option 5	Option 6	Option 7	Option 8	Option 9
If reproduction is:	Equivalent to 1955–80 litter size 2.43 cubs	Adjusted to compensate for mortality	One litter 2.43 cubs at any age	One litter 3 cubs at age 4	One litter 3 cubs at age 5	Two litters 2.43 cubs each at ages 5 and 10	Two litters 2.43 cubs each at ages 4 and 7	Three litters 2.43 cubs each at ages 4, 9, and 12	Litter of 2.43 cubs alternate years
then									
The removal of 0–1-year-olds required for stability is:	46%	0%	This level of reproduction appears insufficient to sustain population with present mortality	Probability 0%	This level of reproduction appears insufficient to sustain population with present mortality	30%	35%	53%	72%
or									
Removal from each age class required for stability is:	7%	0%		Probability 0%		4.85%	7.75%	9.5%	15.5%
Generation time	7.75 years	7 years	Will vary	4 years	5 years	7 years	7 years	7.5 years	7.5 years

397

difficult to translate into actual management. In Seal and Foose (in press) Seal has calculated that the average litter size for this subspecies has been 2.43 cubs with an approximately even sex ratio (Table 5). If reproduction could be regulated to about 1.7 litters per animal evenly distributed over the reproductive ages from 4 to 12, the populations apparently could be stabilized. However, the complexity of managing production of 1.7 litters of 2.43 cubs per animal per lifetime should be obvious. Stabilizing other populations almost certainly will require some removals or harvest of animals.

This type of approach will be applicable to closely managed wild populations, which also may have explosive potentials for increase. Demographic management will be particularly important as perturbation of the ecosystems interferes with the mechanisms that normally regulate these populations. Fowler and Smith (1981) present a diverse collection of studies on demographic management and regulation of wild populations. Owen-Smith (1981) discusses a particularly interesting and pertinent management case of the white rhino where demographic management has been dependent upon interaction with captive facilities and programs.

However, such demographic management must be cognizant of the genetic consequences of these manipulations, as indicated by the data in Table 6 derived from a paper by Ryman et al. (1981). The authors investigated the effect on both generation time and N_e of the various hunting regimes that might be employed to stabilize populations of moose (*Alces alces*). Spectacular differences in both N_e and generation time result, depending on which age and sex classes are removed.

There will be many stationary configurations at the carrying capacity for a particular population in captivity, depending on how the survivorships and fertilities are managed to produce the stability. Captive management might then attempt to produce the best stationary configuration possible, best in this context being defined as optimization of some particular management objective or population property. In terms of preserving genetic diversity, the best configuration will constitute some configuration that optimizes N_e and generation time.

Table 6. *Effect of Removal of N_e and Generation Time in Moose (Alces)*

Removal regime	N_e/N	Generation (years)
Equal removal from all age classes	.426	4.8
Equal removal from adult age classes	.415	4.2
Removal from 0–1 age class	.396	9.9
Removal from 1–2 age class	.289	7.6
A variable removal from each adult age class	.368	6.5
A variable removal from each age class	.302	7.5

Conclusion: A Case Example of Genetic and Demographic Management

Application of this strategy for genetic and demographic management is exemplified by the SSP program for the Siberian tiger (*Panthera tigris altaica*).

1. The carrying capacity has been established at 250 animals with an even sex ratio. This number is equivalent to the minimal population recommended by Franklin (1980) for long-term viability of a closed population. An additional consideration has been an analysis of the capacity of North American zoos for large felids in relation to other taxa in need of captive sanctuary. Table 3 suggests that North American zoos can accommodate four taxa of felids in the lion and tiger class. Although there are 8 subspecies of tiger (*Panthera tigris*) and 11 of lion (*P. leo*) recognized, ecologically and evolutionarily it seems sensible to try to maintain a temperate and a tropical tiger, and an Asian and an African lion.

2. The current population of Siberian tigers in North American zoos is about 250 animals. However, the age and sex structures are still badly distorted (Figure 10). Hence, institutions participating in the program have been advised (a) to resume reproduction to generate a broader base for the age distribution and (b) to remove certain animals from age classes that contain too many individuals for a stable configuration for the population.

3. All animals over the age of 15 are recommended for removal. Such removal almost certainly will entail euthanasia, which is difficult for emotional and political reasons. On the assumption that these problems can be resolved, it is proposed that further truncation of the population occur in a few years by removal of all animals over the age 13. This limit was determined in relation to the proposed production of two litters per lifetime by each tiger in the population. The second, and normally last, litter of cubs would be produced at 7 to 8 years of age. Because of stochastic events, it is important to maintain parents in the population until it is known that their progeny are sufficiently viable and fertile, which will be determined when these offspring are 4 to 5 years old. If the progeny are viable and productive, then their parents can be removed at 12 to 13 years of age. Seven males and seven females have been identified for removal in 1982.

4. Presently the population is represented by 17 founders, but their proportional contributions of offspring are very disparate (Figure 8). Thus the under-represented founders (numbers 11, 13, 19, 20, 21, 22, 23, 410) will be reproduced preferentially.

5. Once founder representation is adjusted, every animal will be permitted to produce two offspring (one male, one female) that will themselves be allowed to survive and reproduce—that is, will be recruited into

the population. This policy will equalize family sizes and sex ratio, thereby maximizing N_e.

6. This equalization of family size will permit each animal to participate in the production of two litters, probably at ages 4 to 5 and 7 to 8. Progeny in these litters in excess of the two progeny per animal to be recruited into the population will have to be removed. It is expected that about 30% of the yearlings will have to be removed each year, at least until the age distribution is stabilized. It then may be possible to restrict reproduction more and thereby reduce the need to remove animals.

7. The opportunity has developed to acquire several new founders from both the Russian and the Chinese populations of this tiger. A strategy has been formulated to incorporate these migrants into the North American population:

 (a) Each new founder should produce five litters in its lifetime. Each litter would be produced with a different mate, selected to distribute the new lineage across the existing bloodlines. The founder would be mated a second time with the same mate only if the first litter fails to produce surviving and reproductive male and female offspring.
 (b) The new founders should not be mated with each other.
 (c) The new founder should not be mated with different mates that are siblings of one another.
 (d) The new founder should not mate with its own offspring.
 (e) Ten progeny (5 males and 5 females) will be recruited to the population from each new founder.

To plagiarize a phrase from the World Wildlife Fund, the zoo ark has embarked. If it can stay afloat it could contribute substantially to the preservation of the earth's vanishing biota.

Summary

With the continuing and accelerating destruction of wildlands and wildlife, captive and cultivated populations will become increasingly important to the strategies for preservation of species. As wild populations and sanctuaries become reduced and fragmented while captive collections and facilities become larger, more naturalistic, and better coordinated, the distinction between wild and captive populations will diminish. Thus, general discussion will continue on how captive populations and their programs can be integrated with wild populations and their programs to maximize the preservation of biotic diversity.

CHAPTER 23 Relevance of Captive Populations to Conservation of Biotic Diversity   **401**

The zoo community in North America is now placing the highest priority on the problems of genetic and demographic management. A discussion of the genetic and demographic objectives and strategies being employed illustrates how zoos are contending with the problems of founding and managing small, fragmented, and frequently isolated populations. Emphasis must be on preservation of genetic diversity and demographic stabilization of populations. The zoological park experience should be helpful to other wildlife managers.

Despite genetic and demographic management, the capacity of zoological parks for viable populations of endangered species is very limited. Hence, zoological parks are engaged in establishing priorities and guidelines concerning the taxonomic diversity and refinement sustainable in their programs. This is another area where close coordination with the "wild" population is vital.

Zoos are attempting to expand their carrying capacity in several ways. One is technological expansion through cryogenics, as mentioned elsewhere in the book. Another is territorial expansion through cooperative programs with private facilities that have vast areas of land and sincere commitments to conservation. A description of these programs provides examples of how the distinction between captive and "wild" populations is diminishing.

CHAPTER
24

The Impact of Research on the Propagation of Endangered Species in Zoos

Kurt Benirschke

Introduction

It may seem contradictory that a chapter on zoo research is included in a book on genetics and conservation. Zoos have their hands full in managing their captive animals, and in the past they have had an "unfortunate record in conservation and research," as noted in a 1967 conference, "The Value of Zoos for Science and Conservation" (Jarvis, 1967). In her presentation Jarvis also said that she anticipated significant improvements to occur in the zoo world. Indeed, much progress has been made since then, and most zoological gardens now recognize the value of research activities and espouse to have in-house programs of investigation.

Other books (Ehrlich and Ehrlich, 1981b; Frankel and Soulé, 1981; Myers, 1979) have given quantitative estimations of the diminishing diversity of natural life and have made recommendations as to what programs might halt or decelerate the tempo of this plunder of our planet. An increasing human population and its ever-increasing expectations are the recognized causes of a diminishing biological diversity, and we concur with Ehrlich and Ehrlich's (1981a) remarks in "The Dangers of Uninformed Optimism" that zero population growth is a more desirable goal than placing unjustified hope in reaching the food production levels necessary to satisfy the expectations of famine-stricken world populations. And this does not take into consideration the fulfillment of the material desires much of this population already has or will have once it emerges from Third World status.

What I wish to argue here is that, without some acceleration of conservation and research activities in zoological gardens, the process of extinction will proceed at an even more accelerated pace. While it is recognized that such research represents but a miniscule aspect in combating a global problem and that its efforts may save but few of the more spectacular species, it is still considered a worthwhile effort to undertake.

Jenkins (1982) has recently stated that "before anything can be done to safeguard endangered species, it is necessary to have a good bit of knowledge about them." Speaking of the then forthcoming reauthorization of the U.S. Endangered Species Act, he concluded that after reauthorization (October, 1982) what cannot wait, even in a tight budget period, is the acquisition of new information. And that is the point I wish to emphasize in this chapter: more new hard data must be acquired on endangered species before many meaningful decisions can be made on their behalf, even for wild populations. As Gibson (1980) aptly summarizes, "medical research should be seen as a social investment and not as a charity." It astonishes me that, as a group, people have accepted the wisdom of medicine and are willing to fund basic medical research, expecting improved health and greater longevity for themselves, but are slow to place emphasis on the wisdom of conservation or to fund similar research directed toward endangered species.

Zoological parks are in a unique position to perform this research function. Indeed many types of research can be accomplished *only* in captive populations. Access to giraffe blood, elephant placentas, or chromosomes of wild species is virtually impossible in free-ranging wild populations. On the other hand, most vertebrates, sooner or later, come through the doors of zoos where meaningful study can be accomplished. In the presentation to follow, some examples will be given of significant advances made by such research. First, however, I should like to review briefly the history of research in zoological parks.

Recognizing the need for closer cooperation between scientists in the biomedical and academic research community and their counterparts in zoos, the National Research Council of the U.S. National Academy of Sciences held a symposium in 1973 (National Research Council, 1975) on all aspects of research in zoos. It summarized the extensive investigative endeavors of many European zoos, notably those in London and Antwerp. It provided insight into the activities of the Penrose Laboratory at the Philadelphia Zoo, and research in the Bronx Zoo, the National Zoological Park in Washington, D.C., and other institutions. More important, though, it provided a first platform for scientific exchange between the zoo community, the American Association of Zoological Parks and Aquariums (AAZPA), and biomedical investigators of academic institutions. In-house research in zoological parks and aquaria was declared a desirable endeavor for the future, and the developments in the intervening decade have shown that these were correct anticipations. Research is now a regular topic at the annual meetings of the AAZPA. Endangered species are not only registered by studbooks and ISIS (Chapter 23), but they are also very rapidly beginning to be managed by knowledgeable geneticists as international pools of animals. Behavioral programs have been established by many organizations, and artifical reproduction, endocrine surveillance of reproductive cycles, genetic assessment, etc., all have suddenly become respectable topics and begun to make significant impacts in captive management. Above all, much more knowledge is rapidly being acquired about the biology of endangered species by zoo investigators through close collaboration among themselves and with outside scientists. The enormous knowledge of human medicine has been applied to help captive reproduction; for instance, currently ultrasonography and amniocentesis are used in gorilla pregnancy, modern antibiotics for therapy against fungus infections are made available, sophisticated chemistry makes its impact in the pursuit of the identification of different steroid pathways in endocrine monitoring, and mitochondrial DNA study is used to investigate subspecific status of selected species. Despite funding difficulties, and the significant philosophical differences that exist among managers of zoo populations, research activities are making major impacts within zoological parks in the care and handling of endangered species. Every prospect exists that zoos will become integral participants in the larger arena of biomedical research. Responsive to much surveillance and some criticism, all ongoing work in zoos is directed toward a better future for the species under consideration. What significance might this work have for the management of wild populations?

1. In the early 1960s it became possible to assess the chromosomal complement of animals more readily, and an assessment of the genetic relationship between species could be undertaken with new tools. An early unexpected finding was the determination that the chromosome complement of Przewalski's horse (*Equus przewalskii*) is $2n = 66$, not 64 as all domestic horses (*E. caballus*) are now known to possess (Benirschke et al., 1965). Although the Przewalski or Mongolian wild horse is presumed to be extinct in the wild and believed to be related ancestrally to domestic horses, this finding and subsequent more detailed genetic assessments of this species (by some considered a subspecies, *E. caballus przewalskii*) provided an important impetus for international conferences to treat it as a discrete genetic pool, lest further loss of genetic variety ensue. Most recently, three Mongolian wild horses were exchanged with the separate

gene pool of horses in Askania Nova, Russia, and other animals have been placed with other institutions for selective, genetically desirable crosses (Ryder and Wedemeyer, 1982). The finding of significant genetic differences between *E. przewalskii* and *E. caballus* also led to an inquiry on the nature of the Mongolian pony, whose chromosomes were found to be the same as those of domestic horses (Schepper and deFrance, 1979). Perhaps more important, the knowledge of "contaminations" of Przewalski's horses through the input by an infamous Halle domestic mare (Mohr, 1959) has been taken more seriously.

This in its turn has led to a yet unresolved controversy over the future management of the two lines of horses, the presumably pure Munich and the Prague lines. The point to be made here is that an earlier effort to gather a greater biological knowledge of specific animals taken into captive management might have averted such contamination. To be sure, however, the technology for such chromosomal distinction did not exist at that time. But this example serves to illustrate that now, at a time when such assessments *can* be made, stocks to be managed should be studied in detail.

A pertinent example of the kind of studies necessary for proper management comes from the orangutan (*Pongo pygmaeus*). Seuanez et al. (1979) were able to chromosomally differentiate Bornean (*P. p. pygmaeus*) from Sumatran (*P. p. abelii*) orangutans. Chromosome 2 of these apes shows consistent differences in structure, which arose by pericentric inversion of a chromosomal segment. If we desire to breed pure stocks in the future, then a chromosomal analysis on breeding partners is indicated *now*, before chromosomal segregation of hybrids in the third generation prevents recognition of the distinctness of an animal by this relatively simple analysis. Chromosome study of apes is easy and can be accomplished by all of the many human cytogenetic laboratories that now exist. Regrettably, the fact that we have been able to recognize these differences only at this late date will lead to our having to maintain a third group of orangutans, the hybrids, because many have been produced inadvertently in the past. Furthermore, if reintroductions to their primeval habitats can be achieved in the future, one would hope that only properly identified consubspecifics will be returned to the respective populations.

One may have different opinions as to whether such minor chromosomal differences play an important role in the future of orangutans or other subspecies. An answer to these questions will come only with the future assessment of the reproductive proclivity of hybrids and possible deleterious effects or fitness that such hybridization may produce. There is a possibility that a resulting heterosis may even turn out to be advantageous, but for the orangutans this remains unanswered for the moment. In other primates, particularly some South American species, the case is more clear-cut. At least three chromosomally distinct subspecies of squirrel monkeys (*Saimiri sciureus*; all $2n = 44$) can be recognized cytogenetically. Each is correlated with origin from different population groups in South America (Jones et al., 1973) In spider monkeys (*Ateles*) many more different chromosomal phenotypes have been identified (Benirschke, 1979). They arise by inversions and translocations and correlate less readily with phenotypic characters allocated to species or subspecies. Indeed, no comprehensive study has correlated cytogenetic constitution of *Ateles* populations or species with locale of origin. Also, nothing is as yet known of the reproductive outcome on fertility or fitness of hybrids with significantly different karyotypes (all have $2n = 34$). In large part this is so because the precise point of origin of most imported spider monkeys is unknown. For zoos the recognition of these cytogenetic parameters is important if the goal of self-perpetuating colonies of primates is to be fulfilled. For the field biologist it may be important to know that such barriers as mountain ranges or streams may be correlated with significant cytogenetic differences of seemingly adjacent troops of animals.

The geneticist is interested in the mechanics of chromosomal evolution and why cytogenetic diversity appears to be more common in some species. Most pronounced among primates perhaps in this respect is the genetic diversity of *Aotus trivirgatus* (owl monkey), a species with very poor long-term reproductive success in captivity. Although taxonomically considered one species with numerous subspecies, *Aotus* has been found to have chromosome numbers varying between $2n = 46$ and $2n = 56$ (Ma, 1981), and hybrids are reported. Cicmanec and Campbell (1977) report that successful reproduction in the owl monkey is enhanced when chromosomally similar subspecies are paired, but Hultsch and Appleby (1980) found this not to be the case. These are recent findings, and a resolution of whether hybrid sterility exists as it occurs in mules (hybrids of a male donkey, *Equus asinus asinus* × a female horse, *E. caballus*), or whether reduced fertility with trisomic losses is to be expected as is true in the chromosomally divergent feral *Mus* (mouse) species from southern Europe (Gropp et al., 1972), remains to be seen. In the latter example, simple Robertsonian translocations account for the cytogenetic diversity, while in South American primates inversions and translocations occur more commonly. These genetic rearrangements have different consequences for reproductive fitness and need to be understood before captive management can be truly successful. Dutrillaux (1979) suggested that each taxon employs its own mechanism for chromosomal evolution for reasons that are not yet understood, and that some taxa apparently are speciating more rapidly at present than others, as judged by their chromosomal heterogeneity. Population surveys of chromosomes have been done for humans, but rarely for wild animals. When Soulie and de Grouchy (1981) surveyed chromosomal structure of 110 *Papio cynocephalus* (yellow baboons) from Kenya, they found no abnormalities and only few polymorphic sites. Perhaps this cytogenetic stability correlates with the remarkable success of this and related cercopithecids (Old World monkeys) in Africa.

The karyotype for some species is difficult to establish, particularly from populations in zoological parks. One reason is that the point of origin of animals is unknown, and the other, that the sample size for any given species is so small. A good example in our experience is the Soemmering's gazelle (*Gazella soemmeringi*). Of the nine specimens to which we have had access, seven had different chromosome numbers; they differed because of both translocations and inversions. It is possible, indeed it is likely, that their very poor reproductive success in the few zoos that have kept them is the result of this chromosomal heterogeneity. Much less likely is the possibility that many different karyotypes exist in the wild. Perhaps only one or two markedly different karyotypes were imported, and, through crossing, the variety now observed was accidentally produced. That such marked diversity in karyotypes exists in nature in relatively similar appearing species is best exemplified by *Muntiacus* (muntjacs or barking deer). The Indian *Muntiacus muntjak* (the Indian muntjac) has $2n = 6$ (female) and 7 (male), while *Muntiacus reevesi* (the Chinese muntjac) has $2n = 46$. Only recently a few single specimens with intermediate chromosome numbers have been discovered in related species. This genus clearly represents a cytogenetic and evolutionary challenge for conservation, which can be resolved only with the help of field biologists, because none of the intermediate species exist in zoos. A similar challenge exists in many other species with the potential findings having similar relevance to conservation management. The peccaries (Tayassuidae) come to mind, with the collared peccary (*Tayassu tajacu*) having $2n = 30$ and the white–lipped peccary (*T. pecari*) in Costa Rica with $2n = 26$ (Hufty et al., 1973). Recently, the thought-to-be-extinct species *Catagonus wagneri* (the Chaco peccary) which is phenotypically similar to the collared peccary, was discovered alive in Paraguay (Wetzel et al., 1975). None of these animals exist in captivity, and their chromosome constitution is unknown. Moreover, in order to avoid the spread of swine diseases, importation of tissue samples to the United States is prohibited by law. *Catagonus* appears to be threatened, and, as others, it may become extinct before its chromosome number and other biologic parameters are assessed. Knowing its chromosomes, for instance, may allow deduction of whether hybrid fertility with the other species is a possibility. Such knowledge obviously has an impact on how to manage a dwindling wild population.

The need to study more than single specimens at a time cannot be overemphasized here. Not only do cytogenetically abnormal individuals exist as natural sporadic events, but apparently inconsequential chromosomal polymorphisms are documented for a few species. Thus, the bongo (*Tragelaphus eurycerus*) has an X chromosome that, in some individuals, is typical of that of other tragelaphines, while in other individuals, a heterochromatic short arm is added without apparent deleterious effect (Benirschke et al., 1982). Because only 14 specimens without precise points of origin have been studied, it is premature to deduce that the different X-bearing individuals come from separate populations, a possibility that clearly would be of genetic and management interest.

Thus, cytogenetic assessment of mammals (the topic most familiar to me) has a significant impact on the establishment of self-sustaining populations of mammals. A priori, one must believe that the same should be true for birds and reptiles, which as a group have been less well studied. For the field biologist, it would appear that cytogenetic assessments can have great benefit in delineating populations of endangered species. Regrettably, the preparation of chromosome karyotypes in the field, usually from bone marrow, is less desirable than that from tissue cultures in the laboratory. The latter results in greater extension of the chromosomes and permits easier detection of differentiation in chromosome banding patterns. Skin biopsies are the hardiest starting material for such studies, and they are readily shipped to laboratories, yet doing so is frequently difficult or impossible because of importation regulations. The other benefit a solid tissue has for analysis is that it can be frozen for perpetuity and made available for future comparative study.

2. Others in this book have addressed the need for population genetic studies and the benefit that electrophoresis and other techniques might have in the management of wild populations and for developing evolutionary insights. I wish to discuss only one case recently investigated in our laboratories by Dr. O. Ryder and his colleagues. The question was posed by Ian Player, a conservationist of Natal, South Africa: are new efforts needed to save the northern subspecies of the white rhinoceros, *Ceratotherium simum cottoni*? Of the five species of rhinoceros (Rhinocerotidae), three are in captivity. It is anticipated that captive propagation might rescue these species while heavy pressure continues to be exerted on the wild populations (Foose, Chapter 23). An estimated 700 remaining northern white rhinoceroses face a more uncertain future because only a few specimens reside in zoos (Anon., 1982a; Foose, Chapter 23). How different is this animal from the southern subspecies, *Ceratotherium simum simum*? Few quantitative data exist, and any successful hybridization is unknown. Because our laboratory makes it a practice to freeze tissue and cell samples of many species, particularly the endangered forms when they come to autopsy, it is possible to compare purified mitochondrial DNA fractions of restriction enzyme digests of these animals. [Comparison of mitochondrial DNA fragments was recently proposed as a tool for the assessment of evolutionary distance (Brown et al., 1979; and Powell, Chapter 14).] Preliminary results from a comparison of individuals of the white rhinoceros show the two subspecies to have marked differences of mitochondrial DNA fragments. An estimation of these differences places the subspecies' divergence at two million years ago, which we feel is enough justification to mount a serious effort for planning a conservation strategy for the northern white rhinoceros. The point to be made for field

biologists is that laboratory techniques can be powerful in assisting the biological characterization of species. For zoos, it is apparent that efforts should be expanded to establish banks of tissues and cells. This will enable future scientists to have materials for study of animals that are either very rare or already extinct.

Zoo research occasionally yields other genetic insights that will be of interest to field biologists. Because the practice of inbreeding is so prevalent in zoos and so difficult to avoid, hereditary anomalies have become evident in carefully monitored species. Hairless offspring (Goodwin, 1980) in the pedigree of a pair of black and white ruffed lemurs (*Varecia variegata variegata*) and red ruffed lemur (*V. v. rubra*) hybrids combined with repeated father × daughter matings suggest that the gene for hairlessness exists in the wild, as does the autosomal dominant gene for pectus excavatum and flat chest (Benirschke, 1980). Likewise, diaphragmatic hernia in *Leontopithecus rosalia* (golden lion tamarin) may represent the expression of a gene whose manifestation was first recognized through captive breeding efforts (Bush et al., 1980). Doubtless, a large number of deleterious genes exist in nature whose phenotypic expressions are either not recognized in the wild because of perinatal loss or are not manifested because of consistent outbreeding. Recognition of such traits—e.g., flat chest in *Varecia variegata*—may provide useful clues for population management; it also makes an important impact in defense of the currently strong recommendation that breeding populations should be started with a sufficiently large founder stock of genetically diverse individuals, as several other authors in this book have emphasized.

3. In his summary of the AAZPA conference previously alluded to (National Research Council, 1975), C.E. Hopla expressed the hope that "a reproductive physiologist will be a standard appointment on a zoological park staff (as pathologists and behaviorists are currently)." The benefits of having modern endocrinologists participate in zoological research are not widely recognized. On the other hand, a perhaps unjustified optimism prevails among some that endocrinology can quickly overcome some of the barriers to captive reproduction.

The captive reproduction of the nine-banded armadillo (*Dasypus novemcinctus*) serves as an example that, despite much new insight and application of modern tools, success in a neglected field of study does not come overnight. Armadillos are doing well in the southern United States; their population is expanding (Humphrey, 1972), yet only rarely have they reproduced in the confinement of captivity. The complex reasons for this failure are not fully understood, but there is some evidence that their behavioral and perhaps nutritional needs are not completely met (Lasley et al., 1979). In the study of their endocrinology in captivity, however, fascinating insights into the fetal–maternal–placental endocrine relationships have been obtained that underscore the diversity of reproductive physiology in mammals (Nakakura et al., in press). Specifically, it was learned that the steroid produced by the fetal adrenal is progesterone and not androgens as in other species endowed with large fetal adrenal glands. Moreover, the fetal adrenal gland appears to maintain pregnancy through its progesterone secretion and perhaps controls the length of gestation. Furthermore, measurements of hormones from wild armadillos differed appreciably from those made after prolonged captivity. The lesson learned from this example underscores the view that insight into normal reproductive endocrine events of exotic species may be achieved more quickly when wild populations are studied with the benefit of laboratory investigations.

Determined efforts in comparative reproductive endocrinology in zoos have been initiated in the United States by B.L. Lasley and his colleagues. Employing urine as starting material has the advantage that it obviates immobilization for bleeding, and it enables shipment of longitudinal samples and characterization of reproductive cycles rapidly by steroid measurements from small samples (Lasley, 1980). Estrogen/testosterone determination in single fecal samples of adult sexually monomorphic birds allows determination of sex as well as an estimation of reproductive status (Bercovitz et al., 1978). These techniques are not only of interest for an understanding of the evolution of reproductive endocrine mechanisms, but are now recognized as necessary for the management of captive populations. Note as well that the techniques also can be used for field research. Stool samples (dried or formalin-fixed to inactivate potential viruses) can be obtained, for instance, from condors (*Gymnogyps* and *Vultur*) or other sexually monomorphic birds and shipped to established laboratories for determination of sex. Specimens collected through the year can be used to detect annual cycles [very pronounced in lemurs (Lemuridae), for example]. They can be of use in diagnosing pregnancy and anticipating musth in elephants (Elephantidae) and in assessing anticipated time of birth. For many species, the endocrine assessment can now be accomplished within hours, and one can anticipate that, ultimately, kits may become available that accomplish these goals in the field much as urine sugar content is now being tested at home by diabetics. It should be understood that only in zoo research laboratories that closely follow the phenomenal advances made in human endocrinology during the last decade can these developments occur. The results soon will have wide applicability for managing livestock as well as for preserving wild populations of exotic species.

While artificial insemination and ovum transfer are easily talked about in the considerations of zoo reproductive physiology, the very fact that they are as yet so rarely practiced indicates that many obstacles still need to be overcome. For successful artificial insemination, not only must semen be available in some quantity, one must well understand the timing of ovulation (as the ovulated egg lives only a few hours). This has not been achieved for most species, and collection and meticulous delineation of ovulatory cycles does not constitute glamorous research, however needed it is. When such studies are undertaken, one is amazed at the diversity displayed by different species in executing their reproductive strategy, although broad principles are similar. The semen successfully stored at the London Zoo for African elephants (*Loxodonta africana*) will have to wait until ovulation can be accurately anticipated in the cows, instrumentation can be developed, and (important) it can be learned what impact anesthesia might have on the endocrine control of ovulation. It is no surprise then that successful insemination and interspecific embryo transfer — e.g., gaur (*Bos frontalis* = *B. gaurus*) to Holstein cow (*B. taurus*: Stover et al., 1979) — has been accomplished almost exclusively for those species for which the basic research foundation has been laid down by countless investigators of domestic stock. It is my opinion that the real success for a wider spectrum of species not closely related to livestock will come only when similar research efforts have been made for these wild species. It would be nice, though, if species not now in sufficient number in zoos or so severely pressed that survival is dubious, and in particular the severely endangered bovine species (e.g., tamaraw, *Bubalus mindorensis*; anoa or dwarf buffalo, *B. depressicornis*; the Sumatran rhinoceros, *Dicerorhinus sumatrensis*; and the giant eland, *Tragelaphus derbianus*) would be saved by collecting their embryos and transferring them to commoner (including domestic) species. Much more must be learned of such species' physiology before such dreams can be realized (Durrant and Benirschke, 1981), and, fortunately, the research is ongoing in several centers. The benefits for management of wild populations are direct. Small refuge populations of endangered species may well benefit from genetic input of other small populations, yet exchange of males may prove impossible or behaviorally unacceptable to the resident animals. It readily can be envisaged that artificial insemination could make a beneficial contribution, say, in remnant populations of large carnivores. Even zoo specimens could become genetic donors for wild stock, as in the reintroduction of species such as the Arabian oryx (*Oryx leucoryx*). The technology that now dominates cattle breeding is being learned for application to exotics and will make an impact on both wild and zoo populations. For the zoological parks the developing technology is desperately needed *now*. Take the case of giraffes (*Giraffa camelopardalis*) kept in Australia and New Zealand, where new importation of even-toed ungulates (Artiodactyla) is currently prohibited because of the fear of introduction of virus diseases. A recent analysis has shown that the present population of 11.10 giraffes derives from 1.2 founders. The population is experiencing continued inbreeding and consequently inbreeding depression (Steele et al., 1981). The similar experience for many zoo species makes the shipment of semen or fertilized eggs among institutions the ultimate answer, rather than the old-fashioned transport of expensive and dangerous animals. "New blood" might become available from gametes of wild populations, but *only* if the basic research is undertaken now.

A final example in which basic research into reproductive physiology not only will benefit captive specimens but will have an impact on the management of wild populations is the recently discovered temperature effect on sex determination of large turtles (Bull, 1981).

Incubation during critical stages of the development of turtles at one temperature yields exclusively or predominantly one sex, while at other temperatures the other sex is produced. What holds for one species (precise temperature and sex) is not necessarily true for another species (Yntema and Mrosovsky, 1980), necessitating that basic studies be conducted for each (Chapter 17 by Carr and Dodd). What is now accepted for turtles may apply also to other reptiles whose sex is not fixed by sex chromosomes. In any event, for the construction of artificial incubation sites in wild populations of endangered reptile species, knowledge of these laboratory data would seem to be essential. (It should be pointed out that this research was conducted in the academic community, not in zoos.)

4. Sophisticated zoo research now includes microbiological, parasitological, and virological studies. Clearly, for the maintenance of zoological park collections a sound knowledge of infectious diseases is important, to both treatment and prevention. For this reason, research on infectious diseases and pathology are investigative fields already firmly established in zoos. They have made important contributions in the past and have led often to identification of diseases in wild populations. As knowledge of therapeutic agents expands and particularly as sophisticated techniques allow preparation of better virus vaccines, the contributions by such research will have ramifications for the management of wild populations. When more species are compressed into smaller reserves, interspecies transfer of infectious agents will become more likely and vaccination against some diseases will become mandatory for some species we seek to preserve. Transfer of such agents (e.g., malignant catarrhal fever virus) from a wild species (e.g., wildebeest, *Connochaetes*) to domestic animals (e.g., cattle, *Bos*) is already documented. If research eventually yields a vaccine, doubtless it will be widely useful. In the meantime, recognition of infectious agents and their infectivity will be important bits of knowledge for the management of wild and captive exotic species.

I have but scratched the surface of what is in need of study. The veterinary profession, having long struggled for a bona fide place in zoological parks, is now firmly entrenched. The knowledge gained by such investigations is brought to bear on the restraint of animals in the wild in the control of parasitic diseases, and in the management, even establishment, of national parks. Ethologists have had a major impact on the management of wild populations, and their knowledge is in steady demand. As a pathologist, I must defend the contributions made by that discipline to our understanding of endangered species. Long espoused by European zoos, the Penrose Laboratories of the Philadelphia Zoo, and the laboratories of the San Diego Zoo, pathology has laid important foundations for the recognition of imported vectors and parasites, of comparative aspects of the structure of the intestinal tract, and of the impact this has had on an understanding of nutritional needs. A comprehensive review of 12,000 consecutive autopsy studies recently published (Griner, 1983) provides details that could not have been amassed by investigations of wild populations, yet the experience detailed therein doubtless will be of great importance to managers of both captive and wild populations.

Conclusions

Scientific studies in zoological parks take many forms. In the past, such studies often asked whether an understanding of "animal models" could not be useful for the benefit of better dealing with human disease. The shoe is now on the other foot. Medicine has made such spectacular advances in biological research that its tools are better applied to the conservation of animals by providing new insights into diagnosis and therapy, into unraveling biochemical mysteries for the benefit of captive and ultimately wild animal management, than just by looking to exotics as tools for human benefits. It is along these lines that I envisage the expanding research effort of zoos will proceed.

Acknowledgments

I wish to express my gratitude to Drs. B.L. Lasley and O.A. Ryder for allowing me to present as yet unpublished observations.

CHAPTER
25

Conclusions: Guidelines to Management: A Beginning Attempt

Christine M. Schonewald-Cox

Introduction

It is stated throughout this book that the survival of species in the long term depends upon their having enough genetic diversity contained both within and between their populations to accommodate new selection pressures brought about by environmental change. Chance events where single or few isolated individuals propagate successfully and produce new healthy populations do occur, but they are relatively rare (Carson, Chapter 11; Selander, Chapter 12). Some individuals, such as those belonging to self-fertilizing species, tend to be uniform genetically in single localities and are prone to faster extinction in the long term (millennia to geologic time). The ability of a species or population to maintain sufficient genetic diversity for long-term survival depends upon a variety of parameters. Foremost among them are the number of existing gene pools, their heterogeneity, and effective population sizes. Since any reserve contains only a finite amount of available habitat, the size of the reserve affects roughly the number and sizes of the gene pools contained therein. Though many other factors affect available habitat size, reserve size is still a useful

CHAPTER 25 Conclusions: Guidelines to Management: A Beginning Attempt

predictor of the potential population sizes and demographic complexity the reserve's resources can sustain. When compared with what is actually existing in the habitat, such estimates can be used to assess the present success or vulnerability of species and to project their needs for one or another form of protection.

Time presses most managers to act quickly in making conservation and management plans for their habitats and species, in the face of rapidly changing landscapes, peripheral land uses, competition for economic resources, and many other problems. Managers need guidelines that serve to develop conservation and management plans. The guidelines and planning method must rely upon data that can be obtained easily in the face of small budgets and inevitably long turn-around times of any detailed ecological surveys.

In this concluding chapter I wish to develop some initial guidelines on how to assess for a park or reserve the projected longevity of focal (or target) species and from this determine how realistic conservation goals are. These guidelines will function to clarify which optional directions management can take, and specifically when *evolutionary genetic considerations* in management and *genetic* management may be useful. The guidelines follow two undercurrents of thought that always should be in the manager's mind during planning: (1) toward what objective is the planning directed, and (2) for what time scale of concern does the manager plan? I approach this task by examining the goal of many conservation and management programs, the promotion of species survival (including ecosystem complexes of species; see Appendix 3). Can the longevity of a species be predicted or assured even roughly? In order to answer this question and accomplish my objectives, I examine influences upon the prediction of species survival from reserve size, from population sizes as they relate to reserve size, and from levels of demographic complexity (that is, apportionment of genetic diversity among the individuals in and between groups or populations).

In this synthesis it becomes apparent that most reserves are too small to maintain certain species in the long term, much as Frankel and Soulé (1981), Soulé, Wilcox and Holtby (1979), Soulé and Wilcox (1980), East (1981), Myers (1979) and others have recently projected. Yet, as they also conclude, many of the intermediate-sized parks still have the capacity to retain the majority of their species diversity if protected from major intrusions, simply because individual space requirements for the majority of species are smaller than they are for the less numerous species such as ungulate and large carnivore species. In addition, other alternatives exist for small and large reserves to increase their functional sizes and capacities for species. The size of the protected area, the complexity of the demographic units contained, and the policy of management in the area all will be major determinants of when "laissez faire" is best and when "intervention" is prescribed (Frankel, Chapter 1).

Reserve Size

The distribution of sizes of the national parks in the United States (Figure 1, Table 1) is consistent with the size distribution of world parks published in the International Union for the Conservation of Nature and Natural Resources (IUCN) survey (1975). Based solely upon difficulties already being reported for parks with respect to declining populations of large vertebrates, one can predict that most of our national parks (90% of which are less than 1 million hectares in size) probably will need some intensive forms of management to support large vertebrates if the choice is made to forestall their disappearances within the parks. For smaller species with smaller space demands, the number of national parks requiring intensive management might decrease to those 55 to 69% that are smaller than 10,000 to 100,000 ha.

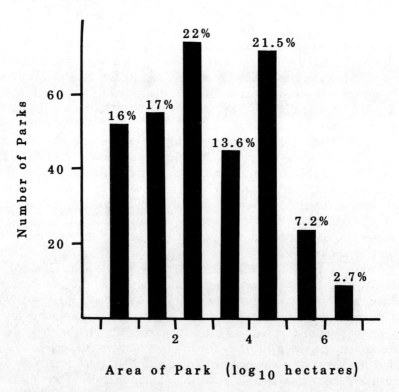

Figure 1. *Histogram of United States park sizes according to $\log_{10}$ of area, for a total of 320 parks. Percentage figures refer to the percentage of parks that belong to the specific size category. Data are taken from Table 1.*

CHAPTER 25 Conclusions: Guidelines to Management: A Beginning Attempt   417

Table 1. *A Sample of Parks and Reserves Listed by Increasing Size*[1]

Unit	Size (ha)	Size (acres)
Larger than 10; smaller than 100 (ha)		
Boston NHP	16	40
Pipe Spring NM	16	40
Fort McHenry NM & HS	17	43
San Juan NHS	21	53
Statue of Liberty NM	23	58
Fort Smith NHS	25	63
Mound City Group NM	28	68
Puukohola Heiau NHS	31	77
Jefferson Memorial NHS	37	91
Sitka NHP	44	108
Fort Clatsop	51	125
Wolftrap Farm Park	53	130
Fort Caroline N MEM	56	139
Larger than 100; smaller than 1000 (ha)		
Pipestone NM	114	282
Navajo NM	146	360
Casa Grande NM	191	473
Oregon Caves NM	197	488
Muir Woods NM	224	554
Lehman Caves	259	640
Devils Postpile NM	323	798
Buck Island Reef NM	356	880
Greenbelt Park	476	1,175
Kaloko-Honolohau NHP	506	1,250
Jewel Cave NM	516	1,275
Mount Rushmore N MEM	517	1,278
Devils Tower NM	545	1,347
Nez Perce NHP	845	2,019
Larger than 1,000; smaller than 10,000 (ha)		
Kennesaw Mountain NBP	1,167	2,884
Scotts Bluff NM	1,213	2,997
Sunset Crater NM	1,230	3,040
Manassas NBP	1,826	4,513
Hot Springs NP	2,358	5,826
Chattahoochee River	2,944	7,274
Natural Bridges NM	3,148	7,779
Rio Grande River	3,885	9,600

Table 1. *(continued)*

Unit	Size (ha)	Size (acres)
Larger than 1,000; smaller than 10,000 (ha) *(continued)*		
Indiana Dunes NL	5,073	12,535
Virgin Islands NP	5,947	14,694
Pinnacles NM	6,565	16,222
Prince William Forest Park	7,516	18,572
Larger than 10,000; smaller than 100,000 (ha)		
Wind Cave NP	11,450	28,292
Cape Lookout NS	11,500	28,414
Haleakala NP	11,597	28,655
Cape Hatteras NS	12,270	30,319
Bryce Canyon NP	14,502	35,835
Bandelier NM	14,962	36,971
Acadia NP	15,805	39,055
Assateague Island NS	16,038	39,631
Cape Cod NS	18,048	44,596
Lava Beds NM	18,843	46,560
Carlsbad Caverns NP	18,921	46,755
Mesa Verde NP	21,079	52,085
Canaveral NS	23,321	57,627
New River Gorge N River	25,101	62,024
Delaware Water Gap NRA	28,178	69,628
Theodore Roosevelt NP	28,497	70,416
Point Reyes NS	28,733	71,000
Guadalupe Mountains NP	30,875	76,293
Blue Ridge Parkway	33,318	82,328
Saguaro NM	33,823	83,576
Big Thicket NM PRES	34,741	85,846
Lassen Volcanic NP	43,048	106,372
Redwood NP	44,215	109,256
Padre Island NS	52,892	130,697
Zion NP	59,308	146,551
Santa Monica Mountains NRA	60,704	150,000
Shenandoah NP	78,983	195,057
Voyaguers NP	88,680	219,128
Hawaii Volcanoes NP	92,747	229,177
Mount Rainier NP	95,267	235,404
Badlands NP	98,463	243,302

Table 1. *(continued)*

Unit	Size (ha)	Size (acres)
Larger than 100,000; smaller than 1,000,000 (ha)		
Channel Islands NP	100,912	249,354
Rocky Mountain NP	108,030	266,943
Grand Teton NP	125,664	310,516
Organ Pipe Cactus NM	133,828	330,689
Canyonlands NP	136,613	337,570
Sequoia NP	162,885	402,488
Kings Canyon NP	186,214	460,136
North Cascades NP	204,281	504,780
Great Smoky Mountains NP	210,550	520,269
Joshua Tree NM	226,613	559,960
Big Cypress N PRES	230,676	570,000
Cape Krusenstern NM	265,757	656,685
Kenai Fjords NM	273,843	676,667
Big Bend NP	286,571	708,118
Yosemite NP	307,939	760,917
Olympic NP	370,468	915,426
Glacier NP	410,196	1,013,595
Grand Canyon NP	493,069	1,218,375
Glen Canyon NRA	500,558	1,236,880
Everglades NP	566,087	1,398,800
Lake Mead NRA	605,666	1,496,601
Kobuk Valley NM	707,826	1,749,037
Death Valley NM	836,825	2,067,627
Yellowstone NP	898,350	2,219,823
Larger than 1,000,000; smaller than 10,000,000 (ha)		
Lake Clark NM	1,065,938	2,516,821
Yukon–Charley Rivers N PRES	1,081,544	2,633,933
Bering Land Bridge NM	1,122,969	2,774,182
Glacier Bay NP and PRES	1,222,338	3,020,396
Katmai NM	1,488,842	3,678,929
Denali NM	1,901,490	4,698,583
Noatak NM	2,653,664	6,557,204
Wrangell–Saint Elias NP	3,012,969	7,445,047
Gates of the Arctic NM	3,034,426	7,498,066

[1] Categories are 10 to 100 ha, less than 1,000 ha, less than 10,000 ha, less than 100,000 ha, less than 1,000,000 ha, and less than 10,000,000 ha. Symbols: B = Battlefield, H = Historical, HS = Historical Site, L = Lakeshore, M = Monument, MEM = Memorial, N = National, P = Park, PRES = Preserve, RA = Recreational Area, S = Seashore.

Most (55%) of the parks are smaller than 1000 ha; they were set aside for their historical or geological attributes and contain very little habitat. Many rodent, insect, and plant populations not requiring protected habitats can survive effectively in areas as small as these. Consider, however, that numerous and successful as individuals in small localized populations may be, they are particularly susceptible to catastrophic events and other causes of population crashes. For most vertebrates and some long-lived plant species, parks in this size range provide little protection beyond the life of individuals or a few dwindling generations that presently constitute the population.

Pinnacles National Monument, at 6565 ha (Table 1), serves as a good example of the 13.6% of parks between 1000 ha and 10,000 ha, our smallest parks recognized for their biological resources. Pinnacles National Monument represents a unique habitat in California, with a highly diverse invertebrate and reptile fauna. The habitat does not sustain large indigenous populations of large bird or mammal species within the park boundaries. The raccoon (*Procyon lotor*) and the California ground squirrel (*Citellus beecheyi*), which are native species in California, are capable of surviving both in and out of the park. They collect within it, being partially supported by food debris left by visitors. These species, and the exotic feral pig (*Sus scrofa*), are considered pests in the park. As is the case with many small parks, the faunal and floral uniqueness of Pinnacles is being threatened by biological imbalance coupled with the small size of the park and by urbanization, which also reduces adjacent habitat spreading toward the park.

Larger parks (10,000 to 100,000 ha) contain more of the species frequently focused upon in management efforts. Canaveral National Seashore has a loggerhead sea turtle (*Caretta caretta*) nesting colony on its beach. Point Reyes National Seashore has a very small population of tule elk (*Cervus elaphus nannodes*) recently founded in the park (and having considerable difficulty: high mortality, slow population growth). Wind Cave National Park has a small population of pronghorn antelope (*Antilocapra americana*) and a population of bison (*Bison bison*) and elk (both of the latter require periodic removal of surplus animals to prevent overexploitation of the habitat). Omnivorous and herbivorous small mammal populations can survive well in protected areas of a few ten thousand hectares, at least in the intermediate term. In this size range (e.g., in Redwoods National Park, Shenandoah National Park, or Mount Ranier National Park), the capacity for large vertebrates or small carnivores is limited, particularly as adjacent habitats disappear.

Parks in the size range of Yellowstone National Park (over 100,000 ha) represent the top 10% of the parks in size, and species in these parks ought not to require much management. Yet Yellowstone, at 898,350 ha, is not large enough to sustain the migrations of its elk population, nor is it able to provide adequate protection for its grizzly bear (*Ursus arctos horribilis*) population of about 180 animals. In addition, this and other very large parks are facing impacts from a variety of exogenous sources that could modify entire species assemblages. A relatively greater abundance of problems have been reported for the larger units of the National Park System (National Park Service, 1980). The habitat available to species is becoming smaller as changes occur so rapidly in air and water and in habitats surrounding the parks that the more inflexible (specialized, long generation time, or low fecundity) species are not able to adapt. Several of the parks in this size range are designated Biosphere Reserves (Risser and Cornelison, 1979), and as such are supposed to be the natural ecological benchmarks by which we are to measure change in surrounding habitats.

The largest parks (over 1,000,000 ha), which are among our most pristine and represent 2.7% of the total number, are those in Alaska (Table 1). However, these represent a limited biotic diversity consisting of only 3 of the 22 total North American biomes (Dasmann, 1972, 1973; Udvardy, 1975).

Maintaining the complete array of species characteristic of a habitat may be impossible in some reserves without sacrificing either the focal species with public appeal or the species which, though less conspicuous, contribute greater stability to the ecosystem.

Can one roughly predict from its area the capacity of a park to maintain a specific population size of a designated species? The practice of determining carrying capacities has certainly been in use for some time. For the purpose of determining management objectives, the capacities of parks to support self-sustaining populations of species might be determined by conducting an extensive synthesis of the existing and historical population sizes of species in areas of known size. The choice of population size analysis is a practical one, because managers of natural resources rarely have much more information to draw from than park size or reserve size, rough estimates of species densities and population sizes for a select few species. In addition this choice of analysis tests whether by other means one can arrive at conclusions on species/area relationships similar to those already published based upon minimum population size and island biogeographic applications. For illustration, I have made an initial review of data that shows the potential usefulness of such an analysis.

Population Size

Effective population size and minimum viable population size have already been discussed by Soulé (Chapter 7) and Foose (Chapter 23), and in Frankel and Soulé (1981) and Soulé and Wilcox (1980). Rather than approaching the question of what population sizes are optimal for long-term survival from theoretical models, I examined censused population sizes of species in areas of known size. This is similar to the approach taken by East (1981) in which he surveyed both species and population densities of large herbivores and carnivores in African reserves.

I have chosen population sizes of mammalian species for this illustration; the data used here are listed in Table 2 together with park or sampling area size. The six mammalian orders represent widely differing reproductive and foraging strategies. Some of these species would be considered K-selected species, while others would be considered r-selected species. However, reproductive and trophic strategies, body size, and behavior of these species are generalizations more pertinent to the issue of optimal population size and long-term survival than the grosser designations of "K" and "r."

Considering that many of the sampled areas have not been determined to be insularized, one might predict that population size shows no direct relationship to reserve or sampling area size but rather to available habitat, socioecological niche, and body size. On the contrary, trends are quite clear. Park or sample area sizes do in fact appear to be good general indicators of the size of populations they are capable of containing, at least for these orders of mammals.

Figure 2 illustrates the relationship between area (or park) size and population size for a selection of mammalian species. These are representative of both temperate and tropical biomes in North America, South America, Europe, and Africa. They are also representative of a variety of trophic niches, reproductive strategies, and levels of complexity in social organization. When historic data were available these were included, and the average for each locality was computed and plotted. It appears, from the sources cited in Table 2, that only a few of the points represent populations declining sufficiently rapidly that their space requirements are serious underestimates (as with grizzly bears in Yellowstone National Park). Estimated population sizes are those reported in the literature, not extrapolations from carrying capacity of reserves or from home range estimates. Three regressions (relationships of population size to area) are plotted:

1. Small herbivores (rodents and lagomorphs) (correlation coefficient = 93; $P < .01$)
2. Large herbivores (cervides, perissodactyls, proboscids, and bovids) (correlation coefficient = .82; $P < .01$)
3. Large carnivores (ursids, canids, and large felids) (correlation coefficient = .88; $P < .01$)

Large carnivores occupy more space overall per individual than do ungulates (similar to the findings of East, 1981), and large herbivores in turn occupy more space per individual than do small herbivores. The

Table 2a. *Locality, Size of Reserve, Population Size, and References for Small Herbivore Species Listed in Figure 2*

Order and species	Locality	Area (ha)	Estimated population size	References
Lagomorpha (pikas, hares, rabbits)				
Lepus americanus (snowshoe hare)	Alberta, Canada	34	59	Keith and Windberg, 1978
	Alberta, Canada	28	52	
Rodentia (rodents)				
Cleithrodontomys (?) (harvest mouse)	Pea Ridge N.M.P., Arkansas, U.S.A.	1,740	8,526	Johnsey and Malinen, 1971
Microtus pinetorum (pine vole)	Pea Ridge N.M.P., Arkansas, U.S.A.	1,740	11,658	Johnsey and Malinen, 1971
Dicrostonyx groenlandicus (collared lemming)	Churchill region, Manitoba, Canada	1	38	Shelford, 1943, 1945; Finerty, 1976
Peromyscus maniculatus (deer mouse)	Pea Ridge N.M.P., Arkansas, U.S.A.	1,740	6,612	Johnsey and Malinen, 1971
Mus musculus (house mouse)	Pea Ridge N.M.P., Arkansas, U.S.A.	1,740	7,482	Johnsey and Malinen, 1971
Peromyscus leucopus (white-footed mouse)	Pea Ridge N.M.P., Arkansas, U.S.A.	1,740	12,006	Johnsey and Malinen, 1971
Sciurus carolinensis (gray squirrel)	Ohio, U.S.A.	162	222	Barkelow et al., 1970
Zapus princeps (western jumping mouse)		4	13.4	Brown, 1970

Table 2b. *Locality, Size of Reserve, Population Size, and References for Large Herbivore Species Listed in Figure 2*

Order and species	Locality	Area (ha)	Estimated population size	References
Proboscidia (elephants)				
Loxodonta africana (African elephant)	Queen Elizabeth N.P., Kenya	505,000	10,487	Petrides et al., 1968
Perissodactyla (odd-toed ungulates)				
Equus burchelli (Burchell's zebra)	Serengeti N.P., Tanzania	2,550,000	17,500	Schaller, 1972
	Manyara N.P., Tanzania	9,100	50	Schaller, 1972
	Ngorongoro crater, Tanzania	26,000	4,500	Kruk, 1972; Schaller, 1972
	Nairobi N.P., Kenya	11,500	488	Foster and Kearny, 1967; Schaller, 1972
	Kruger N.P., Rep. S. Africa	1,908,400	14,400	Pienaar, 1966; Schaller, 1972
Artiodactyla (even-toed ungulates)				
Cervus elaphus nelsoni (Rocky mountain elk)	Wind Cave N.P., North Dakota, U.S.A.	11,450	500	Varland et al., 1978
	Yellowstone N.P., Wyoming, U.S.A.	898,350	14,167	National Park Service, 1967
	Gaellatin N. Forest, Montana, U.S.A.	64,743	1,492	National Park Service, 1967

Table 2b. (continued)

Order and species	Locality	Area (ha)	Estimated population size	References
Cervus elaphus nannodes (tule elk)	Owens Valley, California, U.S.A.	80,929	487	Bureau of Land Management, 1981
	Tupman Reserve, California, U.S.A.	277	40	Bureau of Land Management, 1981
	San Luis Reserve, California, U.S.A.	304	37	Bureau of Land Management, 1981
Cervus elaphus scotticus (red deer)	Scarba Island, Scotland, U.K.	1,530	527	Mitchell and Crisp, 1981
Odocoileus hemionus (mule or black-tail deer)	Beaver River Basin	4,694	624	Mierau and Schmidt, 1981
Odocoileus virginianus borealis (white-tail deer)	Minnesota, U.S.A.	100,999	7,800	Darling, 1969
	Michigan, U.S.A.	5,697	469	Darling, 1969
Alces alces americanus (moose)	Newfoundland, Canada	3,600	152	Bergerud, 1968
Rangifer tarandus (barren-ground caribou)	Northern Canada	107,215	5,796	Parker, 1973
	Northern Canada	58,528	8,904	Parker, 1973
	Northern Canada	26,638	6,494	Parker, 1973
	Northern Canada	20,459	5,412	Parker, 1973
Ovis canadensis canadensis (bighorn sheep)	Canada	11,654	225	Geist, 1971

Table 2b. *(continued)*

Order and species	Locality	Area (ha)	Estimated population size	References
Artiodactyla (continued)				
Ovis dalli kenaiensis (Dahl sheep)	Mt. McKinley N.P. (before conversion to Denali N.P.), Alaska, U.S.A.	784,806	1,000	Murphy and Whitten, 1976
Connochaetes taurinus (wildebeest)	Serengeti N.P., Tanzania	2,550,000	10,000	Schaller, 1972
	Ngorongoro crater, Tanzania	26,000	13,528	Kruk, 1972; Schaller, 1972
	Nairobi N.P., Kenya	11,500	253	Foster and Kearny, 1967; Schaller, 1972
	Kruger N.P., Rep. S. Africa	1,908,400	13,035	Pienaar, 1966; Schaller, 1972
Gazella thomsoni (Thomson's gazelle)	Serengeti N.P., Tanzania	2,550,000	10,000	Schaller, 1972
	Ngorongoro crater, Tanzania	26,000	3,500	Kruk, 1972; Schaller, 1972
	Nairobi N.P., Kenya	11,500	344	Foster and Kearny, 1967; Schaller, 1972
Syncerus caffer (cape buffalo)	Serengeti N.P., Tanzania	2,550,000	50,000	Schaller, 1972
	Manyara N.P., Tanzania	9,100	1,500	Schaller, 1972
Syncerus caffer (cape buffalo)	Ngorongoro crater, Tanzania	26,000	60	Kruk, 1972; Schaller, 1972
	Kruger N.P., Rep. S. Africa	1,908,400	10,614	Pienaar, 1966; Schaller, 1972

Table 2c. Locality, Size of Reserve, Population Size, and References for Large Carnivore Species Listed in Figure 2

Order and species	Locality	Area (ha)	Estimated population size	References
Carnivora (carnivores)				
Ursus americanus (black bear)	Long Island, Washington, U.S.A.	1,953	23	Lindzey and Meslow, 1977a, 1977b
Ursus arctos horribilis (grizzly bear)	Yellowstone N.P. Wyoming, U.S.A.	898,350	180	Craighead, 1974
Panthera leo (African lion)	Ngorongoro crater, Tanzania	26,000	70	Kruk, 1972; Schaller, 1972
	Manyara N.P., Tanzania	9,100	35	Schaller, 1972
	Serengeti N.P., Tanzania	2,550,000	2,250	Schaller, 1972
	Nairobi N.P., Kenya	11,500	25	Foster and Kearny, 1967; Schaller, 1972
	Kruger N.P., Rep. S. Africa	1,908,400	1,120	Pienaar, 1966; Schaller, 1972
Panthera pardus (leopard)	Ngorongoro crater, Tanzania	26,000	20	Kruk, 1972; Schaller, 1972
	Manyara N.P., Tanzania	9,100	10	Schaller, 1972
	Serengeti N.P., Tanzania	2,550,000	900	Schaller, 1972

Table 2c. *(continued)*

Order and species	Locality	Area (ha)	Estimated population size	References
Carnivora (carnivores) *(continued)*				
Panthera pardus (leopard)	Nairobi N.P., Kenya	11,500	10	Foster and Kearny, 1967; Schaller, 1972
	Kruger N.P., Rep. S. Africa	1,908,400	650	Pienaar, 1966; Schaller, 1972
Crocuta crocuta (spotted hyena)	Ngorongoro crater, Tanzania	26,000	479	Kruk, 1972; Schaller, 1972
	Manyara N.P., Tanzania	9,100	10	Schaller, 1972
	Serengeti N.P., Tanzania	2,550,000	3,500	Schaller, 1972
	Nairobi N.P., Kenya	11,500	12	Foster and Kearny, 1967; Schaller, 1972
	Kruger N.P., Rep. S. Africa	1,908,400	1,500	Pienaar, 1972; Schaller, 1972
Canis lupus (wolf)	Northern Canada	107,215	30	Parker, 1973
	Northern Canada	58,528	29	Parker, 1973
	Northern Canada	26,638	11	Parker, 1973
	Northern Canada	20,459	7	Parker, 1973

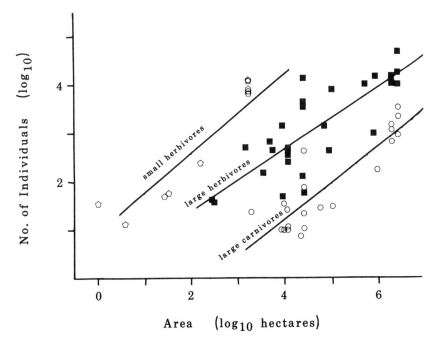

Figure 2. *Regression lines of relationship between $\log_{10}$ of the number of individuals and $\log_{10}$ of the area in hectares for small herbivores (open pentagons) large herbivores (dark squares), and large carnivores (open circles). Data are taken from Table 2.*

regression line for tropical and temperate species overlap, so they have been combined into single regressions for this analysis. The upper size range of a population possibly contained in a reserve of known area is predictable, at least for these species. For example, for an area of 100,000 ha in the temperate zone, a population of 1000 to 2000 individuals of a large native herbivore species may survive within the area for the short term. Interestingly, the population density on the average seems consistent from the tropics to the temperate zone.

In the temperate zone rarely more than 5 species of large herbivores (e.g., 2 to 3 cervid and 1 to 3 bovid species) and rarely more than 1 to 3 carnivores (felids, ursids, and canids) coexist (cf. Hall, 1981). However, as one would expect, in the tropics the species diversity is generally greater per area size than in temperate habitats (more species coexist — between 10 and 21 large herbivore and about 5 large carnivore species: Schaller, 1972; East, 1981). An analysis that classifies species by metabolic rate and trophic niche could certainly refine the descriptions of the population size-to-area and multispecies relationships.

As discussed throughout this book there are known advantages of genetic diversity to the long-term survival of species. This genetic diversity is usually distributed in varying proportions both within and between populations. Can we examine existing conservation practices and protection given to species in terms of a proportion of total natural genetic diversity preserved, measured in terms of fragments and numbers of populations protected? Such an approach can be applied to species as an independent assessment of a population's potential long-term success. This measure is independent because demographic complexity (structure or number of populations protected) is not necessarily related to the size of the population or area protected. While one population can be given optimum protection in one size reserve that accommodates the ideal numbers and sizes of populations for long-term survival, another species with quite a different biology and set of interdependencies may differ in its area and population size requirements for the same "optimum" protection. The actual population size and number of populations necessary to maintain most of the genetic variability may be smaller for one species than for another. These differences in area or population size requirements do not enter into this mode of classification where one is really examining the proportions of genetic diversity preserved. To classify protection according to the complexity of the demographic unit parallels in a fashion what determinations of minimum viable population size do.

Level of Demographic Protection

In practice, what are the levels of demographic protection given to species? Figure 3 suggests nine major types or levels of protection based on current practice ranging from zoological park or botanical garden to large portions of ecosystems that are big relative to the space and resource demands of focal species. These levels are described in Table 3. Each definition is based upon the size and complexity of the demographic unit protected. The highest levels (Levels 8 and 9) have implicit in their definitions the concept of the demographic unit(s)' being capable of remaining self-sustaining with a potential for the species to evolve in the long term. Level 8, in essence, contains the minimum viable population size required for long-term survival. Level 9 does also, and does this simultaneously in several geographic portions of the species' range, giving the species greater immunity to catastrophic and other localized causes of extinction. The multiple use (logging, hunting, fishing, etc.) of species in a specified level of protection can reduce the natural stability of the species, creating dependence upon management or depression, even after the multiple use ceases. As such, multiple use would usually *demote* a level of protection, one to several steps, depending upon the measures taken to enhance, control, or manipulate the population or its habitat. Conservation efforts, it is assumed, are intended to maintain or upgrade the level of protection rather than

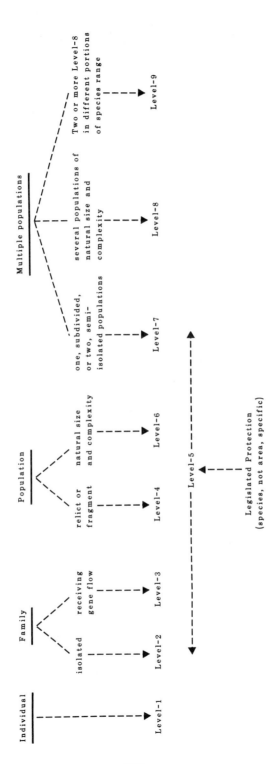

Figure 3. *Method of separation of levels of demographic protection by type of demographic unit and degree of isolation. This forms the basis of the descriptions in Table 3.*

Table 3. *Levels of Protection; A Summary of Current Practice*

Level designation	Description
1	One or a few confined individuals. This includes the provision of survival assistance (of supplementary food, possibly special diet, veterinary/horticultural care, and other management).
2	Family or small group of individuals forming a relict maintained in captivity or semicaptivity or cultivation for educational purposes or for reintroduction to the wild.
3	Several small families, or groups of individuals that are connected demographically, though these may not be close to each other geographically. These are maintained in captivity or semicaptivity or cultivation, or in the wild. The same protection given to Level 2 is also given to Level 3, except that it involves cooperation between geographically separated institutions. Because of the restricted area, supplementary feeding and medical/horticultural attention are necessary. In addition, risks of domestication exist.
4	A reserve that is small relative to the species' body size or space requirements, containing a small population in a fairly uniform habitat. This reserve can be one similar to a semicaptive game reserve, or a natural area in which individuals, though few in number, remain wild in their native habitat. Because space is limited relative to population size, and because the population size that can be sustained by available resources is limited, programs are occasionally necessary to limit population growth beyond the carrying capacity of the reserve. Level 4 can be enhanced by creating very limited gene flow between similar reserves of this sort. This category contains a fraction of the natural within-population heterozygosity.
5	Legislated protection, sometimes not area-restricted but restricted to listed species, to a category of species, or to a taxon. Such levels of protection are usually for especially vulnerable or endangered species. Level 5 is a mobile level, but when it is the only protection given it has a strong area effect similar to that of Levels 4 and 6, hence its placement between them. This category contains a fraction of within-population heterozygosity for each population fraction, and some fraction of between-population diversity.

downgrade it. These levels provide a very useful framework for the manager to plan the type of protection or management to give species.

These levels (Figure 3) are based mostly upon two types of demographic distinction. Within the first there is the individual, the family or equivalent unit, the population, and multiple populations. The second distinguishes between two types of protection given the family unit, based upon whether it is isolated or is exchanging gene flow with other similar

CHAPTER 25 Conclusions: Guidelines to Management: A Beginning Attempt

Table 3. (continued)

Level designation	Description
6	A reserve that is large relative to the species' body size or space requirements, containing one large population with a potential for breeding groups to separate from each other periodically. This level contains most of the natural within-population genetic diversity. This category contains the natural level of within-population heterozygosity, and no between-population diversity if the population is subdivided.
7	A reserve that is large relative to the species' body size or space requirements, containing uniform to heterogeneous habitat and a large population with possibilities of periodic subdivision of the population into different parts of the reserve. Level 7 possesses a limited possibility for maintaining between-population genetic diversity in addition to natural amounts of within-population genetic diversity.
8	A reserve that is very large relative to the needs of a species, containing uniform to heterogeneous habitat and a few very large populations or multiple populations with a possibility of localized adaptation and evolutionary divergence between populations. Natural amounts of both within- and between-population genetic diversity are preserved here. This level accommodates a potential for speciation to occur within the reserve and the greatest probability of stability in the long term for a single reserve.
9	A set of reserves that are each very large relative to the needs of a species, containing heterogeneous habitat and a few very large populations or multiple populations with a possibility of localized adaptation and evolutionary divergence between populations. Natural amounts of both within- and between-population genetic diversity are preserved here, and in addition genetic diversity, that characterizes populations in different geographic portions of the species' range, is also preserved. This level accommodates a potential for speciation to occur within the reserve as well as between reserves. Being composed of several disconnected reserves and consequently less susceptible to species extinction by localized catastrophes, this level has the greatest probability of stability in the long term.

groups (as Foose describes in Chapter 23). This second distinction separates Level 4 and Level 6, based upon whether the population is a small relict or fragment of a population or whether the population is its natural size, and upon the relative proportion of within-population diversity that is protected. Levels 7 and 8 are distinguished on the basis that in Level 7 there is one population, but it is capable of subdividing itself, and there is enough heterogeneity in the habitat to support the semi-isolated subpopulations. In Level 8 there are at least two natural populations that experience natural degrees of isolation, with a potential for recolonization and gene flow to occur. Levels 7 and 8 are distinguished from each other by the proportions of between- as well as within-population diversity they protect. Level 9 is distinguished by the presence of two or more Level-8 systems that exist for a species in different kinds of habitats within the species' geographic range. While the protection in different parts of the species' range could in fact apply to other levels of protection, it is only for Level-8 reserves that it can provide the very long-term value described by Level 9. For reserves protecting species at Levels 1 through 7, the upgrading of protection provided by having reserves in different habitats is comparable to Level-5 protection.

1. Level-1 Protection

The simplest type of protection is given to an individual or to a small confined group of individuals, principally for educational and esthetic purposes. In such a case we are protecting very few genotypes. Our time scale of concern may be as short as a few years, one human generation, or several generations at best. Whether or not the source of populations in the wild is limited, maximum efficiency needs to be made of the initial founding group in captivity or cultivation. The management alternatives depend upon the biology of the species; upon the availability of stock in the wild for addition to the group; the mating system (see Hamrick, Chapter 20; Selander, Chapter 12; and Eisenberg, 1981, pages listed in Appendix 6 of this book); and a manager to maintain breeding records and to control mate selection, or, for plants, to effect pollination (if necessary) or to carry out their artificial propagation. Templeton and Read (Chapter 15) give a good example of Level-1 protection and management, as well as a course of action to take with an animal species for which there is no additional supply from the wild. They have shown that a normally outbreeding species such as Speke's gazelle (*Gazella spekei*) can be adapted to the unavoidable inbreeding within a small captive population. They demonstrate effectively that careful planning can reduce inbreeding depression in the population without losing all the existing genetic diversity in the group. In cases where supplies of additional individuals are available to introduce into the group, alternatives on managing the genetics of the group need to be weighed for their fiscal costs and possible mortality costs. Proceeding in the way that Templeton and Read suggest for their specific type of management problem, versus the traditional method in which very little attention was given to the genetic management of zoo populations, reduces the risk of inbreeding depression and increases the possible longevity of this small demographic unit.

2. Level-2 Protection

In Level 2, protection is extended to a family or group of individuals, preferably. This can be a relict population, or a recently founded population either in captivity/cultivation or in the wild. If the population is a captive/cultivated one the objective is to conserve the potential for restoring the species to its natural habitat at some later date and, in addition, to maintain it for educational purposes. Level-2 protection differs from Level-1 protection in at least three ways: (1) genetic diversity will reflect as much of the original diversity of the parent population as possible and will be maintained throughout the protection period; (2) the imposition of artificial selection on the population is avoided as much as possible; and (3) the inbreeding coefficient is kept as low as possible, especially for naturally outbreeding species (in Level 2, outbreeders need to remain outbreeders particularly because the population is planned for restoration). If the population is already evolutionarily adapted to inbreeding, this latter consideration, of course, is moot. Since for large vertebrates and marine or aquatic organisms, limited facilities exist for Level-2 protection, this level will compete for space and funds with Level 1. In the case of small mammals, insects, and most herbaceous plants, the competition for space depends less on the space needed for each species and more on the large number of species needing protection.

3. Level-3 Protection

Level-3 protection is more stable than Level 2 and accomplishes this by connecting several Level-2 populations. The Species Survival Plan discussed by Foose (Chapter 23) connects zoos by means of controlled migration. This form of management and protection increases gene flow and stabilizes these very small captive or semicaptive populations, which are likely to suffer from loss of genetic diversity and inbreeding. For example, most populations of tule elk (except possibly, the Owens Valley population) are kept somewhere between independently managed Level-1 and Level-2 populations with some recent changes toward creating gene flow, approaching Level 3. A single, small population of a characteristically outbreeding species carries with it the potential dangers of a severe bottleneck in population size: inbreeding depression, potentially slow population growth, and secondary loss of rare alleles through genetic drift (see Carson, Chapter 11; Futuyma, Chapter 22). In these cases it may be particularly beneficial to increase the effective population size by connecting several populations. This can be done in such a way that distinctiveness of groups is preserved rather than homogenized by panmixis, the dangers of which are described by Frankel (Chapter 1) and in Chapters 16, 17 and 18 of this book. As Selander (Chapter 12) clearly shows, not all species find a rough road to successful colonization with small inbred populations. Species that are capable of self-fertilization or are polyploid (Hamrick, Chapter 20; Liu and Godt, Chapter 5; Clegg and Brown, Chapter 13), or that may propagate asexually, have a decided advantage and require little help of the sort described by Foose (Chapter 23) and Chesser (Chapter 4), or as offered by Level 3. Their survival depends more upon multiple, semi-isolated populations, primarily for sources of colonizers if some catastrophe befalls the population.

Level 3 approaches very closely the type of protection that is extended to small parks (under 10,000 ha in Table 1). The forms of management described by Foose (Chapter 23), Ralls and Ballou (Chapter 10), Benirschke (Chapter 24), and Templeton and Read (Chapter 15) all have something to contribute to the critical wildlife management problem of small, restrained, and entirely dependent populations, each of which may presently be managed at Level 2. This type of management will become very much a reality for species such as desert bighorn sheep (*Ovis canadensis*), wapiti or American elk, pronghorn antelope, mountain lions (*Felis concolor*), and grizzly bears. Those who manage small populations must realize that the long-term prognosis is poor for outcrossing species maintained with very small numbers and little potential for growth in numbers. Their survival will be expensive in time, money, and logistics, and their management may end up requiring assistance that is in direct conflict with the established objectives of preserving the habitat, park, or other endemic species, as Frankel points out (Chapter 1).

4. Level-4 Protection

A small reserve containing a small population of a single managed species in a fairly uniform habitat, Level 4, may not experience a loss of individuals due to inbreeding depression but is likely to have a limited life span because of genetic isolation, or isolation of the host species upon which the focal species depends. Small population here means a population that carries only a limited amount of the total genetic diversity of the original population before it was reduced, and that has no other sources of gene flow or direct colonization (if the population disappears). The potential for population crashes resulting from both minor and major catastrophes, disease, overexploitation of the limited habitat, and inflexibility to the changing environment is high in the span of a few hundred years. This level has a small effective population size and is particularly vulnerable. The instability of this level can be improved as it can be with Level 2 — that is, by limited migration between the populations (taking the same precau-

CHAPTER 25 Conclusions: Guidelines to Management: A Beginning Attempt   437

tions to maintain genetic diversity between them), which counters declining fitness in each population. The risks of outbreeding depression (a decrease in fitness due to swamping of the gene pool by new alleles) in this case must be weighed against the total loss. By creating migration or corridors, one might be able to prevent the degeneration of a Level-3 protection to Level 2 or Level 1.

5. Level-5 Protection

Legislated protection (Level 5), such as that provided by the U.S. Endangered Species Act of 1973 as amended, or the Migratory Bird Treaty Act of 1918 as amended, can protect species in a manner that is not always locality-restricted, because protection afforded to the species applies wherever a population may be found. Level-5 protection can be noninterventionist, providing for imprisonment or fines for trade, poaching, and other kinds of grosser impacts on the species. On the other hand, if a species is very rare or highly specialized, as are those qualifying for the Endangered Species Act, Level 5 will convey protection equivalent to Levels 2, 3, or 4, protecting a few isolated or semi-isolated pockets (such as critical habitats), even though legislation has the potential to protect to Level 9. Level-5 protection is the only level that floats depending upon whether it is working only in a noninterventionist manner on the species or in conjunction with other measures. In most cases legislation was initially extended to the species because it did not have unlimited habitat and was in danger of declining in numbers or of becoming extinct. Therefore, the protection given by legislation is generally intermediate, in the range of small demographic unit protection. It usually creates the potential for several small populations to exist in small pockets of available habitat, sometimes with natural opportunities for gene exchange between them to occur, assuming that the species is capable of moving, or that its seeds or pollen can be dispersed. This intermediate level of protection is most stable when it is coupled with the establishment of a few reserves. For some species that are endangered and specifically listed for legal protection, the individual populations may require intense management and a detailed recovery plan, to forestall their extinction. In such a case, the level of protection (initially at least) is most comparable to that of Level 2, or preferably Level 3.

Level 6, Level 7, and Level 8 of demographic protection can be conceptualized as a graded series of reserves with increasing available habitat relative to the needs of individual species, and increasing potential for multiple populations to exist (increasing within- and between-population diversity).

6. Level-6 Protection

Level 6 can sustain a single, large population, but the lack of separation between individuals reduces the opportunity for differentiation into demes and for survival of sudden crises such as disease or nutrient change, or other changes such as new predation or competition, longer term climatic changes. Parks such as Muir Woods National Park may have an endemic population of salamander or plants, but these are highly susceptible to changes in this confined habitat around which urbanization and deforestation have occurred. Hence, these populations would be vulnerable in the short term unless opportunities exist for individuals to move into better habitat or for new individuals from adjacent populations to come into the habitat to help recolonize it should a population crash occur. In Level 6, very little opportunity exists for this sort of recovery. While a city park can hold a sizable population of insect or other invertebrate species, larger species such as elk require very large parks, such as Rocky Mountain National Park. Parks such as Guadalupe Mountains National Park or Redwoods National Park that are intermediate in size protect large ungulate populations at a level closer to Level 4 than Level 6. These intermediate-sized parks may harbor one or two herds, but the herd is susceptible to sudden changes in the habitat and to population declines, with concomitant reduction in allelic diversity within the population. Increasing the available habitat through the use of cooperative agreements with state and county administrations combined with careful management can upgrade these populations (if this is desired) to Level 6. The fact that Level 6 only protects diversity within the population, for a single populations limits its long-term utility for survival of many species.

7. Level-7 Protection

Protection at Level 7 reduces the need for management of gene flow or other aspects of population demography. Level-7 protection can provide habitat and resources for a large population with potential for dispersal and the formation of additional populations. It shows a much greater promise for intermediate-term protection (on the order of hundreds of years) and may be enhanced to include adjacent habitats, via cooperation with states, counties, or the private sector, to approximate Level 8 or Level 9. While for elk, Yellowstone and Glacier National Parks might be considered to be at Level 7 (Glacier National Park possibly at Level 8 because of adjacent Canadian parks), for the grizzly bear these parks are only at Level 3 and teetering on the border of Level 2 or Level 1. The giant sequoia (*Sequoiadendron gigantea*), also found in large protected areas, is protected at a level little above the cultivated situation, Level 2. The fact that this is an

CHAPTER 25 Conclusions: Guidelines to Management: A Beginning Attempt 439

extremely long-lived relict of ancient climatological conditions, long since changed, makes it particularly vulnerable.

8. Level-8 Protection

While Level-7 protection is as close to ideal as is currently possible for the large vertebrate species, and many of the small specialized species too, Level 8 is entirely possible for smaller, rapidly reproducing species including typically inbreeding species, sexual polyploid species and asexually reproducing species. Parks at Level 8 house multiple populations of species and contain significant microhabitat diversity to contain some if not all of the genetic diversity that is characteristic of the species at the locality. Potential exists for continued divergence and long-term evolution. With the exception of a few reserves in the world that are over a few million hectares, the prospect for the large, slowly reproducing and highly specialized species (such as the large carnivores and many of the ungulates) is dim, as Frankel and Soulé (1981) and Soulé and Wilcox (1980) project.

9. Level-9 Protection

With Level-9 protection species have the most possibilities for survival and continued phyletic evolution in geologic time. This increased protection is afforded by the preservation of several large populations in reserves in two or more rather different portions of the species' range. Vulnerability to catastrophic local extinctions still exists, but the presence of other populations in each reserve and in other localities decreases the probability that all populations will vanish catastrophically. No level of protection, not even Level-9 protection, can convey immunity to the extinction resulting from the species' inability to meet changing ecological or climatological requirements that occur globally. For this reason and the reasons stated for Level 8, the prospect of long-term survival for large vertebrates is still dubious.

These rough descriptions of levels of demographic protection can be refined, probably by describing levels in terms of the percentages of the total within- and between-population diversity they contain that are characteristic of the species for specific ecosystems. Nevertheless, even coarse assessments such as this are of value in protection and management planning.

Can the potential of local extinction and the rate of loss be predicted? This is what Terborgh (1974), Terborgh and Winter (1980), Soulé et al. (1979) and Soulé and Wilcox (1980) attempt in their projections of faunal collapse of vertebrates; for example, in Nairobi National Park at the existing rate of species loss a 60% loss in the next 1000 years, or a 40% loss in the next few hundred years, is projected on the basis of the analyses I discussed earlier. Frankel's "time scale of concern" (Frankel, 1974; Frankel and Soulé, 1981) is a projection that planners should reflect, and sometimes itself influences the survival of species or demographic units.

Synthesis

In order to integrate some of the data on park size, population size and level of demographic protection, Figure 4 illustrates how one might project the longevity of a focal species on the basis of this simple information at hand. This model which relies heavily on intuition at this stage is not intended to circumvent the more sophisticated methods but rather to demonstrate to the manager the use of a relatively simple tool for initial thinking on a problem, or a plan. Using Figure 4 one may ask: what is the capacity of a park of specified size to sustain a select species in the long term without the use of intervention other than protection from catastrophic events? While this figure is developed partly from the results of Figure 2, it is not intended to be mammal specific. Rather, each species of convergent niche type defines its own curve. In this figure the longevity of a population in a reserve is the projected length of existence for individuals and their descendents in the reserve as a function of space and level of demographic protection. Species forming curve A may function very well in small areas; however, in time the vulnerability of small protected habitats greatly increases and local extinctions also become more probable than in larger protected areas that contain more than one population. In contrast, species forming curves B and C do not do very well in small protected areas without sustained management; they may not survive well as small populations, and require more space to exist as multiple populations in single reserves. In time these are also vulnerable to local disasters and population extinctions. Ultimately, all curves meet in the span of geological time.

The spatial and protection requirements of small organisms that are rapidly reproducing, that are self-fertilizing, that are sexual polyploids, or that can inbreed are depicted at the extreme left edge of the shading. This is because species that have very small resource, genetic, and spatial demands, even if their populations are numerous in a small protected area, are especially vulnerable to catastrophe to their habitat or host species. The black pineleaf scale insects (*Nuculaspis californica*) mentioned by Liu and Godt (Chapter 5) probably fall significantly above and to the left of curve A. A single population may require only one tree; however, if a catastrophe obliterates the ponderosa pine (*Pinus ponderosa*) population, the local populations of scale insects risk also being extirpated regardless of their genetic health. Thus, the survival of populations in small reserves,

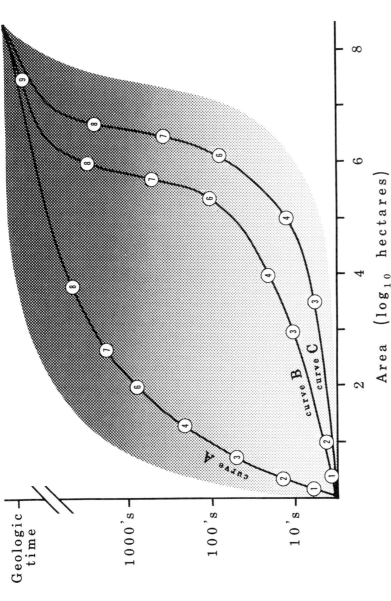

Figure 4. Hypothesized relationship between longevity of demographic unit in relation to the demographic level of protection and the size of protected area. Numbers in circles refer to levels of demographic protection. Curves refer to various hypothetical species (A, B, and C) whose longevity, level of demographic protection, and size of protected area fit the relationships that the curves define.

which fit or fall to the left of curve A, approaches in time the survival curve of their host species. Small mammals such as the pine vole (*Microtus pinetorum*) or deer mouse (*Peromyscus maniculatus*) follow curve A more closely than do the scale insects. They can sustain large populations in areas as small as 2000 ha at about Level 7 of demographic protection. A reserve this size, however, as mentioned earlier, is vulnerable to many deleterious changes caused by activities adjacent to the reserve, and by the loss of keystone species in a reserve so small.

Curve B is the hypothetical relationship for species that reproduce only sexually, that depend upon outbreeding, that are large-bodied, and that have intermediate generation times (1 to 3 years). Most of the large herbivores fit around curve B. The relationship of population size to the size of existing reserves (Figure 2) predicts that space places a major limit on the possibilities of long-term survival for these species, largely because of the obstacles that park boundaries create to migration and population expansion—related, for example, to forage availability or human impact on the species outside the protected area. The ungulate species in Figure 2 would fall into this category. While species such as the white-tailed deer (*Odocoileus virginianus*) fall above and to the left of curve B, elk fall closer to the curve, and African elephants (*Loxodonta africana*) fall at a point well below curve B, closer to curve C.

Some highly specialized and large-bodied species and species at the top of the trophic ladder fall near curve C. The asymptote of curve C is not so much the probability of local extinction of their forage species as it is the patchy distribution of the resources and the limited number of patches contained in the reserves. Even reserves the size of 1 to 2 million hectares contain very little habitat to accommodate a large population of grizzly bear, wolf (*Canis lupus*), mountain lion, and cape hunting dog (*Lycaon pictus*), and some tropical trees that fall on or to the right of curve C. Curve C follows closely the relationship of population size to park size reported in Schaller (1972), Kruk (1972), Foster and Kearny (1967), and Pienaar (1966) for African lions (*Panthera leo*), and by East (1981) for carnivores in general. The grizzly bear, which is nearly extinct in Yellowstone National Park, falls well below and to the right of curve C.

What does Figure 4 say about the longevity of a focal species (taken from curve B, for example) in a reserve of known size? If a manager is about to take action to mitigate the species loss to a reserve of, for example, 10,000 ha, he or she can predict (1) that it will require intense management to prolong the species survival in the reserve (if indeed this is the objective of the program), (2) that it will require intense management to prolong its survival (if this is the objective of the program), and (3) that, while the management that the species will require may not be genetic per se, it will require that population genetics be taken into consideration when developing the management plan. The size and complexity of the demographic

CHAPTER 25   Conclusions: Guidelines to Management: A Beginning Attempt

unit that the habitat is capable of sustaining will determine exactly what considerations need to be made. In addition, a minimum viable population size can be generated, and if this yields a figure that matches a larger reserve size (Figure 4), then the manager should know that the goal of preserving this species in isolation is a goal with a *short-term* time scale of concern. In the short term a species requiring large areas may overcrowd the habitat before its numbers reach a sustainable level. Removal of individuals may be necessary to avoid overexploitation of the habitat and subsequent population crash. In this case the manner in which individuals are selected for removal should be planned to minimize the loss of rare alleles and to preserve the maximum allele diversity in the population. Lastly, the manager will know that the management action taken will forestall the extinction of the demographic unit *only* as long as the continued management effort is invested. This, as Frankel reminds us, is short-lived in the time measured by human interest or political administrations.

If the population of concern were the bighorn sheep (*Ovis canadensis*) population in a park of 10,000 ha, it would have a projected longevity of a few hundred years at best. Even with the best protection and management the population is vulnerable to sudden declines from drought, disease, competition by exotics, or small-scale catastrophic events, with possible concomitant losses of genetic diversity necessary to cope with changing pressures of natural selection or to recolonize after a bottleneck.

Taken in a slightly different way, the question might be asked: given a park of known size, what is the ideal population size or level of protection for the long term, so that the evolutionary potential for the species is preserved, and is this feasible without intensive management of the population? If knowledge of the approximate historical densities of a species, or its density where it is abundant (accounting for any movements and room for several populations), and of the park size is available, one can determine whether the park can accommodate the species and maintain its capacity to evolve. If the park qualifies on the borderline, management can choose either to let nature takes its course, or to increase the stability of the populations by increasing the functional size of the available habitat when the actual size of the reserve is fixed (for example, with the use of cooperative agreements or habitat management).

Frankel and Soulé (1981), Terborgh (1974), East (1981), and Main and Yadov (1971) arrived at conclusions for the space demands of vertebrates similar to those illustrated by Figure 4 (although Main and Yadov's estimates seem low for "long term" and minimum viable population size). East (1981) suggests that for some large herbivores supplementary reserves of a minimum of 50,000 ha may be useful with management assistance to the populations. The prospect for smaller species such as woodrat (*Neotoma*), marmot (*Marmota*), warblers (Parulidae), and flickers (Picidae: *Colaptes*) is quite good in numerous reserves of 50,000 ha, by comparison with the prospect for cape hunting dogs, grizzly bears, pronghorn antelope, or African elephants in reserves less than 1 million to 10 million hectares.

Conclusions

A manager can and needs to determine how realistic his or her objectives are for each focal species, taking into consideration the type of protection, the space available, and the probable longevity of the demographic unit. This approach the manager must also impress upon the policymakers, who are frequently too pressed for time or are reluctant to change existing methods. The levels of demographic protection discussed earlier in this chapter and in Table 3 are descriptions of parks and mammal populations taken from sources easily obtained and are based upon current practice.

All three elements—park size, relationship of species population size to area, and levels of demographic protection—are capable of pointing the manager in the direction of realistic planning. The projections for natural populations derived from effective population size and biogeography lead to similar conclusions. Whichever way one looks at it, when our time scale of concern is truly long term, it is inevitable that some species will become extinct, most likely species from curve C (Figure 4) first and curve A last. As the available habitat decreases, the realistic longevity of species decreases proportionately, even if they are protected. The size of the demographic unit also will act against (if too small) or in support of (if large enough) a species' continued survival.

Managers will find their work increasing in complexity, to managing demographics for declining species, because the probability of increasing available habitat or significantly altering protection is going to decrease with time. The question of how to develop priorities, and whether we should spend time and money fighting inevitable extinctions, is a question of values and foresight, and whether one is willing to concede that some extinctions are and will remain inevitable.

For all the levels of protection identified (Levels 1 to 9) there are cases in which evolutionary or genetic considerations can be utilized in making decisions on whether and how to manage. In some specific cases, with careful consideration genetic management may be necessary or useful. Such cases could be the reduction of isolation between populations across a boundary; the mitigation of undesired or near-term extinction; the founding of populations to reestablish a food web that was disrupted; or the encouragement or discouragement of gene flow between two populations that previously had been isolated. The objective of conserving the natural integrity of populations or taxonomic units and optimally the dynamics of species' evolution, in any case, will require active effort. I hope that this book will assist the manager in this responsibility.

Acknowledgments

I thank Jonathan Bayless for his excellent assistance in preparing this manuscript, and Gary Johnston, Larry E. Morse, Steven M. Chambers, Bruce MacBryde, Larry Thomas, especially O.H. Frankel and Bruce Wilcox for their reviews and suggestions on improving it, and the office of Science and Technology of the National Park Service for their support and cooperation. I should also like to dedicate this to the memory of my fine friend Fred Ludeke, and to my daughter Dominique, born 1981. Statements made in this paper do not necessarily reflect the policies or opinions of The National Park Service, but those of the author who is solely responsible for the paper's content.

List of Contributors

Fred W. Allendorf	Department of Zoology, University of Montana, Missoula, MT 59812
Jonathan Ballou	Division of Biological Research, National Zoological Park, 3001 Connecticut Avenue N.W., Washington, D.C. 20008
Jonathan W. Bayless	Biological Resources Division, National Park Service, Washington, D.C. 20240
John A. Beardmore	Department of Genetics, University of Swansea, Singleton Park, Swansea, Wales SA2 8PP, United Kingdom
Kurt Benirschke	Research Department, Zoological Society of San Diego, P.O. Box 551, San Diego, CA 92112
John W. Bickham	Department of Wildlife and Fisheries Science, Texas A & M University, College Station, TX 77843

List of Contributors

A.H.D. Brown	C.S.I.R.O. (Commonwealth Scientific and Industrial Research Organization) Division of Plant Industry, P.O. Box 1600, A.C.T. 2601, Canberra City, Australia
Tom J. Cade	Division of Biological Sciences, Cornell University, Ithaca, NY 14850
Archie F. Carr III	New York Zoological Society, The Zoological Park, Bronx, NY 10460
Hampton L. Carson	Department of Genetics, University of Hawaii, Honolulu, Hawaii 96822
Steven M. Chambers	Office of Endangered Species, U.S. Fish and Wildlife Service, Department of the Interior, Washington, D.C. 20240; and Department of Biology, George Mason University, Fairfax, VA 22030
Ronald K. Chesser	Department of Biological Sciences, Texas Tech University, Lubbock, TX 79409
Michael T. Clegg	Departments of Botany and Molecular and Population Genetics, University of Georgia, Athens, GA 30603
C. Kenneth Dodd, Jr.	Office of Endangered Species, U.S. Fish and Wildlife Service, Department of the Interior, Washington, D.C. 20240
Paul R. Ehrlich	Department of Biological Sciences, Stanford University, Palo Alto, CA 94305
Thomas J. Foose	American Association of Zoological Parks and Aquariums, Johnny Cake Ridge Road, Apple Valley, MN 55124
O.H. Frankel	C.S.I.R.O. (Commonwealth Scientific and Industrial Research Organization) Division of Plant Industry, P.O. Box 1600, A.C.T. 2601, Canberra City, Australia
Douglas J. Futuyma	Department of Ecology and Evolution, State University of New York, Stony Brook, NY 11794
Mary Jo W. Godt	Savannah River Ecology Laboratory, Aiken, SC 29801
James L. Hamrick	Department of Botany, University of Kansas, Lawrence, KS 66044
Jack R. Harlan	Department of Agronomy, University of Illinois, Urbana, IL 61801
Bruce MacBryde	Office of Endangered Species, U.S. Fish and Wildlife Service, Department of the Interior, Washington, D.C. 20240

Gene Namkoong	Genetics Department, U.S. Department of Agriculture, Forest Service, North Carolina State University, Raleigh, NC 27607
Jeffrey R. Powell	Department of Biology, P.O. Box 6666, Yale University, New Haven, CT 06511
Katherine Ralls	Division of Biological Research, National Zoological Park, 3001 Connecticut Avenue, N.W., Washington, D.C. 20008
Bruce Read	St. Louis Zoo, St. Louis, MO 63166
Christine M. Schonewald-Cox	Biological Resources Division, National Park Service, Department of the Interior, Washington, D.C. 20240; and National Park Service Cooperative Park Studies Unit, University of California, Davis, CA 95616
Robert K. Selander	Department of Biology, University of Rochester, Rochester, NY 14627
Michael E. Soulé	Institute for Transcultural Studies, 901 So. Normandie Ave., Los Angeles, CA 90006
Alan R. Templeton	Department of Biology, Washington University, St. Louis, MO 63130
Larry Thomas	Office of Endangered Species, U.S. Fish and Wildlife Service, Department of the Interior, Washington, D.C. 20240

Glossary

Acrocentric — (adj.) *referring to a chromosome with the centromere close to one end of the chromosome so that one arm of the chromosome is short and the other is long.* Contrast with metacentric and telocentric chromosomes.

Adaptation — *the evolutionary process resulting in an organism's becoming well suited (adapted) to its environment.* See also Natural selection, Evolution, Fitness.

Adaptive value — *the relative fitness of one genotype compared to another.*

Additive genes — *different genes that interact only by enhancement of a trait.* They show no dominance or recessiveness, or if they are nonallelic genes, they show no epistasis.

Agamospermy — *a kind of asexual reproduction in plants in which there is seed formation without fertilization.* The embryo and seed are formed without meiosis or fertilization, in either of two ways. In gametophytic apomixis, the gametophyte is unreduced by meiosis (or $2n$) and usually forms a $2n$ egg and then by parthenogenesis a $2n$ embryo in the seed. In adventitious embryony, there is no gametophyte; the $2n$ embryo develops directly from a $2n$ somatic cell of the ovule. In some agamospermous flowering plants pollination may still be necessary for this asexual seed development to occur (a phenomenon called pseudogamy). In such cases the pollen usually is required for formation of the nutritive tissue (endosperm) for the embryo; the embryo itself is strictly maternal in genotype.

Allele — *one of two or more alternative forms of a gene that determine alternative characteristics in inheritance.* Alleles are situated at the same site (locus) on chromosomes that have genes controlling the same traits (i.e., homologous chromosomes). For example, while a gene is responsible for the characteristic shape, say of a red blood cell, the specific allele that is present on the chromosome will determine whether the shape of the cell will be normal or sickle-shaped (as in sickle-cell anemia in humans).

Allopatric — (adj.) *occurring in different geographic areas or in isolation.* It can refer to populations of the same species that live in different and nonoverlapping portions of the geographic range of the species (allopatric populations), or to related species that are isolated from each other, existing in different geographic locations (allopatric species). It can also refer to a process of speciation in which two allopatric populations of a single species evolve into two distinct species (allopatric speciation).

Allopolyploidy — *a term to describe a condition in which an organism possesses one or more sets of chromosomes that are derived from different species.* This is a special condition of polyploidy. Some of our cultivated crops such as bread wheat are considered to be allopolyploids (contrast with autopolyploidy). Also called amphiploidy and alloploidy. See also Hybrid.

Allozyme — *one of several alternative forms of an enzyme encoded by different alleles at the same locus.* Although the location (locus) on a chromosome of the code specifying an enzyme is the same from individual to individual, different alleles are responsible for determining which forms of the enzyme will be produced.

Amino acid — *one of the building blocks of proteins.* Of the hundreds of amino acids that are known, only 20 are commonly found in proteins. Hundreds to thousands of amino acids are bound together to form protein molecules.

Amphidiploid — *having a complete chromosome set from each of two different diploid parental strains.*

Amphimixis — *sexual reproduction.*

Analogy — *correspondence in function between parts of different structure and origin.*

Anaphase — *the third stage of mitosis or meiosis (both meiosis-I and meiosis-II) during which the chromosomes migrate toward opposite poles of the cell.*

Aneuploidy — *the condition of a cell or organism in which it is either missing one (or a few) whole chromosome(s) or has one (or a few) whole chromosomes too many* — i.e., it has fewer or more than the integral haploid number. Example: the condition in which a human female has three sex chromosomes (XXX) is a case of aneuploidy in which this organism has an extra X (female sex) chromosome.

Angiosperm — *a seed borne in a vessel, thus a member of a group of plants, the flowering plants,* whose seeds are borne within closed ovaries.

Annual — *a plant that completes its life cycle within one year and then dies.*

Apomixis – *reproduction without fertilization in which an individual develops from an unreduced egg or from another cell of specialized generative tissue that has not undergone meiosis.* Parthenogenesis, agamospermy, and vegetative reproduction are kinds of apomixis.

Artificial selection – *the process of choosing parents of the following generation on the basis of one or more heritable traits, or of preventing individuals from reproducing (or eliminating them) on the basis of one or more heritable traits.* Pigs with extra ribs are artificially selected for this trait, which improves their market value. Modern European moose have relatively small antlers when compared with the North American ones because in Europe males with large antlers were overly hunted and effectively eliminated.

Asexual reproduction – *the form of reproduction that occurs without fertilization.* No union of gametes or exchange of genetic material occurs between mating organisms.

Assortative mating – *refers to a selection of mates on the basis of one or more traits.* It is positive when individuals with the same form of a trait mate more often than would be predicted by chance, and is negative when they mate less often.

Autogamy – *self-fertilization.*

Automixis – *self-fertilization.*

Autopolyploid – *a term to describe a condition in which an organism possesses one or more sets of chromosomes that are derived from the same species.* This is a special case of polyploidy (contrast with allopolyploidy).

Autosome – *any chromosome that is not responsible for determining the sex of an organism* – i.e., not a sex chromosome. Autosomes are sometimes called somatic chromosomes.

Backcross – *the mating of a hybrid with its parents or with a genotype identical to that of its parents.*

Biennial – *a plant growing vegetatively during the first year, then fruiting and dying the second year.*

Bimodal distribution – *a statistical distribution having two statistical modes.* A distribution of quantities that shows divergence into two separate distributions.

Bottleneck – *a temporary period when a population becomes reduced to only a few individuals.*

Canonical variables – *a multivariate statistical technique used to analyze many variables simultaneously,* producing a set of transformed axes generated by linear discriminate functions on which individual scores can be plotted. Also called canonical variates.

Centromere – *a chromosomal region that becomes associated with the spindle fibers,* which are involved in the movement of chromosomes to opposite poles of the dividing cell during mitosis and meiosis.

Chiasmata – *an X-shaped configuration of chromosomes during meiosis;* a point at which two chromatids may exchange segments (cross over) during the first prophase of meiosis. See also Recombination.

Chromatid – *one of two filaments making up a chromosome that has recently replicated.* These result from chromosome duplication during mitosis or the second division of meiosis. After a chromosome has duplicated, each original plus replicate are connected. These eventually separate in anaphase (in meiosis this is anaphase II), each going to a different pole of the dividing cell, and become the new chromosomes of the daughter cells. Contrast with homologous chromosomes.

Chromatin – *material that constitutes the chromosomes*, consisting of DNA, histone proteins, non-histone proteins, and some RNA. Species sometimes can be identified by the specific staining properties of chromatin in the nuclei of their cells. Contrast with chromatid.

Chromatography – *a process of separating closely related compounds by allowing a solution of them to seep through an absorbant (as clay or paper)*. Since each compound has characteristic affinities for the solution and the paper or clay, sometimes has a characteristic color, and has a specific molecular weight, it moves at a characteristic rate (is absorbed) on the surface of the absorbant. Dissimilar compounds move at different rates. Thus, this technique can aid in the isolation and identification of different compounds.

Chromosomal aberration (= chromosomal abnormality) – see *Chromosomal mutation*.

Chromosomal mutation – *a change in the structure or number of the chromosomes*. This also is called chromosomal aberration or chromosomal abnormality. The change in the structure of a chromosome can be, for example, in the form of a deletion of a part, an inversion of a part, or the crossing over and trading of a part between two homologous chromosomes.

Chromosomal polymorphism – *the presence in a population of more than one gene sequence for a given chromosome*.

Chromosome – *a threadlike structure, found in the nuclei of cells, that contains the genes arranged in a linear sequence*. The chromosome is composed of DNA (deoxyribonucleic acid), non-histone and histone proteins, and some RNA (ribonucleic acid). See Chromatin.

Chromosome complement – *the total of the chromosomes in a normal cell nucleus of a gamete or zygote*. If the chromosome complement consists of one set of chromosomes, the cell is haploid. If the chromosome complement consists of two sets, it is diploid. If the chromosome complement consists of more than two sets, it is polyploid.

Chromosome set – *the normal gametic complement of chromosomes of a haploid cell*, – i.e., the base chromosome number.

Clade – *the set of species descended from a single ancestral species*. Sometimes used loosely to refer to a set of related species.

Cladistic – *describes branching patterns by which organisms are sorted on the basis of their historical (chronological) sequences of separation (divergence) from their common ancestor* (cladistic analysis or cladistic classification).

Cladogenesis – *the splitting of an evolutionary lineage into two or more lineages* – e.g., the splitting of early ungulates into deer, antelope, giraffes, hippopotami, rhinoceri, tapirs, elephants, hyraxes, camels. The treelike branches used in "evolutionary trees" are intended to show this splitting.

Cleistogamy – *self-fertilizing in self-pollinating flowers that do not open*. Examples occur in violets that develop small, inconspicuous, and nonopening flowers in addition to the usual type.

Cline — *the gradual change in characters of a species that manifests itself along a geographic gradient* — i.e., a gradient in the frequencies of genotypes or phenotypes along a stretch of territory. For example, a species may show a gradual decrease in size from north to south. On the other hand, if the type of change is not continuous between geographic regions, no cline exists. The cause for the cline is usually interpreted as natural selection occurring along an environmental or physical gradient, the characteristics of which also change gradually over the same distance.

Clone — *a collection of organisms that are genetically identical and produced asexually.* The term is used for organisms such as hydras that reproduce by budding, and many plants that propagate vegetatively (e.g., many grasses).

Coadaptation — *the harmonious interaction of genes;* the process of natural selection in which genes that have similarly positive and sometimes synergistic effects become established in a population.

Codominant — (adj.) *describing alternative alleles of a gene that are both equally manifest in the individual* — i.e., both are manifested phenotypically.

Codon — *a group of three, adjacent nucleotides in DNA or messenger RNA* (ribonucleic acid) *that codes for a specific amino acid* (protein building block) *or for a polypeptide chain termination* (also protein building blocks) during the synthesis of proteins. See also Nucleotide, Ribonucleic acid (RNA), Polypeptide, and Protein.

Coefficient of variation — *the standard deviation divided by the mean; the resulting value is sometimes multiplied by 100.*

Conditional lethal mutation — *a mutation that kills the affected organism in one environment but that is not lethal in another environment.*

Consanguinity — *the sharing of at least one recent common ancestor.*

Continuous variation — *variation, with respect to a certain trait, among phenotypes that cannot be classified into clearly distinct classes,* but rather that differ little one from the next.

Crossing over — *the interaction between chromosomes in a homologous pair in which an exchange of parts occurs by physical breakage and reunion.* This usually takes place during prophase of the first meiotic division. Crossing over results in a recombination of genetic material that generates cells or descendants different genetically from those which would have been expected had no crossover occurred.

Cryptogam — *a non-seed plant,* such as a fern or moss.

Cultivar — *a cultivated variant of a plant.*

Darwinian fitness — *the relative fitness of one genotype compared with another,* as determined by its relative contribution to the following generations.

Deficiency — *the absence of one or more genes from a chromosome.*

Deme — *a local population or gene pool of a species.* A clearly defined group of essentially random-mating individuals.

Denatured protein – *a protein that has lost its natural configuration by exposure to a destabilizing agent such as heat.*

Deoxyribonucleic acid (DNA) – *the information-storing portion of the genetic material of the cell consisting of a sequence of nucleotides in which the sugar component is deoxyribose.* Contrast with ribonucleic acid (RNA).

Dihybrid cross – *a cross between two individuals that have different alleles at each of two gene loci.*

Dioecious – (adj.) *describing individuals having either male or female sex organs but not both.*

Diploid – (adj.) *describing a cell or organism that possesses two chromosome sets.* Ploidy refers to the number of chromosome sets. Thus di-ploidy or two sets is termed diploid; higher polyploids are similarly termed triploid, tetraploid, etc. Contrast with haploid, aneuploidy, and polyploidy.

DNA – see Deoxyribonucleic acid.

Dominance – *the property of an allele at a locus to suppress expression of another allele when in the heterozygous condition.* The allele for smooth skin on a pea is dominant over the allele for wrinkled skin. Thus, when a seed carries both characteristics (its genotype is heterozygous) for smooth and wrinkled skin, the seed will appear (its phenotype will be) smooth. See also Heterozygote, Genotype, Phenotype, Epistasis. Contrast with recessive.

Dominant – *an allele, or the corresponding trait, that is manifest and masks the expression of another allele in heterozygotes.* Contrast with recessive.

Duplication – *a chromosomal mutation characterized by the presence of two copies of a chromosome segment in a single chromosome.*

Ecotype – *a genetic race with distinct morphological and/or physiological characteristics found in a particular habitat.* The term is frequently used by botanists to describe the product of genetic responses to a habitat when there is some ecological barrier to free gene exchange with other ecotypes.

Effective population size (N_e) – *the size of an ideal population that would have the same rate of increase in inbreeding or decrease in genetic diversity by genetic drift as the population being studied.* The effective population size of a true population is usually much less than its real size. See also Ideal population, Inbreeding.

Electromorphs – *proteins that can be distinguished by electrophoresis.* Contrast with chromatography. See also Allozyme, Protein, Electrophoresis.

Electrophoresis – *a technique for separating molecules based on their differential mobility in an electric field.* Each type of molecule has a specific electrical charge, shape, a specific attraction to the solution in which it is kept, and a specific molecular weight. All these characteristics "fingerprint" the compounds in a solution so that they can be separated (here, largely on the basis of electrical charge) from other molecules.

Embryo sac – *the female gametophyte of a flowering plant.* See Gametophyte.

Endemic – (adj.) *referring to a plant or animal that is either indigenous (native) in or restricted to a specific geographic locality or other condition.* For example, the Florida woodrat is endemic to the Everglades habitat characteristic of the southern tip of Florida. A plant can be endemic to a particular soil type. Contrasted with native, which is not so restricted, and exotic, which is not native but introduced.

Enzyme — *a protein complex produced in living cells, which, even in very low concentration, speeds up certain chemical reactions but is not used up in the reaction.* See also protein.

Epistasis — *the interaction between genes at different loci such that one of them affects the action (phenotypic expression) of the other.* The one that suppresses is the epistatic gene, and the suppressed gene is the hypostatic gene. See also definitions of Gene, Phenotype, Dominance, Recessive.

Equilibrium — *a condition of stasis, such as population size or genetic composition.* It is also the value of a feature, such as population size or gene frequency, at which stasis occurs.

Eukaryote — *a cell or organism that has a nucleus.*

Evolution — (see Natural selection first) *a cumulative change in the inherited characteristics of groups of organisms, which occurs in the course of successive generations related by descent.* Evolution, a process, is defined as the result of natural selection, and has no predetermined endpoint.

Expressivity — *the intensity with which the effect of a gene is realized in the phenotype.* The expressivity of a gene is influenced by the genotype and environment to varying degrees. See also Gene, Phenotype, Genotype.

Extinction — *the man-induced or natural process whereby a species ceases to exist.* Sometimes used to describe the same process at the population level or other, higher taxonomic levels.

F_1 generation — *first filial generation in a cross (mating) between any two parents.* See also Sexual reproduction, F_2 generation.

F_2 generation — *second filial generation, obtained by crossing two members of the F_1 generation, or by self-pollinating the F_1 (in plants).* See also Sexual reproduction, F_1 generation; see also symbols, Appendix 1.

Fertilization — *the fusion of two gametes of opposite sex to form a zygote.*

Fitness — *the relative survival value and reproductive capability of a given genotype in comparison with others of a population.* That is, an individual that carries genes that ensure greater reproduction and/or survival value for the individual and its descendants is more fit than others not carrying such genes. See also Natural selection, Population, Genotype, Gene.

Fixation — (adj.) *describing an allele for which no alternative alleles at that locus exist in the population.* The population is (or has become) monomorphic (having one form) for the allele; i.e., the allele has attained a frequency of 1 (or 100%) in the population.

Forb — *any herbaceous plant other than a grass, sedge, or rush.*

Founder effect — *when the original founders of a new population or colony contain only a portion of the genetic variability of their parental population.* That is, they are not usually a representative sample of the total genetic variability contained in the parent population.

Frequency-dependent selection — *the case in which the intensity of natural selection depends upon the frequency of genotypes or phenotypes in the population.* See also Natural selection, Genotype, Phenotype, Population.

Gamete—*a mature reproductive cell capable of fusing with another similar cell of the opposite sex to produce a zygote*—i.e., a sex cell. See also Sexual reproduction.

Gametophyte—*the sexual generation of a plant, which produces gametes;* it alternates with an asexual sporophyte generation, which produces spores. See also Sexual reproduction.

Gene—*the unit of heredity (or inheritance) transmitted in the chromosome.* Interacting with other genes, it controls the development of hereditary characters. The gene is a segment of the DNA (deoxyribonucleic acid) molecule that bears the information specifying the amino acid sequence for a particular protein or a major peptide chain (molecule made up of amino acid chains).

Gene flow—*the exchange of genes (in one or both directions) between two populations due to the dispersal of gametes, propagules, or individuals from one population to another.*

Gene frequency—*the proportion of an allele in a population relative to the proportion of other alleles for the same gene in the same population.* If the gene frequency for an allele is 1 (or 100%), the allele is fixed in the population.

Gene pool—*the sum total of genes in a breeding population.*

Genetic drift—*the variation of allele frequency from one generation to the next due to chance fluctuations.*

Genetic marker—*an allele whose inheritance is being observed in a mating* (usually a prescribed mating or cross).

Genetic variance—*the fraction of the phenotypic variance that is due to differences in the genetic constitution of individuals within a population.*

Genome—*in its most specific sense, one haploid chromosome set. In a broader sense it describes the entire genetic endowment of an individual.*

Genotype—*the genetic identity of an individual.* The sum total of the genetic information combined in an organism. Also used to designate the genetic constitution of an organism with respect to one or a few gene loci under consideration. This should be contrasted to phenotype, which refers to the expressed characters only. In examining one or a few gene loci, the genotype refers to all characters carried on the chromosomes, even those that do not appear. Thus if a person carries characters for both blue and brown eyes, his/her genotype is blue-brown for eye color. However, the phenotype of the individual will likely be brown, the dominant color.

Germ cell—*cells in the gonads that give rise to eggs and sperm.* See also Sexual reproduction, Gamete, Zygote.

Grade—*a group of organisms possessing a similar phenotypic level of organization, usually representing a series of closely related branches rather than a single line.* Contrast with clade and cladogenesis.

Haploid—(adj.) *referring to a cell or organism possessing a single chromosome set.* Having a single set of unpaired chromosomes (being haploid) is characteristic of gametes, some sporozoa as well as some parthenogenetically produced individuals (as in male bees), and the spores and gametophytes of many plants.

Hardy-Weinberg law (HWL)—*a principle: the prediction of genotypic frequencies (the frequencies at which a genotype will be observed in the population) on the* basis of the allele frequencies (the frequency of occurrence of specific genes) in the population. This law depends upon a major assumption that mating occurs randomly in the population.

Hemizygous — (adj.) *describing genes that are present only once in the genotype.*

Heritability — in the broad sense, *the fraction of the total phenotypic variance that remains after exclusion of the variance due to the environmental effects.* In the narrow sense, the ratio of the additive genetic variance to the total phenotypic variance.

Hermaphrodite — *an individual capable of producing both male and female gametes (sex cells).* See also Parthenogenesis, Apomixis.

Heterogametic — (adj.) *describing the sex whose gametes possess different types of sex chromosomes* — i.e., human males are heterogametic, and their sex chromosomes are XY, whereas females are homogametic, XX.

Heterosis — *increased vigor of growth or fertility (or other characters influencing survival) in an individual resulting from a cross of two genetically different lines,* as compared with the same characters in each of the individual lines that gave rise to the cross.

Heterozygote — *an individual organism that possesses different alleles (of the same gene) at the same locus on homologous chromosomes.*

Histone — *a type of protein that is part of the molecular construction of the chromosome in eukaryotic organisms* (organisms whose cells have nuclei, not bacteria, viruses, or rickettsia).

Homogametic — (adj.) *describing the sex whose gametes possess the same sex chromosomes* — e.g., human females are homogametic (XX), and human males are heterogametic (XY).

Homologous — (adj.) literally: identical by descent. Genetics: *describes chromosomes that have the same gene loci and the same visible structure.*

Homologous chromosomes — *chromosomes that carry codes for the same functions, but, however, may differ with respect to the alleles they carry.* They are called homologous because they are similar by descent, having the same loci present on each. Two haploid gametes have one full set of chromosomes each. Each gamete joins to the other to form a new zygote (first cell giving rise to the new individual), which possesses both chromosome sets. When a cell undergoes meiosis (the formation of gametes), the homologous chromosomes separate, and each homolog (carrying one of each pair of alleles for the same gene) goes to a different gamete. When a normal cell undergoes ordinary cell division the homologous pairs of chromosomes are duplicated, and each daughter cell ends up with a full diploid set identical to that of the parent cell.

Homology — *possession by two or more species of a trait derived with or without modification, from their common ancestor.* Antlers of wapiti, moose, and other deer are homologous characters; so are the hoofs of antelopes and human nails. Characters not derived from the same ancestor are not homologous but analogous (examples of convergence), though they may appear to be similar — e.g., rhinocerous horn and deer antler. See also Species, Homologous, Analogy.

Homozygote — *a cell or individual organism that has the same alleles at the same locus on each homologous chromosome.* If a cow is born with two recessive alleles for "short-jaw" (a recessive lethal character), she suffers from a disease called "agnathia" in which the lower jaw is 1 to 3 inches shorter than the upper one. Such animals cannot graze effectively. Contrast with heterozygote. See also Locus, Chromosome, Homologous chromosomes, Homologous, Heterozygote, Allele, Phenotype, Genotype.

Hybrid — *an offspring of a cross between two genetically dissimilar individuals.* Such an individual will exhibit a mixture of characteristics of both parents. The resemblance may be stronger to one parent than to the other, depending upon the influence of a variety of allelic interactions, and it may be more or less fit than related nonhybrid individuals.

Hybrid inviability — *the loss or reduction in vigor/fitness of hybrids.* See also Hybrid, Fitness, Heterosis.

Hybrid sterility — *the increased sterility in hybrids.* See also Hybrid, Fitness.

Hybrid vigor — *the increased behavioral or biological performance of a hybrid over the parental strains that produced it.*

Ideal population — a theoretical, diploid, sexually reproducing population that meets the following criteria: individuals in an ideal population mate at random, and their generations do not overlap. In order to make the "ideal population" fit mathematical constructs, the additional restrictions include the following assumptions: there is no migration into or out of the population, there is no selection, and there are no mutations occurring. Such a population definition permits certain manipulations of data and probabilities that assist biologists in determining likely trends and effects in natural populations. This "ideal" population is the basis for much of the small-population demographics and genetics discussed in this book.

Identical by descent — (adj.) *describing two genes that are structurally identical in their nucleotide sequence because they are both derived from a common ancestor.*

Identical in structure — (adj.) *describing two genes that are identical in nucleotide sequence, regardless of their common or noncommon ancestry.*

Imago — *an insect in its final, adult, sexually mature, and typically winged state.*

Inbreeding — *the mating of related individuals, or fusion of gametes in self-fertilization.*

Inbreeding coefficient — *the probability that the two alleles present at a locus are identical by descent* — i.e., that they are derived from the same ancestor. For example, if an offspring exhibits hemophilia, the probability that this is inherited from a sister–brother mating or a cousin or other relative–relative mating is the inbreeding coefficient.

Inbreeding depression — *permanent or temporary reduction in fitness or vigor due to the inbreeding of normally outbreeding organisms.*

Independent assortment — *the random distribution of genes on separate chromosomes during meiosis.* The result is recombination, resulting in a new mixture of genetic material in the offspring. See also Meiosis, Recombination, Chromosomal mutation.

Individual selection — *selection differentially acting upon different genotypes within a population, which affects their contribution to subsequent generations.* See also Natural selection.

Inducer — *an effector molecule responsible for the induction of enzyme synthesis.*

Induction — *the synthesis of new enzyme molecules in response to a specific inducer.*

Inflorescence — *a flower cluster with a morphologically defined arrangement of flowers.*

Interference — *a measure of the degree to which one crossover by a chromatid affects the probability for a second crossover by that same chromatid. Positive (negative) interference indicates that a crossover decreases (increases) the probability of a second crossover.*

Interphase — *the stage of the cell cycle during which metabolism and synthesis occur without any visible evidence of cell division. The functional period in between cell divisions. See also Mitosis, Meiosis.*

Intersex — *an abnormal individual that is intermediate between the two sexes in characteristics, having all its cells of identical genetical composition. This may occur through failure of the sex-determining mechanism of genes, or through hormonal or other influences during development.*

Intrinsic rate of natural increase — *the rate at which a population grows. This assumes that the age and sex composition of the population remain stable and that this intrinsic rate is independent of natural events that normally affect survival.*

Introgression — *the incorporation of genes from one species into the gene pool of another.*

Inversion — *a chromosomal aberration in which the linear sequence of the genes in one segment of the chromosome is reversed:*

Normal configuration: A B C D E F G H I

Inversion: A B G F E D C H I

Isozyme — *one of several forms of an enzyme. See also Allele, Genome, Allozyme.*

Karyotype — *the chromosome complement of a cell or organism, characterized by the number, size, and configuration of chromosomes, usually described during mitotic metaphase. When these are described in the literature, the author has photographed a cell in mitotic metaphase, and cut out and lined up (usually in decreasing size) the outlines of the chromosomes. The number and shape for these often are species-specific.*

Kin selection — *(see Natural selection first) a natural selection process favoring individuals that carry inheritable characteristics that cause them to increase the survival or fitness of their kin. Ants, termites, bees, and lions all provide excellent examples where the care of sisters or other kin may be more advantageous than reproducing one's self.*

Lethal gene or allele — *an allele or gene that causes death when expressed.*

Linkage — *occurrence of two loci (sing. locus) on the same chromosome so that the genes at these loci tend to be inherited as a unit.*

Linkage disequilibrium — *the nonrandom distribution of alleles together at different loci, because the loci are linked.*

Linkage group — *a set of gene loci that can be placed in order according to their relative position along a chromosome segment. See also Linkage.*

Locus — *a site for a specific gene on a chromosome. The specified "locus" can be oc-*

cupied by any one of the alleles of the gene specific to the "locus". See also Gene, Allele, Chromosome.

Macroevolution — *a broad term for great phenotypic change in a lineage usually resulting in the emergence of new genera or other, higher taxa.*

Mass selection — *a form of artificial selection practiced by choosing, in each generation, individuals with maximum (minimum) expressions of a given trait to mate to produce the following generation.*

Mating system — *the pattern or kind(s) of mating or breeding in sexually reproducing organisms.*

Mating type — *the categories of mating compatibility of organisms, used in controlled mating experiments.*

Mean — *the arithmetic mean or average:* the sum of a series of values, divided by the number of values that were summed.

Meiosis — *a process in which a single diploid cell produces four haploid daughter cells by a sequence of two cell divisions.* Genetic recombination takes place by crossing over during the first of these divisions. Gametes may be formed by these haploid cells. In plants, meiosis does not result in the direct production of n gametes but produces n spores. The spores produce multicellular n organisms, which then produce the n gametes. The n or gamete-producing generation (gametophyte, for plants) can consist of few cells, such as in pollen, or many cells, such as the free-living stage found in ferns.

Meiotic drive — *any mechanism that alters meiotic division in a manner such that alternative alleles of the heterozygote are not distributed evenly among the gametes* (especially in the formation of eggs). See also Meiosis.

Mendelian population — *an interbreeding group of organisms* sharing a common gene pool.

Messenger RNA (mRNA) — *the ribonucleic acid that has a base sequence transcribed from DNA.* This base sequence is translated into the amino acid residue sequence of a protein.

Metacentric chromosome — *a chromosome with the centromere near the middle.* Contrast with acrocentric and telocentric chromosomes.

Metaphase — *the second stage of mitosis or meiosis (1 or 2) during which the condensed chromosomes line up on the plane between the two opposite poles of the cell.*

Metric character — *a trait that varies more or less continuously among individuals, which can be measured and assigned to categories according to the measured values of the trait; a quantitative trait requiring measurement.*

Microevolution — *slight evolutionary changes within a species that may, over long stretches of time, be responsible for large changes.* See also definitions of Natural selection, Evolution. Contrast with macroevolution.

Migration — *a movement of any number of individuals or populations from one geographic location to another.* In population genetics: pertains to the dispersal and establishment of reproducing individuals beyond their geographic place of origin. This contributes to the exchange of genetic characteristics between individuals living at different localities—i.e., gene flow. See also Gene flow, Population.

Mitosis—*the process of nuclear and usually cell division with the duplication of chromosomes.* Each of the two daughter cells has the same number of chromosomes as the parent cell.

Modifier gene—*a gene that interacts with other genes by modifying their phenotypic expression.* See also Gene, Phenotype.

Monoecious(adj.) *describing organisms having male and female sex organs in unisexual structures (e.g., unisexual flowers) on the same individual and producing male and female gametes.* Contrast with dioecious.

Monomorphic—(adj.) *having but a single form or structural pattern. Genetics: a population in which all individuals have the same genotype at a given locus* (i.e., identical alleles of a specific gene). A gene locus can also be termed monomorphic if it has no variant alleles. Contrast with polymorphic.

Monozygotic twins—*twins developed from a single fertilized ovum that gives rise to two embryos at an early developmental stage.*

Monte Carlo simulations—*the process of testing a model (formula) by generating solutions using randomly generated values for variables.*

Mosaic—*a mixture of cells in an organism of different genetic composition.* The term is also applied to populations of species that are closely associated but differ genetically.

Multiple alleles—*the occurrence in a population of more than two alleles at a locus.*

Mutator gene—*a gene that increases the mutation rate of other genes in the same organism.* See also Chromosomal mutation, Gene.

Mutualism—*a relationship between two species of organisms in which both derive benefits from the other.* A type of symbiosis.

Natural selection—*the natural process by which organisms leave differentially more/less descendants than other individuals because they possess certain inherited advantages/disadvantages.* Individuals of a species that possess certain inherited advantages that allow them to survive and produce more offspring are more fit than individuals without these advantages. On the other hand, individuals that have inherited disadvantages tend to die earlier and to leave fewer offspring, they may be sterile, and their offspring may be fewer or less likely to survive than offspring of individuals without such disadvantages. An advantage during one time may at a later time become a disadvantage, because of changes in habitat, climate, or other critical parameters. Species that have developed as a result of natural selection and have later become extinct, in the natural course of events, often are examples of organisms whose advantages had transient value, being favored under earlier conditions and then disfavored by natural selection when conditions changed. See also Fitness, Evolution, Species.

Nondisjunction—*the failure of two sister chromatids or homologous chromosomes to separate during cell division, so that both go to the same pole,* thus producing aneuploid nuclei in the daughter cells. See also Homologous chromosome, Aneuploidy.

Nonhomologous—(adj.) *describing chromosomes that contain dissimilar genes and that do not pair during meiosis.* Contrast with homologous chromosomes.

Nonrandom mating—*a mating system in which the frequencies of the various kinds of mating with respect to some trait or traits are different from those expected according to chance.*

Nucleus—*a membrane-enclosed organelle of eukaryotes that contains the chromosomes of the cell and acts as the control center of cell function.*

Ontogeny—*the morphological, physical, etc., development of an individual.* In a specific context the term is used to refer to the development of an individual during the embryonic stages. Contrast with phylogeny, which has to do with the evolutionary development of a species or other lineage.

Oogenesis—*the process of differentiation of a mature egg cell from an undifferentiated germline cell,* including the process of meiosis.

Outbreeding—*a mating system in which matings between close relatives do not usually occur; outcrossing.*

Overdominance—*the condition when the heterozygote exhibits a more extreme manifestation of a trait (usually fitness) than does either homozygote.* Contrast with hybrid vigor.

Ovum—*a female reproductive cell or gamete, an egg.*

Ovule—*a structure in seed plants containing the female gametophyte with its egg cell.* When mature, the (usually) fertilized ovule becomes a seed. See Embryo sac, Gametophyte.

Panmixis—(adj. panmictic) *random mating.*

Paracentric inversion—*a chromosomal inversion that does not include the centromere.*

Parapatry—(adj. parapatric) *populations whose boundaries touch but are nonoverlapping.* Contrast with sympatric and allopatry.

Parental type—*an association of genetic markers, found among the progeny of a cross, that is identical to an association of markers present in a parent.*

Parthenogenesis—*development from an unfertilized egg*—i.e., no contribution of genes from the male parent. Individuals produced parthenogenetically can be haploid or diploid (by the lack of the reduction division in meiosis).

Pedigree—*a diagram showing the ancestral relationships among individuals of a family over two or more generations.*

Penetrance—*the frequency with which a heterozygous dominant or homozygous recessive gene manifests itself in the phenotype of its carriers.* See also Dominance, Recessive.

Pericentric inversion—*a chromosomal inversion that includes the centromere.* See also Inversion.

Phanerogam—*a seed plant.*

Phenetic—(adj.) *having to do with expressed characteristics (phenotype)*—e.g., "phenetic classification," a classification based on the similarity of phenotypes.

Phenocopy—*a manifestation in which a phenotypic character appears to result from an inherited character, but in fact is not genetic and results from a response to environmental stimuli.*

Phenotype – *the visible properties of an individual that are produced by the interaction of the genotype and the environment.*

Phylogeny – *the evolutionary history or genealogy of a group of organisms.* Contrast with ontogeny. See also Evolution, Clade.

Pleiotropy – *a case in which a single gene affects more than one characteristic in the phenotype.*

Polar bodies – *the degenerate smaller cells that are produced during meiosis in oogenesis and that do not develop into functional ova.*

Pollen – *the pollen grains; a pollen grain is a male gametophyte, which produces the sperm.* See also Gametophyte, Sporophyte.

Polygenic – (adj.) *describing traits determined by genes at many loci, usually each gene having only a slight effect on the expression of the trait.* Polygenes are the genes that collectively control a polygenic trait.

Polymorphism – *having more than one form or structural pattern.* Genetics: *the existence within a population of several alternative alleles at a gene locus.* The level of genetic polymorphism is measured by the proportion of gene loci for which alternative alleles exist. A polymorphic gene is one with several alleles rather than one. Contrast with monomorphic. See also Locus.

Polypeptide – *a protein of low molecular weight or a portion of a protein.* These are composed of chains of amino acid residues.

Polyphagous – *an organism that utilizes many kinds of food.*

Polyploid – *a cell or organism possessing three or more whole sets of chromosomes.* Polyploid individuals, subspecies/varieties, and species, though relatively rare in animals, are quite common especially in flowering plants (angiosperms) and ferns. The number of chromosomes in a gamete is n; the number in body (somatic) cells is usually $2n$. Strictly speaking, the number of chromosomes in a single set of chromosomes is x, the haploid (also base) chromosome number. In diploid organisms (most animals and about half the plants) $n = x$, so it is customary also to call n the haploid number, and $2n$ the diploid number. In polyploid organisms, however, n does not equal x but is a multiple of x; for example, in a tetraploid organism $2n = 4x$. Odd polyploids (with odd multiple sets, as $3x$) are usually sterile and reproduce only by asexual methods, while even polyploids (e.g., $4x$) are usually fertile. (See Grant, 1981; Lewis, 1980.)

Polytene chromosome – *a giant chromosome containing many strands of chromatin.*

Polytypy – (adj. polytypic) *the existence of different geographic races or subspecies/varieties within a species, or species within a genus, etc.*

Population – *a group of organisms belonging to the same species that occupy a well-defined locality and exhibit reproductive continuity from generation to generation.* Genetic and ecological interactions are generally more common between members of a population than between members of different populations of the same species. See also Deme, Species.

Prokaryote (procaryote) – *a cell that has no true nucleus and no nuclear membrane.* Bacteria are prokaryotes.

Prophase – *the first stage of meiosis or mitosis (1 or 2) during which the chromosomes condense and become visible as distinct bodies.* See also Mitosis, Meiosis.

Protein — *a large molecule composed of a chain of amino acid residues, which possesses a characteristic three-dimensional shape imposed by the sequence of its component amino acid residues.* See also Amino acid.

Race — *a population or population system within a species that differs statistically in the composition of its gene pool and in its genetically determined phenotypic characters from other populations or population systems within the species. It is useful to think of two arbitrary levels: local races and geographic races. The range extends from local breeding populations to major geographic races (the latter termed subspecies).* See also Deme, Population, Species, Subspecies.

Recessive — *an allele, or the corresponding trait, that is manifest only in the homozygote.* Contrast with dominant.

Recombination — *the result of independent segregation and assortment of genes on separate chromosomes during meiosis. The formation by crossover of gene combinations not present in the parental types is one mechanism.* See also Meiosis.

Relict — *a persistent remnant of an otherwise extinct flora or fauna or kind of organism.*

Reproductive isolating mechanism — *any biological property of an organism that interferes with its interbreeding with individuals of other species.* See also Species.

Reproductive isolation — *the inability to interbreed due to biological differences.*

Rhizome — *a more or less horizontally growing underground stem.*

Ribonucleic acid (RNA) — *a polynucleotide in which the sugar residue is ribose and which has uracil rather than the thymine found in DNA.* See also Deoxyribonucleic acid (DNA).

Ribosomes — *small particles found either free in the cytoplasm or attached to the outer surface of the endoplasmic reticulum in the cell of all eukaryotes and many prokaryotes. Ribosomes contain high concentrations of ribonucleic acid (RNA) and are centers of protein synthesis.* See also Ribonucleic acid (RNA), Protein.

Selection coefficient — *a measure of the intensity of selection. This is measured by the proportional reduction in the gametic contribution of one genotype when compared with another.* See also Natural selection.

Semispecies — *Different groups of individuals that are partially but not entirely reproductively isolated from one another. The genetically controlled mechanisms of behavior, physiology, or anatomy that control reproductive isolation are not entirely effective.* See also Population, Race, Species.

Sex chromosomes — *the chromosomes that are different in the two sexes and that are involved in sex determination.* Contrast with autosome. See also Chromosome.

Sex-linked — (adj.) *describing a gene carried by one of the sex chromosomes. It can be expressed phenotypically in both sexes.*

Sex-related, sex-limited — (adj.) *pertaining to genetically controlled characters that are phenotypically expressed in only one sex.*

Sex ratio — *the number of males divided by the number of females (sometimes ex-*

pressed in percent) at fertilization (primary sex ratio), at birth (secondary sex ratio), and at sexual maturity (tertiary sex ratio).

Sexual reproduction – *the union of two gametic nuclei of parents to produce an individual.* Each parent thus contributes one half of the chromosomes of the offspring in cases where the different sexes have the same chromosome number.

Somatic cells – *all body cells except the gametes and the cells from which these develop.*

Speciation – *any of the processes of species formation.*

Species – There are several functional definitions for species. *A biological species is a group of individuals that can interbreed successfully with one another but not with members of other groups. They are reproductively isolated from members of all other such groups (or species).* Such individuals may mate also with somewhat similar organisms that belong to other species and bear considerable resemblance to them, but these matings either cannot produce offspring, the offspring are sterile, or the offspring have distinct survival disadvantages. In most cases members of different species cannot mate because of morphological, behavioral, or physiological differences. *Microspecies are populations of predominantly uniparental (e.g., asexual or self-fertilizing) organisms that are uniform themselves and are slightly different morphologically from other such populations. Taxonomic species are groups of morphologically similar individuals consistently and persistently distinct from other such groups.* See also Semispecies, Taxon.

Spermatids – *the cells that are produced during meiosis in spermatogenesis and that eventually develop into functional spermatozoa.* See also Meiosis.

Spermatogenesis – *the process of differentiation of a mature sperm cell from an undifferentiated germline cell, including the process of meiosis.*

Spermatozoon – *in animals, a motile sperm cell (male gamete).* Plants can have motile or nonmotile sperm; in flowering plants two nonmotile sperms are developed from each pollen grain.

Spikelet – *the unit of inflorescence in grasses.*

Spore – *a reproductive structure (usually one-celled) capable of developing into a mature organism without fusion with another such cell.* In plants meiosis usually results in the production of spores, which produce gametophytes, which in turn produce gametes.

Sporophyte – *the 2n generation of a plant, which produces spores; it alternates with an n gametophyte generation, which produces gametes.*

Stability – *the propensity to return to equilibrium after having been shifted away from the equilibrium.*

Structural gene – *a gene that codes for a polypeptide: a cistron.*

Subspecies – *a named geographic race; or a set of populations of a species that share one or more distinctive features and occupy a different geographic area from other subspecies.* The edges of subspecies' ranges frequently overlap and show gradual shifting from one subspecies to the other. Some subspecies may be at an early stage of speciation. In botany, the term "variety" is often used in place of "subspecies."

Substitution — *the complete replacement of one allele by another within a population or species;* "fixation" is frequently used instead of "substitution." See also Allele, Population, Gene frequency.

Supergene — *a group of two or more loci between which recombination is so reduced that they remain together, inherited as a single unit.* Linked genes do not necessarily experience this reduced recombination.

Symbiosis — *the living together in more or less intimate association or close union of two dissimilar organisms.*

Sympatric — (adj.) *describing two or more populations that overlap in geographic distribution.*

Synapsis — *the process of the pairing of homologous chromosomes during meiosis.* See also Homologous chromosomes, Meiosis, Crossing over, Recombination.

Systematics — *the study of evolution, including historic, genetic, and phenotypic relationships among organisms.*

Taxon — (pl. taxa) *a group of organisms that form a unit of classification* (e.g., a kingdom, phylum/division, class, order, family, genus or species — including subcategories of these as well). Used for any taxonomic grouping; frequently used at the species level when subspecies/varieties also are included.

Taxonomy — *the science of classification*, of describing, naming, and assigning organisms to taxa. Ideally, the classification is based upon systematic relationships — i.e., of inherited characteristics of anatomy, behavior, chemistry, morphology, physiology, etc. Usually a combination of characteristics and measurements is used.

Telocentric chromosome — *a chromosome with a terminal centromere.* The existence of true, terminal-centromeric chromosomes is being debated. Contrast with metacentric and acrocentric chromosomes.

Telophase — *the fourth and final stage of mitosis or meiosis (1 or 2) in which the chromosomes reorganize to form daughter nuclei, and the cytoplasm divides to form two complete daughter cells.* See also Mitosis, Meiosis.

Tetraploid — *having four sets of chromosomes*, typically as a result of doubling of the unlike single sets in a hybrid; one type of polyploid.

Transcription — *the process of forming messenger RNA, in which the genetic information of the chromosomal DNA is coded (transcribed) in complementary fashion.* See also Messenger RNA (mRNA), Ribonucleic acid (RNA), and Deoxyribonucleic acid (DNA).

Triploid — *a cell, tissue or organism having three chromosome sets.* The food-storage tissue (endosperm) in the seed of a flowering plant is normally triploid.

Vegetative — (adj.) *of, relating to, or involving propagation of plants by asexual methods.*

Zygote — *the cell formed by the union of egg and sperm cells.* Also known as a fertilized egg.

Appendixes

APPENDIX 1

Some Symbols and Abbreviations

The symbols used in this book were selected to maximize consistency from chapter to chapter and with prior literature.

A	A measure of the nongenetic or environmental causes of death.
A_P	Observed number of alleles per polymorphic locus.
B	Number of lethal equivalents per haploid gamete.
β	The impact of inbreeding upon birth weight.
cM	Centimorgan(s): a unit of measure of the distance, and therefore the recombination frequency, between two loci on the same chromosome.
D	Genetic distance (of Nei, 1972) between samples.
D_{ST}	Allelic (or gene) diversity among populations.
F	Inbreeding coefficient.

APPENDIX 1 Some Symbols and Abbreviations

f	Rate of self-fertilization.
F_{ST}	Correlation between two nonrandom gametes from the same subpopulation; used in this book as a measure of allelic differentiation among subpopulations. F_{ST} values are usually similar, if not identical, to G_{ST} values computed for the same subpopulations.
F.N.	Fundamental number: the number of autosomal arms in a karyotype.
F_1	First filial generation: the offspring of a cross between two individuals of a parental (P) generation. The offspring of crosses between F_1 individuals represent an F_2, or second filial generation.
G_{ST}	Proportion of the total allelic (gene) diversity of a population that is distributed among its subdivisions. Usually similar, if not identical, to F_{ST}.
H	Average heterozygosity.
H_S	Allelic (gene) diversity within divisions of a subdivided population.
H_T	Total allelic (gene) diversity.
I	Genetic identity or similarity (of Nei, 1972) between samples. Values for I range from 0 (no similarity) to 1 (absolute identity).
K	Carrying capacity of a habitat, expressed for a particular habitat as number of individuals of a given species.
m	Migration coefficient or migration rate: probability that an individual will breed in a population other than that of its birth.
N	Actual census number of individuals in a population. In theoretical discussions in this book N is equivalent to N_e, although in actual populations N_e is usually less than N.
N_e	Effective population size or number: the number of breeding individuals in a population under ideal conditions. N of theoretical population genetic models is usually equivalent to N_e.
N_f	Number of breeding females in a population.
N_m	Number of breeding males in a population.
n	Number of chromosomes in a gamete; the haploid number of chromosomes.
n_a	Average observed number of alleles per locus over all loci investigated.

APPENDIX 1   Some Symbols and Abbreviations

P	Proportion of loci polymorphic.
P_A	Proportion of observed alleles in a species that are maintained within a given population.
r	Intrinsic rate of natural increase of a population; a measure of potential population growth rate.
s	Reduction in survival probability of homozygous genotypes relative to a superior heterozygote.
σ_p^2	The component of quantitative variation due to genetic differences between populations.
t	Reduction in survival probability of a particular homozygous genotype relative to other genotypes.
V_G	Total quantitative variation in polygenic traits.

APPENDIX 2

Equations and Population Management

Larry Thomas
J. Ballou

The following equations are provided to assist in determining those various parameters that managers must consider in populations management. Refer to the citations in this text for additional references.

I. Effective Population Size
(see Templeton and Read, Chapter 15) for a constant (nongrowing) population of randomly mating individuals.

$$\frac{1}{N_e} = \frac{1}{4N_m} + \frac{1}{4N_f}$$

or, equivalently,

$$N_e = \frac{4N_m N_f}{N_m + N_f}$$

where N_e = effective population size,
N_m = number of potentially breeding males,
N_f = number of potentially breeding females.

Example: The effective population size of one male and three female Speke's gazelle is

$$\frac{1}{N_e} = \frac{1}{(4)(1)} + \frac{1}{(4)(3)}$$

$$\frac{1}{N_e} = \frac{1}{4} + \frac{1}{12}$$

$$\frac{1}{N_e} = \frac{1}{3}$$

$$N_e = 3$$

For more information on calculating and interpreting effective population size, see Crow and Kimura (1970).

II. Measure of Inbreeding Depression

1. *Viability* (see Templeton and Read, Chapter 15)

The effect of inbreeding on viability can be modeled by using the following formula

$$v_i = e^{(-A - BF_i)}$$

where v_i = viability of individuals with inbreeding coefficient F_i,

$= \frac{\text{(number of individuals surviving)}}{\text{(number of individuals born)}}$

for a given level of inbreeding,

A = estimate of percentage survival of noninbred offspring,
B = estimate of the number of lethal equivalents per haploid gamete ($2B$ is the number of lethal equivalents per individual),
e = 2.7183,
F_i = inbreeding coefficient.

Viability and inbreeding coefficients are obtained from observed data, and A and B are obtained from standard regression analysis. If we define $v_i = x_i/n_i$, where x_i is the number of individuals with inbreeding coefficient F_i that are still alive at the index age, and n_i is the total number of births (including stillbirths and abortions) with inbreeding coefficient F_i,

Table 1. *Method for Calculating 30-Day Viability from Data on Births and Deaths of Inbred and Noninbred Young*[1]

Inbreeding coefficient	Number surviving	Number dying	Total number of births	Viability	Viability corrected for small sample size	Natural log of variability
F_i	x_i		n_i	v_i*	v_i**	$\ln v_i$
0	22	6	28	0.786	0.767	−0.266
1/16	2	0	2	1.00	0.750	−0.288
1/8	8	4	12	0.667	0.643	−0.442
3/16	1	0	1	1.00	0.667	−0.405
1/4	7	14	21	0.333	0.348	−1.056

[1] From Templeton and Read, Chapter 15.
*$v_i = x_i/n_i$.
**$v_i = (x_i + 1)/(n_i + 2)$.

then the original equation can be expressed in the format of a standard linear regression by using natural logarithms as

$$\ln\left(\frac{x_i}{n_i}\right) = -A - BF_i$$

As indicated in Templeton and Read, Chapter 15, when one is dealing with a small sample size, a correction factor must be incorporated. The observed equation then becomes the working equation for a logarithmic regression:

$$\ln\left(\frac{x_i + 1}{n_i + 2}\right) = -A - BF_i$$

Table 1 illustrates the results of calculating viability for Speke's gazelle. Using weighted regression techniques, Templeton and Read estimated A

Table 2. *Average Birth Weights for the Offspring of Noninbred Parents as a Function of the Offspring's Inbreeding Coefficient*[1]

Inbreeding coefficient F_i	Average birth weight (kg) w_i	Number of animals
0	1.67	6
1/16	1.39	2
1/8	1.31	7
3/16	1.33	1
1/4	1.17	5

[1] Only live-born offspring are included in this sample.

as 0.22 ± 0.08 and B as 3.09 ± 0.53. Figure 1 in Templeton and Read shows the plot of 30-day viability on inbreeding coefficients for the data in Table 1. The line shown is that obtained from the weighted least-squares regression.

2. *Birth weights* (see Templeton and Read, Chapter 15)

The effects of inbreeding on birth weight can be modeled by using the following formula:

$$w_i = \alpha + \beta F_i$$

where
- w_i = average weight of a newborn animal with inbreeding coefficient F_i,
- α = estimate of birth weight of noninbred offspring,
- β = estimate of the quantitative measure of the effect of inbreeding upon birth weight,
- F_i = inbreeding coefficient.

Both w_i and F_i are obtained from observed data, while α and β are obtained from regression analysis.

Effect of inbreeding on birth weight in Speke's gazelles is given in Table 2. Figure 2 in Templeton and Read shows a graph of the birth weight of liveborn animals plotted against the inbreeding coefficient of the animals obtained by weighted least-squares regression. The regression yields an α of 1.61 kg and a β of -1.93.

III. Genetic Diversity

Table 3 presents some recent data on the frequency of different alleles for the *LAP* locus in voles captured on different grids. The frequencies of the alleles were determined by electrophoresis techniques. These data will be used to illustrate the methods for calculating H_T, H_S, and G_{ST} for this survey.

Table 3. *Allele Frequencies for* LAP *(Leucine Aminopeptidase) Found in* Microtus californicus *Captured at Two Different Grid Sites*[1]

Grid	N	Frequency of *LAP-1.00* ($p_{1.00}$)	Frequency of *LAP-1.15* ($p_{1.15}$)	H
A	33	0.47	0.53	0.50
B	42	0.22	0.22	0.34
	75	$\bar{p}$ = 0.64	$\bar{p}$ = 0.36	H_T = 0.46

[1] From Bowen (1982). N = number of animals captured at each site. Values of H, $\bar{p}_i$, and H_T are calculated below.

1. *Total genetic diversity of each polymorphic locus* (see Hamrick, Chapter 20, and Clegg and Brown, Chapter 13)

$$H_T = 1 - \Sigma \bar{p}_i^2$$

where H_T = total genetic diversity for each polymorphic locus,
$\bar{p}_i$ = mean frequency of the ith allele at its locus as determined from electrophoretic techniques.

Step 1: Calculate the weighted average of the allele frequencies ($\bar{p}_i$) across all populations (grids) for each allele:

$$\bar{p}_i = \frac{(N_A \times p_A) + (N_B \times p_B)}{N_A + N_B}$$

where N_A and N_B are the number of animals sampled in each population (grid), and p_A and p_B are the frequencies of that allele in populations (grids) A and B.

From our example:

$$\bar{p}_{1.00} = \frac{(33 \times 0.47) + (42 \times 0.78)}{75} = 0.64$$

and

$$\bar{p}_{1.15} = \frac{(33 \times 0.53) + (42 \times 0.22)}{75} = 0.36$$

Step 2: Sum the squares of the averages ($\bar{p}_i^2$) and subtract from 1:

$$H_T = 1 - [(0.64)^2 + (0.36)^2] = 0.46$$

2. *Genetic diversity between and among populations* (see Hamrick, Chapter 20)

The total genetic diversity of a population (H_T) can be divided into two components: the component due to variation *within* the subpopulations (H_S), and the component due to variation *between* subpopulations (D_{ST}):

$$H_T = H_S + D_{ST}$$

where H_S = genetic diversity within subpopulations,
$D_{ST} = H_T - H_S$ = genetic diversity between populations.

The method for calculating H_T is shown in Section III.1 of this appendix.

H_s is calculated by determining the weighted average of the heterozygosity of each subpopulation.

The heterozygosity of a subpopulation can be determined by

$$H = 1 - \Sigma p_i^2$$

Calculating H_s for grid A in our example:

Step 1: Calculate p^2 for each allele:

$$p_{1.00}^2 = (0.47)^2 = 0.22$$
$$p_{1.15}^2 = (0.53)^2 = 0.28$$

Step 2: Sum all p^2 values and subtract from 1:

$$H = 1 - (0.22 + 0.28) = 0.50$$

Similar calculations can be made for grid B (see Table 3).

The weighted average of the heterozygosity of each population is calculated by:

$$H_s = \frac{(N_A \times H_A) + (N_B \times H_B)}{N_A + N_B}$$

where N_A and N_B are the number of animals sampled in subpopulations A and B,

H_A and H_B are the heterozygosities of subpopulations A and B.

From our example:

$$H_s = \frac{(33 \times 0.50) + (42 \times 0.34)}{75} = 0.41$$

Now that H_T and H_s are calculated, D_{ST} can be determined:

$$D_{ST} = H_T - H_s = 0.46 - 0.41 = 0.05$$

3. *Proportion of genetic diversity (G_{ST}) due to the among-population component*

G_{ST} is the proportion of the total allelic diversity of a population that is distributed among the subpopulations. It is usually similar, if not identical, to F_{ST}.

One of two methods can be used for calculating G_{ST}:

Method 1: $G_{ST} = D_{ST}/H_T$ (see Hamrick, Chapter 20)

Method 2: $G_{ST} = \dfrac{\Sigma \sigma_i^2}{\Sigma \bar{p}_i(1 - \bar{p}_i)}$ (see Allendorf, Chapter 3)

Calculating G_{ST} using method 1:

The procedure for calculating D_{ST} is illustrated in Section III.2 of this appendix. Thus, for our example, deriving G_{ST} is just a matter of dividing D_{ST} by H_T:

$$G_{ST} = D_{ST}/H_T = 0.05/0.46 = 0.11$$

Calculating G_{ST} using method 2:

Step 1: Calculate the variance, σ_i^2, of each allele between populations:

$$\sigma_i^2 = \overline{p_i^2} - \bar{p}_i^2$$

where $\overline{p_i^2}$ is the weighted average of the squared allele frequencies in each population:

$$\overline{p_i^2} = \dfrac{(N_A \times p_{iA}^2) + (N_B \times p_{iB}^2)}{N_A + N_B}$$

From our example:

$$\overline{p_{1.00}^2} = \dfrac{[33 \times (0.47)^2] + [42 \times (0.78)^2]}{75} = 0.44$$

$$\sigma_{1.00}^2 = 0.44 - (0.64)^2 = 0.03$$

$$\overline{p_{1.15}^2} = \dfrac{[33 \times (0.53)^2] + [42 \times (0.22)^2]}{75} = 0.1507$$

$$\sigma_{1.15}^2 = 0.15 - (0.36)^2 = 0.02$$

Step 2: Sum the variances over all alleles:

$$\Sigma \sigma^2 = 0.03 + 0.02 = 0.05$$

Step 3: Calculate $\bar{p}(1 - \bar{p})$ for each allele and sum the results over all alleles:

$$\bar{p}_{1.00}(1 - \bar{p}_{1.00}) = 0.64(1 - 0.64) = 0.23$$
$$\bar{p}_{1.15}(1 - \bar{p}_{1.15}) = 0.36(1 - 0.36) = 0.23$$
$$\Sigma \bar{p}(1 - \bar{p}) \quad = 0.23 + 0.23 = 0.46$$

Step 4: Divide the results from step 2 by the results from step 3:

$$G_{ST} = \frac{0.05}{0.46} = 0.11$$

IV. Allele Frequency Divergence at Individual Loci, F_{ST}

F_{ST} is the correlation between two nonrandom gametes from the same subpopulation and is used in this book as a measure of allelic differentiation among subpopulations. F_{ST} values are similar, in many cases identical, to G_{ST} values computed for the same population. See Section III.3 of this appendix for methods of calculating G_{ST}.

The equilibrium F_{ST} can be calculated for an island model of migration with the following formula:

$$F_{ST} = \frac{(1-m)^2}{2N - (2N-1)(1-m)^2}$$

where N = the population size of each deme,
        m = the migration rate.

If m is small, this approaches the following (see Allendorf, Chapter 3, and Frankel, Chapter 1):

$$F_{ST} = \frac{1}{4Nm + 1}$$

Example: Two populations containing 100 individuals each exchange 10% of their individuals each generation (i.e., there are 10 migrants each way every generation).

$$N = 100, \, m = 0.10$$

$$F_{ST} = \frac{1}{4(100)(0.10) + 1}$$

$$F_{ST} = \frac{1}{41} = 0.024$$

Thus, with 10 migrants per generation, the allele frequency divergence will stabilize at 0.024.

V. Population Models (see Namkoong, Chapter 19)
1. Exponential growth

$$N_{t+1} = wN_t$$

where N_{t+1} = population number in the next generation,
       N = population size,
       t = time of particular population,
       w = constant rate of growth or decline.

Example: N = 100
          t = time of measurement, year 0
          w = 15% per year (1.15)
          N_{t+1} = wN_t
                   = (1.15)(100)
                   = 115

This relationship can also be expressed by the equation

$$\frac{dN}{dt} = rN$$

where dN/dt is an expression from calculus; here meaning the change of the population size (N) at a particular time (t). The variable r is the growth rate of the population.

In this model, the change of the population's size is a function of the growth rate (r) and the population size. The larger the population, the faster it grows.

2. *Logistic growth model*

$$\frac{dN}{dt} = \left(r - \frac{rN}{K}\right)N$$

where r = rate of increase or decline,
       K = carrying capacity for the population.

In this model, the rate of increase is influenced by how close to carrying capacity the population is. When the population is at carrying capacity ($K = N$), the expression $r - (rN/K)$ equals 0 and the population stops growing.

To predict the size of a population that is growing in a logistic fashion, the following formula is used:

$$N_t = \frac{K}{1 + [e^{-rt}(K - N_0)/N_0]}$$

where K = carrying capacity,
       r = rate of increase,
       N_0 = initial size of population (size at time = 0),
       t = amount of time elapsed since time 0,
       N_t = population size at time t.

Example: Ten white-tailed deer are transplanted to a reserve with a potential carrying capacity of 200 deer. From previous experience, it is known that the maximum intrinsic rate of increase can be as much as 0.80 (George Reserve deer herd, data taken from McCullough, 1979, for illustration purposes only). Assuming the population will increase in a logistic fashion, what will the population size be in 5 years?

Let
$$N_0 = 10$$
$$K = 200$$
$$r = 0.80$$
$$t = 5$$

$$N_5 = \frac{200}{1 + [e^{-(0.8)(5)}(200 - 10)/10]}$$

$$N_5 = 148$$

3. Predator/prey interactions are considered in the following Lotka–Volterra equations (see Namkoong, Chapter 19, for additional references)

$$\frac{dN_a}{dt} = (r_a - x_{ab} N_b) N_a$$

$$\frac{dN_b}{dt} = (r_b - x_{ba} N_a) N_b$$

where N_a = prey or host population size,
       N_b = predator or pathogen population size,
       r_a and r_b are the rates of increase of the prey and host species,
       x_{ab} = the effect that species a has on the survival and reproduction of species b.

In this model, the growth rate of species a is reduced by interactions with species b. The more abundant species b is, the lower the growth rate of species a. Note that if $x_{ab} = 0$ (i.e., there is no interaction between the species), the model becomes the exponential growth model discussed above.

4. *Auto- and allo-regulations are considered in the Gause competition equations* (see Namkoong, Chapter 19, for additional references)

$$\frac{dN_a}{dt} = \left(r_a - \frac{r_a}{K_a} N_a - \frac{r_a}{K_a} x_{ab} N_b \right) N_a$$

$$\frac{dN_b}{dt} = \left(-r_b - \frac{r_b}{K_b} N_b - \frac{r_b}{K_b} x_{ba} N_a \right) N_b$$

where r_a and r_b are the rates of increase of the two populations,
N_a and N_b are the sizes of the prey and predator populations,
K_a and K_b are the carrying capacities of the two populations.
x_{ab} = the effect that species a has on the survival and reproduction of species b.

These models incorporate the effect of both carrying capacity and species interactions on the growth rates. The growth of species a is reduced by how close it is to its carrying capacity as well as the interaction with species b.

These equations could encompass cooperative effects among species if the signs were changed (see Namkoong, Chapter 19).

APPENDIX 3

Questions Posed by Managers

Christine M. Schonewald-Cox
Jonathan W. Bayless

Introduction

As in other fields of science, genetic management generates more questions than can be immediately answered. So many questions were asked by managers at the symposium on the application of genetics to the management of wild plant and animal populations held in Washington, D.C., in August, 1982, that it is not possible in the few months that have passed to provide specific answers to all of them. So, keeping in mind that this examination of questions is open-ended, we should like to lay out an overview of concerns presented by wildlife managers during the symposium. We hope to show the directions open for inquiry for use by students, theoreticians, and practitioners of applying genetics in managing wild plant and animal populations. Questions in this appendix can be used for exercises to stimulate use of the text in formulating answers. Material is included in tables and figures in the appendix for additional reference and clarification of genetic theory.

Concerns of Managers — A Direction

Table 1 is a synopsis of principle concerns for managed species in different state and federal reserves taken from issue papers written for the symposium. The list is not meant to be a thorough survey of state management problems but is we believe a representative sample of organisms mentioned. While one state representative may have emphasized species, another may have concentrated on general conservation strategies. The extent of each states' listing does not necessarily reflect the magnitude of concerns for that state.

Of the organisms cited in Table 1, 34% are plants and 66% are animals. Within plants, 86% are flowering plants (49% dicotyledonous and 37% monocotyledonous), 9% are conifers, with the remaining 6% consisting of ferns. Animals are almost entirely vertebrates with only one exception, an invertebrate sand shrimp. Birds (28%) and mammals (35%) are most often mentioned, totaling 63% of all animals. Fish (15%), reptiles (15%), and amphibians (7%) make up the remaining 37%. So it appears, as is often the case, that organisms higher on the phylogenetic scale receive the most attention in management. There are a number of obvious reasons for this. The majority of the United States' federally listed threatened and endangered species are from vertebrate taxa. Another method of ranking conservation efforts mentioned by a number of managers would be to measure the impact on a community or species of the loss of a species or population.

The principle concerns listed in Table 1 show a diverse array of problems confronting the manager. One concern, the *isolation of populations* (cited 30 times), is mentioned twice as often as any other single category. This is not surprising, because target species are often reduced to isolated and remnant populations — a prerequisite for management concern. The effect of isolation on populations also creates many problems associated with gene flow, genetic drift, inbreeding, and population decline below minimum viable population size. The *founding of populations* (18 mentions) is the next most often cited concern and may be the greatest tool managers have for restoring and buffering habitats against losses in species diversity. Though founding of new populations is a technique that has been used for some time, many genetic aspects of founding are just now being considered. *Preserving genetic diversity* (13 mentions) could cover all aspects of conservation, but here was used to infer the actual genetic constitution of populations and the spatial/temporal patterns of their distributions. Although many managers expressed reservations about using *hybridization* and the *merging of populations* (12 mentions) as management techniques, most believed that there is a very good potential for such processes to be effectively utilized in the future. Finally, as a category, the *viability of small populations* (9 mentions) includes a com-

Table 1. *The Principle Concerns of Some Managers of State and Federal Wildlife Resources*

Organism	Principle concerns	State or province
Plants		
Pteridophytes		
1. Filmy fern	Viability of small populations	Kentucky
2. Cliffbreak fern	Isolated populations	North Carolina
Gymnosperms		
1. Giant sequoias	Preserving genetic diversity	California
2. Douglas-fir	Preserving genetic diversity	Oregon
3. Sitka spruce	Preserving genetic diversity	Oregon
Flowering plants		
Monocotyledons		
1. California Orcutt grass	Extinction of populations	California
2. Crested fringed orchid	Viability of small populations	Kentucky
3. Purple fringed orchid	Viability of small populations	Kentucky
4. White fringless orchid	Viability of small populations	Kentucky
5. Kentucky lady's-slipper	Taxonomic and endangered status	Kentucky
6. Bracted sunnybell	Distinguishing sibling species	Kentucky
7. White sunnybell	Distinguishing sibling species	Kentucky
8. Native lily	Pollution of genes (swamping?)	Kentucky
9. Irises	Influence of hybridization on origin and development of cultivars	Kentucky
10. Small white lady's-slipper	Small population size	Manitoba
11. Side-saddled pitcherplant	Isolation and hybridization of populations	North Carolina
12. Yellow pitcherplant	Isolation and hybridization of populations	North Carolina
13. Texas wild-rice	Identification of endemic habitat	Texas

Table 1. *(continued)*

Organism	Principle concerns	State or province
Dicotyledons		
1. Monterey clover	Genetic distinctiveness	California
2. Lucy Braun's white snakeroot	Viability of small populations	Kentucky
3. French's shooting star	Viability of small populations	Kentucky
4. Wolf willow	Control of exotics	Manitoba
5. Trembling aspen	Control of exotics	Manitoba
6. Virginia bluebells	Isolation of populations	Michigan
7. American lotus	Isolation and merging of populations	Michigan
8. Mead's milkweed	Increasing reproductive efficiency	Missouri
9. Corkwood	Optimal habitat manipulation	Missouri
10. Pondberry	Optimal habitat manipulation	Missouri
11. Frémont clematis	Optimal habitat manipulation	Missouri
12. Queen-of-the-prairie	Maintenance of community integrity	Missouri
13. Wooly meadow foam	Distinguishing sibling subspecies; extinction of populations; founding populations	Oregon
14. Bradshaw's desert parsley	Inbreeding depression	Oregon
15. Tansy ragwort	Control of exotics	Oregon
16. Roses	Influence of hybridization on origin and development of cultivated plants	Oregon
17. Nestronia	Population decline towards extinction; life history extinction-prone?	North Carolina
Animals		
Crustaceans		
1. Sand shrimp	Extinction of population and influence on food chain	Oregon

Table 1. *(continued)*

Organism	Principle concerns	State or province
Fish		
1. Devil's Hole pupfish	Founding of new populations	California
2. Speckled dace	Preserving genetic diversity of populations and taxonomic units	California
3. Native trout	Taxonomic status and isolation of populations	California
4. Blue pike	Declared extinct	Ohio
5. Scioto madtom	Possibly extinct	Ohio
6. Trout	Relationship of genetics to management	Oregon
7. Salmon	Relationship of genetics to management	Oregon
8. Rainbow trout	Genetic comparisons between anadromous and nonanadromous	Washington
9. Steelhead trout	Effect of hatchery stocks on genetic diversity; genetic markers	Washington
10. Cutthroat trout	Isolation of populations	Wyoming
Amphibians		
1. Desert slender salamander	Isolated populations	California
2. Green salamander	Population decline	Kentucky
3. Barking tree frog	Isolation of populations	Kentucky
4. Bird-voiced tree frog	Preserve selection	Kentucky
5. *Plethodon kentuckiae*	Taxonomic and endangered statuses	Kentucky
Reptiles		
1. Blunt-nosed leopard lizard	Isolation of populations; merging of naturally disconnected populations	California
2. Coachella Valley fringe-toed lizard	Extinction of populations	California
3. Desert tortoise	Taxonomic status and genetic variability	California
4. Illinois mud turtle	Isolation of populations; value of preserving subspecies	Illinois, Missouri
5. Corn snake	Isolation of populations	Kentucky

Table 1. *(continued)*

Organism	Principle concerns	State or province
Reptiles		
6. Gopher tortoise	Genetic aspects of relocations	Florida
7. Blanding's turtle	Genetic importance of outliers	Missouri
8. Collared lizard	Habitat manipulation for optimization	Missouri
9. Massasauga rattlesnake	Isolation of populations	Missouri
10. Northern prairie skink	Between-population genetic diversity	Manitoba
Birds		
1. Inyo brown towhee	Isolation of populations	California
2. California condor	Value of preservation	California
3. Elf owl	Preserving genetic diversity	California
4. Great gray owl	Preserving genetic diversity	California
5. Bald eagle	Preserving genetic diversity; restoration, merging populations; isolation of populations	California, Ohio, Wyoming, Michigan
6. Dusky seaside sparrow	Hybridization and preserving genetic diversity	Florida
7. Whooping cranes	Founding of new populations	Florida, Manitoba
8. Osprey	Restoration	Kentucky
9. Snow goose	Understanding genetic variability	Manitoba
10. Prairie chicken	Isolation of populations	Michigan
11. Piping plover	Isolation of populations	Michigan
12. Barn owl	Isolation of populations	Michigan
13. Kirtland's warbler	Restoration	Michigan
14. Trumpeter swans	Restoration of migration routes	Missouri
15. Wild turkey	Founding of new populations	Ohio
16. Ring-necked pheasant	Founding of new populations	Ohio
17. Northern bobwhite	Founding of new populations	Ohio
18. Peregrine falcon	Restoration; isolation of populations; minimum viable population size; genetic variability	Wyoming, Manitoba, Oregon, Michigan
19. Columbia sharp-tailed grouse	Isolation of populations	Wyoming

Table 1. (continued)

Organism	Principle concerns	State or province
Mammals		
1. Salt marsh mouse	Isolation of populations	California
2. Kangaroo rat	Extinction of populations; ecosystem influence	California
3. Amargosa vole	Founding of populations	California
4. San Joaquin kit fox	Merging of populations	California
5. Grizzly bear	Extinction and isolation of populations	California, Wyoming
6. Bighorn sheep	Restoration, founding, and isolation of populations	California, Wyoming
7. White-tailed deer	Preserving genetic diversity	Florida
8. Florida panther	Preserving genetic diversity; restoration	Florida
9. Eastern woodrat	Isolation of populations	Illinois
10. New England cottontail	Hybridization and swamping	Kentucky
11. Eastern cottontail	Hybridization with New England cottontail	Kentucky
12. Long-tailed shrew	Minimum viable population; density effects	Kentucky
13. Otter	Restoration	Kentucky, Ohio, Missouri
14. Wood bison	Restoration	Manitoba
15. Wolf	Captive breeding and restoration	Michigan
16. Marten	Merging of populations	Michigan
17. Allegheny woodrat	Restoration	Ohio
18. Snowshoe hare	Founding of populations	Ohio
19. Columbia white-tailed deer	Minimum viable population; inbreeding depression; isolation of populations	Oregon
20. Moose	Isolation of populations	Wyoming
21. Black-footed ferret	Isolation, restoration, merging of populations	Wyoming
22. Elk	Isolation, restoration of populations	Wyoming
23. Pronghorn antelope	Isolation of populations	Wyoming
24. Mule deer	Isolation of populations	Wyoming

plex variety of genetic mechanisms that affect small populations that are the center of conservation efforts.

Genetic Management Questions

The following lists of questions have been grouped according to the part headings of this book. While these questions do not cover all possible problems, it is believed that they represent a good sample of management concerns. With many of the questions it will be possible to formulate answers through a careful examination of the appropriate chapters. Others are cited to stimulate thought about what managers need in the way of data, interpretation, and methodologies to be able to make decisions about the conservation (including management) of their habitats and species.

Isolation

1. *Determining whether populations are isolated*
(A) How do we determine if a population is isolated? (B) Ernst Mayr (1963) defined a population as "a group of individuals so situated that any two of them have equal probabilities of mating with each other and producing offspring." Some studies have shown gene flow to be restricted to small clusters of individuals. Are these clusters "isolated populations"? (C) At what level of gene flow does a population become isolated?

2. *Manipulated gene flow*
(A) Under what conditions should individuals be transferred between isolated populations? (B) How do we balance the desire to avoid the gene pool impoverishment that can occur with isolation and the differentiation/speciation process that also occurs with isolation? (C) If lack of gene flow causes genetic problems, should we avoid having many small reserves versus a few large reserves? (D) At what point can migration between small reserves offset genetic drift and degradation?

3. *Adaptations to rarity and low population size*
(A) Are populations isolated in disjunct habitats and remnant populations of widespread species different in their genetic variability? (B) Aren't some isolated populations adapted to rarity? (C) Should these rare populations ever have artificial gene flow into them?

Extinction

1. *Minimum viable population size*
(A) Can minimum viable population sizes be determined for different pollination and mating systems? (B) If a species is rare normally, should the estimate of minimum viable population size be applied to it as it would be for more common species?

2. *Long-term survival and risks of extinction*
(A) How many populations of a species are necessary to afford protection from extinction in the long term? (B) Do we know enough about genetics to be able to discuss long-term survival?

3. *Ranking priorities for protective management*
(A) If we use a triage system for endangered species, are there genetic characteristics of species that would help us determine ranks within the triage? (B) For instance, if a species' population cannot reach a minimum viable size, should we decide not to spend effort on its protection? (C) How many generations at some low population size would it take for us to decide not to expend effort on preserving it? (D) Or, instead, should we concentrate on the impact the loss of a particular species would have on the ecosystem? (E) Might common species of plants and animals be more important to the community than naturally rare species? (F) Does every state need a peregrine falcon restoration program, or should some states concentrate more on their endemic flora and fauna?

4. *Captive propagation and cultivation*
(A) When should efforts at artificial propagation, transplantation, and habitat modification begin? (B) Why do botanists seem more reluctant to make use of botanical gardens for conservation by means of artificial propagation of plants than zoologists are with captive propagation of animals in zoos?

Founding and Bottlenecks

1. *Manipulation of bottlenecks in population size*
(A) Some species regularly reach very low population sizes in their cycles. Have these species undergone genetic adaptation to regular bottlenecks? (B) Can this be used as a management strategy where bottlenecks are unavoidable?

2. *Deleterious alleles and fitness*
(A) Is it possible that a population becomes more fit as a bottleneck is ap-

Table 2. *The Minimum Sample Size N of Individuals, Based on Probability Theory, Required to Ensure That All Alleles at a Locus with Frequencies g Are Detected with a Probability of P**

g	N ($P \geq .95$)	N ($P \geq .99$)	N ($P \geq .999$)
.500	6	8	11
.400	7	10	14
.300	11	15	22
.200	21	28	39
.100	51	66	88
.080	65	84	112
.060	92	119	156
.040	152	192	249
.020	341	422	536
.010	754	916	1146
.009	850	1030	1285
.008	972	1174	1462

*After Gregorius, H. 1980. The probability of losing an allele when diploid genotypes are sampled. *Biometrics* **36**: 643–652.

proached? (B) At what level of bottlenecking is the loss of deleterious alleles offset by the loss of genetic variability?

3. *Restoration — source and destination of founders*
(A) How do we determine what sources of founders to use for restoration? (B) Are the taxonomic similarities of the source and restored populations important? (C) Should a species be transplanted outside its normal or known range into suitable habitat?

4. *Restoration — genetic diversity of founders*
(A) How many individuals are necessary to found a population? (See Table 2). (B) Is there a level of genetic diversity that is or should be necessary to successfully found a population?

5. *Restoration planning — baseline data*
What types of genetic analysis should be done prior to restoration and what type of records should be kept?

6. *Restoration planning — monitoring*
(A) What are the major signs of success that can be observed? (B) How many years of observation are necessary?

Hybridization and Merging Populations

1. *Manipulation of merging populations*
(A) Can the genetic consequences of merging populations be predicted? If

not, what risks do we run when we merge populations? (B) If hybridization increases the genetic diversity of populations, though not necessarily their fitness, should hybridization ever be undertaken as a means of preserving genetic diversity? Won't the original genomes be in danger of being lost? (C) Should plant populations threatened with extinction be moved to other sites with preexisting populations of the species? (D) When is it advisable to merge populations?

2. *Swamping and introgression*
(A) Can we protect populations, such as many native fish species, from being swamped by domestic, cultivated, or commercial stocks? (B) Should merging be avoided between populations distinct at the subspecific/varietal level?

Natural Diversity and Taxonomy

1. *Genetic variation and biogeography*
(A) Is there a trend in the geographic distribution of genetic variation within a species that would allow generalizations on whether the central or peripheral populations are most important? (B) How important are outliers to variability? (C) If a species is rare in our state but abundant elsewhere in its range, is it important that these marginal populations be protected?

2. *Genetic variation of species and subspecies/varieties*
(A) Does genetic variation follow subspecies/varieties taxonomy? (B) Should we test for genetic distances between subspecies/varieties versus full species?

3. *Extrapolations — comparability of taxa*
(A) Can the results of one genetic study be applied to other taxa? (B) Are species with different densities, mating systems, or habitat preferences comparable?

4. *Priorities — taxonomic levels for protection*
(A) Are relict taxa more important than taxa with other close relatives with similar genetic material? (B) What should our goal of conservation be: alleles, allele frequencies, heterozygosity, subspecies/varieties, species, or ecosystems? (C) Could standards be developed for determining such priorities? (D) Should we always work to increase diversity and fitness?

5. *Loss of numbers and genetic variability*
(A) What are the genetic results of removal of individuals from a small population as might be done for transplanting or cultivation? (B) Can the harvesting of a species alter the genetic characteristics of a population over

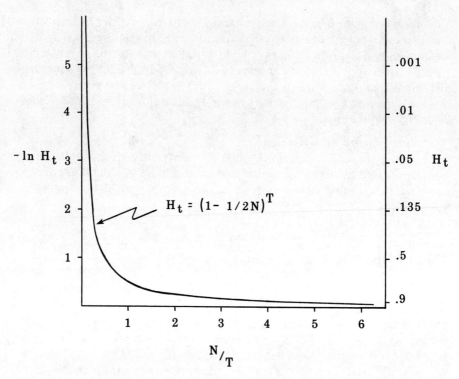

Figure 1. Graph of the heterozygosity (H) remaining after t generations (H_t) and the natural log transform ($-\ln H_t$) as a function of the ratio N/T, the ratio of the effective population size (N) and the number of generations (T). The curve shows the function when both N and T are very large, however, it is accurate to within 1% of the actual value of H_t when N and T are greater than 7. The graph clearly shows that over a few generations the loss of heterozygosity is small unless a very small population size is present. A long-term view of a populations heterozygosity reveals that substantial amounts of heterozygosity can be lost when populations remain at sizes that are small in relation to the number of generations over time.

several decades, and how can this be documented? (C) How quickly is genetic diversity lost in small populations? (See Figure 1).

6. *Mating systems and preserving natural diversity*
(A) Is there information compiled on mammal mating systems as Hamrick has done for plants? (See Table 3). (B) How does one determine the reproductive mode or breeding system for a plant? (See Figure 2).

7. *Adaptation*
(A) How does one determine the maximum amount of habitat alteration in an ecosystem that a species can accommodate and survive? (B) When habi-

Table 3. *Directory to Pertinent Information on Mammalian Mating Systems and Life History Strategies That Can Be Found in Eisenberg (1981).*

Taxon	Subject	Page
Monotremes	Body weight, metabolic rate, maximum captive longevity, encephalization quotient, litter size, litters per year, assumed maximum reproductive life span, estimated maximum lifetime, production of young	15
Monotremes and marsupials	Recent literature on monotremes: ethology, ecology, reproduction, general	26–27
Marsupials	Mean head + body length, weight, home range, density, biomass; social units: mating, parental care, feeding, refuging; defensibility of: mate, feeding site, shelter, temporal stability of resources; reproductive rate	39
Dasypodids (armadillos)	Length of gestation, number of young, number of embryos, time eyes open	50
Edentates and pholidots (anteaters, pangolins)	Head and body length, unit weight, home range, density, biomass 54	
Rodents (general)	Recent life history studies: ecology, behavior, reproduction and ontogeny	89
Rodents (desert-adapted)	Literature references: ecology, behavior, reproduction and ontogeny, 95 physiology	
Muroid rodents	Recent literature: ecology, life history, behavior, reproduction and ontogeny	97
Caviomorph rodents	Recent literature: ecology, behavior, reproduction and ontogeny	108–109
Insectivores	Literature: ecology, behavior, reproduction and ontogeny, physiology	115

Table 3. *(continued)*

Taxon	Subject	Page
Fissiped carnivores	Recent literature: ecology, behavior, reproduction and ontogeny, general review	129–131
Social carnivores	Form of dimorphism, forms of dominance, spacing between groups, social structure, adult relationships	138
Pinnipeds (seals and sea lions)	Selected references: natural history and ecology, behavior, reproduction and ontogeny, physiology	142
Chiroptera (bats)	Ecology, behavior, reproduction, development	155
Primates	Recent literature: behavior and ecology	160–161
Cetaceans (whales)	Ecology, behavior, reproduction	175
Sirenians and proboscideans (dugongs, manatees, elephants)	Natural history, ecology, behavior, reproduction and ontogeny	182
Perrissodactyls (horses and their relatives)	Natural history and ecology, behavior, reproduction and ontogeny	191
Artiodactyls (antelope, deer, and their relatives)	Natural history, ecology, behavior, reproduction and ontogeny	206–207
Mammals (general)	Social system: mating, rearing of young, foraging, refuging, antipredator strategies, size dimorphism, social complexity	423–425

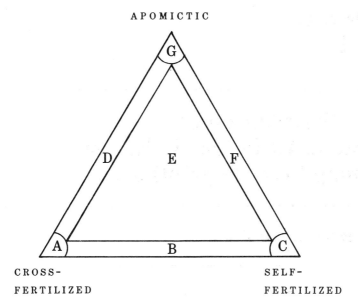

Figure 2. *A simplified "reproductive triangle" with seven general regions representing the kinds of possible reproduction in vascular plants. A) cross-fertilized species; B) partially self-fertilized and partially cross-fertilized species; C) self-fertilized species; D, E, and F) species sometimes reproducing asexually, corresponding to categories A, B, and C respectively when sexual; G) species that reproduce only asexually. All fully sexual species will be in region A-B-C of the triangle; most facultatively apomictic species will be in region A-D-G; few species will be in region C-F-G and the interior, E. (From Fryxell, 1957).*

tats are altered intentionally or unavoidably, do we risk altering species genetically through creating new selection pressures? (C) Can human disturbance of natural processes still preserve evolutionary capacity for species, and for which species?

APPENDIX 4

The Distribution of Genetic Variation Within and Among Plant Populations

James L. Hamrick

Data on the distribution of variation within and among populations of 91 plant species were used to compile the summaries given in Chapter 20, Table 1. H_T = total allelic diversity; H_S = mean allelic diversity within populations; G_{ST} = the ratio of the allelic diversity among populations to the total allelic diversity; A_p = mean number of alleles per polymorphic locus; P_A = the proportion of the total number of alleles found within each population.

Species	Populations	Polymorphic loci	H_T	H_S	G_{ST}	A_p	P_A	Source
Amaranthaceae, Amaranth Family								
Amaranthus spp. (amaranth, pigweed)	4	3	.065	.054	.153	2.33	.619	Jain et al., 1980
Asteraceae, Sunflower Family								
Chrysanthemum leucanthemum (oxeye daisy)	6	7	.385	.367	.045	2.85	.921	Griswold, unpublished
Coreopsis cyclocarpa var. *cyclocarpa*	3	9	.309	.285	.077	2.67	.764	Crawford and Bayer, 1981
Coreopsis cyclocarpa var. *pinnatisecta*	5	11	.325	.261	.186	3.36	.638	Crawford and Bayer, 1981
Helianthus annuus (common sunflower)	5	1	.345	.334	.032	2.00	1.000	Torres et al., 1977
Helianthus annuus	2	1	.338	.336	.006	2.00	1.000	Torres and Diedenhofen, 1979
Hymenopappus artemisiifolius	12	5	.371	.311	.162	4.20	.488	Babbell and Selander, 1974
Hymenopappus scabiosaeus (whitebract hymenopappus)	14	5	.353	.306	.133	4.20	.505	Babbell and Selander, 1974
Liatris cylindracea (cylindric blazing-star)	—	14	.328	.305	.069	2.14	—	Schaal, 1975
Stephanomeria exigua ssp. *carotifera* (a wire-lettuce)	11	8	.331	.300	.077	4.88	.646	Gottlieb, 1975
Tragopogon dubius (yellow salsify)	6	4	.223	.146	.162	2.00	.729	Roose and Gottlieb, 1976
Tragopogon mirus	8	5	.188	.038	.515	2.80	.486	Roose and Gottlieb, 1976
Tragopogon miscellus	6	4	.018	.017	.070	2.00	.667	Roose and Gottlieb, 1976
Tragopogon porrifolius (vegetable-oyster salsify)	3	2	.285	.216	.271	2.00	.832	Roose and Gottlieb, 1976
Xanthium strumarium (rough cocklebur)	12	4	.311	.028	.910	2.25	.481	Moran and Marshall, 1978

Species	Populations	Polymorphic loci	H_T	H_S	G_{ST}	A_P	P_A	Source
Caryophyllaceae, Pink Family								
Silene maritima (sea silene)	4	4	.454	.421	.060	3.75	.724	Baker et al., 1975
Chenopodiaceae, Goosefoot Family								
Chenopodium atrovirens (dark goosefoot)	11	3	.203	.008	.702	2.00	.567	Crawford and Wilson, 1979
Chenopodium desiccatum (desert goosefoot)	8	2	.205	.006	.536	2.00	.563	Crawford and Wilson, 1979
Chenopodium fremontii (Frémont goosefoot)	40	5	.299	.022	.893	2.20	.481	Crawford and Wilson, 1977
Chenopodium hians	12	5	.343	.036	.732	2.00	.567	Crawford and Wilson, 1979
Chenopodium hircinum	10	2	.407	.284	.304	4.00	.450	Wilson, 1981
Chenopodium incognitum	9	5	.438	.023	.951	2.40	.495	Crawford and Wilson, 1979
Chenopodium leptophyllum (slimleaf goosefoot)	11	6	.366	.089	.741	2.16	.555	Crawford and Wilson, 1979
Chenopodium pratericola (narrowleaf goosefoot)	26	6	.143	.034	.477	2.17	.470	Crawford and Wilson, 1979
Chenopodium quinoa (cultivated quinoa)	23	3	.125	.055	.286	2.00	.596	Wilson, 1981
Chenopodium quinoa var. *melanospermum* (highland wild quinoa)	7	2	.122	.103	.296	2.00	.679	Wilson, 1981
Fabaceae, Pea Family								
Baptisia leucophaea (plains wild-indigo)	9	4	.432	.367	.151	2.00	.953	Scogin, 1969
Baptisia sphaerocarpa (yellow wild-indigo)	3	3	.438	.417	.065	2.00	1.000	Scogin, 1969
Desmodium nudiflorum (barestem tick-clover)	5	6	.090	.073	.105	2.33	.600	Schaal and Smith, 1980

Species	Populations	Polymorphic loci	H_T	H_S	G_{ST}	A_P	P_A	Source
Fabaceae, Pea Family (continued)								
Lupinus subcarnosus (Texas bluebonnet)	8	5	.913	.119	.383	3.00	.613	Babbell and Selander, 1974
Lupinus texensis (Texas bluebonnet)	10	5	.525	.428	.183	4.60	.678	Babbell and Selander, 1974
Phaseolus coccineus (scarlet runner bean)	15	1	.319	.140	.561	3.00	.487	Wall and Wall, 1975
Phaseolus vulgaris (garden bean)	24	3	.343	.144	.486	2.67	.528	Wall and Wall, 1975
Trifolium hirtum (rose clover)	17	4	.376	.334	.118	2.00	.912	Jain, 1977
Trifolium hirtum	6	3	.302	.259	.179	2.67	.792	Martins and Jain, 1980
Lamiaceae, Mint Family								
Origanum vulgare (wild marjoram)	2	1	.511	.420	.178	3.00	.833	Elena-Rossello et al., 1976
Lythraceae, Loosestrife Family								
Lythrum tribracteatum (loosestrife)	7	1	.853	.477	.440	11.00	.247	Baker and Baker, 1976
Moraceae, Mulberry Family								
Ficus carica (common fig)	4	2	.534	.532	.004	3.00	.917	Valizadeh, 1977
Myrtaceae, Myrtle Family								
Eucalyptus obliqua (messmate stringybark eucalyptus)	4	3	.489	.369	.243	3.00	.806	Brown et al., 1975
Eucalyptus pauciflora (snow eucalyptus)	3	7	.271	.265	.020	3.43	.944	Phillips and Brown, 1977

Species	Popula-tions	Poly-morphic loci	H_T	H_S	G_{ST}	A_P	P_A	Source
Onagraceae, Evening-primrose Family								
Chamaenerion angustifolium (fireweed)	2	2	.393	.359	.056	2.50	1.000	Verkleij and Koniuszek, 1981
Clarkia biloba (lobed clarkia)	3	5	.345	.304	.121	3.60	.759	Gottlieb, 1974
Clarkia lingulata (Merced clarkia)	2	5	.340	.279	.136	3.20	.844	Gottlieb, 1974
Clarkia rubicunda (farewell-to-spring)	4	6	.271	.238	.106	2.33	.821	Gottlieb, 1973
Gaura demareei (a gaura)	2	5	.198	.186	.060	3.00	.800	Gottlieb and Pilz, 1976
Gaura longiflora (a gaura)	3	6	.270	.252	.054	3.17	.737	Gottlieb and Pilz, 1976
Oenothera biennis (common evening-primrose)	43	4	.377	.154	.588	2.50	.630	Levin, 1975a
Oenothera biennis	3	6	.348	.336	.028	2.33	1.000	Levy and Levin, 1975
Oenothera grandis	26	5	.180	.160	.080	3.17	—	Ellstrand and Levin, 1980
Oenothera laciniata (cutleaf evening-primrose)	26	6	.178	.142	.240	2.80	—	Ellstrand and Levin, 1982
Oenothera parviflora (smallflower evening-primrose)	29	8	.465	.442	.049	2.38	.974	Levy and Levin, 1975
Pinaceae, Pine Family								
Abies lasiocarpa (subalpine fir)	3	1	.394	.388	.015	3.00	1.000	Grant and Mitton, 1977
Picea abies (Norway spruce)	8	4	.469	.462	.021	3.25	.923	Bergmann, 1973
Picea abies	18	1	.488	.362	.258	3.00	.815	Bergmann, 1975a
Picea abies	15	6	.450	.420	.068	3.17	.832	Bergmann, 1975b
Picea abies	18	1	.737	.631	.144	4.00	.972	Bergmann, 1978

Species	Populations	Polymorphic loci	H_T	H_S	G_{ST}	A_p	P_A	Source
Pinaceae, Pine Family (continued)								
Picea abies	4	10	.420	.408	.036	5.30	.764	Lundkvist, 1979
Picea abies	11	4	.372	.352	.052	5.00	.850	Lundkvist and Rudin, 1977
Picea engelmannii (Engelmann spruce)	3	1	.538	.484	.100	3.00	1.000	Grant and Mitton, 1977
Picea sitchensis (Sitka spruce)	10	18	.159	.147	.079	2.56	.713	Yeh and El-Kassaby, 1980
Pinus contorta ssp. latifolia (lodgepole pine)	9	25	.167	.160	.041	3.16	.601	Yeh and Layton, 1979
Pinus longaeva (Great Basin bristlecone pine)	5	11	.484	.465	.038	2.64	.945	Hiebert and Hamrick, 1983
Pinus nigra (Austrian pine)	40	4	.396	.349	.107	3.75	.613	Bonnet-Masimbert and Bikay-Bikay, 1978
Pinus ponderosa (ponderosa pine)	11	12	.289	.284	.015	3.23	.925	Hamrick, unpublished
Pinus ponderosa	2	3	.430	.429	.003	2.00	1.000	Linhart et al., 1981b
Pinus ponderosa	6	7	.300	.289	.041	2.71	.904	Linhart et al., 1981c
Pinus ponderosa	2	1	.449	.426	.051	3.00	1.000	Linhart et al., 1981a
Pinus ponderosa	11	1	.368	.344	.065	2.00	1.000	Mitton et al., 1977
Pinus ponderosa	3	7	.302	.294	.027	2.86	.900	Mitton et al., 1980
Pinus ponderosa	10	12	.123	.108	.122	2.25	—	O'Malley et al., 1979
Pinus pungens (Table Mountain pine)	3	1	.342	.331	.032	2.00	1.000	Feret, 1974
Pinus radiata (Monterey pine)	5	1	.154	.133	.136	2.00	.800	Moran et al., 1980
Pinus rigida (pitch pine)	11	21	.152	.147	.023	3.10	—	Guries and Ledig, 1982
Pinus sylvestris (Scotch pine)	19	3	.364	.338	.088	8.67	.431	Mejnartowicz, 1979
Pinus sylvestris	3	3	.390	.383	.018	5.33	.812	Rudin et al., 1974

Species	Populations	Polymorphic loci	H_T	H_S	G_{ST}	A_P	P_A	Source
Pinaceae, Pine Family (continued)								
Pinus taeda (loblolly pine)	2	12	.250	.245	.022	3.42	.854	Adams and Joly, 1980
Pinus virginiana (Virginia pine)	3	2	.301	.299	.004	3.00	.778	Witter and Feret, 1978
Pseudotsuga menziesii (Douglas-fir)	6	1	.396	.342	.136	3.00	.944	Bergmann, 1975a
Pseudotsuga menziesii	2	1	.320	.312	.025	3.00	1.000	Linhart et al., 1981b
Pseudotsuga menziesii	6	2	.552	.538	.034	6.00	.875	Meinartowicz, 1976
Pseudotsuga menziesii	11	18	.159	.155	.026	3.00	.730	Yeh and O'Malley, 1980
Poaceae, Grass Family								
Anthoxanthum odoratum (sweet vernal grass)	3	3	.531	.507	.045	3.00	.926	Wu and Jain, 1980
Avena barbata (slender wild oat)	16	5	.289	.074	.736	2.00	.606	Clegg and Allard, 1972
Avena barbata	31	1	.829	.397	.521	17.00	.220	Kahler et al., 1980
Bromus mollis (soft brome)	10	6	.211	.205	.030	2.00	.958	Brown et al., 1974
Dactylis glomerata (orchard grass)	4	2	.314	.272	.088	4.50	.722	Lumaret, 1982
Elymus canadensis (Canada wild-rye)	63	4	.210	.094	.499	2.25	.588	Sanders et al., 1979
Hordeum jubatum (foxtail barley)	3	5	.293	.244	.184	2.20	.818	Babbell and Wain, 1977
Hordeum jubatum	11	7	.266	.159	.259	2.86	.555	Shumaker and Babbell, 1980
Hordeum spontaneum (ancestral two-row barley)	28	25	.194	.098	.490	3.88	.381	Brown et al., 1978
Hordeum spontaneum	4	11	.386	.351	.090	3.27	.778	Nevo et al., 1981

Species	Populations	Polymorphic loci	H_T	H_S	G_{ST}	A_p	P_A	Source
Poaceae, Grass Family (continued)								
Hordeum vulgare (six-row barley)	30	4	.572	.536	.036	7.00	.601	Kahler and Allard, 1981
Lolium multiflorum (Italian darnel)	8	1	.661	.589	.109	4.00	.906	Østergaard and Nielsen, 1981
Lolium perenne (perennial darnel)	9	3	.442	.390	.118	2.67	.986	Hayward and McAdam, 1977
Lolium perenne	11	1	.606	.581	.041	4.00	.818	Østergaard and Nielsen, 1981
Panicum maximum (sexual) (Guineagrass panic grass)	3	6	.396	.381	.035	3.29	.985	Usberti and Jain, 1978
Panicum maximum (asexual)	3	8	.172	.159	.080	3.29	.652	Usberti and Jain, 1978
Secale cereale (common rye)	2	1	.652	.616	.055	4.00	.969	Jaaska, 1979
Zea mays (maize)	51	1	.826	.621	.248	21.00	.206	Stuber et al., 1977
Polemoniaceae, Phlox Family								
Gilia achilleifolia (selfing) (yarrow gilia)	3	8	.300	.139	.390	2.77	.527	Schoen, 1982a
Gilia achilleifolia (outcrossing)	4	10	.280	.209	.170	2.77	.527	Schoen, 1982a
Phlox cuspidata	10	7	.197	.100	.455	2.00	.629	Levin, 1975b
Phlox cuspidata	43	5	.074	.046	.299	2.60	.462	Levin, 1978
Phlox drummondii (Drummond phlox)	10	4	.380	.322	.141	2.25	.894	Levin, 1975b
Phlox drummondii (cultivated)	16	7	.323	.100	.601	2.00	.647	Levin, 1976
Phlox drummondii	73	5	.244	.180	.231	2.40	.708	Levin, 1978
Phlox roemeriana (Roemer phlox)	15	4	.356	.278	.244	2.75	.691	Levin, 1975b

Species	Popula-tions	Poly-morphic loci	H_T	H_S	G_{ST}	A_p	P_A	Source
Polygonaceae, Buckwheat Family								
Polygonum pensylvanicum (Pennsylvania smartweed)	6	3	.461	.413	.125	2.00	.889	Kubetin and Schaal, 1979
Rosaceae, Rose Family								
Fragaria chiloensis (Chiloé strawberry)	7	2	.558	.302	.482	4.50	.556	Hancock and Bringhurst, 1979
Fragaria vesca (European strawberry)	3	1	.627	.404	.356	3.00	.667	Arulsekar and Bringhurst, 1981
Fragaria vesca	3	2	.448	.310	.292	3.00	.722	Hancock and Bringhurst, 1978
Fragaria virginiana (Virginia strawberry)	2	1	.579	.577	.003	5.00	.800	Hancock and Bringhurst, 1978
Sarraceniaceae, Pitcherplant Family								
Sarracenia purpurea (common pitcherplant)	11	5	.219	.172	.195	2.40	.659	Schwaegerle and Schaal, 1979
Scrophulariaceae, Snapdragon Family								
Mimulus guttatus (common monkeyflower)	11	9	.492	.375	.213	4.67	.699	McClure, 1973
Solanaceae, Potato Family								
Lycopersicon esculentum (common tomato)	7	5	.113	.094	.161	2.20	.649	Rick and Fobes, 1975
Lycopersicon esculentum var. *cerasiformae* (cherry tomato)	7	5	.081	.042	.312	2.20	.545	Rick and Fobes, 1975
Lycopersicon pimpinellifolium (current tomato)	43	6	.239	.125	.317	2.50	.558	Rick et al., 1977
Solanum grayii	4	1	.428	.187	.563	2.00	.750	Whalen, 1979
Solanum johnstonii	2	4	.350	.295	.120	2.00	.938	Whalen, 1979
Solanum pennellii	23	19	.431	.268	.361	4.58	.468	Rick and Tanksley, 1981
Solanum tenuipes	2	4	.288	.230	.169	2.20	.773	Whalen, 1979
Zosteraceae, Eel-grass Family								
Zostera marina (common eel-grass)	2	2	.396	.396	.001	2.00	1.000	Gagnon et al., 1980

APPENDIX
5

Calculating Inbreeding Coefficients from Pedigrees

J. Ballou

Inbreeding can be loosely defined as any mating between relatives. However, this definition needs to be put in some perspective, since any two individuals in a population can probably be traced back to a common relative—providing their lineages are traced back far enough. We then must be careful to define inbreeding relative to either some point in time or to some "base" population—that is, a population that consists of individuals assumed to be unrelated to each other. Exactly what constitutes the "base" population is important in defining the level of inbreeding in any population.

Determining the base population is usually dependent on how much information is available on the population we are interested in. In the best of cases, the field or zoo manager will have pedigree information available only as far back as the founding of his population. Often times this information is unavailable, and only data on the most recent generations are available. For this reason it is often convenient to define the base population as those individuals in a pedigree beyond which no further information is available. These may be wild-caught founders or those individuals existing in a population when the first pedigree information was recorded. These individuals are assumed to be unrelated (unless it is known otherwise), and any inbreeding occurring in the population is relative to this group of individuals.

The genetic effect of inbreeding is to increase the average homozygosity of the population, and inbred offspring (offspring resulting from consanguineous matings) have a higher probability of being homozygous for more loci than noninbred offspring. This genetic effect is measured by the "inbreeding coefficient," designated as F, and defined as the probability that the two alleles present at a given locus are "identical by descent"—that is to say, are derived by replication of a single allele from a common ancestor. The F value ranges from 0 (for noninbred individuals) to 1 (for totally homozygous individuals), and the degree of inbreeding is dependent on how closely related the parents of the individual are. (For a full discussion of inbreeding, see Hartl, 1980; Crow and Kimura, 1970; Lasley, 1978; or Falconer, 1981.)

Various methods have been used to calculate inbreeding coefficients (for examples, see the above references). We shall present two common methods: One using Path Analysis techniques and the other using an Additive Relationship method.

Path Analysis Technique

Once a pedigree of an individual (say X) is obtained, we can calculate its inbreeding coefficient, F_x. The steps involved can be summarized as follows:

Step 1: *Draw the pedigree so that common ancestors appear only once.* A common ancestor is any individual related to both parents of X, the individual for whom we wish to determine F_x. If there are no common ancestors, then $F_x = 0$, and X is noninbred.

Step 2: *Determine the inbreeding coefficients of all common ancestors.* If there is no pedigree information on the common ancestor, it is often assumed to be noninbred. If the common ancestor is inbred, then its inbreeding coefficient, F_{CA}, must be calculated before calculating F_x. Calculate F_{CA} as you would F_x as described in the following steps. Once F_{CA} is determined, F_x can then be calculated by returning to step 3 (see example 4 below).

Step 3: *Look for loops in the pedigree.* A loop is a path that runs from X, through one parent, to the common ancestor, through the other parent, and back to X without going through any individual more than once. Determine the number of steps in each loop.

APPENDIX 5 Calculating Inbreeding Coefficients from Pedigrees   511

Step 4: *Calculate the contribution of each loop to the inbreeding coefficient.* The contribution of each loop to the F_x is determined by using the following formula:

$$\left(\frac{1}{2}\right)^{i-1} \times (1 + F_{CA})$$

where: F_{CA} = inbreeding coefficient of the common ancestor,
   i = number of steps in each loop as defined in step 3.

Step 5: *Sum the contributions of each loop.* The summation of all the contributions will be the inbreeding coefficient of individual X.

Several examples follow to illustrate the principles involved.

Example 1: Half-sib mating
The pedigree of individual X is shown in Figure 1A. X's mother and father are D and E, respectively, and they share the same mother, A, but have

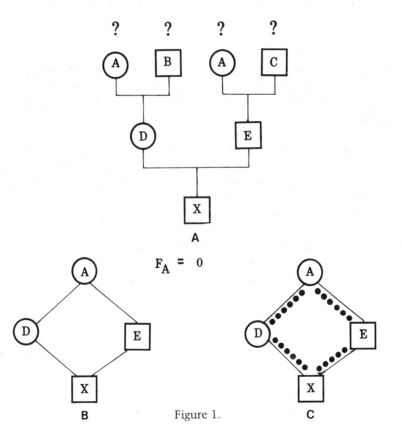

Figure 1.

different fathers. We wish to calculate the inbreeding coefficient of X, the offspring of two half-sibs. In this example, A is the common ancestor, since it is related to both parents of X.

Step 1: Figure 1B shows the pedigree redrawn so that the common ancestor, A, appears only once.

Step 2: Since the pedigree of A is unknown, we assume that it is noninbred and let F_A, the inbreeding coefficient of A, equal 0.

Step 3: After examining the pedigree, it is seen that there is only one four-step loop, shown in Figure 1C (dotted line). Therefore, $i = 4$.

Step 4: The contribution of this loop is calculated:

Loop	F_{CA}	i	Contribution to F_X
X–D–A–E–X	0.0	4	$(1/2)^3 \times (1 + 0.0) = 0.125$

Step 5: Since only one loop is involved: $\qquad F_X = 0.125$
Individual X therefore has an inbreeding coefficient of 0.125.

Example 2: Multiple-loop pedigree

The pedigree of individual X is shown in Figure 2A. H and G are the parents of X; E and F are the parents of H. A is the mother of E, F, and G; however, they all have different fathers. We wish to calculate the inbreeding coefficient of X.

Step 1: The abbreviated pedigree is shown in Figure 2B, with A, the only common ancestor, appearing once. Only individuals involved in the pathways from the common ancestor to X are shown (that is, B, C, and D need not be included).

Step 2: Since the pedigree of A is unknown, we assume that A is noninbred and let $F_A = 0$.

Step 3: Two separate five-step loops can be drawn through the pedigree as shown by the dashed and dotted lines in Figure 2C.

Step 4: The contribution of each loop to F_X is calculated:

Loop	F_{CA}	i	Contributions to F_X
X–H–E–A–G–X	0.0	5	$(1/2)^4 \times (1 + 0.0) = 0.0625$
X–H–F–A–G–X	0.0	5	$(1/2)^4 \times (1 + 0.0) = 0.0625$

Step 5: Summing over all loops: $\qquad F_X = 0.125$
The inbreeding coefficient of X is 0.125.

APPENDIX 5 Calculating Inbreeding Coefficients from Pedigrees

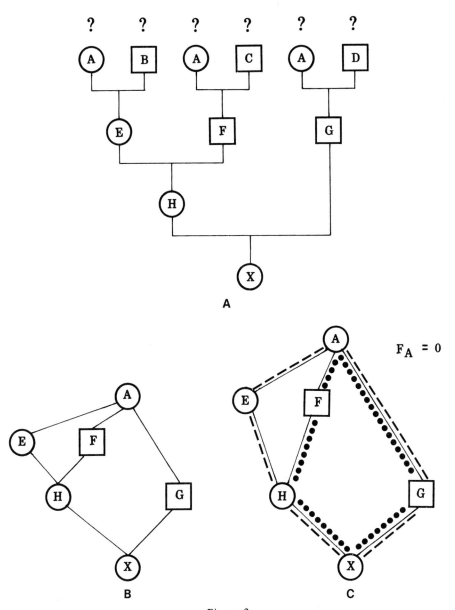

Figure 2.

Example 3: Full-sib mating

The complete pedigree of X is shown in Figure 3A. X is the offspring of two full sibs, C and D, who are offspring of A and B. The parents of A and B are unknown. We wish to calculate the inbreeding coefficient of X.

APPENDIX 5 Calculating Inbreeding Coefficients from Pedigrees

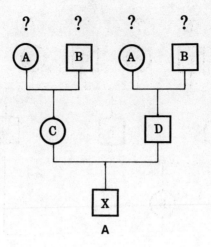

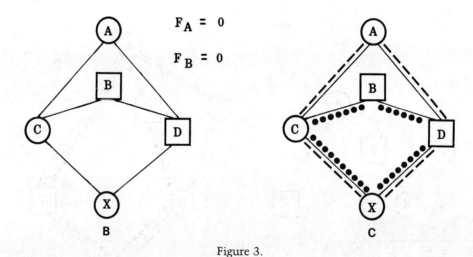

Figure 3.

Step 1: A and B are both common ancestors, since they are related to both parents of X. The abbreviated pedigree is shown in Figure 3B, with A and B appearing only once.

Step 2: Since neither of the common ancestor's pedigrees is known, they are assumed to be noninbred, and F_A and F_B are 0.

Step 3: Only one loop can be traced through each of the common ancestors, as shown in Figure 3C (dashed and dotted lines).

Step 4: The contribution of each loop to F_X is calculated:

Loop	F_{CA}	i	Contribution to F_x
X-C-$\underline{A}$-D-X	0.0	4	$(1/2)^3 \times (1 + 0.0) = 0.125$
X-C-$\underline{B}$-D-X	0.0	4	$(1/2)^3 \times (1 + 0.0) = 0.125$

Step 5: Summing over all loops: $\qquad F_x = 0.25$
The inbreeding coefficient of X is 0.25.

Example 4: Complex pedigree

Step 1: Figure 4A shows the already abbreviated pedigree of individual X. A, C, and E are all common ancestors, since they are related to both B and D, the parents of X. We wish to determine the inbreeding coefficient of X.

Step 2: In calculating the inbreeding coefficient of A, C, and E, the common ancestors, we assume that both A and E are noninbred. However, C is inbred, and we must first calculate F_C before determining F_x.

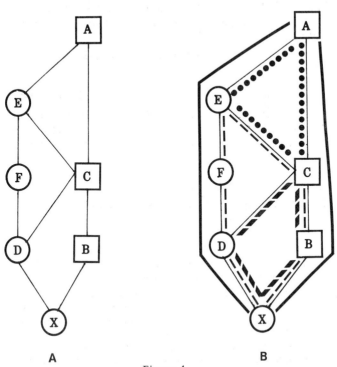

Figure 4.

Calculating F_c — Path Analysis

We determine F_C just as we would any other inbreeding coefficient. However, since the abbreviated pedigree is already drawn and we know that the inbreeding coefficient of A, the common ancestor of C, is 0, we can skip steps 1 and 2. We proceed directly to step 3 and look for loops involving A and C.*

Step 3: Figure 4B shows the only three-step loop from C to A (dotted line).

Step 4: The contribution of the loop is calculated:

Loop	F_{CA}	i	Contribution of F_C
C-E-$\underline{A}$-C	0.0	3	$(1/2)^2 \times (1 + 0.0) = 0.25$
			$F_C = 0.25$

Step 5: Since there is only one loop, and we conclude that the inbreeding coefficient of the common ancestor, C, is 0.25.†

Proceeding with the determination of the inbreeding coefficient of X, starting at step 3:

Step 3: Figure 4B shows the three loops that can be traced between X and its common ancestors, C, E, and A (bold dashed, light dashed, and solid lines).

Step 4: The contribution of each loop to F_X is calculated:

Loop	F_{CA}	i	Contribution to F_X
X-D-F-E-$\underline{A}$-C-B-X	0.0	7	$(1/2)^6 \times (1 + 0.0) = 0.0156$
X-D-F-$\underline{E}$-C-B-X	0.0	6	$(1/2)^5 \times (1 + 0.0) = 0.0313$
X-D-$\underline{C}$-B-X	0.25	4	$(1/2)^3 \times (1 + 0.25) = 0.1563$

Step 5: Summing over all loops: $F_X = 0.2032$
The inbreeding coefficient of X is 0.2032.

---

*Note that A is both a parent and a common ancestor of C, because A is related to both of C's parents — it is the father of E and is itself the father of C.

†Two important points can be made here: (1) C is the offspring of a father–daughter mating, and all such matings, like all mother–son matings, have inbreeding coefficients of 0.25; (2) example 3 as well as the above example shows different mating patterns that result in inbreeding coefficients of 0.25. Examples 1 and 2 also show different breeding patterns that result in the same inbreeding coefficient. Different breeding schemes can obviously result in similar inbreeding coefficients; therefore one cannot draw conclusions about the mating structure based solely on the inbreeding coefficients of the offspring produced.

APPENDIX 5 Calculating Inbreeding Coefficients from Pedigrees   517

The advantages in using Path Analysis are that it is fast and does not require the use of computers. The disadvantages, however, are apparent if inbreeding coefficients of individuals in complex pedigrees are desired or if many inbreeding coefficients need to be calculated. This method then becomes extremely tedious. Calculating inbreeding coefficients from complex pedigrees is considerably simplified, however, by using the additive relationship method. The method is easily incorporated into computer language and is extremely useful for breeding plans since inbreeding coefficients for large numbers of hypothetical offspring and "founder representation" (the proportion of founder genes present in descendents) can be calculated.

Calculating F_c – Additive Relationship Matrix Method

Probably the simplest way to describe this method is to work through an example. Since the technique is the same regardless of how complex or long a pedigree is, we will use one relatively simple example, that already presented in Example 3 of the Path Analysis (See Figure 3).

The steps for using the Additive relationship method are as follows:

Step 1: Construct a table by placing the names of the individuals in the pedigree across the top and down the left side in order of their births (chronological order). Above each individual in the top row, place the names of its parents (If only one parent of an individual is known, place that parent above the individual.):

	A	B	AB v C	AB v D	CD v X
A					
B					
C					
D					
X					

Step 2: Place a "1 +" in each of the boxes in the diagonal:

	A	B	A B v C	A B v D	C D v X
A	1+				
B		1+			
C			1+		
D				1+	
X					1+

Step 3: Work across the top row, from left to right. For the diagonal (first) box, do nothing. For all other boxes on the top row, locate the values in this row for each parent of the individual under consideration (the individual under consideration is the individual listed at the top of that column). Add the values of the two parents (or use single value if only one parent is known), divide by two and enter the result in the individual's box. In our example, B has no known parents, so it gets a "0." C's parents are A and B. The values in A and B's boxes (1 & 0) are therefore added, divided by two, and entered as C's value (1/2). Calculate the values for each individual in the top row. When the top row is completed, copy the same values down the first column on the table:

	A	B	A B v C	A B v D	C D v X
A	1+0	0	1/2	1/2	1/2
B	0	1+			
C	1/2		1+		
D	1/2			1+	
X	1/2				1+

Step 4: Complete successive rows one at a time in the same way:

For each box in the row: 1) identify the parents of the individual at the top of that column; 2) identify the parent's columns (or parent's column if only one is known); 3) from the row on which you are working, find and add the two values in the parent's columns (or use single value if only one parent is known); 4) divide by two and enter the result in the box under consideration. For example, to fill the box for individual X on the third row, we see the parents of X are C and D. The values in columns C and D on the third row are 1 and 1/2, their sum is 1½. Halving that results in the value 3/4, which is entered in box X on the third row.

For each diagonal box: 1) identify the parents of the individual at the top of that column (If only one parent is known, add 0 to the 1 in the diagonal box); 2) identify the box where the row labeled by one parent intercepts with the column labeled by the other parent; 3) divide the value in the box by 2 and add it to the 1 in the diagonal box. For example, to calculate the value for box X in row 5 (the diagonal box), we see that X's parents are C and D. The value in the box where row C intersects column D is 1/2. Therefore 1/4 is added to the 1 already in the diagonal box.

Copy the values in each row, as completed, into the corresponding columns. Therefore, the second row read from left to right should read the same as the second column from top to bottom, etc. The completed table is:

	A	B	A B v C	A B v D	C D v X
A	1 + 0	0	1/2	1/2	1/2
B	0	1 + 0	1/2	1/2	1/2
C	1/2	1/2	1 + 0	1/2	3/4
D	1/2	1/2	1/2	1 + 0	3/4
X	1/2	1/2	3/4	3/4	1 + 1/4

Several valuable pieces of information can be obtained from the table. Of primary interest are the inbreeding coefficients. These are calculated from the values in the diagonal boxes. One subtracted from this value is the inbreeding coefficient of the individual at the top of that column. Therefore, the inbreeding coefficient of individual D is "0" while X's is $1/4$ (0.25).

Inbreeding coefficients of all possible pairings between individuals in the population can also be calculated. One-half the value in any box (except for the diagonals) will be the inbreeding coefficient of the offspring resulting from matings between the individual labeling the row and the individual labeling the column of that box. For example, the inbreeding coefficient of any offspring of C and D will be one-half of $1/2$, or $1/4$. This is confirmed above when it was shown that the inbreeding coefficient of X (the offspring of C & D) was indeed $1/4$. As another example, the inbreeding coefficient of offspring resulting from a mating between X and D will be $3/8$.

In addition, the "founder representation", (the proportion of a founder's genes in a descendent) can also be calculated. The boxes in rows labeled by founders indicate the proportion of genes from the founder in the individual labeling the column. Therefore, $1/2$ of C, D, and X's genes come from founder A and $1/2$ from founder B.

Acknowledgments

I would like to thank T. Foose and N. Flesness for permission to reproduce their original description of this technique, and Jonathan Bayless for his comments on the manuscript.

APPENDIX 6

Monogamy and Polygyny Models

Ronald K. Chesser

Monogamous Mating

Symbols used for the variables involved in the monogamy model are as follows:

F = inbreeding coefficient,
N = number of individuals in the population,
m = migration rate,
P_o = number of original natives,
Q = number of immigrants,
R = number of relatives (produced by $Q \times P_o$ matings).

Assuming a steady influx of immigrant individuals each generation, the first generation would consist of $Q = Nm$, $P_o = N - Nm$, and $R = N - (Q + P_o) = 0$. Subsequently, the probabilities of matings among the different types of individuals and the associated inbreeding coefficients in their offspring can be calculated as follows:

Mating type	Probability	F-statistic
Natives × natives	$(P_o/N)^2$	$F_{t+1} = 2F_t - F_t^2$
Natives × immigrants	$2P_oQ/N^2$	0
Natives × relatives	$2P_oR/N^2$	$F_{t+1} = 0.5F_t$
Immigrants × immigrants	$(Q/N)^2$	0
Immigrants × relatives	$2QR/N^2$	0
Relatives × relatives	$(R/N)^2$	$F_{t+1} = 0.25F_t$

Thus, the weighted average inbreeding coefficient of the population for each generation is approximated by

$$\bar{F}_{t+1} = P_o^2 (2F_t - F_t^2) + 2P_oR (0.5F_t) + R^2 (0.25F_t)$$

Polygynous Mating

Symbols for variables involved in the polygyny model are the same as those used in the monogamy model except for the addition of the number of females in each harem, n. This model assumes only male dispersal, and it assumes that each incoming male will be successful in attaining a harem.

In the first generation $Q = Nm$, $P_o = N - Nm$, and $R = N - (Q + P_o) = 0$. The probabilities of mating types are somewhat more complicated than with monogamy because the number of remaining available mates must be considered, and the probabilities of matched pair combinations assume a multinomial rather than binomial distribution. For example, a native male may mate with 1, 2, or n native females or with various combinations of relatives and natives. Probabilities of mating types were calculated as follows:

Mating Type	Probability
Native male × native females	$\sum_{i=1}^{n}\{[P_o/(N - Nm)]^2\}^i$
Native male × relative females	$\sum_i[2P_oR/(N - Nm)^2]^i$
Relative male × native females	
Immigrant male × native females	$\sum_i[P_o/(N - Nm)]^i$
Immigrant male × relative females	$\sum_i[R/(N - Nm)]^i$
Relative male × relative females	$\sum_i\{[R/(N - Nm)]^2\}^i$

Inbreeding coefficients are calculated as in monogamy. The probability of each mating type must be multiplied by the relative frequency of each male type and the number of remaining available harems to obtain the probability that the males of a given type will mate at all (this is not a panmictic model). Thus, the weighted average of inbreeding coefficients of each generation is approximated by

APPENDIX 6 Monogamy and Polygyny Models

$$F_{t+1} = \Big[\Sigma_i\big(\{[P_o/(N-Nm)]^2\}^i\big)\,(P_o/N)(N/2n - Nm)(2F_t - F_t^2)\,2 \;+$$
$$\Sigma_i\{[2P_oR/(N-Nm)^2]^i\}\,(2P_oR/N)(N/2n - Nm)(0.5F_t)\,2 \;+$$
$$\Sigma_i\big(\{[R/N-Nm)]^2\}^i\big)\,(R/N)(N/2n - Nm)(0.25F_t)^2\Big]\frac{1}{N}$$

Each of the above values was multiplied by 2 because each mating is assumed to produce 2 offspring (equal replacement of parents). An arbitrary N value of 100 was chosen. However, since the equations use proportions, population size is irrelevant.

References

Abplanalp, H.A. 1974. Inbreeding as a tool for poultry improvement, pp. 897–908. In *First World Congress on Genetics Applied to Livestock Production*. Graficas Orbe, Madrid, Spain.

Adams, W.T., and R.W. Allard. 1982. Mating system variation in *Festuca microstachys*. *Evolution* **36**:591–595.

Adams, W.T., and R.J. Joly. 1980. Allozyme studies in loblolly pine seed orchards: clonal variation and frequency of progeny due to self-fertilization. *Silvae Genetica* **29**:1–4.

Adler, S. 1948. Origin of the golden hamster *Cricetus auratus* as a laboratory animal. *Nature* **162**:256–257.

Alatalo, R.V., L. Gustafsson, and A. Lundberg. 1982. Hybridization and breeding success of Collared and Pied Flycatchers on the Island of Gotland. *Auk* **99**: 285–291.

Alerstam, T., B. Ebenman, M. Sylven, S. Tamm, and S. Ulfstrand. 1978. Hybridization as an agent of competition between two bird aleospecies: *Ficedula albicollis* and *F. hypoleuca* on the Island of Gotland in the Baltic. *Oikos* **31**:326–331.

Allard, R.W. 1975. The mating system and microevolution. *Genetics* **79** (Suppl.):115–126.

Allard, R.W., S.K. Jain, and P.L. Workman. 1968. The genetics of inbreeding populations. *Advances in Genetics* **14**:55–131.

Allard, R.W., G.R. Babbel, M.T. Clegg, and A.L. Kahler. 1972. Evidence for coadaptation in *Avena barbata*. *Proceedings of the National Academy of Sciences of the United States of America* **69**:3043–3048.

Allen, W.R., and P.M. Sheppard. 1971. Copper tolerance in some California populations of the monkey flower, *Mimulus guttatus*. *Proceedings of the Royal Society of London Series B-Biological Sciences* **177**:177–196.

Allendorf, F.W., and S.R. Phelps. 1981. Use of allelic frequencies to describe population structure. *Canadian Journal of Fisheries and Aquatic Sciences* **38**: 1507–1514.

American Ornithologists' Union. 1980. Resolution 1: Conservation of gene pools. In. A.O.U. Proceedings of the 97th stated meeting of the American Ornithologists' Union, August 1979, College Station, Texas. *Auk* **97**(1,Suppl.):10AA.

Anderson, E. 1949. *Introgressive Hybridization*. Wiley and Sons, New York.

Anderson, R.F.V. 1977. Ethological isolation and competition of allospecies in secondary contact. *American Naturalist* **111**:939–949.

Anonymous. 1975. *Research in Zoos and Aquariums*. National Academy of Sciences, Washington, D.C.

Anonymous. 1978. U.S.–Mexico restoration efforts may be only hope for Kemp's ridley. *Endangered Species Technical Bulletin* **3**(10):6–8.

Anonymous. 1979. Ban upheld on Cayman turtle products. *Endangered Species Technical Bulletin* **4**(6):5.

Anonymous, 1982a. Elephants and rhinos in Africa. *Oryx* **16**:274.

Anonymous. 1982b. Mariculture operation exemption again sought. *Endangered Species Technical Bulletin* **7**(5):4.

Antonovics, J. 1971. The effects of a heterogeneous environment on the genetics of natural populations. *American Scientist* **59**:593–599.

Antonovics, J., and A.D. Bradshaw. 1970. Evolution in closely adjacent plant populations. VIII. Clinal patterns at a mine boundary. *Heredity* **25**:349–362.

Aquadro, C.F., and J.C. Avise. 1982. Evolutionary genetics of birds. VI. A reexamination of protein divergence using varied electrophoretic conditions. *Evolution* **36**:1003–1019.

Arnold, R.A. 1980. The endangered Lepidoptera of San Bruno Mountain: their status, habitats, ecology, and prospects for survival. Unpublished manuscript.

Arnold, R.A. 1981. Distribution, life history, and status of three California Lepidoptera proposed as endangered or threatened species. California Department of Fish and Game, Sacramento, California.

Arnold, R.A. 1982. *Ecological Studies of Six Endangered Butterflies (Lep :Lycaenidae): Island Biogeography, Patch Dynamics, and the Design of Nature Preserves*. University of California Publications in Entomology. Berkeley, California. In press.

Arnold, S.J. 1981. Behavioral variation in natural populations. I. Phenotypic, genetic, and environmental correlations between chemoreceptive responses to prey in the garter snake, *Thamnophis elegans*. *Evolution* **35**:489–509.

Arulsekar, S., and R.S. Bringhurst. 1981. Genetic model for the enzyme marker PHI in diploid California *Fragaria vesca* L. *Journal of Heredity* **72**:117–120.

Aston, J.L., and A.D. Bradshaw. 1966. Evolution in closely adjacent plant populations II. *Agrostis stolonifera* in maritime habitats. *Heredity* **21**:649–664.

Averhoff, W.W., and R.H. Richardson. 1976. Multiple pheromone system controlling mating in *Drosophila melanogaster*. *Proceedings of the National Academy of Sciences of the United States of America* **73**:591–593.

Avise, J.C. 1976. Genetic differentiation during speciation, pp. 106–122. In F.J. Ayala (ed.), *Molecular Evolution*. Sinauer Associates, Sunderland, Massachusetts.

Avise, J.C. 1977. Genic heterozygosity and rate of speciation. *Paleobiology* **3**:422–432.

Avise, J.C., and M.H. Smith. 1974a. Biochemical genetics of sunfish. I. Geographic variation and subspecific intergradation in the bluegill, *Lepomis macrochirus*. *Evolution* **28**:42–56.

Avise, J.C., and M.H. Smith. 1974b. Biochemical genetics of sunfish. II. Genic similarity between hybridizing species. *American Naturalist* **108**:458–472.

Avise, J.C., R.A. Lansman, and R.O. Shade. 1979. The use of restriction endonucleases to measure mitochondrial DNA sequence relatedness in natural populations. I. Population structure and evolution in the genus *Peromyscus*. *Genetics* **92**:279–295.

Ayala, F.J. 1968. Genotype, environment, and population numbers. *Science* **162**:1453–1459.

Ayala, F.J., and J.W. Valentine. 1978. Genetic variation and resource stability in marine invertebrates. In B. Battaglia and J.A. Beardmore (eds.), *Marine Organisms: Genetics, Ecology and Evolution*. Plenum Press, New York.

Ayala, F.J., M.L. Tracey, D. Hedgecock, and R.C. Richmond. 1974. Genetic differentiation during the speciation process in *Drosophila*. *Evolution* **28**:576–592.

Babbell, G.R., and R.K. Selander. 1974. Genetic variability in edaphically restricted and widespread plant species. *Evolution* **28**:619–630.

Babbell, G.R., and R.P. Wain. 1977. Genetic structure of *Hordeum jubatum*. I. Outcrossing rates and heterozygosity levels. *Canadian Journal of Genetics and Cytology* **19**:143–152.

Bacon, P.R. 1975. Review on research, exploitation, and the management of the stocks of sea turtles. *FAO Fisheries Circular* **334**:1–19.

Baker, A.J.M. 1978. Ecophysiological aspects of zinc tolerance in *Silene maritima* With. *New Phytology* **80**:635–642.

Baker, H.G. 1959. Reproductive methods as factors in speciation in flowering plants. *Cold Spring Harbor Symposia on Quantitative Biology* **24**:177–191.

Baker, H.G. 1965. Characteristics and modes of origin of weeds, pp. 147–168. In H.G. Baker and G.L. Stebbins (eds.), *The Genetics of Colonizing Species*. Academic Press, New York.

Baker, H.G. 1972. Migration of weeds, pp. 327–347. In D.H. Valentine (ed.), *Taxonomy, Phytogeography and Evolution*. Academic Press, New York.

Baker, I., and H.G. Baker. 1976. Variation in an introduced *Lythrum* species in California vernal pools, pp. 3–68. In S. Jain (ed.), *Vernal Pools: Their Ecology and Conservation*. Institute of Ecology Publication 6. University of California, Davis, California.

Baker, J., J.M. Smith, and C. Strobeck. 1975. Genetic polymorphism in the bladder campion, *Silene maritima*. *Biochemical Genetics* **13**:393–410.

Baker, R.J., S.L. Williams, and J.C. Patton. 1973. Chromosomal variation in the plains pocket gopher, *Geomys bursarius major*. *Journal of Mammalogy* **54**:765-769.

Ballou, J., and K. Ralls. 1982. Inbreeding and juvenile mortality in small populations of ungulates: a detailed analysis. *Biological Conservation* **24**:239-272.

Ballou, J., and J. Seidensticker. 1983. Demographic and genetic status of the captive Sumatran tigers (*Panthera tigris sumatrae*). *International Tiger Studbook*. In press.

Bantock, C.R., and W.C. Cockayne. 1975. Chromosome polymorphism in *Nucella lapillus*. *Heredity* **34**:231-245.

Barber, H.S. 1951. North American fireflies of the genus *Phorturis*. *Smithsonian Miscellaneous Collections* **117**:1-58.

Barclay, J.H., and T.J. Cade. 1983. Restoration of the Peregrine Falcon in the eastern United States. Annual Report of the International Council for Bird Preservation, United States Section. In press.

Barkelow, F.S., R.B. Hamilton, and R.F. Soots, 1970. The vital statistics of an unexploited gray squirrel population. *J. Wildlife Management* **34**(3):489-500.

Barrett, J.A. 1981. The evolutionary consequences of monoculture, pp. 209-248. *In* J.A. Bishop and L.M. Cook (eds.), *Genetic Consequences of Man-Made Change*. Academic Press, London.

Barrett, J.A. 1983. Plant-fungus symbioses. *In* D.J. Futuyma and M. Slatin (eds.), *Coevolution*. Sinauer Associates, Sunderland, Massachusetts. In press.

Bateson, P. 1982. Preferences for cousins in Japanese quail. *Nature* **295**:236-237.

Baudoin, M. 1975. Host castration as a parasitic strategy. *Evolution* **29**:335-352.

Beardmore, J.A. 1960. Developmental stability in constant and fluctuating temperatures. *Heredity* **14**:411-422.

Beardmore, J.A. 1966. Genetic information in populations. *Advances in Sciences* **23**:109, 128-132.

Beardmore, J.A. 1970. Ecological factors and the variability of gene pools in *Drosophila*. *Evolutionary Biology* (Suppl.):299-313. (Chapter 2 in *Essays in Evolution and Genetics* in honor of Th. Dobzhansky.)

Beardmore, J.A. 1980. Genetical considerations in monitoring effects of pollution, pp. 258-266. *In* A.D. McIntyre and J.B. Pearce (eds.), *Biological Effects of Marine Pollution and the Problems of Monitoring. Rapports et Proces-Verbaux des Réunions.* Conseil International pour l'Exploration de la Mer.

Beardmore J.A., and F.B. Abreu Grobois. 1983. Taxonomy and evolution in the brine shrimp *Artemia*. *In* G.S. Oxford and D. Rollinson (eds.), *Adaptation and Taxonomic Significance of Protein Variation*. Academic Press, New York. In press.

Beardmore, J.A., and L. Levine. 1963. Fitness and environmental variation. 1. A study of some polymorphic populations of *Drosophila pseudoobscura*. *Evolution* **17**:121-129.

Beardmore, J.A., and S.A. Shami. 1979. Heterozygosity, fitness and the optimum phenotype. *Aquilo Serie Zoologica* **20**:100-110.

Beardmore, J.A., and R.D. Ward. 1977. Polymorphism, selection and multilocus heterozygosity in the plaice *Pleuronectes platessa* L., pp. 208-221. *In* F.B. Christiansen and T. Fenchel (eds.), *Measuring Selection in Natural Populations*. Springer-Verlag, Berlin.

Beardmore, J.A., Th. Dobzhansky, and D. Pavlovsky. 1960. An attempt to measure the fitness of monomorphic and polymorphic populations of *Drosophila pseudoobscura*. *Heredity* **14**:49-55.
Beattie, A.J., and D.C. Culver. 1979. Neighborhood size in *Viola*. *Evolution* **33**:1226-1229.
Bell, G. 1982. *The Masterpiece of Nature. The Evolution and Genetics of Sexuality*. University of California Press, Berkeley, California.
Belyaev, D.R., and V.I. Evisiksov. 1967. Genetics of fertility of animals. The effect of the fur color on the fertility of mink (*Lutreola vison* Brisson). *Genetika (Moscow)* **2**:21-34.
Bengtsson, B.O. 1978. Avoid inbreeding: at what cost? *Journal of Theoretical Biology* **73**:439-444.
Benirschke, K. 1977. Genetic management. *International Zoo Yearbook* **17**:50-55.
Benirschke, K. 1979. Chromosomes of spider monkeys. *ZooNooz (San Diego)* **52**(10):11.
Benirschke, K. 1980. Pectus excavatum in ruffed lemurs *Lemur (Varecia) variegatus*, pp. 169-172. In *XII Internationades Symposium uber Erkrankugnen Zootieren*. Akademie Verlag, Berlin.
Benirschke, K., A.T. Kumamoto, G.N. Esra, and K.B. Crocker. 1982. The chromosomes of the bongo, *Taurotragus (Boocerus) eurycerus*. *Cytogenetics and Cell Genetics*. **34**:10-18.
Benirschke, K., N. Malouf, R.L. Low, and H. Heck. 1965. Chromosome complement: differences between *Equus caballus* and *Equus przewalskii*, Poliakoff. *Science* **148**:382-383.
Benirschke, K., B. Lasley, and O. Ryder. 1980. The technology of captive propagation, pp. 225-242. In M.E. Soulé and B.A. Wilcox (eds.), *Conservation Biology*. Sinauer, Sunderland, Mass.
Benirschke, K., A.T. Kumamoto, G.N. Esra, and K.B. Crocker. 1982. The chromosomes of the bongo, *Taurotragus (Boocerus) eurycerus*. *Cytogenetics and Cell Genetics* **34**:10-18.
Benoit, R., and E. Boesiger, 1974. Comparison des moyennes et de la variabilité des poids de douze muscles bilatoraux de la caille japonaise (*Coturnix c. japonica*) au cours de quatre générations de croisements consanguins. *Acta Anatomica* **91**:612-630.
Bercovitz, A.B., N.M. Czekala, and B.L. Lasley. 1978. A new method of sex determination in monomorphic birds. *Journal of Zoo Animal Medicine* **9**:114-124.
Bergerud, A.T. 1968. Numbers and densities, pp. 21-42. In Golley, F.B. and Buechner, H.K. (eds.) *A Practical Guide to the Study of the Productivity of Large Herbivores*. International Biology Programme Handbook No. 7. Blackwell Scientific Publications, Oxford, Edinburgh.
Bergmann, F. 1973. Genetische Untersuchungen bei *Picea abies* mit Hilfe der Isoenzym-Identifizierung. III. Geographische Variation an 2 Esterase- und 2 Leucine-aminopeptidase-Loci in der Schwedischen Fichtenpopulation. *Silvae Genetica* **22**:63-66.
Bergmann, F. 1975a. Adaptive acid phosphatase polymorphism in conifer seeds. *Silvae Genetica* **24**:175-177.
Bergmann, F. 1975b. Herkunfts-Identifizierung von Vorstsaatgut auf der basis von Isoenzym-Genhaufigkeiten. *Allg. Forst aud Jagd.-Zeitung* **146**:191-195.

Bergmann, F. 1978. The allelic distribution at an acid phosphatase locus in Norway spruce (*Picea abies*) along similar climatic gradients. *Theoretical Applied Genetics* **52**:57–64.

Berry, R.J. 1977. *Inheritance and Natural History*. Collins, London.

Bickham, J.W., and R.J. Baker. 1977. Implications of chromosomal variation in *Rhogeesa* (Chiroptera: Vespertilionidae). *Journal of Mammalogy* **58**:448–453.

Bickham, J.W., and R.J. Baker. 1979. Canalization model of chromosomal evolution. *Bulletin of the Carnegie Museum of Natural History* **13**:70–84.

Bickham, J.W., and R.J. Baker. 1980. Reassessment of the nature of chromosomal evolution in *Mus musculus*. *Systematic Zoology* **29**:159–162.

Bickham, J.W., K.A. Bjorndal, N.W. Haiduk, and W.E. Rainey. 1980. The karyotype and chromosomal banding patterns of the green turtle (*Chelonia mydas*). *Copeia* **1980**(3):540–543.

Bielewicz, M. 1967. The causes of the vanishing butterfly *Papilio podalirius* L. in Poland. *Chronmy Przyrode Ojczysta [Protection of Nature in Poland]* **23**:21–29.

Bishop, J.A. 1981. A neo-Darwinian approach to resistance: examples from mammals, pp. 37–51. *In* J.A. Bishop and L.M. Cook (eds.), *Genetic Consequences of Man-Made Change*. Academic Press, London.

Bishop, J.A., and D.J. Hartley. 1976. The size and age structure of rural populations of *Rattus norvegicus* containing individuals resistant to the anticoagulant poison warfarin. *Journal of Animal Ecology* **45**:623–646.

Bishop, J.A., D.J. Hartley, and G.G. Partridge. 1977. The population dynamics of genetically determined resistance to warfarin in *Rattus norvegicus* from mid-Wales. *Heredity* **39**:389–398.

Bjorndal, K. (ed.). 1982. *The Biology and Conservation of Sea Turtles*. Smithsonian Institution Press, Washington, D.C.

Boesiger, E. 1974. Le maintien des polymorphisms et de la polygenotypie par l'avantage selectif des heterozygotes. *Memoires Société Zoologique de France* **37**: 363–416.

Bogart, R. 1966. Inbreeding of zoo animals. *Parks and Recreation* **31**:254–257.

Bohlin, R.G., and E.G. Zimmerman. 1982. Genic differentiation of two chromosome races of the *Geomys bursarius* complex. *Journal of Mammalogy* **63**:218–228.

Bollengier, R., Jr. (ed.). 1979. *Eastern Peregrine Falcon Recovery Plan*. U.S. Fish and Wildlife Service, Department of the Interior. Washington, D.C. p. 147.

Bonnel, M.L., and R.K. Selander. 1974. Elephant seals: genetic variation and near extinction. *Science* **184**:908–909.

Bovee, K.C., M. Bush, J. Dietz, P. Jezk, and S. Segal. 1981. Cystinuria in the maned wolf of South America. *Science* **212**:919–920.

Bowen, B.S. 1982. Temporal dynamics of microgeographic structure of genetic variation in *Microtus californicus*. *Journal of Mammalogy*. **63**(4):625–638.

Bowman, J.C. 1977. The future of Przewalski horses in captivity. *International Zoo Yearbook* **17**:62–68.

Bowman, J.C., and D.S. Falconer, 1960. Inbreeding depression and heterosis of litter sizes in mice. *Genetical Research* **1**:262–274.

Bradley, B.P. 1980. Developmental stability of *Drosophila melanogaster* under artificial and natural selection in constant and fluctuating environments. *Genetics* **95**:1033–1042.

Bradshaw, A.D. 1965. Evolutionary significance of phenotypic plasticity in plants. *Advances in Genetics* **13**:115-155.

Bradshaw, A.D. 1972. Some of the evolutionary consequences of being a plant, pp. 25-47. *In* T. Dobzhansky, M.K. Hecht, and W.C. Steere (eds.), *Evolutionary Biology*, Vol. 5. Appleton-Century-Croft, New York.

Braithwaite, L.W., and B. Miller. 1975. The mallard, *Anas platyrhynchos*, and mallard-black duck, *Anas superciliosa rogersi*, hybridization. *Australian Wildlife Research* **2**(1):47-61.

Breedlove, D., and P.R. Ehrlich. 1972. Coevolution: patterns of legume predation by a lycaenid butterfly. *Oecologia* **10**:99-104.

Britten, R.J., A. Cetta, and E.H. Davidson. 1978. The single copy DNA sequence polymorphism of the sea urchin *Strongylocentiotus purpuratus*. *Cell* 15:1175-1186.

Brittnacker, J.G., S.R. Sims, and F.J. Ayala. 1978. Genetic differentiation between species of the genus *Speyeria* (Lepidoptera:Nymphalidae). *Evolution* **32**:199-210.

Brongersma, L.D. 1980. Turtle farming and ranching. *British Herpetological Society Bulletin* **2**:15-19.

Brown, A.H.D. 1979. Enzyme polymorphism in plant populations. *Theoretical Population Biology* **15**:1-42.

Brown, A.H.D., and D.R. Marshall. 1981. Evolutionary changes accompanying colonization in plants. *In* G.G.E. Scudder and J.L. Reveal (eds.), *Evolution Today*. Proceedings of the Second International Congress of Systematic and Evolutionary Biology. Carnegie-Mellon University, Pittsburgh.

Brown, A.H.D., and G.F. Moran. 1981. Isozymes and genetic resources in forest trees, pp. 1-10. *In* M.T. Conkle (ed.), *Proceedings of the Symposium on Isozymes of North American Forest Trees and Forest Insects*. Pacific Southern Forest and Range Experiment Station. General Technical Report PSW-48.

Brown, A.H.D., A.C. Matheson, and K.G. Eldridge. 1975. Estimation of the mating system of *Eucalyptus obliqua* L'Hérit. by using allozyme polymorphisms. *Australian Journal of Botany* **23**:931-949.

Brown, A.H.D., E. Nevo, and D. Zohary. 1977. Association of alleles at esterase loci in wild barley *Hordeum spontaneum* L. *Nature* **268**:430-431.

Brown, A.H.D., E. Nevo, D. Zohary, and O. Dagan. 1978. Genetic variation in natural populations of wild barley *(Hordeum spontaneum)*. *Genetica* **49**:97-108.

Brown, B.A. and M.T. Clegg. 1983. Influence of flower color polymorphism on genetic transmission in a natural population of the common morning glory, *Ipomoea purpurea*. *Evolution* (submitted).

Brown, I.L., and P.R. Ehrlich. 1980. Population biology of the checkerspot butterfly, *Euphydryas chalcedona*: structure of the Jasper Ridge colony. *Oecologia* **47**: 239-251.

Brown, K.S. 1970. Rediscovery of *Heliconius nattereri* in eastern Brazil. *Entomology News* **81**:129-140.

Brown, L.N. (1970) Population dynamics of the western jumping mouse (*Zapus princeps*) during a four-year study. *J. Mammalogy.* **51**(4):651-658.

Brown, W., M. George, and A.C. Wilson. 1979. Rapid evolution of animal mitochondrial DNA. *Proceedings of the National Academy of Sciences of the United States of America* **76**(4):1967-1971.

Brumback, W.E. 1981. Endangered plant species programs for botanic gardens with examples from North American institutions. M.S. thesis, University of Delaware, Newark.

Brussard, P.F., and P.R. Ehrlich. 1970a. The population structure of *Erebia epipsodea* (Lepidoptera: Satyrinae). *Ecology* **51**:119-129.

Brussard, P.F., and P.R. Ehrlich. 1970b. Adult behavior and population structure in *Erebia epipsodea* (Lepidoptera: Satyrinae). *Ecology* **51**:880-885.

Brussard, P.F., and P.R. Ehrlich. 1970c. Contrasting population biology of two species of butterfly. *Nature* **227**:91-92.

Bryant, E.H. 1974. On the adaptive significance of enzyme polymorphisms in relation to environmental variability. *American Naturalist* **108**:1-19.

Bryant, E.H., H. van Dyk, and W. van Delden. 1981. Genetic variability of the face fly, *Musca autumnalis* De Geer, in relation to a population bottleneck. *Evolution* **35**:872-881.

Bull, J.J. 1981. Sex determination in reptiles. *Quarterly Review of Biology* **55**:3-21.

Bumpus, H.E. 1896. The variations and mutations of the introduced sparrow, *Passer domesticus*, pp. 1-5. *In Lectures.* Marine Biology Laboratory, Wood's Hole, Massachusetts (1896-1897).

Burdon, J.J. and D.R. Marshall. 1981. Biological control and the reproductive mode of weeds. *J. Applied Ecology* **18**:649-658.

Bureau of Land Management. 1981. The Tule Elk in California; 5th annual report to Congress. Report prepared by the Bureau of Land Management for the Secretary of the Interior under P.L.94.398. 50 pp.

Burnet, B. 1972. Enzyme protein polymorphism in the slug *Arion ater*. *Genetical Research* **20**:161-173.

Bush, G.L. 1975. Modes of animal speciation. *Annual Review of Ecology and Systematics* **6**:339-361.

Bush, G.L., S.M. Case, A.C. Wilson, and J.L. Patton. 1977. Rapid speciation and chromosomal evolution in mammals. *Proceedings of the National Academy of Sciences of the United States of America* **74**:3942-3946.

Bush, G.L., R.W. Neck, and G.B. Kitto. 1976. Screwworm eradication: Inadvertent selection for noncompetitive ecotypes during mass rearing. *Science* **193**:491-493.

Bush, M., R.J. Montali, D.G. Kleiman, J. Randolph, M.D. Abramowitz, and R.F. Evans. 1980. Diagnosis and repair of familial diaphragmatic defects in golden lion tamarins. *Journal of the American Veterinary Medicine Association* **77**:858-862.

Cade, T.J. 1955. Variation of the Common Rough-legged Hawk in North America. *Condor* **57**:313-346.

Cade, T.J. 1980. The husbandry of falcons for return to the wild. *International Zoo Yearbook* **20**:23-35.

Capen, D.E., W.J. Crenshaw, and M.W. Coulter. 1974. Establishing breeding populations of Wood Ducks by relocating wild broods. *Journal of Wildlife Management* **38**:253-256.

Carlquist, S.J. 1974. *Island Biology*. Columbia University Press, New York.

Carnahan, H.L. 1947. Combining ability in flax (*Linum usitatissimum*). M.S. thesis, University of Minnesota. Quoted in J.W. Gowen (ed.), *Heterosis*. 1950. Iowa State College Press, p. 52. Ames, Iowa.

Carr, A.F., Jr. 1956. *The Windward Road*. Knopf, New York.
Carr, A.F., Jr. 1979. Preface, pp. xiii–xxxii. *In The Windward Road*. University Presses of Florida, Tallahassee, Florida.
Carr, A.F., Jr. 1980. Some problems of sea turtle ecology. *American Zoologist* **20**: 489–498.
Carr, A.F., Jr., H. Hirth, and L. Ogren. 1966. The ecology and migrations of sea turtles. 6. The hawksbill turtle in the Caribbean Sea. *American Museum Novitates* **2248**:1–29.
Carr, A.F., Jr., M.H. Carr, and A. B. Meylan. 1978. The ecology and migrations of sea turtles. 7. The West Caribbean green turtle colony. *Bulletin of the American Museum of Natural History* **162**:1–46.
Carson, H.L. 1970. Chromosome tracers of the origin of species. *Science* **193**:1414–1418.
Carson, H.L. 1971. Speciation and the founder principle. *Stadler Genetics Symposia, University of Missouri* **3**:51–70.
Carson, H.L. 1975. The genetics of speciation at the diploid level. *American Naturalist* **109**:83–92.
Carson, H.L. 1983. Speciation and the founder effect on a new oceanic island. *In* P.H. Raven (ed.), *Biogeography of the Pacific Islands*. B.P. Bishop Museum, Honolulu. In press.
Carson, H.L., and J.S. Yoon. 1982. Genetics and evolution of Hawaiian Drosophila, pp. 298–344. *In* M. Ashburner, H.L. Carson, and J.N. Thompson, Jr. (eds.), *The Genetics and Biology of Drosophila*, Vol. 3b. Academic Press, New York.
Cates, R.G. 1975. The interface between slugs and wild ginger: some evolutionary aspects. *Ecology* **56**:391–400.
Cavalli-Sforza, L.L. 1966. Population structure and human evolution. *Proceedings of the Royal Society of London Series B-Biological Sciences* **164**:362–379.
Cavalli-Sforza, L.L., and W.F. Bodmer. 1971. *The Genetics of Human Populations*. Freeman, San Francisco, California.
Chai, C.K. 1966. Characteristics in inbred mouse populations plateaued by directional selection. *Genetics* **54**:743–753.
Chambers, S.M. 1980. Genetic divergence between populations of *Goniobasis* (Pleuoceridae) occupying different drainage systems. *Malacologia* **20**:63–81.
Chapin, F.S., III, and M.C. Chapin. 1981. Ecotypic differentiation of growth processes in *Carex aquatilis* along latitudinal and local gradients. *Ecology* **62**:1000–1009.
Chapman, R.W., J.C. Stephens, R.A. Lansman, and J.C. Avise. 1982. Models of mitochondrial DNA transmission genetics and evolution in higher eucaryotes. *Genetic Research, Cambridge* **40**:41–57.
Charlesworth, B. 1982. Review of *The Masterpiece of Nature: The Evolution and Genetics of Sexuality*, by G. Bell. *New Scientist* **94**:725.
Chesser, R.K. 1981. Genetic and morphologic variation within and among populations of the blacktailed prairie dog. Ph.D. dissertation, University of Oklahoma, Norman, Oklahoma.
Chesser, R.K. 1983. Genetic variability within and among populations of the black-tailed prairie dog. *Evolution* **37**(2):320–331.
Chesser, R.K., M.H. Smith, and I.L. Brisbin, Jr. 1980. Management and maintenance of genetic variability in endangered species. *International Zoo Yearbook* **20**:146–154.

Chesser, R.K., C. Reuterwall, and N. Ryman. 1982a. Genetic differentiation among moose populations in Sweden over short geographic distances. *Oikos*. In press.

Chesser, R.K., M.H. Smith, P.E. Johns, M.N. Manlove, D.O. Straney, and R. Baccus. 1982b. Spatial, temporal and age-dependent heterozygosity of beta-hemoglobin in white-tailed deer. *Journal of Wildlife Management* **46**(4): 983–990.

Chew, F.S. 1981. Coexistence and local extinction in two pierid butterflies. *American Naturalist* **118**:655–672.

Chew, F.S., and J.E. Rodman. 1979. Plant resources for chemical defense, pp. 271–307. *In* G.A. Rosenthal and D.H. Janzen (eds.), *Herbivores: Their Interaction with Secondary Plant Metabolites*. Academic Press, New York.

Chichester, L.F., and L.L. Getz. 1973. The terrestrial slugs of northeastern North America. *Sterkiana* **51**:11–42.

Chinnici, J.P. 1971a. Modification of recombination frequency in *Drosophila*. I. Selection for increased and decreased crossing-over. *Genetics* **69**:71–83.

Chinnici, J.P. 1971b. Modification of recombination frequency in *Drosophila*. II. The polygenic control of crossing-over. *Genetics* **69**:85–96.

Christiansen, F.B. 1974. Sufficient conditions for protected polymorphism in a subdivided population. *American Naturalist* **108**:157–166.

Christiansen, F.B. 1975. Hard and soft selection in a subdivided population. *American Naturalist* **109**:11–16.

Christie, W.J. 1974. Changes in the fish species composition of the Great Lakes. *Journal of the Fisheries Research Board of Canada* **31**:827–854.

Cicmanec, J.C., and A.K. Campbell. 1977. Breeding the owl monkey (*Aotus trivirgatus*) in a laboratory environment. *Laboratory Animal Science* **27**:517.

Clarke, B.C. 1976. The ecological genetics of host–parasite relationships, pp. 87–103. *In* A.E.R. Taylor and R. Muller (eds.), *Genetic Aspects of Host–Parasite Relationships*. Blackwell, Oxford.

Clarke, C.A., and P.M. Sheppard. 1960. Super genes and mimicry. *Heredity* **14**:175–185.

Clarke, C.A., C.G.C. Dickson, and P.M. Sheppard. 1963. Larval color pattern in *Papilio demodocus*. *Evolution* **17**:130–137.

Clayton, W.D. 1977. New grasses from eastern Africa. Studies in the Gramineae: XLII. *Kew Bulletin* **32**:1–4.

Clegg, M.T. 1980. Measuring plant mating systems. *BioScience* **30**:814–818.

Clegg, M.T., and R.W. Allard. 1972. Patterns of genetic differentiation in the slender wild oat species *Avena barbata*. *Proceedings of the National Academy of Sciences of the United States of America* **69**:1820–1824.

Clegg, M.T., and R.W. Allard. 1973. Viability versus fecundity selection in the slender wild oat species *Avena barbata*. *Science* **181**:667–668.

Clegg, M.T., R.W. Allard, and A.L. Kahler. 1972. Is the gene the unit of selection? Evidence from two experimental plant populations. *Proceedings of the National Academy of Sciences of the United States of America* **69**:2474–2478.

Cocks, P.S., and J.R. Phillips. 1979. Evolution of subterranean clover in South Australia. I. The strains and their distribution. *Australian Journal of Agricultural Research* **30**:1035–1052.

Confer, J.L., and K. Knapp. 1981. Golden-winged warblers and blue-winged warblers: the relative success of a habitat specialist and a generalist. *Auk* **98**:108–114.

Conley, W. 1978. Population modelling. *In* J.L. Schmidt and D.L. Gilbert (eds.),

Big Game of North America: Ecology and Management. Stackpole Books, Harrisburg, Pennsylvania.

Conner, J.L. 1975. Genetic mechanisms controlling the domestication of a wild house mouse population (*Mus musculus* L). *Journal of Comparative Psychology and Physiology* **89**:118ff.

Conterio, F., and L.L. Cavalli-Sforza. 1960. Selezione per caratteri quantitative nell'uomo. *Atti Associanione Genetica Italaliana* **5**:295–304. Quoted in L.L. Cavalli-Sforza and W.F. Bodmer. 1971. *The Genetics of Human Populations.* Freeman, San Francisco.

Conway, W. 1967. The opportunity for zoos to save vanishing species. *Oryx* **9**:154–160.

Cook, S.A. 1962. Genetic system, variation, and adaptation in *Eschscholzia californica*. *Evolution* **16**:278–299.

Copeman, D.G., and D.E. McAllister. 1978. Analysis of the effect of transplantation on morphometric and meristic characters in lake populations of the rainbow smelt, *Osmerus mordax* (Mitchell). *Environmental Biology of Fishes* **3**:253–260.

Corbin, K.W., C.G. Sibley, and A. Ferguson. 1979. Genetic changes associated with the establishment of sympatry in orioles of the genus *Icterus*. *Evolution* **33**:624–633.

Coyne, J.A. 1976. Lack of genetic similarity between two sibling species as revealed by varied techniques. *Genetics* **84**:593–607.

Craighead, J.J., J.R. Varney, and J.P. Craighead. 1974. A population analysis of the Yellowstone grizzly bears. *Bulletin* **40**. Montana Forest and Conservation Experiment Station, School of Forestry, University of Montana, Missoula, Montana.

Crawford, D.J., and R.J. Bayer. 1981. Allozyme divergence in *Coreopsis cyclocarpa* (Compositae). *Systematic Botany* **6**:373–379.

Crawford, D.J., and H.D. Wilson. 1977. Allozyme variation in *Chenopodium fremontii*. *Systematic Botany* **2**:180–190.

Crawford, D.J., and H.D. Wilson. 1979. Allozyme variation in several closely related diploid species of *Chenopodium* of the western United States. *American Journal of Botany* **66**:237–244.

Crick, F. 1979. Split genes and RNA splicing. *Science* **204**:264–271.

Crook, J.H. 1960. *The Present Status of Certain Rare Land Birds of the Seychelles Islands.* Seychelles Government Bulletin.

Crow, J.F., and M. Kimura. 1970. *An Introduction to Population Genetics Theory.* Harper and Row, New York.

Daly, J.C. 1981. Effects of social organization and environmental diversity on determining the genetic structure of a population of the wild rabbit *(Oryctolagus cunicalus)*. *Evolution* **35**:698–706.

Darling, F.F. 1969. *A Herd of Red Deer; a Study in Animal Behavior.* Oxford University Press. Reprinted Lowe and Brydone (Printers) Ltd. London.

Darwin, C. 1859. *The Origin of Species.* Harvard Facsimile 1st. ed. 1964.

Darwin, C. 1868. *The Variation of Animals and Plants under Domestication.* John Murray, London.

Dasmann, R.F. 1972. Towards a system for classifying natural regions of the world and their representation by natural parks and reserves. *Biological Conservation* **4**:247–255.

Dasmann, R.F. 1973. A system for defining and classifying natural regions for purposes of conservation: A progress report. Morges, *IUCN Occasional Paper* No. 7.

Davies, M.S., and R.W. Snaydon. 1973. Physiological differences among populations of *Anthoxanthum odoratum* L. collected from the Park areas experiment, Rothamsted. I. Response to calcium. *Journal of Applied Ecology* **10**:33-45.

Davies, M.S., and R.W. Snaydon. 1974. Physiological differences among populations of *Anthoxanthum odoratum* L. collected from the Park grass experiment, Rothamsted. III. Response to phosphate. *Journal of Applied Ecology* **11**:699-707.

Davies, S.M. 1977. The *Arion hortensis* complex, with notes on *A. intermedius* Normand (Pulmonata:Arionidae). *Journal of Conchology* **30**:123-127.

de Jong, G., A.J.W. Hoorn, G.E.W. Thorig, and W. Scharloo. 1972. Frequencies of amylase variants in *Drosophila melanogaster*. *Nature* **238**:453-454.

de Vos, V., and H.H. Braack. 1980. Castration of a black rhinoceros *Diceros bicornis minor*. *Koedoe Monograph* **23**:185-187.

de Wet, J.M.J., and J.R. Harlan. 1966. Morphology of the compilospecies *Bothriochloa intermedia*. *American Journal of Botany* **53**:94-98.

de Wet, J.M.J., and J.R. Harlan, 1970. *Bothriochloa intermedia* — a taxonomic dilemma. *Taxon* **19**:339-340.

de Wet, J.M.J., J.R. Harlan, and E.G. Price. 1976. Variability in *Sorghum bicolor*, pp. 453-464. *In* J.R. Harlan, J.M.J. de Wet, and A.B.L. Stemler (eds.), *Origins of African Plant Domestication*. Mouton, The Hague.

Deakin, M.A.B. 1966. Sufficient conditions for genetic polymorphism. *American Naturalist* **100**:690-692.

Deakin, M.A.B. 1968. Genetic polymorphism in a subdivided population. *Australian Journal of Biological Sciences* **21**:165-168.

DeBach, P. 1966. The competitive displacement and coexistence principles. *Annual Review of Entomology* **11**:183-212.

Dempster, J.P., and M.L. Hall. 1980. An attempt at reestablishing the swallowtail butterfly at Wicken Fen. *Ecological Entomology* **5**:327-334.

Dempster, J.P., M.L. King, and K.H. Lakhani. 1976. The status of the swallowtail butterfly in Britain. *Ecological Entomology* **1**:71-84.

Denniston, C. 1978. Small population size and genetic diversity. Implications for endangered species, pp. 281-290. *In* S.A. Temple (ed.), *Endangered Birds: Management Techniques for Preserving Threatened Species*. University of Wisconsin, Madison, Wisconsin.

Diamond, J.M. 1975. The island dilemma: lessons of modern biogeographic studies for the design of natural preserves. *Biological Conservation* **7**:129-146.

Diamond, J.M. 1982. Evolution of bowerbirds' bowers: animal origins of the aesthetic sense. *Nature* **297**:99-102.

Dobzhansky, Th. 1943. Genetics of natural populations. IX. Temporal changes in the composition of populations of *Drosophila pseudoobscura*. *Genetics* **28**:162-166.

Dobzhansky, Th. 1955. A review of some fundamental concepts and problems of population genetics. *Cold Spring Harbor Symposia on Quantitative Biology* **20**:1-15.

Dobzhansky, Th. 1970. *Genetics of the Evolutionary Process*. Columbia University Press, New York.

Dobzhansky, Th. 1974. Genetic analysis of hybrid sterility within the species *Drosophila pseudoobscura*. *Hereditas* **77**:81-88.

Dobzhansky, Th., F.J. Ayala, G.L. Stebbins, and J.W. Valentine. 1977. *Evolution.* W.H. Freeman, San Francisco, California.

Dodd, C.K., Jr. 1982. Does sea turtle aquaculture benefit conservation? *In* K. Bjorndal (ed.), *The Biology and conservation of Sea Turtles.* Smithsonian Institution Press, Washington, D.C.

Doty, H.A., and A.D. Kruse. 1972. Techniques of establishing local breeding populations of Wood Ducks. *Journal of Wildlife Management* **36**:428-435.

Drew, A., and C. Reilly. 1972. Observations on copper tolerance in the vegetation of a Zambian copper clearing. *Journal of Ecology* **60**:439-444.

Drury, W.H. 1974. Rare species. *Biological Conservation* **6**:162-169.

Duffey, E. 1977. The re-establishment of the large copper butterfly *Lycaena dispar batava* Obth. on Woodwalton Fen National Nature Reserve, Cambridgeshire, England, 1969-73. *Biological Conservation* **12**:143-158.

Duncan, C.J. 1975. Reproduction, pp. 309-366. *In* V. Fretter and J. Peake (eds.), Pulmonates. Vol.1. *Functional Anatomy and Physiology.* Academic Press, London.

Dungan, M.L., T.E. Miller, and D.A. Thompson. 1982. Catastrophic decline of a top carnivore in the Gulf of California rocky intertidal zone. *Science* **216**: 989-990.

Durrant, B. and K. Benirschke. 1981. Embryo transfer in exotic animals. *Theriogenology* **15**:77-83.

Dutrillaux, B. 1979. Chromosomal evolution in primates: tentative phylogeny from *Microcebus murinus* (prosimian) to man. *Human Genetics* **48**:251-314.

Eanes, W.F. 1978. Morphological variance and enzyme heterozygosity in the monarch butterfly. *Nature* **276**:263-264.

East, R. 1981. Species—area curves and populations of large mammals in African savanna reserves. *Biological Conservation*, **21**:111-126.

Eanes, W.F. 1978. Morphological variance and enzyme heterozygosity in the monarch butterfly. *Nature* **276**:263-264.

Edmunds, G.F., Jr., and D.N. Alstad. 1978. Coevolution in insect herbivores and conifers. *Science* **199**:941-945.

Ehrendorfer, F. 1959. Differentiation-hybridization cycles and polyploidy in *Achillea. Cold Spring Harbor Symposia on Quantitative Biology* **24**:141-152.

Ehrendorfer, F. 1964. Cytologie, Taxonomie und Evolution bei Samenpflanzen. *Vistas in Botany* **4**:99-186.

Ehrenfeld, D.W. 1974. Conserving the edible sea turtle: Can mariculture help? *American Scientist* **62**:23-31.

Ehrenfeld, D.W. 1976. The conservation of non-resources. *American Scientist* **64**: 648-656.

Ehrenfeld, D.W. 1980. Commercial breeding of captive sea turtles: status and prospects, pp. 93-96. *In* J.B. Murphy and J.T. Collins (eds.), *Reproductive Biology and Diseases of Captive Reptiles, SSAR Contributions to Herpetology,* No. 1.

Ehrenfeld, D.W. 1982. Options and limitations in the conservation of sea turtles. *In* K. Bjorndal (ed.), *The Biology and conservation of Sea Turtles.* Smithsonian Institution Press, Washington, D.C.

Ehrlich, A.H., and P.R. Ehrlich. 1981a. Dangers of uninformed optimism. *Environmental Conservation* **8**:173-175.

Ehrlich, P.R. 1965. The population biology of the butterfly, *Euphydryas editha.* II. The structure of the Jasper Ridge colony. *Evolution* **19**:327-336.

Ehrlich, P.R. 1980. The strategy of conservation, 1980-2000, pp. 329-344. *In* M.E.

Soulé and B.A. Wilcox (eds.), *Conservation Biology: An Evolutionary-Perspective*. Sinauer Associates, Sunderland, Massachusets.

Ehrlich, P.R. 1983. The structure and dynamics of butterfly populations. *Proceedings of the Royal Entomological Society on the Biology of Butterflies*. In press.

Ehrlich, P.R. and L.C. Birch, 1967, The "balance of nature" and "population control." *American Naturalist* **101**:97–107.

Ehrlich, P.R. and A.H. Ehrlich. 1981. *Extinction; the Causes and Consequences of the Disappearance of Species*. Random House, New York. 305 pp.

Ehrlich, P.R., and L.G. Mason. 1966. The population biology of the butterfly, *Euphydryas editha*. III. Selection and the phenetics of the Jasper Ridge colony. *Evolution* **20**:165–173.

Ehrlich, P.R., and D.D. Murphy. 1981. The population biology of checkerspot butterflies (*Euphydryas*). *Biologiches Zentralblatt* **100**:613–629.

Ehrlich, P.R., and P.H. Raven. 1969. Differentiation of populations. *Science* **165**: 1227–1232.

Ehrlich, P.R., and R.R. White. 1980. Colorado checkerspot butterflies: isolation, neutrality, and the biospecies. *American Naturalist* **115**:328–341.

Ehrlich, P.R., D.E. Breedlove, P.F. Brussard, and M.A. Sharp, 1972. Weather and the "regulation" of subalpine populations. *Ecology* **53**:243–247.

Ehrlich, P.R., R.R. White, M.C. Singer, S.W. McKechnie, and L.E. Gilbert. 1975. Checkerspot butterflies: a historical perspective. *Science* **188**:221–228.

Ehrlich, P.R., D.D. Murphy, M.C. Singer, C.B. Sherwood, R.R. White, and I.L. Brown. 1980. Extinction, reduction, stability, and increase: the responses of checkerspot butterfly populations to the California drought. *Oecologia* **46**: 101–105.

Ehrlich, P.R., B.A. Wilcox, P.R. Brussard, D.D. Murphy, and C.E. Holdren. 1983. Constancy and change in gene frequencies of checkerspot butterfly (*Euphydryas*) populations. In preparation.

Eisenberg, John F. 1981. *The Mammalian Radiations*. The University of Chicago Press.

Elena-Rossello, J.A., A. Kheyr-Pour, and G. Valdeyron. 1976. La structure génétique et le régime de la fécondation chez *Origanum vulgare* L.; répartition d'un marqueur enzymatique dans deux populations naturelles. *Comptes Rendus Hebdomadaires des Séances de l'Academie de Sciences, Série D* **283**:1587–1589.

Ellstrand, N.C., and D.A. Levin. 1980. Recombination system and population structure in *Oenothera*. *Evolution* **34**:923–933.

Ellstrand, N.C., and D.A. Levin. 1982. Genotypic diversity in *Oenothera laciniata* (Onagraceae), a permanent translocation heterozygote. *Evolution* **36**:63–69.

Ellstrand, N.C., A.M. Torres, and D.A. Levin. 1978. Density and the rate of apparent outcrossing in *Helianthus annuus* (Asteraceae). *Systematic Botany* **3**: 403–407.

Emets, V.M. 1977. Rare and disappearing species of diurnal butterflies in the Usmanskii Forest and possible measures for their protection. *Zoologichesky Zhurnal (U.S.S.R.)* **56**:1889–1890.

Ennos, R.A. 1981. Quantitative studies of the mating system in two sympatric species of *Ipomoea* (Convolvulaceae). *Genetica* **57**:93–98.

Eshed, N., and A. Dinoor. 1981. Genetics of pathogenicity in *Puccinia coronata*: the host range among grasses. *Phytopathology* **71**:156–163.

Ewens, W.J. 1972. The sampling theory of selectively neutral alleles. *Theoretical*

Population Biology **3**:87-112.

Falconer, D.S. 1960. *Introduction to Quantitative Genetics*. Ronald Press, New York.

Falconer, D.S. 1981. *Introduction to Quantitative Genetics*. 2nd ed. Longman, London and New York.

Falconer, D.S., and J.W.B. King. 1953. A study of selection limits in the mouse. *Journal of Genetics* **51**:561-581.

Feldman, M.W., and R.C. Lewontin. 1975. The heritability hang-up. *Science* **190**: 1163-1168.

Feret, P.P. 1974. Genetic differences among three small stands of *Pinus pungens*. *Theoretical Applied Genetics* **44**:173-177.

Festing, M.F.W. 1976. Effects of marginal malnutrition on the breeding performance of inbred and F_1-hybrid mice—a diallele study, pp. 99-114. *In* T. Antikatzites, S. Erickson, and A.G. Spiegel (eds.), *The Laboratory Animal in the Study of Reproduction*. Fisher, Stuttgart.

Finerty, J.P. 1976. *The Population Ecology of Cycles in Small Mammals*. Yale University Press. New Haven.

Finnley, D. (ed.). 1977a. *Endangered Species Technical Bulletin* **2**(9):1-8.

Finnley, D. (ed.). 1977b. *Endangered Species Technical Bulletin* **2**(12):1-8.

Finnley, D. (ed.). 1978a. *Endangered Species Technical Bulletin* **3**(4):1-12.

Finnley, D. (ed.). 1978b. *Endangered Species Technical Bulletin* **3**(8):1-12.

Fishbein, L. 1976. pp. 555-603. *In* C.F. Wilkinson (ed.), *Insecticide Biochemistry and Physiology*. Heyden, London.

Fisher, J. 1966. *Zoos of the World*. Aldus Books, London.

Fitch, W.M. 1976. Molecular evolutionary clocks, pp. 160-178. *In* F.J. Ayala (ed.), *Molecular Evolution*. Sinauer, Sunderland, Mass.

Fitch, W.M., and E. Margoliash. 1967. Construction of phylogenetic trees. *Science* **155**:279-284.

Flack, J.A.D. 1976. Hybrid parakeets on the Mangere Islands, Chatham group. *Notornis* **23**:253-255.

Flemming, C.A. 1939. Birds of the Chatham Islands, Part 2. *Emu* **38**:492-509.

Flesness, N.R. 1977. Gene pool conservation and computer analysis. *International Zoo Yearbook* **17**:62-68.

Flesness, N.R. 1978. Gene pool conservation and computer analysis. *International Zoo Yearbook* **17**:77-81.

Flesness, N., L. Grahm, and K. Hastings. 1982. *General Information: International Species Inventory System (ISIS)*. ISIS, Apple Valley, Minnesota.

Flor, H.H. 1956. The complementary genic systems in flax and flax rust. *Advances in Genetics* **8**:29-54.

Foltz, D.W., and J.L. Hoogland. 1981. Analysis of the mating system in the black-tailed prairie dog (*Cynomys ludovicianus*) by likelihood of paternity. *Journal of Mammalogy* **62**:706-712.

Foltz, D.W., B.M. Schaitkin, and R.K. Selander. 1982a. Gametic disequilibrium in the self-fertilizing slug *Deroceras laeve*. *Evolution* **36**:80-85.

Foltz, D.W., H. Ochman, J.S. Jones, S.M. Evangelisti, and R.K. Selander. 1982b. Genetic population structure and breeding systems in arionid slugs (Mollusca: Pulmonata). *Biological Journal of the Linnean Society* **17**:225-241.

Foltz, D.W., H. Ochman, and R.K. Selander. 1982c. Genetic diversity and breeding systems in terrestrial slugs of the families Limacidae and Arionidae. Proceedings of the Congress on Molluscan Genetics, American Malacologists Union. *Malacologia.* In press.

Foose, T. 1977. Demographic models for management of captive populations. *International Zoo Yearbook* **17**:70–76.

Foose, T. 1980. Demographic management of endangered species in captivity. *International Zoo Yearbook* **20**:154–165.

Foose, T. 1981. Demographic problems and management in captive populations. *AAZPA 1980 Annual Conference Proceedings.* American Association of Zoological Parks and Aquariums, Wheeling, West Virginia.

Foose, T., and E. Foose. Demographic and genetic status and management. *In* B. Beck and C. Wemmer (eds.), *Père David's Deer: The Biology and Conservation of an Extinct Species.* Noyes, Park Ridge, New Jersey. In press.

Foose, T., and U.S. Seal. 1981. *A Species Survival Plan for Siberian Tiger (Panthera tigris altaica) in North America.* American Association of Zoological Parks and Aquariums, Wheeling, West Virginia.

Ford, E.B. 1956. Rapid evolution and the conditions which make it possible. *Cold Spring Harbor Symposia on Quantitative Biology* **20**:230–238.

Ford, E.B. 1964. *Ecological Genetics.* Wiley and Sons, New York.

Ford, H.D., and E.B. Ford. 1930. Fluctuation in numbers and its influence on variation in *Melitaea aurinia* Rott. (Lepidoptera). *Transactions of the Royal Entomological Society of London* **78**:345–351.

Foster, J. and D. Kearny. 1967. Nairobi National Park game census, 1966. *East African Wildlife J.* **5**:112–120.

Fowler, C., and T. Smith. 1981. *Dynamics of Large Mammal Populations.* Wiley and Sons, New York.

Frankel, O.H. 1954. Invasion and evolution of plants in Australia and New Zealand. *Caryolgia* **4** (Suppl.):600–619.

Frankel, O.H. 1970. Variation, the essence of life. Sir William Macleay Memorial Lecture. *Proceedings of the Linnean Society* **95**:158–169.

Frankel, O.H. 1974. Genetic conservation: our evolutionary responsibility. *Genetics* **78**:53–65.

Frankel, O.H. 1981. Evolution in jeopardy: the role of nature reserves. *In* W.R. Atchley and D. Woodruff (eds.), *Evolution and Speciation.* Cambridge University Press, Cambridge, England.

Frankel, O.H., and M.E. Soulé. 1981. *Conservation and Evolution.* Cambridge University Press, Cambridge, England.

Franklin, I.R. 1980. Evolutionary change in small populations, pp. 135–149. *In* M. Soulé and B. Wilcox (eds.), *Conservation Biology: An Evolutionary–Ecological Perspective.* Sinauer Associates, Sunderland, Massachusetts.

Frazier, J.G. 1980. Marine turtles and problems in coastal management, pp. 2395–2411. *In* B.L. Edge (ed.), *Coastal Zone '80*, Vol. 3, *Proceedings of the Second Symposium on Coastal and Ocean Management.* American Society of Civil Engineers, New York, New York.

Frick, J. 1976. Orientation and behavior of hatchling green turtles (*Chelonia mydas*) in the sea. *Animal Behavior* **24**:849–857.

Fryxell, P.A. 1957. Mode of reproduction of higher plants. *Botanical Review* **23**(3): 135–233.

Futuyma, D.J. 1979. *Evolutionary Biology.* Sinauer, Sunderland, Massachusetts.

Futuyma, D.J. 1983. Evolutionary interactions among herbivorous insects and plants. *In* D.J. Futuyma and M. Slatkin (eds.), *Coevolution*. Sinauer Associates, Sunderland, Massachusetts.

Futuyma, D.J., and G.C. Mayer. 1980. Non-allopatric speciation in animals. *Systematic Zoology* **29**:254–271.

Futuyma, D.J., S.L. Leipertz, and C. Mitter. 1981. Selective factors affecting clonal variation in the fall cankerworm *Alsophila pometaria* (Lepidoptera: Geometridae). *Heredity* **47**:161–172.

Gagnon, P.S., R.L. Vadas, D.B. Burdick, and B. May. 1980. Genetic identity of annual and perennial forms of *Zostera marina* L. *Aquatic Botany* **8**:157–162.

Galbraith, I.C.J. 1956. Variation, relationships and evolution in the *Pachycephala pectoralis* superspecies (Aves, Muscicapidae). *Bulletin of the British Museum (Natural History) Zoology* **4**:133–222.

Gardner, A.L. 1977. Chromosomal variation in *Vampyressa* and a review of chromosome evolution in the Phyllostomidae. *Systematic Zoology* **26**:300–318.

Gartside, D.W., and T. McNeilly. 1974. Genetic studies in heavy metal tolerant plants. II. Zinc tolerance in *Agrostis tenuis*. *Heredity* **33**:303–308.

Gaymer, R., R.H.H. Blackman, D.G. Dawson, M. Penny, and C.M. Penny. 1969. The endemic birds of Seychelles. *Ibis* **111**:157–176.

Geist, V. 1971. *Mountain Sheep; a Study in Behavior and Evolution*. University of Chicago Press, Chicago.

Ghiselin, M.T. 1969. The evolution of hermaphroditism among animals. *Quarterly Review of Biology* **44**:189–208.

Gibson, J.B., and B.P. Bradley. 1974. Stabilizing selection in constant and fluctuating environments. *Heredity* **33**:293–302.

Gibson, W.C. 1980. The cost of *not* doing medical research. *Journal of the American Medical Association* **244**:1817–1819.

Gilbert, L.E. 1980. Food web organization and the conservation of neotropical diversity, pp. 11–33. *In* M.E. Soulé and B.A. Wilcox (eds.), *Conservation Biology: An Evolutionary-Ecological Perspective*. Sinauer Associates, Sunderland, Massachusetts.

Gilbert, L.E., and M.C. Singer. 1973. Dispersal and gene-flow in a butterfly species. *American Naturalist* **107**:58–72.

Giles, R.H. 1978. *Wildlife Management*. Freeman, San Francisco, California.

Gill, F.B. 1980. Historical aspects of hybridization between blue-winged and golden-winged warblers. *Auk* **97**:1–18.

Gillespie, J. 1977. A general model to account for enzyme variation in natural populations. III. Multiple alleles. *Evolution* **31**:85–90.

Goel, N.S., S.C. Maitra, and E.W. Montroll. 1971. *Nonlinear Models of Interacting Populations*. Academic Press, New York.

Goodman, D. 1980. Demographic intervention for closely managed populations, *In* M.E. Soulé and B.A. Wilcox (eds.), *Conservation Biology: An Evolutionary–Ecological Perspective*. Sinauer Associates, Sunderland, Massachusetts.

Goodman, M. 1976. Protein sequences in phylogeny, pp. 149–159. *In* F.J. Ayala (ed.), *Molecular Evolution*, Sinauer Associates, Sunderland, Massachusetts.

Goodman, P.J. 1973. Physiological and ecotypic adaptations of plants to salt desert conditions in Utah. *Journal of Heredity* **61**:473–494.

Goodwin, L.G. 1980. Scientific Report: Zoological Society. Nuffield Laboratories of Comparative Medicine. *Journal of Zoology* **190**:553–571.

Gorman, G.C., M. Soulé, S.Y. Yang, and E. Nevo. 1975. Evolutionary genetics of insular Adriatic lizards. *Evolution* **9**:52-71.

Gottlieb, L.D. 1973. Enzyme differentiation and phylogeny in *Clarkia franciscana, C. rubicunda* and *C. amoena. Evolution* **27**:205-214.

Gottlieb, L.D. 1974. Genetic confirmation of the origin of *Clarkia lingulata. Evolution* **28**:244-250.

Gottlieb, L.D. 1975. Allelic diversity in the outcrossing annual plant *Stephanomeria exigua* ssp. *carotifera* (Compositae). *Evolution* **29**:213-225.

Gottlieb, L.D. 1981. Electrophoretic evidence and plant populations. *Progress in Phytochemistry* **7**:1-46.

Gottlieb, L.D., and G. Pilz. 1976. Genetic similarity between *Gaura longiflora* and its obligately outcrossing derivative *G. demareei. Systematic Botany* **1**: 181-187.

Gould, F. 1979. Rapid host range evolution in a population of the phytophagous mite *Tetranychus urticae* Koch. *Evolution* **33**:791-802.

Gould, F. 1983. Genetics of plant-herbivore systems: interactions between applied and basic study. *In* R.F. Denno and M.S. McClure (eds.), *Impact of Variable Host Quality on Herbivorous Insects.* Academic Press, New York.

Gould, S.J., and D.S. Woodruff. 1978. Natural History of Cerion VIII: Little Bahama Bank-a revision based on genetics, morphometrics, and geographical distribution. *Bulletin, Museum of Comparative Zoology, Harvard University.* **148**: 372-415.

Grant, M.C., and J.B. Mitton. 1977. Genetic differentiation among growth forms of Engelmann spruce and subalpine fir at tree line. *Arctic and Alpine Research* **3**: 259-263.

Grant, V. 1958. The regulation of recombination in plants. *Cold Spring Harbor Symposia on Quantitative Biology* **23**:337-363.

Grant, V. 1975. *Genetics of Flowering Plants.* Columbia University Press, New York.

Grant, V. 1981. *Plant Speciation.* 2nd ed. Columbia University Press, New York.

Greenway, J.C., Jr. 1967. *Extinction and Vanishing Birds of the World.* Dover Publications, New York.

Greenwell, G.A., C. Emerick, and B. Biben. 1982. Inbreeding depression in mandarin ducks: a preliminary report on some continuing experiments. *Avicultural Magazine.* **88**(3):145-148.

Greenwood, P.J., and P.H. Harvey. 1978. Inbreeding and dispersal in the great tit. *Nature* **27**:52-54.

Gregorius, H.R. 1978. The concept of genetic diversity and its formal relationship to heterozygosity and genetic distance. *Mathematical Biosciences* **41**:253-271.

Gregorius, H.R. 1980. The probability of losing an allele when diploid gametes are sampled. *Biometrics* **36**:643-652.

Gregory, R.P.G., and A.D. Bradshaw. 1965. Heavy metal tolerance in populations of *Agrostis tenuis* Sibth and other grasses. *New Phytologist* **64**:131-143.

Greig, J.C. 1979. Principles of genetic conservation in relation to wildlife management in southern Africa. *South African Journal of Wildlife Research* **9**:57-78.

Griner, L.A. 1982. *The Pathology of Zoo Animals.* Zoological Society of San Diego, San Diego.

Gropp, A., H. Winking, L. Zech, and H.J. Muller. 1972. Robertsonian chromosomal variation and identification of metacentric chromosomes in feral mice. *Chromosoma* **39**:265-288.

Grula, J.W., T.J. Hall, J.A. Hunt, T.D. Giuni, G.J. Graham, E.H. Davison, and R.J. Britten. 1982. Sea urchin DNA sequence variation and reduced interspecies differences of the less variable DNA sequences. *Evolution* **36**:665-676.

Guillery, R.W., and J.H. Kaas. 1973. Genetic abnormality of the visual pathways in a "white" tiger. *Science* **180**:1287-1289.

Guries, R.P., and F.T. Ledig. 1982. Genetic diversity and population structure in pitch pine (*Pinus rigida* Mill.). *Evolution* **36**:387-402.

Haldane, G.B.S., and S.D. Jayakar. 1963. Polymorphism due to selection of varying direction. *Journal of Genetics* **58**:237-242.

Hall, E.R. 1981. *The Mammals of North America*, Vols. 1 and 2. Wiley and Sons, New York.

Hall, W.P. 1973. Comparative population cytogenetics, speciation, and evolution of the iguanid lizard genus *Sceloporus*. Ph.D. dissertation, Harvard University, Cambridge, Massachusetts.

Hall, W.P., and R.K. Selander. 1973. Hybridization of karyotypically differentiated populations in the *Sceloporus grammicus* complex (Iguanidae). *Evolution* **27**:226-242.

Halliday, T. 1978. *Vanishing Birds*. Holt, Rinehart and Winston, New York.

Hamrick, J.L. 1970. An index of selection intensity based on patterns of variation in natural plant populations. Ph.D. dissertation, University of California, Berkeley, California.

Hamrick, J.L. 1976. Variation and selection in western montane species. II. Variation within and between populations of white fir on an elevational transect. *Theoretical Applied Genetics* **47**:27-34.

Hamrick, J.L., and R.W. Allard. 1972. Microgeographical variation in allozyme frequencies in *Avena barbata*. *Proceedings of the National Academy of Sciences of the United States of America* **69**:2100-2104.

Hamrick, J.L., and R.W. Allard. 1975. Correlations between quantitative characters and enzyme genotypes in *Avena barbata*. *Evolution* **29**:438-442.

Hamrick, J.L., and L.R. Holden. 1979. Influence of microhabitat heterogeneity on gene frequency distribution and gametic phase disequilibrium in *Avena barbata*. *Evolution* **33**:521-533.

Hamrick, J.L., and W.J. Libby. 1972. Variation and selection in western U.S. montane species. I. White fir. *Silvae Genetica* **21**:29-35.

Hamrick, J.L., Y.B. Linhart, and J.B. Mitton. 1979. Relationships between life history characteristics and electrophoretically-detectable genetic variation in plants. *Annual Review of Ecology and Systematics* **10**:173-200.

Hamrick, J.L., J.B. Mitton, and Y.B. Linhart. 1981. Levels of genetic variation in trees: Influence of life history characteristics, pp. 35-41. *In* M.T. Conlisle (ed.), *Proceedings of the Symposium on Isozymes of North American Forest Trees and Forest Insects*. U.S. Department of Agriculture Forest Service. Pacific Southern Forest and Range Experimental Station General Technical Report PSW-48.

Hancock, J.F., and R.S. Bringhurst. 1978. Interpopulational differentiation and adaptation in the perennial, diploid species *Fragaria vesca* L. *American Journal of Botany* **65**:795-803.

Hancock, J.F., and R.S. Bringhurst. 1979. Hermaphroditism in predominately dioecious populations of *Fragaria chiloensis* (L.) Duchn. *Bulletin of the Torrey Botanical Club* **106**:229-231.

Handford, P. 1980. Heterozygosity at enzyme loci and morphological variation. *Nature* **286**:261-262.

Harborne, J.B. 1973. *Phytochemical Methods*. Chapman and Hall, London.

Hare, J.D., and D.J. Futuyma. 1978. Different effects of variation in *Xanthium strumarium* L. (Compositae) on two insect seed predators. *Oecologia* **39**:109-120.

Harlan, J.R. 1963a. Natural introgression between *Bothriochloa ischaemum* and *B. intermedia* in West Pakistan. *Botanical Gazette* **124**:294-300.

Harlan, J.R. 1963b. Two kinds of gene centers in Bothriocholininae. *American Naturalist* **97**:91-98.

Harlan, J.R. 1966. Plant introduction and biosystematics, pp. 55-83. *In* K. Frey (ed.), *Plant Breeding*. Iowa State University Press, Ames, Iowa.

Harlan, J.R. 1982. Human interference with grass systematics, pp. 37-50. *In* J.R. Estes, R.J. Tyrl, and J.N. Brunken (eds.), *Grasses and Grasslands: Systematics and Ecology*. University of Oklahoma Press, Norman, Oklahoma.

Harlan, J.R., and J.M.J. de Wet. 1963. The compilospecies concept. *Evolution* **17**:497-501.

Harlan, J.R., and J.M.J. de Wet. 1971. Toward a rational classification of cultivated plants. *Taxon* **20**:509-517.

Harlan, J.R., and J.M.J. de Wet. 1974. Sympatric evolution in *Sorghum*. *Genetics* **78**:473-474.

Harlan, J.R., M.H. Brooks, D.S. Borgaonkar, and J.M.J. de Wet. 1964. Nature and inheritance of apomixis in *Bothriochloa* and *Dichanthium*. *Botanical Gazette* **125**:41-46.

Harper, J.L. 1977. *Population Biology of Plants*. Academic Press, London.

Hart, E.B. 1978. Karyology and evolution of the plains pocket gopher, *Geomys bursarius*. *Occasional Papers*. *Museum of Natural History*. *University of Kansas* **71**:1-20.

Hartl, D.L. 1980. *Principles of Population Genetics*. Sinauer Associates, Sunderland, Massachusetts.

Hartl, D.L. 1981. *A Primer of Population Genetics*. Sinauer Associates, Sunderland, Massachusetts.

Hatchett, J.H., and R.L. Gallun. 1970. Genetics of the ability of the Hessian fly, *Mayetiola destructor*, to survive on wheats having different genes for resistance. *Annals of the Entomological Society of America* **63**:1400-1407.

Hayward, M.D., and N.J. McAdam. 1977. Isozyme polymorphism as a measure of distinctiveness and stability in cultivars of *Lolium perenne*. *Zeitschrift für Pflanzenernahrung und Bodenkunde* **79**:59-68.

Hedrick, P.W., M.E. Ginevan, and E.P. Ewing. 1976. Genetic polymorphism in heterogeneous environments. *Annual Review of Ecology and Systematics* **7**:1-32.

Hendrickson, J.R. 1979. Chemical discrimination of tortoise-shell materials and reptilian leathers. *Final Report* (Contract 14-16-0002-3701). U.S. Fish and Wildlife Service, Albuquerque, New Mexico.

Heusmann, H.W. 1974. Mallard-black duck relationships in the Northeast. *Wildlife Society Bulletin* **2**(4):171-177.

Hiebert, R.D., and J.L. Hamrick. 1983. Patterns and levels of genetic variation in Great Basin bristlecone pine, *Pinus longaeva*. *Evolution* **37**(2):302-311.

Highton, R., and A. Larson. 1979. The genetic relationships of the salamander of the genus *Plethodon*. *Systematic Zoology* **28**:579-599.

Hillis, D.M. 1981. Premating isolating mechanisms among three species of the *Rana pipiens* complex in Texas and southern Oklahoma. *Copeia* **1981**:312-319.

Hinegardner, R. 1976. Evolution of genome size, pp. 179-199. In F.J. Ayala (ed.), *Molecular Evolution*. Sinauer, Sunderland, Massachusetts.

Hirth, H.F., and W.M. Schaffer. 1974. Survival rate of the green turtle, *Chelonia mydas*, necessary to maintain stable populations. *Copeia* **1974**:544-546.

Holdren, C.E., and P.R. Ehrlich. 1981. Long-range dispersal in checkerspot butterflies: transplant experiments with *Euphydryas gillettii*. *Oecologia* **50**:125-129.

Hollingsworth, M.J. and J. Maynard Smith. 1955. The effects of inbreeding on rate of development and on fertility in *Drosophila subobscura*. *Journal of Genetics* **53**:295-314.

Honeycutt, R.L., and D.J. Schmidly. 1979. Chromosomal and morphological variation in the plains pocket gopher, *Geomys bursarius*, in Texas and adjacent states. *Occasional Papers. Museum of Texas Tech University* **58**:1-54.

Honeycutt, R.L., R.J. Baker, and H.H. Genoways. 1980. Results of the Alcoa Foundation-Suriname expeditions. III. Chromosomal data for bats (Mammalia: Chiroptera) from Suriname. *Annals of the Carnegie Museum* **49**:237-250.

Hoogland, J.L. 1982. Prairie dogs avoid extreme inbreeding. *Science* **215**: 1639-1641.

Hsu, K.J., Q. He, J.A. McKenzie, H. Weissert, K. Perch-Nielsen, H. Oberhänsli, K. Kelts, J. LaBrecque, L. Tauxe, U. Krähenbühl, S.F. Percival, Jr., R. Wright, A.M. Karpoff, N. Petersen, P. Tucker, R.Z. Poore, A.M. Gombos, K. Pisciotto, M.F. Carman, Jr., E. Schreiber. 1982. Mass mortality and its environmental and evolutionary consequences. *Science* **216**:249-256.

Hubbard, J.P. 1977. The biological and taxonomic status of the Mexican duck. *New Mexico Department of Game and Fish Bulletin* No. 16.

Hufty, M.P., C.J. Sedgwick, and K. Benirschke. 1973. The karyotypes of the white-lipped and collared peccaries. Aspects of their chromosomal evolution. *Genen en Phaenen* **16**:81-86.

Hultsch, E., and J.W. Appleby. 1980. Laboratory breeding of the owl monkey (*Aotus trivirgatus*) with special reference to management and behavioral patterns. *Abstract 15.6 American Society of Primatologists*, p. 55.

Humphrey, S.R. 1972. Zoogeography of the nine-banded armadillo (*Dasypus novemcinctus*) in the United States. *BioScience*. **24**:457-462.

IUCN. 1966 et seq. *Survival Service Commission Red Data Book*, Vol. 1. *Mammals*. International Union for the Conservation of Nature and Natural Resources, Morges, Switzerland.

IUCN. 1971. *Marine Turtles. Proceedings of the 2nd Working Group of Marine Turtle Specialists. IUCN Publications New Series, Supplementary Paper* No. **31**: 1-109. International Union for the Conservation of Nature and Natural Resources, Morges, Switzerland.

IUCN. 1975. *United Nations List of National Parks and Equivalent Reserves*. International Union for Conservation of Nature and Natural Resources. Morges, Switzerland.

IUCN. 1980. *World Conservation Strategy*. International Union for the Conservation of Nature and Natural Resources/United Nations Environmental Program/World Wildlife Fund. Gland, Switzerland.

Jaaska, V. 1979. Genetic polymorphism of acid phosphatase in populations of rye, *Secale cereale* L. s.1. *Eesti Nov Teaduste Akadeemia Toimetised* 28 Koide Bioloogia B-3, pp. 185–193.

Jacguard, A. 1974. *The Genetic Structure of Populations.* Springer-Verlag, New York.

Jain, S.K. 1976. Patterns of survival and microevolution in plant populations. In S. Karlin and E. Nevo (eds.), *Population Genetics and Ecology.* Academic Press, New York.

Jain, S.K. 1977. Inheritance and population genetics of four marker traits in rose clover. *Journal of Heredity* **68**:48–52.

Jain, S.K., and A.D. Bradshaw. 1966. Evolutionary divergence among adjacent plant populations. I. The evidence and its theoretical analysis. *Heredity* **21**:407–441.

Jain, S.K., and D.R. Marshall. 1967. Population studies in predominantly self-pollinating species. X. Variation in natural populations of *Avena fatua* and *A. barbata. American Naturalist* **101**:19–33.

Jain, S.K., and P.S. Martins. 1979. Ecological genetics of the colonizing ability of rose clover (*Trifolium hirtum* All.). *American Journal of Botany* **66**:361–366.

Jain, S.K., and K.N. Rai. 1980. Population biology of *Avena.* VIII. Colonization experiment as a test of the role of natural selection in population divergence. *American Journal of Botany* **67**:1342–1346.

Jain, S.K., L. Wu, and K.R. Vaidya. 1980. Levels of morphological and allozyme variation in Indian amaranths: a striking contrast. *Journal of Heredity* **71**:283–285.

James, F.C. 1980. Miscegenation in the Dusky Seaside Sparrow. *BioScience* **30**: 800–801.

James, S.H. 1982. The relevance of genetic systems to conservation practice. In R.H. Groves and W.D.L. Ride (eds.), *Species at Risk: Research in Australia.* Australian Academy of Science, Canberra. Springer-Verlag.

Jarvis, C. 1967. The value of zoos for science and conservation. *Oryx* **9**:127–136.

Jefferies, R.L., A.J. Davy, and T. Rudmik. 1981. Population biology of the salt marsh annual *Salicornia europaea* agg. *Journal of Ecology* **69**:17–31.

Jenkins, R.E. 1982. Endangered species and the primacy of information. *The Nature Conservancy News* **32**:23–30.

Jennrich, R.I. 1977. p. 76. In K. Enstein, A. Ralston, and H.S. Wild (eds.), *Statistical Methods for Digital Computers.* Wiley and Sons, New York.

Jewell, P., and S. Holt. 1982. *Problems in Management of Locally Abundant Wild Animals.* Academic Press, New York.

John, B. 1981. Chromosome change and evolutionary change, pp. 23–51. In W.R. Atchley and D.S. Woodruff (eds.), *Evolution and Speciation: Essays in Honor of M.J.D. White.* Cambridge University Press, Cambridge, England.

Johnsey, P.G., and M.O. Malinen. 1971. *Final report on population densities of small mammals in relation to specific habitat in Pea Ridge National Military Park, Barton County, Arkansas.* Unpublished science report to the National Park Service, U.S. Department of the Interior, Washington, D.C.

Johnsgard, P. 1961. Evolutionary relationships among the North American mallards. *Auk* **78**:3–43.

Johnsgard, P. 1967. Sympatric changes and hybridization incidence in mallards and black ducks. *American Midland Naturalist* **77**:51–63.

Johnson, M.S., B. Clarke, and J. Murray. 1977. Genetic variation and reproductive isolation in *Partula. Evolution* **31**:116–126.

Johnston, R.F., and R.K. Selander. 1964. House sparrows: rapid evolution of races in North America. *Science* **144**:548-550.
Jones, D.A. 1973. Co-evolution and cyanogenesis, pp. 213-242. *In* V. Heywood (ed.), *Taxonomy and Ecology*. Academic Press, London.
Jones, J.S. 1982. Of cannibals and kin. *Nature* **299**:202-203.
Jones, J.S., B.H. Leith, and P. Rowling. 1977. Polymorphism in *Cepaea*: A problem with too many solutions? *Annual Review of Ecology and Systematics* **8**: 109-143.
Jones, M. 1983. History in captivity. *In* B. Beck and C. Wemmer (eds.), *Père David's Deer: The Biology and Conservation of an Extinct Species*. Noyes, Park Ridge, New Jersey. In press.
Jones, T.C., R.W. Thorington, M.M. Hu, E. Adams, and R.W. Cooper. 1973. Karyotypes of squirrel monkeys *(Saimiri sciureus)* from different geographic regions. *American Journal of Physical Anthropology* **38**:269-278.
Jowett, D. 1958. Populations of *Agrostis* spp. tolerant of heavy metals. *Nature* **182**: 816-817.
Jowett, D. 1964. Population studies on lead-tolerant *Agrostis tenuis*. *Evolution* **18**: 70-80.
Kahler, A.L., and R.W. Allard. 1981. Worldwide patterns of genetic variation among four esterase loci in barley *(Hordeum vulgare* L.). *Theoretical Applied Genetics* **59**:101-111.
Kahler, A.L., R.W. Allard, M. Krzakowa, C.F. Wehrhahn, and E. Nevo. 1980. Associations between isozyme phenotypes and environment in the slender wild oat *(Avena barbata)* in Israel. *Theoretical Applied Genetics* **56**:31-47.
Karlin, S. 1976. Population subdivision and selection migration interation, pp. 617-657. *In* S. Karlin and E. Nevo (eds.), *Population Genetics and Ecology*. Academic Press, New York.
Karlin, S. 1977. Gene frequency patterns in the Levene subdivided population model. *Theoretical Population Biology* **11**:356-385.
Karlin, S., and R.B. Campbell. 1980. Polymorphism in subdivided populations characterized by a major and subordinate demes. *Heredity* **44**:51-168.
Karr, J.K. 1982. Avian extinction on Barro Colorado Island, Panama: a reassessment. *American Naturalist* **119**:228-239.
Kear, J., and A.J. Berger. 1980. *The Hawaiian Goose: An Experiment in Conservation*. Buteo Press, Vermillion, South Dakota.
Keith, L.B., and L.A. Windberg. 1978. A demographic analysis of the snowshoe hare cycle. *Wildlife Monographs* **No. 58**.
Kerney, M.P., and R.A.D. Cameron. 1979. *A Field Guide to the Land Snails of Britain and Northwest Europe*. Collins, London.
Kettlewell, H.B.D. 1973. *The Evolution of Melanism*. Clarendon Press, Oxford.
Kim, Y.J. 1972. Studies of biochemical genetics and karyotypes in pocket gophers (family Geomydiae). Ph.D. dissertation, University of Texas, Austin, Texas.
Kimura, M., and J.F. Crow. 1963. The measurement of effective population number. *Evolution* **17**:279-288.
Kimura, M., and T. Ohta. 1971. *Theoretical Aspects of Population Genetics*. Princeton University Press, Princeton, New Jersey.
King, J.A. 1955. Social behavior, social organization, and population dynamics in a black-tailed prairie dog town in the Black Hills of South Dakota. *University of Michigan Contributions from the Laboratory of Vertebrate Biology* No. 67, Ann Arbor, Michigan.

King, M.C., and A.C. Wilson. 1975. Evolution at two levels: molecular similarities and biological differences between humans and chimpanzees. *Science* **188**: 107–116.

King, W.B. 1981. *Endangered Species of the World. The ICBP Bird Red Data Book.* Smithsonian Institution Press and International Council for Bird Preservation, Washington, D.C.

Kinloch, B.B., and J.L. Littlefield. 1977. White pine blister rust: hypersensitive resistance in sugar pine. *Canadian Journal of Botany* **55**:1148–1155.

Kleiman, D.G. 1982. Cooperative research and management for the golden lion tamarins. *American Association of Zoological Parks and Aquariums Newsletter* **23**(2):16.

Kloppers, J.J. 1976. Butterflies—a word of warning. *Fauna and Flora (Transvaal)* **34**: 8–9.

Knerer, G., and C.E. Atwood. 1973. Diprinoid sawflies: Polymorphism and speciation. *Science* **179**:1090–1099.

Koch, A.L., A.F. Carr, and D.W. Ehrenfeld. 1968. The problem of open-sea navigation: the migration of the green turtle to Ascension Island. *Journal of Theoretical Biology* **22**:163–179.

Koehn, R. 1969. Esterase heterogeneity: dynamics of a polymorphism. *Science* **163**: 943–944.

Koenig, W.D., and F.A. Pitelka. 1979. Relatedness and inbreeding avoidance: counterploys in the community nesting Acorn Woodpecker. *Science* **206**: 1103–1105.

Kruuk, H. 1972. *The Spotted Hyena.* University of Chicago Press. Chicago.

Kubetin, W.R., and B.A. Schaal. 1979. Apportionment of isozyme variability in *Polygonum pensylvanicum* (Polygonaceae). *Systematic Botany* **4**:148–156.

Kuris, A.M. 1974. Trophic interactions: similarity of parasitic castrators to parasitoids. *Quarterly Review of Biology* **49**:129–148.

Laerm, J., J.C. Avise, J.C. Patton, and R.A. Lansman. 1982. Genetic determination of the status of an endangered species of pocket gopher in Georgia. *Journal of Wildlife Management* **46**:513–518.

Lamas Mueller, G. 1974. Supuesta extinción de una mariposa en Lima, Perú (Lepidoptera, Rhopalocera). *Revista Peruana Entomologica* **17**:119–120.

Lang, E.M. 1977. What are endangered species? *International Zoo Yearbook* **17**:2–5.

Langley, C.H., Y.N. Tobari, and K. Kojima. 1974. Linkage disequilibrium in natural populations of *Drosophila melanogaster. Genetics* **78**:921–936.

Lansman, R.A., R.O. Shade, J.F. Shapira, and J.C. Avise. 1981. The use of restriction endonucleases to measure mitochondrial DNA sequence relatedness in natural populations. III. Techniques and potential applications. *Journal of Molecular Evolution* **17**:214–229.

Larson, A., and R. Highton. 1978. Geographic variation and divergence in the salamanders of the *Plethodon welleri* group (Amphibia, Plethodontidae). *Systematic Zoologist* **27**:431–448.

Lasley, B.L. 1980. Recent techniques in endocrine monitoring of zoo animals. *International Zoo Yearbook* **20**:166–170.

Lasley, B.L., N.M. Czekala, K.C. Nakakura, S.G. Amara, and K. Benirschke. 1979. Armadillos for studies on delayed implantation, quadruplets, uterus simplex, and fetal adrenal physiology, pp. 447–451. *In* N.J. Alexander (ed.), *Animal Models for Research on Contraception and Fertility.* Harper and Row, Hagerstown, Maryland.

Lasley, J.F. 1978. *Genetics of Livestock Improvement*, 3rd ed. Prentice-Hall, Englewood Cliffs, New Jersey.

LaVal, R.K. 1973. Systematics of the genus *Rhogeessa* (Chiroptera: Vespertilionidae). *Occasional Papers, Museum of Natural History, University of Kansas* **19**: 1-47.

Lee, F.B., and H.K. Nelson. 1966. The role of artificial propagation in wood duck management, pp. 140-150. In J.B. Trefethen (ed.), *Wood Duck Management and Research: A Symposium*. Wildlife Management Institute, Washington, D.C.

Lenarz, M., and W. Conley. 1980. Demographic considerations in reintroduction programs of bighorn sheep. *Acta Theriologica* **25**:71-80.

Lerner, I.M. 1954. *Genetic Homeostasis*. Wiley and Sons, New York.

Levene, H. 1953. Genetic equilibrium when more than one ecological niche is available. *American Naturalist* **87**:331-333.

Levin, D.A. 1975a. Genic heterozygosity and protein polymorphism among local populations of *Oenothera biennis*. *Genetics* **79**:477-491.

Levin, D.A. 1975b. Interspecific hybridization, heterozygosity and gene exchange in *Phlox*. *Evolution* **29**:37-51.

Levin, D.A. 1976. Consequences of long-term artificial selection, inbreeding and isolation in *Phlox*. II. The organization of allozymic variability. *Evolution* **30**: 463-472.

Levin, D.A. 1978. Genetic variation in annual *Phlox*: self-compatible versus self-incompatible species. *Evolution* **32**:245-263.

Levin, D.A. 1979. The nature of plant species. *Science* **204**:381-384.

Levin, D.A., and H.W. Kerster. 1968. Local gene dispersal in *Phlox*. *Evolution* **22**: 130-139.

Levin, S.A., and J.D. Udovic. 1977. A mathematical model of coevolutionary populations. *American Naturalist* **111**:657-675.

Levins, R. 1968. *Evolution in Changing Environments*. Princeton University Press, Princeton, New Jersey.

Levy, M., and D.A. Levin. 1975. Genic heterozygosity and variation in permanent translocation heterozygotes of the *Oenothera biennis* complex. *Genetics* **79**: 493-512.

Lewis, W.H. (ed.). 1980. *Polyploidy: Biological Relevance*. Basic Life Sciences, Vol. 13. Plenum Press, New York.

Lewontin, R.C. 1972. The apportionment of human diversity. *Evolutionary Biology* **6**:381.

Lewontin, R.C. 1974. *The Genetic Basis of Evolutionary Change*. Columbia University Press, New York.

Lewontin, R.C., and L.C. Birch, 1966. Hybridization as a source of variation for adaptation to new environments. *Evolution* **20**:315-335.

Lewontin, R.C., and J. Krakauer. 1973. Distribution of gene frequency as a test of the theory of the selective neutrality of polymorphisms. *Genetics* **74**:175-195.

Libby, W.J., R.F. Stettler, and F.W. Setz. 1969. Forest genetics and forest tree breeding. *Annual Review of Genetics* **3**:469-494.

Lindemann, H. 1982. *African Rhinoceroses in Captivity*. University of Copenhagen, Denmark.

Lindzey, F.G., and E.C. Mesow. 1977a. Population characteristics of black bears on an island in Washington. *J. Wildlife Management* **41**(3):408-412.

Lindzey, F.G., and E.C. Mesow. 1977b. Home range and habitat use by black bears in southwestern Washington. *J. Wildlife Management* **41**(3):413–425.

Linhart, Y.B., J.B. Mitton, K.B. Sturgeon, and M.L. Davis. 1981a. An analysis of genetic architecture in populations of ponderosa pine, pp. 53–59. *In* M.T. Conkle (ed.), *Proceedings of the Symposium on Isozymes of North American Forest Trees and Forest Insects*. U.S. Department of Agriculture, Forest Service. Pacific Southern Forest and Range Experiment Station. General Technical Report PSW-48.

Linhart, Y.B., M.L. Davis, and J.B. Mitton. 1981b. Genetic control of allozymes of shikimate dehydrogenase in ponderosa pine. *Biochemical Genetics* **19**:641–646.

Linhart, Y.B., J.B. Mitton, K.B. Sturgeon, and M.L. Davis. 1981c. Genetic variation in space and time in a population of ponderosa pine. *Heredity* **46**:407–426.

Long, T. 1970. Genetic effects of fluctuating temperature in populations of *Drosophila melanogaster*. *Genetics* **66**:401–416.

Loudenslager, E.J., and G.A.E. Gall. 1980. Geographic patterns of protein variation and subspeciation in cutthroat trout, *Salmo clarki*. *Systematic Zoology* **29**:27–42.

Lovejoy, T.E. 1978. Genetic aspects of dwindling populations: a review, pp. 275–280. *In* S.A. Temple (ed.), *Endangered Birds: Management Techniques for Preserving Threatened Species*. University of Wisconsin Press, Madison, Wisconsin.

Lumaret, R. 1982. Protein variation in diploid and tetraploid orchard grass (*Dactylis glomerata* L.): formal genetics and population polymorphism of peroxidases and malate dehydrogenases. *Genetica* **57**:207–215.

Lundkvist, K. 1979. Allozyme frequency distributions in four Swedish populations of Norway spruce (*Picea abies* K.). I. Estimations of genetic variation within and among populations, genetic linkage, and a mating system parameter. *Hereditas* **90**:127–143.

Lundkvist, K., and D. Rudin. 1977. Genetic variation in eleven populations of *Picea abies* as determined by isozyme analysis. *Hereditas* **85**:67–74.

Luria, S.E., S.J. Gould, and S. Singer. 1981. *A View of Life*. Benjamin/Cummings, Menlo Park, California.

Lynch, C.B. 1977. Inbreeding effects upon animals derived from a wild population of *Mus musculus*. *Evolution* **31**:526–537.

Ma, N.S.F. 1981. Chromosome evolution in the owl monkey, *Aotus*. *American Journal of Physical Anthropology* **54**:293–303.

MacArthur, R.H., and E.O. Wilson. 1967. *The Theory of Island Biogeography*. Princeton University Press, Princeton, New Jersey.

Mackay, T.F.C. 1981. Genetic variation in varying environments. *Genetical Research* **37**:79–93.

Macnair, M.R. 1977. Major genes for copper tolerance in *Mimulus guttatus*. *Nature* **268**:428–430.

Macnair, M.R. 1982. Tolerance of higher plants to toxic materials, pp. 177–207. *In* J.A. Bishop and L.H. Cook (eds.), *Genetic Consequences of Man-Made Change*. Academic Press, London.

Main, A.R., and M. Yadov. 1971. Conservation of macropods in reserves in Western Australia. *Biological Conservation* **3**:123–132.

Manlove, M.N., M.H. Smith, H.O. Hillestad, S.E. Fuller, P.E. Johns, and D.O. Straney. 1976. Genetic subdivision in a herd of white-tailed deer as demonstrated by spatial shifts in gene frequencies. *Proceedings of the Conference of the Southeastern Association for Game and Fish Commissioners* **30**:487-492.

Marshall, D.R., and P.W. Weiss. 1982. Isozyme variation within and among Australian populations of *Emex spinosa* (L.) Campd. *Australian Journal of Biological Sciences* **35**(3):327-332.

Marshall, D.R., and R.W. Allard. 1970a. Maintenance of isozyme polymorphisms in natural populations of *Avena barbata*. *Genetics* **66**:393-399.

Marshall, D.R., and R.W. Allard. 1970b. Isozyme polymorphisms in natural populations of *A. fatua* and *A. barbata*. *Heredity* **25**:373-382.

Marshall, D.R., and S.K. Jain. 1968. Phenotypic plasticity of *Avena fatua* and *A. barbata*. *American Naturalist* **102**:457-467.

Marshall, D.R., and S.K. Jain. 1969. Genetic polymorphism in natural populations of *Avena fatua* and *A. barbata*. *Nature* **221**:276-278.

Martin, R.D. (ed.). 1975. *Breeding Endangered Species in Captivity*. Academic Press, London.

Martins, P.S., and S.K. Jain. 1980. Interpopulation variation in rose clover: recently introduced species in California rangelands. *Journal of Heredity* **71**:29-32.

Mason, L.G., P.R. Ehrlich, and T.C. Emmel. 1968. The population biology of the butterfly *Euphydryas editha*. VI. Phenetics of the Jasper Ridge colony, 1965-1966. *Evolution* **22**:46-54.

Mather, K. 1950. The genetical architecture of heterostyly in *Primula sinensis*. *Evolution* **4**:340-342.

Mather, K.M. 1973. *Genetical Structure of Populations*. Chapman and Hall, London.

May, R.M. 1980. Inbreeding among zoo animals. *Nature* **283**:430-431.

May, R.M., and R.M. Anderson. 1983. Parasite-host coevolution. In D.J. Futuyma and M. Slatkin (eds.). *Coevolution*. Sinauer Associates, Sunderland, Massachusetts. In press.

May, R.M., and G.F. Oster. 1976. Bifurcations and dynamic complexity in simple ecological models. *American Naturalist* **110**:573-599.

Maynard Smith, J. 1976a. What determines the rate of evolution? *American Naturalist* **110**:331-338.

Maynard Smith, J. 1976b. A comment on the Red Queen. *American Naturalist* **110**:325-330.

Maynard Smith, J. 1977. The sex habit in plants and animals, pp. 315-331. In F.B. Christiansen and T.M. Fenchel (eds.), *Measuring Selection in Natural Populations*. Springer-Verlag, Berlin.

Maynard Smith, J. 1978. *The Evolution of Sex*. Cambridge University Press, Cambridge, England.

Mayr, E. 1963. *Animal Species and Evolution*. Harvard University Press, Cambridge, Massachusetts.

Mayr, E. 1969. *Principles of Systematic Zoology*. McGraw-Hill, New York.

Mayr, E. 1970. *Populations, Species, and Evolution*. Belknap Press, Cambridge, Massachusetts.

Mayr, E., and L.L. Short. 1970. Species taxa of North American birds, and contribution to comparative systematics. Publication 9, *Nuttall Ornithological Club*.

McAndrew, B.J., R.D. Ward, and J.A. Beardmore. 1982. Lack of relationship between morphological variance and enzyme heterozygosity in the plaice *Pleuronectes platessa*. *Heredity* **48**:117–125.

McCabe, R.A. 1947. The homing of transplanted young wood ducks. *Wilson Bulletin* **59**:104–109.

McClure, S. 1973. Allozyme variability in natural populations of the yellow monkeyflower, *Mimulus guttatus*, located in the North Yuba River drainage. Ph.D. Dissertation, University of California, Berkeley, California.

McCracken, G.F., and J.W. Bradbury. 1977. Paternity and genetic heterogeneity in the polygynous bat, *Phyllostomus hastatus*. *Science* **198**:303–306.

McCracken, G.F., and R.K. Selander. 1980. Self-fertilization and monogenic strains in natural populations of terrestrial slugs. *Proceedings of the National Academy of Sciences of the United States of America* **77**:684–688.

McCullough, D.R. 1969. *The Tule Elk: Its History, Behavior, and Ecology*. University of California Publications in Zoology, Berkeley, California.

McCullough, D. 1979. *The George Reserve Deer Herd*. University of Michigan Press, Ann Arbor, Michigan.

McDonald, J.F., and F.J. Ayala. 1974. Genetic response to environmental heterogeneity. *Nature* **250**:572–574.

McKechnie, S.W., P.R. Ehrlich, and R.R. White. 1975. Population genetics of *Euphydryas* butterflies. I. Genetic variation and the neutrality hypothesis. *Genetics* **81**:571–594.

McMichael, D. 1982. What species, what risk? In R.H. Groves and W.D.L. Ride (eds.), *Species at Risk: Research in Australia*. Springer-Verlag.

McNeilly, T. 1967. Evolution in closely adjacent plant populations. III. *Agrostis tenuis* on a small copper mine. *Heredity* **23**:99–108.

McNeilly, T., and A.D. Bradshaw. 1967. Evolutionary processes in populations of copper tolerant *Agrostis tenuis* Sibth. *Evolution* **22**:108–118.

Meagher, T.R., J. Antonovics, and R. Primack. 1978. Experimental ecological genetics in *Plantago*. II. Genetic variation and demography in relation to survival of *Plantago cordata*, a rare species. *Biological Conservation* **14**:243–257.

Mecham, J.S. 1969. New information from experimental crosses on genetic relationships within the *Rana pipiens* species group. *Journal of Experimental Zoology* **170**:169–180.

Meise, W. 1936. Zur Systematik und Verbreitungsgeschichte der Haus- und Weidensperlinge, *Passer domesticus* (L.) und *hispaniolensis* (T.) *Journal of Ornithology* **84**:631–672.

Mejnartowicz, L. 1976. Genetic investigations on Douglas-fir (*Pseudotsuga menziesii* [Mirb.] Franco) populations. *Arboretum Kornickie* **21**:126–187.

Mejnartowicz, L. 1979. Genetic variation in some isoenzyme loci in Scots pine (*Pinus sylvestris* L.) populations. *Arboretum Kornickie* **24**:91–104.

Mierau, G.W., and J.L. Schmidt. 1981. The Mule deer of Mesa Verde National Park. *Mesa Verde Research Series*. Mesa Verde Museum Association Inc., Mesa Verde National Park, Colorado.

Miller, R.M. 1982. Zinc-tolerant ecotypes in *Melilotus alba* (Leguminosae). *American Journal of Botany* **69**:363–368.

Minawa, A., and A.J. Birley. 1975. Genetical and environmental diversity in *Drosophila melanogaster*. *Nature* **255**:702–704.

Mitchell, B., and J.M. Crisp. 1981. Some properties of red deer (*Cervus elaphus*) at

exceptionally high population density in Scotland. *Journal of Zoology* **193**: 157-169.
Mitter, C., D.J. Futuyma, J.C. Schneider, and J.D. Hare. 1979. Genetic variation and host plant relations in a parthenogenetic moth. *Evolution* **33**:777-790.
Mitton, J.B. 1978. Relationship between heterozygosity for enzyme loci and variation of morphological characters in natural populations. *Nature* **273**:661-662.
Mitton, J.B., Y.B. Linhart, J.L. Hamrick, and J.S. Beckman. 1977. Observations on the genetic structure and mating system of ponderosa pine in the Colorado Front Range. *Theoretical Applied Genetics* **51**:5-13.
Mitton, J.B., K.B. Sturgeon, and M.L. Davis. 1980. Genetic differentiation in ponderosa pine along a steep elevational transect. *Silvae Genetica* **29**:100-103.
Mohr, E. 1959. *Das Urwildpferd*. Ziemsen Verlag, Wittenberg.
Moore, J.A. 1944. Geographic variation in *Rana pipiens* Schreber of eastern North America. *Bulletin of the American Museum of Natural History* **82**:345-370.
Moore, J.A. 1946. Hybridization between *Rana palustris* and different forms of *Rana pipiens*. *Proceedings of the National Academy of Sciences of the United States of America* **32**:209-212.
Moore, J.A. 1949. Patterns of evolution in the genus *Rana*, pp. 315-338. In G.L. Jepson, G.G. Simpson, and E. Mayer (eds.), *Genetics, Paleontology and Evolution*. Princeton University Press, Princeton, New Jersey.
Moran, G.F., and D.R. Marshall. 1978. Allozyme uniformity within and variation between races of the colonizing species *Xanthium strumarium* L. (Noogoora Burr). *Australian Journal of Biological Science* **31**:283-291.
Moran, G.F., J.C. Bell, and A.C. Matheson. 1980. The genetic structure and levels of inbreeding in a *Pinus radiata* D. Don seed orchard. *Silvae Genetica* **29**: 190-193.
Morreale, S.J., G.J. Ruiz, J.R. Spotila, and E.A. Standora. 1982. Temperature-dependent sex determination: current practices threaten conservation of sea turtles. *Science* **216**:1245-1247.
Morton, A. 1982. The importance of farming butterflies. *New Scientist* **94**:503-511.
Morton, N.E., J.F. Crow, and J.J. Muller. 1956. An estimate of the mutational damage in man from data on consanguinous marriages. *Proceedings of the National Academy of Sciences of the United States of America* **42**:855-863.
Mrakovcic, M., and L.E. Haley. 1979. Inbreeding depression in the Zebra fish *Brachytanio rerio* (Hamilton Buchanon). *Journal of Fish Biology* **15**:323-327.
Mrosovsky, N., and C.L. Yntema. 1980. Temperature dependence of sexual differentiation in sea turtles: implications for conservation practices. *Biological Conservation* **18**:271-280.
Mueller, L.D., and F.J. Ayala. 1981. Dynamics of single species population growth: stability or chaos? *Ecology* **62**:1148-1154.
Muller, H.J. 1950. Our load of mutations. *American Journal of Human Genetics* **2**: 111-176.
Murphy, D.D., and P.R. Ehrlich. 1980. Two California checkerspot butterfly subspecies: one new, one on the verge of extinction. *Journal of the Lepidopterists' Society* **34**:316-320.
Murphy, D.D., A.E. Launer, and P.R. Ehrlich. 1983. The role of adult feeding in egg production and population dynamics of the checkerspot butterfly *Euphydryas editha*. *Oecologia* **56**:257-263.
Murphy, E.C., and K.R. Whitten. 1976. Dall sheep demography in McKinley Park

and a reevaluation of Murie's data. *Journal of Wildlife Management* **40**(4): 597–609.

Murphy, M.R. 1971. Natural history of the Syrian golden hamster. *American Zoologist* **11**:632.

Murray, N.D. 1982. Ecology and evolution of the *Opuntia–Cactoblastis* ecosystem in Australia. *In* J.S.F. Baker and W.T. Starmer (eds.), *Ecological Genetics and Evolution: The Cactus-Yeast-Drosophila Model System*. Academic Press, Sydney.

Myers, N. 1979. *The Sinking Ark: A New Look at the Problem of Disappearing Species*. Pergamon Press, Oxford.

Nakajima A., Fuiiki K., N. Yasuada, and K. Kabasawa. 1980. Population genetics of eye diseases among the Japanese, pp. 409–430. *In* A.W. Erickson, H. Forsius, H.R. Nevarlinna, P.L. Workman, and R.K. Norris (eds.), *Population Structure and Genetic Disorders*. Academic Press, London.

Nakakura, K., N.M. Czekala, B.L. Lasley, and K. Benirschke. 1982. Fetal-maternal gradients of steroid hormones in the nine-banded armadillo (*Dasypus novemcinctus*). *Journal of Reproduction and Fertility* **66**(2):635–643.

Namkoong, G., J.H. Roberds, L.B. Nunnally, and H.A. Thomas. 1979. Isozyme variations in populations of southern pine beetles. *Forest Science* **25**:197–203.

National Park Service. 1967. Hearings before a subcommittee of the Committee on Appropriations United States Senate, Ninetieth Congress, first session on control of elk population, Yellowstone National Park. U.S. Government Printing Office, Washington, D.C.

National Park Service. 1980. State of the Parks 1980; a report to the Congress. Report prepared by Office of Science and Technology, National Park Service, U.S. Department of the Interior, Washington, D.C.

National Research Council. 1975. *Research in Zoos and Aquariums*. National Academy of Sciences, Washington, D.C.

Nei, M. 1972. Genetic distance between populations. *American Naturalist* **106**: 283–292.

Nei, M. 1973. Analysis of gene diversity in subdivided populations. *Proceedings of the National Academy of Sciences of the United States of America* **70**: 3321–3323.

Nei, M. 1975. *Molecular Population Genetics and Evolution*. American Elsevier, New York.

Nei, M., and A. Chakravarti. 1977. Drift variances of F_{st} and G_{st} statistics obtained from a finite number of isolated populations. *Theoretical Population Biology* **11**:307–325.

Nei, M., and W.H. Li. 1979. Mathematical model for studying genetic variation in terms of restriction endonucleases. *Proceedings of the National Academy of Sciences of the United States of America* **76**:5269–5273.

Nei, M., and T. Maruyama. 1975. Lewontin-Krakauer test for neutral genes. *Genetics* **80**:395.

Nei, M., and A. Roychoudhury. 1974. Sampling variances of heterozygosity and genetic distance. *Genetics* **76**:379.

Nei, M., T. Maruyama, and R. Chakraborty. 1975. The bottleneck effect and genetic variability in populations. *Evolution* **29**:1–10.

Nei, M., A. Chakravarti, and Y. Tateno. 1977. Mean and variance of F_{st} in a finite number of incompletely isolated populations. *Theoretical Population Biology* **11**:291–306.

Nelson, D.A. 1980. A mallard × mottled duck hybrid. *Wilson Bulletin* **92**:527–529.

Nelson, K., and D. Hedgecock. 1980. Enzyme polymorphism and adaptive strategy in the decapod Crustacea. *American Naturalist* **116**:238–280.

Nevo, E. 1976. Adaptive strategies of genetic systems in constant and varying environments. *In* S. Karlin and E. Nevo (eds.), *Population Genetics and Ecology*. Academic Press, New York.

Nevo, E. 1978. Genetic variation in natural populations: patterns and theory. *Theoretical Population Biology* **13**:121–177.

Nevo, E., T.S. Shimony, and M. Lilni. 1977. Thermal selection of allozymes in barnacles. *Nature* **267**:699–701.

Nevo, E., A.H.D. Brown, D. Zohary, H. Storch, and A. Beiles. 1981. Microgeographic edaphic differentiation in allozyme polymorphisms of wild barley (*Hordeum spontaneum*, Poaceae). *Plant Systematics and Evolution* **138**:287–292.

Newton, E. 1867. On the land birds of the Seychelles Archipelago. *Ibis* **3**:335–360.

Nietschmann, B. 1979. Ecological change, inflation, and migration in the far western Caribbean. *Geographical Review* **69**:1–24.

Niles, G.A. 1980. Breeding cotton for resistance to insect pests, pp. 337–369. *In* F.G. Maxwell and P.R. Jennings (eds.), *Breeding Plants Resistant to Insects*. Wiley and Sons, New York.

Ochman, H., and R.K. Selander. 1982. Genetic interaction between self-fertilizing strains of the land snail *Rumina*. In preparation.

Odum, E.P. 1971. *Fundamentals of Ecology*. 3rd ed. Saunders, Philadelphia.

Oliver, C.G. 1972. Genetic and phenotypic differentiation and geographic distance in four species of Lepidoptera. *Evolution* **26**:221–241.

Olney, P.J.S. (ed.). 1977. Section I. Breeding endangered species in captivity. *International Zoo Yearbook* **17**:1–122.

Olney, P.J.S. (ed.). 1980. Section I. Breeding endangered species in captivity. *International Zoo Yearbook* **20**:1–189.

O'Malley, D.M., F.W. Allendorf, and G.M. Blake. 1979. Inheritance of isozyme variation and heterozygosity in *Pinus ponderosa*. *Biochemical Genetics* **17**:233–250.

Østergaard, H., and G. Nielsen. 1981. Cultivar identification by means of isoenzymes. I. Genotypic survey of the Pgi-2 locus in tetraploid ryegrass. *Zeitschrift für Pflanzenzüchtung* **87**:121–132.

Owen-Smith, N. 1982. The white rhino overpopulation problem and a proposed solution, pp. 129–150. *In* P.A. Jewell and S. Holt (eds.), *Problems in Management of Locally Abundant Wild Animals*. Academic Press, New York.

Pace, A.E. 1974. Systematic and biological studies of the leopard frogs (*Rana pipiens* complex) of the United States. *Miscellaneous Publications. Museum of Zoology, University of Michigan* **148**:1–140.

Packer, C. 1979. Inter-troop transfer and inbreeding avoidance in *Papio anubis*. *Animal Behavior* **27**:1–36.

Panov, E.N. 1972. Interspecific hybridization and fate of hybrid populations as exemplified by two species of shrikes: *Lanius collurio* L., and *L. phoenicuroides*

Schalow. (In Russian with English summary.) *Zhurnal Obshchei Biologii* **33**(4): 409–426.

Parker, G.R. 1973. Distribution and densities of wolves within barren-ground caribou range in northern mainland Canada. *Journal of Mammalogy* **54**(2): 341–348.

Parsons, P.A. 1973. *Behavioural and Ecological Genetics: A Study in Drosophila.* Clarendon Press, Oxford.

Parsons, P.A. 1975. The comparative evolutionary biology of the sibling species, *Drosophila melanogaster* and *D. simulans. Quarterly Review of Biology* **50**: 151–169.

Partridge, G.G. 1980. The vitamin K requirements of wild brown rats (*Rattus norvegicus*) resistant to warfarin. *Comparative Biochemistry and Physiology* **66A**: 83–87.

Passmore, J. 1974. *Man's Responsibility for Nature.* Duckworth, London.

Pathak, M.D. 1970. Genetics of plants in pest management, pp. 137–157. In R.L. Rabb and R.E. Guthrie (eds.), *Concepts of Pest Management.* North Carolina State University Press, Raleigh, North Carolina.

Patton, J.L. 1972. Patterns for geographic variation in karyotype in the pocket gopher, *Thomomys bottae* (Eydoux and Gervais). *Evolution* **26**:574–586.

Patton, J.L., and S.Y. Yang. 1977. Genetic variation in *Thomomys bottae* pocket gophers: macrogeographic patterns. *Evolution* **31**:697–720.

Patton, J.L., R.J. Baker, and H.H. Genoways. 1980. Apparent chromosomal heterosis in a fossorial mammal. *American Naturalist* **116**:143–146.

Patton, J.L., S.Y. Yang, and P. Myers. 1975. Genetic and morphological divergence among introduced rat populations (*Rattus rattus*) of the Galapagos Archipelago, Ecuador. *Systematic Zoology* **25**:296–310.

Penny, M. 1974. *The Birds of Seychelles and Outlying Islands.* Taplinger Publishing Company, New York.

Person, C. 1966. Genetic polymorphism in parasitic systems. *Nature* **212**:266–267.

Petrides, G.A., F.B. Golley, and I.L. Brisbin. 1968. Energy flow and secondary productivity. In F.B. Golley and H.K. Buechner (eds.), *A Practical Guide to the Study of the Productivity of Large Herbivores.* International Biological Programme, Handbook No. 7. Blackwell Scientific Publications, Oxford, Edinburgh.

Phillips, M.A., and A.H.D. Brown. 1977. Mating system and hybridity in *Eucalyptus pauciflora. Australian Journal of Biological Sciences* **30**:337–344.

Phillips, P.A., and M.M. Barnes. 1975. Host race formation among sympatric apple, walnut, and plum populations of the codling moth, *Laspeyresia pomonella. Annals of the Entomological Society of America* **68**:1053–1060.

Pickett, S.T.A., and J.N. Thompson. 1978. Patch dynamics and the design of nature reserves. *Biological Conservation* **13**:27–37.

Pienaar, U. de. 1966. An aerial census of elephants and buffalo in the Kruger National Park. *Koedoe* **9**:40–107.

Pimentel, D. 1968. Population regulation and genetic feedback. *Science* **159**: 1432–1437.

Pinder, N.J., and J.P. Barkham. 1978. An assessment of the contribution of captive breeding to the conservation of rare mammals. *Biological Conservation* **13**: 187–245.

Platz, J.E. 1976. Biochemical and morphological variation of leopard frogs in Arizona. *Copeia* **1976**:660–672.
Platz, J.E., and A.L. Platz. 1973. *Rana pipiens* complex: hemoglobin phenotypes of sympatric and allopatric populations in Arizona. *Science* **179**:1334–1336.
Portères, R. 1955. Les céréales mineures du genre *Digitaria* en Afrique et Europe. *Journal d'Agriculture Tropical et de Botanique Appliquée* **2**:349–386, 477–510, 620–675.
Powell, J.R. 1975a. Protein variation in natural populations of animals. *Evolutionary Biology* **8**:79–119.
Powell, J.R. 1975b. Isozymes and non-Darwinian evolution: a re-evaluation, pp. 9–26. In C. Markert (ed.), *Isozymes IV*. Academic Press, New York.
Powell, J.R. 1979. Population genetics of *Drosophila* amylase. II. Geographic patterns in *Drosophila pseudoobscura*. *Genetics* **92**:613–622.
Powell, J.R., W.J. Tabachnick, and J. Arnold. 1980. Genetics and the origin of a vector population: *Aedes aegypti*; a case study. *Science* **298**:1385–1387.
Powell, J.R., W.J. Tabachnick, and G. Wallis. 1982. *Aedes aegypti* as a model of the usefulness of population genetics of vectors, pp. 396–413. In W.M.M. Steiner (ed.), *Recent Developments in the Genetics of Mosquito Disease Vectors*, Stipes.
Powell, J.R., and H. Wistrand. 1978. The effect of heterogeneous environments and a competitor on genetic variation in *Drosophila*. *American Naturalist* **112**:935–947.
Prakash, S. 1972. Origin of reproductive isolation in the absence of apparent genetic differentiation in a genetic isolate of *Drosophila pseudoobscura*. *Genetics* **72**:143–145.
Prakash, S., R. Lewontin, and J. Hubby, 1969. A molecular approach to the study of genic heterozygosity in natural populations. IV. Patterns of genic variation in central, marginal and isolated populations of *Drosophila pseudoobscura*. *Genetics* **61**:841–858.
Prescott-Allen, R., and C. Prescott-Allen. 1981. Wild plants and crop improvement. *World Conservation Strategy, Occasional Paper No. 1*. World Wildlife Fund, U.K.-Godalming, Surrey, England.
Price, M.V., and N.M. Waser. 1979. Pollen dispersal and optimal outcrossing in *Delphinium nelsonii*. *Nature* **277**:294–297.
Price, P.W. 1980. *Evolutionary Biology of Parasites*. Princeton University Press, Princeton, New Jersey.
Pritchard, P.C.H. 1979. "Head-starting" and other conservation techniques for marine turtles Cheloniidae and Dermochelyidae. *International Zoo Yearbook* **19**:38–42.
Pritchard, P.C.H. 1980. The conservation of sea turtles: practices and problems. *American Zoologist* **20**:609–617.
Prout, T. 1968. Sufficient conditions for multiple niche polymorphism. *American Naturalist* **102**:493–496.
Pyle, R.M. 1976. Conservation of Lepidoptera in the United States. *Biological Conservation* **9**:55–75.
Pyle, R.M., M. Bentzien, and P. Opler. 1981. Insect conservation. *Annual Review of Entomology* **26**:233–258.
Quick, H.E. 1960. British slugs (Pulmonata: Testacellaidae, Arionidae, Limacidae). *Bulletin of the British Museum (Natural History) Entomology* **6**:105–226.

Ralls, K., and J. Ballou. 1982a. Effects of inbreeding on juvenile mortality in some small mammal species. *Laboratory Animals* **16**:159–166.

Ralls, K., and J. Ballou. 1982b. Effects of inbreeding on infant mortality in captive primates. *International Journal of Primatology* **3**:491–505.

Ralls, K., K. Brugger, and J. Ballou. 1979. Inbreeding and juvenile mortality in small populations of ungulates. *Science* **206**:1101–1103.

Ralls, K., K. Brugger, and A. Glick. 1980. Deleterious effects of inbreeding in a herd of captive Dorcas gazelle. *International Zoo Yearbook* **20**:138–146.

Ramshaw, R.A.M., J.A. Coyne, and R.C. Lewontin. 1979. The sensitivity of gel electrophoresis as a detector of genetic variation. *Genetics* **93**:1019–1037.

Rao, P.S.S., and S.G. Inbaraj. 1980. Inbreeding effects on fetal growth and development. *Journal of Medical Genetics* **17**:27–33.

Rausher, M.D. 1982. Population differentiation in *Euphydryas editha* butterflies: larval adaptation to different hosts. *Evolution* **36**(3):581–590.

Read, B., and R.J. Fruch. 1980. Management and breeding of Speke's gazelle (*Gazella spekei*) at the St. Louis Zoo with a note on artificial insemination. *International Zoo Yearbook* **20**:99–104.

Rees, H., and J.B. Thompson. 1956. Genotypic control of chromosome behavior in rye. III. Chiasma frequency in homozygotes and heterozygotes. *Heredity* **3**:409–424.

Rick, C.M., and J.F. Fobes. 1975. Allozyme variation in the cultivated tomato and closely related species. *Bulletin of the Torrey Botanical Club* **102**:376–384.

Rick, C.M., and S.D. Tanksley. 1981. Genetic variation in *Solanum pennellii*: comparisons with two other sympatric tomato species. *Plant Systematics and Evolution* **139**:11–45.

Rick, C.M., J.F. Fobes, and M. Holle. 1977. Genetic variation in *Lycopersicon pimpinellifolium*: evidence of evolutionary change in mating systems. *Plant Systematics and Evolution* **127**:139–170.

Ricklefs, R.E. 1973. *Ecology*. Chiron Press, Newton, Massachusetts.

Ripley, D. 1957. *A Paddling of Ducks*. Harcourt Brace, New York.

Risser, P.G. and K.D. Cornelison. 1979. *Man and the Biosphere; U.S. Information Synthesis Project, MAB-8 Biosphere Reserves*. Oklahoma Biological Survey. Norman, Oklahoma.

Ritzema-Bos, J. 1894. Untersuchungen uber die folgen der Zucht in engster Blutverwandschaft. *Biologisches Centralblatt* **14**:75–81.

Roberts, D.F. 1980. Genetic structures and the pathology of an isolate population, pp. 7–26. In A.W. Eriksson, H. Forsius, H.R. Nevanlinna, P.L. Workman, and R.K. Norris (eds.), *Population Structure and Genetic Disorders*. Academic Press, London.

Roberts, D.R., W.D. Alecrim, J.M. Heller, S.R. Ehrhardt, and J.B. Lima. 1982. Male *Eufriesia purpurata* a DDT collecting euglossine bee in Brazil. *Nature* **297**:62–63.

Robertson, A. 1975. Remarks on the Lewontin-Krakauer test. *Genetics* **80**:396.

Rocovich, S.E., and D.A. West. 1975. Arsenic tolerance in a population of the grass *Andropogon scoparius* Michx. *Science* **188**:263–264.

Rodman, J.E., and F.S. Chew. 1980. Phytochemical correlates of herbivory in a community of native and naturalized Cruciferae. *Biochemical Systematics and Ecology* **8**:43–50.

Rogers, J.S. 1972. Measure of genetic similarity and genetic distance. *Studies on Genetics. VII. University of Texas Publication No. 7213*.

Romer, A. 1949. Time series and trends in animal evolution, pp. 103–120. *In* G.L. Jepson, E. Mayr, and G.G. Simpson (eds.), *Genetics, Palaeontology and Evolution*. Princeton University Press, Princeton, New Jersey.
Roose, M.L., and L.D. Gottlieb. 1976. Genetic and biochemical consequences of polyploidy in *Tragopogon*. *Evolution* **30**:818–830.
Rosenthal, G.A., and D.H. Janzen (eds.). 1979. *Herbivores: Their Interaction with Secondary Plant Metabolites*. Academic Press, New York.
Rosenzweig, M.L. 1973. Evolution of the predator isocline. *Evolution* **27**:84–94.
Roughgarden, J. 1971. Density-dependent natural selection. *Ecology* **52**:453–474.
Roughgarden, J. 1979. *Theory of Population Genetics and Evolutionary Ecology: An Introduction*. Macmillan, New York.
Roughgarden, J. 1983. The theory of coevolution. *In* D.J. Futuyma and M. Slatkin (eds.), *Coevolution*. Sinauer Associates, Sunderland, Massachusetts.
Rudin, D., G. Eriksson, I. Ekberg, and M. Rasmuson. 1974. Studies of allele frequencies and inbreeding in Scots pine populations by the aid of the isozyme technique. *Silvae Genetica* **23**:10–13.
Runham, N.W., and P.J. Hunter. 1970. *Terrestrial Slugs*. Hutchinson, London.
Rupp, R.S., and M.A. Redmond. 1966. Transfer studies of ecologic and genetic variation in the American smelt. *Ecology* **47**:253–259.
Ryder, O., and E. Wedemeyer. 1982. A cooperative breeding program for the Mongolian wild horse *Equus przewalskii* in the United States. *Biological Conservation* **22**:259–272.
Ryder, O., P. Brisban, A. Bowling, and E. Wedemeyer. 1981. Monitoring genetic variation in endangered species, pp. 417–424. *In* G. Scudder and J. Reveal (eds.), *Proceedings of the Second International Congress of Systematics and Evolutionary Biology*. Hunt Institute for Botanical Documentation, Pittsburgh, Pennsylvania.
Ryman, N., and G. Ståhl. 1981. Genetic perspectives of the identification and conservation of Scandinavian stocks of fish. *Canadian Journal of Fisheries and Aquatic Sciences* **38**:1562–1575.
Ryman, N., F.W. Allendorf, and G. Ståhl. 1979. Reproductive isolation with little genetic divergence in sympatric populations of brown trout (*Salmo trutta*). *Genetics* **92**:247–262.
Ryman, N., C. Reuterwall, K. Nygren, and T. Nygren. 1980. Genetic variation and differentiation in Scandinavian moose (*Alces alces*): are large mammals monomorphic? *Evolution* **34**:1037–1049.
Ryman, N., R. Baccus, C. Reuterwall, and M. Smith. 1981. Effective population size, genetic interval, and potential loss of genetic variability in game species under various hunting regimes. *Oikos* **26**:257–266.
Salthe, S.N. 1969. Geographic variation in the lactate dehydrogenases of *Rana pipiens* and *Rana palustris*. *Biochemical Genetics* **2**:271–303.
Salthe, S.N., and J.S. Mecham. 1974. Reproductive and courtship patterns. *In* B. Lofts (ed.), *Physiology of the Amphibia*, Vol. 2. Academic Press, New York.
Sanders, T.B., J.L. Hamrick, and L.R. Holden. 1979. Allozyme variation in *Elymus canadensis* from the tallgrass prairie region: geographic variation. *American Midland Naturalist* **101**:1–12.
Schaal, B.A. 1975. Population structure and local differentiation in *Liatris cylindracea*. *American Naturalist* **109**:511–528.
Schaal, B.A., and D.A. Levin. 1978. Morphological differentiation and neighborhood size in *Liatris cylindracea*. *American Journal of Botany* **65**:923–928.

Schaal, B.A., and W.G. Smith. 1980. The apportionment of genetic variation within and among populations of *Desmodium nudiflorum*. *Evolution* **34**:214-221.

Schaffer, W.M., and M.L. Rosenzweig. 1978. Homage to the Red Queen. I. Coevolution of predators and their victims. *Theoretical Population Biology* **14**:135-157.

Schaller, G. 1972. *The Serengeti Lion*. University of Chicago Press, Chicago.

Schepper, G.G. de, and H.F. deFrance. 1979. Chromosome analysis of the Mongolian domestic horse: A preliminary study, pp. 85-86. *In* L.E.M. de Boer and J.E. Bowman (eds.), *Genetics and Hereditary Diseases of the Przewalski Horse. Proceedings of the Arnhem Conference, October 1978.* Foundation for the Preservation of the Przewalski Horse, Rotterdam.

Schnell, G.D., and R.K. Selander. 1981. Environmental and morphological correlates of gentic variation in mammals, pp. 60-99. *In* N.H. Smith and J. Jould (eds.), *Mammalian Population Genetics*. University of Georgia Press, Athens, Georgia.

Schoen, D.J. 1982a. Genetic variation and the breeding system of *Gilia achilleifolia*. *Evolution* **36**:361-370.

Schoen, D.J. 1982b. The breeding system of *Gilia achilleifolia*: variation in floral characteristics and outcrossing rates. *Evolution* **36**:352-360.

Schoen, D.J. 1983. Relative fitness of selfed and outcrossed progeny in *Gilia achilleifolia* (Polemoniaceae). *Evolution* **37**(2):292-302.

Schonewald-Cox, C., J.W. Bayless, and J. Schonewald. *Cervus elaphus roosevelti* and *C.e. nannodes*; a cranial morphometric analysis. In preparation.

Schwaegerle, K.E., and B.A. Schaal. 1979. Genetic variability and founder effect in the pitcher plant *Sarracenia purpurea* L. Evolution **33**:1210-1218.

Schwartz, O.A., and K.B. Armitage. 1980. Genetic variation in social mammals: the marmot model. *Science* **207**:665-667.

Scogin, R. 1969. Isoenzyme polymorphism in natural populations of the genus *Baptisia* (Leguminosae). *Phytochemistry* **8**:1733-1737.

Seal, U.S. 1978. The Noah's Ark problem: multigeneration management of wild species in captivity, pp. 303-319. *In* S.A. Temple (ed.), *Endangered Birds: Management Techniques for Preserving Threatened Species*. University of Wisconsin Press, Madison, Wisconsin.

Seal, U.S., and N. Flesness. 1979. Noah's Ark—sex and survival, pp. 214-228. *In AAZPA 1978 Annual Conference*. American Association of Zoological Parks and Aquariums, Wheeling, West Virginia.

Seal, U.S., and T. Foose. 1983. Genetics and demography of Siberian tigers in North America with evidence for inbreeding depression. *Zoo Biology*. In press.

Seidel, M.E., and R.V. Lucchino. 1981. Allozymic and morphological variation among musk turtles *Sternotherus carinatus*, *S. depressus*, and *S. minor*. *Copeia* **1981**:119-128.

Selander, R.K. 1970. Behavior and genetic variation in natural populations. *American Zoologist* **10**:53-66.

Selander, R.K. 1976. Genic variation in natural populations, pp. 32-45. *In* F.J. Ayala (ed.), *Molecular Evolution*. Sinauer Associates, Sunderland, Massachusetts.

Selander, R.K., and R.O. Hudson. 1976. Animal population structure under close inbreeding: the land snail *Rumina* in southern France. *American Naturalist* **110**:695-718.

Selander, R.K., and D.W. Kaufman. 1973. Genetic variability and strategies of adap-

tation in animals. *Proceedings of the National Academy of Sciences of the United States of America* **70**:1875-1877.
Selander, R.K., and H. Ochman. 1983. The genetic structure of populations as illustrated by molluscs. In M.C. Rattazzi, J.G. Scandalios, and G.S. Whitt (eds.), *Isozymes: Current Topics in Biological and Medical Research*. Volume 10. Alan R. Liss, New York. In press.
Selander, R.K., S.Y. Yang, R.C. Lewontin, and W.E. Johnson. 1970. Genetic variation in the horseshoe crab (*Limulus polyphemus*), a phylogenetic relic. *Evolution* **24**:402-414.
Sene, F.M., and H.L. Carson. 1977a. Genetic variation in Hawaiian *Drosophila pseudoobscura*. *Genetics* **84**:609-629.
Sene, F.M., and H.L. Carson. 1977b. Genetic variation in Hawaiian *Drosophila*. IV. Allozymic similarity between *D. silvestis* and *D. heteroneura* from the island of Hawaii. *Genetics* **86**:187-198.
Senner, J.W. 1980. Inbreeding depression and the survival of zoo populations, pp. 209-224. In M.E. Soulé and B.A. Wilcox (eds.), *Conservation Biology; An Evolutionary-Ecological Perspective*. Sinauer Associates, Sunderland, Massachusetts.
Seuanez, H.N., H.J. Evans, D.E. Martin, and J. Fletcher. 1979. An inversion of chromosome 2 that distinguishes between Bornean and Sumatran orangutans. *Cytogenetics and Cell Genetics* **23**:137-140.
Shabica, S.V. 1982. Planning for protection of sea turtle habitat, pp. 513-518. In K. Bjorndal (ed.), *The Biology and conservation of Sea Turtles*. Smithsonian Institution Press, Washington, D.C.
Shank, D.B., and M.W. Adams. 1960. Environmental variability within inbred lines and single crosses of maize. *Journal of Genetics* **57**:119-126.
Shapiro, A.M. 1978. Weather and the lability of breeding populations of the checkered white butterfly, *Pieris protodice* Boisduval and LeConte. *Journal of Research on the Lepidoptera* **17**:1-23.
Shelford, V.E. 1943. The abundance of the collared lemming (*Dicrostonyx groenlandicus* (Tr.) var. *richardsoni* Mer.) in the Churchill area, 1929 to 1940. *Ecology* **24**:472-484.
Shelford, V.E. 1945. The relation of snowy owl migration to the abundance of the collard lemming. *Auk* **62**:592-596.
Shields, W.M. 1982a. Inbreeding and the paradox of sex: a resolution? *Evolutionary Theory* **5**:245-279.
Shields, W.M. 1982b. Optimal inbreeding and the evolution of philopatry. In I.R. Swingland and P.J. Greenwood (eds.), *The Ecology of Animal Movement*. Oxford University Press, Oxford.
Shields, W.M. 1982. *Philopatry, Inbreeding, and the Evolution of Sex*. State University of New York Press, Albany.
Short, L.L. 1963. Hybridization in the wood warblers *Vermivora pinus* and *V. chrysaetos*, pp. 147-160. *Proceedings of the 13th International Ornithological Congress*.
Short, L.L. 1969. Taxonomic aspects of avian hybridization. *Auk* **86**:84-105.
Short, L.L. 1972. Hybridization, taxonomy and avian evolution. *Annals of the Missouri Botanical Garden* **59**:447-453.
Shultz, F.T. 1952. Concurrent inbreeding and selection in the domestic fowl. *Heredity* **7**:1-21.

Schultz, F.T., and W.E. Briles. 1953. The adaptive value of blood group genes in chickens. *Genetics* **38**:34-50.

Shumaker, K.M., and G.R. Babbell. 1980. Patterns of allozyme similarity in ecologically central and marginal populations of *Hordeum jubatum* in Utah. *Evolution* **34**:110-116.

Shumaker, K.M., R.W. Allard, and A.L. Kahler. 1982. Cryptic variability at enzyme loci in three plant species, *Avena barbata, Hordeum vulgare* and *Zea mays*. *Journal of Heredity* **73**:86-90.

Sibley, C.G. 1957. The evolutionary and taxonomic significance of sexual dimorphism and hybridization in birds. *Condor* **59**:166-191.

Siegel, B.Z., and S.M. Siegel. 1982. Mercury content of *Equisetum* plants around Mt. St. Helens one year after the major eruption. *Science* **215**:292-293.

Silander, J.A. 1979. Microevolution and clone structure in *Spartina patens*. *Science* **203**:658-660.

Silvers, W.K. 1979. *The Coat Colors of Mice: A Model for Mammalian Gene Action and Interaction*. Springer-Verlag, New York.

Simberloff, D., and L.G. Abele. 1982. Refuge design and island biographic theory: effects of fragmentation. *American Naturalist* 120:41-50.

Simmons, J.B., R.I. Beyer, P.E. Brandham, G. Lucas, and V.T.H. Parry. 1976. *Conservation of Threatened Plants*. NATO Conference Series I. Ecology, Vol. 1. Plenum Press, New York.

Simpson, G. G. 1980. *Splendid isolation*. New Haven: Yale University Press.

Singer, M.C. 1972. Complex components of habitat suitability within a butterfly colony. *Science* **176**:75-77.

Singer, M.C., and P.R. Ehrlich. 1979. Population dynamics of the checkerspot butterfly *Euphydryas editha*. *Fortschritte Zoologie* **25**:53-60.

Singh, R., R. Lewontin, and A. Felton. 1976. Genetic heterogeneity within electrophoretic "alleles" of xanthine dehydrogenase in *Drosophila pseudoobscura*. *Genetics* **84**:609-629.

Singh, S.M., and E. Zouros. 1978. Genetic variation associated with growth rate in the American oyster (*Crassostrea virginica*). *Evolution* **32**:343-353.

Sites, J.W., Jr. 1983a. Chromosome evolution in the iguanid lizard *Sceloporus grammicus*. I. Chromosome polymorphisms. *Evolution*. In press.

Sites, J.W., Jr. 1983b. Morphological variation within and among three chromosome races of *Sceloporus grammicus* (Sauria, Iguanidae) in the northcentral part of its range. *Copeia*. In press.

Sites, J.W., Jr., and J.R. Dixon. 1981. A new subspecies of iguanid lizard, *Sceloporus grammicus*, from northeastern Mexico, with comments on its evolutionary implications and the status of *S.g. disparilis*. *Journal of Herpetology* **15**:59-69.

Sites, J.W., Jr., and I.F. Greenbaum. 1983. Chromosome evolution in the iguanid lizard *Sceloporus grammicus*. II. Allozyme variation. *Evolution*. In press.

Sittman, K., H. Abplanalp, and R.A. Fraser. 1966. Inbreeding depression in Japanese quail. *Genetics* **54**:371-379.

Skibinski, D.O.F., and R.D. Ward. 1982. Correlation between heterozygosity and evolutionary rate of proteins. *Nature* **289**:490-492.

Slatis, H.M. 1960. An analysis of inbreeding in the European bison. *Genetics* **45**:275-287.

Slatkin, M. 1980. The distribution of mutant alleles in a subdivided population. *Genetics* **98**:503-523.

Slatkin, M. 1981. Estimating levels of gene flow in natural populations. *Genetics* **95**:323–335.

Slatkin, M. 1982. Testing neutrality in subdivided populations. *Genetics* **100**: 533–545.

Smith, M.F., and J.L. Patton. 1980. Relationships of pocket gopher (*Thomomys bottae*) populations of the lower Colorado River. *Journal of Mammalogy* **61**: 681–696.

Smith, M.H., C.T. Garten, and P.E. Ramsay. 1975. Genetic heterozygosity and population dynamics in small mammals, pp. 85–102. *In* C.L. Markert (ed.), *Isozymes, Genetics and Evolution*. Academic Press, New York.

Smith, M.H., H.O. Hilstad, M.N. Manlove, D.O. Straney, and J.M. Dean. 1978a. Management implications of genetic variability in loggerhead and green sea turtles, pp. 302–312. *In Proceedings of the 13th Congress of Game Biologists*.

Smith, M.H., M.N. Manlove, and J. Joule. 1978b. Spatial and temporal dynamics of the genetic organization of small mammal populations, pp. 99–113. *In* D.P. Snyder (ed.), Small Mammals under Natural Conditions. *Special Publication Series Pymatuning Laboratory of Ecology*. Vol. 5. Pittsburgh Press, Pittsburg, Pennsylvania.

Smith, M.H., M.W. Smith, S.L. Scott, E.H. Liu, and J.C. Jones. 1983. Rapid evolution in a post thermal environment. *Copeia*. **1983**(1):193–197.

Smith, M.W., M.H. Smith, and R.K. Chesser. 1983. Biochemical genetics of mosquitofish I. Environmental correlates, and temporal and spatial heterogeneity of allele frequencies within a river drainage. *Copeia*. In press.

Smith, R.H. 1979. On selection for inbreeding in polygynous animals. *Heredity* **43**: 205–211.

Snaydon, R.W., and M.S. Davies. 1972. Rapid population differentiation in a mosaic environment. II. Morphological variation in *Anthoxanthum odoratum*. *Evolution* **26**:390–405.

Snaydon, R.W., and T.M. Davies. 1982. Rapid divergence of plant populations in response to recent changes in soil and conditions. *Evolution* **36**:289–297.

Sokal, R.R., and T.J. Crovello. 1970. The biological species concept: a critical evaluation. *American Naturalist* **104**:127–153.

Solbrig, O.T. 1976. On the relative advantages of cross- and self-fertilization. *Annals of the Missouri Botanical Garden* **63**:262–276.

Somero, G.N., and M. Soulé. 1974. Genetic variations in fishes as a test of the niche variation hypothesis. *Nature* **249**:670–672.

Sosa, O., Jr. 1981. Biotypes J and L of the Hessian fly discovered in an Indiana wheat field. *Journal of Economic Entomology* **74**:180–182.

Soulé, M.E. 1967. Phenetics of natural populations. II. Asymmetry and evolution in a lizard. *American Naturalist* **101**:141–160.

Soulé, M. 1976. Allozyme variation: its determinants in time and space, pp. 60–77. *In* F.J. Ayala (ed.), *Molecular Evolution*. Sinauer Associates, Sunderland, Massachusetts.

Soulé, M.E. 1979. Heterozygosity and developmental stability: another look. *Evolution* **33**:396–401.

Soulé, M. 1980. Thresholds for survival: maintaining fitness and evolutionary potential, pp. 151–169. *In* M.E. Soulé and B.A. Wilcox (eds.), *Conservation Biology: An Evolutionary–Ecological Perspective*. Sinauer Associates, Sunderland, Massachusetts.

Soulé, M.E., and B.A. Wilcox (eds.). 1980. *Conservation Biology: An Evolutionary–*

Ecological Perspective. Sinauer Associates, Sunderland, Massachusetts.
Soulé, M. and S.Y. Yang. 1973. Genetic variation inside blotched lizards on islands in the Gulf of California. *Evolution* **27**:593-600.
Soulé, M.E., B.A. Wilcox, and C. Holtby. 1979. Benign neglect: a model of faunal collapse in the game reserves of East Africa. *Biological Conservation* **15**: 259-272.
Soulie, J., and J. de Grouchy. 1981. A cytogenetic survey of 110 baboons (*Papio cynocephalus*). *American Journal of Anthropology* **56**:107-113.
Spieth, P.T. 1974. Gene flow and genetic differentiation. *Genetics* **78**:961-965.
Squillace, A.E. 1976. Analysis of monoterpenes of conifers by gas-liquid chromatography, pp. 120-157. In J.P. Mikesche (ed.), *Modern Methods in Forest Genetics*. Springer-Verlag, New York.
Stace, C.A. 1980. *Plant Taxonomy and Biosystematics*. University Park Press, Baltimore, Maryland.
Staiger, H. 1957. Genetical and morphological variation in *Purpura lapillus* with respect to local and regional differentiation of population groups. *Annals of Applied Biology* **33**:251-258.
Stebbins, G.L. 1957. Self-fertilization and population variability in the higher plants. *American Naturalist* **91**:337-354.
Stebbins, G.L. 1959. The role of hybridization in evolution. *Proceedings of the American Philosophical Society* **103**:231-251.
Stebbins, G.L. 1950. *Variation and Evolution in Plants*. Columbia University Press, New York.
Stebbins, G.L. 1971. *Chromosomal Evolution in Higher Plants*. Edward Arnold, London.
Stebbins, G.L. 1977. *Processes of Organic Evolution*, 3rd ed. Prentice-Hall, Englewood Cliffs, New Jersey.
Stebbins, R.C. 1966. *A Field Guide to Western Reptiles and Amphibians*. Houghton Mifflin Co., Boston, Massachusetts.
Steele, F., D. Langdon, and J. McKenzie. 1981. The current levels of inbreeding in Australian and New Zealand populations of the giraffe *Giraffa camelopardalis* with special reference to the Melbourne herd. *Bulletin of Zoo Management* **20**: 23-28.
Stine, G.J. 1977. *Biosocial Genetics*. Macmillian, New York.
Stoneburner, D.L., M.N. Nicora, and E.R. Blood. 1980. Heavy metals in loggerhead sea turtle eggs (*Caretta caretta*): evidence to support the hypothesis that demes exist in the western Atlantic population. *Journal of Herpetology* **14**(2): 171-175.
Stover, J., J. Evans, and E.P. Dolensek. 1979. Interspecies embryo transfer from the gaur to domestic Holstein, pp. 122-124. In M.E. Fowler (ed.), *Proceedings of American Zoo Veterinarians*. Seattle, Washington.
Strong, L.C. 1978. Inbred mice in science, pp. 45-67. In H.C. Morse, III (ed.), *Origins of Inbred Mice*. Academic Press, New York.
Stuber, C.W., M.M. Goodman, and F.M. Johnson. 1977. Genetic control and racial variation of c-glucosidase isozymes in maize (*Zea mays* L.). *Biochemical Genetics* **15**:383-394.
Sturgeon, K.B. 1979. Monoterpene variation in ponderosa pine xylem resin related to western pine beetle predation. *Evolution* **33**:803-814.
Summer, F.B. 1924. The stability of subspecific characters under changed conditions of environment. *American Naturalist* **58**:481-505.

Svejgaard, A., M. Hauge, C. Jelsid, P. Platt, L.P. Ryder, L. Staub-Nielsen, and M. Thomseni. 1979. *The HLA System*, 2nd ed. Karger, Basel.

Synge, H., and H. Townsend. 1979. *Survival or Extinction*. The Bantham-Moxon Trust, Royal Botanic Gardens, Kew, England.

Tabashnik, B.E. 1983. Host range evolution: the shift from native legume hosts to alfalfa by the butterfly *Colias philodice eriphyle*. *Evolution*. In press.

Tauber, C.A., and M.J. Tauber. 1977a. Sympatric speciation based on allelic changes at three loci: evidence from natural population in two habitats. *Science* **197**:1298–1299.

Tauber, C.A., and M.J. Tauber. 1977b. A genetic model for sympatric speciation through habitat diversification and seasonal isolation. *Nature* **268**:702–705.

Tauber, C.A., M.J. Tauber, and J.R. Tauber. 1977. Two genes control seasonal isolation in sibling species. *Science* **197**:592–593.

Tavormina, S.J. 1982. Sympatric genetic divergence in the leaf-mining insect *Liriomyza brassicae* (Diptera: Agromyzidae). *Evolution* **36**:523–534.

Taylor, R.H. 1975. Some ideas on speciation in New Zealand parakeets. *Notornis* **22**:110–121.

Templeton, A.R. 1979. The unit of selection in *Drosophila mercatorum*. II. Genetic revolutions and the origin of coadapted genomes in parthenogenetic strains. *Genetics* **92**:1265–1282.

Templeton, A.R. 1980. The theory of speciation via the founder principle. *Genetics* **94**:1011–1038.

Templeton, A.R. 1982. Adaptation and the integration of evolutionary forces, pp. 15–31. *In* R. Milkman (ed.), *Perspectives on Evolution*. Sinauer Associates, Sunderland, Massachusetts.

Templeton, A.R., and B. Reed. A breeding program to eliminate inbreeding depression in a captive herd of Speke's gazelle (*Gazella spekei*). Submitted to *Zoo Biology*.

Terborgh, J.W. 1974. Preservation of natural diversity: The problem of extinction prone species. *BioScience* **24**:715–722.

Terborgh, J. 1975. Faunal equilibria and the design of wildlife preserves, pp. 369–380. *In* F. Golley and E. Medina (eds.), *Tropical Ecological Systems: Trends in Terrestrial and Aquatic Research*. Springer-Verlag, New York.

Terborgh, J.W. 1976. Island biogeography and conservation: Strategy and limitations. *Science* **193**:29–30.

Terborgh, J.W. and B. Winter. 1980. Some causes of extinction, pp. 119–134. *In* M.E. Soulé and B.A. Wilcox (eds.), *Conservation Biology: An Evolutionary-Ecological Perspective*. Sinauer Associates, Sunderland, Massachusetts.

Thoday, J.M. 1953. Components of fitness. *Symposia of the Society for Experimental Biology* **7**:96–113.

Thomas, J.A. 1980a. Why did the large blue become extinct in Britain? *Oryx* **15**:243–247.

Thomas, J.A. 1980b. The extinction of the large blue and the conservation of the black hairstreak butterflies (a contrast of failure and success), pp. 19–23. *In Annual Report 1979*. Institute for Terrestrial Ecology.

Thornton, I.W.B. 1978. White tiger genetics—further evidence. *Journal of Zoology* **185**:389–394.

Timm, R.M., and R.D. Price. 1980. The taxonomy of *Geomydoecus* (Mallophaga: Trichodectidae) from the *Geomys bursarius* complex (Rodentia: Geomyidae). *Journal of Medical Entomology* **17**:126–145.

Timofeeff-Ressovsky, N.W. 1935. Uber geographische Temperaturrassen bei *Drosophila funebris*. *Archiv fuer Naturschutz und Landschaftsforschung* **4**: 245–257.

Tomlinson, J. 1966. The advantage of hermaphroditism and parthenogenesis. *Journal of Theoretical Biology* **11**:54–58.

Torres, A.M., and U. Diedenhofen. 1979. Baker sunflower populations revisited. *Journal of Heredity* **70**:275–276.

Torres, A.M., U. Diedenhofen, and I.M. Johnstone. 1977. The early allele of alcohol dehydrogenase in sunflower populations. *Journal of Heredity* **68**:11–16.

Treus, V.D., and N.V. Lobanov. 1971. Acclimatization and domestication of the eland at Askenya-Nova Zoo. *International Zoo Yearbook* **11**:147–156.

Tucker, P.K., and D.J. Schmidly. 1981. Studies of a contact zone among three chromosomal races of *Geomys bursarius* in eastern Texas. *Journal of Mammalogy* **62**:258–272.

Turkington, R., and J.L. Harper. 1979. The growth, distribution and neighbor relationships of *Trifolium repens* in a permanent pasture. IV. Fine scale biotic differentiation. *Journal of Ecology* **67**:245–254.

Turner, M.E., L.C. Stephens, and W.W. Anderson. 1982. Homozygosity and patch structure in plant populations as a result of nearest-neighbor pollination. *Proceedings of the National Academy of Sciences of the United States of America* **79**:203–207.

Udvardy, M.D.F. 1975. *A Classification of the Biogeographical Provinces of the World.* IUCN Occasional Paper No. 18. International Union for the Conservation of Nature and Natural Resources. Morges, Switzerland.

UNESCO. 1979. *Programme on Man and the Biosphere (MAB). MAB Information System: Biosphere Reserves.* UNESCO, Paris.

Urquhart, C. 1971. Genetics of lead tolerance in *Festuca ovina*. *Heredity* **26**:19–33.

Usberti, J.A., and S.K. Jain. 1978. Variation in *Panicum maximum*: a comparison of sexual and asexual populations. *Botanical Gazette* **139**:112–116.

U.S. Fish and Wildlife Service. 1981. Amendments to appendices of the Convention on International Trade in Endangered Species of Wild Fauna and Flora (CITES). *Federal Register* **46**(172):44660–44674.

Valentine, J.W., and F.J. Ayala. 1976. Genetic variation in krill. *Proceedings of the National Academy of Sciences of the United States of America* **73**:658–660.

Valizadeh, M. 1977. Esterase and acid phosphatase polymorphism in the fig tree (*Ficus carica* L.). *Biochemical Genetics* **15**:1037–1048.

van der Plank, J.E. 1963. *Plant Diseases: Epidemics and Control.* Academic Press, New York.

van der Plank, J.E. 1968. *Disease Resistance in Plants.* Academic Press, New York.

Van Valen, L. 1973. A new evolutionary law. *Evolutionary Theory* **1**:1–30.

Varland, K.L., A.L. Lovaas, and R.B. Dahlgren. 1978. *Herd Organization and Movements of Elk in Wind Cave National Park, South Dakota.* Natural Resources Report No. 13. National Park Service, Department of the Interior. U.S. Government Printing Office, Washington, D.C.

Verkleij, J.A.C., and J.W.J. Koniuszek. 1981. Genetic variability in *Chamaenerion angustifolium* (L.) Scop. (Onagraceae) occurring on contrasting soils. *Genetics* **55**:151–159.

Via, S. 1982. Genetic covariance of female host plant preference and larval performance in an insect herbivore. Address delivered to Society for the Study of Evolution, Stony Brook, New York.

Vogel, H., and A.G. Motulsky. 1979. *Human Genetics.* Springer-Verlag, Heidelberg.
Wade, M.J. 1978. A critical review of the models of group selection. *Quarterly Review of Biology* **53**:101–114.
Wade, N. 1978. Bird lovers and bureaucrats at loggerheads over peregrine falcons. *Science* **199**:1053–1055.
Wahl, I., N. Eshed, A. Segal, and Z. Sobel. 1978. Significance of wild relatives of small grains and other wild grasses in cereal powdery mildews, pp. 83–100. *In* D.M. Spencer (ed.), *The Powdery Mildews.* Academic Press, London.
Wall, J.R., and S.W. Wall. 1975. Isozyme polymorphisms in the study of evolution in the *Phaseolus vulgaris-P. coccineus* complex of Mexico, pp. 287–305. *In* C.L. Markert (ed.), *Isozymes IV.* Academic Press, New York.
Wallace, B. 1981. *Basic Population Genetics.* Columbia University Press, New York.
Wallace, R. 1923. *Farm Livestock of Great Britain*, 5th ed. Oliver and Boyd, Edinburgh.
Walley, K.A., M.S. Khan, and A.D. Bradshaw. 1974. The potential for evolution of heavy metal tolerance in plants. I. Copper and zinc tolerance in *Agrostis tenuis. Heredity* **32**:309–319.
Wallis, G.P. 1981. Biochemical genetics of gobies. Ph.D. dissertation, University of Wales.
Wasserman, S.S., and D.J. Futuyma. 1981. Evolution of host plant utilization in laboratory populations of the southern cowpea weevil, *Callosobruchus maculatus* Fabricius (Coleoptera: Bruchidae). *Evolution* **35**:605–617.
Watson, J. 1970. *The Molecular Biology of the Gene*, 2nd ed. W.A. Benjamin, New York.
Watt, W.B. 1977. Adaptation at specific loci. I. Natural selection on phosphoglucose isomerase of *Colias* butterflies: biochemical and population aspects. *Genetics* **87**:177–194.
Wetzel, R.M., R.E. Dubois, R.E. Martin, and P. Myers. 1975. *Catagonus*, and "extinct" peccary alive in Paraguay. *Science* **189**:379–381.
Whalen, M.D. 1979. Allozyme variation and evolution in *Solanum* section *Androceras. Systematic Botany* **4**:203–222.
White, M.J.D. 1973. *Animal Cytology and Evolution*, 3rd. ed. Cambridge University Press, Cambridge, Massachusetts.
White, M.J.D. 1978a. *Modes of Speciation.* W.H. Freeman, San Francisco, California.
White, M.J.D. 1978b. Chain processes in chromosomal speciation. *Systematic Zoology* **27**:285–293.
White, R.R., and M.C. Singer. 1974. Geographical distribution of hostplant choice in *Euphydryas editha* (Nymphalidae). *Journal of the Lepidopterists' Society* **28**:103–107.
Whitehead, G.K. 1980. Captive breeding as a practical aid to preventing extinction and providing animals for reintroduction. *Deer* **5**:7–13.
Whitemore, T.C. 1973. Frequency and habitat of tree species in the rainforest of Ulu Kelantan. *Gardens' Bulletin (Straits Settlement)* **26**(2):195–210.
Wiener, G., and S. Hayter. 1974. Crossbreeding and inbreeding in sheep, pp. 19–26. *In Report of the Animal Breeding Research Organization.*
Wiener, G., and J.A. Woolliams. 1980. The effects of crossbreeding and inbreeding on the performance of three breeds of hill sheep in Scotland. *Proceedings World Congress on Sheep and Beef Cattle Breeding*, Massey University, New Zealand.

Wiens, J.A. 1976. Population responses to patchy environments. *Annual Review of Ecology and Systematics* **7**:81-120.
Wiens, J.A. 1982. Forum: avian subspecies in the 1980's. *Auk* **99**:593.
Wilcox, B.A. 1980. Insular ecology and conservation, pp. 95-117. *In* M.E. Soulé and B.A. Wilcox (eds.), *Conservation Biology; an Evolutionary-Ecological Perspective*. Sinauer Associates, Inc. Sunderland, Massachusetts.
Williams, G.C. 1966. *Adaptation and Natural Selection*. Princeton University Press, Princeton, New Jersey.
Williams, G.C. 1975. *Sex and Evolution*. Princeton University Press, Princeton, New Jersey.
Williams, J.E. 1977. *Observations on the Status of the Devils Hole Pupfish in the Hoover Dam Refugium*. Bureau of Reclamation Report REC-ERC-77-11, Washington, D.C.
Williams, M. 1970. Report on mallard-grey duck hybridization in New Zealand. *In Wildlife—A Review*. New Zealand Department of Internal Affairs, Wildlife Series 43-44.
Williams, M., and C. Roderick. 1973. Breeding performance of grey duck (*Anas superciliosa*), mallard (*Anas platyrhynchos*) and their hybrids in captivity. *International Zoo Yearbook* **13**:62-69.
Williamson, M. 1959. Studies on the colour and genetics of the black slug. *Proceedings of the Royal Physical Society*, Edinburgh **27**:87-93.
Willis, E.O. 1974. Populations and local extinction of birds on Barro Colorado Islands, Panama. *Ecological Monographs* **44**:153-169.
Willis, E.O. 1980. Ecological roles of migratory and resident birds on Barro Colorado Islands, Panama, pp. 205-225. *In* A. Keast and E.S. Morton (eds.), *Migrant Birds in the Neotropics: Ecology, Behavior, Distribution, and Conservation*. Smithsonian Institution, Washington, D.C.
Wilson, A.C. 1976. Gene regulation in evolution, pp. 225-234. *In* F.J. Ayala (ed.), *Molecular Evolution*. Sinauer Associates, Sunderland, Massachusetts.
Wilson, A.C., G.L. Bush, S.M. Case, and M.C. King. 1975. Social structure of mammalian populations and rate of chromosomal evolution. *Proceedings of the National Academy of Sciences of the United States of America* **72**:5061-5065.
Wilson, D.S. 1980. *The Natural Selection of Populations and Communities*. Benjamin Cummings, Menlo Park, California.
Wilson, E.O., and W.L. Brown, Jr. 1953. The subspecies concept and its taxonomic application. *Systematic Zoology* **2**:97-111.
Wilson, H.D. 1981. Genetic variation among South American populations of tetraploid *Chenopodium* sect. *Chenopodium* subsect. *Cellulata*. *Systematic Botany* **6**:380-398.
Witter, M.S., and P.P. Feret. 1978. Inheritance of glutamate oxalo-acetate transaminase isozymes in Virginia pine megagametophytes. *Silvae Genetica* **27**:129-134.
Wood, J.R. 1982. Release of captive green sea turtles by Cayman Turtle Farm Ltd. *Marine Turtle Newsletter No. 20*:6-7.
World Conference on Sea Turtle Conservation. 1979. *Sea Turtle Conservation Strategy*. International Union for the Conservation of Nature and Natural Resources (IUCN), Gland, Switzerland.
Wright, S. 1921. Systems of mating. *Genetics* **6**:111-178.
Wright, S. 1922a. Coefficients of inbreeding and relationship. *American Naturalist* **56**:330-338.

Wright, S. 1922b. The effects of inbreeding and crossbreeding on guinea pigs. I. Decline in vigor. *Bulletin of the U.S. Department of Agriculture* **1093**:37–63.
Wright, S. 1922c. The effects of inbreeding and crossbreeding on guinea pigs. III. Crosses between highly inbred families. *U.S. Department of Agriculture Technical Bulletin* 1121.
Wright, S. 1923. Mendelian analysis of the pure breeds of livestock: II. The Duchess family of shorthorns as bred by Thomas Bates. *Journal of Heredity* **14**:405–422.
Wright, S. 1931. Evolution in Mendelian populations. *Genetics* **16**:97–159.
Wright, S. 1932. The roles of mutation, inbreeding, crossbreeding and selection in evolution. *Proceedings of the International Conference on Genetics* **1**:356–366.
Wright, S. 1943. Isolation by distance. *Genetics* **28**:114–138.
Wright, S. 1951. The genetical structure of populations. *Annals of Eugenics* **15**:323–354.
Wright, S. 1969. *Evolution and the Genetics of Populations*, Vol. 2. *The Theory of Gene Frequencies*. University of Chicago Press, Chicago, Illinois.
Wright, S. 1977. Inbreeding in animals: differentiation and depression, pp. 44–96. In *Evolution and the Genetics of Populations*, Vol. 3. University of Chicago Press, Chicago, Illinois.
Wright, S. 1978. *Evolution and the Genetics of Populations*, Vol. 4. *Variability within and among Natural Populations*. University of Chicago Press, Chicago, Illinois.
Wright, S. 1980. Genic and organismic selection. *Evolution* **34**:825–842.
Wright, S. 1982. Character change, speciation, and the higher taxa. *Evolution* **36**:427–443.
Wu, L., and S.K. Jain. 1980. Self-fertility and seed set in natural populations of *Anthoxanthum odoratum* L. *Botanical Gazette* **141**:300–304.
Wu, L., A.D. Bradshaw, and D.A. Thurman. 1975. The potential for evolution of heavy metal tolerance in plants, III. The rapid evolution of copper tolerance in *Agrostis stolonifera*. *Heredity* **34**:165–187.
Yamashima, Y. 1947. Notes on the Marianas mallard. *Pacific Science* **11**:121–124.
Yeh, F.C., and Y.A. El-Kassaby. 1980. Enzyme variation in natural populations of Sitka spruce (*Picea sitchensis*). I. Genetic variation patterns among trees from 10 IUFRO provenances. *Canadian Journal of Forest Research* **10**:415–422.
Yeh, F.C., and C. Layton. 1979. The organization of genetic variability in central and marginal populations of lodgepole pine *Pinus contorta* ssp. *latifolia*. *Canadian Journal of Cytology* **21**:487–503.
Yeh, F.C., and D. O'Malley. 1980. Enzyme variation in natural populations of Douglas fir, *Pseudotsuga menziesii* (Mirb.) Franco, from British Columbia. 1. Genetic variation patterns in coastal populations. *Silvae Genetica* **29**:83–92.
Yntema, C.L. and N. Mrosovsky. 1980. Sexual differentiation in hatchling loggerheads (*Caretta caretta*) incubated at different controlled temperatures. *Journal of Herpetology* **36**:33–36.
Young, C.W., W.J. Typler, A.E. Freeman, H.H. Voelker, L.D. McGilliard, and T.M. Ludwick. 1969. Inbreeding investigations with dairy cattle in the north central region of the United States. *Minnesota University Agricultural Experiment Station Technical Bulletin* **266**:3–15.
Zaret, T.M., and R.T. Paine. 1973. Species introduction in a tropical lake. *Science* **182**:449–455.
Zeedyk, W.D., R.E. Farmer, Jr., B. MacBryde, and G.S. Baker. 1978. Endangered plant species and wildland management. *Journal of Forestry* **76**(1):31–36.

Zimmerman, E.G., and N.A. Gayden. 1981. Analysis of genic heterozygosity among local populations of the pocket gopher, *Geomys bursarius*, pp. 272–378. *In* M.H. Smith and J. Joule (eds.), *Mammalian Population Genetics*. University of Georgia Press, Athens, Georgia.

Zimmerman, E.G., C.W. Kilpatrick, and B.J. Hart. 1978. The genetics of speciation in the rodent genus *Peromyscus*. *Evolution* **32**:565–579.

Zukowski, R. 1959. Extinction and decrease of the butterfly *Parnassius apollo* in Polish territories. *Sylwan* **103**:15–30.

AUTHOR INDEX

Abplanalp, H.A., 138, 178
Abele, L.G., 332
Abreu Grobois, F.B., 144
Adams, M.W., 146
Adams, W.T., 221, Appendix 4
Adler, S., 175
Alatalo, R.V., 304
Alerstam, T., 304
Allard, R.W., 80, 82, 214, 218, 219, 221–223, 342, 345, Appendix 4
Allen, W.R., 84
Allendorf, F.W., 52, 55
Alstad, D.M., 80, 86, 92, 368, 369
American Ornithologists Union, 307
Anderson, R.F.V., 291
Anderson, R.M., 367
Anonymous, 282, 283, 408
Antonovics, J., 92
Appleby, J.W., 406
Aquadro, C.F., 354
Armitage, K.B., 179
Arnold, R.A., 161
Arulsekar, S., Appendix 4
Aston, J.L., 80
Atwood, C.E., 86
Averhoff, W.W., 253
Avise, J.C., 97, 144, 237, 354, 360
Ayala, F.J., 97, 139, 141, 143, 358

Babbell, G.R., 339, Appendix 4
Bacon, P.R., 281
Baker, A.J.M., 81
Baker, H.G., 205, 210, 214, Appendix 4
Baker, I., Appendix 4

Baker, R.J., 100, 104
Ballou, J., 171, 172, 175, 179, 184
Bantock, C.R., 128
Barber, H.S., 99
Barkelow, F.S., 423
Barkham, J.P., 164
Barnes, M.M., 369, 370
Barrett, J.A., 367
Bateson, P., 304
Baudoin, M., 366
Bayer, R.J., Appendix 4
Beardmore, J.A., 129, 135, 137, 139, 143, 144, 149
Beattie, A.J., 336
Bell, G., 202, 204
Belyaev, D.R., 38
Bengtsson, B.O., 76, 178
Benirschke, K., 35, 139, 406, 407, 409, 411
Benoit, R., 147
Bercovitz, A.B., 410
Berger, A.J., 178
Bergerud, A.T., 425
Bergmann, F., 341, Appendix 4
Berry, R.J., 128, 134, 135
Bickham, J.W., 104, 280
Bielewicz, M., 161
Bikay-Bikay, Appendix 4
Birch, L.C., 159, 300
Birley, A.J., 143
Bishop, J.A., 88, 89
Bjorndal, K., 278
Bodmer, W.F., 171
Boesiger, E., 147
Bogart R., 165, 178

Bohlin, R.G., 100
Bollengier, R., Jr., 307
Bonnel, M.L., 149, 209, 364
Bonnet-Masimbert, Appendix 4
Bouman, J.C., 165, 171
Bowman, J.C., 138, 175, 242
Braack, H.H., 139
Bradbury, J.W., 179
Bradley, B.P., 135, 149
Bradley, B.P., 135, 149
Bradshaw, A.D., 78, 80, 81, 83–85, 219
Braithwaite, L.W., 293
Breedlove, D., 157
Briles, W.E., 150
Bringhurst, R.S., Appendix 4
Britten, R.J., 352
Brittnacker, J.G., 360
Brongersma, L.D., 281
Brussard, P.F., 162
Brown, A.H.D., 4, 216, 220, 222–224, 226, 335, 338, 340, 343, 347, Appendix 4
Brown, B.A., 223
Brown, I.L., 162
Brown, K.S., 161
Brown, L.N., 423
Brown, W., 238, 408
Brown, W.L., Jr., 358
Bryant, E.H., 142, 354
Bull, J.J., 412
Bumpus, H.E., 134
Burnet, B., 208
Bush, G.L., 355
Bush, M., 179, 359, 409

Cade, T.J., 291, 307
Cameron, R.A.D., 209
Campbell, R.B., 69, 406
Capen, D.E., 306
Carlquist, S.J., 111
Carnahan, H.L., 138
Carr, A.F., 279, 280, 283
Carr, A.F., Jr., 282, 283
Carson, H.L., 192, 193, 238, 358
Cates, R.G., 368
Cavalli-Sforza, L.L., 61, 134, 171
Chai, C.K., 147
Chakravarti, A., 54
Chambers, S.M., 361
Chapin, F.S., III, 83
Chapin, M.C., 83
Chapman, R.W., 352
Charlesworth, B., 204

Chesser, R.K., 9, 45, 49, 63, 66–76, 82, 87, 180
Chew, F.S., 161, 368, 371
Chichester, L.F., 209
Chinnici, J.P., 128
Christiansen, F.B., 69, 321
Christie, W.J., 366
Cicmanec, J.C., 406
Clarke, C.A., 129, 369
Clayton, W.D., 271
Clegg, M.T., 9, 82, 218, 219, 221–223, 342, 345, 346, Appendix 4
Cockayne, W.C., 128
Cocks, P.S., 219
Confer, J.L., 296, 297, 300, 305
Conley, W., 384
Conner, J.L., 172, 175
Conterio, F., 134
Conway, W., 164
Cook, S.A., 83
Copeman, D.G., 359
Corbin, K.W., 292, 360
Cornelison, K.D., 421
Coyne, J.A., 354, 357
Craighead, J.J., 427
Crawford, D.J., Appendix 4
Crick, F., 229
Crisp, J.M., 425
Crook, J.H., 301
Crovello, T.J., 96
Crow, J.F., 68, 70, 201
Culver, D.C., 336

Daly, J.C., 67, 179
Darling, F.F., 425
Darwin, C., 38, 137, 174
Dasmann, R.F., 421
Davies, M.S., 80, 84, 343
Davies, S.M., 210
Davies, T.M., 78, 79, 84
Deakin, M.A.B., 69
DeBach, P., 365
De Grouchy, J., 406
De Jong, G., 136
De Vos, V., 139
De Wet, J.M.J., 3, 271, 272, 273, 276
Dempster, J.P., 160, 162
Denniston, C., 67, 68, 118, 165
Diamond, J.M., 121, 198
Diedenhofen, U., Appendix 4
Dinoor, A., 369
Dobzhansky, Th., 32–34, 137, 194, 238, 267
Dodd, C.K., Jr., 282

Doty, H.A., 306
Drew, A., 80, 84
Drury, W.H., 305, 307
Duffey, E., 159
Duncan, C.J., 205
Dungan, M.L., 126
Durrant, B., 411
Dutrillaux, B., 406

Eanes, W.F.,
East, R., 415, 422, 429, 422-444
Edmunds, G.F., Jr., 80, 86, 88, 92, 368, 369
Ehrendorfer, F., 267
Ehrenfeld, D.W., 278, 282, 286
Ehrlich, A.H., 109, 111, 153, 158, 161, 286, 402, 403
Ehrlich, P.R., 79, 85, 109, 111, 121, 153-159, 161, 162, 286, 402, 403
Eisenberg, J.F., 434, Appendix 3
Elena-Rossello, J.A., Appendix 4
El-Kassaby, Y.A., Appendix 4
Ellstrand, N.C., 340, 343
Emets, V.M., 161
Ennos, R.A., 223
Eshed, N., 369
Evsikov, V.I., 38
Ewens, W.J., 232, 354

Falconer, D.S., 67, 73, 138, 146, 174, 175, 178, 253
Feldman, M.W., 16
Feret, P.P., Appendix 4
Festing, M.F.W., 179
Finerty, J.P., 423
Finnley, D., 293, 294, 295
Fishbein, L., 130
Fisher, J., 165
Fitch, W.M., 353, 354
Flack, J.A.D., 301, 303
Flemming, C.A., 302, 303
Flesness, N., 118, 383, 389
Flesness, N.R., 67, 68, 165, 171, 180, 242, 386, 389, 392
Flor, H.H., 367
Fobes, J.F., 339, Appendix 4
Foltz, D.W., 202, 205-207
Foose, E., 384, 387-389
Foose, T., 67, 68, 376, 384, 386-389, 392-394
Ford, E.B., 91, 162
Ford, H.D., 162
Foster, J., 424, 426-428, 442

Fowler, C., 398
Frankel, O.H., 2-4, 8-10, 13, 45, 55, 63, 103, 111, 112, 115, 118, 120, 125-127, 161, 175, 178, 179, 183, 189, 190, 218, 227, 241-243, 250, 259, 260, 353, 402, 415, 421, 439, 440, 443
Franklin, I.R., 5, 63, 118, 165, 179, 183, 242, 243, 379, 386, 389, 392, 394, 399
Frazier, J.G., 278
Frick, J., 283
Fruch, R.J., 242
Futuyma, D.J., 355, 358, 368, 369

Gagnon, P.S., Appendix 4
Galbraith, I.C.J., 292, 300
Gall, G.A.E., 361
Gallun, R.L., 366, 367
Gardner, A.L., 18
Gartside, D.W., 84
Gayden, N.A., 100
Gaymer, R., 301
Geist, V., 425
Getz, L.L., 209
Ghiselin, M.T., 210
Gibson, W.C., 135, 403
Gilbert, L.E., 10, 121, 155, 351
Giles, R.H., 183
Gill, F.B., 296, 298-300
Gillespie, J., 155
Godt, M.J.W., 49, 78-95
Goel, N.S., 326
Goodman, D., 384, 392
Goodman, M., 353
Goodman, P.J., 80, 92
Goodwin, L.G., 409
Gottlieb, L.D., 335, 338, 339, 340, 341, Appendix 4
Gould, F., 369
Gould, S.J., 20, 31, 357, 371
Grant, M.C., Appendix 4
Grant, V., 29, 35, 96, 205
Greenway, J.C., Jr., 112
Greenwood, P.J., 179
Gregorius, H.R., 318, 353, 354, 355
Gregory, R.P.G., 81, 83, 84
Grieg, J.C., 183
Griner, L.A., 413
Gropp, A., 406
Grula, J.W., 352
Guillery, R.W., 179
Guries, R.P., 341, Appendix 4

Haldane, G.B.S., 141
Haley, L.E., 178
Hall, E.R., 100, 104, 429
Hall, W.P., 160
Halliday, T., 118
Hamrick, J.L., 80, 82, 92, 223, 335, 338–345, Appendix 4
Hancock, J.F., Appendix 4
Harborne, J.B., 86
Hare, J.D., 368
Harlan, J.R., 3, 267, 270, 271–273, 276
Harper, J.L., 205, 343
Hart, E.B., 100
Hartl, D.L., 52, 53, 70
Hartley, D.J., 88
Harvey, P.H., 179
Hatchett, J.H., 366, 367
Hayter, S., 176, 177
Hedgecock, D., 140
Hedrick, P.W., 140, 155
Heingardner, R., 35
Hendrickson, J.R., 280
Heusmann, H.W., 294
Hiebert, R.D., 341, 342, Appendix 4
Highton, R., 357
Hirth, H.F., 284
Holden, L.R., 82, 343, 345
Holdren, C.E., 156
Hollingsworth, M.J., 204
Holtby, C., 415
Honeycutt, R.L., 100, 104
Hoogland, J.L., 139, 179, 202
Hsu, K.J., 127
Hubbard, J.P., 293, 294, 295
Hudson, R.O., 211
Hufty, M.P., 407
Hultsch, E., 406
Humphrey, S.R., 409
Hunter, P.J., 205

Inbaraj, S.G., 244
IUCN, 5, 6, 164, 281, 282, 374, 416

Jaaska, V., Appendix 4
Jacquard, A., 73
Jain, S.K., 78, 82, 218, 219, 223, 339, Appendix 4
James, F.C., 306
James, S.H., 8, 9
Janzen, D.H., 368
Jarvis, C., 402
Jayakar, S.D., 141
Jefferies, R.L., 80

Jenkins, R.E., 403
Jennrich, R.I., 233
Johnsey, P.G., 423
Johnsgard, P., 293, 294
Johnson, M.S., 358
Johnston, R.F., 358
Joly, R.J., Appendix 4
Jones, D.A., 368
Jones, J.S., 79, 215, 387
Jones, T.C., 406
Jowett, D., 79, 84

Kaas, J.H., 179
Kahler, A.L., 218, 223, Appendix 4
Karlin, S., 69, 321, 322
Karr, J.K., 113, 119
Kaufman, D.W., 141
Kear, J., 178
Kearny, D., 424, 426–428, 442
Kerney, M.P., 209
Keith, L.B., 423
Kerster, H.W., 336
Kettlewell, H.B.D., 89, 162
Kim, Y.J., 100
Kimura, M., 43, 55, 63, 68, 70, 201
King, J.A.,
King, J.W.B., 146, 300
King, M.C., 357
Kinloch, B.B., 347
Kleiman, D.G., 165
Kloppers, J.J., 161
Knapp, K., 296, 297, 300, 305
Knerer, G., 86
Koch, A.L., 285
Koehn, R., 136
Koenig, W.D., 304
Koniuszek, J.W.J., Appendix 4
Krakauer, J., 61
Kruuk, H., 424, 426–428, 442
Kruse, A.D., 306
Kubetin, W.R., Appendix 4
Kuris, A.M., 366

Laerm, J., 350
Lamas Mueller, G., 161
Lang, E.M., 67
Langley, C.H., 129
Lansman, R.A., 237, 352
Larson, A., 357
Lasley, B.L.,
Lasley, J.F., 174, 177, 409, 410
LaVal, R.K., 104
Layton, C., 341, Appendix 4
Ledig, F.T., 341, Appendix 4

Lee, F.B., 306
Lenarz, M., 384
Lerner, I.M., 67, 146
Levene, H., 69, 320
Levin, D.A., 85, 336, 340, Appendix 4
Levin, S.A., 367
Levine, L., 143
Levins, R., 219
Levy, M., Appendix 4
Lewontin, R.C., 7, 16, 61, 125, 130, 232, 238, 300, 318, 343, 344, 355
Li, W.H., 352
Libby, W.J., 345
Lindemann, H.,
Lindzey, F.G., 427
Linhart, Y.B., Appendix 4
Littlefield, J.L., 347
Liu, E.H., 49
Lobanov, N.V., 242
Long, T., 143–145
Lovejoy, T.E., 165
Lucchino, R.V., 357
Lumaret, R., Appendix 4
Luria, S.E., 17, 20, 31
Lynch, C.B., 172, 175

Ma, N.S.F., 406
McAllister, D.E., 359
McAndrew, B.J., 149
McCabe, R.A., 306
McClure, S., 345, Appendix 4
McCracken, G.F., 179, 205–207
McCullough, D.R., 359
McDonald, J.F., 143
Mackay, T.F.C., 143
McKechnie, S.W., 156
McMichael, D., 9, 10
Macnair, M.R., 79, 81
McNeilly, T., 80, 84, 85
Main, A.R., 443
Malinen, M.O., 423
Manlove, M.N., 66
Margoliash, E., 353
Marshall, D.R., 216, 217, 219, 220, 222, 224, 226, Appendix 4
Martin, R.D., 164
Martins, P.S., 218, 223, Appendix 4
Maruyama, T., 61
Mather, K., 146
Mather, K.M., 133
May, R.M., 242, 324, 367

Maynard Smith, J., 112, 202–204, 365, 370
Mayr, E., 66, 67, 96, 267, 290, 291, 292, 351, 355, 357, 358, 365
Meagher, T.R., 224
Meise, W., 292
Mejnartowicz, L., Appendix 4
Mendel, G., 21–27, 38
Meslow, E.C., 427
Mierau, G.W., 425
Miller, B., 293
Miller, R.M., 84
Minawa, A., 143
Mitchell, B., 425
Mitter, C., 369
Mitton, J.B., Appendix 4
Mohr, E., 405
Moran, G.F., 219, 343, 347, Appendix 4
Morreale, S.J., 278
Morton, A., 161
Morton, N.E., 244
Motulsky, 136
Mrakovcic, M., 178
Mrosovsky, N., 278, 412
Mueller, L.D., 324
Muller, H.J., 195
Murphy, D.D., 153, 154, 155, 156, 158
Murphy, E.C., 426
Murphy, M.R., 175
Murray, N.D., 226
Myers, N., 109, 111, 161, 402, 415

Nakajima, A., 138
Namkoong, G., 319
National Research Council, 403, 409
Nei, M., 34, 40, 51, 54, 55, 61, 97, 126, 175, 209, 217, 231, 232, 336, 347, 352–354, 360, 362
Nelson, D.A., 294
Nelson, H.K., 306
Nelson, K., 140
Nevo, E., 131–133, 136, 140, 208, 335, Appendix 4
Newton, E., 301
Nielsen, G., Appendix 4
Niles, G.A., 371

Ochman, H., 209
Odum, E.P., 350
Ohta, T., 43, 55, 63
Olney, P.J.S., 164
O'Malley, D.M., Appendix 4

Oster, G.F., 324
Ostergaard, H., Appendix 4
Owen-Smith, N., 398

Packer, C., 179
Paine, R.T., 366
Panov, E.N., 292
Parker, G.R., 425, 428
Parsons, P.A., 98, 139
Partridge, G.G., 88
Pathak, M.D., 366
Patton, J.L., 100, 101, 357, 360
Penny, M., 301
Person, C., 367
Petrides, G.A., 424
Phelps, S.R., 52, 55
Phillips, M.A., Appendix 4
Phillips, P.A., 370
Phillips, J.R., 218, 369
Pickett, S.T.A., 113
Pienaar, U., (de) 424, 426–428, 442
Pilz, G., 339, Appendix 4
Pimentel, D., 367
Pinder, N.J., 164
Pitelka, F.A., 304
Portères, R., 275
Powell, J., 143, 208
Powell, J.R., 130, 234, 235
Prakash, S., 238
Prescott-Allen, C., 5
Prescott-Allen, R., 5
Price, M.V., 304, 369
Price, R.D., 101
Pritchard, P.C.H., 278
Prout, T., 69
Pyle, R.M., 161

Quick, H.E., 208

Rai, K.N., 82
Ralls, K., 67, 171, 172, 175, 179, 184, 242, 304
Ramshaw, R.A.M., 357
Rao, P.S.S., 244
Rausher, M.D., 370
Raven, P.H., 79, 85, 121, 351
Read, B., 242, 320
Redmond, M.A., 359
Reilly, C., 80, 84
Richardson, R.H., 253
Rick, C.M., 339, Appendix 4
Ricklefs, R.E., 159
Ripley, D., 306
Risser, P.G., 421

Ritzema-Bos, J., 174
Roberts, D.F., 138
Roberts, D.R., 127
Robertson, A., 61
Rocovich, S.E., 81
Rodman, J.E., 368, 371
Rogers, J.S., 354
Romer, A., 126
Roose, M.L., 341, Appendix 4
Rosenthal, G.A., 368
Rosenzweig, M.L., 112, 370
Roughgarden, J., 326, 367
Roychoudhury, A., 231, 354
Rudin, D., Appendix 4
Runham, N.W., 205
Rupp, R.S., 359
Ryder, O., 165, 387, 392, 405
Ryman, N., 61, 66, 358, 398

Sanders, T.B., Appendix 4
Schaal, B.A., 336, 343, Appendix 4
Schaffer, W.M., 284, 370
Schaller, G., 424, 426–428, 429, 442
Schmidly, D.J., 100
Schmidt, J.L., 425
Schnell, G.D., 209
Schoen, D.J., 222, 340, Appendix 4
Schonewald-Cox, C., 359
Schwaegerle, K.E., Appendix 4
Schwartz, O.A., 179
Scogin, R., Appendix 4
Seal, U.S., 165, 171, 376, 386, 392, 393
Seidel, M.E., 357
Selander, R.K., 66, 67, 87, 141, 146, 149, 205–209, 211, 339, 354, 358, 364, Appendix 4
Sene, F.M., 358
Senner, J.W., 118, 165, 172, 180, 242, 384
Shabica, S.V., 278
Shami, S.A., 135
Shank, D.B., 146
Shelford, V.E., 423
Sheppard, P.M., 84, 129
Shields, W.M., 202, 203, 304
Short, L.L., 290–292, 297, 307
Shultz, F.T., 146, 150
Shumaker, K.M., Appendix 4
Sibley, C.G., 292, 293
Siegel, B.Z., 126
Siegel, S.M., 126
Silander, J.A., 78, 80, 343

Silvers, W.K., 38
Simberloff, D., 332
Simpson, G.G., 49
Singer, M.C., 154, 155
Singer, S., 17, 20
Singh, R., 139, 238, 354, 357
Sittman, K., 149
Skibinski, D.O.F., 146
Slatis, H.M., 244
Slatkin, M., 52
Smith, M.F., 101
Smith, M.W., 89
Smith, M.H., 66, 68, 79, 80, 89, 90, 97, 139, 280, 360
Smith, R.H., 178
Smith, T., 398
Smith, W.G., Appendix 4
Snaydon, R.W., 78-80, 84, 343
Sokal, R.R., 96
Solbrig, O.T., 205
Somero, G.N., 142
Sosa, O., Jr., 366
Soulé, M.E., 3, 4, 6, 8, 9, 10, 55, 63, 111, 112, 115, 117, 120, 125, 126, 127, 141, 142, 149, 161, 165, 175, 178, 179, 183, 189, 190, 227, 241, 242, 250, 259, 260, 353, 402, 415, 421, 439, 440, 443
Soulie, J., 406
Spieth, P.T., 55
Squillace, A.E., 86
Ståhl, G., 61, 62
Staiger, H., 128
Stebbins, G.L., 205, 210, 224, 267
Steele, F., 411
Stine, G.J., 244
Stoneburner, D.L., 280
Stover, J., 411
Strong, L.C., 175
Stuber, C.W., Appendix 4
Sturgeon, K.B., 368
Summer, F.B., 199, 359
Svejgaard, A., 129

Tanksley, S.D., Appendix 4
Tauber, C.A., 86, 98
Tauber, M.J., 86, 98
Tauber, J.R., 98
Tavormina, S.J., 88, 369
Taylor, R.H., 301-303
Templeton, A.R., 197, 243, 245, 250, 253, 254, 259, 320, 354

Terborgh, J.W., 118, 119, 121, 439, 443
Thoday, J.M., 141
Thomas, J.A., 160
Thompson, J.N., 113
Thornton, I.W.B., 179
Timm, R.M., 101
Timofeeff-Ressovsky, N.W., 142
Torres, A.M., Appendix 4
Treus, V.D., 242
Tucker, P.K., 100
Turkington, R., 343
Turner, M.E., 90

Udovic, J.D., 367
Udvardy, M.D.F., 421
UNESCO, 6
Urquhart, C., 84
Usberti, J.A., 339, Appendix 4
U.S. Dept. of the Interior, Bureau of Land Management, 425
U.S. Dept. of the Interior, Fish and Wildlife Service, 289, 295, 296
U.S. Dept. of the Interior, National Park Service, 421, 424

Valentine, J.W., 141
Valizadeh, M., Appendix 4
Van der Plank, J.E., 366
Van Valen, L., 112, 367
Varland, K.L., 424
Verkleij, J.A.C., Appendix 4
Via, S., 369
Vogel, H., 136

Wade, N., 307, 366
Wahl, I., 369
Wain, R.P., Appendix 4
Wall, J.R., Appendix 4
Wall, S.W., Appendix 4
Wallace, B., 134, 139, 177
Walley, K.A., 84
Wallis, G.P., 142
Ward, R.D., 139, 146
Waser, N.M., 304
Wasserman, S.S., 369
Watson, J., 128
Watt, W.B., 155
Wedemeyer, E., 165, 405
Weiss, P.W., 217
West, D.A., 81
Wetzel, R.M., 407
Whalen, M.D., Appendix 4
White, M.J.D., 35, 355

White, R.R., 155, 156
Whitehead, G.K., 183
Whitemore, T.C., 6
Whitten, K.R., 426
Wiener, G., 176, 177
Wiens, J.A., 85, 358
Wilcox, B.A., 125, 415, 421, 439
Williams, G.C., 365, 366
Williams, M., 293, 295
Williams, J.E., 359
Williamson, M., 209
Willis, E.O., 113, 118, 119
Wilson, A.C., 33, 179, 358
Wilson, D.S., 366
Wilson, E.O., 358
Wilson, H.D., Appendix 4
Windberg, L.A., 423
Winter, B., 118, 119, 439
Wistrand, H., 143
Witter, M.S., Appendix 4
Wood, J.R., 282
Woolliams, J.A., 177

Woodruff, D.S., 357
World Conference on Sea Turtle Conservation, 278
Wright, S., 9, 51, 53–55, 66, 68, 69, 73, 88, 133, 174, 175, 177, 178, 191, 204, 267, 336
Wu, L., 84, Appendix 4

Yadov, M., 443
Yamashima, Y., 293
Yang, S.Y., 101
Yeh, F.C., 341, Appendix 4
Yntema, C.L., 278, 412
Yoon, J.S., 193
Young, C.W., 175–177

Zaret, T.M., 366
Zeedyk, W.D.,
Zimmerman, E.G., 100
Zouros, E., 139
Zukowski, R., 161

SPECIES INDEX

Abies concolor. See White fir (*Abies concolor*)
Acanthoscelides. See Bean weevils (*Acanthoscelides obtectus*)
Acer. See Maple (*Acer*)
Achillea lanulosa. See Western yarrow (*Achillea lanulosa*)
Acouchi (*Myoprocta pratti*), inbreeding effects on juvenile mortality, 170
Aedes aegypti. See Yellow fever mosquito (*Aedes aegypti*)
African chiclid fish, extinction of native fishes in Gatún Lake, Panama, 366
African elephant (*Loxondonta africana*), area size, 424; artificial insemination, 441; future prospects, 444; locality, 424; population size, 424; population-size to park-size relationship, 441, 442; reserve size for, 444; space and survival of, 442
African lion (*Panthera leo*), area size for, 427; captive facilities, 381; habitat patchiness, 442; locality, 427; maintenance limitation, 381; population longevity prospects, 441, 442; population size, 427; population-size to park-size relationship, 441, 442; subspecies, 381
Agrostis stolonifera. See Carpent bent grass (*Agrostis stolonifera*)
Agrostis tenuis = *A. capillaris.* See Colonial bent grass (*Agrostis tenuis* = *A. capillaris*),
Aix sponsa. See North American Wood duck (*Aix sponsa*)
Alces alces americanus. See Moose (*Alces alces americanus*)
Alfalfa (*Medicago sativa*), sulphur butterfly adaptation to, 369
Alocasia (*Alocasia*), weed, 274
Alocasia. See Alocasia (*Alocasia*)
Alsophila pometaria. See Fall cankerworm (*Alsophila pometaria*)
Ambystoma tigrinum. See Tiger salamander (*Ambystoma tigrinum*)
American Black duck (*Anas rubripes*), competitive exclusion effects, 308; contact with Mallard duck, 293; gene pool assimilation, Mallard duck, 294; habitat preference, 294; hunting, 294, bag, 294, other duck comparisons, 294; hybrids, 294, fitness of, 294, frequency of, 294, survival, 294; hybridization with Mallard duck, 294, 308, interactions, 294, 295, 308; population size, 294, other duck comparisons, 294; range expansions, 294; response to environmental change, 294, 308
American chestnut (*Castanea dentata*),

Note to reader: References to individual species are not usually cross-referenced under higher taxonomic categories.

threatened extinction, 366, by introduced fungi, 366
American elm (*Ulmus americana*), threatened extinction, 366, by introduced fungi, 366
Ammospiza maritima nigrescens and *Ammospiza maritima peninsulae*. *See* Dusky Seaside sparrow and Scott's Seaside sparrow (*Ammospiza maritima nigrescens* and *Ammospiza maritima peninsulae*)
Ammospiza maritima nigrescens. *See* Dusky Seaside sparrow (*Ammospiza maritima nigrescens*)
Ammospiza maritima peninsulae. *See* Scott's Seaside sparrow (*Ammospiza maritima peninsulae*)
Anas fulvigula. *See* Mottled duck (*Anas fulvigula*)
Anas "oustaleti". *See* Marianas Mallard duck (*Anas "oustaleti"*), 293
Anas diazi. *See* Mexican duck (*Anas diazi*)
Anas platyrhynchos and *Anas diazi*. *See* Mallard duck and Mexican duck (*Anas platyrhynchos* and *Anas diazi*)
Anas platyrhynchos and *Anas fulvigula*. *See* Mallard duck and Mottled duck (*Anas platyrhynchos* and *Anas fulvigula*)
Anas platyrhynchos and *Anas rubripes*. *See* Mallard duck and American Black duck (*Anas platyrhynchos* and *Anas rubripes*)
Anas platyrhynchos and *Anas superciliosa "rogersi"*. *See* Mallard duck and Australian Black duck (*Ana platyrhynchos* and *Anas superciliosa "rogersi"*)
Anas platyrhynchos and *Anas superciliosa*. *See* Mallard duck and Grey duck (*Anas platyrhynchos* and *Anas superciliosa*)
Anas platyrhynchos. *See* Mallard duck (*Anas platyrhynchos*)
Anas poecilorhyncha. *See* Spotbill duck (*Anas poecilorhyncha*)
Anas rubripes. *See* American Black duck (*Anas rubripes*)
Anas superciliosa. *See* Grey duck (*Anas superciliosa*)
Anas superciliosa "rogersi". *See* Australian Black duck (*Anas superciliosa "rogersi"*)
Anoa (*Bubalus depressicornis*), endangered species, 411; interspecific embryo transfer proposal, 411; survival, 411
Anole lizards (*Anolis*), heterozygosity and morphological asymmetry, 149
Anolis. *See* Anole lizards (*Anolis*)
Ant (*Myrmica sabuleti*), butterfly symbiotic relationship, 160
Ant (*Myrmica scabrinodis*), replacement of *M. sabuleti*, 160
Anthonomus grandis. *See* Boll weevil (*Anthonomus grandis*)
Anthoxanthum odoratum. *See* Sweet vernal grass (*Anthoxanthum odoratum*)
Antilocapra americana. *See* Pronghorn antelope (*Antilocapra americana*)
Antirrhinum majus. *See* Snapdragon (*Antirrhinum majus*)
Anura. *See* Toads, Frogs (*Anura*)
Aotus trivirgatus. *See* Owl monkey (*Aotus trivirgatus*)
Aphids (*Homoptera: Aphidoidae*), damage reduction in cotton, 371, with sesquiterpene, gossypol, 371
Apis mellifera. *See* Honey bee (*Apis mellifera*)
Apple (*Malus domestica*), codling moth pest, 369, 370
Apus apus. *See* Common swift (*Apus apus*)
Arabian oryx (*Oryx leucoryx*), genetic donors for wild stock, 411
Araceae. *See* Edible aroids (Araceae)
Arctic sedge (*Carex aquatilis*), ecotype variation of, 83, along latitudinal, local gradients, 83
Arionidae. *See* Terrestrial slugs (Arionidae)
Arionid slug (*Arion ater ater*), distribution, 210; genetic variation and breeding system, 206; heterozygosity (H), 206, 208; homozygosity, 206, 208; mixed species, 206; polymorphism (P), 206; populations, number, 206; region, 206

Arionid slug (*Arion ater rufus*), genetic variation and breeding system, 206; heterozygosity (H), 206; mixed species, 206, 208; polymorphism (P), 206; populations number, 206; region, 206, 210; selfing species, 206, 208

Arionid slug (*Arion circumscriptus*), genetic variation and breeding system, 206; heterozygosity (H), 205, 206; homozygous strains, 208, 209; polymorphism (P), 206; populations, number, 206; region, 206, 209; self-fertilizing species, 206

Arionid slug (*Arion distinctus*), distribution, 208; genetic variation and breeding system, 206; heterozygosity (H), 206, 208; polymorphism (P), 206; populations, number, 206; outcrossing species, 206; region, 206

Arionid slug (*Arion fasciatus*), genetic variation and breeding system, 206; heterozygosity (H), 206, 208; homozygous strains, 208; polymorphism (P), 206; populations, number, 206; region, 206; self-fertilizing species, 206

Arionid slug (*Arion hortensis*), distribution, 208; genetic variation and breeding system, 206; heterozygosity (H), 206, 208; outcrossing species, 206, 208; polymorphism (P), 206; populations, number, 206; region, 206

Arionid slug (*Arion intermedius*), genetic variation and breeding system, 206; heterozygosity (H), 206; homozygous strains, 208–210; outcrossing, none, 210; polymorphism (P), 206; population number, 206; region, 206; self-fertilizing species, 206, 210

Arionid slug (*Arion lusitanicus*), genetic variation and breeding system, 206; heterozygosity (H), 206; outcrossing species, 206; polymorphism (P), 206; populations, number, 206; region, 206

Arionid slug (*Arion oweni*), genetic variation and breeding system, 206; heterozygosity (H), 206; polymorphism (P), 206; populations, number, 206; outcrossing species, 206; region, 206

Arionid slug (*Arion silvaticus*), genetic variation and breeding system, 206; heterozygosity (H), 205, 206, 208; homozygous strains, 208, 209; polymorphism (P), 206; populations, number, 206; region, 206; self-fertilizing species, 206

Arionid slug (*Arion subfuscus*), distribution, 208, 209; genetic variation and breeding system, 206; heterozygosity (H), 206, 208, 209; homozygosity, 208, 209; mixed species, 206; polymorphism (P), 206; outcrossing species, 206, 208; populations, number, 206; self-fertilizing species, 206; region, 206

Artemia. See Brine shrimp (*Artemia*)
Artemia franciscana. See Brine shrimp (*Artemia*)
Artemia persimilis. See Brine shrimp (*Artemia*)
Artemia salina. See Brine shrimp (*Artemia*)
Artemia urmiana. See Brine shrimp (*Artemia*)
Asarum canadense. See Wild ginger (*Asarum canadense*)
Asian wild horse. See Przewalski's wild horse (*Equus przewalski*)
Asteraceae. See Sunflower Family (Asteraceae)
Ateles. See Spider monkeys (*Ateles*)
Ateles fusciceps robustus. See Black spider monkey (*Ateles fusciceps robustus*)
Attwater's prairie-chicken (*Tympanuchus cupido attwateri*), 104; clutches, 118; extinction, 118; ground nesting, 118
Australian Black duck (*Anas superciliosa "rogersi"*), hybridization with Mallard duck, 293; See also Spotbill duck (*Anas poecilorhyncha*)
Australian flatback turtle (*Chelonia depressa*), occurrence, 277

Austrian pine (*Pinus nigra*), Appendix 4
Avena. See Oats (*Avena*)
Avena barbata. See Slender wild oat (*Avena barbata*)
Avena fatua. See Common wild oat (*Avena fatua*)

Baboons (*Papio*), adaptation capacity, 115; in environmental change, 115
Baboons (*Theropithecus*), adaptation capacity, 115; in environmental change, 115
Bald eagle, Appendix 3
Banana (*Musa*), weed races, 272
Baptisia leucophaea. See Plains wild-indigo (*Baptisia leucophaea*)
Baptisia sphaerocarpa. See Yellow wild-indigo (*Baptisia sphaerocarpa*)
Barestem tick-clover (*Desmodium nudiflorum*), Appendix 4
Bark beetles (Coleoptera: Scolytidae), immunization agents against extinction, 373
Barking deer (*Muntiacus*). See Muntjacs (*Muntiacus*)
Barking tree frog, species of interest to managers, Appendix 3
Barley (*Hordeum*), gene flow, 274; mimetic races, 274; selection pressures, 274
Barren-ground caribou (*Rangifer tarandus*), area size, 425; locality, 425; population size, 425
Bean weevils (*Acanthoscelides obtectus*), artificial selection on laboratory, 369; improvement in adaptation to inferior hosts, 369
Beet (*Beta*), weed races, 274
Bengal tiger (*Panthera tigris altaica*), population maintenance difficulty, 380
Beta. See Beet (*Beta*)
Bighorn sheep (*Ovis canadensis canadensis*), area size, 425; bottleneck in population, 443; competition, 443; genetic diversity loss, 443; locality, 425; longevity, 443, in reserve, 443; park area, 443; population management, 443; population size, 425; recolonization, 443; selection pressure, 443; small managed populations, 436

Bird-voiced tree frog, species of interest to managers, Appendix 3
Bird's-foot trefoil (*Lotus corniculatus*), geographic variation in cyanide-producing genotypes, 368; response to herbivory, 368
Birds of paradise (Paradisaeidae), displays, 198; population alteration, 199; reproductive characters under balanced polygenic control, 198; secondary sexual characteristics under genetic control, 198, 199
Bison (*Bison bison*), habitat overexploitation, 420; population in park, 420; removal of animals
Bison bison. See Bison (*Bison bison*)
Bison bonasus. See European bison (*Bison bonasus*)
Black and white ruffed lemur (*Varecia variegata variegata*), flat chest, 409; hairless offspring, inbreeding effect, 409
Black bear (*Ursus americanus*), area size, 427; locality, 427; population size, 427
Black duck. See American Black duck (*Anas rubripes*); Australian Black duck (*Anas superciliosa "rogersi"*)
Black lemur (*Lemur macaco*), inbreeding effects on juvenile mortality, 168
Black pineleaf scale insect (*Nuculaspis californica*), adaptation to chemical heterogeneity in host trees, 86, 87; adaptation to specific trees, 86, 87, 95; adaptive genetic differentiation in, 80, 86, 87, 95; differentiation of populations, 86, 87; divergence causes, 80; extinction with ponderosa pine extinction, 441; gene flow restriction, 87; genetic differentiation of host trees, 86; plant toxins, 80; ponderosa pine host, 86; populations adjacent, 80; selection pressure, 87; selection pressure by ponderosa pine, 87, 88; semi-isolation of, 87; survival with host species, 442
Black rhinoceros (*Diceros bicornis*), distribution, 377; endangered, 376–377; population, 376–378; recessive sex-linked gene, 139
Black spider monkey (*Ateles fusci-*

ceps robustus), inbreeding effects on juvenile mortality, 168
Black turtle (*Chelonia agassizii*), distinction from *C. mydas*, 278; distribution, 278
Black-tailed prairie dog (*Cynomys ludovicianus*), rarity of inbreeding in, 179
Blanding's turtle, species of interest to managers, Appendix 3
Blue pike, Appendix 3
Blue-winged warbler (*Vermivora pinus*), breeding distribution, 298; competitive exclusion effects, 299; distribution, 296, changes, 296; ecological capabilities of, 300; ecological competition, 300; habitat change, 299; habitat generalist, 297; history, 296, 299; hybrid swarms, 300; hybridization, 296, 299, 300; nesting, 296; numbers in zones of contact, 296; phenotype persistence, 299; physiological capabilities, 300; range extension, 296, and Golden-winged genes, 300; response to environmental change, 297; spread, 298; species persistence, 299; stabilizing selection, 300; sympatry with Golden-winged warbler, 300
Blue-winged warbler and Golden-winged warbler (*Vermivora chrysoptera*), breeding distribution, 297; contact zones, 297, 299; competitive exclusion effects, 299; genetic assimilation, 299; Golden-wing genotype destruction, 300; habitat, 299, change, 296, 297; hybrid swarms, 300; hybridization, 296, 299, 300; nesting, 296; numbers in zones of contact, 296; phenotype, elimination, 300; range, 297; secondary contact, 299; selection against, 300; sympatry to hybridization, 295
Bluegill, subspecies classification by electrophoresis, 360
Blunt-nosed leopard lizard, species of interest to managers, Appendix 3
Boll weevil (*Anthonomus grandis*), damage effect on cotton, 371; damage effect minimization, 371; and cotton frego bract genetic alteration, 371

Bollworms (*Pectinophora* and *Heliothis*), damage reduction on cotton, 371; with absence of extrafloral nectaries, 371; with sesquiterpene gossypol, 371
Bongo (*Tragelaphus eurycerus*); heterochromatic short arm, 407; X chromosome, 407; comparison with other tragelaphines, 407
Boris (*Octodontomys gliroides*), inbreeding effects on juvenile mortality, 170
Bos frontalis = *Bos gaurus*. See Gaur (*Bos frontalis* = *B. gaurus*)
Bos. See Cattle (*Bos*)
Botflies (Diptera: Oestridae), host life extension, 366
Bothriochloa bladhii. See Grass (*Bothriochloa bladhii*)
Bothriochloa ewartiana. See Desert blue grass (*Bothriochloa ewartiana*)
Bothriochloa grahamii. See Grass (*Bothriochloa grahamii*)
Bothriochloa insculpta. See Grass (*Bothriochloa insculpta*)
Bothriochloa intermedia. See Grass (*Bothriochloa intermedia*)
Bothriochloa ischaemum. See Grass (*Bothriochloa ischaemum*)
Bothriochloa longifolia. See Grass (*Bothriochloa longifolia*)
Bothriochloa-Dichanthium complex. See Old World bluestem grasses (*Bothriochloa-Dichanthium* complex)
Bovidae. See Cattle, sheep, antelope, etc. (Bovidae), 382
Bovinae. See Cattle, etc. (Bovinae), 190, 382
Bowerbirds (Ptilonorhynchidae), displays, 198; population alteration, 199; reproductive characters under balanced polygenic control, 298; secondary sexual characteristics, 198; under genetic control, 198, 199
Bracted sunnybell, Appendix 3
Bradshaw's desert parsley, Appendix 3
Brassica. See Mustard (*Brassica*)
Brine shrimp (*Artemia*), average heterozygosity, 144; distribution, 144; genetic variation levels, 144; species age and abundance, 144;

heterozygosity in bisexual species, 144
Bromus mollis. See Soft brome (*Bromus mollis*)
Bromus rubens. See Foxtail brome (*Bromus rubens*)
Brown lemur (*Lemur fulvus*), inbreeding effects on juvenile mortality, 168
Brown planthopper (*Nilaparvata lugens*), evolution of strains, 366, against resistant rice strains, 366
Brown rat (*Rattus norvegicus*), alleles, 88, 89, fixation of, 89, rodenticide resistance and, 88; food, 88, 89, vitamin K in, 88, 89; genetic differentiation factors, 88; genetic structure of adjacent populations, 88, 89; heterogeneous condition, 88; homozygous condition, 88; migration between breeding units, 89; population size, 88; population structure, 88; populations adjacent, 88, 95, random event effects, 88; selective pressures, 88; semi-isolation of populations and stochastic events, 88, effect in genetic structure of, 88, and population extinction, 88; survival factors, 88; vitamin K and rodenticide resistance in, 88, 89; warfarin resistance in, 88, 89, 95, dominant allele, 88
Brown trout (*Salmo trutta*), allele frequency divergence at loci in local demes, 61, 62
Bubalus depressicornis. See Anoa (*Bubalus depressicornis*)
Bubalus minorensis. See Tamaraw (*Bubalus minorensis*)
Burchell's zebra (*Equus burchelli*), area size, 424; locality, 424; population size, 424
Burrowing shearwater (*Puffinus*), parakeet habitat degradement from, 303
Buteo lagopus. See Rough-legged hawk (*Buteo lagopus*)

Cactoblastis cactorum. See Phycitid moth (*Cactoblastis cactorum*)
California condor, Appendix 3

California ground squirrel (*Citellus beecheyi*), survival in park, 420; survival out of park, 420
California Orcutt grass, Appendix 3
California poppy (*Eschscholzia californica*), adaptive genetic differentiation in, 80, 83; annual or perennial growth forms, 83; character variation over climatic regions, 83; differentiation with sharp environmental gradients, 83; stamen number, 83
Caligo butterflies (*Caligo*), pest of banana, 371
Camel Family: Camels, llamas (Camelidae), 382
Camelidae. *See* Camel Family (Camelidae)
Canis. See Dogs (*Canis*)
Canis lupus. See Wolf (*Canis lupus*)
Cape buffalo (*Syncerus caffer*), area size, 426; locality, 426; population size, 426
Cape hunting dog (*Lycaon pictus*), future prospects, 444; habitat patchiness, 442; population-size to park-size relationship, 442; population longevity prospects, 441, 442; reserve size for, 444
Capillipedium parviflorum. See Grass (*Capillipedium parviflorum*)
Capillipedium spicigerum. See Scented top grass (*Capillipedium spicigerum*)
Caprinae. *See* Sheep and Goat Subfamily (Caprinae), 382
Capra. See Goats (*Capra*)
Capricornis crispis. See Japanese serow (*Capricornis crispis*)
Caretta caretta. See Loggerhead sea turtle (*Caretta caretta*)
Carex aquatilis. See Arctic sedge (*Carex aquatilis*)
Carica. See Papaya (*Carica*)
Carolina parakeet (*Conuropsis carolinensis*), social behavior and extinction of, 117, 118
Carpet bent grass (*Agrostis stolonifera*), adaptive genetic differentiation in, 80, 81, 83, 95; copper tolerance in populations of, 83, 84; nickel tolerance in populations of, 83, 84; populations, 80, 81, 83, ad-

jacent, 81, distance between, 80; rapidity of genetic changes, 84; wind exposure and population differentiation, 80; zinc tolerance in populations of, 81
Carps and minnows (Cyprinidae), heterozygosity levels, 146
Carrot (*Daucus*), weed races, 274
Caryophyllaceae. *See* Pink Family (Caryophyllaceae)
Castanea dentata. *See* American chestnut (*Castanea dentata*)
Castor (*Ricinus*), weed races, 274
Cat Family: Lions, tigers, etc. (Felidae), capacity of ISIS facilities for, 381
Cats (*Felis*), in species extinction, oceanic islands, 111
Cat (*Felis catus*), predation on Yellow-crowned parakeet, 302
Catagonus wagneri. *See* Chaco peccary (*Catagonus wagneri*)
Cattle (*Bos*), breeding of, 190, in recombination and selection of existing genetic variability, 190
Cavia porcellus. *See* Guinea pig (*Cavia porcellus*)
Celebes black ape (*Macaca nigra*), inbreeding effect on juvenile mortality, 169
Centrachidea. *See* Sunfishes and freshwater basses (Centrachidae)
Cerastomella ulmi. *See* Dutch elm disease (*Cerastomella ulmi*)
Ceratitis capitata. *See* Mediterranean fruit fly (*Ceratitis capitata*)
Ceratotherium simum cottoni. *See* Northern white rhinoceros (*Ceratotherium simum cottoni*)
Ceratotherium simum simum. *See* Southern white rhinoceros (*Ceratotherium simum simum*)
Ceratotherium simum. *See* White rhinoceros (*Ceratotherium simum*)
Cercomys cunicularis. *See* Punare (*Cercomys cunicularis*)
Cercyonis sthenele sthenele. *See* Sthenele brown butterfly (*Cercyonis sthenele sthenele*)
Cervidae. *See* Deer Family (Cervidae)
Cervus elaphus nelsoni. *See* Elk (*Cervus elaphus*)

Cervus elaphus nannodes. *See* Tule elk (*Cervus elaphus nannodes*)
Cervus elaphus nelsoni. *See* Elk (*Cervus elaphus*)
Cervus elaphus roosevelti. *See* Roosevelt elk (*Cervus elaphus roosevelti*)
Cervus elaphus scotticus. *See* Red deer (*Cervus elaphus scotticus*)
Cervus eldi thamin. *See* Eld's deer (*Cervus eldi thamin*)
Chaco peccary (*Catagonus wagneri*), chromosome complement unknown, 407; hybrid possibilities with other peccaries, 407; threatened species, 407
Chalcedon checkerspot butterfly (*Euphydryas chalcedona*), characteristics by ecological distance of populations, 155; ecotype similarity, 155, and geographical distance, 155; monitoring, 155
Chamaenerion angustifolium. *See* Fireweed (*Chamaenerion angustifolium*)
Chatam Islands Yellow-crowned or Forbes parakeet (*Cyanoramphus auriceps forbesi*), phenetic differences from close congeners, 301
Checkered white butterfly (*Pieris protodice*), population regulation, "mosaic" pattern of habitat, 159
Checkerspot butterfly *See* Nearctic checkerspot butterflies and Editha cheskerspot butterfly (*Euphydryas* and *Euphydryas editha*)
Cheetah (*Acinoyx jubatus*), capacity of ISIS for, 381
Chelonia agassizii. *See* Black turtle (*Chelonia agassizii*)
Chelonia depressa. *See* Australian flatback turtle (*Chelonia depressa*)
Chelonia mydas. *See* Green turtle (*Chelonia mydas*)
Chenopodiaceae. *See* Goosefoot Family (Chenopodiaceae)
Chenopodium (*Chenopodium hians*), Appendix 4
Chenopodium (*Chenopodium hircinum*), Appendix 4
Chenopodium (*Chenopodium incognitum*), Appendix 4

Chenopodium atrovirens. See Dark goosefoot (*Chenopodium atrovirens*)
Chenopodium desiccatum. See Desert goosefoot (*Chenopodium desiccatum*)
Chenopodium fremontii. See Frémont goosefoot (*Chenopodium fremontii*)
Chenopodium hians. See Chenopodium (*Chenopodium hians*)
Chenopodium hircinum. See Chenopodium (*Chenopodium hircinum*)
Chenopodium incognitum. See Chenopodium (*Chenopodium incognitum*)
Chenopodium leptophyllum. See Slimleaf goosefoot (*Chenopodium leptophyllum*)
Chenopodium pratericola. See Narrowleaf goosefoot (*Chenopodium pratericola*)
Chenopodium quinoa var. *melanospermum.* See Highland wild quinoa (*Chenopodium quinoa* var. *melanospermum*)
Chenopodium quinoa. See Cultivated quinoa (*Chenopodium quinoa*)
Cherry tomato (*Lycopersicon esculentum* var. *cerasiformae*), Appendix 4
Chicken (*Gallus gallus*), breeding of, 190, in recombination and selection of existing genetic variability, 190; inbreeding depression in, 150; polymorphism level and inbreeding, 150; segregation at B locus, 150
Chimpanzee (*Pan troglodytes*), comparison with humans, 357; DNA hybridization technique, 357; electrophoresis, 357; immunology technique, 357; inbreeding effects on juvenile mortality, 169
Chinese muntjac (*Mutiacus reevesi*), karyotype, 407
Chloridion cameronii. See Grass (*Chloridion cameronii*)
Choeropsis liberiensis. See Pygmy hippopotamus (*Choeropsis liberiensis*)
Chrysanthemum leucanthemum. See Oxeye daisy (*Chrysanthemum leucanthemum*)
Chrysolina gemellata. See Chrysomelid beetles (*Chrysolina gemellata*)

Chrysomelid beetles (*Chrysolina gemellata*), introduction for control of Klamath weed or St. John's-wort, 366
Chrysopa carnea. See Green lacewing (*Chrysopa carnea*)
Chrysopa downesi. See Green lacewing (*Chrysopa downesi*)
Citellus beecheyi. See California ground squirrel (*Citellus beechyi*)
Citrus. See Oranges (*Citrus*) or Lemons (*Citrus*)
Clarkia. See Clarkia (*Clarkia*)
Clarkia (*Clarkia*), species distribution and Gst values, 339; Appendix 4
Clarkia biloba. See lobed clarkia (*Calrkia biloba*)
Clarkia lingulata. See Merced clarkia (*Clarkia lingulata*)
Clarkia rubicunda. See farewell-to-spring (*Clarkia rubicunda*)
Cleithrodontomys. See Harvest mouse (*Cleithrondontomys*)
Cliffbreak fern, Appendix 3
Coachella Valley fringe-toed lizard, Appendix 3
Coatimundi (*Nasua nasua*, near-ground predator on birds, 119
Codling moth (*Laspeyresia pomonella*), adaptation to English walnut, 369, 370; pest of apple, 369, 370; pest of plum, 369, 370; tree cultivation areas and abundance, 370
Coleoptera: Scolytidae. See Bark beetles (Coleoptera: Scolytidae)
Colias philodice eriphyle. See Sulphur butterfly (*Colias philodice eriphyle*)
Collared flycatcher and Pied flycatcher (*Ficedula hypoleuca* and *Ficedula albicollis*), breeding success, 304; competitive influence of hybridization, 304; hybridization, 304; island populations, 304; population replacement, 304
Collared lemming (*Dicrostonyx groenlandicus*), area size for, 423; locality, 423; population size, 423
Collared lizard, Appendix 3
Collared peccary (*Tayassu tajacu*), karyotype, 407
Colobus monkey, environmental change and survival of, 114, 115

Colonial bent grass (*Agrostis tenuis* [= *A. capillaris*]) adaptive genetic differentiation of, 80, 83; heavy metal tolerance in populations of, 80, 83; populations, 80; distance between, 80

Colonial bent grass (*Agrostis tenuis*) and Carpet bent grass (*Agrostis stolonifera*), adaptive genetic differentiation in, 84, 95; hybrid, 84; zinc tolerance of populations of, 84

Colorado potato beetle (*Leptinotarsa decemlineata*), movement from wild to cultivated potato, 371

Common eel-grass (*Zostera marina*), Appendix 4

Common evening-primrose (*Oenothera biennis*), Appendix 4

Common monkeyflower (*Mimulus guttatus*), adaptive genetic variation in, 81, 84; allozyme variation, 345; between population variance/total genetic variance, 344, 345; Gst value, 345; life history influence on quantitative traits, 344, 345; morphometric study, 345; populations adjacent, 81; reproduction by vegetative and sexual means, 344; tolerance of populations of, 81, 84; zinc tolerance of populations of, 81, 84; Appendix 4

Common morning-glory (*Ipomoea purpurea*), allozyme variation, 219; anther-stigma-distance variation, 223; flower color polymorphisms, 219; genetic determination of mating system, 223; introduction, 219; pollinator behavior, 223, and flower color, 223, and outcrossing rate, 223; mating structure shifts, 223, from habitat change, 223

Common pitcherplant (*Sarracenia purpurea*), Appendix 4

Common sunflower (*Helianthus annuus*), genetic structure of populations, 343; family structure, 343; Appendix 4

Common swift (*Apus apus*), clutch sizes and progeny number, 134, 135

Common tomato (*Lycopersicon esculentum*), Appendix 4

Common wild oat (*Avena fatua*), allozyme variation, single gene morphological trait, quantitatively inherited trait associations, 344; slender wild oat comparison in genetic variability, and phenotypic plasticity, 219

Common rye (*Secale cereale*), Appendix 4

Condors (*Gymnogyps* and *Vultur*), sex determination, 410; sexual monomorphism, 410; stool samples, 410

Connochaetes taurinus. See Wildebeest (*Connochaetes taurinus*)

Conuropsis carolinensis. See Carolina parakeet (*Conuropsis carolinensis*)

Coreopsis (*Coreopsis cyclocarpa* var. *cyclocarpa*), Appendix 4

Coreopsis (*Coreopsis cyclocarpa* var. *pinnatisecta*), Appendix 4

Coreopsis cyclocarpa var. *cyclocarpa*. See Coreopsis (*Coreopsis cyclocarpa* var. *cyclocarpa*)

Coreopsis cyclocarpa var. *pinnatisecta*. See Coreopsis (*Coreopsis cyclocarpa* var. *pinnatisecta*)

Corkwood, Appendix 3

Corn (*Zea mays*), decrease in yield, by new race of southern corn leaf blight fungus, 367. See also Maize.

Cotton (*Gossypium*), bract alteration and resistance to boll weevil, 371; damage increase, thrips, 371, white flies, 371; leaf hairiness and resistance to leafhoppers, boll weevils, 371; leaf hairlessness and resistance to mirid bugs, pink boll worm, 371; nectaries absence and protection against mirids, boll worms, 371; sesquiterpene, gossypol, and resistance to leafhoppers, boll worms, aphids, 371; traits for resistance or linked genes, 371; weed races, 274

Coturnix japonica. See Japanese quail (*Coturnix japonica*)

Cowpea (*Vigna*), weed races, 274

Crabgrass (*Digitaria sanguinalis*), cultivated as finger millet, 275; weed, 275

Crocuta crocuta. See Spotted hyena (*Crocuta crocuta*)

Crustacea. See Sand shrimp

Cucumber (*Cucumis sativus*), spider mite adaptation to cultivar of, 371; toxicity of cucurbitacin, 371

Cucumis sativus. See Cucumber (*Cucumis sativus*)
Cucurbita. See Gourd (*Cucurbita*)
Cultivated barley (*Hordeum vulgare*), inbreeding, 223; multilocus associations, 223; Appendix 4
Cutthroat trout (*Salmo clarki*), allelic diversity analysis, between subspecies, 361, within localities, 361, from inland drainages, 361; subspecies classification, 363, at isozyme level, 363; Appendix 3
Cyanoramphus auriceps. See Yellow-crowned parakeet (*Cyanoramphus auriceps*)
Cyanoramphus auriceps forbesi. See Chatam Islands Yellow-crowned or Forbes parakeet (*Cyanoramphus auriceps forbesi*)
Cyanoramphus novaezelandiae. See Red-fronted parakeet (*Cyanoramphus novaezelandiae*)
Cyanoramphus novaezelandiae and *Cyanoramphus auriceps*. See Red-fronted parakeet and Yellow-crowned parakeet (*Cyanorhamphus novaezelandiae* and *Cyanorhamphus auriceps*)
Cylindric blazing-star (*Liatris cylindracea*), allelic diversity analyses, 342; long-lived perennial outcrossing, 342; morphological and allozyme variation association, 344; outcrossing perennial, 342, 343; population analysis, genetic diversity, 342; subpopulation analysis, genetic diversity, 342
Cynomys ludovicianus. See Prairie dog (*Cynomys lucovicianus*)
Cyprinidae. See Carps and minnows (Cyprinidae)
Cyprinodon diabolis. See Devil's Hole pupfish (*Cyprinodon diabolis*)
Cyrtosperma (*Cyrtosperma*), weedy, 274

Dactylis glomerata. See Orchard grass (*Dactylis glomerata*)
Dahl sheep (*Ovis dalli kenaiensis*), area size, 426; locality, 426; population size, 426
Dark goosefoot (*Chenopodium atrovirens*), Appendix 4

Dasypus novemcinctus. See Nine-banded armadillo (*Dasypus novemcinctus*)
Daucus. See Carrot (*Daucus*)
Decapoda. See Decapods (Decapoda)
Decapods (Decapoda), correlations of genetic variation with trophic generality, 140, and number of species per genus, 140; patterns of environmental diversity, 140; trophic instability, 140
Deer Family (Cervidae), capacity of ISIS facilities for ungulates, 382
Deer mice (*Peromyscus*), measured character changes in subspecies of, 359; subspecific characteristics, 199; difficulty of maintenance, 199
Deer mouse (*Peromyscus maniculatus*); area size, 423; locality, 423; population size, 423; population-size to park-size relationship, 441, 442; survival, 441, 442
Dendroica kirtlandii. See Kirtland's warbler (*Dendroica kirtlandii*)
Dermochelys coriacea. See Leather-back turtle (*Dermochelys coriacea*)
Deroceras agreste. See Limacid slug (*Deroceras agreste*)
Deroceras caruanae. See Limacid slug (*Deroceras caruanae*)
Deroceras laeve. See Limacid slug (*Deroceras laeve*)
Deroceras reticulatum. See Limacid slug (*Deroceras reticulatum*)
Desert blue grass. (*Bothriochloa ewartiana*), hybridization, 271, with *B. bladhii*, 271
Desert goosefoot (*Chenopodium desiccatum*), Appendix 4
Desert slender salamander, species of interest to managers, Appendix 3
Desert tortoise, species of interest to managers, Appendix 3
Desmodium nudiflorum. See Bare-stem tick-clover (*Desmodium nudiflorum*)
Devil's Hole pupfish (*Cyprinodon diabolis*), color change, 359; measured character increase, 359; transplanted population, 359; species of interest to managers, Appendix 3
Dicerorhinus sumatrensis. See Suma-

tran rhinoceros (*Dicerorhinus sumatrensis*)
Diceros bicornis minor. See Black rhinoceros (*Diceros bicornis*)
Diceros bicornis. See Black rhinoceros (*Diceros bicornis*)
Dichanthium annulatum. See Grass (*Dichanthium annulatum*)
Dichanthium bladhii. See Grass (*Bothriochloa*)
Dichanthium fecundum. See Grass (*Dichanthium fecundum*)
Dichanthium insculptum. See Grass (*Bothriochloa insculpta*)
Dichanthium ischaemum. See Grass (*Bothriochloa ischaemum*)
Dicrostonyx groenlandicus. See Collared lemming (*Dicrostonyx groenlandicus*)
Digitaria sanguinalis. See Crabgrass (*Digitaria sanguinalis*)
Dik-dik (*Madoqua kirki*), inbreeding effects on juvenile mortality, 167
Dipodomys ordii. See Ord kangaroo rat (*Dipodomys ordii*)
Dog welk (*Nucellus lapillus*), chromosome number variability, 128
Dogs (*Canis*), in species extinction, oceanic islands, 111
Dolichotis salincola. See Salt-desert cavy (*Dolichotis salincola*)
Domestic horse (*Equus caballus*), chromosome complement, 404; comparison with Przewalski's horse, 404, 405
Dorcas gazelle (*Gazella dorcas*) NZP successful breeding group, 184; inbreeding effects on juvenile mortality, 167
Douglas-fir (*Pseudotsuga menziesii*), Appendix 3; Appendix 4
Drosophila (*Drosophila*), behavioral pattern differences in sibling species of, 99, chromosome analysis, 355; chromosome variation in sibling species of 99; cosmopolitan distribution, 98; courtship behavior differences, 98; crucial phase in founding, 197; darkness response differences in mating behavior, 98, 99; early generations after founder event, 197; eye color mutants in, 34; frequency of chromosome polymorphism in sibling species of, 99; genetic diversity at taxonomic level, 355; habitats, 98; homosequential species of Hawaiian, 355; individual differences and uniqueness, 196; inversion polymorphisms in, 36; meiotic chromosome differences, 99; mutation rates, 37; nutritional requirement differences, 99; polymorphism of chromosomes in, 355; predictable mutation rates, 37; quantitative differences in sibling species of, 99; reproductive organ differences, 98; seasonal variation in population size, sibling species of, 99; taxonomic information, 355, by chromosomal markers, 355; temperature tolerance differences, 99; temperature tolerance variation among sibling species of, 99
Drosophila (*Drosophila melanogaster*), alcohol tolerance of larvae, 99; chromosome rearrangements, 99; comparison with *Drosophila simulans*, 98, 99; high chromosome polymorphism, 99; high resistance to physiological stresses, 99; high temperature tolerance, 99; hybrid sterility, 98; inversion in chromosome III, 99; peak population size, January, 99
Drosophila (*Drosophila simulans*), alcohol tolerance of larvae, 99; chromosomal monomorphism, 99; comparison with *Drosophila melanogaster* 98, 99; hybrid sterility, 98; low resistance to physiological stresses, 99; low temperature tolerance, 99; peak population size, May, 99
Drosophila melanogaster. See Drosophila (*Drosophila melanogaster*)
Drosophila simulans. See Drosophila (*Drosophila simulans*)
Drummond phlox (*Phlox drummondii*), Appendix 4
Dusky Seaside sparrow (*Ammospiza maritima nigrescens*), Endangered Species Act, 306; gene pool of five males 306, 309; hybridization and preservation of alleles, 306, 309; hybridization with Scott's Seaside sparrow, 306, 309, F1 hybrids, 306,

backcross, 306, 309; sex ratio extinction, 117; U.S. Fish and Wildlife decisions, 306; wild bird genotypes, 306, 309, Appendix 3

Dusky Seaside sparrow and Scott's Seaside sparrow (*Ammospiza maritima nigrescens* and *Ammospiza maritima peninsulae*), hybridization to save Dusky Seaside sparrow genotype, 306, 309; U.S. Fish and Wildlife legal decision, 306

Dutch elm disease (*Ceratostomella ulmi*), elm and, 127

Dwarf buffalo (*Bulbalus depressicornis*). See Anoa (*Bulbalus depressicornis*)

Ectopistes migratorius. See Passenger pigeon (*Ectopistes migratorius*)

Edible aroids (Araceae), weediness, 274

Editha checkerspot butterfly (*Euphydryas editha*), adaptation to local abundance of host, 370; allele frequencies, 155, 157; anthropogenic extinction, 155; bottlenecks 155, 156; demographic units, 155, stress response, 155, dynamics, 156; ecotypes, 155, difference between, 155, environmental heterogeneity and, 155, 159; extinction factors, 155, 156; gene flow, 114, 155; gene frequency stability, 154; genetic variation, 155, 156, loss, 155, restoration, 155; habitat change, 155; heterozygosity, 155, 156

Elaphurus davidianus. See Père David's deer (*Elaphurus davidianus*)

Elaphus maximus. See Indian elephant (*Elaphus maximus*)

Eld's deer (*Cervus eldi thamin*), inbreeding effects on juvenile mortality, 166

Elephant Family (Proboscidia), area size, 424; population size, 424

Elephantidae. See Elephants (Elephantidae)

Elephants (Elephantidae) musth determination, 410; pregnancy diagnosis, 410; stool samples, 410

Elephantulus rufescens. See Elephant shrew (*Elephantulus rufescens*)

Elephant shrew (*Elephantulus rufescens*), inbreeding effects on juvenile mortality, 170

Eleusine coracana. See Finger millet (*Eleusine coracana*)

Elf owl, Appendix 3

Elk (*Cervus elaphus*), habitat overexploitation, 410; management level, 436; migration, 420; population decline, 438; population in park, removal of surplus animals, 420; protection measures, 438; reduction of allelic diversity, 438; small dependent populations, 436; space and survival, 420, 442; subspecies classification, 359; subspecies extinction, 359; subspecies relationships, 359; vulnerability, 438

Elm (*Ulmus*), and Dutch elm disease, 127

Elymus canadensis. See Canada wildrye (*Elymus canadensis*)

Emex spinosa. See Spring emex (*Emex spinosa*)

Engelmann spruce (*Picea engelmannii*), Appendix 4

English walnut (*Juglans regia*), and codling moth pest, 369, 370

Equidae. See Horse Family: horses, zebras, etc. (Equidae)

Equus burchelli. See Burchell's zebra (*Equus burchelli*)

Equus caballus przewalski. See Przewalski's (or Asian) wild horse (*Equus przewalski*)

Equus caballus. See Domestic horse (*Equus caballus*)

Equus grevyi. See Grevy's zebra (*Equus grevyi*)

Equus przewalski. See Przewalski's (or Asian) wild horse (*Equus przewalski*)

Eretmochelys and *Chelonia*. See Sea turtles (*Eretmochelys* and *Chelonia*)

Eretmochelys imbricata. See Hawksbill turtle (*Eretmochelys imbricata*)

Eschscholzia californica. See California poppy (*Eschscholzia californica*)

Eucalyptus obliqua. See Messmate stringybark eucalyptus (*Eucalyptus obliqua*)

Eucalyptus pauciflora. See Snow eucalyptus (*Eucalyptus pauciflora*)
Eufriesia purpurata. See Euglossine bee (*Eufriesia purpurata*)
Euglossine bee (*Eufriesia purpurata*), DDT tolerance, 127
Euphydryas. See Nearctic checkerspot butterflies (*Euphydryas*)
Euphydryas editha bayensis. See San Francisco Bay region checkerspot butterfly (*Euphydras editha bayensis*)
Euphydryas aurinia. See Marsh fritillary butterfly (*Euphydryas aurinia*)
Euphydryas chalcedona. See Chalcedon checkerspot butterfly (*Euphydryas chalcedona*)
Euphydryas editha. See Editha checkerspot butterfly (*Euphydryas editha*)
Euphydryas gillettii. See Gillett's checkerspot butterfly (*Euphydryas gillettii*)
European bison (*Bison bonasus*), early inbreeding, 175; lack of inbreeding effects in, 175
European strawberry (*Fragaria vesca*), Appendix 4
Evening-primrose (*Oenothera*), selfing and genetic variation, 340; among populations, 340; Appendix 4

Fabaceae. See Pea Family (Fabaceae)
Falco peregrinus. See Peregrine falcon (*Falco peregrinus*)
Fall cankerwork (*Alsophila pometaria*), abundance in maple, 369, oak, 369; clonal reproducing genotypes, 369; differences in phenological attributes, 369, behavioral attributes, 369
Farewell-to-spring (*Clarkia rubicunda*), Appendix 4
Felidae. See Cat Family: lions, tigers, etc. (Felidae); see also Large felids (Felidae)
Felis. Cats (*Felis*)
Felis catus. See Cat (*Felis catus*)
Felis concolor. See Mountain lion (*Felis concolor*)
Festuca microstachys = *Vulpia microstachys.* See Small fescue (*Festuca microstachys* = *Vulpia microstachys*)
Festuca ovina. See Sheep fescue (*Festuca ovina*)
Festuca rubra. See Red fescue (*Festuca rubra*)
Ficedula hypoleuca and *Ficedula albicollis.* See Collard flycatcher and Pied flycatcher (*Ficedula hypoleuca* and *Ficidula albicollis*)
Ficus carica. See Common fig (*Ficus carica*)
Filmy fern, Appendix 3
Finger millet (*Eleusine coracana*), gene flow, 274; mimetic races, 274; selection pressures, 274
Fireflies (*Photurus*), light flashes, 99; sibling species, 99
Fireweed (*Chamaenerion angustifolium*), Appendix 3; Appendix 4
Flickers (Picidae: *Colaptes*), future prospects, 443; reserve size for, 443
Four-striped rat, inbreeding effects on juvenile mortality, 170
Foxtail barley (*Hordeum jubatum*), Appendix 4
Foxtail brome (*Bromus rubens*), soft brome comparison with, 219; in genetic variability, 219; phenotypic plasticity, 219
Fragaria chiloensis. See Chiloé strawberry (*Fragaria chiloensis*)
Fragaria vesca. See European strawberry (*Fragaria vesca*)
Fragaria virginiana. See Virginia strawberry (*Fragaria virginiana*)
Frémont clematis, Appendix 3
Frémont goosefoot (*Chenopodium fremontii*), Appendix 4
French's shooting star, Appendix 3
Freshwater snails (*Goniobasis floridensis*), complex allelic diversity between drainage systems, 361
Frogs (Anura), genetic diversity gradients, 139; habitat type and genetic variation, 140
Frogs, song or call in sibling species identification of, 99

Galago crassicaudatus crassicaudatus. See Greater galago (*Galago crassicaudatus crassicaudatus*)

Gall aphids (Homoptera: Eriosomatidae), immunization agents against extinction, 373
Galliforms and Hummingbirds (Galliformes and Trochilidae), hybrid, 291; sympatry, 291
Galliformes and Trochilidae. See Galliforms and Hummingbirds (Galliformes and Trochilidae)
Gallus gallus. See Chicken (*Gallus gallus*)
Gallus. See Chicken (*Gallus*)
Garden bean (*Phaseolus vulgaris*), Appendix 4
Gaur (*Bos frontalis* = *B. gaurus*), in cross with Holstein cow, 411
Gaura (*Gaura demareei*), Appendix 4
Gaura (*Gaura longiflora*), Appendix 4
Gaura (*Gaura*), species distribution limitation and Gst values, 339
Gaura demareei. See Gaura (*Gaura demareei*)
Gaura longiflora. See Gaura (*Gaura longiflora*)
Gazella dorcas. See Dorcas gazelle (*Gazella dorcas*)
Gazella soemmeringi. See Soemmering's gazelle (*Gazella soemmeringi*)
Gazella spekei. See Speke's gazelle (*Gazella spekei*)
Gazella thomsoni. See Thomson's gazelle (*Gazella thomsoni*)
Geomys. See Pocket gophers (*Geomys*)
Geomys breviceps. See Pocket gopher (*Geomys breviceps*)
Geomys bursarius and *Geomys attwateri.* See Pocket gopher (*Geomys bursarius* and *Geomys attwateri*)
Geomys bursarius complex. See Pocket gophers (*Geomys bursarius*) complex
Geomydoecus ewingi. See Louse (*Geomydoecus ewingi*)
Geomydoecus subgeomydis. See Louse (*Geomydoecus subgeomydis*)
Giant eland (*Tragelaphus derbianus*), endangered species, 411; interspecific embryo transfer proposal, 411; survival, 411
Giant sable antelope (*Hippotragus niger variani*), captive habitat for, 383

Giant sequoia (*Sequoiadendron gigantea*), climatology and, 439; in cultivated situation, 438; in large protected areas, 438; long-lived species, 439; relict species, 439; reserves, 2; vulnerability of, 439, Appendix 3
Gilia (outcrossing) (*Gilia achilleifolia*), 222, Appendix 4
Gilia achilleifolia. See Yarrow gilia (*Gilia achilleifolia*) and Gilia (outcrossing) (*Gilia achilleifolia*)
Gillett's checkerspot butterfly (*Euphydryas gillettii*), eggs and larvae, 156; population extinction, 156; population fluctuation, 156; transplant populations, 156
Giraffe and Okapi Family (Giraffidae), capacity of ISIS facilities, for, 382
Giraffidae. See Giraffe and Okapi Family (Giraffidae)
Giraffa camelopardalis. See Giraffe (*Giraffa camelopardalis*)
Giraffe (*Giraffa camelopardalis*), importation, 165; inbreeding depression, 411; inbreeding effects on juvenile mortality, 166; introduction prohibition, 411; reproduction in zoos, 411; shipment of semen or fertilized eggs, 411
Glaucopsyche lygdamus. See Silver blue butterfly (*Glaucopsyche lygdamus*)
Glaucopsyche xerces. See Xerxes blue butterfly (*Glaucopsyche xerces*)
Glycine. See Soybean (*Glycine*)
Goat's beard (*Tragopogon*), gene flow between populations, 341; Gst values, 341; selfed and plumose-seeded, 341; Appendix 4
Goats (*Capra*), in species extinction, oceanic islands, 111
Gobiidae. See Goby (Gobiidae)
Goby (Gobiidae), genetic variation and environmental variation, 142
Golden hamster (*Mesocricetus auratus*), genetic variation present in; effects of inbreeding in, 175
Golden lion tamarin (*Leontopothecus rosalia rosalia*), inbreeding effects on juvenile mortality, 169; zoo breeding plans for, 165
Golden whistler (*Pachycephala pec-*

toralis), phenetic differences from close congeners, 300
Golden-winged warbler (*Vermivora chrysoptera*), adaptation of genotype to environment, 300; breeding grounds, 296; competition, 299; destruction, 300, disappearance, 299; distribution, 296, 297; habitat 296, 297, 299, 300, change, 297, 299, manipulation, 300; habitat specialist, 296; history, 296, 297; hybrid phenotypes, 299; hybridization, 299, 300; nesting, 296; numbers, 297; persistence, 300, in latitude, 300, in high altitude, 300; preservation, 300, long-term, 300; range, 296, expansion, 296; response to environmental changes, 297; sympatry, 300; with Blue-winged warbler, 300
Goniobasis floridensis complex. *See* freshwater snails (*Goniobasis floridensis*) complex
Goosefoot Family (Chenopodiaceae), Appendix 4
Gopher tortoise, Appendix 3
Gossypium. *See* Cotton (*Gossypium*)
Gourd (*Cucurbita*), weed races, 274
Grass (*Bothriochloa bladhii*), adaptation to climate, 270; apoximis, 270; aromatic compounds, 270; clones, 270; compilospecies, 271, 276; contact zones, 270, along suture, 270; crosses, 270; facultative sexual plant intercrosses, 270; genera integrity and, 271; genetic aggressor, 269–272; habitat, 270, disturbance, 270, 272, suture, 270; history, 270, taxonomy, 270; human transport of, hybrid, 276; hybrid swarms, 270; hybridization, 270, 276, with *B. ewartiana*, 271, 276, with *B. insculpta*, 271, 276, with *B. ischaemum*, 270, 276, with *Capillipedium*, 271, 276, with *Dichanthium*, 270, 271, 276; intergrades, 271, 276; interspecies contacts, 200, 271, 276; introgression products, 271, 276; model, compilospecies, 271, 276; morphological characters, 270, differentiation 270; name changed to *Dichanthium*, 271; ploidy, 269, 276; races, 270; sexual potential, 270; taxonomic dilemma, 271

Grass (*Bothriochloa insculpta*), genetic interaction with *B. bladhii*, 271; name changed to *Dichanthium*, 271

Grass (*Bothriochloa ischaemum*), apoximis, 270; contact zones, 269, 270, 276; distribution, 270, 275; habitat, 269, 270, 275, suture, 276; history, 270; hybridization with *B. bladhii*, 270; morphological characters, 270, 276, differentiation, 270, 276; ploidy, 269, 275; sexual potential, 270; variability, 270; weed, 271, 272

Grass (*Bothriochloa longifolia*), like possible diploid progenitor of *B. bladhii*, 269

Grass (*Capillipedium parviflorum*), hybridization, 271, with *B. bladhii*, 271

Grass (*Chloridion cameronii*), adaptive genetic differentiation, 80, 84, 95; copper tolerance in populations of, 80, 84; populations distance between, 80

Grass (*Dichanthium annulatum*), distribution, 271; hybridization, 270, 271; ploidy, 271; sexual potential, 271; sibling species, 271

Grass (*Dichanthium fecundum*), distribution, 271; sibling species, 271

Grass (*Trachypogon spicatus*), adaptive genetic differentiation in, 80, 84; copper tolerance of populations, 84; populations, distance between, 80; zinc tolerance of populations, 86

Gray squirrel (*Sciurus carolinensis*), area size, 423; locality, 423, population size, 423

Great Basin bristlecone pine (*Pinus longaeva*), long-lived tree, 342; outcrossing species, 342; population analysis, genetic diversity, 342; subpopulation analysis, genetic diversity, 342; succession stage and within population variation, 341; Appendix 4

Great gray owl, Appendix 3

Great tit (*Parus major*), deleterious effects of inbreeding in, 179; variation in clutch size, 133

Greater galago (*Galago crassicaudatus crassicaudatus*), inbreeding effect on juvenile mortality, 168
Green lacewing (*Chrysopa carnea*) comparison with *Chrysopa downesi*, 98; color, 98; description, 98; habitat, 98; reproduction rate, 98; *See also* Green lacewings (*Chrysopa*)
Green lacewing (*Chrysopa downesi*) breeding season, 98; color, 98; comparison with *Chrysopa carnea*, 98; habitat, 98; *See also* Green lacewings (*Chrysopa*)
Green lacewings (*Chrysopa*) color differences, 98; crossing experiments with, 98; fixation of alternate alleles, 98; genetic differences, 98; habitat differences, 98; interfertile, under laboratory conditions, 98; reproductive isolation and gene loci, 98; reproductive isolation by differences of three loci, 98; reproductive isolation under natural conditions, 98; reproductive season differences, 98; reproductive timing differences, 98; sibling species, 98
Green salamander, Appendix 3
Green turtle (*Chelonia mydas*), 104; distribution, 277; electrophoresis studies, 280; heterozygosity, 280; mark recapture programs, 279; populations as functional units, 280; sibling species, 281
Grevy's zebra (*Equus grevyi*), bloodlines, 385; captive habitat on Texas ranch, 383, 385; cooperation of institutions, 385; facilities, 385; founder stock, 385; population establishment (SSP), 383, 385; stock selection, 385; zoos, 383, 385
Grey duck (*Anas superciliosa*), hybridization, 293, with Australian Black duck, 293, with Mallard duck, 293
Griffon vulture (*Gyps fulvus*), extinction threat, 133; one egg, 133
Grizzly bear (*Ursus arctos horribilis*), area size, 427; future prospects, 444; habitat, 421; locality, 427; management level, 436; populations, 420; protection, 420, reserve size, 444; underestimation, space requirements, 422; vulnerability, 420, 421
Grus americana. See Whooping crane (*Grus americana*)
Guava (*Psidium*), cultivated/weed race interactions, 272, 275; naturalized escape, 275; weed races, 272, 275
Guineagrass panic grass (*Panicum maximum*), genetic variation in asexual and sexual races, 339; genetic variation within and between populations, 339; Appendix 4
Guinea pig (*Cavia porcellus*), inbreeding studies with, 175; inbred young and survival, 175
Guppy (*Poecilia reticulata*), stabilizing selection on caudal fin ray number, 135
Gymnogyps and *Vultur. See* Condors (*Gymnogyps* and *Vultur*)
Gymnosperms, Appendix 3, 4
Gyps fulvus. See Griffon vulture (*Gyps fulvus*)

Harvest mouse (*Cleithrodontomys*), area size, 423; locality, 423; population size, 423
Hawksbill sea turtle (*Eretmochelys imbricata*), diffuse nesting, 279; distribution, 279; mark and recapture problems, 279
Heart-leaved plantain (*Plantago cordata*), genetic investigations, 224; primacy of ecological factors, 224; rare species, 224
Heath hen (*Tympanuchus cupido cupido*), clutches, ground nesting of, and extinction, 118
Helianthus. See Sunflowers (*Helianthus*)
Helianthus annuus. See Common sunflower (*Helianthus annuus*)
Helianthus kubiniji. See Sun star (*Helianthus kubiniji*)
Helminthosporium maydis. See Southern corn leaf blight fungus (*Helminthosporium maydis*)
Hemiptera: Miridae. *See* Mirid bugs (Hemiptera: Miridae)
Herpestes. See Mongooses (*Herpestes*)
Hessian fly (*Mayetiola* [= *Phytoph-*

aga] destructor), gene for gene system, 367; virulent strains for wheat breeders, 366

Hibiscus. See Okra (*Hibiscus*)

Highland wild quinoa (*Chenopodium quinoa* var. *melanospermum*), Appendix 4

Hippopotamidae. See Hippopotamus Family (Hippopotamidae)

Hippopotamus Family (Hippopotamidae), capacity of ISIS for, 382

Hippotragus niger variani. See Giant sable antelope (*Hippotragus niger variani*)

Hippotragus niger. See Sable antelope (*Hippotragus niger*)

Homo sapiens. See Humans (*Homo sapiens*)

Homoptera: Cicadellidae. See Leafhoppers (Homoptera: Cicadellidae)

Homoptera: Aleyrodidae. See White flies (Homoptera: Aleyrodidae)

Homoptera: Eriosomatidae. See Gall aphids (Homoptera: Eriosomatidae)

Honey bee (*Apis mellifera*), DDT lethal dose, 127

Hordeum. See Barley (*Hordeum*)

Hordeum jubatum. See Foxtail barley (*Hordeum jubatum*)

Hordeum spontaneum. See Wild barley (*Hordeum spontaneum*), and ancestral two-row barley (*Hordeum spontaneum*)

Hordeum vulgare. See Cultivated barley (*Hordeum vulgare*), and six-row barley (*Hordeum vulgare*)

Horse Family: Horses, zebras, etc. (Equidae), capacity of ISIS for, 382

Horseshoe or King crab (*Limulus polyphemus*), genetic variation level, 146

House mouse (*Mus musculus*), area size, 423; locality, 423; population size, 423; breeding groups, 87; gene flow between demes, 87; genetic subdivision of, 90; in a single barn, 90; inbreeding effects, 172; mating patterns and differentiation of, 90; movement patterns and differentiation of, 90; pleiotropic effects of recessive allele in, 38

House sparrow (*Passer domesticus*), evolution of racial differences, 358; 359; storm survival by medium size phenotype, 134

House sparrow and Willow sparrow (*Passer domesticus* and *Passer hispaniolensis*), environmental adaptation of hybrid swarms, 292

Humans (*Homo sapiens*), cause of extinctions, 111, 134; comparison with chimpanzees, 357; DNA hybridization technique, 357; electrophoresis, 357; height and marriage, 134; immunology technique, 357

Hymenopappus (*Hymenopappus*), species distribution limitation and Gst values, 339

Hymenopappus (*Hymenopappus artemisiifolius*), Appendix 4

Hymenopappus artemisiifolius. See Hymenopappus (*Hymenopappus artemisiifolius*)

Hymenopappus scabiosaeus. See Whitebract hymenopappus (*Hymenopappus scabiosaeus*)

Hymenoptera (Hymenoptera), parasitoid biological control agent of scale insects, 365; competitive exclusion by new introduction, 365

Hypericum perforatum. See Klamath weed or St. John's-wort (*Hypericum perforatum*)

Icterus. See Midwestern oriole (*Icterus*)

Illinois mud turtle, Appendix 3

Indian elephant (*Elaphas maximus*), inbreeding effects on juvenile mortality, 166

Indian muntjac (*Muntiacus muntjak*) karyotype, 407, male and female differences, 407

Indian rhinoceros (*Rhinoceros unicornis*), distribution, 377; endangered, 376; numbers, 377; population trends, 377

Inyo brown towhee, Appendix 3

Ipomoea purpurea. See Common morning-glory (*Ipomoea purpurea*)

Irises, Appendix 3

Italian darnel (*Lolium multiflorum*), Appendix 4

Japanese quail (*Coturnix japonica*), egg weight variation from inbred 149, greater than outbred, 149; inbreeding and egg weight variance, 149; inbreeding and body weight, muscle weight, 147; homeostatic capacity at intraindividual level, 149; mate preference, 304; relatedness, 304

Japanese serow (*Capricornis crispis*), inbreeding effects on juvenile mortality, 167

Javan rhinoceros (*Rhinoceros sondaicus*), capacity of ISIS for, 377; endangered, 376; numbers, 377; population trends, 377; rare species, 376

Johnson grass (*Sorghum halepense*), adaptation to climate, 274; aggression, 274; distribution, 274; evolution, 274; introgression, 274, of cultivated diploid sorghum, 274; origin, U.S.A., 274; populations, 274, merging of, 274; prussic acid content, 274

Juglans regia. See English walnut (*Juglans regia*)

Kangaroo rat, Appendix 3

Katsuwonis pelamis. See Skipjack tuna (*Katsuwonis pelamis*)

Kemp's ridley sea turtle (*Lepidochelys kempii*), colonies, 284; distribution, 277; endangered species, 277; extinction threat, 279; human activities, 284, egg taking, 284, international cooperation, 284, mortality in shrimp trawls, 284; nesting, 279, grounds, 279; population establishment, 284, after crash, 284; research problems, 279; U.S. and Mexican government cooperation, 284

Kentucky lady's-slipper, Appendix 3

Kirtland's warbler (*Dendroica kirtlandii*), endangered species management, 300; habitat manipulation for, 300; species elimination, 300, with negative biological impact, 300; Appendix 3

Klamath weed or St. John's-wort (*Hypericum perforatum*), control by chrysomelid beetles, 366

Lacerta mellisellensis. See Lizard (*Lacerta mellisellensis*)

Lactuca. See Lettuce (*Lactuca*)

Lagomorpha. See Pika, Rabbit, Hare (Lagomorpha)

Lampropeltris triangulum. See Milk snake (*Lampropeltris triangulum*)

Lanius collurio and *Lanius phoenicuroides*. See Red-backed shrike and Red-tailed shrike (*Lanius collurio* and *Lanius phoenicuroides*)

Large blue butterfly (*Maculinea arion*), ant-host symbiosis, 160; environmental change, 160; extinction, 160; habitat, sheep relationships, 160, ant relationships, 160; larval food, wild thyme, 160

Large copper butterfly (*Lycaena dispar*), collecting 159; ecotypes 160; extinction, 159; genetic differentiation of stocks, 159; habitat, 159; management 159; mosaic population dynamic pattern, 260; reintroduction, 159; survival, 159, with human aid, 159

Large felids (Felidae), ISIS captive facilities, 381

Largemouth bass (*Micropterus salmoides*), allele frequency, 90, divergence of, 90; breeding programs, 105; catchability, 105; genetic divergence, 90, with thermal effluents, 90; growth rate, 105; populations, 90, adjacent but isolated, 90; return to normal, 90; selection pressure, 90, man-imposed, 90; thermal stability and enzymes, 90, in populations of, 90; trait selection, 105, economic importance, 105

Laspeyresia pomonella. See Codling moth (*Laspeyresia pomonella*)

Leaf-nosed bats (Phyllostomidae), chromosomes, 16, 17

Leafhoppers (Homoptera: Cicadellidae), damage reduction on cotton, 371, with cotton hairiness, 371, with sesquiterpene, gossypol, 371

Leatherback sea turtle (*Dermochelys coriacea*), distribution, 277; feeding grounds, 278; mark and recapture programs, 279; migration, 277, 278; nesting grounds, 279; size, 277

Lemons (*Citrus*), cultivated/weed race interactions, 272, 275
Lemur catta. See Ring-tailed lemur (*Lemur catta*)
Lemur fulvus. See Brown lemur (*Lemur fulvus*)
Lemur macaco. See Black lemur (*Lemur macaco*)
Lemuridae. See Lemurs (Lemuridae)
Lemurs (Lemuridae), annual cycles, 410; stool samples, 410
Leontopithecus rosalia rosalia. See Golden lion tamarin (*Leontopithecus rosalia rosalia*)
Leopard (*Panthera onca*), capacity of ISIS for, 381
Leopard (*Panthera pardus*), area size, 427, 428; locality, 427, 428; population size, 427, 428
Lepidochelys kempii. See Kemp's ridley sea turtle (*Lepidochelys kempii*)
Lepidochelys olivacea. See Olive ridley sea turtle (*Lepidochelys olivacea*)
Lepidoptera: Pieridae. See Pierid butterfly (Lepidoptera: Pieridae)
Lepidopteran caterpillars (Lepidoptera), viral death of, 366
Lepomis. See Sunfish (*Lepomis*)
Leptinotarsa decemlineata. See Colorado potato beetle (*Leptinotarsa decemlineata*)
Lepus americanus. See Snowshoe hare (*Lepus americanus*)
Lettuce (*Lactuca*), weed races, 274
Liatris cylindracea. See Cylindric blazing-star (*Liatris cylindracea*)
Lice (Trichodectidae) association with pocket gophers, 101, 103
Limacidae. See Terrestrial slugs (Limacidae)
Limacid slug (*Deroceras agreste*), distribution, 210; heterozygosity (H), 207, 208; homozygous strains, 207, 208; outcrossing species, 207; polymorphism (P), 207, 208; populations, number, 207; region, 207
Limacid slug (*Deroceras caruanae*), heterozygosity (H), 207, 210; outcrossing species, 207; polymorphism (P), 207, 208; populations, number, 207; region, 207
Limacid slug (*Deroceras laeve*), heterozygosity (H), 207, 208; mixed system, 207; outcrossing species, 207; polymorphism (P), 207, 208; populations, number, 207; region, 207
Limacid slug (*Deroceras reticulatum*), distribution, 208; heterozygosity (H), 207, 208; outcrossing species, 207; polymorphism (P), 207; populations, number, 207; region, 207
Limacid slug (*Limax marginatus*), genetic variability and breeding system, 207.
Limacid slug (*Limax maximus*), distribution, 208; genetic variability and breeding system, 207; heterozygosity (H), 207, 208; outcrossing species, 207, 208; polymorphism (P), 207; populations, number, 207; region, 207
Limacid slug (*Limax pseudoflavus*), heterozygosity (H), 207; genetic variability and breeding system, 207; heterozygosity (H), 207; outcrossing species, 207; polymorphism (P), 207; populations, number, 207; region, 207
Limacid slug (*Limax tenellus*) heterozygosity (H), 207; genetic variability and breeding system, 207; heterozygosity (H), 207; outcrossing species, 207; polymorphism (P), 207; populations, number, 207; region, 207
Limacid slug (*Limax valentianus*), genetic variability and breeding system, 207; heterozygosity (H), 207; outcrossing species, 207; polymorphism (P), 207; populations, number, 207; region, 207
Limacid slug (*Milax budapestensis*), heterozygosity (H), 207; outcrossing species, 207; polymorphism (P), 207; populations, number, 207; region, 207
Limacid slug (*Milax gagates*), heterozygosity (H), 207; outcrossing species, 207; polymorphism (P), 207; populations, number, 207; region, 207
Limacid slug (*Milax sowerbyi*), heterozygosity (H), 207; outcrossing

species, 207; polymorphism (P), 207; populations, number, 207; region, 207
Limax marginatus. See Limacid slug (*Limax marginatus*), and Mollusks
Limax maximus. See Limacid slug (*Limax maximus*), and Mollusks
Limax pseudoflavus. See Limacid slug (*Limax pseudoflavus*), and Mollusks
Limax tenellus. See Limacid slug (*Limax tenellus*), and Mollusks
Limax valentianus. See Limacid slug (*Limax valentianus*), and Mollusks
Limulus polphemus. See Horseshoe or King crab (*Limulus polyphemus*)
Lion (*Panthera leo*), *See* African lion (*Panthera leo*)
Lizard (*Lacerta mellisellensis*), lack of congruence in subspecies, 360, classification, 360
Lizards, displays, 99; sibling species, 99
Lobed clarkia (*Clarkia biloba*), Appendix 4
Loblolly pine (*Pinus taeda*), Appendix 4
Lodgepole pine (*Pinus contorta* spp. *latifolia*), succession stage and within population variation, 341; Appendix 4
Loggerhead sea turtle (*Caretta caretta*), distribution, 277; electrophoresis studies, 280; heterozygosity, 280; mark and recapture results, 279; park beach nesting, 420; populations as independent units, 280
Lolium multiflorum. See Italian darnel (*Lolium multiflorum*)
Lolium perenne. See Perennial darnel (*Lolium perenne*)
Loosestrife (*Lythrum tribracteatum*), Appendix 4
Lotus corniculatus. See Bird's-foot trefoil (*Lotus corniculatus*)
Louse (*Geomydoecus ewingi*), infestation of pocket gopher (*Geomys breviceps*), 101
Louse (*Geomydoecus subgeomydis*), in interspecific contact of pocket gophers 101, 103; infestation of pocket gopher (*Geomys attwateri*), 101

Loxodonta africana. See African elephant (*Loxodonta africana*)
Lucy Braun's white snakeroot, Appendix 3
Lupines (*Lupinus*), butterfly impact on buds, 157; early flowering, 157; snowstorm effects, 157, and extinction of butterflies, 157; species distribution limitation and Gst values, 339
Lupinus subcamosus. See Texas bluebonnet (*Lupinus subcamosus*)
Lupinus texensis. See Texas bluebonnet (*Lupinus texensis*)
Lupinus. See Lupines (*Lupinus*)
Lycaena dispar. See Large copper butterfly (*Lycaena dispar*)
Lycaon pictus. See Cape hunting dog (*Lycaon pictus*)
Lycopersicon esculentum var. *cerasiformae. See* Cherry tomato (*Lycopersicon esculentum* var. *cerasiformae*)
Lycopersicon esculentum. See Common tomato (*Lycopersicon esculentum*)
Lycopersicon. See Tomato (*Lycopersicon*)
Lythrum tribracteatum. See Loosestrife (*Lythrum tribracteatum*)

Macaca fascicularis. See Crab-eating macaque (*Macaca fascicularis*)
Macaca mulatta. See Rhesus macaque (*Macaca mulatta*)
Macaca nemestrina. See Pig-tailed macaque (*Macaca nemestrina*)
Macaca nigra. See Celebes black ape (*Macaca nigra*)
Maculinea arion. See Large blue butterfly (*Maculinea arion*)
Madagasgar Turtle dove (*Streptopelia picturata picturata*), characters, 301, plumage, 301, wings, 301; contact zone, 301, islands, with Seychelle Turtle dove, 301; history, 301; hybridization, 301; multiple introductions, 301; stabilization of hybrid swarms, 301
Madoqua kirki. See Dik-dik (*Madoqua kirki*)
Maize (*Zea mays*), gene flow, 274;

mimetic races, 274; phenotypic variance ratio, inbred to outbred, 146; selection pressures, 274, breeding of individuals genetically unique, 196; with environmentally unique experiences 196; Appendix 4. *See also* Corn.

Mallard duck (*Anas platyrhynchos*), adaptability, 293; competitive exclusion, 308; domestication, 293, hybridization in, 293; distribution, 293, 308; environmental change response, 293; gene exchange, 295; genetic assimilation capacity, 293, 294; habitat, 293, modification by human activities, 293; hunting, 294, 295, bag, 294, 295; hybrid swarm with American Black duck, 293; hybrids, 293–295, 308, fitness, 294, fertility, 294–295, hybridization, 293–295, frequency with other *Anas*, 293; hybridizations, 293–295, 308, with American Black duck, 293, 308, with Australian Black duck (Spotbill duck), 293, with Grey duck, 293, 294, with Mexican duck, 293, 295, 308, with Mottled duck, 293, 294, 308; introductions, 293, 294; invasive capacity, 293; numbers increase, 294, 308; range expansion, 293, 308, duck contact zones, 293, 294; relative population sizes, ducks at contact zones, 294, 308

Mallard duck and American Black duck (*Anas platyrhynchos* and *Anas rubripes*), decline, 294; habitat, 294; hybrid, 293, 294, 309; range extension, 294; survival, 294

Mallard duck and Australian Black duck (*Anas platyrhynchos* and *Anas superciliosa "rogersi"*), hybrid, 293

Mallard duck and Grey duck (*Anas platyrhynchos* and *Anas superciliosa*), hybrid, 293

Mallard duck and Mexican duck (*Anas platyrhynchos* and *Anas diazi*), distribution, 295; hybrid, 293, 295, 308; hunting pressure, 295; populations, 295, 296; stability, 296; range, 295; taxonomic status, 295; U.S. population legal action, 295

Mallard duck and Mottled duck (*Anas platyrhynchos* and *Anas fulvigula*), infrequent hybrid, 293, 294, 308

Malus domestica. *See* Apple (*Malus domestica*)

Man (*Homo sapiens*), *See* Humans (*Homo sapiens*)

Mandrill (*Mandrillus sphinx*), inbreeding effects on juvenile mortality, 169

Mandrillus sphinx. *See* Mandrill (*Mandrillus sphinx*)

Mangifera. *See* Mango (*Mangifera*)

Mango (*Mangifera*), cultivated/weed race interactions, 272, 275; naturalized escape, 275; weed races, 272, 275

Maple (*Acer*), fall cankerwork in, 369

Marmota flaviventris. *See* Yellow-bellied marmot (*Marmota flaviventris*)

Marmota. *See* Marmots (*Marmota*)

Marmots (*Marmota*), future prospects, 443; reserve size for, 443

Marsh fritillary butterfly (*Euphydryas aurinia*), population size fluctuation, 162; phenotypic changes in variability 162

Massasauga rattlesnake, Appendix 3

Mead's milkweed, Appendix 3

Medicago sativa. *See* Alfalfa (*Medicago sativa*)

Mediterranean fruit fly (*Ceratitis capitata*), malathion spraying control, 158, 159; effect on butterfly population, 158

Mediterranean land snail (*Ruminia decollata*), adaptation of mixed genomes, 211; allele fixation in strains of, 211; backcrossing effects, 214; body color 211, 212; colonies, 212; dispersal, 214; facultative self-fertilization, 211; fecundity, 211, 212, 214; genetic diversity, 211, 213; genetic structure, 213; genomes, mixed 211; geographic distribution, 214; habitat, 214; hybrids, 212, 213; individuals, electrophoresis, 212; model, selfing strains of molluscs, 212–214; population structure, 211, 214, change, 211; reproductive age

differential, 211; selfing rate, 211; strains (two local), 211; strains (many), 214; survival, 214; temporal changes, 211
Melilotus alba. See White sweet clover (*Melilotus alba*)
Merced clarkia (*Clarkia lingulata*), Appendix 4
Mesocricetus auratus. See Golden hamster (*Mesocricetus auratus*)
Messmate stringybark eucalyptus (*Eucalyptus obliqua*), Appendix 4
Mexican duck (*Anas diazi*), competitive exclusion effects, 308; distribution, 295; endangered population, 295; 296; endangered species, 295; federal action, 295; hunting, 295; hybridization, 293, 295, 296, with Mallard, 293, 295, 296; interactions, 295, 296, 308; numbers, 295; population stability, 296; pressure, 295; response to environmental changes, 295; taxonomic status, 295
Mice (*Mus*), chromosomal divergence, 406; individuals genetically unique, 196; with environmentally unique experiences, 196
Micropterus salmoides. See Largemouth bass (*Micropterus salmoides*)
Microtus pinetorum. See Pine vole (*Microtus pinetorum*)
Midwestern oriole (*Icterus*), assortative mating to sympatry and speciation, 292
Milax budapestensis. See Limacid slug (*Milax budapestensis*), and Mollusks
Milax gagates. See Limacid slug (*Milax gagates*), and Mollusks
Milax sowerbyi. See Limacid slug (*Milax sowerbyi*), and Mollusks
Milk snake (*Lampropeltis triangulum*), biological significance, 363; esthetics of subspecies, 363
Mimulus guttatus. See Common monkeyflower (*Mimulus guttatus*)
Mink (*Mustela vison*), alleles for coat color controlling fertility (pleiotropic effects), 37, 38
Mirid bugs (Hemiptera: Miridae), damage increase on cotton with frego bract, 371; damage reduction on hairless cotton, 371, with extrafloral nectaries absence, cotton, 371
Mirounga angustirostris. See Northern elephant seal (*Mirounga angustirostris*)
Mollusks (*Arion ater ater*), genetic variance and breeding system, 206
Mollusks (*Arion ater rufus*), genetic variance and breeding system, 206
Mollusks (*Arion circumscriptus*), genetic variance and breeding system, 206
Mollusks (*Arion distinctus*), genetic variance and breeding system, 206
Mollusks (*Arion hortensis*), genetic variance and breeding system, 206
Mollusks (*Arion intermedius*), genetic variance and breeding system, 206
Mollusks (*Arion lusitanicus*), genetic variance and breeding system, 206
Mollusks (*Arion oweni*), genetic variance and breeding system, 206
Mollusks (*Arion silvaticus*), genetic variance and breeding system, 206
Mollusks (*Arion subfuscus*), genetic variance and breeding system, 206
Mollusks (*Deroceras agreste*), genetic variance and breeding system, 207
Mollusks (*Deroceras caruanae*), genetic variance and breeding system 207
Mollusks (*Deroceras laeve*), genetic variance and breeding system, 207
Mollusks (*Deroceras reticulatum*), genetic variance and breeding system, 207
Mollusks (*Limax marginatus*), genetic variance and breeding system, 207
Mollusks (*Limax maximus*), genetic variance and breeding system, 207
Mollusks (*Limax pseudoflavus*), genetic variance and breeding system, 207
Mollusks (*Limax tenellus*), genetic variance and breeding system, 207
Mollusks (*Limax valentianus*), genetic variance and breeding system, 207
Mollusks (*Milax budapestensis*), genetic variance and breeding system, 207
Mollusks (*Milax gagates*), genetic variance and breeding system, 207

Mollusks (*Milax sowerbyi*), genetic variance and breeding system, 207
Mongolian wild horse. *See* Przewalski's horse (*Equus przewalski*)
Mongooses (*Herpestes*), in species extinction, oceanic islands, 111
Monterey clover, Appendix 3
Monterey pine (*Pinus radiata*), Appendix 4
Moose (*Alces alces americanus*), area size, 425; hunting-age class, 398; locality, 425, population size, 425; removal effect on effective population size, 398; removal effect on generation time, 398; variation, organization in, 77, isolation by distance and, 77
Mottled duck (*Anas fulvigula*), competitive exclusion effects, 308; contact with Mallard duck, 294; hybridization with Mallard duck, 294, 308, interactions, 308; population size, 294, other duck comparisons, 294; response to environmental change, 308
Mountain lion (*Felis concolor*), habitat available, 442, patchiness, 442; management level, 436; population longevity prospects, 441, 442; population-size to park-size relationship, 441, 442; small dependent populations, 436
Mouse (*Mus*), *See* Mice (*Mus*)
Mule = black-tail deer (*Odocoileus hemionus*), area size, 425; locality, 425; population size, 425
Muntiacus muntjak. *See* Indian muntjac (*Muntiacus muntjak*)
Muntiacus reevesi. *See* Chinese muntjac (*Muntiacus reevesi*); *See also* Muntjac (*Muntiacus reevesi*)
Muntiacus. *See* Muntjac (*Muntiacus*)
Muntjac (*Muntiacus*), karyotypic diversity, 407
Mus. *See* Mice (*Mus*)
Mus musculus. *See* House mouse (*Mus musculus*)
Musa. *See* Banana (*Musa*)
Muscicapid (Muscicapidae) flycatcher and Golden whistler (*Pachycephala pectoralis*), environmental adaptation of hybrid swarms, 292

Muscicapidae and *Pachycephala pectoralis*. *See* Muscicapid (Muscicapidae) flycatcher and Golden whistler (*Pachycephala pectoralis*)
Mustard (*Brassica*), weed races, 274
Mustela vison. *See* Mink (*Mustela vison*)
Mayetiola (= *Phytophaga*) *destructor*. *See* Hessian fly (*Mayetiola* [= *Phytophaga*] *destructor*)
Myoprocta pratti. *See* Acouchi (*Myoprocta pratti*)
Myrmica sabuleti. *See* Ant (*Myrmica sabuleti*)
Myrmica scabrinodis. *See* Ant (*Myrmica scabrinodis*)

Narrowleaf goosefoot (*Chenopodium pratericola*), Appendix 4
Nasua nasua. *See* Coatimundi (*Nasua nasua*)
Native lily, Appendix 3
Nearctic checkerspot butterflies (*Euphydryas*), allozyme frequencies, 153; demographic unit changes, 153, 156; density-independence, 156; dynamics, 153, 156; inbreeding depression, lack, 156; population extinction, 153, 156; population size and gene frequency interaction, 153, 156; reestablishment, 156; role of genetics, 156, in extinction, 156; stochastic (random) extinction, 156; transplants, 156
Neotoma. *See* Wood rats (*Neotoma*)
Nestronia, Appendix 3
Nicotiana tabacum. *See* Tobacco (*Nicotiana tabacum*)
Nilaparvata lugens. *See* Brown planthopper (*Nilaparvata lugens*)
Nine-banded armadillo (*Dasypus novemcinctus*), behavioral needs, 409; endocrine relationships, 409; nutritional needs, 409; progesterone secretion, 410, after captivity, 410; reproductive physiology, 409, problems in captivity, 409
North American elk. *See* Elk (*Cervus elaphus*)
North American Wood duck (*Aix sponsa*), captive propagation and release, 306; endangered species

history, 306; hatchling translocations, 306, habitat aid, 306; protection from hunting, 306; reestablishment, 306; response to management, 306
Northern elephant seal (*Mirounga angustirostris*), electrophoretic assay, 149; genetic variation in, 149, 354, five breeding colony samples, 354; niche competition free, 150; population bottleneck, 149; variation unrevealed by electrophoresis, 354; within population selection, 150
Northern oriole, subspecies by electrophoresis, 360
Northern white rhinoceros (*Ceratotherium simum cottoni*), See also White rhinoceros (*Ceratotherium simum*), captive habitat, 379; captive populations, 378, 379; captive propagation, 408; distribution, 377; effective population size, 379; endangered species, 380; facilities, 379; institutions with, 378; in the wild, 377; number in wild, 408; number in zoos, 408; number of males/number of females, 378; population trends, 377; reproduction in captivity, 378; research at San Diego Zoo, 380; SSP program, 380; strategy for conservation, 408; systematics, 380; white rhino dilemma, 380
Norway spruce (*Picea abies*), succession stage and within population variation, 341; Appendix 4
Nucellus lapillus. See Dog welk (*Nucellus lapillus*)
Nuculaspis californica. See Black pineleaf scale insect (*Nuculaspis californica*)

Oak (*Quercus*), fall cankerworm in, 369
Oats (*Avena*), gene flow, 274; mimetic races, 274; selection pressures, 274
Octodontomys gliroides. See Boris (*Octodontomys gliroides*)
Odocoileus hemionus. See Mule = black-tail deer (*Odocoileus hemionus*)

Odocoileus virginianus borealis. See White-tailed deer (*Odocoileus virginianus*)
Odocoileus virginianus. See White-tailed deer (*Odocoileus virginianus*)
Oenothera (*Oenothera grandis*), Appendix 4
Oenothera biennis. See Common evening-primrose (*Oenothera biennis*)
Oenothera grandis. See *Oenothera* (*Oenothera grandis*)
Oenothera laciniata. See Cutleaf evening-primrose (*Oenothera laciniata*)
Oenothera parviflora. See Smallflower evening-primrose (*Oenothera parviflora*)
Oenothera. See Evening-primrose (*Oenothera*)
Okapi (*Okapia johnstoni*), founder representation, 392; zoo exchanges, 392
Okapia johnstoni. See Okapi (*Okapia johnstoni*)
Okra (*Hibiscus*), weed races, 274
Old world wild rabbit (*Oryctolagus cuniculus*), rarity of inbreeding in, 179
Old-world bluestem grasses (*Bothriochloa-Dichanthium*) complex, apomixis, 269; sterility escape, 269
Olive baboon (*Papio anubis*), deleterious effects of inbreeding in, 179
Olive ridley sea turtle (*Lepidochelys olivacea*), distribution, 277
Opuntia stricta. See Prickly-pear (*Opuntia stricta*)
Oranges (*Citrus*), cultivated/weed race interactions, 272, 275; naturalized escape, 272; weed races, 272, 275
Orangutan (*Pongo pygmaeus*), chromosome structure differences, 405, with other apes, 405
Orchard grass (*Dactylis glomerata*), Appendix 4
Ord kangaroo rat (*Dipodomys ordii*), allelic diversity between populations, 360
Origanum vulgare. See Wild majoram (*Origanum vulgare*)
Oryx dammah. See Scimitar-horned oryx (*Oryx dammah*)
Oryx leucoryx. See Arabian oryx (*Oryx leucoryx*)

Oryza sativa. See Rice (*Oryza sativa*)
Oryza. See Rice (*Oryza*)
Osmerus mordax. See Rainbow smelt (*Osmerus mordax*)
Owl monkey (*Aotus trivirgatus*), chromosome complement, 406, variability, 406; genetic diversity, 406; hybrids, 406; reduced fertility, 406, sterility, 406; subspecies, 406; success in captivity, 406
Ovis aries. See Sheep (*Ovis aries*)
Ovis canadensis canadensis. See Bighorn sheep (*Ovis canadensis canadensis*)
Ovis dalli kenaiensis. See Dahl sheep (*Ovis dalli kenaiensis*)
Oxeye daisy (*Chrysanthemum leucanthemum*), differentiation at geographic levels, 343; differentiation at microgeographic levels, 343; habitat, 343; outcrossing species, 342, 343, variation within population subdivisions, 342, 343; population analysis, genetic diversity, 342; short-lived perennial subpopulation analysis, genetic diversity 342; Appendix 4

Pachycephala pectoralis. See Golden whistler (*Pachycephala pectoralis*)
Pan troglodytes. See Chimpanzee (*Pan troglodytes*)
Panicum maximum. See Guineagrass panic grass (*Panicum maximum*)
Panthera leo. See African lion (*Panthera leo*)
Panthera onca. See Leopard (*Panthera onca*)
Panthera pardus. See Leopard (*Panthera pardus*)
Panthera tigris altaica. See Bengal tiger (*Panthera tigris altaica*) and Siberian tiger (*Panthera tigris altaica*)
Panthera tigris sumatrae. See Sumatran tiger (*Panthera tigris sumatrae*)
Panthera tigris tigris. See White tiger (*Panthera tigris tigris*)
Panthera tigris. See Tiger (*Panthera tigris*)
Panthera uncia. See Snow leopard (*Panthera uncia*)

Papilio dardanus. See Swallowtail butterfly (*Papilio dardanus*)
Papilio machaon. See Swallowtail butterfly (*Papilio machaon*)
Papio anubis. See Olive baboon (*Papio anubis*)
Papio cynocephalus. See Yellow baboon (*Papio cynocephalus*)
Papio. See Baboons (*Papio*)
Papaya (*Carica*), cultivated/weed race interactions, 275, naturalized escape, 275
Paradisaeidae. See Birds of paradise (Paradisaeidae)
Parulidae. See Warblers (Parulidae)
Parus major. See Great tit (*Parus major*)
Passenger pigeon (*Ectopistes migratorius*), breeding behavior and extinction of, 118; K-selected species, 118
Passer domesticus. See House sparrow (*Passer domesticus*)
Passer domesticus and *Passer hispaniolensis.* See House sparrow and Willow sparrow (*Passer domesticus* and *Passer hispaniolensis*)
Pea Family (Fabaceae), color inheritance, 26; Mendelian inheritance and, 22–24; Appendix 4
Pearl millet (*Pennisetum glaucum*), gene flow, 274, mimetic races, 274; selection pressures, 274
Pectinophora and *Heliothis.* See Bollworms (*Pectinophora* and *Heliothis*)
Pectinophora gossypiella. See Pink bollworm (*Pectinophora gossypiella*)
Pennisetum glaucum. See Pearl millet (*Pennisetum glaucum*)
Pennsylvania smartweed (*Polygonum pensylvanicum*), Appendix 4
Père David's deer (*Elaphurus davidianus*), early inbreeding, 175; founder number, 387; inbreeding effects on juvenile mortality, 166; lack of inbreeding effects in, 175; success of captive preservation, 387; variability by electrophoresis, 387
Peregrine falcon (*Falco peregrinus*), artificial restocking, 307; environmental changes, 307; genetic

heterogeneity maximization, 307; individuals, 307, hybridized in captivity, 307, from other parts of range, 307; legal status, 307, exotics, 307, intraspecific hybrids, 307; population recovery, 307
Peromyscus. See Deer mice (*Peromyscus*)
Peromyscus maniculatus. See Deer mouse (*Peromyscus maniculatus*)
Petromyzon marinus. See Sea lamprey (*Petromyzon marinus*)
Phaseolus coccineus. See Scarlet runner bean (*Phaseolus coccineus*)
Phaseolus vulgaris. See Garden bean (*Phaseolus vulgaris*)
Phlox (*Phlox cuspidata*), Appendix 4
Phlox (*Phlox*), selfing and genetic variation, 340, among populations, 340
Phlox cuspidata. See Phlox (*Phlox cuspidata*)
Phlox drummondii. See Drummond phlox (*Phlox drummondii*)
Phlox roemeriana. See Roemer phlox (*Phlox roemeriana*)
Photurus. See Fireflies (*Photurus*)
Phycitid moth (*Cactoblastis cactorum*), prickly-pear control by, 226
Phyllostomidae. See Leaf-nosed bats (Phyllostomidae)
Phyllostomus hastatus. See Spearnose bat (*Phyllostomus hastatus*)
Phytophthora infestans. See Potato blight (*Phytophthora infestans*)
Picea abies. See Norway spruce (*Picea abies*)
Picea engelmannii. See Engelmann spruce (*Picea engelmannii*)
Picea sitchensis. See Sitka spruce (*Picea sitchensis*)
Picidae: *Colaptes*. See Flickers (Picidae: *Colaptes*)
Picture-winged *Drosophila*, cross-fertilizing species, 192; founder event on oceanic islands, 192; gene pool transported to new locality, 192, population bottleneck, 192, population characteristics, 192, 193, lack genetic change in transportation, 192, 193; sexually reproducing species, 192
Pied flycatcher. See Collared flycatcher and Pied flycatcher

Pierid butterfly (Lepidoptera: Pieridae), egg laying on introduced crucifers, 371, larval survival failure, 371; glucosinolate sharing with species, 371
Pieris protodice. See Checkered white butterfly (*Pieris protodice*)
Pig Family (Suidae), capacity of ISIS for, 382
Pig-tailed macaque (*Macaca nemestrina*), inbreeding effects on juvenile mortality, 169
Pigs (*Sus*), in species extinction, oceanic islands, 111; pest, 420
Pigweed (*Amaranthus* spp.), Appendix 4
Pika, Rabbit and Hare Family (Lagomorpha), area size, 423; population size, 423
Pine vole (*Microtus pinetorum*), area size, 423; locality, 423; population size, 423; population-size to park-size relationship, 441, 442; survival, 441, 442
Pink bollworm (*Pectinophora gossypiella*), damage reduction on hairless cotton, 371
Pink Family (Caryophyllaceae), Appendix 4
Pinus contorta ssp. *latifolia*. See Lodgepole pine (*Pinus contorta* ssp. *latifolia*)
Pinus contorta. See Lodgepole pine (*Pinus contorta* ssp. *latifolia*)
Pinus lambertiana. See Sugar pine (*Pinus labertiana*)
Pinus longaeva. See Great Basin bristlecone pine (*Pinus longaeva*)
Pinus nigra. See Austrian pine (*Pinus nigra*)
Pinus ponderosa. See Ponderosa pine (*Pinus ponderosa*)
Pinus pungens. See Table Mountain pine (*Pinus pungens*)
Pinus radiata. See Monterey pine (*Pinus radiata*)
Pinus rigida. See Pitch pine (*Pinus rigida*)
Pinus sylvestris. See Scotch pine (*Pinus sylvestris*)
Pinus taeda. See Loblolly pine (*Pinus taeda*)
Pinus virginiana. See Virginia pine (*Pinus virginiana*)
Pitch pine (*Pinus rigida*), succes-

sional stage and within population variation, 341; Appendix 4
Plagiodontia. See zagouti (*Plagiodontia*)
Plains wild-indigo (*Baptisia leucophaea*), Appendix 4
Plantago cordata. See Heart-leaved plantain (*Plantago cordata*)
Plethodon kentuckiae. See Salamander (*Plethodon kentuckiae*)
Plethodon. See Salamander (*Plethodon*)
Plum (*Prunus*), codling moth pest, 369, 370; weed races, 272
Pocket gophers (*Geomys*), allopatric or parapatric species of, 101; clay soil barriers between species of, 101, 102; contact zones between species of, 100–103; contact zones, lack of gene flow across, 101; distribution and population differentiation, 101, 102; fixation of chromosomal rearrangements, 100; founder events among populations of, 101; gene flow and species of, 100, 101; genetic drift, 100; historical factors in population genetics, 101; inbreeding among, 100; inviable backcrosses among species of, 100, 101; lice as tracers in interspecific contact of, 101, 103; limited dispersal ability, 100; morphological similarity among, 101; sand habitat, 101, 102; soil habitat distribution patterns and differentiation of, 101, 102
Pocket gopher (*Geomys attwateri*), contact zone with *G. breviceps*, 100, 101; contact zone with other species, 100, 101; distribution pattern, 101; founder effect in population differentiation, 101; karyotype, 101; lice species associated with, 101; louse infestation *Geomydoecus subgeomydis*, 101, 103
Pocket gopher (*Geomys breviceps*) complex, chromosomal variability due to polymorphisms, 100; contact zone with *G. attwateri*, 100; contact zones with other species, 100, 101; distribution pattern, 101, 102; founder effect in population differentiation, 101; karyotype, 100, 101, chromosomal arms, 100, 101; karyotypically and reproductively isolated sibling species, 100, lice species associated with, 101, 103; louse infestation *Geomydoecus ewingi*, 101; louse infestation *Geomydoecus subgeomydis*, 103; site of hybridization with *Geomys attwateri*, 101
Pocket gopher (*Geomys breviceps* and *Geomys attwateri*), contact zone gene flow, 101; fixed alleles at enzymatic gene loci, 101; genic differentiation subsequent to differentiation event, 101; F_1 embryos with backcross karyotypes, 100; F_1 hybrids found, 100; F_2 progeny inviable, 100, 101; hybrids in small area, 100; species chromosomal differences effective in hybrid breakdown, 101
Pocket gopher (*Geomys bursarius*) complex, 21, subspecies, 100; contact zones with other species, 100, 101; distribution, 100; F_{ST} values in, 100; genetic drift in, 100; heterozygosity in, 100; hybridization in, 100; karyotypic diversity in, 100; range, 100; reproductive isolation among, 100; sibling species in, 100–103
Pocket gopher (*Thomomys bottae alenus*), karyotype congruence, lack of, 360; subspecies classification, 360; composite of chromosomal races, 360
Poecilia reticulata. See Guppy (*Poecilia reticulata*)
Polygonum pensylvanicum. See Pennsylvania smartweed (*Polygonum pensylvanicum*)
Pomegranate (*Punica*), weed races, 272
Pondberry, Appendix 3
Ponderosa pine (*Pinus ponderosa*), allelic diversity analyses, 342; allozyme variation among populations, 344; beetles and frequency of defense terpinoids, 368; chemical heterogeneity of, 86; family structure, 343; genetic control, 86; genetic variation distribution, 342, 343; genetic variation, environmental gradients, 344; host tree to Black pineleaf scale insect, 86; microgeographic variation, 343; morphometric variation with

environmental variation, 344; outcrossing tree, 342; Black pineleaf scale insect response to chemicals in, 86; population analysis, genetic diversity, 342; population extinction by catastrophic event, 440, and population of scale insects, 440; population genetic structure, 342–344, selection pressure on Black pineleaf scale insect by, 87; subpopulation analysis, genetic diversity, 342; succession stage and within population variation, 341; Appendix 4

Pongo pygmaeus. See Orangutan (*Pongo pygmaeus*)

Potato (*Solanum tuberosum*), Colorado potato beetle adaptation to, 371; weed races, 274; wild relatives, 367, in genetic variability to resist potato blight, 367, in genetic variability maintenance of resistant cultivars, 367

Potato blight (*Phytophthora infestans*), potato resistance to, 367

Prairie dog (*Cynomys ludovicianus*), gene flow between demes, 87; social structure, 87

Prickly-pear (*Opuntia stricta*), genetic variability, 225, low levels, 225, 226; insect control, 226; intercontinental colonization, 225, 226; introduction, 226; population expansion, 226; range expansion, 225; release from environmental constraints, 225, 226; weedy species, 225

Primrose (*Primula*), heterostyly, 129; linkage disequilibrium, 129; stamen length 129; style length, 129

Primula. See Primrose (*Primula*)

Proboscidia. See Elephants (Proboscidia)

Procyon lotor. See Racoon (*Procyon lotor*)

Proechimys semispinosus. See Spiny rat (*Proechimys semispinosus*)

Pronghorn antelope (*Antilocapra americana*), future prospects, 444; management level, 436, in park, 420; reserve size, 444; small dependent populations, 436

Pronghonr Antelope Family (Antilocapridae), capacity of ISIS for, 382

Prunus. See Plum (*Prunus*)

Przewalski's (or Asian) wild horse (*Equus przewalski*), captive populations, 384; chromosome complement, 404; domestic horse relationship, 404, 405; extinct in the wild, 404; genetic variability loss, 404; inbreeding effects, 170, 171; input by Halle domestic mare, 405; lines for management, 405; reproductive value for age class, 384; selective genetically desirable crosses, 405; systematics, 404; zoo breeding plans for, 165; zoo exchanges, 389

Pseudotsuga menziesii. See Douglas-fir (*Pseudotsuga menziesii*)

Psidium. See Guava (*Psidium*)

Pteridophytes. See Filmy fern (Pteridophytes)

Ptilonorhynchidae. See Bowerbirds (Ptilonorhynchidae)

Puccinia graminis var. *tritici.* See Wheat stem rust (*Puccinia graminis* var. *tritici*)

Puffinus. See Burrowing shearwaters (*Puffinus*)

Punare (*Cercomys cunicularis*), inbreeding effects on juvenile mortality, 170

Punica. See Pomegranate (*Punica*)

Purple fringed orchid, Appendix 3

Pygmy hippopotamus (*Choeropsis liberiensis*), inbreeding effects on juvenile mortality, 166

Radish (*Raphanus*), weed races, 274

Rainbow smelt (*Osmerus mordax*), differences in transplanted stock, 359

Rainbow trout, Appendix 3

Rangifer tarandus. See Barren-ground caribou (*Rangifer tarandus*), and Reindeer (*Rangifer tarandus*)

Raphanus. See Radish (*Raphanus*)

Rat (*Rattus rattus*), morphometric and electrophoretic data, 357; congruence, 357; effects of inbreeding in, 174; pre-Mendelian experiment with, 174; in species extinction, oceanic islands, 111

Rattus norvegicus. See Brown rat (*Rattus norvegicus*)

Rattus rattus. See Rat (*Rattus rattus*)

Rattus. See Rat (*Rattus*)
Red deer (*Cervus elaphus scotticus*), area size, 425; locality, 425; population size 425; variation organization in, 77; isolation by distance and, 77
Red fescue (*Festuca rubra*), adaptive genetic differentiation in, 81, 84
Red ruffed lemur (*Varecia variegata rubra*), father/daughter matings, 409; genes in wild, 409; hybrids, 409
Red-backed shrike and Red-tailed shrike (*Lanius collurio* and *Lanius phoenicuroides*), environmental adaptation of hybrid swarms, 292
Red-fronted parakeet (*Cyanoramphus novaezelandiae*), comparison with Yellow-crowned parakeet, 303; control, 305; distribution, 302; environmental changes, 303; food supply increase, 303; habitat, 301; human intervention, 302; hybridization, 301, 303, 308; mixed matings, 303; island recolonization, 302; nesting, 301; numbers, 303, hybrid, 303; reproductive compatibility, 303, factors, selection, 303; shooting of, 303; sympatry, to hybridization, 301
Red-fronted parakeet and Yellow-crowned parakeet (*Cyanoramphus novaezelandiae* and *Cyanoramphus auriceps*), control, 302, 305; environmental change, 303; hybrid, 303; mixed matings, 303; number, 303; numbers, 303; random genetic shifts for reproductive compatibility, 303; reproductive isolation breakdown, 303; shooting, 303; sympatry to hybridization sequence, 301
Reindeer (*Rangifer tarandus*), inbreeding effects on juvenile mortality, 166. See Barren-ground caribou.
Rhabdomys pumilio. See Four-striped rat (*Rhabdomys pumilio*)
Rhesus macaque (*Macaca mulatta*), inbreeding effects on juvenile mortality, 169
Rhinocerotidae. See Rhinoceros (Rhinocerotidae)
Rhinoceros (Rhinocerotidae), captive program for propagation, 408; African, 377–379; Asian 377–379; captive habitat, 379; captive populations, 378, 379, for taxon, 379; distribution, 377; effective population size, 379; endangered, 376; estimated numbers, 377; facilities for, 379, 382; generation time, 376, 379; managed population numbers, 378, 379; management, 379, ISIS, 382; number of males/number of females, 378; population trends, 377; populations in captivity, 378; preservation time, 379, for genetic diversity maintenance, 379; reproduction in captivity, 378; species, 377; zoos, 383
Rhinoceros sondaicus. See Javan rhino (*Rhinoceros sondaicus*)
Rhinoceros unicornis. See Indian rhino (*Rhinoceros unicornis*)
Rhogeessa tumida. See Yellow bat, Central American (*Rhogeessa tumida*)
Rhynchosia (*Rhynchosia monophylla*), adaptive genetic differentiation in, 80, 84; copper tolerance in populations of, 80, 84; populations, 80, 84, distance between, 80, 84; zinc tolerance in populations of, 80
Rhynchosia monophylla. See Rhynchosia (*Rhynchosia monophylla*)
Roosevelt elk (*Cervus elaphus roosevelti*), of morphologically different populations, 359

Saccharum. See Sugarcane (*Saccharum*)
Saddle-back tamarin (*Saguinus fuscicollis*), inbreeding effects on juvenile mortality, 168
Saguinus fuscicollis. See Saddle-back tamarin (*Saguinus fuscicollis*)
Saimiri sciureus. See Squirrel monkey (*Saimiri sciureus*)
Salamanders (*Plethodon*), genetic variation measurement techniques, 375, congruence in results, 357, DNA hybridization data, 357, electrophoresis, 357, immunological data, 357
Salmo clarki. See Cutthroat trout (*Salmo clarki*)
Salmo trutta. See Brown trout (*Salmo trutta*)

608 Species Index

Salmon, Appendix 3
Salt meadow cord grass (*Spartina patens*), genetic structure of populations, 343; vegetative reproduction, 343
Salt-desert cavy (*Dolichotis salinicola*), inbreeding effect on juvenile mortality, 170
San Francisco Bay region checkerspot butterfly (*Euphydryas editha bayensis*), adults, 154; anthropogenic extinction, 158; area G demographic unit, 153–155; bottlenecking, 154; climatic relationships, 154; disappearance under San Francisco, 158; eggs, 154; environmental stress, 155; extinction 153–155
Sand shrimp (Phyllum: Crustacea), Appendix 3
Sarracenia purpurea. See Common pitcherplant (*Sarracenia purpurea*)
Scale insects, adaptation to individual pine trees of same species, 369
Scarlet runner bean (*Phaseolus coccineus*), Appendix 4
Scented top grass (*Capillipedium spicigerum*), hybridization with *B. bladhii*, 271
Scimitar-horned oryx (*Oryx dammah*), inbreeding effects on juvenile mortality, 167
Scioto madtom, Appendix 3
Sciurus carolinensis. See Gray squirrel (*Sciurus carolinensis*)
Scotch pine (*Pinus sylvestris*), Appendix 4
Scott's Seaside sparrow (*Ammospiza maritima peninsulae*), alleles of distinctive female sex chromosome, 306; F_1 hybrids, 306; hybridization with Dusky Seaside sparrow, 306, 309; U.S. Fish and Wildlife Service decisions, 306
Sea lamprey (*Petromyzon marinus*), impact of introduced predator, 365, 366, on fishes of North America Great Lakes, 366
Sea silene (*Silene maritima*), Appendix 4
Sea turtles (*Eretmochelys*) and Sea turtles (*Chelonia*), amino acid composition of keratins, 280; beach colonization, 285; behavior, 283;

biology, 278, 279, 284, 286; breeding, 281, 282; calipee, 281; captive release, 283; conservation, 278, 281, 282; controlled exploitation, 278; egg removal, 282; egg/hatchling beach transfer, 278; electrophoresis studies, 280; extinction threat, 279; habitat protection, 286; hatchlings, 284; homing instinct, 283; human activities, 284; hybridization, 278, in captivity, 291, through human action, 278, 280, 281, 284, threats, 286, 287; international cooperation, 282, 284; laws and agreements, 278; life cycle, 279, 280; management, 278, 279; migration, 285; nest site, 286, fixity, 286, poaching at, 279, protection, 279, 286; ranching, 281, 282; return to wild, 281, 282; shell, 281; survival, 278, 285; turtle excluder device, 286, 279, 280
Sea urchins, laboratory techniques, 352; nucleotide sequence differences, 352; within and between species diversity, 352, in DNA sequences, 352
Secale. See Rye (*Secale*)
Sesame (*Sesamum*), weed races, 274
Sesamum. See Seasame (*Sesamum*)
Seychelles Turtle dove (*Streptopelia picturata rostrata*), island populations, 301, and Madagascar Turtle dove, 301
Sheep, domestic (*Ovis aries*), in environmental effect on Yellow-crowned parakeet, 302; inbreeding and crossbreeding experiment in, 177
Sheep fescue (*Festuca ovina*), adaptive genetic differentiation in, 81, 84, 95; lead tolerance in populations of, 81, 84; populations adjacent, 81, 84
Siberian tiger (*Panthera tigris altaica*), age class, 395, 396; age/sex structure, 395, 399; age distribution, North American zoos, 395, 396, 399; bloodlines, 400; captive habitat, 383; carrying capacity for, 397, 399; crosses, 400; demographic management, 395–398; demographic stabilization, 395, 397; family size equalization, 400;

fertility, 395, 399; founders, 391, 392, 399, 400, equalization of representation of, 391, 392, 399, 400; genetic diversity loss management, 383–400; mates, 400; migrants, 400; numbers, 399; offspring, 399, survival, 399; parents, 395, 399; populations, 395, 399; recruitment, 399, 400; removal, 395, 399, 400; reproduction, 399, regulation, 399, senescence, 395; sex class, 395, 399; sex ratio equalization, 391, 399, SSP program, 399, 400; stock, 395; subspecies, 380; survival, 395, 399; viability, 395, 399; zoos, 400
Silver blue butterfly (*Glaucopsyche lygdamus*), climatic event and extinction, 157; food plant relationships, 157
Sitka spruce (*Picea sitchensis*), Appendix 3, Appendix 4
Six-row barley (*Horedum vulgare*), Appendix 4
Skipjack tuna (*Katsuwonis pelamis*), lack of gene exchange, 114
Slender wild oat (*Avena barbata*), adaptation to physical environmental factors 344; adaptive genetic differentiation in, 80, 82, 95; allele diversity analyses, 342, 343; alleles, 218; allozyme and polygenic variation relationship, genotype distribution, nonrandom, 343; allozyme marker loci in differentiation of, 82, genotypes of, 82; between population variance/total genetic variance, 344, 345; climatic zones, colonial populations vs. source populations, 217, 218; common wild oat comparisons, 218; monomorphism, 219; comparison, 219; genetic variation comparisons, 217, 218; genotypes of, 82, and climatic variation, 82, and habitat, 82; Gst value, 345; intrapopulation differentiation, 342, 343; microhabitat size, 82; phenotypic plasticity, 219; polymorphism, 219; population analysis, 342, 343; quantitative trait differentiation among populations, 344, 345; quantitative trait distribution, 344, 345; reproduction mode, 344; selection mode, 82, 83; water stress and, 80, 95; selfing annual, 342, 343, 344; specific multilocus genotype-environment correlations, 223; subpopulation analysis, 342, 343; transitional polymorphism, 82

Slimleaf goosefoot (*Chenopodium leptophyllum*), Appendix 4
Small fescue (*Festuca microstachys* = *Vulpia microstachys*) outcrossing rate, 221
Small white lady's-slipper, Appendix 3
Smallflower evening-primrose (*Oenothera parviflora*), Appendix 4
Snapdragon (*Antirrhinum majus*), incomplete dominance in, 25
Snow eucalyptus (*Eucalyptus pauciflora*), Appendix 4
Snow leopard (*Panthera uncia*) captive facilities, 381; captive history, 387
Snowshoe hare (*Lepus americanus*), area size, 423; locality, 423; population size, 423
Soemmering's gazelle (*Gazella soemmeringi*), chromosomal number variation, 407; inversion in chromosomes, 407; karyotypes in wild, 407; translocation of chromosomes, 407
Soft brome (*Bromus mollis*), alleles, 218; colonial populations vs. source populations, 217, 218; foxtail brome comparison, 219; in genetic variability, 219; phenotypic plasticity, 219; genetic variation comparisons, 217, 218
Solanum (*Solanum grayii*), Appendix 4
Solanum (*Solanum johnstonii*), Appendix 4
Solanum (*Solanum pennellii*), Appendix 4
Solanum (*Solanum tenuipes*), Appendix 4
Solanum grayii. See Solanum (*Solanum grayii*)
Solanum johnstonii. See Solanum (*Solanum johnstonii*)
Solanum pennellii. See Solanum (*Solanum pennellii*)
Solanum tenuipes. See Solanum (*Solanum tenuipes*)
Solanum tuberosum. See Potato (*Solanum tuberosum*)

Solanum. See Potato (*Solanum*)
Sorghum (*Sorghum bicolor*), adaptation, 272; crossing with cultivars, 274, with wild races, 273, 274; cultivated races, 272-274; distribution, 272; diversity 273; domestication, 272, 273; gene flow, 273; hybridization, 273; mimicry, 273; races, 272-274; seed dispersal, 273; selection pressure, 273; shattering genotype, 273, "shattercane", 273; single recessive allele, 273; spikelets, 273; weeds, 272-274

Sorghum bicolor. See Sorghum (*Sorghum bicolor*)

Sorghum halepense. See Johnson grass (*Sorghum halepense*)

Southern corn leaf blight fungus (*Helminthosporium maydis*), epidemic on genetically uniform cultivar of corn, 367

Southern white rhino (*Ceratotherium simum simum*), capacity of ISIS for, 381; captive habitat, 379, 380; differences with northern white rhinoceros 408; distribution, 377; effective population size, 379; endangered status, 379; existence or elimination in zoos, 380; facilities, 379, 380, institutions with, 378; management, 379, 380; mtDNA analysis, 408; numbers, 377; population trends, 377; populations in captivity, 378; relocation in captivity, 379, 380; reproduction in captivity, 378, 379; San Diego Zoo research, 380; social system effects, 379; SSP program, 380; systematic relationships, 380; white rhino dilemma, 380; wild status, 380

Soybean (*Glycine*), weed races, 274

Spartina patens. See Salt meadow cord grass (*Spartina patens*)

Spearnose bat (*Phyllostomus hastatus*), rarity of inbreeding in, 179

Speckled dace, Appendix 3

Speke's gazelle (*Gazella spekei*), adaptation to inbreeding, 260; breeding history, 245-250; breeding of, 434; breeding plan for, 180; breeding program, elimination of inbreeding depression, 250-259; founders, 242, 251; gene pools, 251, in small captive populations, 434, inbreeding depression reduction in, 434; habitat, 242, 260; inbreeding coefficients, 248, 249, 255-258; management of introductions, 434; offspring, 254, birth weights, 249, 258; outbreeding species, 243, 244, 434; parentage, 246; survival, 241-260, viability data, 246, 255, zoos, 259, 260

Speyeria, subspecies classification by electrophoresis, 360

Spider mites (*Tetranychus urticae*), improvement in adaptation to inferior hosts, 369; damage against cotton, 371, with absence of extrafloral nectaries, 371; artificial selection in laboratory, 369; sesquiterpene gossypol, 371

Spider monkeys (*Ateles*), cytogenetic parameters, 406; different chromosomal phenotypes, 406; inversion chromosome, 406; karyotype, 406; origin unknown, 406; translocation of chromosome, 406

Spiny rat (*Proechimys semispinosus*), inbreeding effect on juvenile mortality, 170

Spotbill duck (*Anas poecilorhyncha*), hybridization with Mallard duck, 293. See also Australian Black duck (*Anas superciliosa "rogersi"*)

Spotted hyena (*Crocuta crocuta*), area size, 428; locality, 428; population size, 428

Spring emex (*Emex spinosa*), genetic variation, loss in colonization, 217

Squash, epistatic influence of alleles in fruit color of, 37. See Gourd.

Squirrel monkey (*Saimiri sciureus*), chromosomally distinct subspecies, 405, 406; inbreeding effect on juvenile mortality, 168

Steelhead trout, Appendix 3

Stephanomeria exigua ssp. *carotifera.* See Wire-lettuce (*Stephanomeria exigua* ssp. *carotifera*)

Sthenele brown butterfly (*Cercyonis sthenele sthenele*), disappearance under San Francisco, 157, 158

Streptopelia picturata picturata. See Madagascar Turtle dove (*Streptopelia picturata picturata*)

Streptopelia picturata rostrata. See

Seychelles Turtle dove (*Streptopelia picturata rostrata*)
Subalpine fir (*Abies lasiocarpa*), Appendix 4
Subterranean clover (*Trifolium subterraneum*), adaptation, 218; collections, 218; flowering time, 218; genetic changes, 216, and founder events, 226; genotypes (unique), 218; introduction, 218; outcrossing and recombination, 218; self-fertilizing species, 218
Sugar pine (*Pinus lambertiana*), resistance to white pine blister rust, 347, by rare dominant allele, 347
Sugarcane (*Saccharum*), chromosome numbers, 269; clones, 269; complex hybrid derivatives, 269; differentiation to generic level, 269; genera, 269; hybridization, 269
Suidae. *See* Pig Family
Sulphur butterfly (*Colias philodice eriphyle*), adaptation to alfalfa, 369; adaptation to wild legumes, 370; local host in abundance, 370; members of the order, 371; pest of banana, 371
Sumatran rhino (*Dicerorhinus sumatrensis*), distribution, 377; endangered species, 411; endangered, 376, 377; interspecific embryo transfer proposal, 411; numbers, 377; population proposed, 379; population trends, 377; populations, 376–379; rare species, 376; survival, 411
Sumatran tiger (*Panthera tigris sumatrae*), breeding analysis of, 180; genetic differences in three populations of, 180; population maintenance difficulty, 380
Sunfish (*Lepomis*), D values, 97
Sunfishes and freshwater basses (*Centrachidae*), heterozygosity levels, 146
Sunflower Family (Asteraceae), Appendix 4
Sunflower Family (Asteraceae: *Coreopsis cyclocarpa* var. *pinnatisecta*), Appendix 4
Sunflower Family (Asteraceae: *Coreopsis cyclocarpa* var. *cyclocarpa*), Appendix 4
Sunflower Family (Asteraceae: *Hymenoppapus artemisiifolius*), Appendix 4
Sunflower Family (Asteraceae: *Tragopogon mirus*), Appendix 4
Sunflower Family (Asteraceae: *Tragopogon miscellus*), Appendix 4
Sunflowers (*Helianthus*), weed races, 274; Appendix 4
Sun star (*Helianthus kubiniji*), decline in numbers by sea temperature change, 126
Sus. See Pigs (*Sus*)
Swallowtail butterfly (*Papilio dardanus*), body color, form 129; polymorphism, 129
Swallowtail butterfly (*Papilio machaon*), environmental change, 160; extinction 160; habitat loss, 160; inbreeding depression lack, 160; low mobility phenotype, 160; monitoring, 160; reestablishment, 160, failure, 160
Sweet vernal grass (*Anthoxanthum odoratum*), adaptation to physical environmental factors, 343; adaptive genetic differentiation in, 80, 84, 95; copper tolerance in populations of, 80; mineral nutrients in differentiation of, 84; populations, 80, distance between, 80; rapidity of genetic changes, 84; Appendix 4
Syncerus caffer. See Cape buffalo (*Syncerus caffer*)

Table Mountain pine (*Pinus pungens*), Appendix 4
Tamaraw (*Bubalus minorensis*), endangered species, 411; interspecific embryo transfer proposal, 411; survival, 411
Tamarind (*Tamarindus*), cultivated/weed race interactions, 272, 275; naturalized escape, 275; weed races, 272, 275
Tamarindus. See Tamarind (*Tamarindus*)
Tansy ragwort, Appendix 3
Tayassu tajacu. See Collared peccary (*Tayassu tajacu*) or White-lipped peccary (*Tayassu tajacu*)
Teleost fishes (Teleostei), genetic variation and environmental variation lack of correlation, 142

Teleostei. *See* Teleost fishes (Teleostei)
Terrestrial slugs (Limacidae), breeding systems, 205–210
Tetranychus urticae. *See* Spider mite (*Tetranychus urticae*)
Texas bluebonnet (*Lupinus subcamosus* and *Lupinus texensis*), Appendix 4
Texas wild-rice, Appendix 3
Thamnophis elegans. *See* Western terrestrial garter snake (*Thamnophis elegans*)
Theropithecus. *See* Baboons (*Theropitchecus*)
Thrips (Thysanoptera), damage increase on cotton, 371, with sesquiterpene, gossypol, 371
Thomomys bottae alenus. *See* Pocket gopher (*Thomomys bottae alenus*)
Thomson's gazelle (*Gazella thomsoni*), area size, 426; locality, 426, population size, 426
Thymus praecox. *See* Wild thyme (*Thymus praecox*)
Thysanoptera. *See* Thrips (Thysanoptera)
Tiger (*Panther tigris*), captive facilities, 381; competition for habitat, zoos, 383; institutions, 381; maintenance limitation, 381; management, 383; subspecies, 381, 399; temperate, 380, 399; tropical, 380, 399; zoos, 380, capacity, 380, capacity of ISIS for, 381
Tiger salamanders (*Ambystoma tigrinum*), biological significance, 363; esthetics of subspecies, 363
Toads (Anura), average heterozygosity ($\bar{H}$), 140; average polymorphism ($\bar{P}$), 140; habitat type and genetic variation, 140
Tobacco (*Nicotiana tabacum*), heterozygosity, 221; negative assortative mating, 221; selectivity for mutants, 221; self-sterility system, 221
Tomato (*Lycopersicon*), species distribution limitation and Gst values, 339; weed races, 274
Trachypogon spicatus. *See* Grass (*Trachypogon spicatus*)
Tragelaphus derbianus. *See* Giant eland (*Tragelaphus derbianus*)
Tragelaphus eurycerus. *See* Bongo (*Tragelaphus eurycerus*)
Tragopogon. *See* Goat's beard (*Tragopogon*)
Tragopogon (*Tragopogon mirus*), Appendix 4
Tragopogon (*Tragopogon miscellus*), Appendix 4
Tragopogon dubius. *See* Yellow salsify (*Tragopogon dubius*)
Tragopogon mirus. *See* Tragopogon (*Tragopogon mirus*)
Tragopogon miscellus. *See* Tragopogon (*Tragopogon miscellus*)
Tragopogon porrifolius. *See* Vegetable-oyster salsify (*Tragopogon porrifolius*)
Trembling aspen, Appendix 3
Trichodectidae. *See* Lice (Trichodectidae)
Trifolium hirtum. *See* Rose clover (*Trifolium hirtum*)
Trifolium repens. *See* White clover (*Trifolium repens*)
Trifolium subterraneum. *See* Subterranean clover (*Trifolium subterraneum*)
Triticum aestivum. *See* Wheat (*Triticum aestivum*)
Tule elk (*Cervus elaphus nannodes*), area size, 425; gene flow between populations, 435; locality, 425; morphometric studies, 359, subspecies and, 359; mortality, 420, in park, 420; population, 420, growth, 420, size, 425; morphometric variation, 363, among populations, 363; subspecies classification, 363
Tympanuchus cupido attwateri. *See* Attwater's prairie-chicken (*Tympanuchus cupido attwateri*)
Tympanuchus cupido cupido. *See* Heath hen (*Tympanuchus cupido cupido*)
Tympanuchus cupido. *See* Attwater's prairie-chicken (*Tympanuchus cupido attwateri*) and Heath hen (*T. c. cupido*)

Ulmus americana. *See* American elm (*Ulmus americana*)
Ulmus. *See* Elm (*Ulmus*)
Uredinales. *See* Rusts (Uredinales)

Ursus americanus. See Black bear (*Ursus americanus*)
Ursus arctos horribilis. See Grizzly bear (*Ursus arctos horribilis*)

Varecia variegata rubra. See Red ruffed lemur (*Varecia variegata rubra*)
Varecia variegata variegata. See Black and white ruffed lemur (*Varecia variegata variegata*)
Vegetable-oyster salsify (*Tragopogon porrifolius*), Appendix 4
Vermivora chrysoptera. See Golden-winged warbler (*Vermivora chrysoptera*)
Vermivora pinus and *Vermivora chrysoptera*. See Blue-winged warbler and Golden-winged warbler (*Vermivora pinus* and *Vermivora chrysoptera*)
Vermivora pinus. See Blue-winged warbler (*Vermivora pinus*)
Vigna. See Cowpea (*Vigna*)
Virginia bluebells, Appendix 3
Virginia pine (*Pinus virginiana*), Appendix 4

Warblers (Parulidae), reserve size and, 443
Western jumping mouse (*Zapus princeps*), area size, 423; locality, 423; population size, 423
Western terrestrial garter snake (*Thamnophis elegans*), Arnold's model, 370, 371; behavioral responses to different prey items, slugs, leeches, 371; genetic correlation, 371
Western yarrow (*Achillea lanulosa*), between population variance/total genetic variance, 344, 345; life history trait influence on quantitative traits, 344, 345; outcrossing species, 344; quantitative trait distribution, 344, 345; reproduction mode, 344; reproduction also by rhizomes, 344
Wheat (*Triticum aestivum*), against evolution of new strains of Hessian fly, 366; breeding of, 190; differentiation-hybridization cycle, 268; gene-for-gene system, 367; inbred, 5; in recombination and selection of existing genetic variability, 190
Wheat stem rust (*Puccinia graminis*), attack on resistant wheats, 366
White clover (*Trifolium repens*), adaptation to biotic factors, 343; genetic structure of populations, 343; geographic variation in cyanide-producing genotypes, 368; response to herbivory, 368
White fir (*Abies concolor*), between population variance/total genetic variance, 344, 345; life history trait influence on quantitative traits, 344, 345; quantitative trait distribution, 344, 345; reproduction mode, 344; wind–pollinated tree, 344
White flies (Homoptera: Aleyrodidae), damage increase on cotton with sesquiterpene gossypol, 371
White fringeless orchid, Appendix 3
White rhinoceros (*Ceratotherium simum*), distribution, 377; endangered, 376; numbers, 377; population trends, 377
White sweet-clover (*Melilotus alba*), adaptive genetic differentiation in, 81, 84; populations adjacent, 81; zinc tolerance in populations of, 81, 84
White tiger (*Panthera tigris tigris*), in competition for captive tiger habitat, 380, 383
White-footed mouse (*Peromyscus maniculatus*). See Deer mouse (*Peromyscus leucopus*)
White-lipped peccary (*Tayassu tajacu*), karyotype, 407
White-tailed deer (*Odocoileus virginianus*), area size, 425, locality, 425, population size, 425; cross-breeding programs, 105; economic importance, 105; population-size to park-size relationship, 442; selection, 105, variation organization in, 77, isolation by distance and, 77; space and survival, 442
Whitebract hymenopappus (*Hymenopappus scabiosaeus*), Appendix 4
Whooping crane (*Grus americana*), 104
Wied's red-nosed mouse (*Wiedomys pyrrhorhinos*), inbreeding effects on juvenile mortality, 170

Wiedomys pyrrhorhinos. See Wied's red-nosed mouse (*Wiedomys pyrrhorhinos*)
Wild barley (*Hordeum spontaneum*), allozyme and morphological traits, 344; distribution of variation, 344, within and among populations, 344; inbreeding, 223; multilocus associations, 223; Appendix 4
Wild ginger (*Asarum canadense*), growth rate and slugs, 368
Wild marjoram (*Origanum vulgare*), Appendix 4
Wild mouse (*Mus musculus*), See House mouse (*Mus musculus*)
Wild thyme (*Thymus praecox*), butterfly larval food plant, 160
Wildebeest (*Connochaetes taurinus*), area size, 426, locality, 426, population size, 426; inbreeding effects on juvenile mortality, 167; infectious agent transfer, 412, to domestic animals, 412; models of survival for herd of, 172–174
Willow sparrow (*Passer hispaniolensis*), in hybrid swarm, 292
Wire–lettuce (*Stephanomeria exigua* ssp. *carotifera*), Appendix 4
Wolf (*Canis lupus*), area size, 428; habitat availability, 442; patchiness, 442; locality, 428; longevity prospects, 441, 442; population size, 428; population-size to park-size relationship, 441, 442
Wolf willow, Appendix 3
Woodrats (*Neotoma*), future prospects, 443; reserve size for, 443
Wooly meadow foam, Appendix 3

Xanthium strumarium. See Rough cocklebur (*Xanthium strumarium*)
Xanthosoma. See Yautia (*Xanthosoma*)
Xerxes blue butterfly (*Glaucopsyche xerces*), disappearance under San Francisco, 158

Yarrow gilia (*Gilia achilleifolia*), fecundity, 222; heterozygous advantage, 222; inbreeding depression, 222; loss in fitness, 222, generations, 222; outcrossing, 222, rates, 222, generations, 222; selfed generations, 222; survival, 222; Appendix 4
Yautia (*Xanthosoma*), on disturbed land, 274; weediness, 274
Yellow baboon (*Papio cynocephalus*), chromosome structure, 406, no abnormalities, 406, few polymorphic sites, 406; cytogenetic stability, 406; success, 406
Yellow bat (*Rhogeessa tumida*), chromosomal variability, 104, morphological variability, 104; cytotype, 104; genetic diversity in, 104; habitat destruction and, 104; karyotype, 104; sibling species, 104; two chromosomal races, 104, distribution of, 104
Yellow-bellied marmot (*Marmota flaviventris*), rarity of inbreeding in, 179
Yellow-crowned parakeet (*Cyanoramphus auriceps*), comparison with Red-fronted parakeet, 301, 303; distribution, 302; environmental change, genetic assimilation of, 303; habitat, 301, change, 303, disadvantage for, 303; human intervention, 302; hybridization, 301, 303, 308; island recolonization, 302; nesting, 301, numbers, 302, 303; reproductive compatibility, 303; selection factors, 303; sympatry to hybridization, 301
Yellow fever mosquito (*Aedes aegypti*), allele frequencies, five loci, 234; allozyme variation, 234; compared with humans, 234; canonical variables, isozyme data, 234
Yellow pitcherplant, Appendix 3
Yellow salsify (*Tragopogon dubius*), Appendix 4
Yellow wild–indigo (*Baptisia sphaerocarpa*), Appendix 4

Zagoutis (*Plagiodontia*), inbreeding effects on juvenile mortality, 170
Zapus princeps. See Western jumping mouse (*Zapus princeps*)
Zea mays. See Corn and Maize (*Zea mays*)
Zebra, inbreeding effects on juvenile mortality, 166
Zostera marina. See Common eelgrass (*Zostera marina*)

SUBJECT INDEX

Adaptation
  to climate, 292
  cross-adaptation to another species, 373
  differentiation of populations over short distances, 94
  to ecological diversity, 320, 365
  to environment, 38
  to environmental shift, 217
  to external environment, 243
  and extinction, 133
  extremes for survival, 331
  genetic variation and, 149, 218, 219
  hybridization and, 307
  with hybridization, 289
  hybridization and exchange of genes, 289
  to inbreeding, 243–245, 250, 254, 258, 259, 320, 434, 435
  increase, 38
  influences on, 289
  to insect pollinator, 225
  to internal environment, 243
  on islands, 111
  local populations, 351
  to marine environment, 285
  monomorphic population, 137
  mutation and, 37
  to new selection regime, 93
  to one species, 373
  by phenotypic plasticity, 219
  polymorphic population, 137
  preadaptation to heavy metal stress, 84
  by preadapted genetic variants, 93
  rapid, in founding populations, 217, 218
  to reduction of gene pool size, 187
  in restricted areas, 331
  sea turtle hatchlings
    demes to local conditions, 285
    to marine environment, 285
    to new migration routes, 285
    to new nesting beaches, 285
    to new nesting partners, 285
  to shifting environment, 292
  to steady state, 333
Adaptation to inbreeding
  rapidity of, 245, 250, 258, 259
  small population size and, 259
  for survival, 320
Adaptation rate
  to inbreeding, 253
  selective pressures and, 253
Adaptive genetic differentiation
  over small geographic scale, 92
Adaptive genetic divergence and fitness between adjacent populations, 90, 91
Adaptive norm
  heterozygous genotypes, 194
  homozygous genotypes, 194
Adaptive strategies
  morphological, 90
  physiological, 90
Adjacent areas
  effect of activities in, 442
Adjacent populations

Subject Index

barrier separation of, 89
  with genetic divergence, 79
  biology of, 79
  genetic structure, 89
  interactions, 89
Africa, 115
  biomes of, 422
  grass distribution, 269
  Grevy's zebra imports, 384, 385
  rhinoceros, 139, 376, 377
  weed rices, 274
African reserves, 183
  carnivores, 422
    densities of, 422
  herbivores, 422
    densities of, 422
Age, stable distribution of, 395
Age class
  distribution and population size, 117
  in fertility, 384
  in survival, 384
Age distribution, of Siberian tiger, North American zoos, 396
Age structures, differential in captive populations, 180
Aggressive races (plants), adaptation to present state of disturbance, 276
Agricultural plants
  extinction from insects, 373
  extinction from pathogens, 373
  mortality from insects, 373
  mortality from pathogens, 373
Agricultural systems, parasite variability and homogeneous host populations, 368, 369
Agriculturalists, observation of extinction, 109
Alaska
  northwestern birds, 291
  park size in, 421
Allele coding
  for human sickle cell hemoglobin, 36
  lethals, 195
  semilethals, 195
  subvitals, 195
Allele combinations
  with high inbreeding coefficients, 252
  of inbred animals, 252
  for inbreeding, 252

Allele frequencies, 40–43
  alteration of, 90
  differences of male and female, 321
  divergence, 64, 65
  equability, 341
  genetic divergence at individual loci, 61, 62
  genetic variability and dynamic changes, 155
  genotype frequencies from, 40
  migrants per generation and, 62, 63
  natural selection modes simulation, 57
  patterns of divergence, 62
  patterns of divergence and deme distance, 61
  in populations, 40
  relationship to population size, 331
  similarity in different demes with heterozygous advantage, 56, 57
  stability, 155
Allele frequency analysis, 232–236
  stepwise linear discriminant analysis, 233
Allele frequency divergence, 62
  computer simulations, 64
  at loci, 61
  occurrence, 64
  among semi-isolated demes, 63–64
Allele frequency divergence interactions
  gene flow, 64
  genetic drift, 64
  natural selection, 64
Alleles
  adaptive combinations of, 211
  alternative, 253
  beneficial in heterozygous condition, 197
  best possible at any gene locus, 129
  calculation of, from genotype frequencies, 40
  causes of variation, 61
  change of frequencies in, 40
  changes in frequencies, Monte Carlo simulations, 52
  coding of, 33, 90
  codominant, 25
  combinations of, 195
    in diploid species, 195
  combinations in particular chromosomes, 355
  cryptic for polymorphic loci, 346
  definition of, 22

Subject Index 617

deleterious, 175, 184, 194, 196, 244
deleterious number in lethality, 244
desirable for breeder, 31
in disease resistance, 347
divergence in frequency, 319
dominant, 23, 45
   harmful, 45
   in rat resistance to warfarin, 88
elimination of deleterious, 178
fixation, 68, 197
   in separate populations, 320
fixation of, 89
   in warfarin resistance, 89
fixation of alternate, 320
   at same locus, 98
founder contribution in herd, 252
from founders, 253
frequencies between, 7
heterozygous, 244
homozygous, 244
in host/pathogen relationships, 367
independently acting lethal, 244
inheritance at a single locus, 37, 38
inheritance at different loci, 37, 38
lethal, 9, 195
   translocation heterozygotes, 9
loss of, 67, 197, 313, 320
loss to genetic drift, 64
loss rate, 392
low frequency, 321
maintenance in natural populations, 33
maintenance in semi-isolation, 49
movement from gene pool, 292
multiple, 25
multiple character expression from one locus, 37
natural selection maintenance, 33
number in colonial populations vs. number in source populations, 217
one character expression from separate loci, 37
in pest resistance, 347
in populations, 7
preservation in heterozygous individuals, 67
random genetic drift in frequency, 197
random loss of by genetic drift, 68
rare, 244, 347, 443

recessive, 9, 23, 38, 45, 137, 139, 244, 305
reduction of number of, 69
selection against, 320
selectivity equivalent (neutral value), 33
single recessive, 197
substitution effects at phenotypic level, 346, 347
transmittal by parent in zygote formation (model), 52
uncommon, in heterozygous condition, 45
undesirable, 31
Allelic diversity. *See also* Genetic diversity
absolute amount, 362
allele success in phenotypes, 347
colonial populations vs. source populations, 217, 218
comparisons, plant, 336–338, 342
conservation and, 362
   subspecies in, 362, 363
degree of genetic difference, 347
degree of phenotypic difference, 347
in differences between populations (G_{ST}), 360
frequency of, 319
maintenance of high frequency of heterozygous individuals, 67
maintenance of population subunits homozygous for an allele, 67
maintenance of other subunits homozygous for other alleles, 67
partitioning, 362
reduction of, 438
relative amount, 362
substitutable alleles, 346, 347
total (H_T), 360
Allelic polymorphism
with environmental variance increase, 322
extinction and, 321
increase, 322
low frequencies, 321
maintenance, 326
protectedness in, 321, 326
Allelic systems
changes in, 332
maintenance of high levels of heterozygosity in, 332

Subject Index

Allelic variation, for multiple loci, 318
Allozyme(s), 33, 34
  in differentiation of genotypes, 82
  as genetic markers, 33
  marker loci, 82
  natural selection maintenance, 33
  selectively equivalent (neutral value), 33
  in a single environmental gradient, 345
Allozyme frequencies
  bottlenecks and, 155
  butterflies, 153, 155
Allozyme frequency evaluation
  insects, 156
Allozyme genetic variation
  among species within taxonomic groups, 335
  among taxonomic groups, 335
Allozyme heterozygosity, summary indices of genetic variation, 365
Allozyme loci
  codominance, 343
  plant monomorphism over, 219
Allozyme surveys, partner selection, 8
Allozyme studies. *See* Protein electrophoresis
Allozyme traits, patterns, 344
Allozyme variation distribution
  evolution factor effects, 345
  quantitative variation distribution and, 344, 345
  selection pressure effects, 345
Allozyme variation distribution, plant populations
  allelic diversity among population subdivisions (H_{sp}), 342
  allelic diversity among populations (D_{st}), 337, 338
  allelic diversity among populations/total allelic diversity, ratio (G_{st}), 337–342, 344–346
  allelic diversity within populations (H_s), 337, 338
  allelic diversity total (H_T), 337, 338, 340, 342, 346
Allozyme variation levels, at polymorphic loci
  endemic species, 338
  widespread species, 338
American Association of Zoological Parks and Aquariums (AAZPA), 164, 404, 409
  NZP captive breeding of Dorcas gazelles, 184
  Species Survival Plan (SSP), 375
American Ornithologists' Union, 295
  contamination of native gene pool prevention, 307
  intraspecific hybrid status, 307
American zoos, number of species in, 164, 376, 403
Amino acid composition
  of keratins, 280
  overlapping ranges, 280
  of turtles, 280
  variation within ratios, 280
Amino acid residues, 32, 33
Amino acid sequences
  compared with DNA sequencing, 352
  of cytochrome c, 353
  of globins, 353
  individual differences at a site, average phylogenetic relationships, 353
  probabilities, 353
Amino acids
  glutamic acid, 36
  valine, 36
Amphibians, larger, migration, 9
Analysis methods, sibling species relationships
  electrophoresis, 103
  karyology, 103
  meiotic analysis of hybrids, 103
  parasite distribution, 103
Ancestral population, heterozygote fitness for new population, 194
Andes Mountains, potatoes, 274
Aneuploidy
  definition, 35
  Down's syndrome, 35
  homologue segregation failure, 35
  nondisjunction, 35
  trisomy, 35
Animal evolution, polyploidy in, 35
Animal management, genetic diversity maintenance, 52
Animal numbers
  for effective population size, 386
  reduction for maintenance, 386
Animal populations, allelic heterozygosity in, 194
Animal removal

age class, 398
  dominant males, 391
  sex class, 398
  in zoos, 394
Animals
  adaptation on islands, 111
  adverse environmental condition
    response of, 85
  diploid, cross-fertilized, 199
  equal exchange in slowing of inbreeding, 76
  extinction by man, 111, 112
  game management, 278
  genotype reserves in management
    of, 7
  homeostatic mechanisms mitigate
    effects of environmental
    change, 85
  intensity of selection pressures on,
    85
  inversion in, 36
  isolation by distance, 77
  migration success with, 9
  most valued, 199
  motility of, 85
  periodicity, 76
  in physical environment, 85
  polyploid occurrence in, 35
  of restricted mobility, 94
  sex chromosome systems in, 27–29
  sex determination systems in, 27–29
  spatial selection pressures of, 85
  variation in, 77
Animals, larger
  evolution, 9, 439
  population size for survival, 6
Animal species
  asexual, 365
  and catastrophic mortality, 117
  chromosome complements,
    404–407
  demographic stochasticity, 117
  extinction and breeding activities,
    118
  extinction and dispersal
    mechanisms, 119, 120
  extinction and foraging activities,
    118
  extinction in island-dwelling
    populations, 120
  extinction and nesting activities,
    118
  extinction by intrinsic factors,
    117–120

extinction in small population size,
  117–120
extinction by vulnerability interactions, 120
genetic deterioration, 118–120
large breeding congregations of, 117
outcrossing, 209
sexual divergence in migration, 321
wide-ranging, 209
Annual crops, weed races, 272
Annuals
  diploid, 225
  selfing, 205
Aphids, biotypes with different
  cultivars, 369
Apomixis, dominant in hybrid
  swarms, 270
Area(s)
  and longevity, 441
  multiple small, 332
  of populations, 423–428
  single large, 332
  of U.S. parks and reserves,
    417–419, 443, 444
Area, scale
  large, 443
  mutualists, 121
  small and treefall habitats, 121
Area size. See also Park and Reserve
  size
  extinction and, 118–120
  large, 332
  and population size, 49
  small, 332, 440
Area size decrease
  alteration of ecological interactions,
    120
  as barriers to recolonization, 120
  catastrophe, 120
  and competition increase, 120
  decrease in N_e, 120
  demographic, 120
  disease, 120
  in extinction of ground/near
    ground species, 120
  in extinction of key resource/habitat dependent species, 120
  in extinction of large animals, 120
  in extinction of large predators, 120
  in extinction of mutualist dependent species, 120
  in extinction of plants, 120
  in extinction of rare species, 120
  extrinsic factors and, 120

genetic, 120
habitat/resources disappearance, 120
with increase of overuse effects, 120
intrinsic factors and, 120
isolation (area effects), 120
isolation between fragments, 119
by numerical increase of some herbivores, 120
by numerical increase of some vertebrates, 120
small predator increase and, 120
social, 120
species abundance alteration and distribution, 120
species diversity changes, 120
Argentina, 144
Arizona, ducks, 295, 296
Artificial founder events, natural analogs of, 192
Artificial inbreeding program
fixation of genes, 150
natural selection operative, 150
Artificial sample, propagation of, 191
Artificial selection, 196
genetically novel population production, 199
for a phenotype, 288
Ascension Island, 285
Asexual forms of life
evolution comparison with sexual forms, 365
relation to sexual forms, 365
Asexual species, genetic homogeneity, 226
Asexuality, 204
Asia
rhinoceros, 376, 377
weed rices, 274
Asia, southern, grass distribution, 269
Associations, allozyme traits/quantitative traits, 345
Assortative mating. *See also*, Mating of parental forms in hybrid zone
self-fertilization, 221
speciation from, 292
sympatry from, 292
Atlantic Coast, warblers, 296
Atlantic Coast, France
grass distribution, 269
Atlantic Coast, Iberia, grass distribution, 269

Atlantic Ocean, sea turtles, 285
Attributes
offspring, 252
parental, 252
Australia
European colonization and plant spread, 228
grass distribution, 269, 271, 276
plant colonization, 218
plant genetic variation, 217
plant introduction, 218
plant races monomorphic, 219
species for preferential preservation, 9
turtles, 277
Autosomal linkage, 29–31. *See also* Crossing over
Autosomes, genes on, 29
Average heterozygosity
decrease after founder event, 231
decrease continuance after founder effect, 231
population bottleneck and, 231
population bottleneck in measurement of, 231
statistical problems in measurement of, 231
Avian diseases, in species extinction, oceanic islands, 111
Avian hybrid situations, North America
hybrid swarms, 291
hybrid zones, 291
zones of overlap, 291
Avian populations
assortative mating, 294
behavioral interactions, 289, 308
conspecific, 289
ecological interactions, 289, 308
fitness, 294
genetic interactions, 289, 308
geographic differentiation, 289
hybrid progeny, 294, 295
hybridization degree, 290
mixed flocks, 294, 295
pairing, 294
parental, 294
reproductive isolation, 289
secondary contact, 289, 291
sympatry, 289
Avian species. *See also* Birds
climatic adaptation, 292
hybridization in evolution of, 292
Pleistocene conditions, 292

range extension by hybrid swarms, 292
social behavior and inbreeding avoidance, 304

Backcrossing, to parental forms, 292
Bacteria, extinction question, 126
Balanced polygenic control and reproductive characters, 198
Balanced system, new one to replace old, 197
Balanced system of inheritance
  high fitness heterozygote, 196
  retained by natural selection, 196
Balancing selection
  complement of species in maintenance of, 372
  polymorphism for host utilization and, 370
Barriers. See also Human activities structures
  chemical, 86
  creation for vulnerable species, 305
  ecological, 305
  geographical, 305
  to mating, 319
  natural
    deforested terrain, 121
    rivers, 121
  park and reserve boundaries, 442
  physical, 94
    genetic variability partition by, 94
  soil type, 101
  thermal, 86
Barriers to dispersal
  habitat type variance, 67
  physical barriers, 67
  unfavorable terrain, 119
  water, 119
Barro Colorado Island, Lake Gatún, Panama, vulnerability of bird species on, 119
Behavior
  competition, 325
  cooperation, 325
  linear models, 325
  sibling species differences in, 99
Behavioral responses, *Drosophila*
  female, 198
  male, 198
Bermuda, sea turtles, 284
Biochemical techniques, electrophoresis, 335

Biogeography, effective population size and, 444
Biological control agents, asexual species effectiveness, 226
Biological diversity, 322–330
  from habitat destruction, 308
  hybridization in conservation, 309
  increase by new environments, 292
  losses from genetic assimilation, 307, 308
  losses of communities, 308
  populations, 308
  species of animals and plants, 308
Biological diversity measurements
  conservation criteria, 350
  measure of average differences between individuals, 350
  number of species in community, 350
  quantification of diversity in populations, 350
  relative abundance of species, 350
  scope of field, 350
  Shannon information index, 350
Biological factors
  host-parasite relationships, 126
  and extinction, 126
Biological imbalance, in parks and reserves, 420
Biological interactions, cause and effect mechanisms, 326
Biological properties, sibling species
  speciation by allopatric divergence, 106
  speciation by rapid divergence with founder events, 106
Biological resources, noncontinuous distribution of, 66
Biological species concept, 96
Biology
  of adjacent populations, 79
  with genetic divergence, 79
Biomass, strains of *Drosophila* and production of, 144
Biomes
  in Alaskan parks, 421
  number of, 421
  temperate, 422, 429
  tropical, 422, 429
Biosphere reserves, ecological benchmarks in habitat-change measurements, 421. See also Park and reserve size, large
Biosystematic studies

genetic diversity measurement, 357
  concordance with, 357
Biota
  in Alaska, 421
  diversity of, 421
  in ecological chain reactions, 7
  management intervention, 7
  on North American continent, 421
  in self-regulation, 6
Biotic diversity, 322–330, 400
Biotic factors
  differences in fitness, 322
  in extinction, 365
  in fitness, 322
  selective forces, 322
  variations among populations, 322
Bird populations, inbreeding depression in, 178
Bird species
  adaptation and natural selection, 133
  clutches, 133
  dispersal and water barriers, 119
  eggs, 133
    losses, 133
  extinction threat, 133
  failure rates by resource climatic fluctuation, 119
  food supply for young, 133
  numbers of young, 133
  population decline, 161
  population extinction, 161
  random matings among hybrids, parental forms, 291
  rarity and extinction in habitat patches, 118
  recolonization, 119
  resource dependence of, 118, 119
  taxa in hybridization, 290
  vulnerability of, to near ground predators, 119
Birds. See also Avian species
  adaptation in absence of predation, 111
  adaptation on islands, 111
  behavioral patterns, 198
  clutch sizes and live progeny per female, 135
  courtship systems, 198
  display effects, 198
  extinction of, 111
  extinction of by social behavior factors, 117
  flight loss of, on islands, 111
  game management, 278
  ground foraging of, 111
  ground nesting of, 111
  K-selected species, 118
  loss of fear, 111
  loss of flight, 111
  males, 198
  natural sexual selection, 198
  nesting behavior, 118
  plumages, 198
  reproductive potential, 118
  sociability, 117
  song or call in sibling species identification of, 99
  territorial defense, 198
  vocalization, 198
Birds, larger, migration, 9
Birds, tropical
  factors in extinction of, 118, 119, 121
  natural barriers and dispersal of, 121
Birth
  in elimination of inbreeding depression, 255
  with inbreeding, 246, 247
Birth weight
  depression with inbreeding, 243
  inbreeding and, 258
  in nonselected animals, 259
  reduction with inbreeding, 248–250
  regression vs. inbreeding coefficient, 258
  in selected animals, 259
  survival for 30 days, 256–258
Blood groups, codominance in phenotypes of, 25
Body size
  genetic effect of several loci, 37
  level of demographic protection, 438
  park and reserve size and, 438
  and rarity of species, 115
Bogotá, Colombia, *Drosophila* studies, 238
Botanical gardens, xix
  allelic and species components, 333
  breeding designs for, 63
  construction, 331
  management of, 333
  natural areas, 333
  population management for, 63
  relict populations, 333
  source of new genotypes, 7

stability of management system, 333
Bottlenecks, effects of, 191
Brazil, 104
Brazil Coast, sea turtles, 285
Brazos River, Texas, 101-103
Breeders
  animal, 196
  change in characters, 190
  culling methods, 199
  genetic variability manipulation, 190
  in maintenance of genetic diversity, 179
  plant, 196
  and selection, 196
  selection against deleterious alleles, 197
  use of selection, 190, 191
Breeders' success
  based on variability, 191
  in retention of inherent values of natural species, 191
  selective regimes, 191
Breeding
  for adaptation to small population size, 331
  contact prevention, 192
  controlled, 76
  crossover individuals, 30, 31
  early studies on laboratory animals, 174
  history of, 174
  import animals, 250
  inbreeding, 137, 433
  inbreeding avoidance, 250
  of large, rare, and endangered animals, 192
  linebreeding, 176
  of livestock, 174
  outbreeding, 137, 196, 435
  physiology, 409-412
  rate of, 117
  selective, 190, 198
  to separate desirable from undesirable alleles, 30, 31
  typically inbreeding, 439
Breeding activity, catastrophic events during, 117
Breeding colonies, sea turtle
  genetic groupings, 280
  reproductive limits, 280
  separate demes, 280
  systematic separation by nesting beach, 280
Breeding design, inbreeding depression elimination, 250
Breeding efforts, relict populations, 313
Breeding plans
  inbreeding of Speke's gazelle, 187
  outbreeding maximization, 196, 197
  for small captive populations, 180
  for white tigers, 179, 180
  for zoos
    inbreeding to no inbreeding, generation of survival, 172, 173
Breeding pool, selection time before entry into, 151
Breeding populations, expansion, 150
Breeding programs. *See also* Captive species
  for captive animals, 68, 69, 242
  captive populations, 180
  choice of parents, 254
  cooperative efforts, 180
  for demographic stability, 180
  design, 242
  design for Speke's gazelle, 245-259
  economic feasibility, 69
  effective population size, 68
  founder number, 242
  generations of preservation, 133, 134
  genetic diversity maintenance, 180
  genetic diversity preservation, 133, 134
  genetic input of founders, 253
  genetic variability maximization, 252, 254
  high genetic variance between populations, 68
  impact on genetic mortality, 255
  inbred animals as parents, 255
  inbreeding avoidance, 259
  inbreeding minimization, 242
  inbreeding nonavoidance, 245
  increase of, 68
  low genetic variance within populations, 68
  number of lethal equivalents reduction, 255
  options in breeding, 259
  pairing of parental animals by offspring attributes, 254
  pedigree judgments, 254

preservation of founder heterozygosity, 259
reduction of selection differential, 133, 134
use of founder genetic contribution, 180, 182
use of numerous isolated breeding units, 69
Breeding program design (Speke's gazelle)
adaptation to inbreeding, 260
breeding loans between zoos, 260
elimination of inbreeding depression, 260
equalizing genetic ancestry from founders, 260
herd size expansion during selection, 260
inbred parent selection, 260
maximization of genetic variability, 260
parent viability maximization, 260
selection for inbreeding (genes), 260
zoo cooperation, 260, 261
Breeding program success, inbreeding depression elimination, 261
Breeding records
mammals, 171
primates, 171
species, 171
ungulates, 171
Breeding schemes
captive species propagation, 68
for isolated populations, 68–70
for maintenance of genetic polymorphisms, 76
for small populations, 68
for zoo populations, 67–70
Breeding strategies
genetic change limitation, 67
inbreeding depression avoidance, 243
inbreeding depression nonavoidance, 243
for isolated populations of normally outbreeding species, 67
maintenance of genetic variability, 67
maintenance of viable individuals, 67
Breeding strategies, small populations
crosses, number, and sex, 245
founder ancestry equalization, 245
genetic variation maximization, 245

inbreeding coefficients, 245
for effective selection, 245
inbreeding depression minimization, 245
offspring inbred, 245
parents selection, 245
population size increase, 245
Breeding studies
adjunct to electrophoretic studies, 356
in evaluation of protein variation, 356
evolutionary divergence estimates, 356
genetic divergence estimates, 356
laboratory, 356
Mendelian inheritance of morphological traits, 356
Mendelian inheritance of physiological traits, 356
in species with behavioral isolating mechanisms, 356
Breeding success, 304
Breeding systems. See also Effective population size (N_e)
asexual, 203
inbreeding, 118, 139, 203
inbreeding in normally outcrossed species, 118
linkage disequilibrium effects, 129
in maintenance of genetic variation, 76
meiosis, 203
mutational load, 203
outbreeding, 118, 205
population genetic structures and, 204
recombinational load, 203
relative advantages of different, 204
segregational load, 203
self-fertilizing, 205
in terrestrial slugs, 205–210
variety in molluscs, 215
Breeding types, 499
dispersal within semi-isolated populations, 73, 74, 75
inbreeding, 109
outbreeding, 109
Breeding units of individuals mating at random, small, 89. See also Demes
British Columbia, Canada, *Drosophila* studies, 238
British Isles, slugs, 205, 209

Buffering (homeostatic capacity) at intraindividual level, 149
Burleson County, Texas, 101
Butterflies
  collections, 161
  dynamic and genetic events, 162
  ecological factors and extinction, 109
  extinction by larval food plant stresses, 161, 162
  factors for extinction in, 109
  factors in extinction, 121
  genetic adaptiveness in, 109
  key indicator organisms, 161
  life history characteristics of, 109
  loss, 161
  model-mimic resemblances, 162
  monitoring, 155, 161, 162
  natural barriers and dispersal of, 121
  rates of extinction of populations, 162
  studies, 161
Butterfly larvae, dependence on food plants, 155
Butterfly populations, allozyme frequencies, 153
  biology of, 162
  buffering, 155
  climatic relationships, 154, 155
  density-independent regulation, 154, 156
  dependence on other biota, 154
  dynamic history, 156
  dynamics, 155, 161
  extinction, 154, 156, 162
  fluctuation in size, 156
  food plant relationships, 154, 155
  genetic variability loss, 155
    restoration, 155
  genetics, 154, 161
  geographical proximity, 153, 155
  habitat changes, 155
  host-plant relationships, 161
  human disturbances and, 162
  inbreeding depression resistance, 156
  index to plant community status, 161
  isolation, 156
  larval food plants and, 154, 155, 161, 162
  levels of variability, 156
  mortality, 154
  mosaic pattern in regulation, 159
  persistence lack in small numbers, 156
  pollution and, 162
  reestablishment, 156
  selective pressures, 162
Butterfly transplants
  eggs and larvae, 156
  population extinction, 156
  population fluctuation time, 156
  reproductive output, 156

California, 153, 155
  butterfly populations, 156
  climate, 82
  correlations in slender wild oat, 223
  plant colonization, 218
  plant genetic variation, 218
  slender wild oat, 344
  specific multilocus genotype-environment, 223
California, Central Valley, 159
California, Edgewood, butterflies, 158, 159
Cambridgeshire, England, butterflies, 160
Canadian parks, 438
  habitat adjacent to U.S. parks, 438
Canaveral National Seashore, Florida
  loggerhead sea turtle in, 420
  size, 420
Captive animal introduction, 242
Captive animals
  breeding programs, 242, 254
  inbreeding depression, 242
  numbers, 242
Captive breeding programs. *See also* Breeding program
  crossbreeding tolerance evaluation, 104
  in generations, 103
  genetic and karyotypic data on founders, 104
  genetic alterations induced, 259
  inbreeding and population maintenance, 241
  long-term, 241
  management, 103, 104
  populations of different karyotypes, 103
    treatment as different species, 103

populations with fixed differences at many gene loci, 103
  treatment as different species, 103
  preservation of rare or endangered species, 259
Captive-bred animals, elimination of inbreeding depression, 245
Captive collections, 375
  enlargement, 400
Captive group, division into numerous populations, 69
Captive habitat
  competition for, 376
  for endangered taxa, 376
  need for increase, 383–387
Captive herds
  genetic variation maximization, 250
  inbreeding depression and, 260
Captive management
  animal numbers limitation, 386
  artificial reproduction, 404, 410, 411
  behavioral programs, 404
  effective population size, 388
  endocrine surveillance, reproductive cycles, 404, 409, 410
  family size, 383
  founders, 384
  genetic assessment, 404
  genetics of species, 387–392
  medical technology, 404
  sex ratio, 384, 386
  stationary configuration management, 398
Captive population management
  demographic regulation, 392–398
  effective population size, 398
  generation time, 398
  stable age distribution, 395
Captive population problems
  population management, 375
  taxa selection, 375
Captive population programs
  preservation of biotic diversity, 400
  wild population programs and, 400
Captive populations, 103. *See also* Genetic populations
  age structure, 394, 395
  bottlenecks in population size, 187
  breeding schemes for preservation of genetic diversity, 67–70
  controlled gene migration, 314
  deleterious effects of inbreeding, 165–174
  demographic management, 392
  differences in age structures, 180
  differences in fecundity, 180
  differences in sex ratios, 180
  differences in sizes, 180
  differences in survivorship, 180
  dispersal and, 76
  distinction from wild diminishing, 401
  early development of characteristics in, 180
  effective population size management, 389–392
  founder effects in, 180
  genetic diversity preservation, 387–392
  genetic management of, 179–182
  genetic variability, 241
  inbreeding accumulation in, 76, 180, 183, 241
  inbreeding depression in, 109, 172
  instability and senescence in, 395
  isolation by distance, 69
  isolation by ecological distance of, 76
  isolation by geographic distance of, 76
  loss of fitness in, 183
  maintenance of all polymorphisms, 76
  manipulated matchmaking efforts by man, 76
  mtDNA analysis in lineage studies, 238
  natural population enrichment, 314
  numerous isolated breeding units, 69
  of rare and endangered species, 164
  small, management of, 177
  stationary, 395
  strategy for incorporation of founders, 388
  studbooks and, 180
  successful adaptation to inbreeding, 244
  survival time of, 172
  total for taxon, 379
Captive species propagation. *See also* Breeding programs
  factors for, 68, 69

Subject Index 627

in numerous isolated populations, 68
Captive stock
  failure of introductions, 384
  genetic diversity deficiency, 384
  inbred lineage, 384
  new population establishment, 384
Captive zoo populations
  breeding program for, 259
  genetics, 241
  inbreeding depression elimination, 259
  inbred populations, 259
  management, 241
Caribbean, 144
Caribbean Ocean, sea turtles, 277, 283
Carnivore species, 415
Carnivores
  highly specialized, 439
  population-size to park-size relationship, 442
Carnivores, large
  number of species, 429
  number of individuals of a given species, 429
  space requirements of, 422, 427, 428
Carrying capacity
  for captive populations, 386, 392
  in parks, 421
  soft selection type and, 134
Castle-Hardy-Weinberg equilibrium. See Hardy-Weinberg equilibrium
Cat, color markings and extra X chromosome, 28
Catastrophes
  climatic, 121
  colonization success following, 436
  disease, 436, 438
  environmental change, 436
  fires, 121
  habitat overexploitation, 436, 440
  inflexibility of species to, 436
  landslides, 121
  overexploitation of limited habitat, 436
Catastrophic events
  cometary, 126
  competition by exotics, 443
  disease, 443
  drought, 443
  extinction and, 117
  recolonization after, 436
  small organism vulnerability to, 440
  small scale, 443
  volcanic eruption, 126
Cayman Islands, sea turtle farming and ranching, 282
Cayman Turtle Farms, farmed turtles release, 282
Cell, genotype of two sets, 18
Cells
  chromosomes in single, 19
  diploid, 19–21
  division, 19–21. See also Meiosis, Mitosis
  haploid, 19–21
  meiotic, 101
Cellular reproduction
  meiosis, 19–21
  mitosis, 19
Census data, of rare mammal species in zoos, 164
Central America, 367
  plant introduction from, 219
Cereals, 274
Ceratostomella ulmi. See Dutch elm disease (*Ceratostomella ulmi*)
Chance events, individuals in propagation, 414
Character
  gene effect
  gene effects of single locus on, 37, 38
  gene effects at different loci on, 37, 38
  polygenic effect on, 198
Character variance
  decrease within groups, 202
  inbreeding effect on, 202
  increase between groups, 202
  total increase in population, 202
Characteristics
  of demes, 199
  economically important, 184
  effects on genetic variation distribution, 336
  influenced by environment, 15
  influenced by heredity, 15
  of populations, 193
  of species, 199
  of subspecies, 199
Characteristics (species, subspecies)

Subject Index

Characters (cont.)
  maintenance of, 198, 199
  and polygenic balances, 197
Characters
  in adjacent populations, 78
  alteration, 359
  analysis of individual, 362
  change in laboratory populations, 359
  change in managed populations, 359
  change in transplanted populations, 359
  components and selection action, 133
  concordance in subspecies, 359, 360
  discrete differences in sibling species, 99
  distance from Darwinian fitness characters, 133
  factor loading, 362
  genetic variance affecting, 133
  with higher additive components, 133
  increase in advantages, 38
  increase in number, 359
  mating behavior, 133
  with lower dominance and epistatic components, 133
  morphological, 133
  nested analysis of variance, 362
  polygenic, disruption of, 178
  in proportion of total variance, 362
  reproductive, 198
  for reproductive adaptation, 38
  selection affecting, 133
  special, of taxonomic groups, in sibling species identification, 99
  stabilizing selection and, 149
  in subspecies, 358, 359, 362
  for survival of individuals, 38
  variation ability, 133
  variations of, in individual survival, 38
Character modification
  morphological, 84
  physiological, 84
Chromatids
  breaks, 30, 31
  crossing over, 31
  in gametes, 31
  gene linkage, 31
  rejoining, 31
  in second meiotic division
Chromosomal analysis
  chromosomal marker behavior, 355
  genetic diversity at taxonomic level, 355
  karyotypic identity, 355
  meiosis studies in species differentiation, 355
  polymorphism, 355
Chromosomal banding studies, sea turtles, 280
Chromosomal incompatibility
  deleterious effects, 8
  parents, 8
Chromosomal mutations
  aneuploidy, 35
  deletions, 35, 36
  duplications, 35, 36
  inversions, 35, 36
  polyploidy, 34, 35
  translocations, 35, 36
Chromosomal races, 360
Chromosomal studies
  karyotypic variation, 101
  lower cost, 357
Chromosome(s)
  animal, 404–408
  arrangements, 129
  assortment into gametes, 21
  assortment of homologous pairs, 18, 128
  autosomes, 27
  in cell division, 18
  centromere position, 17
  chromatids, 19, 30
  crossing over, 30, 31, 128
  diploid number, 128
  duplication, 19
  fusion of arms, 128
  gene loci, 22, 130
  haploid set, 18
  inversion, 99, 128
  karyotype, 17
  linear sequence of genes, 128
  linkage disequilibrium, 128
  maps, 30
  maternal, 18, 21
  in meiosis, 19–21, 99
  in mitosis, 19, 99
  monomorphism, 99
  number and sex, 16

Subject Index 629

number species-specific, 16, 128
number variability, 128
pairing, 30, 31, 128
paternal, 18, 21
ploidy, 18
polymorphism, 99
rearrangements, 99
recombination, 21
recombination within pairs, 128
sequence of gene loci, 22
sets, 16
sex, 18, 27–29
in sibling species, 99–101
sibling species differences, 99
species differences, 101
species-specific, 127
structure alteration, 35
synapsis, 21
Chromosome maps and linkage disequilibrium evidence, 129
Chromosome number
diploid, 19–21, 128
haploid, 19–21, 128
species-specific, 128
Chromosome sets. See Genomes
Chromosome structure, 35, 36
Chromosome techniques, mapping, 355
Chromosome type
autosome, 27
in humans, 129
X chromosome, 28
Y chromosome, 28
Z chromosome, 29
City parks, level of protection in, 438
Classical system of population structure, 197
Classification. See also Systematics
outdated concepts, 350
techniques, 357
Climate
affecting genotype formation, 82
butterfly phase relationships, 154
deterioration in and decline in fitness of species, 114
fluctuation of, 119
rainfall, 119
seasonal fluctuations and larval phases, 154
seasonal fluctuations and stress, 119
Climatic change
long-term, 438

population survival, 438
species decline effects, 126
Climatic effect
drought, 154, 155
extinction, 157
rain, 154, 155
snow, 157
temperature, survival and species abundance, 126
Climatic factors, host-parasite relationships and extinction by, 126
Climatic variation and gene arrangements, 137
Climatology
extinction from changes in, 439
of giant sequoia, 438, 439
Clones, 193, 269
aromatic compounds in, 270
complex hybrid derivatives, 269
Clubs in species extinction, oceanic islands, 111
Clutch, eggs per, 133
Clutch size
per female and, 135
live progeny and, 135
and natural selection, 133
variation for food supply, 133
Codominance, alleles expressed independently in, 25
Coevolution effects
distribution, 314
genetic variation amounts, 314
interacting species, 314
within ecosystems, 318
Collected populations. See Captive populations
Collecting, 116, 120
butterflies, 161
habitat information, 103
locality information, 103
Collections, 349
Colonies
definition, 40
inbreeding effects in, 205
Colonization
arrival ability, 192
between islands, 192
breeding type and success, 436
genetic systems in, 193
by invariant clones, 193
loss of genetic variation in, 217
of oceanic islands, 193

repeated, 192
    by single colonist (propagule), 193
    success, 192
Colonizers, following catastrophe, 436
Colonizing ability, 212–215
Colonizing events. *See also* Founding
    adaptive shifts, 226
    genetic changes, 226
Colonizing success, genetic variability and, 219
Colorado, butterfly populations, 156
Color pattern inheritance, 22, 23, 25, 28
Commercial investors, sea turtle farming and ranching, 281
Communities
    against novel enemies, 373
    conservation of, 372
    destabilization by evolution of parasites, 367
    in genetic defenses, 373
    species-rich/species-poor, 373
Community biology, comprehension, 12
Comparisons
    allozyme data/quantitative genetic variation distribution, 344, 345
    plant trait distribution/genetic variation distribution, 344, 345
Competition, 438
    herbaceous plants, 435
    insects, 435
    interspecific, 299
    lack of, 154
    mammals, 435
    for space, 435
Competitive exclusion, avian populations, 308
Competitors, target species genetic diversity and, 332
Compilospecies capacity for genetic absorption, 271
Compilospecies concept, 276
Computer simulations. *See also* Models
    genetic diversity maintenance, 52
Concept of diversity
    analysis of structure, 314
    elements measured, 314
    measured characteristics, 314
Connecticut warblers, 296
Conservation
    allelic diversity and, 362
    apportionment of effort, 362

areas for population subdivisions, 372, 373
biological factors in, 363
commercial factors in, 363
criteria, 350
of the dynamics of species evolution, 444
of ecological communities, 372
of ecosystems, 2
endangered species research, 402, 403
esthetic factors in, 363
esthetic ideals and unpopular organisms, 372
and evolution, 189
existing practices, 430
experimental procedures (sea turtles), 278
focus on small populations, 187
in fragments of populations, 430
of genes, 347
genetic diversity, 359, 430
genetic diversity measures, 332
genetic input, 125
genetic limitations in, 200
of genetic resources, 336, 337, 348
goals, 190
of habitat, 2
hybridization, 309
hybridization influences, 289
integrated strategies, 375
interaction of species and, 372
international cooperation in, 283
interventionist, 14
laissez-faire, 14
large tropical trees, 6
locally adapted plant alleles, 347
management, 2
manager, 313
natural integrity of populations, 444
in numbers of populations, 430
for operation of natural selection, 190
in perpetuity, 3, 4
preservation of ecosystem, 2
priorities in our time, 10
promotion of level-of-protection, 430, 431
retention of natural breeding communities, 190
role of genetic management, 13
and self-regulation in nature, 6
survival enhancement, 2

systematic confusion, sea turtles, 286
of taxonomic units, 444
time factor of, 2
time scale in, 2
time scale relaxed, 13
of valuable species, 189
of variation, 46
Conservation biology, 46
Conservation of ecosystems, 161
Conservation ethic, 189, 286, 287
  interventionist, 12
  laissez faire, 12
  and maintenance of natural characteristics, 199
  and preservation, 199
Conservation genetics
  "genetics of scarcity", 117
  mating system information, 8
  and success rate of animal reintroductions, 183
Conservation goals
  classification and, 350
  genetic diversity and, 350
Conservation guidelines, avian populations, 292
Conservation vs. improvement, 189, 191
Conservation management
  objective (target) in, 2
  procedures, 2
  programs, 2
  of rare and endangered species, 2
  sibling species, 103–105
Conservation manager
  avoidance of selective breeding, 198
  utilization of natural selection, 198
Conservation methods
  limits of knowledge, 278
  sea turtles, 281
Conservation philosophies, 12
Conservation planning
  in perpetuity, 13
  relaxed time scale, 13
Conservation programs
  for esthetics, 369
  importance of mating system in design, 8, 9
  to save species from extinction, 313
  sibling species, 103, 104
  systematic relationship research, 376
Conservation of species
  commercial species, 351

ecosystem preservation and, 350, 351
esthetic species, 351
intervention for survival, 351
keystone species, 351
Conservation strategies
  microenvironments, 6
  for plants, 6, 224
Conservation type
  long term, 2, 3
  short term, 2, 3
Conservationists
  concern about human activities, 126
  concern about inbreeding, 165
  observation of extinction, 109
Construct of five species, 271
Contact zones
  pocket gophers, 100–102
  species of grasses, 270
Continental populations, avian species, 293–300
Continents, habitat patches, mosaic, and survival on, 112
Convention on International Trade in Endangered Species of Wild Fauna and Flora (CITES), 164
Cooperative agreements
  with counties, 438
  for increase of habitat size, 443
  with private sector, 438
  with state, 438
Cosmopolitan species
  and genetic variation, 131, 132
  heterozygosity, 132
Cost of resistance, plants/insects, 368
Crises
  climate changes, 438
  competition, 438
  habitat destruction, 438
  new predation, 438
  nutrient change, 438
Critical age for reproduction, 44
Crop plants
  analogs of wild populations, 272
  gene-for-gene system, with pathogens, 367
Crops, 272–276
  resistance overcome, 366
    to fungi, 366
    to pathogens, 366
  resistant strains, 366
Crop species, 3

improvement, 275
  by genetic interactions, 275
  mimicry in, 273
  weed forms, 273
  weed races, 272
Crop-weed relationships, 269
Crosses
  dihybrid, 25–27
  genotypic ratios, 23
  interspecific, of pocket gopher, 103
  with linked loci, 29
  one parent inbred, 252
  phenotypic ratios, 23
  two parents inbred, 252
Crossing
  backcrosses, 294
  in peas, 22–24
Crossing experiments
  color in green lacewings, 98
  sibling species, 98
Crossing over, 30, 31
Crossmating, generations of and loss of alleles, in small populations, 69
Crossover, unequal, 36
Cryptic species. See Sibling species
Culling
  of individuals, 139
  reduction in heterogeneity by, 139
Culling methods from breeders, 199
Cultivars
  biotype association with, 369
  genetic uniformity, 367
  susceptibility to pathogens, 367
Cultivated populations, 103
  genotype reserves, 7

Dairy cows, autosomes of, and milk production, 29
Dallas, Texas, Speke's gazelle, 251
Darwin, Charles, 38, 174
  genetic variety is adaptive, 137
  inbreeding depression in natural plant populations, 222
Darwinian fitness, 135
  reproduction, 135
Darwinian selection, 202
Darwin-Wallace theory of natural selection, 38
Data analysis
  electrophoresis of protein index, 97
  genetic distance (D), 97
Death
  in elimination of inbreeding depression, 255
  with inbreeding, 246
Decapods
  heterozygosity and polymorphism in, 140
  trophic generalism, 140
  trophic instability, 140
Deleterious genes, segregation in captive/cultivated populations, 196
Deletion, definition of, 36
Deme conservation (sea turtles)
  right to evolution, 286
  right to existence, 286
Deme model
  allelic frequency divergence, 54, 55
  divergence by genetic drift, 55
  divergence by migration, 55
  exchange rate per generation, 55
  F_{ST} values, 54, 55
  m values, 54, 55
  mN values, 54, 55
  N values, 54, 55
  qualitative genetic diversity exchange, 55
  quantitative genetic diversity exchange, 55
Demes. See also Local populations
  adaptedness to local conditions, 63
  allele exchange amount, 55
  allele exchange rate, 55
  allele frequencies among, simulation model, 53–55
  allele frequency divergence among, 53
  breeding unit(s) of individuals mating at random, 39, 40
  definition of, 39
  differentiation between, 438
  differentiation into, 438
  distance between (isolation), 63
  divergence among, 56, 57
  emigration amount, 61
  fixation of chromosomal or genetic changes in inbred, 106
  gene flow between, 64
  genetic differentiation of animal populations in, 85
    contrasted with plants, 85
  genetic divergence in, causes of, 87
  genetic drift in, 63
  immigrant individuals mating (model), 52
  immigration amount, 61
  inbreeding among, 305

interbreeding, 77
local individuals mating (model), 52
localized gene flow in plants, 85
loss of alleles in, 63
migration (m) (model), 53
numbers of individuals (N) in (model), 53
with patterns of selection, 56, 57
population size (N) (model), 52, 53
population size effect on genetic divergence (model), 55, 56
reduced genetic variability by population subdivision in, 93
reproduction between, 52
reproduction within, 52
similarity among, 55
size
  drift, 55
  effects, 53, 55, 56
  heterozygous advantage effects, 53
  reduced divergence, 56, 57
social structure in, 87
Demes, isolated (plant), counteraction of inbreeding depression, 9
Demes, semi-isolated local populations, 87
allele frequency divergence among, 64
differential selection and, 64
heterozygous advantage and, 64
migrant exchange in management of, 64
size function in, 64
Demographic complexity levels
among individuals, 415
between groups or populations, 415
genetic diversity apportionment, 415
in groups or populations, 415
Demographic factors (new populations)
age and sex in fertility and survival, 384
population age characteristics, 384
population sex characteristics, 384
reproductive value, 384
Demographic management
age and sex class, fertilities, and survivorship, 394
age specificity, 394
before inbreeding, 250
case example, Siberian tiger, 399, 400
expansion to carrying capacity, 394
objectives, 400, 401

regulation for carrying capacity, 394
stability of captive populations, 392
stabilization at carrying capacity, 394, 398
stationary configuration management, 398
strategies, 400, 401
Demographic protection
concepts of, 437
graded series of reserve sizes, 437
Demographic protection levels, 1-9, 430-440. *See also* Protection
descriptions of, 439
level 1, individual, 432
level 2, family, isolated, 432
level 3, family receiving gene flow, 432
level 4, population, relict or fragment, 432
level 5, legislated protection, species-specific, 432
level 6, population, natural size and complexity, 433
level 7, multiple populations, one subdivided or two isolated, 433
level 8, multiple populations, several of natural size and complexity, 433
level 9, multiple populations, two or more level 8, in different portions of species' range, 433
percentages of total within and between, 439
population diversity, 439
Demographic regulation
of managed populations, 392
captive, 392
wild, 392
Demographic stability
fertility management, 394-398
goal of captive management, 392
survivorship management, 394-398
Demographic stabilization
of populations, 401
of Siberian tiger, 397
Demographic stochasticity
age class, 117
random fluctuation in population size, effects and extinction, 117
random fluctuation in population variables, 117
sex ratio, 117
Demographic unit
adaptation to stress, 155

complexity of, 442, 443
dynamics of, 152
longevity of, 444
in mosaic habitats, 159
protection, 437
size of, 442–444
small, 434
Demographic unit area G (Stanford U.)
fluctuation, 153
genetic variability decay in small, 153
population extinction, 153
reestablishment, 153
stochastic (random) extinction, 153
Demography
little potential for growth, 436
numbers, small, 436
species/area relationships, 422
units of, 415
Deoxyribonucleic acid, 16, 32, 33, 36. *See also* DNA
amount in genome coding for polypeptide products, 128
in chromosomes of eukaryotic organisms, 16
in control of protein synthesis, 32, 33
in *Drosophila*, 128
haploid amount in eukaryotic species, 128
in mammals, 128
nucleotide pairs in fungi, 128
in vascular plants, 128
Design models, for breeding structures, 64
Development
habitat preservation, 279
sea turtles, 279
Developmental stability, 149. *See also* Individuals homeostatic capacity (buffering)
morphological asymmetry, 149
Differential directional selection, 60–63
in allele frequency divergence among demes, 60
different alleles in different demes (model), 53, 57
divergence increase among demes, 53, 57
relative divergence and simulation, 59
simulation steady state F_{ST} values for demes with, 60

Differential sex migration, genetic diversity maintenance, 321
Differentiation,
adaptive, over short distances, 94
patterns of
on geographical scale, 79–83
on microhabitat scale, 79–83
in steady evolutionary advance, 9
Differentiation-hybridization cycles
buffered genotypes, 268
closure by hybridization, 269, 271
differentiation to speciation, 268
length, 268
mode of evolution in plants, 267
systems of evolutionary advance, 267
systems for evolutionary change, 275
systems for speciation, 275
Dihybrid crosses, 25–27
Dinosaurs
extinction of, 126
thermal stress and, 127
Diploid cross-fertilizing species
genetic loads, 193
genetic structure of populations, 193
reproductive mode, 193
Directional evolution
mutation and, 39
natural selection and, 39
Directional selection. *See also* Differential directional selection
in every tree patch (insects), 87
genetic variation reduction, 133
Disassortative mating
level of genetic variation increase, 253
new gene combination increase, 253
Disassortative mating by pedigree
adaptation rate to inbreeding slow down, 253
amelioration of inbreeding effects, 253
selective pressure reduction and, 253
Disassortative mating system(s)
genetic variability preservation, 253
pedigree, 253
transiliences and, 253
Disease resistance, 150
decrease with inbreeding, 179
Dispersal
behavioral distance and, 67

birthplace to birthplace of offspring, 70
  ecological distance and, 66, 67
  of forest birds, 119
  geographic distance and, 66, 67
  intrinsic inhibitions to, 121
  of pocket gophers across clay soil barrier, 101
  of seeds, 119
  sibling species differences in, 99, 100
  within populations, 70-73
Dispersal capacity, 106. See also Vagility
  sibling species, 106
Dispersal distance, 71
Dispersal rates
  constant per generation, 71
  effects on inbreeding coefficients, 70
  within semi-isolated populations, 70
Distance, types of, between populations
  behavioral, 67, 69
  ecological, 66, 67, 69
  geographic, 66, 69
Distance barriers
  artificial, 76
  ecological, 76
  geographic, 76
  man's assistance in traversing, 76
Distance scale(s)
  large, 91
  microgeographical, 91
  short, 90, 92
  small, 89
Distribution
  of allozymes
    allozyme traits, 344
    quantitative (morphometric) traits, 344
    variation, 344
    variation associations, 344
  breeding, 297, 298
  dispersed, 126
  and extinction, 126
  inbreeding, 201
  localized, 126
  narrow, 126
  species decline and, 126
  of quantitative variation, 344
  of warblers, 296
Distributional data, gophers, 102

Divergence. See also F_{ST}, Genetic differences
  among demes, 53-64
  among populations, environmental effects, 320
  continued, of species, 439
  local, 91
Diversity, genetic. See also Genetic diversity
  of organisms, 349
Diversity measures. See also Multiple population preservation, Species architecture
  weighted counts of allelic variations, 318
  for multiple loci, 318
DNA analysis, 351-354
  DNA cleavage at recognition sites, 236
  enzyme restriction endonucleases, 236
  length studied, 236
  pieces separated by electrophoresis in agarose gel, 236
  and taxonomy at family or genus level, 237
  variation capability tested at sites, 236
DNA sequence diversity
  between species, 352
  within species, 352
DNA technology, potential of, 236, 237
Dobzhansky, T., 137
Dolphins, in aid of conspecifics, extinction, 117
Domesticated populations, selection pressures in, 3
Domestication, links with, 4-7
Dominance. See also Incomplete inheritance
  codominance, 25
  incomplete, 25
Duplication, of chromosome, definition, 36
Dutch elm disease, 127, 366
Dynamic changes
  allele frequencies and, 155
  genetic variability and, 155

Earthworms, polyploidy in, 35
East Africa
  grass distribution, 269
  grass species, 271, 276

Eastern Pacific Ocean, 114
Ecological associations, subdivided populations, 372
Ecological change and extinction, 439
Ecological compatibility
  genetic differentiation of partners, 8
  two avian species, 308
Ecological disintegration, forked-path domino theory, 120–122
Ecological disintegration rate, dependence on habitat fragment size, 124
Ecological diversity, adaptation to, 320
Ecological environment
  changes affecting fitness, 364, 365
  changes and extinction, 365
Ecological factors
  biotic, 224
  in habitat patch decline, 114
  physical, 224
  in plant population founding, 224
  in population extinction, 109
Ecological interaction effects
  ecological overload, 324
  self-damping effects, 324
Ecological models, 322, 323
  alloregulator, 329
  autoregulator, 329
  of species interactions, 324–330
  for use in management, 325
Ecological niches, nonadjustment of hybrids to parents, 103
Ecological races. *See* Ecotypes
Ecological surveys
  long turn-around-time for, 415
  for management, 415
Ecological systems management, 313
Ecological threats
  in comparison with traits in ensemble of populations, 372
  in different habitats, 372
  introduction of new species, 372
  nonadaptation, 372
  traits in populations under local selection, 372
Ecological variability
  human effects, 331
  natural events, 331
  in species evolution, 330
Ecology
  air, 421
  genetic diversity, 350
  number of species in community, 350
  relative abundance of species in community, 350
  sibling species and, 99
  water, 421
Economic resources, 415
  competition for, 415
Economically important species, 105
  crossbreeding programs for desired traits, 105
Ecosystems. *See also* Environmental impact
  agro- vs. natural, 7
  balance, 11
  butterflies, indicators of health, 152, 162
  chain reactions, 10
  collapse, 10, 12
  coevolutionary effects, 318
  component diversity, 7
  composition, 3
  conservation goal, 350, 351
  environmental arrays, 334
  environmental selective changes, 334
  extinction of species, 333, 334
  gene flow, 12
  genetic diversity in, 350
  genetic diversity of species components in, 318
  genetic management of, 10, 11
  vs. genetic system, 329, 330
  and genetic system stability, 3, 329, 330
  genetic tuning to local environments, 305
  genotype migration effects, 11
  health of, 161
  hybrids vs. parental forms, 307
  impact of population attrition, 11
  impact of population enrichment, 11
  impoverishment, 10
  interactions (chain reactions), 7
  interdependent species in, 127
  interpopulational structures, 318
  introduction of genetic materials, 11
  introduction of genetic variants, 333, 334
  large effect, 10
  large herbivore effect, 10
  management of multiple, 334
  meta-stability, 334

natural selection, 11
in one-locus, two-allele system, 333
population variability, 334
regulation species, 10
resource impact, 11
species interdependence, 127
stability, 13, 329, 330, 333
supplemental resources, 11
target species, 10
temporal sequence of site disturbance, 326
time scale, 13
for two species, 333
unstable, 333
Ecotypes
close genetic timing, 305
formation of, with genetic variation, 82
morphological characteristics, 82
physiological characteristics, 82
population dissimilarity between, 155
population similarity within, 155
Edible aroids, 274
Education, use of species in, 435
Effective population number. See also Population size
breeding individuals in population, 44
for carrying capacity, 391
census population number and, 44
effective population size and, 44
N_e/N ratio, 44, 45
nonrandom distribution of offspring effects, 44
nonrandom mating effects, 44
overlapping generation effects, 44
unequal sex ratio effects, 44
Effective population size, 421. See also Breeding system
actual size and, 242
and dispersal distances, 71
equalizing family size, method of increase, 68
European bison (*Bison bonasus*), 244
family size, 390
gametic individuals per generation, 242
genetic drift and, 392
genetic variability and, 68
for inbreeding, 245
increase with migration, 8
influences on, 336

of large mammal populations, 183
long-term maintenance, 242, 243
maintenance, 242, 243, 384, 386
mating system effects, 336
minimum, 242, 243, 392
number, 390
plants, 336
population size effects, 336
reproduction mode effects, 336
rhinos, 376
seed dispersal effects, 336
sex ratio, 390
for short-term maintenance, 183, 242, 243
small, 6, 43, 245
and stochastic events, 130
total number of animals and, 390, 391
in wild animals, 391
in zoo animals, 391
Egg, 41
Eggs (sea turtle), translocation (*in extremis*), 286
Egg viability
at approach of extinction, 160
monitoring, 160
mortality, 196
Electrophoresis, 101, 103, 387
allozyme variation, 335
average values for heterozygosity, 353
genetic diversity measurements, 353
limitations, 230
at lower taxonomic levels, 354, 357
majority of variation in genome undetected, 230
multiple locus data sets, 353
overall diversity analysis limitations, 353
protein difference studies, 353
sea turtles, 280, 286, 287
snails, 212
sequential technique, comparison with protein sequencing, 357
among species within taxonomic groups, 335
statistical problems, 353
among taxonomic groups, 335
value below genetic level, 354
value in evaluation of genetic resources, 347, 348
variation at loci coding for proteins, 230

widespread use, 239
within and between sample measurements, 353
Electrophoresis data
  alleles at locus indistinguishable, 346
  allelic diversity underestimation, 346
  limitations in species range, 346
  rare allele omission, 346
  variation in one category, 346
  weightings, 346
Electrophoretic analysis, 357, 365
Electrophoretic assay
  limitations in survival studies, 149
  limits in population survival analysis, 149
  in population size analysis, 150
  reflection of population size, 150, and time, since the last restriction of size, 150
Electrophoretic assay of heterozygosity
  fitness components, 139
  genetic diversity limits of electrophoretic assay, 150
  genetic variation and survival, 149, 150
  growth rate components, 139
Electrophoretic methods, 25
  at higher taxonomic levels, 357
  at lower taxonomic levels, 357
Electrophoretic surveys
  of geographic populations of a range of species, 142
  in relation to climatic variables, 142
  variation in heterozygosity, 142
Electrophoretic techniques, gene pool studies, 130
Endangered species, 104, 117, 118
  biology, 404
  of birds, 150
  intense management of, 437
  legislative acts, 437
  of mammals, 150
  preservation, 190
  recovery plan for, 437
Endangered Species Act of 1973, 289, 306
Endemic characters, 301
Endemic species preservation, 436
England, 159
Environment

adaptation to external, 243
adaptation to internal, 243
adaptation to new, by hybrid swarms, 292
biological heterogeneity of hosts, 86
biotic, 224
biotic changes and extinction, 365
biotic diversity analysis, 326
biotic element evolution, 318
breeding system advantage in, 204
chemically extreme, 85
components and genetic variability, 139
disruptive selective pressures of, 92
diversity, 315
future, 331
geographical dispersal of populations in, 142
heavy metals on spoil sites, 95
and heredity, 15, 16
heterogeneity, 155, 159, 321
heterogenous (patchy), 92
heterogenous, and genetic variation preservation, 69
heterogeneous, over short distances, 94
insect scale fitness, 86
intermediate conditions, 82
isolation of genotypes, 85
with large environmental gradients, 94
man-modified, 307
microgeographical, 95
microhabitat differentiation, 92
past, 331
physical, 224
physical changes and extinction, 365
physical factors, 343
poor autocorrelation of, between localities, 205
population adaptation to, 38
rapid change over small distances, 79
reproduction in new, 306
rodenticides, 95
selection in a constant, 39
selective forces in, 82
utilization, 144
Environmental change, 155, 159, 160
  adaptation to, 95, 217–219
  avian population reactions, 308
  ant species replacement, 160
  biotic, 214

by chemicals in soil, 126
cometary events, 126
fitness and, 144
flora change, 160
genetic options, 217–219
grazing sheep, 160
heterozygous advantage and, 139
high temperature input and, 79
and human activity, 79, 92, 313
inflexibility of species to, 436
life span in hundreds of years, 436
man-induced, 293, 307, 308
metals and, 79
by natural causes, 313
by natural heterogeneity of habitat, 79
physical, 214
plant communities in, 217
plant phenotypic plasticity and, 219
pollutants and, 79
population instability and, 436
rate, 49, 115
role in extinction of species, 365
survival and, 49
temperature increase effects, 127
toxins and, 79
volcanic eruption and, 126
Environmental change effects, ocean productivity, 127
Environmental change in habitat/resources
catastrophe, 120
disturbance, 120
environmental variation, 120
long-term trends, 120
succession, 120
Environmental change by other species
competition, 120
disease, 120
hunting, 120
mutualism, 120
predation, 120
Environmental change rate, 49, 115
and generation time, 115
and genetic substitution in species, 115
Environmental conditions, lack of genetic resources to adapt to, 93
Environmental diversity
genetic diversity and, 319
indices in decapod species, 140
masking by high migration rates, 319
selective agent for genotypic diversity, 319
Environmental factors
allele frequencies and, 136
in plant breeding, 222
interaction with genetic factors, 222
Environmental vs. genetic intervention, 11
Environmental gradients and human activity, 94
Environmental heterogeneity, protector of genetic diversity, 321
Environmental impact. *See also* Ecosystem
balance of ecosystem effect, 11
ecosystem composition effect, 11
of human activity, 94
Environmental management
areas for management, 331
controlled fires, 332
cost, 10
effect on genetic diversity, 331
effect on species structure, 331
harvesting, 332
natural state maintenance, 332
nutrients, 11
pesticides, 11
protection from herbivores, 11
varying levels of selecting wilderness or natural areas, 331
water supply, 11
Environmental patches
genotypes in competition in, 320
genotype survival in, 320
individual genotypes in, 320
Environmental pressures, species change through, 38
Environmental relationships with genotype formation, 82
Environmental shift, response to, 217
Environmental tolerance, 224
Environmental variability and genetic variability, 142, 143
Environmental variables
competition coefficients, 113
disease, 113
microclimate, 113
temperature, 142
Environmental variance, increase with polymorphism increase, 322
Environmental variation

640 Subject Index

effects on genetic variation, 140, 141
genetic variation and, 137, 140–143
and genetic variation in temporal, 140, 141, 143
individual genotypes and, 330
man-made, 92
natural, 92
spatial, 140, 141, 143
strong selection pressures with, 92
Environmental variation type
  effect on gene pool, 142
  long-term variation, 141
  multiple, spatial niche, 141
  short-term, 141
  trophic variation, 141
Enzyme variation
  and codominance, 25
  neutrality question, 365
Enzymes, 33, 368
Enzymes (restriction endonucleases), 351, 352
  maps, 351
Epidemic population, single subsample in species diversity preservation, 319
Epistasis
  definition of, 37
  phenotype by, 37
Ethiopia
  Speke's gazelle, 242, 260
  weedy sorghum, 273
Eukaryotic genome, evolution of, 229
Eurasia, grass distribution, 269
Europe
  biomes of, 422
  slugs, 210
  Sumatran tigers in, 180
European frog species, habitat type and genetic variation in, 140
Evolution. *See also* Speciation
  advance by local differentiation, 9
  breeding systems, 203–205
  change in population genetic composition, 39
  coevolution, 370
  and conservation, 189
  continued phyletic, 439
  definition, 39
  dependence on random genetic drift, 9
  direction of, over generations, 39
  ecological complexity, 314
  ecological variability effects, 330

by environmental crises, 313
of eukaryotic genome, 229
and extinction, 125
fate of inbreeders, 214, 215
fauna, 49
flora, 49
future, and population size, 319
genetic divergence speed in, 84
genetic structure of species in, 51
genetic variation and, 144, 217, 323, 330
of genetic variation for adaptation, 370
  to new hosts, 370
  new parasites, 370
  new predators, 370
  new prey, 370
geologic time, 439
historical evidence, 9
inbreeding avoidance, 139
of interactions of species, 370
long-term considerations, 372
long-term, of species, 439
man's part in, 12, 13
of metal tolerant grass populations, 95
mutation and, 37
mutation effecting, 34
of parasites, 367
population change rate by genotype, 323
population variability effects, 330
rapid, 144
role of isolation in, 49
selection pressures and, 93
selective factors, 314
of selfing, 203–205
short distance environmental gradients and, 94
of sibling species, 97, 98
slow, 146
small populations, 9
species in semi-managed populations, 332
species viability, 387
speed of, 370
subpopulation structure, 314
of superficial morphological characters, 357
  convergent, 357
  parallel, 357
time scale, 39
variations in biotic environment, 314

Subject Index   641

Evolutionary adaptation
  to inbreeding, 435
  organism adaptation to environment, 243
  organism adaptation to inbreeding, 243
Evolutionary biologists' use of genetic markers, 33
Evolutionary biology, 46
Evolutionary change, 39
  domination of genetic drift in, 91
  domination of selection in, 91
  genetic variation and, 146
  human activities, 93
  over large distances, 91
  rapid, 146. *See also* Speciation
  rate, 146
  regulatory genes in, 33
  traits in, 253
Evolutionary dynamics, genetic variations, 313, 314
Evolutionary genetics
  in management, 415
  and population management, 45, 46
Evolutionary origin
  selfing strains from outcrossers (model), 215
Evolutionary potential
  long-term superiority for outcrossing, 215
  short-term superiority for selfing, 215
Evolutionary record, island size, 9
Exotics, introduction of, 293, 307, 308
Extinction
  adaptation and natural selection, 133
  allelic polymorphism and, 321
  aspect of evolution, 125
  biological basis for extinction, 151
  biology, 161
  by biotic factors, 365, 366
  breeding techniques and, 250
  butterfly populations, 152–163
  by catastrophe, 69, 157
  catastrophe (climatic), 157
  causes of, 109
  chain reactions in ecosystem, 10, 121
  by collecting, 159
  on continents, 112
  danger of, 437
  by decrease of habitat, 444
  definition of, 114

  demise of geographic units, 114
  demise of systematic units, 114
  differential susceptibility to, 157
  of dominant, key species, 10
  dry season stress, tropical forest birds, 119
  environmental change and, 119, 120, 126, 365
  by epidemic, 69
  epidemics in mammalian species, 126
  evolution and, 125
  as evolutionary process, 109
  exacerbation, 109
  by factors intrinsic for species, 119, 120
  by failure to adapt to environmental changes, 364
    that lower fitness, 364, 365
  failure to adapt to genetic changes in other species, 367
  by flood, 159
  gene frequencies in populations at, 329
  genetic factors, 49
  genetic variation and, 127
  genetic variation in two species, 334
  genetics of, 152, 153
  in geologic time, 126, 367
  by habitat change, 159
  at the hands of man, 111, 112
  heterozygosity and, 127
  host populations, 366, 367
  from human activities, 109, 111
  inevitability, 126, 444
  by insects, 366
  interactions between causes of, 120
  interpretation of observed, 161
  by introductions, 111, 366
  isolation and, 49
  of large animals, 112
  levels of, 114
  life norm, 126
  local, by human activity, 93
  of local geographic groups, 114
  of local populations, 113, 114, 313, 314, 439
  long-term studies of, 115
  by loss of fitness, 112
  loss rate of species prediction, 439
  by man's hand, 125
  minimization, 250
  natural, 112, 124

of natural populations, 241
natural process, 109
near, 444
one species every day, 109
with park protection levels, 439
by pathogens, 366
pathways of, 120
of plants, 112
political decisions on, 151
of population, 364
population size expansion and, 250
prediction, 439
prediction of vulnerability to, 119, 120
probabilities for birds, 183
probabilities for large mammals, 183
prognosis, 103
rarity and vulnerability to, 119
rates, 205, 403
by recent introductions, 365, 366
reproduction and, 118
risk minimization, 150
with rodenticides, 88
seasonal fluctuation of resources, 119
self-fertilizing species and, 414
senescence in zoos, 395
in short periods of time, 126
from single plant founder, 221
in small populations, 109
of species, 114, 313, 314
species grouping factors in, 117, 118
at species level, 313
in species, predisposition, 109
by stochastic events, 115, 156
of subspecies, 114
of taxa, 313
taxa endangered, 333
threat, sea turtles, 278
threat of, 313
threat to key species, 10
time scales of, 183
undesired, 444
of varieties, 114
Extinction, analysis of vulnerability by interaction
environmental changes in habitat/resource, 120
environmental changes by other species, 120
exacerbating qualities (species), 120
Extinction avoidance
breeding population expansion, 150
maximum population size, 150
reproductive capacity expansion, 150
room for selection, 150
Extinction factors
climatic events, 157, 158
drought on habitat factors, 113
human activities, 157, 158
Extinction factors, oceanic islands
cats, 111
clubs, 111
diseases, 111
dogs, 111
exotic birds, 111
goats, 111
guns, 111
mongooses, 111
pigs, 111
weedy plants, 111
Extinction by interaction of species and environment
fundamental characteristics of species, 124
physical and environmental changes in, 124
Extinction, intrinsic factors
behavioral dysfunction, 124
demographic stochastity, 124
genetic deterioration, 124
social dysfunction, 124
Extinction of local populations
catastrophe, 116
competition, 116
disease, 116
environmental variation, 116
extinction or reduction of mutualist populations, 116
founder effects, 116
habitat destruction, 116
habitat disturbance, 116
hunting and collecting, 116
hybridization, 116
inbreeding, 116
limited dispersal ability, 116
predation, 116
long-term environmental trends, 116
loss of heterozygosity, 116
rarity (low density), 116
rarity (small infrequent patches), 116
successional loss of habitat, 116
Extinction of local populations, conclusions
by competition, 116

by habitat destruction, 116
by isolation, 116
by man, 116
by predation, 116
sensitivity of island forms, 115
    to environmental change, 115
Extinction process
    by decline in numbers, 112
    time scale of, 112
Extinction rate
    accelerating, 109
    anthropogenic, 162
    environmental change, 313, 314
    natural, 162
Extinction type
    anthropogenic, 124, 157, 160
    natural, 153–157

F_{ST} (measure of divergence in individual loci)
    allele frequency divergence amount among demes (model), 53–64
    in genetic diversity simulations, 53–55
    steady state (model), 55
F_{ST} values
    for demes with differential directional selection, 60
    distribution, 62
F_1 generation, 23, 24, 41, 42
    phenotypic ratios, 24
F_2 generation, 23, 24, 41, 42
    mortality of hybrids of sibling species, 103
    phenotypic ratios (peas), 24
Family size
    effective population size and, 390
    management, 390, 391, 400
    regulation, 390, 391, 400
Fecundity
    difference in snails, 211
    differential in captive populations, 180
    high lifetime, 338
    level, 211
    low, 225
    number of offspring during lifetime, 44
    retention of, 191
    variability of, 211–213
Feeding
    habitat preservation, 279
    sea turtles, 279
Females, 28, 29

Fertility
    age class, 394
    Darwinian fitness, 133
    demographic stability and, 398
    depression with inbreeding, 243
    loss, 150
    low rate of, 205
    reduction, 36
    retention of high, 191
    sex class, 394
Fertilization
    by inbreeding (selfing), 204
    by outbreeding, 204
Fertilized egg. See Zygote
Fish populations, inbreeding depression in, 178
Fisheries managers, observation of extinction, 109
Fishes
    polyploidy in, 35
    population decline, 161
    population extinction, 161
Fitness, 204. See also Heterozygous advantage
    average, 90, 91
    B chromosomes and plant, 128
    correlations in Drosophila, 145
    decrease of, 112, 205
    environmental change and, 144
    fecundity, 44
    and female gametes, 322
    and genetic variation in Drosophila, 137
    genetic variation simulation, 54
    genotypes and, 137
    heavy metal tolerance and, 84, 85
    heterozygosity and loss of, 45
    in heterozygotes, 322
    in host, 366
    in host/pathogen relationships, 367
    with hybridization, 289
    improvement by mutagens, in homozygous population, 139
    inbreeding depression and, 434, 435
    loss compensation by outcrossing, 68
    loss of, 194, 195
        by inbreeding depression, 45, 68
    and male gametes, 322
    negative correlations of, 205
    outbreeding depression and, 437
    in parasite, 366
    plant mating system shift and, 222
    of populations, 77, 139
        when new, 194

reduction by hybridization, 285
reduction with inbreeding, 67
and response to environmental conditions, 90, 91
restoration of, 195
viability, 44
Fitness index
competitive ability, 144
performance, 144
productivity, 144
Fixation of genes and inbreeding programs, 150
Flower colors or forms
genetic variability system and, 196
inheritance of, in peas, 22, 23, 25
selection by plant breeder, 196
Focal (target) species, 415
longevity of, 440
in reserve of known size, 442
Food plants, for butterfly larvae, 157
Food webs, species to reestablish, 444
Forage, outside protected area, 442
Foraging activity, catastrophic events during, 117
Forest habitats (tropical), timing of treefall and stages of mobile mutualists, pollinators, 121
Forest trees
allozyme variation among populations of, 344
behavior change, 319
from endemic to epidemic populations, 319
morphometric variation levels with environmental variation in, 344
successional stage and genetic variation, 341
Foresters, observation of extinction, 109
Forests
ecological processes and population extinction, 120
successional stages and isolated patch catastrophes, 120
Forked-path domino theory of ecological disintegration, 120–122
Founder(s)
artificial, 191
avoidance of small group size, 242
in captive breeding program, 250
extinction risk, 193
genetic diversity, 384
genetic input, 253

genetic resources of species, 217
heterozygous fitness and, 194
individual impact on gene pool, 252
information on original habitat of, 103
information on original locality of, 103
interfertile, 388
management, 384, 386
from natural populations, 191
noninbred, 388
number, 204, 205, 242, 384, 386
for preservation, 388
and number of lethal equivalents per individual, 247
numbers constraint, 252
on oceanic islands, 192
offspring contributions, 400
one or very few, 192
outbred, 259
pairs, 384
population size, 388
reproductive value, 384
single fertilized (gravid) female, 192, 197
small number of, 192, 245
success of, 192, 193
unrelated, 388
Founder acquisition, Siberian tiger, 400
Founder ancestry equalization in individual animals, 252, 253
Founder effect, 209
after founder events, 197
in artificially worked populations, 191
constraints imposed by man, 197
deleterious result, 197
effect on average heterozygosity measurement, 231
genetic variation reduction and, 231
genetics of, 187
and inbreeding, 200
loss or fixation of alleles, 197
in natural island situations, 197
on oceanic islands, 197
question of, 154
random genetic drift of allele frequencies, 197
from small number of individuals, 191
in small populations, 209
Founder (founding) event. See also Population bottlenecks

artificial, 192
average heterozygosity decrease
 after, 231
 genetic events associated with, in
  plants, 228
 and inbreeding, 197
 man-induced, 187
 natural, 187
 numbers of individuals after, 197
 shift in balance of, 198, 199
 by single pregnant female, 187
Founder population
 mitochondrial DNA studies, 352
Founder population(s), small, 242, 245
 gene pool variability influence, 261
 individual variability influence, 261
Founder population(s), zoos
 additional wild collection lack, 259
 inbreeding avoidance impracticality, 259
 initial inbreeding depression, 259
 maintenance, 259
 nonavoidance of inbreeding, 250
Founder population size
 at least five founders, 180
 mitochondrial DNA analysis for estimation of, 187
 molecular techniques in estimation of, 187
Founder representation, equalization of, 392
Founding. See also Founder event
 by *Drosophila* flies of Hawaii, 192
 by human agents, 187
 management of, 401
 size of group, 192
Founding populations, 444
 demographic factors, 384
 genetic factors, 384
 introductions, 331
 model for, 384
 plant
  adaptive potential, 227
  demographic structure, 227
  new environmental change, 227
  release from environmental constraints, 227
France
 lethal equivalents in human populations in, 244
 snails, 211
Frequency distribution
 features for preservation, 318
 variance, 317

Front Range, Rocky Mountains, Colorado, allozyme variation in ponderosa pine, 344
Fungi
 evolution of new forms against resistant crops, 366
 genetic variability in virulence, 369

Gametes, 41, 43
 crossing over and, 31
 with crossovers, 30
 eggs, 253
 genetic random combination at fertilization, 22
 genomic constitutions from the union of, 128
 plant
  female, 321
  male, 321
  migration of, 321
 pollen, dispersal of, 79
 sperm, 253
 waste, 304
Gangetic plain, India, grass species, 271
Gatún Lake, Panama, African cichlid fish introduction, 366
Gause competition equations, 325, 327
Genes, 319
 bizarre elements (transposons), 128
 chemical substance (DNA), 16
 coded segments of polypeptide products, 128
 coding function, 33, 128
 definition, 22
 deleterious, 196
 dispersal in genome, 128
 evolution into gene complexes, 330
 exchange, 114
 frequencies, 137
 hereditary units in genotype, 16
 independent assortment, 22–24
 linear distribution on chromosome, 22
 linkage of, 30
 linked, 371
 at loci, 22, 95, 128
 migration effects, 11
 neutral, 7
 noncoded segments (introns), 128
 numbers, 128
 overdominance, 155

pairs, 22
question of organization of, 45
random loss of, 130
rate of mutation, 37
recessive, 37, 137, 139, 178, 244
regulatory loci, 130
repeated sequences, 128
reshuffling, 128
segregation, 22
sex-linked, 139
single protein (definition), 33
structural loci, 130
variation (cyclic), 137
Gene action
  type, 321
  and genetic diversity location, 321
Gene arrangements
  climatic variation and, 137
Gene coding, 33
Gene combinations between gametes from different populations, 289
Gene complexes, recombinational load, 304
Gene conservation strategies
  electrophoretic studies, 335, 346, 347, 348
  morphometric studies, 344, 345, 347
  physiological studies, 347
Gene exchange
  between populations, 292
  by geographic entity, 114
  by groups, 114
  by populations, 114
  by species, 114
  by subspecies, 114
  through two populations, 292
Gene flow, 156
  amount of, 64, 91
  in animals, 85
  between adjacent populations, 94
  between populations on different trees, 87
  between two populations, 444
  destabilization of some plant breeding systems, 8
  factors in reduction of, 66, 67
  by gamete, 91
  genetic patchiness by limitation of, 94
  geographic distance interference in, 66
  human activities, barriers to, 78, 79
  interpopulation, 8

lack of, 66, 67
lack of continuous, 66
limited, in motile animals, 85
in maintenance of same alleles in semi-isolation, 49
management of, 438
man-made barriers and, 86
natural selection, 11
natural selection under conditions of limited, 82
by offspring dispersal, 91
over large geographical distances, 91
patterns of differentiation, 82
plant genetic restructuring, 9
population without, 436
from populations of same ecotype, 156
in rapid environmental changes, 82
rare interpopulation, 8
in restoration of genetic variability, 156
restriction by gametic dispersal mechanism, 90
restriction by specific behavior mechanism, 90
selection and, 373
selection pressure of insects in heterogeneous trees, 86
in slender wild oats on hillside, 82
over small distances, 82
stabilization of, 435
from wild to captivity, 388
Gene flow increase (within captive population)
  redistribute individuals periodically, 383
  regulate family size of reproducing animals, 383
  regulate sex ratio of reproducing animals, 383
Gene flow reduction, by distance type
  behavioral, 67
  ecological, 66, 67
  geographic, 66
Gene frequency
  bidirectional behavior, 329
  initial, 331
  at loci samples, 152
  monotonic behavior, 329
  population size and, 153
Gene levels, study techniques, 351–356

Gene loci. *See also* Loci
  body color in snails, 211
  polymorphic, 63
Gene movement and genetic variation among populations, 346
Gene pool
  alleles away from, 292
  diminishing, 374
  diversity of, 415
  elimination of original, 289
  environmental effects on, 141
  environmental variation effects on, 142
  existing, 414
  formerly connected, 313
  gene frequency destabilization, 370
  genetic variability in, 252
  genetic variability maximization, 254
  heterogeneity of, 414
  individual elimination from (culling, 139, 196
  mutations in, 370
  in new locality, 192
  number of, 414
    in parks and reserves, 414, 415
  population isolation, 313
  populations of *Drosophila* with seasonal variation, 142
  by random genetic drift, 370
  reduction in size, 187
  reservoir of genetic diversity, 139
  sizes of, in parks and reserves, 414, 415
  swamping of, with new alleles, 437
  transportation of, between islands, 192
  variability in, 194
  zygotic diversity in, 193
Gene pool shift, ancestral genetic input equalization, 252
Gene types
  regulatory, 33
  structural, 33
Gene-for-gene system, host/pathogen, 367
Genera
  breakdown of integrity of, 271
  by genetic aggression, 271
  merging, 271
Generation time
  duration of, 94
  and environmental change, 115
  genetic drift and, 392

intermediate, 442
long, 338
rapid genetic divergence and scale of, 93
Generations
  after founder effect, 192
  after founder event, 197
  arrays of disparate genotypic combinations, 190
  to bottleneck, 191
  in breeding programs, 133, 134
  critical, transitional, 253
  genetic diversity preservation through, 133, 134
  importance for breeders and managers, 197
  of inbreeding, 71, 250
  without inbreeding, 250
  and inbreeding coefficients in small populations, 69
  of inbreeding depression, 250
  individuals after founder event, 197
  initial, 245, 246, 250
  of intense inbreeding, 245, 246
  and loss of alleles by crossmating in small populations, 69
  of migrants, 320
  overlap, 5, 6
  and phenotypic variance, 146
  selection reduction, 133, 134
  selective removal by age classes and, 398
Genetic absorption
  capacity for, 271
  taxonomic consequences, 271
Genetic adaptation, speed of, 93
Genetic aggression (plants), 269-272
Genetic aspects of multispecies associations, 372
Genetic assimilation, 271
  conservation problem, 308
  island endemics, 308
  relict populations, 308
  warblers, 299
Genetic associations
  quantitative, 346
  and allozyme traits, 346
Genetic buffering
  in outcrossing species, 269
  ploidy and, 269
  in self-fertilizing species, 269
Genetic change
  in few or many loci, 131, 133
  linear or nonlinear, 133

of pests, 373
with population bottleneck occurrence, 193
in populations, 193
rapid or slow, 133
time lag, 11
Genetic code, sequence of bases in DNA molecules, 33
Genetic coding, 33, 34
for allozymes, 33
for proteins, 33
Genetic composition of populations, 45
Genetic consequences of plant mating systems, 221
Genetic conservation
management objective, 313
objective, 313
target species, 13
time scale of concern, 13
Genetic control of chromosome recombination, 128
Genetic correlations in plant resistance to enemies, 371
Genetic demands, small, 440
Genetic deterioration
genetic variability decrease in populations, 118
inbreeding depression, 118
Genetic differences
among natural populations, 53
between individuals and groups of individuals, 351
between individuals in local populations, 51
between other local populations, 51
between species of sea turtles, 280
Chrysopa, 98
determinations (model), 53–55
differential reproduction and, 180
Drosophila, 98, 99
of founding animals, 180
reproductive isolation, sibling species, 97, 98
Genetic differences (in captive populations)
age structures, 180
fecundity rates, 180
sex ratios, 180
sizes, 180
survivorship, 180
Genetic differentiation
adaptive, 95
of adjacent populations, 94

among subpopulations, 267
degree, 268
in different tree habitats, 95
distance effect, 49
and environments heterogeneous or patchy over short distances, 94
in heterogeneous environments over short distances with strong selective forces, 95
before hybridization, 267
in insects, 86
in motile animal populations, 85
in rats, 88
with strong selection forces, 94
from thermal effluents, 90
Genetic distance (D). *See also* Locus
definition, 97
in morphologically distinct species, 97
in sibling species, 97
Genetic divergence, 49
adaptive in animal vs. plant populations, 85
constancy of, 91
on either side of barriers, 89
gene flow and, 91
genetic drift and, 91
under influence of strong directional selection, 89
magnitude of forces exerted, 91
mating patterns and, 91
microgeographic scale, 79
migration and, 91
patchiness and, 91
patterns of, 79, 91
rate dependence on number of migrants exchanged (mN) (model), 55
in reproductive isolation, 357, 358
selection factors in, 91
under selection pressures, 79
of semi-isolated groups, 87
with small population sizes, 78
speed of, 79
Genetic divergence (single locus)
between populations
by migration rates, 49
by population sizes, 49
by selection patterns, 49
with given migration rates, 49
population size affecting, 49
selection pattern affecting, 49
Genetic diversity. *See also* Allelic diversity

Subject Index 649

advantages to long-term species survival, 429
average heterozygosity (H), 140
average polymorphism (P), 140
below species level, 332
between populations, 321, 360, 414
comparison of measurement methods, 356
conservation, 349, 350, 359
deficiency, 384
description of scope, 349-351
differences of individual or population at loci, 351
  or other defined amount of material, 351
distribution, 430
  between populations, 430
  within populations, 430
distribution parameters, 52
elements for measurement, 350, 351
environmental diversity, 319
  selective agent for, 319
gene flow loss and reduction, 436
in gene pools, 139
genetic drift effect, 51
in genotype, 356
heterozygosity within a population, 314
initial level loss, 319
initial level maintenance, 319
interspecific interactions and, 364-373
large population size, 314
loss by random drift, 376, 379
  for various population sizes, 376
losses of, 443
maintenance, 52, 151, 321
  of minimized inbreeding by mild inbreeding and outcrossing (zoos), 183
  in populations, 436, 437
  in selection process, 394
management by allowing genetic drift and divergence, 333
maximization of, 320
  at intraspecies level, 333
  at species level, 333
measures in conservation, 331-333
migration (gene flow) effect, 52
by mutation, 34
natural selection effect, 52
in numbers of strains, 214
overdominance in, 320

parent population diversity maintenance, 435
in phenotype, 356
plant introductions, 217
population size variables (model), 52
in populations, 196
preservation of, 401, 434
  in a single species, 438
  protection of all of a species, 439
  in multiple populations, 439
  protection through female line, 322
rate of dissipation, 319
reduction in small population, 436
reflection of population size, 150
retention by selection type, 134
in selfing populations, 212, 214, 215
self-regulation and, 326
with single parameter of genetic variation, 325
statute for preservation, 280, 281
subspecies indicators of, 359
survival and, 429, 430
within and between sample measurements, 352
within populations, 321, 360, 414
  majority, 360
Genetic diversity distribution parameters
  migration rates (model), 52
  natural selection intensity (model), 52
  natural selection mode (model), 52
  population size variables (model), 52
Genetic diversity loss
  effective population sizes, 379
  generation time, 379
  heterozygosity decline per generation (random drift), 379
Genetic diversity maintenance, 325
  conservation in wild, 183
  dynamics, 331
  experience in zoos, 183
  in geographically subdivided populations, 49
  in multiple populations, 330
  in one population, 330
  structures, 331
Genetic diversity maintenance model
  genetic drift and population size, 52
  migration (gene flow), 53

natural selection (genotype
success), 53
Genetic diversity maintenance patterns model, 52
Genetic diversity management
enhancing, 333
losing nonessential, 333
preserving, 333
Genetic diversity measurement, 350
average difference between individuals, 350
comparison of methods, 356, 357
cost, 357
Genetic diversity measurement techniques
allelic combination analysis (statistical), 355
amino acid residue sequencing, 352, 353
breeding studies, 356
chromosome analysis (maps), 355
cost, 351, 352
DNA sequencing, 352, 353
electrophoresis, 353, 354
enzyme studies, 351, 352
immunological methods, 354
morphometric analysis, 355, 356
mitochondrial DNA, 352
nucleic acid hybridization studies, 352
nucleic acids, 352, 353
protein sequencing, 352, 353
Genetic diversity preservation
effective population size, 398
equalization of founders through time, 392
generation time, 398
Genetic diversity protection
by females, 322
by males, 322
Genetic diversity relationship
to species evolutionary capacity, 325
to species survival, 325
Genetic diversity simulations
parameters, 53
Genetic drift, 191, 209
allele frequency change, 52, 53
allele frequency effects over generations, 155
allele loss by, 392
allelic variation of proteins and, 61
chance change in allele frequencies, 52

deme size and drift-induced divergence, 55
effect on small populations, 53
effective population size and, 392
generation time and, 392
genetic variation decrease by, 76
with isolation, 49
loss of alleles by, 64
loss of genetic variation by, 68
migration and, 55
migration opposing force, 55
population size (model) and, 52, 53
populations, 313
with populations separated by barriers, 89
slowing of, by increase of effective population size, 76
small populations, 9, 232
subpopulation loss of genetic variability by, 93
Genetic effects of natural stabilizing selection, 135
Genetic erosion
by frequent founder events, 217
by intense inbreeding, 217
Genetic equivalent of death (culling), by castration, 139
Genetic factors
in extinction, 49
flower color polymorphism, 223
isolation, population extinction, 49
at low population sizes, 116, 118
in plant breeding, 222
in plant mating system, antherstigma distance, 223
reaction with environmental factors, 222
segregational load and, 205
Genetic factors assessment
behavioral interactions, 363
biological interactions, 363
ecological interactions, 363
Genetic heterogeneity
result of limited mating, 94
result of limited movement, 90
Genetic information redundancy, 269
Genetic input founders, female and male, 252
Genetic input (into herd gene pool)
founder, 252
individual contribution of alleles, 252
Genetic interaction

between differentiated populations, 269
between selfing strains, 210–214
cultivated vs. wild populations, production of weedy races, 276
cultivated/weed/wild populations or species, 274, 275
economic effect, 269
Genetic intervention, time for, 14
Genetic load
  balanced lethals, 196
  in balancing selection, 199
  in captive mammal populations, 195
  deleterious genes, 195
  in diploid cross-fertilizing species, 199
  in *Drosophila* flies, 195
  in humans, 195
  inbreeding and, 195
  lethal genes, 195
  outbreeding and occurrence, 195
  removal by culling, 199
  semilethal, 195
  in small population size, 139
  subvital, 195
  wild species and, 195
Genetic locus, number of, involved in heavy metal tolerance, 84
Genetic management. *See also* Population management, Wildlife management, Management
  allelic diversity analysis, 360–362
    by species classification in, 360–362
  animal reintroductions, 183
  animal species, 12
    of ecological significance, 12
    of great human interest, 12
  arguments, 12, 13
  breeding for adaptation to new population size, 331
  of captive populations, 179–182
  case for intervention, 10
  chain reaction effects in ecosystems, 3
  characteristics for host-migrant relationship, 8
  culling, 12
  cost, 10
  diversity, 49
  ecological compatibility, 8
  ecosystem, 10, 11

extinction mitigation, 444
focus on population size disastrous, 68
founding new populations, 331
future events, 12, 13
gene flow between two isolated populations, 444
gene pools, 444
genetic diversity of subspecies, 362
history of, in captive populations, 180
host-migrant relationship, 8
of inbred stock, 178
increased reproductive rate, 7
induced gene flow, 8
induced migration, 7–10, 12
introducing new genotypes into breeding pool, 331
isolation reduction, 444
karyotic data, 103
large animal community impact, 10
lessons from animal breeders, 184
  of captive species, 184
  of small populations in wild, 184
  of zoo species, 184
limitation of scope, 14
loci in measured divergence, 356
mating system (plant), 8, 9, 225
migration, 7–9
mitigation of near-term extinction, 444
"mobile links" in plant species life processes, 10, 120
objectives, 400, 401
options, 415
  in parks and reserves, 415
  of particular species, 3
planned matings, 178
plant populations, 217
population genetic dynamics, 314
population sources, 7
preservation of diversity from wild gene pools, 387–392
reduction of isolation across boundaries, 444
reproductive rate, 7
research needs in evolutionary genetics, 46
  on target species, 9
reserve size and viable populations, 183
role of in conservation, 13
for short-term maintenance, 183

stability of community, 10
stability of ecosystem, 10
strategies, 400, 401
vigor, 7
wild plant species mating structure, 221
of zoo populations, 434
in zoos, 165
   history of, 165
Genetic management, captive populations
   acquire adequate number of founders, 388
   equalize founder representation, 392
   expand population to carrying capacity, 389
   manage inbreeding coefficients, 392
   maximize effective population size, 389–392
   subdivide population, 388
Genetic management goal(s), genotype introduction, 7
Genetic maps, linkage disequilibrium, 129
Genetic markers
   in evolutionary history studies, 33
   in genetic divergence studies, 33
   in population structure studies, 33
Genetic material
   chemical differences, 351
   in expression of genome, 351
   functionally heterogeneous, 351
   taxonomic distance between samples, 351
Genetic materials, introduction of host materials, 11
Genetic mechanisms
   of extinction, 125
   of speciation, 125
Genetic options, 139
   in environmental shift, 217–219
   for species, 217
Genetic patchiness. See also Genetic subdivision
   and gene flow, 94
   patchiness of genetic structure, 90
Genetic plasticity
   extinction and, 116
   loss of, 116
Genetic polymorphism
   adaptive significance, 137
      in space, 137
      in time, 137

color, 129
differences of fitness to different genotypes, 135
   by nature of environment, 135, 137
form, 129
maintenance, 321
   in environmental heterogeneity, 321
   by homozygosity in numerous populations, 68
   of stable, 320
phenotypic plasticity, negative correlation, 219
protection at species level, 320
scales of time and space, 137
stability in ecosystems, 329
stability protection, 320
Genetic population management
   family size regulation, 390, 391
   number offspring produced in lifetime, 390, 391
   sex ratio regulation, 390, 391
Genetic population structures
   breeding system differences, 205
   evolutionary consequences, 214
   extinction rates, 205
Genetic populations
   levels of inbreeding in, 180
   spatial integration of various, 180
   unequal genetic contributions of founding animals, 180
      in various countries, 180
      in zoos, 180
Genetic principles, application to natural populations, 45
Genetic recombination
   diverse genetic types, 194
   from field of variability, 194
   in gene pool, 194
   mating system control, 220
   mating system influence on, 223, 224
Genetic resources
   conservation of, 337
   crops, 14
   gene flow limitation on, 67
   livestock, 14
   wild relatives of domesticates, 14
Genetic resources preservation. See Conservation
Genetic restructuring
   drastic, 9
   rapid, 9

Genetic selection, 139
Genetic structure
  nonrandom distribution, 343
  patchiness of, 90
  of populations, 8
  stochastic event effects, 89
  strong selection effects, 89
Genetic structure of plant populations
  biotic factors and, 343
  family structures and, 343
  mating systems and, 343
  nonrandom, 342
  vegetative reproduction and, 343
Genetic subdivision. See also Genetic patchiness
  by limited movement and mating, 90
  over short distances, 90
Genetic systems
  diploid, cross-fertilized, 193, 200
  dynamic, 199
  vs. ecosystem, 329, 330
  natural selection and, 193
  in oceanic island colonization, 193
Genetic theory, growth rate, 323, 324
Genetic uniformity in self-fertilizing species, 414
Genetic variability, 155, 156, 206, 207
  and area size, 429, 430
  as balanced hybrid vigor, 194
  in balanced state, in populations, 193
  chance production of phenotypes, 190
  decrease in populations, 118
  in demes, 93
  of diploid cross-fertilized species, 193
  in *Drosophila*, 142, 143
  environmental variability and, 142, 143
  factors promoting, 142, 143
  in gene pool, 193
  with genetic drift, 93
  held in sexual populations, 199
  in hybridization, 292
  with inbreeding, 93
  in laboratories, 142, 143
  level in plant populations, 227
  levels, 156
  loss of, 68, 155, 156, 165
    by catastrophic effects, 67
    by inbreeding in small populations, 165
  maintenance by outcrossing, 68
  mating structure manipulation by level of, 225
  maximum in herds at zoos, 261
  number of unique phenotypes possible, 190
  from observed heterozygosity, 42
  organization in plant populations, 227
  potential for, 190
  preservation of, in captivity, 179–182.
  preservation by population size insufficient, 68
  recombination of, 190, 194
  reductions, 208
  and relative fitness, 190, 191
  in reproductive event, 199
  restoration, 155
  retention, 144
  role in colonizing success, 219
  selection of, 190
  shift without reduction in, 198
  slugs, 209
  survival and, 190
  in variable environments, 142, 143
Genetic variability (average heterozygosity) and environmental components, 139
Genetic variability distribution (plant), ploidy level, 224
Genetic variability maximization
  at herd level, 245
  at individual level, 245
Genetic variability structure
  homeostasis location and, 330
  mating pattern and, 330
  reproduction biology and, 330
Genetic variance
  with phenotypic variance, 146
  rate of erosion, 7
Genetic variants
  origin of, 93
  preexistence in population, 93
Genetic variation
  abundance levels, 151
  adaptation, 137
    to environmental shifts, 219
    to familiar species, 370
    a necessity, 149
    to unfamiliar species, 370
  alleles in, 151

alternate genotype energy allocation, 325
as alternative alleles in genetic combinations, 253
among local populations, 370
amount detected, 354
amount in plant or animal population, 129
arrangement in gamete formation, 253
available to populations, 141
average heterozygosity over loci (H), 205–208
average polymorphism over loci (P), 205–208
between species, 155
by bottleneck time lapse, 142
canonical variables on isozyme data, 234, 235
comparison of levels of, 131–133
controls on distribution among individuals, 220
correlation of effects, 370
decay, 156
decline within populations, 76
dependence on balance between selection and gene flow, 372
depression of, 130
determination by population size, 142
distribution between populations, 51
distribution within populations, 51
and ecologically stable equilibrium, 333
effects of depression of, 130
by electrophoretic assay, 149
detection, 51
elimination of, 333
environmental effects, 140, 141
  in gene pools of populations, 141
  spatial, 140, 141
  temporal, 140, 141
environmental variation and, 137, 140–143
erosion in absence of balancing selection, 372
estimates, 335
  in animal populations, 335
  in plant populations, 335
evolution and, 144–146, 217
evolutionary dynamics, 313, 314
exhaustion by evolution in species, 145
existence among rather than within local populations, 370

extinction and, 127, 149–151
factors in amount of, 149, 150
and fitness in *Drosophila*, 137
function, 135–137
in fungi, 369
in gene pools of populations, 141
growth rate and, 323
in habitat generalists, 133
in habitat specialists, 133
habitat variation and, 137, 151
hosts, 368–370
human interaction and, 51
inbreeding effects, 130, 146, 151
insects, 368–370
  in host response to environmental change, 372
  new hosts, 372
in interacting species of insects and plants, 370
in interacting species of parasites and hosts, 370
in interacting species of predator and prey, 370
at interpopulational level, 313
isolation by distance and, 77
level for selection, 253
level found in allozyme studies, 231
level increase in disassortative mating, 253
levels, 127, 131–133, 146, 149–151
life history traits and, 337, 338
location in species structure, 330
loci in, 151
long-term adaptive values, 109
long-term survival necessity, 149
loss, 313
  by colonization, 217
  by genetic drift, 68
low levels, 210
  colonizing success, 219
magnitude of, 194
maintenance, 67–70, 151
  by interaction of species, 372
  in restricted captive populations, 67
measurement techniques, 304
from migration, 135
monomorphism, 219
mutation source of, 128, 135
nature of, in populations, 127, 129, 130
neutral variation, 231
neutrality, 155
organization of, 127
origin in mutation, 128

in outcrossing populations, 209
in parasites, 368, 369
in pathogens, 368
phenotypic variation and, 146, 149
in phytophagous arthropods, 369
plant control agents and asexual reproduction, 226
in plant founding populations, 219
in plant mating systems, 220, 221
  in heterozygosity level, 221
  in recombination level, 221
plants/phytophagous arthropods, 368
plant polyploids, 227
  buffer against genetic erosion, 227, 269
  founder events, 227
polymorphism, 219
  at many loci, 130
in polyploid species, 227
population bottlenecks and, 142
population size and, 142, 150
preadapted, 93
by preservation of heterogeneous environment, 69
production of genomic constituents and, 128
promotion by interspecific interaction, 373
of protein loci, 51
random genetic drift and, 76
recombination and, 194
reduction due to founder effect, 231
reproductive capacity effects, 151
response to new species
  parasites, 372
  predators, 372
  of prey, 372
scale, 130–133
selection in action on, 127
selection modes operative on, 127
selection pressure responses, 319
selection types, 133–135
selective control, 155
selective pressure operative on preexisting, 90
short-term adaptive values, 109
significance for extinction, 127
slow rundown, 150
in small population, size, 209
source populations vs. founder populations, 217–219
in space, 137, 140, 141
in species evolution, 330

stochastic accumulation of among populations, 76
from stochastic causes, 135
in study case, 151
summary indices, 365
  allozyme heterozygosity, 365
  superficial artifacts, 319
survival and, 127, 144
in time, 137, 140, 141
in traits adaptable to ecological variation, 365
with tropic instability in decapods, 140
in two species with one-locus two-allele system, 333
types, 129
unique patterns in colonizing populations, 218, 219
within and between populations, 234
within populations, 319, 373
within species, 155
Genetic variation conservation, subspecies appropriateness, 363
Genetic variation decline
  with fitness reduction, 313
  by loss of alleles, 313
  by loss of heterozygosity, 313
Genetic variation distribution
  among populations, 336
  predictive rules, 314
  within populations, 336
Genetic variation estimates
  effective number of alleles per locus, 335
  mean number of loci heterozygous per individual, 335
  number of loci polymorphic per population, 335
Genetic variation indices, allozyme heterozygosity, 365
Genetic variation levels
  between populations, 336
  evolutionary survival and, 144
  fast-evolving species, 146
  higher, 150, 151
  increase/decrease of phenotypic variance and, 146
  low and phenotypic variance, 146
  lower, 149
  slow-evolving species, 146
  within populations, 336
Genetic variation measurements
  average number of alleles per locus, 232

average heterozygosity, 231
  decrease with bottleneck, 231
  decrease dependence on population growth rate, 231, 232
Genetic variation measures (levels)
  alleles per locus, average number, 217
  allelic frequency distribution over loci, fraction of loci polymorphic, 217
Genetic variation structuring (within species) advantage with environmental variation, 330
Genetic variation studies, genetic variation within populations or species, 335
Genetic variation types
  by genetic drift (stochastic events), 129
  by inflow of genes (migration), 129
  by recurrent mutational changes, 129
  by selection, 129
Genetic variety
  adaptive, 137
  in habitat utilization, 137
  in time and space, 137
Geneticists use of genetic markers, 33
Genetics
  of captive zoo populations, 241
  definition of, 15
  of extinction, 152, 153
  history of evolutionary theory and, 38
  population dynamics and, 161
Genetics for zoos
  role of diminishing genetic variability, 241
  role of increased inbreeding, 241
  role of small population size, 241
Genetics of reproduction, a dynamic variability system, 190, 191
Genome
  agressive weed, 272
  alteration of (sea turtles), 286
  complete set of genes, 18
  differences in, 272
  genetic material of organism, 18
  haploid set of chromosomes, 18
  heterozygous, 211, 213, 214
  homozygous, 211, 213
  lack, 214
  measurement of expression of, 351, 357
  mixed, 211
  techniques, congruence of, in measurement of, in population, 357
Genome (haploid), fraction coding for polypeptides, 128
Genotype, 22–27
  of adaptively differentiated populations, 95
  with altered ecology, 137
  altering differentially, 137
  changed over time, 317
  comparison with phenotype, 15, 16
  definition of, 15, 16, 22
  fitness and extinction of, 365
  fitness factors, 137
  fitness retention through generations, 364
  fitness under selective pressure, 43, 44
  fitness values, 137
  heterozygous, 23, 40–42, 45, 60, 194, 195, 213, 214
  heterozygous overdominance, 221
  homozygous, 23, 41, 42, 45, 60, 194, 213
  inbred (homozygous), 137
  insects, 370
    adaptation to different hosts, 370
    coexistence within a population, 370
      or larger spatial scale, 370
  introduction into breeding pool, 331
  isolation by heavy metal soils, 85
  moisture gradients and, 95
  new introductions, 331
  numbers in study, 151
  outbred (heterozygous), 137
  preservation, 313
  rare, 130
  recombination and, 130
  relative advantages of outbred (heterozygous), 137, 138
  segregation by moisture conditions, 82
  transitional polymorphic, 82
  two sets of chromosomes, 18
Genotype destruction, warblers, 299
Genotype exploitation and room for selection, 151
Genotype frequencies, 40–43
  after single generation of random mating, 41, 42
  allele frequencies from, 40

Subject Index   657

generations in Hardy-Weinberg equilibrium frequencies, 41
monitoring, 313
parental allele frequency dependence in second generation, 41
Genotype relationships with environment, 82
Genotype replacement
  time scale of, 365
  extinction and, 365
Genotype-habitat associations, segregation of, 95
Genotypic combination range (plants), function of mating system, 220
Genotypic variation
  by outcrossing strains, 213
  purging by backcrossing of hybrids, to strain of higher fitness/lower outcrossing, 213
Genus, 318. *See also* Genera
Geographic area
  large, 115
  small, 115
Geographic boundaries in preservation of allelic diversity, 362
Geographic distance
  in genetic divergence of populations, 66, 67
  isolation by, 66, 67
Geographic distribution, 101, 102
Geographic division of species and gene flow, 130
Geographic isolation, generation of complex strain diversity, 214
Geographic populations
  climatic variables and heterozygosity variation, 142
  of a range of species, 142
Geographic range
  dispersal in, 142
  restriction, 209
Geographic scales
  large regions, 83
  macro, 79, 83
  micro, 79, 83
  small size, adaptive generation over, 92
Geographic separations
  long distance, 66
  period of, 292
  short distance, 66
Geographic subdivisions (of populations)

inbreeding increase, 201
minimization of disease risk, 150
minimization of genetic variance loss, 150
population number restriction, 201
Geographic variation in subspecies, 359
Geologic time, Cretaceous period extinctions, 126
Georgia, U.S.A.
  flower color polymorphisms, 223
  plant colonizing success and allozyme variation, 219
Glacier National Park, Mont.
  elk management in, 438
  grizzly bear management in, 438
  level of demographic protection, 438
  park and reserve size, 438
Gladys Porter Zoo, Brownsville, Texas, 251
Gotland, Baltic Sea, flycatcher hybrids, 304
Great Basin butterfly populations, 156
Great Britain, rodenticide, 88
Great Lakes region, warblers, 296
Group(s)
  differentiation in, 269
  distinctiveness of, 435
  diversity between, 360
  genetic diversity maintenance, 49
  preservation of, 435
  size
    extinction and, 118
    reproduction and, 118
Growth
  in numbers, 436
  potential for, 436
Growth rate
  genetic variation and, 323
  parameter of fitness, 323
  in populations, 323
  in transplanted fish populations,) 359
Guadalupe Mountains National Park, Texas, preservation levels in, 438
Guatemala, northern, *Drosophila* studies, 238
Gulf of California, climatic changes, 126
Gulf of Mexico, turtles, 277
Guns in species extinction, oceanic islands, 111

Habitat
  adjacent, 438
  altered by man, 93
  artificial, 272
  boundaries and, 134
  captive, 376
  carrying capacity, 117, 134
  confined, 438
  deforestation around, 438
  ecological diversity, 140
  exploitation by chemically preadapted genotypes, 85
  extinction and, 158
  field, 272
  fine genetic adjustment to, 163
  forage availability, 442
  garden, 272
  gene exchange between pockets of, 437
  gene exchange between populations, 437
  human activities and, 242, 260
  hybridization, 307
  increase with cooperative agreements, 438
  individual exploitation strategies for, 90
  island of, 121
  limitations in large parks and reserves, 442
  limited, 436
  loss, 160, 161
  manipulation for vulnerable populations, 305
  microdifferentiation of populations in each tree, 87
  minimization of mutagens in, 130
  modification, 49, 307
  modified by human activities, 293
  mosaic of patches, 159
  mosaic population dynamic pattern, 160
  in mosaics on continents, 112
  needs of individual species, 437
  new for species, 292
  nutritional advantage of, in selective, 88
  orchard, 272
  overcrowding, 443
  overexploitation of, 420, 436, 443
  patchiness, 442. See also Habitat patches
  pollen dispersal between, 437
  population fluctuation, 117
  population movement between, 437
  preservation, 436
  protection of adjacent, 438
    with cooperation, 438
    with counties, 438
    with private sector, 438
    with state, 438
  recolonization of, 284, 438
  removal of individuals, 443
  seed dispersal between, 437
  size of demographic unit sustained by, 443
  small pockets of, 437
  species exploitation of, 94
  stable boundary, 134
  unfavorable barrier crossing, 121
  uniform, 436
  urbanization around, 438
  weeds, 272
Habitat area, 153
Habitat assessment vulnerability of species, 415
Habitat availability, 444
  increase of, 437
  restriction on population size, 150
Habitat change
  by human activities, 155
  mating structure shift, 223
  measurement of, 421
  rate of extinction and, 109
Habitat classification, microhabitat conditions, 82
Habitat decrease and small inbreeding populations, 183
Habitat destruction, 104, 159, 161, 163, 442
  extinction on islands and, 111, 301
  by human activities, 241
  man at war with wild nature, 189
  sources of, 2, 3
Habitat differences, sibling species, 98
Habitat disturbance, 271
Habitat diversity and evolutionary survival, 151
Habitat factors, in survival of patches, 113
Habitat generalists
  genetic variation and, 131, 132
  heterozygosity and, 133
  invertebrates, 133
  plants, 133
  vertebrates, 133

Habitat generalists (genetically diverse) to inherit the earth, 151
Habitat limitations
  by park size, 420
  by urbanization of adjacent areas, 420
Habitat management
  declining plants and, 2, 224, 569
  increase by cooperative agreements, 443
Habitat modification by human populations, 49
Habitat patches
  birth, 113
  death, 113
  die out, 113
  ecological extinction in isolated, 121
  extinction and, 112
  extinction and patch death rate, 113
  extinction by drought, 113
  extinction by herbivore increase, 113
  extinction by overgrazing, 113
  factors in, 113
  local population extinction and, 113, 118, 119
  mosaic pattern, 162
  population survival and, 113
  populations and, 113, 114, 118, 119
  regional trends and, 114
  in relation to species decline, 113
  size and species extinction, 113
  survival of species and, 113
  tropical forest, 118, 119
  variability and population, 113
Habitat preservation, sea turtles, 279
Habitat requirements, gophers, 101, 102
Habitat/resource dependence, animals to treefall in tropics, 120
Habitat size decrease. See also Area size decrease
Habitat size decrease, in initiation of extinction
  rare species extinction by decrease of population sizes, 124
  resource/habitat disappearance and extinction of dependent species, 124
Habitat specialists
  genetic variation and, 131, 132
  heterozygosity, 132
    invertebrates, 133
    plants, 133
    vertebrates, 133
Habitat type
  average heterozygosity, 133
  generalists, 132
  genetic variation and, 140
  habitat generalists, 133
  habitat specialists, 133
  specialists, 132
Habitat variety
  ecosystem components and, 150
  genetic variation levels, 150
  species effects, 150
Hardy-Weinberg equilibrium, 210, 321
  behavior of allele and genotype frequencies, 41
  two alleles at a single locus, 45
Hardy-Weinberg expectation deviation (sampling error) causes
  allele frequency change (random genetic drift), 43
  migration, 44
  mutation, 43
  selection, 43
Harem size, inbreeding rate increase with larger, 71, 72
Hawaii, 192
  bird colonization of, 192
  insect colonization of, 192
Hawaiian *Drosophila*, founding event and divergence, 187
Heavy metals, from mining activities, 92
Heavy metals effects
  evolutionary distinctions in adjacent populations, 95
  genetically distinct populations from, 92
Heavy metal stress, 94
  in plants, 84
Heavy metal tolerance, adaptation of species in, 84
Heavy metals type
  copper, 83
  lead, 83
  nickel, 83
  zinc, 83
Herbaceous plants. See also Plants
  competition for space, 435
Herbivores, large
  in known area, 429
  level of demographic protection, 441, 442
  longevity, 441, 442

number of individuals for a given
    species, 429
number of species, 429
size of protected area for, 441, 442
space requirements of, 422, 424-426
Herbivores, small
    space requirements of, 422, 423
Herd
    adaptation to inbreeding, 245
    alteration of genetic makeup, 253
    in short time, 253
    extinction minimization, 250
    gene pool composition, 252
    manipulation of breeding structure, 245
    single founding male and gene pool of, 252
    size, 242
Hereditary characteristics and population size, 152
Heredity and environment, 15, 16
Heritability, definition, 16
Heterogeneity
    environmental, 321
    niche, 321
    sources of, 255
Heterosis
    and fitness, at loci affecting growth rate, 325, 326
    fitness dependence on, 205
    mean fitness and, 205
Heterozygosity, 68
    advantage, 137
    in allele complexes, 194
    amount in populations, 194
    average, 231
        calculation, 232
        decrease, 231, 232
        founder event, 231
    average correlations and evolutionary rate of change, 146
    of an average individual, 131
    average taken as measure of genetic variation, 231
    in balanced population structure, 194
    between and within individuals, 146-149
    biochemical flexibility (homeostatic capacity, 149
    in butterflies, 155, 156
    characters under stabilizing selection, 149
    comparisons, 132

in cosmopolitan species, 132
decrease, 231, 232
deleterious recessive alleles and, 244
dependence on rate of population growth, 231, 232
desirable, 313
electrophoretic assay, 149
electrophoretic detection at loci, 148, 149
evolutionary rate of change, 146
in expanding small mammal populations, decrease/increase, 139
extinction and, 116, 127
fitness and, 139, 144, 145, 197
in frogs by habitat type, 140
genetic diversity, 314
in habitat generalists, 133
in habitat specialists, 133
increase with migration, 8
indicator of genetic diversity, 313
indicator of genetic variation status, 313
indicator of population quality, 313
by interspecific allelic combinations, 313
level determination by population size, 365
levels, 137, 146, 156
in life forms, 131
loss, 116, 154, 313
loss by inbreeding, 150
loss in populations, 68
loss with selfing, 221
maintenance by crossmating from different populations, 68
measure of population quality, 313
measurement of genetic variability, 42
morphological asymmetry and, 149
of outbred populations, 149
paradox
    heterozygote expectations, inbreeding/outbreeding species, 222
phenotypic variance of morphological characters, 146-149
at plant self sterility locus, 221
in plants, 221
at polymorphic loci, 338
polymorphism and, 140
polymorphism correlation, 131
population size and, 139, 142
protection by sexes in species, 322

regained by crossmating with individuals from other populations, 68
relationships, 148, 149
  with polymorphism in decapods, 140
reproductive performance and, 137
sea turtles, 280
structure, 365
survival and, 127
in temperate species, 132
in tropical species, 132
undesirable, 313
use of mutagen (ionizing radiation) and, 139
in vertebrates, 133
Heterozygote
  balancing selection in, 199
  biochemical flexibility, compared with homozygotes, 149
    internal environment and, 149
  elimination, 221
    in plants, 221
  fitness
    artificially managed, 322
    compared with homozygote, 320
    female, 322
    higher, 194, 195, 197
    naturally managed, 322
    over all environments, 320
    superiority of, 199
  overdominance, 322
  superiority
    of fitness, 199
    and heterosis in genetic diversity maintenance, 326
  vitamin K requirement, 88, 89
  for warfarin-resistant allele, 88
Heterozygous advantage, 194, 322
  by female, 322
  growth rate, 139
  by male, 322
  outbred/inbred, 138
  plants and, 222
  populations and, 137-139
  relative divergence simulation and, 56-59
  selection for similar allele frequencies in different demes, 56, 57
  under severe environmental conditions, 139
  stable equilibrium maintenance (model), 53, 57

Heterozygous strains, 205-209
Himalaya-Karakoram mountain system, grass hybridization along, 269
Holarctic, duck hybrids, 293
Holland, butterflies, 159, 160
Homeostasis
  individual genetic, 320
  single populational genetic, 320
Homozygosity. *See also* Inbreeding
  in classical population structure, 194
  decreased fertility with, 178
  decreased viability with, 178
  deleterious allele, appearance with, 197
  deleterious recessive alleles and, 244
  fitness level, 204
  harmful recessive alleles and, 45
  increase, 205
  less fit segregants, 194
  levels and morphological asymmetry, 149
  with localized mating, 321
  recessive deleterious alleles with, 178
  in reduced expression of heterozygote superiority, 178
  species avoidance, 139
Homozygote
  in balanced genetic load, 196
  biochemical flexibility
    compared with heterozygotes, 149
    internal environment and, 149
  fitness compared with heterozygote, 320
  mortality, 196
    in egg, 196
  segregation of, 196
Homozygous strains, 203-214
  persistence of, 213
Host
  change in, 160
  dependent parasitic species and, 126
  in environmental change, 126
  extinction, 160, 440
  extinction by virus, 366, 367
  genotypes, 367
  life extension by parasites, 366
  parasite dependence on, 440
Host trees, heterogeneity in defensive compounds, 86
Human activities. *See also* Barriers, Structures

Subject Index

axe, 2
building construction, 78
bulldozer, 2
canal digging, 119
dam construction, 78
egg taking, 284
environmental impact, 94
extinction and, 112, 125, 157, 313
fence construction, 78
fire, 2
   preventive, 3
fishing, 284
habitat destruction, 161, 162, 286.
   See also Habitat destruction
hunting, 119
hunting and animal extinction, 117
killing, 284
local extinction by, 93
logging, 3
man as pathogen of nature, 111
man-made barriers, 94
Mediterranean fruit fly spraying, 158, 159
microhabitat differentiation and, 78
mine sites, 84
mortality in shrimp trawls, 284
mutagenic substances and, 129
pesticides, 114
plant transport by, 271
road construction, 78
shotgun, 2
in species extinction, 109
strong selective pressures following, 93
Human impact
   on species outside protected area, 442
   species response to, 94
Human interest
   of demographic units and, 443
   extinction, 443
   of species and, 13
Human leucocyte antigen system (HLA)
   in Caucasian populations, 129
   chromosome type frequencies, 129
   linkage disequilibrium, 129
Human populations
   adaptation to inbreeding, 244
   inbreeding depression, 244
   juvenile mortality, 248
   outcrossing species, 244
   severe inbreeding depression, 244
Human pressures, 109

Humans
   bleeder's disease, sex-linked inheritance, 28
   blood groups, 25
   color blindness, sex-linked inheritance, 28
   comparison with chimpanzees, 357
   by DNA hybridization technique, 357
   Down's syndrome (mongolism), 35
   electrophoresis, 357
   gene loci number in genomes of, 32
   genetic load, 195
   genetic variation, between/within populations, 233
   immunology, 357
   impact on species, 437
   numbers of, 403
   selective pressures, 195
   skin color, 16
Hunters, duck bag, 293, 294
Hybrid genotypes (recombinants)
   adaptation for new environments, 305
   adaptation potential increase, 305
   creation of new populations, 305
   for fitness increase, 305
   novel genetic diversity source, 305
   for reproduction in old range, 305
   source of novel genetic diversity, 305
   tool for population restructuring, 305
   tool for creating new populations, 305
Hybrid populations
   adaptation response vs. parental forms, 307
   change by catastrophe, 307
   courtship behavior, 295
   of *Drosophila* in laboratories, 139
   embryonic death and, 294
   exclusive area of occupation, 292
   existence in modified habitat, 307
   fitness, 294
   gene exchange restriction, 294, 295
   heterozygosity and fitness in, 139
   increase in biological diversity, 307
   infertility rate, 294
   interpopulational, 139
   man-induced, 307
   in modified habitats, 307
   new gene complexes, 307
   progeny, 294

reproduction timing, 294
single locality, 139
species recognition, 295
Hybrid swarms, 270
  adaptation to new environments, 292
  in avian species, 291, 292, 300
  populations with new gene complexes, 307
  stability, 308
  warblers, 300
Hybrid zones, stability, 292
Hybridization
  adaptation and, 289, 307
  adaptation and differentiation closed by cycles of, 275
  alteration of genome, 286
  between allopatric populations, 308
  between parapatric allospecies, 291
  between parapatric populations, 292
  biological consequences, 289
  bird taxa in, 291
  conservation and, 295
  degree, 289
  effect on small populations, 289
  evolution and, in plants vs. animals, 267
  evolutionary consequences, 291
  fitness and, 289
  gamete waste, 304
  genetic buffering and, 267
  human induced, 283
  influence on ecological adaptation, 289
  influence on speciation, 289
  intraspecies, 276
  introgression and, 270, 276
  introgression vs., 308
  limited (between sympatric species), 291
  loss of phenotypic diversity, 289
  outbreeding extreme, 304
  phase of differentiation-hybridization cycles, 275, 276
  political decisions on, 295
  polymorphic color phases, 291
  preservation and, 295
  preservation of distinctive alleles, 309
  rare or endangered population and, 305
  recombinational load, 304
  reproductively vigorous population and, 305

  in secondary contact, 308
  sibling species, 98, 100, 101
  stabilizing selection, 300
  strain persistence and, 211
  U.S. endangered species list and, 295
  zones of contact, 289, 290
  zones of overlap, 291
Hybridization (sea turtles)
  adaptation with, 287
  alteration of genome, 286, 287
  breeding and, 287
  effect on fertilization, 287
  effect on fitness, 287
  farming and ranching dangers, 287
  hatchlings, 287
  interspecific, 281, 287
  intraspecific, 287
  localized or population-wide, 287
  survival capacity and, 287
  translocation dangers for, 287
Hybridization control
  artificial restocking of habitat, 305
  conservation of biological diversity, 305
  reduction of species in zone of contact, 305
  regulation of habitat features, 305
  translocation of species from zone of contact, 305
Hybrids, 211
  gene exchange, 295
  mates, 299
  phenotypic disabilities of, 8
  purebred distinctions from, 289
  selection against, 292

Illinois, warblers, 296
Immigrants
  inbreeding reduction, 74
  males with large harems, 74
  offspring production, 74
Immunological assays, for higher taxonomic levels, 357
Immunological techniques
  costs, 354
  in gene pool studies, 130
  immunological distances, 353
  microcomplement fixation, 353
  phylogenies of higher taxa, 353
Import animals, founding individuals, 250
Imprinting, hatchlings (turtle), 283

Inbred animals, as parents, 256
Inbred line(s)
  comparison with noninbred, 179
  genetic variability preservation in, 179
  heterosis level (hybrid vigor), 204, 205
    with self-fertilization, 204, 205
  loss by random extinction, 202
  loss by strain interactions, 202
  phenotype variation with wild phenotypes, 179
  survival, 202
    of one line, 202
  wild species, 179
Inbred parents, effectiveness of use, 257, 258
Inbred populations
  adaptation to inbreeding, 243, 244
  from outcrossing species, 243, 244
Inbreeders, evolutionary fate of, 214, 215
Inbreeding. *See also* Homozygosity
  accumulation for monogamous matings within populations, 71
  adaptation and, 304
  adaptation to, 242, 245, 254
  adaptive responses in, 243
  advantages among small family demes, 305
  alleviation measures, 76
  analysis by standard regression, 246–250
  average heterozygosity reduction, 201
  avoidance, 139, 250, 254
  avoidance impossibility, 260
  avoidance in outbreeding species, 150
  breeding success and, 171
  in captive populations, 76
  changes in phenotypic variance, 146
  costs, 178, 179
  dangers questioned, 183
  death of animals and, 171
    death in calves, 184
  delayed puberty with, 177
  deleterious consequences of, 246
  deleterious effects of, 165
    in captive exotic animals, 165
    predicted, 178
  deleterious homozygous allele appearance and, 197
  depression, 304, 434, 435. *See also* Inbreeding depression

for development of gene complexes, 215
with different patterns of dispersal, 71
with different patterns of mating, 71
disease resistance and, 179
domestic animals, 175–178
in *Drosophila* flies, 197
early experiments, 174
ecotype development, 305
effective sizes, 254
effects, 45, 137, 138, 150, 165–174, 176, 177, 255
egg weight variance, Japanese quail, 149
elimination of deleterious alleles, 178
evidence of, 171
expression of deleterious recessives, 178
fertility reduction, 177
fitness production, 178
fitness reduction, 67
founder effect and, 200
gene sharing by descent, 45
generation numbers with, 178
generations, 146
  genetic variation change with, 146
  phenotypic variation change with, 146
generations after and, 197
genetic consequences, 243, 244
by genetic occurrence relationship, 201
genetic structure of populations and, 8
growth and, 176
heterozygosity reduction with, 150, 205
high intensity, 205
homozygosity in populations and, 178
homozygosity with, 205
homozygosity increase with, 221
inbred lines, 177, 179, 204, 205
increase, 241
  within populations, 67
intense, 227
intensity, 205
intensive, 197, 244
  inbreeding depression and, 244
  elimination of, 244, 245
juvenile mortality increase, 176
levels, 176, 247, 254, 260

in captive populations, 180
in cattle, 184
change in, 254
individuals alive at index age, 247
life span and, 179
linear effects, 177
loss of alleles, 67
loss of reproductive performance, 137
low fecundity and, 304
maintenance of multilocus associations in plants, 223
in milk production, 176
nonavoidance, 180, 245, 250
noninbred lines, 177
in normally outbreeding species, 118
  rate of, 118
outbreeding vs., 304
outcrossed progeny and, 205
outcrossing alternative, 243, 244
part of selection procedure, 245
by phenotype (assortative mating), 201
for a phenotype, 288
philopatry well-developed, 304
plants, 149, 223
  effective rate of recombination reduction, 223
  genetic diversity levels, 149
  heterozygote frequency reduction, 149
polygenic character disruption with, 178
polymorphism after, 150
population critical point, 76
population maintenance with increase in, 241
population number restriction, 201
for preservation of gene complexes, 215
rare problems, 157
rate of, 117, 178
  with a skewed sex ratio, 117
reproductive effects, 176, 177
reproductive success and, 179
severe, 249, 250
small effective population size, 304
in small populations, 250
subpopulation loss of genetic variability, 93
success with laboratory animals, 174, 175
survival and, 175
undomesticated animals, 178, 179

use, 178, 254
  to eliminate deleterious alleles, 178
  to fix desirable traits, 177
usefulness for zoos, 180
vigor, 175
  of young, 175
weight and, 175
in wild, 178, 179
within geographic subdivisions, 201
within populations, 70–73
young born, 175
Inbreeding coefficients
  accumulation of inbreeding (F) (model), 70
  avoidance of extreme, 254
  birth weight change, 249, 258
  change in, 137
  change in limitations per generation, 241, 242, 248
  in chicken (*Gallus gallus*), 150
  effective selection and, 254
  increase of, 245
  infant mortality and, 245
  intensity of inbreeding in offspring, 246
  low, for outbreeders, 435
  management of, 392
  reproductive performance, 137
  survivorship change with noninbred parents, 248
Inbreeding depression
  accumulation, 76
  adaptation to, 260
  alleles loss, 67
  alleviation, 70, 73–76
  avoidance of, 227
  birthweight decrease, 241, 248–250
  breeding types and, 74
  comparison with maladaptive syndrome to climatic changes, 243
  disease resistance loss, 150
  dispersal and, 74
  dominant heterozygous genotype decrease, 221
  effects, 304
  elimination
    in captive herd of Speke's gazelle, 241–261
    by use of inbred parents, 257, 258
  theory, 250
  evidence of, 175
  extinction and, 116

factors
  deleterious alleles, 244
  rare alleles, 244
  recessive alleles, 244
fat yield, 176
fecundity decrease, 118
fertility loss, 150, 241, 384
fitness loss, 45, 67, 70, 73, 74
heterozygosity decrease, 45
as inbreeding coefficient rises, 246
index, 244
  number of lethal equivalents, 244, 255
juvenile mortality, 248
lack, 156, 160, 175, 243
loss of disease resistance, 150
maladaptive syndrome, 245
matings between closely related members of same family, 304
model, 172
in monogamous species, 74
in normally outbreeding species, 67
occurrence, 245
planning and reduction of, 434, 435
plant counteraction, 9
in polygynous species, 74, 75
population bottleneck and, 175
qualified by number of lethal equivalents, 244
recessive homozygous genotype increase, 221
reduction from mutational loads, 205
reduction from segregational loads, 205
resistance to, 156
reversal, 67, 69
risk, 434
severe, 243
severity reduction, 259
in small captive populations of mammals, 109
survival and, 384
viability loss, 118, 241, 248
vigor loss, 118, 150
Inbreeding effects
  abnormalities, 45
  alleles, 45
  in artificial breeding programs, 150
  by breeding system, 137
  deleterious, 45, 137, 138
  on early life characters, 174
  on fecundity, 173
  on fertility, 174, 177, 184
  on fitness, 45, 67
  fixation of genes, 150
  heterozygosity loss, 150
    prediction, 150
  on juvenile mortality, 184
  on levels on genetic variation, 150
  in livestock, 176, 177
  natural selection and, 150
  nonselected animals, 255
  on population fitness, 137
  on populations with no history of inbreeding, 137
  on reproductive performance, 137
  selected animals, 255
  severity in various species, 137, 138
  sexual maturity delay, 184
  on variance, 202
    between/within groups, 202
    population total, 202
  on viability, 174, 246
  on vigor, 184
  in zoos, 165–174
Inbreeding programs
  fixation of genes, 150
  natural selection operative, 150
Inbreeding studies
  age criterion, 171
  animals' survival time, 171
  inbred animals, 171
  juvenile mortality, 171
  noninbred animals, 171
Inbreeding survey, juvenile mortality effect in zoo animals, 178
Indiana, warblers, 296
Individual, offspring during life of, 44
Individual genetic variability by crosses favorable to founder genetic material, 251
Individual genotypes
  environmental endurance, 330
  in migration, 320
Individual numbers, 242
Individual sex
  female, 242
  male, 242
Individuals
  alleles from founders, 253
  average distance between, 350
  culling, 196
  developmental stability and, 149
  elimination of extreme, 39
  environmentally unique experiences, 196
  exchange of, among populations within neighborhoods, 69
  exchange rates within populations, 69

exchange rates within neighborhoods, 69, 70
fitness, 70
founder ancestry, 252
genetic diversity within, 351
  measurements, 351
genetic differences in, 351
genetic similarities in, 351
genetic units of, mating at random, 39
genetic variability, 292
  level, 252
heterozgosity, 149
homeostatic capacity (buffering), 149
intraindividual level variations, 149
inbred, 243
long-term advantage of, 139
  in gene pool, 139
loss of, 436
  by inbreeding depression, 436
maintenance of unrelated, 76
number of
  breeding, 70
  founders in new populations, 182
  in a species, 38
regulating numbers of, 394
outbred/inbred advantage, 138
phenetic variability, 292
plants, 223
  heterozygous at many loci, 223
  recombination occurrence, 223
recruitment, 67
related, 69, 76
removal, 394, 443
of self-fertilizing species, 414
  extinction of, 414
  genetic uniformity, 414
  in long term, 414
sexual, and fertilization, 204
single or few isolated, 414
  in population establishment, 414
  in propagation, 414
in small populations, 69
symmetrical exchange of
  among populations, 69
  in relatedness increase, 69
transplantation, 305
type specimens and, 196
uniqueness, 195, 196
variant, 38
Individuals, wild-caught founders
  American population, 180
  European population, 180
  studbook numbers of, 180

Infant mortality rate, increase with rapid inbreeding, 245
Inheritance
  Medelian, 37
  quantitative (polygenic), 37, 197
  sex-linked, 28, 29
  at two loci, 25–27
Insect pests
  resistance to, 371
    by linked genes, 371
    by trait, 371
Insect populations
  allozyme frequency, 157
  of different ecotypes, 157
  dynamics, 157
  ecological factors, 157
  extinction susceptibility, 157, 160
  inbreeding depression, 178
  regulation, "mosaic" pattern, 158
  small population size (temporary), 156
  survival, 420
  symbiotic relationships, 160
Insect species
  on plants, 372
  predation on crop pests, 152, 153
  selection pressures, 372
  in trait polymorphism maintenance in plants, 372
Insect vectors. See Pollinator(s)
Insects
  area for survival, 420
  B chromosomes in, 128
  behavior change from endemic to epidemic populations, 319
  chemical response to hosts chemicals, 371
  competition for space, 435
  cost of resistance to, 368
  genetic differentiation of, 86
  genetic variation in adaptation to different species of host plants, 370
  genotypes of, 370
    in adaptation, 370
    coexistence within a population or larger spatial scale, 370
    to different hosts, 370
  host-specific factors of genetic differentiation in, 86, 87
  mutagenic insecticides and resistence development, 130
  natural herbivory and variation maintenance in plants, 373
  park and reserve size and, 438
  in plant extinction, 373

668 Subject Index

in plant mortality, 373
polyploidy in, 35
reaction of plants to, 368
response to related plants, 371
song or call in sibling species, 99
   identification of, 99
trait variability, 368
   for plants as hosts, 368
variation by variation in different plant species, 373
Insular founding, 192
Insular populations, phenetic differences, 300
Interaction of two species (hypothetical) in population size
   critical point, movement toward, movement away (saddle point), 329
   critical point stability, movement toward, 328
   critical point unstable, deflection toward extinction, 328
   oscillation in phase, 327
   oscillation out of phase, mutual cycle regulation, 327
Interactions
   dynamic/genetic in butterflies, 161, 162
   genetic and environmental factors, plant breeding systems, 222
   population size and gene frequency, 153
   population size and hereditary characteristics, 152
Interactions of species
   between animals and pathogenic microorganisms, 372
   between crop plants and pathogens, 367
   between plants and pathogenic microorganisms, 372
   between plants and phytophagous insects, 372
   between populations in recent secondary contact, 293–304
   between prey and predators, 314, 372
   conservation and, 372
   evolution, 370
   in genetic equilibrium, 370
   variation patterns in host-pathogen, 314
Intercontinental colonization events, release from native parasites, 226

Intermediate inheritance. *See* Incomplete dominance
International cooperation, Mexico-U.S. translocation and head start program, sea turtle, 284
International Species Inventory System (ISIS), vital statistics on zoo animals, 383. *See also* ISIS
International Union for Conservation of Nature and Natural Resources (IUCN), 374
   survey of park sizes, 416
   Red Data Book, 164
International Zoo Yearbook rare mammal species, 164
Interpopulational differences
   environmental diversity, 319
   species' reproductive biology dependence, 319
   specific traits, 319
   variability, 319
Interspecific interactions, genetic diversity and, 364–373
Intervention
   capacity of a park to protect without, 440
   case for, 10–12
   gene exchange, 289
   hybridization, 289
   by management, 391, 392
   in management techniques, 331
   migration in, 14
   philosophy, 11, 12
   species diversity preservation, 289
   species survival with, 351
Intervention type
   direct, 331
   indirect, 331
Intraspecific hybrids
   biological status, 307
   legal status, 307
Introduction(s)
   exotics, 307, 308
   in extinction, 366
   failure of, 384
   survival in host environment, 8
Introgressions, 271, 274
   hybridization vs., 308
Inversions, loci order reversal, 35
Invertebrates
   collections, 161
   density independent factors and extinction of, 117
   diversity of mating systems in, 187
   fauna, 420

genetic variation, 132, 133
heterozygosity (H), 131, 133
park and reserve size and, 438
polymorphism (P), 131
studies, 161
Ireland, slugs, 208
ISIS, 381, 382, 404
    captive facilities
        capacity for ungulates, 382
        for large felids, 381
Island endemics
    conservation problems, 308
    genetic assimilation, 308
Island model of migration, 54, 61, 63
    allele frequency changes, Monte
        Carlo simulations, 52
    diploid outbreeding species, 52
    generations, 52
    locus with two alleles, 52
    natural selection modes, 52
    parent allele transmittal, 52
    population size, 52
    random mating groups (demes, sub-
        populations), 52
    zygotes for next generation, 52
Island model pattern of gene flow, 61
Island populations
    extinction factors and, 120
    genetic assimilation in hybridiza-
        tion, 300
    genetic variation and, 131
Island size, evolutionary record, 9
Island species
    anthropogenic extinction of, 112
    vulnerability of, 300
Islands
    colonization and, 192
    gene pools from, 192
    older, 192
    species isolation of, 119
    with similar environments, 192
Isoenzymes, 33. See also Allozymes
Isolated populations
    breeding schemes for preservation
        of genetic diversity, 67-70
    breeding strategies for genetic vari-
        ability maintenance, 67-70
    extinction, 49
    genetic differences between, 67
    genetic differences within species,
        67
    genetic variation loss, 313
    heterozygosity loss, 313
    loss of fitness compensated by out-
        crossing, 68

same allele maintenance in semi-
    isolation, 49
speciation, 49
Isolation
    adaptation to, 86, 111
        with lack of herbivory/predation,
            111
    area effects of, 112
    by barriers, 89, 90
    between island populations, 305
    breeding, 88
    by breeding habits, 87
    in cascades of extinction, 124
    consequences, 49
    control among populations, 314
    demes, 9
    by distance, 77
        behavioral type, 67-69
        ecological type, 66, 68, 69
        genetic variation maintenance
            and, 77
        geographical, 66, 68, 69
        in management areas, 68
        in zoological parks, 68
    ecotypes in, 156
    evolution and, 49
    extinction and, 49, 50
    extinction rate, 313
    fitness decline, 313
    founding and, 187
    gene flow and, 49
    gene pools of adjacent populations,
        91
    genetic, 436
    genetic patchiness and, 94
    genetic variation loss, 313
    by geographical distance, 66, 68, 69
    by heterogeneous environments, 87
    higher taxa and, 49
    human activities and, 93
    of insects, 86
    island, 119
    less intrapopulational variability
        with, 319
    of local populations, 87
    loss of genetic variation, 313
    in management areas, 77
    mechanisms, 49
    merging and, 50
    park populations and, 49
    patch extermination, 112
    population diversity and, 49
    of populations, 444
    of populations adapted to chemi-
        cals/heavy metals, 85

populations by distance, 66–69, 77, 82
reduction between populations across a boundary, 444
reproductive, 300
reserve populations and, 49
role of land masses in, 49
semi-, 88
  in insects, 87
by social structure, 87
speciation effect, 49
stochastic (random) effects and, 87
by strong selective pressures, 93
subpopulations by barriers, 93
survival with, 49
by transplantation of genetically different individuals, 305
trigger for species diversity erosion, 121
by water barriers, 119
Isozyme studies. See Electrophoresis, Isoenzymes, Allozymes
Israel
  specific multilocus genotype-environment correlation, absence in, vs. California, 223

Jakarta, Sumatran tigers in, 180
Jasper Ridge, California, 154
  butterflies, 153, 154, 158, 159
Jasper Ridge Biological Reserve, Stanford University, California, 153
Java, rhinoceros, 376, 377
Juveniles
  mortality, 172
    before 30 days, 247
    environmental factors (nongenetic), 247
    genetic factors, 247
  survival, 172

Karyological studies, sea turtles, 280
Karyotypes, 404–408
  definition of, 17
Kazakhstan, birds, 292
Kentucky, warblers, 296
Kenya, weedy sorghum, 273
Kenya national parks, 383
Key species. See also Species
  extinction, 10, 11
  identification, 10
Keystone mutualists (plant), support of mobile links, 10

Keystone role
  competitor removal, 11
  seedling protection, 11
Keystone species, other species dependence on, 351
Khyber Pass, sorghum races, 274

Laboratories, breeding studies, 356
Laboratory facilities, for DNA sequencing techniques, 352
Laboratory methods, quinacrine chromosome staining, 99
Laboratory populations
  adaptation to new environments, 149
  Drosophila and asymmetry decrease, 149
Laboratory strains
  extinction of, 241, 245
  inbred, 241, 245
  success of, 241
Laboratory studies
  differential sex migration, 321
  fruit flies (Drosophila), 259
  variables from natural populations, 137
Laissez faire, 11, 12
Lake Gatún, Panama, 119
Lake Lulejaure, Sweden, trout, 61
Lamarkism
  in evolutionary theory, 39
  inheritance of acquired characteristics, 39
Land management plan
  founding population introduction, 331
  founding populations in, 331
Land use, peripheral, 415
Large mammals
  effective population size of, 183
  population decline, 161
  population extinction, 161
Large vertebrates, habitat capacity for, 420
Largemouth bass, enzymes and allele coding in, 90
Larval food plants, butterflies, 157
Legislation
  Migratory Bird Treaty, Act of 1918, 437
  U.S. Endangered Species Act of 1973, 289, 437
Lethal equivalents, 242
  in human life span, 244
  per inbred animal, 257

of inbred parent, 256
inbred populations, 248
per individual, 244
of noninbred parent, 256
Levels of demographic protection, 443, 444
for endangered species, 437
Levels of diversity
intraspecies, 333
species, 333
Levels of protection
degeneration factors in, 437
instability of population in, 437
intermediate, 437
legislation, 437
migration corridors between populations, 437
Life cycles of sexually reproducing organisms, 21
Life diversity, views of evolutionary biologists, 38
Life forms
heterozygosity, (H), 131
polymorphism, (P), 131
Life histories
butterflies, 154
extinction and, 109
sea turtles, 279, 280
Life history characteristics and their interactions, 109
Life history traits
distribution of quantitative variation, 344, 345
genetic variation and, 337, 338
plants
allozyme variation distribution and, 335–337, 343–345
geographic range and, 335–339
lifetime fecundity, 335
mating system, 335–337, 340
population structure and, 336, 341–343
reproduction mode, 336, 337, 339
seed dispersal mechanism, 336, 337, 340
stage of succession, 336, 337, 341
Life span, decrease with inbreeding, 179
Lineages, maternal, mitochondrial DNA analysis for tracing of, 187
Linkage disequilibrium, 128, 129
breeding system effects, 129
stamen length, 129
style length, 129

HLA (human leucocyte antigen) system, 129
mimetic polymorphism
of body color, 129
body form, 129
Linnaeus, 349
Little Mangere Island, Chatham Islands, parakeets, 301, 302
Little River, Texas, 101, 102
Livestock, breeding, 174
Livestock populations
deleterious alleles in, 178
inbreeding rates in, 178
Livestock species, 3
Local extinction
harbinger of species extinction, 113, 313, 314
by introduction of competitors, 365
by introduction of parasites, 365
by introduction of predators, 365
ordinary evolutionary process, 313
Local heterogeneous environments, plant differentiation among, 347
Local populations. *See also* Demes
breeding units of individuals mating at random, 39, 40
extinction, 113, 114
extinction vulnerability prediction, 119, 120
factors in extinction, 115–119
genetic differences, 51
Loci. *See also* Genetic distance
difference at single, 208
distance in crossing over, 30
effective number of alleles per locus, 335
electrophoretically assayed, 139
evolutionarily trivial variations, 319
evolutionary variation commonly shared, 319
fixation at, 320
in genetic changes, 133
genetic invariability in homozygous organism, 194
genetic variability at, 194
genetic variation of protein, 51
genotype incomplete expression, 38
heterotic effects, 320
heterozygotes with higher efficiency (overdominance), 45
heterozygous per individual, 335, 338
in host/pathogen relationships, 367
interactions between, 355
linking at, 30

mean allele frequencies at five most different, 234
measurements of differences between, 358
missing, 36
multi-, 223
mutation rates at, 37
nonrandom assortment, 355
number measured, 142, 144
  as genetic variation, 142, 144
  in vertebrates, 146
numbers in mammals, 128
numbers for study, 150
numbers studied, 136, 148, 150, 151
  in brine shrimp, 144
  in decapods, 140
  in *Drosophila*, 136
  in frogs, 140
  in vertebrates, 146
overdominance, 326
pleiotropic, 205
polymorphic, 155
  predominant allele maintenance and, 155
  per population, 335
  within a population, 335, 338
  between sea turtle populations, 280
polymorphism at, 130, 320
proportion, 338
random assortment, 355
recessive allele, 179
segregation at B locus, 150
selection and, 133
  few loci, 133
  many loci, 133
selective effects of the environment on, 320
self-sterility (plant), 221
significant numbers, 150
statistical relationships between, 355
in subsets of yellow fever mosquitoes, 234
tracking changes in large samples of, 332
trait sampling at, 346
single
  calculation of allele frequency at, 40
  calculation of genotype frequency at, 40
type
  regulatory, 130
  structural, 130

Longevity
  length of existence of individuals, 440
    their descendents and, 440
    in time, 440
  species, 415
Los Angeles, California, butterflies, 158
Lotka-Volterra equations, predator-prey, 324

Mahe, Seychelles Islands, doves, 301
Mainland populations, genetic variation and, 131
Maintenance
  realization of subspecies characters, 200
  of small founder populations, 259
Maladaptive syndrome, inbreeding depression, 243, 245
Males, 28, 29
  rotation in producing groups, 386
  single founding, 252
Mammals
  adaptation in absence of predation, 111
    loss of fear, 111
  behavior of, 422
  body size, 422
  cytogenetic assessment, 404–408
  endemic radiation, 49
  epidemics in extinction of some, 126
  extinction of, 111
  K-selected, 422
  large, migration, 9
  loci numbers in, 128
  polygynous and inbreeding, 178, 179
  rapid rate of chromosomal evolution in some, 178, 179
  rare species, 164
  reproductive strategies of, 422
  r-selected, 422
  in sample area sizes, 422
  small
    competition for space, 435
    effects of inbreeding in, 184
    juvenile mortality in, 184
  social organizations of, 422
  trophic strategies of, 422
Managed populations, 63, 64
  changes in characters, 200, 227, 359
  ecological threats to, 372
  effective population size, 70

Subject Index 673

rate of accumulation of inbreeding, 70
rate of exchange of individuals, 70
migrant (founder) introduction, 388, 389
numbers after founder event, 197
management of, 197
rate of exchange of individuals, 70
successful survival in, 197
trait preservation under local selection, 372
in different habitats, 372
in ensemble of populations, 372
Management. *See also* Captive management, Genetic management, Population management
adaptive potential variables, 227
allelic diversity maximum, 443
areas
fixation of alleles in, 68
isolation by distance, 77
single large, 332
areas (small)
breeding schemes for isolated and small populations, 67
isolated, 67
movement modes of individuals between, 67
biological systems in operation, 2
biotic factor effects on fitness, 322
botanic gardens, xix, 331
breeding, 63, 179
captive breeding, 103, 104
captive populations, 375
captive zoo populations, 241
concepts, 314
conservation of sibling species, 103, 104
continuous in preservation, 443
coordinated environmental and genetic, 331
cost for target species, 10
criteria for choosing mates, 253, 254
criteria for population subdivision, 363
criteria for target species selection, 9, 10
crucial problem of numbers, 6
data, spotty around world, sea turtle, 280
decision-making, 444
determination of park accommodations for a species, 443
determinations of species capacity to evolve, 443
decision roles, 314

demes
exchange, 63
in local selective pressures, 64
number restraints, 63
size restraints, 63
demographic structure variables, 227
demographic unit size, 416–430, 444
design
biotic variables, 332
breeding structures, 63
choice of options, 332
level of demographic protection, 440–444
management unit size variables, 332
physical variables, 332
population size variables, 332
ecological systems, 313
ecosystem, 14
effective population size, 63
techniques, 389–392
effort, 332
elk populations, 438
by cooperative agreements with state and county, 438
upgrading, 438
environment, 14, 331
evolutionary considerations, 444
evolutionary genetic considerations, 46, 415
evolutionary potential of species, 63
extinction of biota, 2
mitigation, 444
observation, 109
family size, 390, 391
field units, localized ecosystems, 318
financial restraints, 63
focal species, 444
focus on small populations, 187
founder origins, 103
founders, 384, 386
founding of populations, 444
functional size of habitat, 443
funds, 332, 435
gene flow, 438
between two isolated populations, 444
genetic and karyologic data, 103
genetic conservation and, 13
genetic considerations, 444
genetic diversity, 46, 49, 52, 362
analysis, 360–362
maintenance, 52

protection, 320
  of species, 362
  of subspecies, 359, 362
genetic diversity increase, 333
  within a species' population, 333
  within a few populations, 333
  within multiple populations, 333
genetic intervention, 10, 11
genetic resources, 336
  distribution of, 336
genetics of group, 434
  fiscal costs, 434
  mortality costs, 434
goals
  control breeding, 179
  control gene flow between two isolated populations, 444
  enhance diversity, 331
  ensure species buffering, 333
  ensure species continued evolution, 200, 318, 331, 333
  ensure species ecological adaptation, 333
  ensure species survival, 331
  forestall extinction of demographic units, 443, 444
  handle variability realistically, 199
  increase functional size of habitat, 443
  increase stability of populations, 443
  minimize loss of rare alleles, 443
  preserve capacity of species to survive/evolve, 200, 317
  preserve dynamic capacity of species, 200, 318
  preserve frequency of (specific) genotypes, 317, 319
  preserve genetic diversity, 331, 362
  preserve genetic diversity in species, 318
  preserve genetic diversity in subspecies, 362
  preserve operation of natural selection, 200
  preserve species' structures, 319
  preserve survival of taxa, 318
  protect rare, unique alleles, 347, 362
  provide populations for exhibit, 179, 375
  reduce isolation, 444
  reintroduce biota to wild, 179, 374
  stabilize coexistence in population unit, each component saved in each controlled population, 329
  stabilize species for variable future, 331
guidelines, 414–444
  aesthetics and conservation, 198
  in conservation planning, 415
  founding new plant populations, 187
  multiple-populations management (diversify selective pressures in additional units), 322
  one-sex management (double net fitness of heterozygotes), 322
  one-unit management (heterozygotes favored), 322
  for populational ecological diversity, 322
  use of diversity in multiple populations, 315
  use of multiple management units and diversity, 315
habitat availability, 437
habitat restraints, 63
hybridization, 285, 286
'ideal' amount of exchange among demes, 63
inbreeding, 49
  coefficients, 392
  minimization, 49
individual demes to local conditions, 63
intensity, 332, 442, 443
intervention, 10, 331, 391, 392
  breeding structure analysis, 9, 499
  constructive, 9
  destructive, 9
  environment modification for carrying capacity, 6
  environmental, 11
  genetic intervention, 6
  to preserve genetic diversity, 333
isolation of populations, 49, 67–69, 77, 314, 444
large herbivores, 443
large population maintenance, 319
levels of demographic protection (no. 1–9), 430–440
limitations on habitat, 63
longevity of demographic unit, 440–444

Subject Index 675

measurement, 317
measures in population bottlenecks, 6
methods
 indirect, by reservation areas, 331
 intensive, by direct intervention, 331
 simple monitoring, 331
migration between parks and reserves, 437
monitoring, 332
morphometric studies, 359
multiple ecosystems, 334
multiple populations, 314, 315, 320, 332, 437
multiple populations for diversity, 320
multiple small areas, 332
natural levels, 331
number of offspring produced in a lifetime, 390, 391
numerical relations between species, 2
objectives
 conflicts in, in preservation of outcrossing species, 436
 conserve dynamics of species evolution, 444
 conserve natural integrity of populations or taxon units, 444
 defined, 314
 diversity enhancement, 318
 diversity maintenance, 318
 dynamics of species evolution conservation, 444
 goals with short-term time scale, 13, 443
 heterozygosity control, 313, 314
 isolation among populations control, 314
 migration control, 314
 population conservation, 444
 population genetic dynamics control, 314
 preserve diversity structure, 314
 preserve endemic species, 436
 preserve habitat, 436
 preserve natural diversity, 313, 317
 preserve park, 436
 prolong species survival in reserve, 442
 taxonomic unit conservation, 444
one migrant individual among demes per generation, 63

operation difficulties/operation diversity, 8
options
 basic cytogenetic data, 105
 basic genetic data, 105
 breeding structure manipulation, 63, 227
 criteria in survival of species, 332
 population relationships, 105
 practical applications, systematic studies, 105
 reproductive relationships, populations, 105
 sibling species, 105
park size, 444
particular genetic characteristics of species, 63
philosophy
 interventionist, 11, 12, 415
 laissez faire, 11, 12, 334, 415
planning
 animals, for restoration to wild, 374, 435
 demography and, 430
 for endangered species, 437, 569
 to forestall extinction, 437
 founding of new populations, 180–182, 224
 human action effects on genetic variation, 51
 for individuals selected for removal, 139, 196, 443
 intervention for natural diversity preservation, 314
 levels of demographic protection (no. 1–9), 430–440
 matings, 178
 multiple use of species, 430
 objectives, 415
 park size evaluation, 441, 444
 population genetics in formulation of, 442
 to reduce inbreeding depression, 434
 reduction of inbreeding depression in populations, 434, 435
 relationships of species population size to area, 421–430, 444
 research needs in evolutionary genetics, 46
 research on biology of target species, 9
 time scale of concern, 415, 440
planning parameters

levels of demographic protection, 444
park and reserve size, 444
population size to area size relationships, 444
policy, in levels of demographic protection, 444
policy determination
  intervention, 415
  laissez faire, 415
population(s), 313, 443
  demography, 438
  size, 326
  subdivision, 320
  subdivision criteria, 362
practices on genetic diversity, 326
preservation and, 362
preservation of variation, 46
  recommendations, 46
priorities, 444
problems
  budgets, 415
  long turnaround times of ecological surveys, 415
programs
  alleviate inbreeding depression, 76
  biological constraints of particular species, 313
  breeding efforts with relict populations, 313
  breeding structure design, 63
  captive populations, 67
  cost effectiveness of alternate programs, 313
  dependence on genetic conservation objectives, 313
  exchange of animals, 76
  genetics studies, 362, 363
  genotypic frequency monitoring, 313
  maintenance of genetic variation, 67–70
  maintenance of unrelated individuals, 76
  mating systems in design of, 8, 9
  monitoring genotypic arrays, 332
  monogamous species, 76
  polygynous species, 76
  population size increase, 250
  preservation of natural diversity, 319
  in restricted biological resources, 67
  segregating population subunits, 76
  slowing inbreeding accumulation, 76
  success evaluation, 318
  systematic studies, 362, 363
  tracking changes in sample of genetic loci, 332
protection type, 444
racial differences, 358
rare allele loss, 443
representation of target species, 13
reproductive isolation, 357, 358
research on founder events, 187
reticent, 14
role
  maintain genetic variance, 14
  maintain population size, 14
scope of conservation, 2, 3
sex differences for genetic diversity, 322
sex ratio, 390, 391
sibling species, 103, 104
small captive populations, 177
smaller species, 443
space available, 444
in space and time, 8
species, expensive in time, money, logistics, 436
species population size to area, 421–430, 444
species structures, in natural variability, 319
species subdivision, 362
stability of populations, 443
strategy
  in genetic diversity maintenance, 49
  for preserving natural diversity, 318
strategy (plants)
  genetic diversification increase, 226
  genetic structuring of founding populations, 226
subspecies, 362
subspecies classification, 361, 363
success, 317
survival of biota, 2
system, non-steady-state, 319
targets (species/subspecies), 363
target species
  environment costs, 10
  genetic costs, 10
taxonomy, 358
techniques
  in controlled gene migration, 314

measurement of genetic diversity, 196, 350–357
in sibling species detection, 49
use of captive, collected populations, 314
time scale in, 2
type
  direct, 332
  indirect, 332
unit type
  game park, 70
  zoological park, 70
units
  population reduction schemes, 332
  size, 332
  steady-state maintenance, 332
  successional state maintenance, 332
  temporal variations, 332
useful genetic variation, 63
  between demes, 63
  in a deme, 63
variability of genetic expression, 190, 191
wild populations, 227, 375
wilderness levels, 331
within and between population diversity, 437
zoological parks, 331
Managerial information
  estimates of population sizes for select few species, 422
  estimates of species diversity, 422
  park or reserve size, 422
  population size analysis, 422
Managers
  action, speed required for conservation, 415
  allowing self-regulation to preserve genetic regulation, 326
  apportionment of diversity, 360
    methods of choice, 360
  choice of areas to manage, 318
  choice of populations, 318
  concept of desirable diversity, 314
  concerns
    basic genetic and cytogenetic data, 105
    population relationships, 105
    practical applications, systematic studies, 105
    reproductive relationships of populations, sibling species, 105

evaluation of diversity in collected species, 318
expense of technique, 318
genetic diversity measurement technique selection, 356, 357
genotypes worthy of preservation, 94
intervention, 49, 318
  in gene flow, 49
  use of, 318, 415
judgment in culling, 139, 196
knowledge of factors in extinction, 114
knowledge of founder events in nature, 187
knowledge of variability of characteristics, 199
maintenance of genetic diversity among populations, 320
migrant introduction for genetic variation within populations, 320
objectives for types of protection, 444
options for collective measure of diversity, 318
population thinking, 200
populations to manage, 318
preservation goals, 200
problem solving in conservation management, 415
problem solving in evolutionary genetics, 46
removal of genetic load, 199, 200
rescue attempts on doomed populations, 114
responsibilities of, in conservation, 444
retention of a shifting variability system, 200
selection against deleterious alleles, 197
self-regulation in preservation of genetic diversity, 326
taxonomic evaluation of populations, 350
typological thinking, 200
understanding of genetic differentiation, 94
  species capability for adaptation, 94
understanding of physical barrier effects, 94
understanding of population structure, 78

understanding of process of genetic
differentiation, 93
of populations over short distances, 93
use of population diversity, 321
to protect genetic diversity,
321
use of research on founder events,
187
Mangere Island, Chatham Islands,
parakeets, 301
Manipulation (plant)
constructive, 9, 227
destructive, 9
Manitoba, Canada, 100
Marine mammals
harem breeding, 117
skewed sex ratio, 117
Marine organisms, limited facilities
for, 435
Mates
founders, 253
inbred animals, 253, 254
noninbred animals, 254
Mating. *See also* Assortative mating
brother-sister, 197
between closely related members of
same family, 304
between genetically different individuals, 304
between individuals of intermediate
genetic similarity, 304
coadapted gene complexes, 304
consanguineous, 73, 74
disruption of coadapted gene complexes (genotypes), 304
father-daughter, 178, 179, 245
father-grandaughter, 245
first cousin, 246
half sib, 246
harems, 74
immigrant males, 74
inbreeding statistics, 73
individual probability of, 71
monogamous, 70, 71, 73–75
mother-son, 197
in normally outbreeding species,
304
patterns of, 70–72
polygynous, 70, 71, 74, 75
recombinational load and, 304
relatedness between individuals, 70
rate, 69
restriction of, 5, 6
type, 201
assortative (by phenotype), 201

by genetic relationship (inbreeding), 201, 243
random (outbreeding), 201, 243
waste of gametes, 304
Mating behavior
Darwinian fitness, 133
monogamous, 70, 71, 73–75
polygynous, 70, 71, 74, 75
Mating displays, sibling species differences in, 99
Mating system
association with seed dispersal
mechanisms, 346
potential for seed movement, 346
control of genetic recombination,
220
destabilization, 9
dispersal rates, 70
diversity of, 187
in invertebrates, 187
in plants, 8, 9, 499
genetic variability in plants, 220,
221
influence on genetic recombination,
223, 224
interpopulational gene flow, 9
polygyny and monogamy, 73–75
effects on inbreeding, 70, 71
shift, 218
with habitat change in plants,
223
in roadside plant populations,
218
toward outcrossing, 218
type, 499
asexual, 219, 220
inbreeding, 201, 220, 243
outbreeding, 201, 220, 243
sexual, 219, 220
within semi-isolated populations,
70, 71
Mating systems and dispersal rates in
semi-isolated populations
(model)
effects on inbreeding coefficients, 70
variables
exchange rates among units, 70
number of animals in each management unit, 70
number of females in each
harem, 70
type of mating (monogamous,
polygynous), 70
Measures of diversity
relationship to changes within
taxa, 314

relationship to species-level differences, 314
Mediterranean region
   evolution of brine shrimp, 144
   hybrid swarms of birds, 292
   plant genetic variation, 217, 218
   sorghum races, 274
Meiosis
   I and II, 19-21
   alleles, 244
   chromatid pairing during synapsis, 30, 31
   chromosome numbers, 193
   crossover frequencies, 193
   recombination, 190, 193, 194, 197
   segregation of heterozygous gametes at, 197
   synapsis, 19
   zygote formation, 193
Melbourne, Australia, 99
Mendel, 21
   genetic consequences of self-fertilization, 221
   heterozygosity loss and self-fertilization, 221
   inheritance in plants, 221
Mendelian characters, deviation from, 38
Mendelian expectation deviations
   expressivity, 38
   incomplete penetrance, 38
Mendelian genetics, 21-24
Mendelian heredity, 174
Mendelian inheritance, alleles at a single locus, 37
Mendelian laws of heredity, 21, 22, 38
Mexico
   Drosophila studies, 238
   ducks, 295
   sea turtles, 283
Mice, inbreeding depression, 172
Microenvironmental differences, selection effects on genotypes, 319
Microenvironments, conservation, 6
Microevolutionary change, 199
Microgeographic scale
   ability of populations to differentiate on, 93
      under strong selection, 93
   adaptive divergence on a, 92
   genetic divergence at, 79
   plant vs. animal differentiation in, 85
Microhabitat

adaptation, 343
   to physical environment, 343
   differentiation of populations in, 92
   diversity of, 439
   effective size of, 82
   genetic structure and, 343
Microhabitat differentiation
   application of strong selective pressures over short distances, 93
   preservation of genetic variability, 93
Microhabitat type
   mesic, 82
   xeric, 82
Microorganisms
   persistence and, 365
   rapid mutation rate, 365
Migrant(s)
   in deme exchanges, 64
   effect in retarding divergence, 63
   number and allelic frequency, 62-64
   number needed, 14
   one per generation, 44, 63, 64
   reproduction in similar demes, 63
Migration. *See also* Founding
   barriers to, 442
   between populations, 319
   between small populations, 89
   definition, 44
   effective population size increase effect on, 8
   exchange of alleles, 321
   genetic diversity location and patterns, 321
   genetic drift and, 55
   global, 322
   habitat preservation, 279
      sea turtles, 279
   heterozygosity increase, 3, 8
   high, 192
   induced, 7-10
   island model, 53
   in larger mammals, 9
   limitation on ability for, 321
   low, 192
   in management, 14
   movement of breeding individuals, gametes, 44
   number of individuals involved, 7, 9
   in plant populations, 44
   pollen, 44, 79
   in prevention of protection-level degeneration, 437
   restriction by gametic dispersal mechanism, 90

restriction by specific behavior mechanism, 90
restriction types, 321
risk in plant species, 9
sea turtles, 277, 278
seeds, 44
sex difference in plants and, 321
sexual divergence in, 321
spores, 44
success, 9
terminology, 44
Migration rate (m)
between populations, 322
control, 314
exchange proportion among demes (model), 55
in genetic diversity simulations, 53–55
island model, 53
to maintain alleles, 320
in separate populations, 320
number per generation, 44 62–64
one individual per generation, 44
population genetic divergence patterns by, 49
Migrators, male, 321
Migratory Bird Treaty Act, 288, 289 437
Milam County, Texas, 101
Mississippi River, warblers, 296
Mitochondrial DNA, 404, 408
Mitochondrial DNA studies. *See also* DNA
availability, 239
finer distinction than nuclear DNA analysis, 238
markers of genetically different populations, 240
nucleotide substitution rate, 237, 238
number of males and females in successive generations, 240
technology, 237–239
use in founder event studies, 237
use in maternal lineage studies, 237, 238, 239
Mimicry
due to gene flow, 273
due to selection pressures, 273
Missouri, warblers, 296
Mobile links, 120
ants, 10, 160
bees, 10
hummingbirds, 10

Model(s). *See also* Computer simulations
Arnold's model of western terrestrial garter snake, 370, 371
behavioral responses to different prey items, 371
genetic correlation, 371
autoselection in outcrossing hermaphrodite populations, 203
fate of rare recessive mutation for selfing, 203
balance between selection and gene flow, 370
demes, number of individuals (N), 53
ecological, of species interactions, 324–330
effect of inbreeding depression on probable survival time of closed captive populations, 172–174
gene-for-gene system, host/pathogen, 367
genetic diversity distribution parameters, population size variables, 52
impact of inbreeding on birthweight, 248–250
of inbreeding depression, 172
juvenile survival, 172
island model, pattern of gene flow, 61
island model of migration, 52
natural selection modes, 52
number of individuals (N), 53
panmixis, 8, 41
population growth rate, 323, 324
selfing species from outcrossers, 212, 214, 215
standard population genetic model of inbreeding depression, 246, 247
Models of extinction paths, by reduction in area
extinction I, extinction of key species, 120–124
extinction of other species, 120–124
loss of habitat, 120–124
extinction II, habitat/resource disappearances, 120–124
extinction of dependent species, 120–124
with subsequent isolation, 120–124

Subject Index   **681**

Models of population conditions, mating systems, 71
  constant dispersal rate per generation, 71
  constant population size, 71
  equal probability of mating, 71
  equal sex ratio, 71
Models, theoretical, of ecological genetics
  sympatric speciation, 79
  Wright's shifting balance model, 79
Models of variance, mating systems
  exchange rate among units, 70
  number of animals in each management unit, 70
  number of females in a harem, 70, 71
  type of mating, 70
Molecular biology, in study of genetics, 229
Molecular genetics, 32–34
  DNA control of protein synthesis, 32
  enzymes, isoenzymes and allozymes, 33, 34
  genetic code, 33
Molecular technology
  DNA analysis, 236, 237
  mitochondrial DNA analysis, 237–239
Molluscs
  breeding systems, 205–208, 214, 215
  genetic diversity, 205
  hermaphrodites, 210
  homozygous strains, 208
  hybridization, 206
  mating behavior, 210
  outcrossing species, 214
  polymorphic and heterozygous populations, 208
Monitoring
  butterflies, 155, 161, 162
  genotypic frequencies, 313
Monogamy
  accumulation of inbreeding depression, 71
  comparison with polygyny, 75, 77
  genetic variation maintenance, 75
  immigrants and, 74
  rate of consanguineous mating, 74
  relatedness in breeding units, 74
Monomorphism, 137, 219, 365
Monte Carlo simulations, 52
Montpelier, France, snails, 211

Morocco, grass species, 271
Morphology
  characteristics in adjacent populations, 78
  convergent evolution in superficial characters, 357
  environmental gradients and trait variation, 344
  parallel evolution in superficial characters, 357
  plant
    among geographic regions, 347
    among local heterogeneous environments, 347
    sibling species character comparisons, 357
  trait variation among populations, 345
Morphometric analysis
  character measurements, 355
  cost, 356, 357
  computer programs, 356
  effects of gene loci measurements, 355, 356
  extant vs. extinct population comparison, 356
  multivariate statistical analysis, 355
  overall phenotypic distance, 355
  problem of nongenetic variation measurement, 356
Morphometric studies, computer applications, 105
Morphometric traits, environmental gradient association, 344
Mortality
  butterfly larval phases and plant senescence in, 154
  from deleterious alleles, 244
  early genetic death, 248
  environmental, 254
  estimate of environmental component, 248
  from father-daughter matings, 184
  genetic, 248, 254
  from inanition, 184
  inbred rats, 184
  inbreeding and, 171
  infant, 245
  juvenile, 184
  juvenile, of inbred young, 171
  from lethal alleles, 244
  nongenetic, 248, 254
  of noninbred young, 184
  from prematurity, 184
  sibling species, 103

Mosaic pattern of population regulation, 162
Mount Rainier National Park, Washington
 carrying capacity, 420
 size, 420
Mount St. Helens, Washington, volcanic eruption, 126
 effect on fauna, 126
 effect on flora, 126
 long-term soil effects, mercury in soil, 126
Movement. *See also* Vagility
 active transport by man, 67
 animals among isolated breeding units, 69
 animals over distance, 66, 67
 pollen over distance, 66
 populations, 66, 67
 seeds over distance, 66
 sweepstake dispersal, 67
Muir Woods National Park, Calif. 438
Multilocus associations, in inbreeding species, 223
Multilocus system
 selection, 197, 198
 shift, 197, 198
 trait control, 197, 198
Multiple alleles. *See also* Alleles
 in populations, 25, 194
Multiple factor inheritance, data from electrophoresis, 194
Multiple management units
 biotic diversity, 315
 environmental diversity, 315
 population size diversity, 315
Multiple populations
 arrangement of interbreeding units among, 69
  for control of relatedness, 69
 creation of diversity in different, 315
 division into groups, 69
 division into neighborhoods, 69
 enhancing effect in genetic preservation, 69
 genetic diversity preservation, 318
 genetic polymorphism maintenance, 321
 heterozygotes by female selection, 321
 maintenance of genetic variation in units, 69
 in management of diversity, 320
 preservation, 314
 relatedness of individuals between, 69
 in maintenance of genetic variation, 69
 single species in, 332
Multiple use of species
 demotion of level of protection, 430
 reduction of natural stability, 430
Multispecies associations, genetic aspects, 372
Mutagens
 insecticides, 130
 ionizing radiation, 139
  and fitness increase, 139
Mutants
 plant selectivity for, 221
 preadapted, 94
Mutation. *See also* Chromosomal mutations
 allelic variation by selectively neutral, 61
 definition, 34
 deleterious, 129
 directional evolution and, 39
 fitness variation, 93
 frequency of, 34
 genetic variants and, 93
 harmful, 37
 probability by effective population size, 93
 probability by mutation rate, 93
 probability by time span, 93
 purebred distinctions from, 289
 recurrent, 93
 source of genetic diversity, 34, 128
Mutation rate
 with chemicals, 37
 population genetic change and, 43
 with ultraviolet light, 37
 with x-rays, 37
Mutation type classification
 chromosomal, 34–36
 point, 36, 37
Mutual assimilation
 desirability, 8
 inevitability, 8

N (census population number), 45
N (deme sizes), 58
N (number of individuals), in genetic diversity simulations, 53–55
N_e (effective population size or number)
 census population number comparison, 44, 45

Subject Index   **683**

genetic drift rate as in ideal population, 44
for useful genetic preservation, 63
Nairobi National Park, Kenya, rate of species loss in, 439
Napa Valley, California, slender wild oat, 344
National Park Service (NPS), 421
National Zoological Park, Washington, D.C., 165
  effects of inbreeding in wildebeest herd in, 172
Native plants
  extinction from insects, 373
  extinction from pathogens, 373
  mortality from insects, 373
  mortality from pathogens, 373
Natural areas
  adaptation extremes for survival in, 331
  species restriction to less hospitable, 331
Natural catastrophes and extinction, 112
Natural colonization, origins of species, islands, 192
Natural colonizing event, 187
Natural diversity. *See also* Biological diversity
  diminishing of, 402
  preservation, 313
    intervention, 314
    strategy, 318
  structure
    ecological complexity, 314
    subpopulation structure, 314
Natural ecosystems, alterations, 307
Natural environments
  moisture levels and, 79
  salinity levels and, 79
  sunlight levels and, 79
  temperature levels and, 79
  wind levels and, 79
Natural extinction
  competition, 112
  disease, 112
  lack of information in, 112
  natural environmental disturbances, 112
Natural factors in environmental change, 313
Natural forces, pressure on natural populations, 126
Natural plant populations. *See also* Plant populations

fitness loss with mating system shifts, 222
heterozygote expectation with mating shift, 222
heterozygous advantage, 222
inbreeding depression, 222
Natural populations
  adaptation to man-made environments, 127
  allele frequencies and, 61
  allozyme maintenance in, 33
  computer simulations of situations in, 150, 151
  dynamics and genetics, 152
  enrichment, 314
    by captive population supplements, 314
  evolutionary pressures on, 94
  extinction of, 241
  fitness and heterozygosity, 139
  founder effect in, 197
  founder events, 197
  founders from, 191
  generations after founder event, 197
  genetic diversity in, by mutation, 34
  genetic diversity maintenance, 52
  human activity effects, 78
    on dynamics of, 78
    on genetic structure of, 78
  isolation by distance, 69
  man's activities, and extinction of, 126
  natural forces and, 126
  natural selection potential in genetic divergence (model), 53
  projections for, 444
  resources for, 14
  selection in, 134
    soft type, 134
  short distance environmental gradients and, 94
  single founder in, 193
  small number of individuals derived from, 191
  surveys on environmental variation with genetic variation, 137
  survival, 14
  variation and, 77
Natural selection, 64
  absence (model), 53
  adaptation
    to habitat, 285
    to inbreeding, 256
    local conditions, 285
    of organisms, 38

in allozyme maintenance, 33
biological factors in natural environment and, 86
changing pressures of, 443
under conditions of limited gene flow, 82
creative force, 38, 39
differential directional selection (model), 53
differential survival of genotypes (model), 52
differential survival probabilities (model), 53
ecotypic variation and, 83
evidence for action of, 61
evolution and, 38–40
exploitation of heterozygosity, 194
genetic drift in absence of, 63
genetic interchange, 7
in genetic systems, 193
genetic variation under, 155
genetically novel population production, 199
heavy metals and, 83
heterozygous advantage (model), 53
history of evolutionary theory and, 38
inbreeding programs and, 150
at individual gene loci, 61
individuals of intermediate genetic similarity, 304
intensity, 61
intensity (model simulations), 52
intensity of genetic divergence (model), for value m, for value N, 53
on juvenile stage, 91
mode, 61
mode and intensity in genetic diversity simulations, 53
model, 53
polygenic balances and, 197
population differentiation and, 94
selective neutrality (model), 53
stabilization
 genetic variation decrease, 135
 intensity effect on characters, 135
 long-term fitness decrease, 135
 in retention of heterozygosity, 135
stage of postzygotic, 91
theory, 38
through time, 285
trait selection and, 198
Natural selection modes (models)
heterozygous advantage, 52

opposing directional selection in different demes, 52
selective neutrality, 52
Natural sexual selection, naturally occurring shift, 199
Natural species, inherent values in, 191
Natural systems
chain reactions, 7
managerial intervention, 7
targets of impact, 7
Naturalists
butterfly collections, 161
population terminology and, 40
use of term population, 40
Nature conservation, evolutionary process continuance under, 9
Nature reserves
breeding designs, 63
continuing evolution capability of, 12
inadequacies, 12
large, for space-demanding animal species, 6
"life expectancy" limitations, 14
management, 13
political vigilance, 12
population management for, 63
protection, 12
Near East countries, sorghum races, 274
Neighborhood
boundaries to relatedness, 70
establishment from populations, 69, 70
exchange rate of individuals in, 69, 70
interbreeding demes within, 77
maintenance of segregated, 75
 for pools of unrelated individuals, 75
 for cross-mating, 75
segregation, 76
size, number of breeding individuals and dispersal, 70
Nesting
grounds (sea turtles), 279
habitat preservation (sea turtles), 279
populations, mitochondrial DNA studies, 352
Netherlands, slugs, 208
New Mexico, ducks, 295, 296
New population establishment, demographic factors, 384
New Zealand

duck hybrids, 293
flora, 12
large vertebrate lack, 12
Niche(s)
  competition-free, and within population selection, 150
  heterogeneity, 321
  zygote dispersal to different, 321
Nondisjunction, 35
Nonselected animals
  birth weight, 258, 259
    inbreeding coefficient and, 258, 259
  outcrossed, 255
Norfolk Island, 192
  bird colonization, 192
  insect colonization, 192
North America, 144
  biomes of, 422
  birds, 291
  captive rhinos, 378, 379
  Drosophila studies, 238
  ducks, 306
  European colonization and introduced plant spread, 227, 228
  fishes, 144
  slugs, or snails, 205, 209
  tigers, 395
  zoo exchanges, 389
  zoo geneticists, 387
North American biota, in extinction of South American mammals, 365
North American coast, turtles, 277
North American continent, biotic diversity on, 421
North Central states, U.S., 175
Nucleic acids. *See* DNA, RNA
Nucleotide pairs
  in fungi, 128
  in mammals, 128
  in some vascular plants, 128
Number
  effective population size and, 390
  of lethal equivalents, 244
  of offspring produced in lifetime, 390, 391
Numbers, increase by transplantation, 305
Nutrition, sibling species and, 99

Oceanic islands
  extinction on, 111
  founder effect in natural populations, 197

habitat patches and extinction on, 112
lack of herbivory effects in, 111
lack of predation effects in, 111
Offspring
  abnormalities in, 45
  attribute level for individual genetic variability, 252
  attributes
    founder ancestry, 254
    genetic variability, 254
    by pairing, 254
  birth weight, 249, 250
  inbreeding coefficients, 246, 248, 249, 254
  noninbred parents, 246, 248, 249, 254
  number in next generation, 323
  number of by an individual, 44
  one year survival, 246, 248, 249
  percentage of founder contribution, 252
  thirty day survival, 246, 248, 249
Ohio, warblers, 296
Oklahoma, 100
Operation Green Turtle, eggs and hatchlings to Caribbean beaches, 283
Organisms
  adaptation to breeding system, 243
  economically important, 152, 272
Outbreeding
  adaptation and, 304
  high fecundity, 304
  inbreeding vs., 304
  random dispersal in large populations, 304
  random or vagrant dispersal in large populations, 304
Outbreeding depression, 205, 437
Outbreeding organisms
  genetic variance, 146
  inbreeding, 146
  phenotypic variance, 146
Outbreeding species
  inbred, 137, 146, 150
  inbreeding depression, 137
Outcrossed populations, adaptation to inbreeding in, 243, 244
Outcrossing
  in fertility restoration, 177
  fertilization risks, 205
  in genetic variability maintenance, 69, 205, 213
  heterotic condition and, 69
  inbred populations, 68, 69

to maintain genetic variability, 69
marginal benefit, 203
outbreeding depression, 205
perennial habit, 205
plants by pollinators, 90
rate, variability in, 221
in vigor restoration, 177
Overdominance
to ensure genetic diversity, 320
heterozygosity and fitness, 45

Pacific coast, Chiapas, Mexico, 104
Pacific coast, China, grass distribution, 269
Pacific coast, Taiwan, grass distribution, 269
Pacific coast, U.S.A., 363
elk populations, 359
Pacific islands, sea turtle farming and ranching, 282
Pacific Ocean, founding populations, 192
Pacific tectonic plate, 192
Padre Island, Texas, sea turtle nesting population establishment, 283
Pakistani plains, grass hybrid, 271
Palearctic, duck hybrids, 293
Panama Canal, Panama, 119
Pairing
gene separation into gametes, 22
offspring attributes and, 254
Panmixis
adaptation reduction, 8
dangers of, 435
effective population size doubling, 8
joining of two populations, 8
model, 8
once every generation, 7
small number exchange, 7
between two populations, 7
Par Pond, South Carolina, 89
Parakeets, Chatham Islands, 301–303
Parasites
adaptability to alternate hosts, 369
definition, 369
fitness dependence on parasite fitness, 366
genetic variability, 368
and virulence, 368
in identification of sibling species, 103
life extension of host by, 366
persistence in ecological change affecting hosts, 369
plant release from, 226
selection for virulence, 366

Parasite species
dependence on host species, 126
environmental factors affecting host and, 126
evolutionary prospects, 126
extinction with host species extinction, 440
survival time with host species, 442
host-specific in sibling species identification, 99–101, 103
Parental forms
assortative mating, males and females of, 292
backcrossing ability to, 292
Parent selection, inbred individuals in good health, 252
Parents
in breeding, 198
closely related, 246
criteria in selection, 252
offspring desired, 252
different founders
in ancestry, 253
disassortative mating, pedigree, 253
good health, 252
inbred, 254, 260
inbred animals, 254, 255
in meiosis, 23
by natural selection, 198
one parent inbred vs. both parents inbred, 256
selection criteria, 246
Park management
carrying capacity determinations, 421
surplus animal removal, 420
Park and reserve size
acreage in, 420
habitat in, 420
in Alaska, 421
and demographic complexity of resources within, 415
histories of, 443
ideal population size for long term, 443
increase potential, 415
for species functional size and capacity, 415
intensity of management requirements, 415, 416
intermediate, 415, 416, 418, 420, 438
large, 415, 416, 419, 420, 421
large-bodied species in, 442

Subject Index   687

level of demographic protection and, 440–443
management decisions in a fixed, 443
multiple populations and, 439, 440
populations of plants and, 438
populations of salamanders and, 438
population size realtionships, 415, 441, 442
in prediction of population size, 415
relationship to population size, 442
relationship to species population size, 444
small, 415–417, 420
  focal species dependence on, 436
  genetic isolation, 436
  of single species, 436
  isolation of host species, 436
  maintenance of large populations, 440–442
  maintenance of small organisms, 440–442
  species diversity maintenance relationships, 415
  in U.S.A., 416, 417
  by increasing size, 417–419
  in world, 416
Parks and reserves
  adjacent area effects, 442
  adjacent areas, 438
  available habitat, 414, 415
  biological imbalance in, 420
  biosphere reserves, 421
  controlled migration between, 435
  design, 104
  faunal uniqueness of, 420
  floral uniqueness of, 420
  graded series of, 437
  guidelines on projected longevity of focal (target) species, 415
  habitat change, 421
    in surrounding areas, 421
  habitat overexploitation in small, 436
  habitat reduction around, 421
  levels of demographic protection in, 430, 444
  longevity of a population in, 440
  multiple populations in, 439
  "optimum" protection of species, 430
  patchiness of resources in, 442
  prediction of population size, 429, 430
  protection in, 1, 415
  reduction in available habitat in, 421
  refuge lines and species boundaries, 104
  resources and demographic complexity, 415
  resources and potential population sizes, 415
  restrictions in available habitat, 420
  single populations in, 438
  size effects on gene pools, 414, 415
  supplementary, 443
  urbanization of adjacent areas, 420
Partners, change, 8
Patches. See Habitat patches
Patchiness, genetic characteristics of, 91
Parthenogenetic Homoptera, biotypes with different cultivars, 369
  evolution of new forms, 366
    against crop resistance, 366
  extinction by, 366
  gene-for-gene system, with hosts, 367
  genetic impoverishment with host species, 370
  in genetic drift, 370
  genotypes, 367
  in plant extinction, 373
  in plant mortality, 373
Pelagic species
  mass reduction in, 127
  ocean productivity and, 127
  reduction from reduced ocean productivity, 127
  reduction from increased temperature from CO_2, 127
Perennial crops, weed races, 272
Perpetuation of sameness, 198, 199
Phenetic variability
  in hybridization, 292
  random mating variability, 292
Phenotype, 22–27
  advantage, 133
  behavior, 16
  central, 134, 135
  change over time, 317
  comparison with genotype, 15, 16
  definition, 15, 16
  diversity of, 133, 134
    for next generation, 133, 134
    through selection, 133, 134
  extreme, 134, 135
  homozygosity and fitness in genetic variation simulation, 54
  low mobility, 160

optimums in selection type, for genetic diversity, 134
physiology, 16
point mutation effect, 36
skin color, 16
sum of effects of genes at many different loci, 37
typological thinking and, 194–196
Phenotypic changes, dynamic events and, 161, 162
Phenotypic characters
genetic architecture of, 133
variation, 133
Phenotypic plasticity
adaptation to environmental shifts, 219
genetic polymorphism negative correlation, 219
level, 219, 225
range inadequacy in new conditions, 219
reproduction and, 219
survival and, 219
Phenotypic variance
with genetic variance, 146
of morphological characters, 146–149
ratio of inbred to outbred, 146
Phenotypic variation
change by genetic variation change, 146
change with inbreeding generations, 146
of inbred animal stocks, 146
of inbred plant stocks, 146
of inbred stocks greater than outbred, 146
Philippines, grass species, 271
Physical factors
extinction and, 126
host-parasite relationships, 126
Physiological cost, in host/pathogen relationships, 367
Physiological stresses, sibling species differences in response to, 99
Physiology, characteristics in adjacent populations, 78
Physiology, plant
among geographic regions, 347
among local heterogeneous environments, 347
Phytophagous arthropods, genetic variation in, 369
Pinnacles National Monument, Calif.
invertebrate fauna in, 420

reptiles in, 420
size, 420
Planning methods. *See also* Management, Wildlife management
Plant(s)
adaptation on islands, 111
adaptation to reproduction, 9
affected by physical environments, 85
area for survival, 420
breeding systems, 8, 499
characteristics of response to disruptive selection pressures, 92
chemical/heavy metal tolerance, 85
selective forces of, 85
chemical preadaptation and location of, 85
cold tolerance, 269
colonization, 217
complexity, 8
density independent factors and extinction of, 117
diploid cross-fertilized, 199
dispersal limitations, 85
diverse breeding systems, 9, 499
diversity, 8
dormant state in adverse conditions, 85
extinction of, 111
extinction by man, 111, 112
in food webs, 10
frequency of tropical tree species, 6
gene distribution, 224
genetic distinction among local populations of, 82
genetic variation of, 82, 133
genotype reserves in management, 7
heterozygosity, 131, 133
hybrid(s)
apomixis, 270
introgression products, 270
maintenance, 8, 9
seed without fertilization (agamospermy), 270
sexual plant intercrosses, 270
sexual plant sterility, 270
swarms, 307
vigor, 270
inbred, 223
inversion in, 36
larval food for butterflies, 153–155, 161, 162
limited seed dispersal mechanisms, 90
loss of defenses against herbivores, 111

male gametes, 321
microenvironment selective, 6
migration of, 44
mobile links in plant processes, 10, 120
morphology, 347
  among geographic regions, 347
  local heterogeneous environments, 347
most valued, 199
nonvagile, 6
in parks and reserves, 2, 438
physiology, 347
  among geographic regions, 347
  local heterogeneous environments, 347
ploidy level, 220, 269
pollen dispersal, 85
pollination (gene flow) dependence on vectors, 6
polyploid occurrence in, 34, 465
reciprocal transplant experiments, 82
reproductive mode, 226, 499
sedentary nature of, 85
seed dispersal, 85
seed zygotes, 321
self-fertilization, 90
sex chomosome systems in, 28, 29
sex determination systems in, 28, 29
small external gene flow, 9
subvariation, 314
tolerance to metals, 83, 84
weedy species, 217–219
wind-pollinated, 8, 340
Plant assortative mating
  negative (self-sterility), 221
  positive (self-fertilization), 221
Plant colonization
  demographic correlates, 225
  factors in success of, 187
  genetic correlates, 225
  intercontinental
    bottlenecks by long distance founding, 217
    novel selective pressures, 217
    repeated migration absence, 217
  monomorphism of, 219
  responses
    asexual reproduction, 225
    fecundity (high), 225
    genetic change, 225
    genetic differentiation among populations, 225
    genetic variation loss, 225
    inbreeding depression, 225
    phenotypic plasticity, 225
    pollination agent shift, 225
    self-fertilization, 225
    survivorship (high), 225
  source populations vs. founder populations, 217–219
  success, 227
    and genetic variation, 217–219
Plant communities
  butterflies, index to status of, 161
  spatial mosaics of populations, 216
Plant crosses, introgression products, 269
Plant domestication
  domesticated races, 272
  plant adaptation, 272
  reproductive habits, 272
  weed races, 272
  wild plants in artificial habitat, 272
Plant evolution, polyploidy in, 34, 271, 465
Plant genetic correlates
  fixed, 225
  manipulated, 225
Plant genetic organization manipulation
  introduction of polyploidy, 227
  maintenance of genetic variation level, 227
  return to healthier state, 227
Plant genetic structure
  microhabitat adaptation and, 343
  physical environmental factors and, 343
Plant genotype, with range of phenotypes, 219
Plant geographic range, genetic distribution effect
  endemic, 337, 338
  narrow, 337, 339
  regional, 337–339
  widespread, 337–339
Plants, high genetic variation levels
  high lifetime fecundity, 338
  late succession occurrences, 338
  long generation time, 338
  outcrossed mating system, 338
  wide range, 338
  wind pollination, 338
Plant intergrades, 276
Plant introductions, genetic diversity of, 217
Plant management
  genetic diversity maintenance, 52
  nutrients, 11

pesticides, 11
protection from herbivores, 11
seedling planting, 11
water, 11
Plant management strategies, allozyme plus quantitative genetic variation distribution studies, 348
Plant management techniques
chromosomal complement doubling, 226
hybridization, 226
mates available, 226
migration from source, 226
pollinators available, 226
subdivided populations, 226
wide founding sample, 226
Plant manipulation
genetic variability, 225, 226
mating structure, 225, 226
Plant mating structure shifts, habitat change, 223
Plant mating systems, 499
environmental influences, 222, 223
genetic distribution effect
mixed–animal, 337, 340
mixed–wind, 337, 340
outcrossed–animal, 337, 340
outcrossed–wind, 337, 338, 340, 346
selfed, 337, 340, 346
genetic factors, 222, 223
genetic recombination and, 223
genotypic frequency distribution, 223
influence on genetic variation partitioning among/within populations, 346
outbreeding, 225
plastic, 220
shifts in, 222
subject to evolutionary change, 220
Plant migration, by seeds, pollen, spores, 44
Plant pollination, 6
adapted to local environments, 85
Plant population(s)
adaptation to soil moisture changes and to other changes (of competitors, pathogens, insects) of biotic environment, 365
adjacent, differing in resistance to heavy metals, 95
allelic diversities, 217
allelic heterozygosity in, 194
allozyme variation distribution, 337
analysis, 342
biotic environment, 224
causes of patchiness in, 90
chemical compound effects on, 92
colonial, 217
differentiation by heavy metals in, 84
differentiation in, compared with animal, 85
dispersal localized in, 85
dispersed in space, 216
ecological factors and, 224
environmental change and, 216
fertilizer, cause of differentiation in, 84
founding, 216, 217, 224, 227, 228
genetic differentiation of, 216
genetic diversity, 337
analysis, 342
between, 342
proportion among components, 342
within, 342
genetic management, 225
genetic recombination, 223
genetic structure
breeding system, 227
multilocus organization of alleles, 227
ploidy level, 227
genetic subdivision, 90
heavy metal stress in, 83
heterozygote elimination, 221
inbreeding depression in, 178
induced selective changes, 220
liming treatments, cause of differentiation, 84
management, genetics for wild populations, 224
merging, 269
metal-tolerant, 83
multilocus association maintenance, 223
number of alleles per locus, 217
patchiness in, 90
physical environment, 224
rare interpopulation gene flow, 8
reduction in size, 187
short distance environmental gradients and, 94
small demes in, 85
small subpopulations, 9
source, 217
source vs. founder, 217–219
in spatial mosaics, 216
spread, 227, 228

Subject Index 691

subdivisions, 342
subpopulations
  genetic diversity analysis, 342
  genetic diversity between populations, 342
  genetic diversity proportion among components, 342
  genetic diversity within populations, 342
survival, 420
weedy, 216
Plant reproduction mode, genetic distribution effect
  asexual, 337, 339
  asexual/sexual, 337, 339
  sexual, 337, 339
Plant species
  adaptation to insect pollinator, 225
  animal mutualists and extinction, 120
  annual, 222, 225
  asexual, 225, 365
  crops, 269
  defenses against insects, 368
  defensive role of plant secondary compounds, 368
  distribution, 225
  domestic, 272
  environmental tolerance, 222, 223, 225
  extra (B) chromosomes, 128
  fecundity, 225
  genetic correlations to enemy resistance, 371
  genetic differentiation among species, 225
  genetic diversity analysis, 336, 337
  genetic organization plasticity, 227
  genetic variation loss with inbreeding, 221, 222, 225
    mating system, 220, 222
    outbreeding, 221
    phenotypic plasticity, 219
    self-fertilized, 222
    survivorship, 225
  genetic variation, 132
    in plant mating systems, 220, 221, 225
  heterogeneous, 347
  island, and extinction factors, 120
  life forms, 342
    in subpopulations, 342
  life history traits, 336
    and allozyme distribution, 337
  life history types, 342
    in populations, 342

  long-lived, 420
  mating system, 220, 499
  migration and sexes, 321
  outcrossing, 209
  ploidy level, 220, 465
  reproduction mode, 336, 337, 499
  resistance to pests, 371, 372
    by linked genes, 371
    and not other pests, 371, 372
    by observed trait, 371
  seeds dispersed by animals, and vulnerability, 120
  self-fertilizing, 218
  stigma length, genetic control of, 129
  style length, genetic control of, 129
  success, 347
  variability in outcrossing rate, 221
  weeds, 271–275
  weedy annual, 227
  wide-ranging, 209
  wild, 272
  woody, 227
Plant succession stage, genetic distribution effect
  absence of, 120
  early (also weedy), 120, 337, 441
  late, 337, 338, 441
  middle, 337, 441
Pleiotropic effects
  coat color, 37
  on fertility in mink, 38
  single locus effects on different characters, 37, 38
Ploidy
  buffering and, 269
  diploid, 190, 220, 224, 269, 271, 275, 465
  hexaploid, 269, 271, 275, 276
  high ploid, 269
  level (plants)
    in gene duplication, 220
    in heterozygosity, 220
    in segregation, 220
    in selection response, 220
  pentaploid, 269, 275, 276
  in plants, 34, 271, 465
  polyploid, 220, 224, 436, 465
  tetraploid, 269, 271, 275, 276, 465
Point Reyes National Seashore
  removal of surplus animals, 420
  size, 420
  tule elk in, 420
Point mutations, 36, 37
  adaptation and, 36
  allele frequency change and, 43

Drosophila studies, 37
human sickle cell hemoglobin, 36
lethality, 37
rate increase factors, 37
rates, 37
survival and, 37
Political action, on probability of extinction, 12, 151
Political administrations, extinction of demographic units and, 443
Policial boundaries, in preservation of allelic diversity, 362
Pollen, 44, 465
Pollination
by animals, 343
differentiation and, 343
gravity, 220
by insects, 220
outcrossing rate correlation, 223
wind, 220
Pollinator(s)
behavior, 223
color preference, 223
insect vectors, 321
lack, and change in breeding system, 222
wind vectors, 321
Pollutants, 86
mercury in soils, 127
mutagenic insecticides, 130
thermal, 86, 127
Pollution
acid rain, 162
pesticide drift, 162
Polygenic balance
founder event and, 197
inbreeding and, 197
system, founder event in destabilization of, 197
Polygynous mating systems
cross-mating, 75
dispersal rates and, 74
fitness and, 75
genetic variability maintenance, 75
harem sizes and, 75
immigrant unrelated males, 74
and inbreeding depression alleviation, 74
loss of genetic polymorphism, 75
segregated neighborhoods, 75
viability maintenance, 75
Polygynous species
accumulation of inbreeding, 72
compared with monogamous, 75, 77
Polymorphic loci (plant)
distribution and, 338, 339
number of alleles per, 337, 338
number vs. heterozygosity, 338
species variability in, 338
Polymorphic populations, size and/or biomass maintenance, 137
Polymorphism(s), 209, 367
allele frequencies in, 136
environmental factors and, 136
breeding schemes and for maintenance of, 76
in *Drosophila*, 137
by environmental heterogeneity, 155
enzyme, 368
flower color, 218, 219, 223
frequency dependent selection, 367
in polymorphism maintenance, 367, 368
in frogs by habitat type, 140
gene-for-gene system, 368
heterozygosity correlation, 131
heterozygosity and, in decapods, 140
of host, association with asexuality, 369
in life forms, 131
in loci, 130, 131
loss in populations, 68
maintained by marginal overdominance, 155
maintenance, 367
as chaotic genetic oscillations, 367
stable cycles, 367
stable gene frequencies, 367
protection of, 320
reduction by loss of alleles, 67
Polypeptide chains, 36
Polyphagous insects
different genotypes and different hosts, 369
population genetic subdivision, 369
Polyploid species
buffered, 227
level of demographic protection, 439
Polyploids, colonization success by, 436
Polyploidy
chromosome sets in, 34
in new plant species, 34, 269
Population(s). *See also* Species
adaptation to environment, 38
adapted to heavy metal stress, 84
adaptive differentiation of, 94
adjacent, 90

local, 93
   on microgeographical scale, 95
adjacent, 91, 438
after founder event, genetically altered, 198, 199
age class distribution and extinction, 117
allele frequencies, in migration between, 44
   differences among, 321
allele maintenance among semi-isolated, 49
allelic diversity by hierarchy level, 361
   for analysis of variation in species, 40
ancestral, 192, 194
arrays of genetic difference, 196
artificial, 198
attenuated, 197, 198, 199
attrition, 11
between population variance, 345
bottlenecks, 6
captive, 103, 198, 434
in captivity/cultivation, 435
causes of extinction, 115–119
change in gene frequency, 326
change in genetic composition of, over time, 39
creation of genetically distinct, 91
cultivated, 103, 198
cycles, 327
donor/recipient, 7
decline, 160
   in habitat and range, 113
   of variation within, 76
decrease, 328
definition, 39, 40
   and subgroups of, 40
   local, 114
   Pennsylvania, 114
   South Atlantic, 114
demographic stabilization, 394, 401
densities, 422. *See also* Population size
dependent upon management, 436
difference in allele frequencies between, 7
differences of phenotype, 196
   of individuals in, 196
of different ecotype, 155
with different genotype arrays, 313
dimensions of adapted patches and, 95
   constant over space and time, 95
direction change, 327

dispersal within, 70–73
disruptive selection imposed on, 94
distances between, 88
distribution scale, 91
   of allele frequencies at loci in, 61
   effects, 126
diversity of, 320
   in isolation, 49
dynamic interactions between, 293
ecological isolation, 8
ecosystem enrichment, 11
ecosystem impact, 11
effect of selection on, 151
effective dispersal rates and, 67
effective population size maximization, 389–392
effective size, 336
endemic, 319, 331, 438
   of plants, 438
enrichment of restricted, 11, 314
environmental variation effects, 141
epidemic, 319, 331
establishment of new, 384
evolutionary adaptability of, 118
exchange of genes between different gene pools, 289
exchange of individuals among small, 69
   and inbreeding effects after generations, 69
exchange rate of individuals in, 69
expansion, 250, 251, 389, 394
extinction, 109, 156, 160, 313, 328, 364, 440
   in zoos, 172
few, 126
   of some mammalian species, 126
   and extinction, 126
fitness of, 94, 139, 313, 436
   decrease of, with inbreeding, 178, 313
   heterozygosity and, 139
   levels of management, 437
formation of additional, 438
   in large parks and reserves, 438, 439
founder establishment, 192
founding of, 180–182
fragmentation of, 78
frequency distribution
   genotypes, 317
   phenotypes, 317
gene flow
   between, 66, 130
   characteristics of, 92
   in, 94

generation time and adaptation of, 93
generations of inbreeding depression, 250
genetic change in characteristics, 193
genetic composition, 289
genetic differences in, 180
genetic differentiation of, 93
  over short distances, 93
genetic distinction in ranges of two, 292
genetic divergence of, 66
  single locus, 49
genetic diversity
  between, 351
  heterozygous fitness by sexes, 322
  in, 436, 437
  maintenance, 330
  measurements, 351
  in a single population, 360
  within, 351
genetic drift, 49, 155, 313
genetic exchange between, 66
genetic factors in extinctions of, 49
genetic interchanges between, 7, 8
genetic patchiness in, 94
genetic structure, 8, 194
genetic variability, 118, 142, 143, 193
  for adaptation of, 93
  decrease in, 118
  environmental variability, 142, 143
  spatial, 142, 143
  temporal, 142, 143
genetic variation loss, 313
genetically novel, 199
genetically uniform, 367
  and insects, 373
  and pathogens, 373
geographically divided, 130
geographically subdivided, 49
growth rate and heterozygosity, 231, 232
heterozygosity comparisons, 137
heterozygous, 139
of high genetic variability, 92
  and adaptation, 92
homozygous, 178
hybridization between allopatric, 308
  in secondary contact, 308
hybridization effects, 270, 308
increase, 326, 327, 328

and death rates, 38
  potential for, 395
increased proportion of homozygous individuals, 178
  with inbreeding, 178
independent functional units, 280
  sea turtles, 280
individual hybridity maintenance, 8, 9
intensive management of, 443
interaction of selective forces with different gene flow, 92
interbreeding, 282, 283, 319, 351
interpopulational differentiation, 319
intrapopulational variability, 319
introduction of different genotypes, 305, 331
introgression effects, 270, 308
isolation, 314
insular, 289
large, park and reserve size and, 438, 439
length of existence, 142
less fit individuals, 194
life span increase, 172
  with introduction of unrelated animals, 172
long-distance dispersal of, 85
longevity, 435
long-term genetic variability of, 75
long-term genetic viability of, 75
loss of alleles, 313
loss and genetic variation loss, 313
loss of fitness, 313
loss of heterozygosity, 313
low density, 10
maintenance, 241
  of genetic diversity in, 414
male gametes from different, 321
man-made selection effects, 151
managed, allele frequencies and, 61
management, 61, 63, 64, 313
management upgrading of, 438
manipulation of by human activities, 93, 225
maximum fitness, 194
merging of genetically different, 269
microhabitat differentiation of, 92
  from human activities, 92
migration and, 320
migration connection and, 319
migration rate for neutral allele, 44
  composition for similar, 44
mobility of, 94

monitoring, 161, 162
monomorphic for alleles, 365
  for local adaptation to ecological variables, 365
multiple, 439, 440
  in colonizing success, 436
  in protected units, 437
  in a reserve, 439
  semi-isolated, 436
multiplicity, 315
natural, 198
natural effects, 151
in nature reserves, 14
new, 194
nontolerant to heavy metals, 83, 84
numerical strength, 308
numerous, 440
  by division of captive group, 69
  to lessen likelihood of extinction, 69
  in small protected area, 440
offspring and growth rate, 323
outcrossing of inbred, 68, 69
panmixis, 8
partner selection, 8
patterns of divergence in adjacent, 92
population crash, 443
with preadapted genetic variants, 93
preadapted genotypes of low frequency in, 85
with preadapted mutants, 94
preexisting genetic variants in, 93
preservation of genetic variability in captive, 179–182
proliferation and reproduction curtailment, 395
quantification of diversity amounts, 350, 351
random mating within, 8, 321
rate of exchange of individuals among semi-isolated, 70
  to alleviate inbreeding depression, 70
recently founded, 434, 435
recruitment in, 67
reestablishment, 153, 159
relict, 289, 330, 434, 435
reproductive isolation between, 292
response to density effects, 326
retention of genetic diversity in, 134
  selection, 134
of salamanders, 438
  in parks and reserves, 438

of same ecotype, 155, 156
sea turtles
  localized breeding patterns, 286
  nest site fixity, 286
  organization, 286
secondary contact, 293
security
  isolated from formerly connected gene pools, 313
  newly founded, 313
segregation into neighborhoods, 76
selection pressures and, 127, 162
selective differences, 321
selective pressure, 127, 162
self-regulating, density dependent effect, 327
semicaptive, 435
semi-isolated and colonizing success, 436
senescence, 395
separate, allele fixation in, 320
separation over distance, 66–70
sex ratio and extinction, 117
sexual, 199
in short term, 443
short-term survival, 438
sibling species distribution, 104
single founder in, 193
single gravid female establishment of, 192
swamped by another species, 313
two interbreeding, 305
single
  heterozygote fitness, 322
  homozygote fitness, 322
  in species diversity preservation, 320
  species management, 333
  species survival, 333
size, 156
size effects (model), 53. See also Population size
slugs
  heterozygous strains, 205–209
  homozygous strains, 203–214
  introduced, 205, 209, 210
small scale differentiation of, 93
social structuring in individual dispersal, 67
soluble protein polymorphism in, 194
speciation in, 49
species diversity within, 320
specific environmental conditions and, 94

speed of genetic divergence between, 89
stability of, 329
stabilization, 394
stochastic accumulation of genetic variation among, 76
stochastic effects in, 88
stochastic event importance, 130
stochastic genetic divergence of groups within, 90
subdivided, 93
subdivided, genetic characteristics in patchiness, 94
subdivision, 151, 201, 320, 389
  for allele frequencies, 320
  and fixations, 320
subpopulations from barriers, 93
success with little or no genetic variation, 149
survival, 241, 442, 443
  in small reserves, 440–442
  variation at loci and, 149
  in zoos, with inbreeding (generations), 172, 173
  in zoos, with no inbreeding (generations), 172, 173
susceptibility to new strains of pathogens, 367
swamped by another species, 313
temporal changes in distribution, 308
tolerant to heavy metals, 83, 84
two species
  interactions, 327–329
  phases, 327
  saddle point, 329
  stable critical point, 328
  unstable critical point, 328
types and genetic behavior variation, 326
undomesticated
  deleterious alleles in, 178
  inbreeding depression in, 178
  inbreeding rate in, 178
  planned matings in, 178
upgrading of, 438
under strong directional selection, 89
variability in ecosystems, 334
in variable environments, 142
viable, 118
viability of, 118
  in evolutionary sense, 387
  selection in different, 321
vitality, in evolutionary sense, 387

vulnerability, 436
  to extinction, 155
well-adapted, 331
  to steady-state environment, 331
  in wild, 435
Population(s), animal
  adjacent, differing in resistance to toxins, 88, 89
  differentiation by population structure, 88
  differentiation by selective pressure of warfarin, 88
  differentiation in, compared with plant, 85
  dispersal in gene flow of, 85
  genetic differentiation factors, 88, 94, 95
  genetic differentiation with limited gene flow in motile, 85
  genetic structure of, 89
    of adjacent, 88, 89
    elements in establishing, 89
  genetic subdivision of, 90
  limited migration between, 88
  man-made barriers to movements of, 86
  man's alteration of selective forces in gene flow of, 86
  migration between, 89
  short distance environmental gradients and, 94
Population behavior
  oscillatory, 326, 327
  stable, 326, 327
Population biologists, 375
Population bottleneck
  constraints imposed by man, 197
  and founder event effects, 197
  extinction and, 244
  extreme, 191
  founder events, 101
  genetic drift, 191
  level of genetic variation and, 142
  molecular techniques in diagnosis of, 187
  in natural situations, 197
  on oceanic islands, 197
  severity of, 232
    and electrophoretic assay, 149
  Sewall Wright effect, 191
  sudden genetic change and, 192
  survey of heterozygosity, 232
  random drift, 191
  recolonization after, 443
Population(s), butterfly
  allozyme frequency evaluation, 157

of different ecotypes, 157
  differentially susceptible to extinction, 157
  ecological factors, 157
  egg laying in plants, 154
  phase relationships with climate, 154
    mortality and, 154
    small population size, temporary, 156
  plant relationships, 154
Population conditions (model)
  constant dispersal rate per generation, 71
  constant population size, 71
  equal probability of mating (monogamy), 71
  equal sex ratio, 71
Population crashes, 436
  ecological overload, 324
  of insect populations, 420
  of plant populations, 420
  of rodent populations, 420
  vulnerability to, 438
Population decline, elk, 438
Population decrease, 326, 327
Population density
  in temperate zones, 429
  in tropics, 429
  vagility of individuals, selfing and, 204
Population differentiation, 74, 94
  of adjacent populations, 78
  human impact and, 94
  of insects, 86
  on microgeographic scale, 78
  of plants, 82
  on a small scale, 87
  speed of, 94
  successional stage and alleles in populations, 337, 341, 347
Population divergence, 86, 97
  between adjacent populations, 78
  by random genetic drift, 66
  by restriction in gene flow, 78
  speed factors, 93
  by structures, 78
Population diversity, 319–332
  within and between populations, 363
  within species, 319
Population dynamics, 155. *See also* Social behavior
  adult sex ratio, 172
  butterflies, 154, 155, 161
  ducks, 293

dynamic/genetic interactions, 153
effective population size (N_e), 172
extinction, 155
genetics and, 161
group vulnerability to extinction, 117
juvenile survival, 172
migration patterns by sex, 321
mortality, 117, 118, 172
numbers in breeding activities, 118
numbers in foraging activities, 118
population size (actual) (N), 172
population survival, 20 yrs., 173
reproduction behavior and extinction, 118
species development, 153
young
  inbred, 172
  noninbred, 172
Population expansion, barriers to, 442
Population extinction, 114. *See also* Species extinction, Species vulnerability analysis
  compared with species extinction, 113
  dispersal mechanisms and, 124
  ecological interactions and, 109, 124
  and extinction of rare species by decrease in population size, 124
  genetic factors in, 109, 153
  habitat size (area) and, 124
  on islands, 124
  isolation or resource/habitat disappearance and species extinction, 124
  population size and, 109
  by stochastic ecological events, 113
Population, fish
  adjacent populations, 89
  selective pressures within, 90
  separation by barriers, 89
Population founding, from single individuals, 199
Population genetic variation, 127–131, 250, 330
  abundance, 151
  among local populations, 370
  evolution in adaptation to new hosts, new predators, new parasites, and new prey, 370
  in gene pools, 141
  levels, 373
  little, 149
  from mutation, 128
  and selection, 133–135

and survival, 49
within, 319
Population geneticists
  population terminology and, 39, 40
  use of term population, 39
Population genetics
  allelic heterozygosity, 194
  control of dynamics, 314
  effective population number, 44, 45
  experimental, 191
  extinction and deterioration in, 118, 119
  genotype and allele frequencies, 40
  Hardy-Weinberg equilibrium, 41–43
    causes of deviations (sampling error), 43
  inbreeding, 45
  to maintain genetic variance for genetic adaptation, 14
  to maintain minimum population size to preserve fitness, 14
  in management planning, 442
  natural, 191
  parameters, 101
  selection and gene flow, 370
    model, 370
  theoretical, 191
  theory, 40, 51, 52, 64, 260
    influence by individual genetic variability, by gene pool variability, 261
    selection operative in small founder populations, 261
Population growth rate
  carrying capacity overshoot, 323, 324
  equations, 323, 324
  equilibrium, 324
  few numbers in new niches, 324
  self-damping behavior and, 324
  slow, 435
Population growth regulation
  pathogens, 324
  predators, 324
  self-damping behavior, 324
Population inbreeding, 49, 68, 70–73, 201, 215
  by geographic subdivision, 201
  inbreeding depression in, 241, 434
Population(s), insect
  biological interaction between, 86
  differentiation in, 86
  divergence of, 86
  from selective factors, 86
  genetic divergence in, 86, 87
  isolated, 87
  microdifferentiation of, 87
    adaptive modes in each habitat, 87
    every patch (tree), 87
    in limited gene flow between, 87
    with strong directional selection in, 87
  semi-isolation of, 86
  small, 86
Population, isolated, 49, 401, 319
  consequences, 49
  by distance, 82
  by human activities, 78
  mechanisms, 49
  by structures, 82
Population longevity
  few hundred years, 443
  function of space, 440, 441
  level of demographic protection and, 440, 441
Population loss
  genetic significance of a population portion, 343
  of an entire, 343
  genetic variation loss and, 313
    evolution by environmental crises and species-level extinction from, 313
Population maintenance, endangered species, 14
Population management
  artificial selection avoidance, 435
  of biological properties affecting speciation mode, 106
  breeding systems, 139
  breed to maximize heterozygosity, 196
  cytogenetic methods, 196
  economically important species, 105
  effects of inbreeding depression, 137, 139
    mitigation, 139
  effective population size increase, 68
  electrophoretic methods, 196
  by environmental manipulations, 331
  evolutionary genetics and, 45, 46
  family size equalization, 68
  of fecundity, 199
  of fertility, 199
  fisheries stocks, 105
  founding populations, 331
    inadvertent introductions, 331
  game animals, 105
  of genetically altered populations, 199

Subject Index   699

geographic diversity, 150
habitat diversity, 150
heterozygosity, 196
heterozygous advantage (fitness), 139
inbreeding, 151
  coefficient level in, 435
individual migration control, 331
individual population control, 331
individuals culled, 196
isolation affecting speciation and genetic divergence, 106
maintenance of valued natural characteristics, 199
minimization of mutagens, 130
multiple starts, 331
monitoring, 331
outbreeding maximization, 196
population size and structure, 130, 150, 151
population size maximization, 198
removal of genetic load, 199
reproductive capacity, 137, 150, 151
role in conservation, 2
selection against deleterious alleles, 197
selection types, 133–135
self-regulation, 331
sibling species prediction, 106
of vigor, 199
Population management goals
  conserve operative natural selection, 189, 190
  preserve intrinsic biological state, 190
Population manager
  conservation ethic, 199, 200
  in preservation of characteristics of species, 199
  retention of field of genetic variability, in fecundity, 191
    in fertility, 191
    in somatic vigor, 191
Population number. *See* Population size
Population recovery, 438
Population reduction schemes, 332
Population regulation
  density-independent, 154
  "mosaic" pattern, 158, 159
Population size, 38
  allelic frequency divergence, 55
  allelic frequency relationship, 331
  animals reproducing, 390
  area limitations, 49

area size, multispecies relationships and, 429
in artificial populations, 198
avoidance of fluctuations in, 182
biomass and heterozygosity levels in, 139
bottleneck factors, 435
bottleneck time, 155
  underestimation, 155
bottlenecks, 154, 155, 187, 435
buffering, 155
in captive populations, 198
catastrophic events and, 126, 127, 187
by census in areas of known size, 422
change over time, 323
change rate, 322
with competitive interactions, 327
constant, 71
with cooperative interactions, 327
critical size, 389
in cultivated populations, 198
death rate and, 38
decline to extinction, 163
decline over time, 187
declining, 153, 443
degree of reproductive isolation and, 64
diversity, 315
effect on heterozygosity level, 365
effective, 172, 389–392, 421
  inbreeding rate and extinction, 118
effective for mutation, 93
estimates of, 422–427
expansion, 250
extinction and, 109, 153, 155, 157
extinction in reduction of, 364
final size in new, 182
fluctuation, 117, 153, 155, 162, 163, 392
forest trees, 6
founders, 388
founding by human agents and, 187
gene frequency interactions, 153
in generation overlap, 5,6
genetic assimilation by large, 289
  in hybridization, 289
and genetic differentiation, 88
genetic factor and, 118
genetic variability and, 155
genotype exploitation and, 151
  reproductive capacity and, 151
  zygote production, selection and, 151

geometric increase, 38
groups of equal size in neighborhood, 69
growth of, 324
  and new neighborhood establishment, 70
  hereditary characteristics and, 152
heterosis and, 326
heterozygosity and, 142
heterozygotes with fitness heterosis and, 326
heterozygous advantage and, 139
historical, in areas of known size, 421
ideal, for long term, 443
and inbreeding effects, 175
increase, 7, 252
  by migration use, 14
  before selection, 254
  during selection, 254
individuals (N), 53, 56, 389
initial number, 150
initial zygote number, 150
large. *See* Population size–large
limitations, 70
maintenance, 384
maximum possible, 150
  maintenance, 150
minimum, 150, 421
  for social animals, 117, 118
  viable, 443
modest, 151
and mutagens, 130
natural colonizing event and, 187
in neighborhoods, 69
number restriction, 201
numbers, 6, 153
  butterflies, 153
  of individuals in new, 182
optimum, 421
oscillation of, 327
in park areas
  large carnivores, 422, 427, 428
  large herbivores, 422, 424–426
  small herbivores, 422, 423
in park and range size, 415
peaks in sibling species, 99
plant distribution restrictions, 6
plant mating restrictions, 6
random fluctuation in, 117
  extinction and, 117
  small, 117
ratio of, to census population size, 45

recovery in park and reserve sizes, 438
reduction in, 187, 209
reduction effects, 209
reduction of outbreeding species to inbreeding depression, 364
  extinction and, 364
  loss of fitness, 364
relationship to park and reserve size, 440–445
reserve size and viable, 183
restricted mating, 5, 6
restriction by habitat availability, 150
sex ratio effects and extinction, 117
for short-term maintenance, 183
significant factors in determination of, 390
small, 116–118, 390. *See also* Population size–small
of small mammals, 139
  heterozygosity and, 139
species variation in area requirements and, 430
and stochastic events, 130
subpopulation size, 93
sustained by habitat, 443
temporal sequence of site disturbance and, 326
underestimation, 155
variation, 209
  behavioral complexes and, 329
  by weather pattern, 157
variability, 333
Population size–large, 93
in adaptation, 93
  genetic diversity, 93
assimilation of small, 289
characteristics from a previous bottleneck, 191
  managed populations, 191
genetic diversity, 314
genetic diversity loss and, 389
habitat for, 438, 439
high biomass and heterozygosity levels, 139
longevity of species in, 441, 442
park and reserve level of protection for, 438
with polymorphism, 137
protection levels for, 438, 439
in reduction, 319
resources for, 438, 439
small-bodied animals, 440–442
  in small reserves, 440–442

small-bodied mammals, 442
species diversity distribution and, 319
without barriers to mating, 319
without genetic distinctions, 319
Population size—small
  breeding in zoos and, 172
  captive, 435
  captive, inbreeding, 434
  change in genetic composition of, 187
  colonization success and, 436
  difficulty of maintaining, in zoos, 183
  exchange of individuals and panmixis in, 69
  expansion, 390
  fixation for a single allele at loci, 68
  founder effects, 209
  gene flow lack and, 436
  genetic assimilation in hybridization, 300
  genetic diversity maintenance, 49, 436
  genetic drift, 9, 209
  genetic variability loss, 68
  heterozygosity low with, 68
  inbreeding, 165, 183, 436
  inbreeding effects in, 250
  insular, 304
  long-term prognosis, 436
    outcrossing species, 436
  loss of, 436, 437
  mammal, park/reserve size and, 420
  management of, 401, 436
  management strategies, 68
  minimization of mutagens, 130
  park and reserve size and, 440
  polymorphism high overall, 68
  probable lifespan, 173
  recessive genetic load manifested, 139
  recolonization success and, 436
  remnants, 187
  reproduction in, 191
  semi-isolated, 88
  size decline over time, 187
  small numbers of individuals, 205
  species diversity distribution, 319
  survival, 436
    and expense in maintenance, 436
  time and genetic drift in, 232
  in zoos, 184
Population stabilization
  carrying capacity in demographic management, 394
  prevention by regulating reproduction, 394
  removal by regulating age class for survival, 394
Population structure
  balance system vs. classical system, 197
  balance view, 194, 195
  change in, 211
  classical view, 194
  effect on heterozygosity level, 365
  natural populations and, 130
  plant
    clumped distribution, 341
    genetic structure nonrandom, 341
  selfing with mate unavailable, 204, 205
Population subdivision
  different selection pressures, 320
  effective population size maximization, 389
  fixation of alternate alleles, 320
  genetic drift minimization, 389–392
  genetic material exchange, 389
  inbreeding depression minimization, 389
  "neighborhood," 389
Population units
  family or equivalent, 432
  individual, 432
  multiple population, 432
  population, 432
Population vulnerability, in short term, 438
Predation
  human activities and, 119
  new, 438
  rates, physiological limitations, 325
Predators
  animal, loss of fear in absence of, 111
  in extermination of prey patches, oceanic islands, 112
  ground, 119
  large, 119
  loss of flight in absence of, 111
  near-ground, 119
Preservation
  of adaptively differentiated populations, 95
  biotic diversity, 374, 400

captive breeding programs, 241
distinctiveness of groups, 435
dynamics of species evolution, 330
of ecosystems, 350, 351
ecotype, 2
of endangered species, 190
of endangered subspecies, 199
entire ecosystems, 350, 351
genetic changes and, 199
genetic diversity, 281, 374, 401
  in multiple populations, 318
  in sea turtles, 281
  of single species, 438
  of subspecies, 362
genetic variability, 253
genotypes, 317, 318
of a genotype array, 313
global strategies, 374
goals, 190
of habitat, 436
improvements in populations and, 191
levels of, 318, 432
multiple managed populations, 314
natural diversity, 313, 314, 318, 330
  dynamics of species evolution and its structures, 330
objectives for manager, 313
of parks, 436
of particular characters, 198, 199
phenotypes, 317, 318
political action, 12, 150
  on suitable habitat, 150
population diversity, 50
pure races, 307
rare alleles, 347
species, 9
  characteristics, 199
  diversity, 319
  structures, 330
static, of a species, 2
target species, 13, 199
time scale, 13
of traits, 200, 372
  in different habitats, 372
  in ensemble of populations, 372
  under local selection, 372
useful genetic variation in outcrossing species, 63
Primates, 405, 406
effects of inbreeding in, 184
juvenile mortality in, 184
Productivity, strains of *Drosophila*, 144

to produce biomass, 144
in utilization of environment, 144
Progeny
inbred in outcrossed species, 205
outcrossed, 205
Propagation
of artificial sample, 191
asexual, 436
in botanical gardens, 103
of captive species in numerous isolated populations, 68
in zoos, 103
Protected areas
habitats, 13
size of, 415–419
small
  cultural role, 13
  educational role, 13
  recreational role, 13
  scientific role, 13
Protection
amount, 362
assessment of success in, 415
in botanical gardens, xix, 2
from catastrophic events, 440
faunal assessments for demographic planning, 439
in fishing reserves, 1, 2
floral assessments for demographic planning, 439
in individual breeding populations, 287
  limitations, sea turtles, 287
legislation for, 437
  noninterventionist, 437
  on species, 437
  provisions, 437
by level of complexity of demographic unit, 430
long-term, 14, 443
management levels of, in parks and reserves, 437
nature reserves, 14
in park planning, 430, 431. *See also* Demographic levels of protection
period, genetic diversity and, 435
portion of existing populations, 362
plants and animals. *See also* Conservation
sea turtle statutes, 278
short-term, 443
in small parks, 436
of species, 415
type of, 444

in wildlife parks, 2
with time, 444
in zoos, 2
Protection levels. *See also* Demographic protection levels
degeneration of, 437
for evolution, 439
in geologic time, 439
intermediate sized parks and, 438, 439
large sized parks and, 438, 439
legislation provisions and, 437
for long-term survival, 439
population diversity in, 438, 439
for short-term survival, 438
for a single population, 438
for survival, 432–439
Protein electrophoresis, 97
for detection of protein variability, 194
limitations of, 230, 231
theoretical considerations, 231, 232
Protein polymorphism
detection by electrophoresis, 194
in wide-ranging outcrossing animals and plants, 209
Protein synthesis, 32, 33
amino acid residues, 32
DNA, 32
mRNA, 32
polypeptides, 32
proteins, 32
Pungence
indicator of genetic information movement across suture, 270
Punjab, India, grass species, 271
Purebreds, distinctions, 289

Quantitative monitoring, factors in extinction, 112, 113
Quantitative traits
of different life forms, 344, 345
ratios of between population variance, 344, 345

Rabbits (Leporidae), myxamatosis, 127
Rama, Nicaragua, 104
Rancho Nuevo beach, Gulf coast of Mexico, sea turtles, 284
Random events affecting genetic divergence of populations, 66
Random genetic drift, 43, 191
theory, 365
Random mating. *See also* Mating, random
model, 52
panmixis and, 8
survival of genotypes to, 320
Range
changes
with competition, 115
with long-term climatic trends, 115
with predation, 115
with short-term climatic trends, 115
decline, 114
die-off, 114
expansion, warblers, 296
mosaic of habitat patches, 112, 113, 118, 119
patch death, 114
populations and, 114
restricted geographic, 209
shrinking, 114
Rare and endangered species, 164, 242, 295
captive habitat for, 376
captive reproduction of taxa, 386
habitat patches and extinction of, 118, 119
plant, 224, 225, 569
research on biology of, cost, 9, 10
selection of, for preservation, 9
of small geographic area, 115
taxa, 333
temporal persistence, 115
Rare and endangered species preservation. *See also* Captive breeding programs, Conservation, Zoo breeding programs
adaptation to captive environment, 259
genetic alteration from wild state, 259
individuals selected for domestic traits, 259
reintroduction capabilities, 259, 260
survival chances, 260
Rare genes, in gene pool shift, 252
Rare genotypes
recombination from, 130
survival under environmental change and, 130
Rarity (of species)
extinction vulnerability, 119
to other species of similar body size, 115

patchiness of resource and, 119
taxonomic group, 115
trophic level, 115
Recolonization
capabilities of tropical forest birds, 119
of habitat patches, 162
after population bottleneck, 443
repeated, 205
small genetic differences and, 159
Recombination
genotypes for survival through, 130
patterns of partial inbreeding, 193
patterns of self-fertilization, 193
polyploidy and, 193
rare genotypes by, 130
restriction on, 193
Recombinational load, gene complexes, 304
Recruitment
increase of inbreeding without, 67
new genetic material, 67
Redwoods National Park, Calif.
carrying capacity, 420
elk in, 438
levels of protection in, 438
size, 420
Reestablishment, 160
factors preventing, 159
Refuges
fauna survival, 13
flora survival, 13
Refuges and parks boundaries and species areas, 104
Region of contact. *See* Zone of contact
Regions of monomorphism in plants, 219
Reintroduction
to former habitat, 14
habitat requirements, 159
survival chances, 159
Relatedness
among mates, 69, 70
slowing of, 69
Relative advantage, outbred/inbred species, 137
Relict populations
breeding efforts, 313
conservation problems, 308
Relict populations, genetic assimilation, 308
Relict wild populations, existence of progenitors, 190
Replication, no true biological, 195, 196

Reproduction. *See also* Breeding, Inbreeding
age at, 384
average litter size, 398
critical age for, 44
differential, 135
diploid species, 190
eggs, 118, 190
episode of (generation), 190
extinction and, 118
fast, 439, 440
fertilization, 191
gametes, 190
genetic load, 199
inbred, 440
K-selected species, 118
low capacity, 137
meiotic recombination, 190
Mendelian heredity, 190
phenotypic plasticity, 219
prevention, 394
rate of, 7
seasons for, in sibling species, 98
self-fertilized, 436
sexual polyploids, 440
slow, 439
sperm, 190
Reproduction mode. *See also* Mating system
asexual, 226, 365, 439
cross-fertilization, 190, 193
diploid, 193
inbreeding, 304
outbreeding, 304
outcrossed, 205, 209, 218
outcrossing, 220, 269
parthenogenesis, 193
in plants, 220, 221, 499
selfed, 205
self-fertilization, 215, 218, 219, 220, 269, 440
sexual, 193, 219, 365, 442
vegetative, 193
Reproduction rates, physiological limitations, 325
Reproductive barriers
amounts of genetic divergence in reproductive isolation, 357
behavioral incompatibilities, 357
crossability, 318
developmental incompatibilities, 357
electrophoretic studies, 358
gene, chromosome rearrangement analysis, 358

Reproductive biology, 319
Reproductive capacity
  expansion, 150
  genotype exploitation and, 151
  low, and exploitation of desirable genotypes, 151
  natural populations and, 150, 151
  population size and, 151
  scale of zygote production, 130
  two avian species, 308
Reproductive characters
  balanced polygenic control, 198
  natural selection and, 198
  shift after founder event, 198, 199
Reproductive compatibility
  hybridization, 308
  sympatry and, 308
Reproductive effort, net survival to breeding stage, 150
Reproductive event, genetic variability in, 199
Reproductive isolation
  degree of, from allele frequency data, 64
  genetic divergence and, 357, 358
  in green lacewings, 98
  numbers of gene differences, 97, 98
  partial, 292
  in sibling species, 97, 98
  techniques of identification, sibling species, 99
Reproductive output. See Fecundity
Reproductive season differences, sibling species, 98
Reproductive success
  decrease with inbreeding, 179
  reliance on passive agents, 220
  in plants, 220
Reproductive systems
  diploid cross-fertilizing, 199
  outbreeding, 225
Reproductive value
  age specific patterns in survival and fertility, 384
  in new population establishment, 384
  sex-specific patterns in survival and fertility, 384
Reptiles
  adaptation in absence of predation, 111
  fauna, 420
  loss of fear, 111
  migration of larger, 9

population decline, 161
population extinction, 161
Research
  on endangered plants (lists), 10
  on key species, 10
Research problems, sea turtles, 279
Research studies, by genetic markers, 33
Reserve size. See Park and reserve size
  extinction and, 183
Reserves
  area for animals, 6
  area for plants, 6
  large tropical trees, 6
Reserves, large, conservation capabilities of, 2, 432
Reserves, small, conservation capabilities of, 2, 432
Resistance to natural enemies
  genetic variance among populations in, 373
  genetic variance within populations in, 373
Resource demands, small, 440
Resource exploitation, extinction and, 109
Resources
  competition, 11
  fruit, 119
  natural populations, 14
  nectar, 119
  nectar and reproduction in butterflies, 154, 155, 161, 162
  patchy distribution, 118, 119, 442
  pressure increase, 11
  seasonal fluctuation of and extinction, 119
  supplementation, 11
  vulnerable to climatic fluctuation, 119
  wild relatives of domesticates, 14
Reunion and Seychelles, Indian Ocean, sea turtle farming and ranching, 282
Ribosomes, amino acid residues and, 32
Rio Grande River, duck hybrids, 294, 295
RNA (ribonucleic acid), 32
Roadside populations
  demographic attributes, 218
  genetic attributes, 218
  mating system shifts, 218
  weedy plants, 218

Rodenticides, warfarin, 88
Rodents
  area for survival, 420
  survival of populations of, 420

Saint Catherine's Island Survival Center of the New York Zoological Society, facility enlargement, 383
Sampling drift, in small populations, 209
San Antonio Zoo, Texas, 251
San Diego Wild Animal Park, facility enlargement, 383
San Francisco, Calif., butterflies, 158
São Paulo, Brazil, birds, 118
Savannah River Nuclear Plant, Aiken, South Carolina, 89
Scale, genetic variation, 130
Scale, space, 137
  adaptive genetic divergence, 91
  microgeographic, 93
Scale, time, 137
  few hundred years, 443
Scientists, agreement on priorities, 10
Sea turtle aquaculture
  eggs from natural nesting grounds, 282
  farming and ranching, 282
  hatchling release compensation, 282
  hybridization dangers, 282
  markets for products, 282
  poaching encouragement, 282
  wild stock relationships, 282
Sea turtle biology, knowledge for conservation, 278
Sea turtle conservation
  beach colonization, 286
  captive breeding, 278
  commercial market elimination, 279, 286
  cost-effective methods, 279
  egg hatcheries, 278
  eggs and hatchlings to beaches, 283
  egg translocation, 286
  extinction threat, 279
  farming or ranching, 286
    questionable value, 286
  habitat protection, 286
  hatchling translocation, 286
  head starting, 278, 279
  hybridization, 278, 279
    dangers for turtles, 282
  definition, 278, 279
  human induced, 283
  with local stocks, 287
  with wild stocks, 287
  international agreements, 284
  mark and recapture program, 278, 279
  market control, 286
  nesting population establishment, 284
  nesting grounds protection, 279
  poaching elimination, 279
  predator control, 278, 279
  promote sea turtle excluder device, 279
    for shrimp trawls, 279, 286
  protection of habitat, 279
    for development, 279
    for feeding, 279
    for migration, 279
    for nesting, 279
  relocation of eggs/hatchlings, 278, 279
Sea turtle management, 285
Sea Turtle Specialist Group, International Union for the Conservation of Nature and Natural Resources (IUCN), 281, 282
Sea turtle systematics, hybridization and, 280
Seasonal variation and gene pools of *Drosophila*, 142
Secondary sexual characters, female behavioral responses, 198
Seed(s), 44
  dispersal, 321
  without fertilization (agamospermy), 269
Seed crops
  cultivated races and, 274
  genetic interaction, 274
  weed, 274
  wild, 274
Seed dispersal mechanisms
  association with mating systems, 346
  gene movement and, 346
  genetic distribution effect
    animal-attached, 337, 340, 341
    animal-ingested, 337, 340
    large, 337
    plumose, 337, 340, 341, 346
    small, 337, 340, 341
    winged, 337, 340, 341, 346
Segregational load, 204, 205

Selected animals
  birth weight, 258, 259
    inbreeding coefficient, 258, 259
  inbred, 255
  one parent inbred, 255
  trends, 256
  two parents inbred, 255
Selection
  allele frequency alteration by, 43
  of animals, 394
    preservation of genetic diversity and, 394
  animals for removal, 394
  animals for reproduction, 394
  artificial, 135, 198, 199
    avoidance of, 435
  balancing, 199, 372
  coefficient, 43
  on complex mating behavior, 133
  correlation effects, 371
  against deleterious alleles, 197
  density-dependent, 134
  density-independent, 134
  different genotypic arrays, 313
  differential directional, 60
  directional, 38, 39, 133
  disruptive, 134
    and gene flow, 94
  efficiency, 250
    for inbred, healthy animals, 250
  extinction and, 133
  fitness for survival by, 43
  founder effects in multilocus systems and, 197, 198
  frequency-dependent, 134, 368
  frequency-independent, 134
  gene flow and, 373
  generation by recombination, 194
    of superior heterozygotes, 194
  genotype contribution to next generation, 43
  of harmful alleles, 45
  against hybrids, 292
  inbred, healthy animals, 253
  maintenance by natural enemies, 368
  man-made, 126, 127, 151
  multiple-niche, 134
  natural, 126, 133, 135, 151, 198, 199
  in promotion of local polymorphism, 370
    for genotypes in best adaptation, 370
    to local conditions, 370
  reduction for generations of genetic diversity preservation, 133, 134
  scope for, in catastrophic events, 126, 127
  on simple morphological characters, 133
  speed of, 133
  stabilizing, 38, 39, 134
  strong, 89
  traits for adaptation to ecological environment, 365
  within population, 150
  zygote production, and entry into breeding pool, 151
Selection effects
  in different sexes, 321
  with localized mating, 321
Selection by inbreeding
  intensity, 253
  population size during, 254
Selection intensity
  deme size interactions simulation, 56, 58, 59
  migration rate interactions simulation, 56, 58, 59
  mN of 5 on 3 deme sizes, 56, 59
  relative amount of divergence and, 56, 58, 59
Selection-migration model, 320
Selection modes, operative on genetic variation, 127
Selection patterns, 49
  forms of, 321
Selection pressures, 162
  degree of, in genetic divergence, 88
  different for each local population, 86–88
  in diploid species, 194
  disruptive, 87, 92
  genetic differentiation and, 64
  in genetic variability use, 253
  of heavy metals in plants, 84
  in insects in heterogeneous host trees, 86
  insect species on plant species, 372
  large, 89
  in largemouth bass populations, 89
  man-imposed, 90
  metal tolerance in plants and, 85
  new, by environmental change, 414
  rodenticides in animals, 88, 89
  strong, 88, 90, 93
  thermal effluents, 89, 90
  variation of response to, 46, 127
  warfarin, 88, 89

708 Subject Index

Selection regime, system of higher fitness of heterozygotes, 194
Selection regime classification, 134
  classification I
    frequency-dependent, 134
    heterozygote superiority, 134
    multiple-niche, 134
  classification II
    disruptive, 134
    stabilizing, 134
Selection for removal
  age class, 398
  sex class, 398
Selection stabilization
  average phenotype, 134
  genetic consequences, 135
  reproduction, 134
  survival, 134
Selection type
  directional, 133
  frequency-dependent, density-independent, 134
  density-dependent, frequency-independent, 134
  natural populations and, 133–135
Selective equivalence (of neutral value) in allozyme maintenance, 33
Selective neutrality, 61
  advantages for models, 61
  in deme genetic variation, 54, 55
  divergence amount expected with, (model), 53, 57
  null hypothesis, 61
Self-fertilization, 208
  in colonization of new habitats, 187
Selfing
  alleles to fixation, 204
  annual habit, 205
  autoselection model, 203
  colonizing ability, 210
  evolution of, 203–205
  evolutionary correlations, 203
  fate, 214, 215
  evolutionary model, 212, 214, 215
  fitness, 204
  founder numbers and, 204
  genetic consequences, 204
  genetic interactions, 204, 210–214
  heterozygosity loss, 221
  homozygosity, 204
  hybrid vigor, 204
  in inbreeding species, 204
  individual vagility (population density) and, 204
  in low vagility species, 210
  mate availability and, 205
  mate finding and, 210
  in outbreeding species, 204
  outcrossing generations rates, 222
  outcrossing with, 221
  recessive mutation, 204
  reproduction assured, 210
  segregational load, 204, 205
Selfing rate, variability, 211–213
Selfing species
  consist of few strains, 211, 213
    despite hybridization, 211, 213
    mutation, 211, 213
  from outcrossers (model), 212, 214, 215
Selfing strains
  genetic interactions of, 210–214
Semispecies, partial reproductive isolation, 358
Senegal, grass species, 271
Separate populations (of species), allele maintenance by migration between, 320
Sequence data (DNA) (proteins)
  costly, 357
  for higher taxonomic levels, 357
Sequential electrophoretic procedure, 354, 357. See also Electrophoresis
Sewell Wright effect, 191
  genetic drift, 191
  random drift, 191
  from outcrossers (model), 214, 215
Self-regulation, genetic diversity and, 326
Sex
  chromosomes, 27–29
  determination, 27–29
  female, 322, 386
  male, 322, 386
  one in genetic diversity control, 322
  one in polymorphism protection, 322
Sex-limited autosomal traits, secondary sexual characteristics, 29
Sex-linked traits, 28, 29
  recessive traits, 28
Sex class
  in fertility, 384
  in survival, 384
Sex chromosomes, complements, 28, 29

Subject Index   709

females, number of, 28, 29
genes on, 27
inheritance in, 27, 29
males, number of, 28, 29
Sex determination
  in birds, 29
  in dioecious plants, 29
  in insects, 28, 29
  in mammals, 28
  in nonmammalian vertebrates, 29
  temperature, in sea turtles, 278
Sex differences
  in allelic polymorphism, 322
  among population migration rates, 322
  in genetic diversity, 322
  management, 322
  measurement, 322
  use in management, 322
Sex ratio(s)
  differential in captive populations, 180
  distribution and population size, 117
  effective population size and, 390
  equal, 71
  founders, 242
  management, 386, 390, 391, 400
  regulation, 391, 400
  skewed, and breeding rate, 117
  unequal, 242
Shannon information index, 350
Shenandoah National Park, Va.
  carrying capacity, 420
  size, 420
Sibling species, 96–106
  chromosomal races, 104
  for conservation, 97
  conservation boundaries, 104
  distribution, 104
  contact zones, 100–103
  distinction, 97
  divergence between, 49
  Drosophila, 98, 99
  ecological characteristics overlap, 99
  electrophoretic studies, 97
  endangered through habitat destruction, 104
  evaluation techniques, 105
  fireflies, 99
  fruit flies, 98, 99
  genetic differences, 97, 98, 99
  genetic distance and morphological distance, comparison, 97
  green lacewings, 98
  identification, 99, 100, 106
  lizards, 99
  for management, 97
  morphological differences, 97, 98
  morphological similarity, 98
  pocket gophers, 100–103
  prediction, 106
  separate evolutionary paths, 97
  small natural distributions, 104
  taxonomic evaluation, 350
  technical methods of identification, 105
  yellow bat, Central American, 104
Sibling species identification
  behavioral differences, 99
  biochemical analysis, 99
  discrete character differences (electrophoresis or karyotype analysis), 99
  mating displays, 99
  morphological differences, 99
  parasite species, host-specific, 100
  species-specific songs or calls, 99
  taxonomic groups, special characters in, 99
Sibling species management, 103, 104
Sibling species morphological differences, 99
Simulation experiments, on population size, 150
Simulations, Monte Carlo, in allele frequencies, 52
Single population
  heterozygote fitness vs. homozygote, 322
  samples, loss of initial levels of diversity, 319
  single species in, 332
Single species
  effect in ecosystem, 127
  introduction, 127
  removal, 127
  target, 127
Size
  of founding group, 192
  gene pool, 187
  in transplanted fish populations, 359
Sizes, differential in captive populations, 180
Skin color, genic effect at several loci, 37
Slugs. *See* Terrestrial slugs
  outcrossing, 210

Small animal populations, reduced in size, 187
Small animals, butterflies representative of, 161
Small carnivores, habitat capacity for, 420
Small founder group sizes, avoidance of inbreeding, 242
Small mammal populations, 420
  heterozygosity, 139
  heterozygous advantage in, 139
Small organisms. *See also* Species, small-bodied
  protection requirements of, 440
  spatial requirements of, 440
Snakes, predators on birds, 119
Social behavior, 117
  dysfunction of, and catastrophic mortality, 117
  extinction and, 117, 118
  and fatal decline in numbers, 117
Social organization, levels of complexity in, 422
Soil chemicals, mercury, 127
Soil type
  clay, 101, 102
  sand, 101, 102
Soils
  gopher populations and, 101, 102
  heavy metals in, 83
    affecting genetic differentiation, 83
Somalia, Speke's gazelle, 242, 260
Somatic vigor, retention of, 191
Songs or calls, species-specific in sibling species identification, 99
South America, 367
  biomes of, 422
  mammals, 49
    extinction by North American biota, 365
South Australia, plant flowering-time shift, 218
South Pacific islands, birds, 292
Space
  availability of, 444
  competition for, 435
  demands of vertebrates, 443
  function of, 440
    in longevity of species, 440
  for large-bodied species, 442
  and long-term species survival limit, 442
  and species survival limit, 441
Space requirements

for large carnivores, 415, 422, 427, 428
for large herbivores, 422, 424–426, 429
for large vertebrates, 416
mammals, 422
for small-bodied species, 415
small herbivore species, 422, 423
for ungulates, 415, 422
Spatial demands, small, 440
Spatial limitations, reserve life expectancy, 13
Speciation. *See also* Evolution
  by allopatric divergence, 106
    by rapid divergence with founder events, 106
  among avian species, 292
  from divergence, 187
  hybridization and exchange of genes, 289
  influence on, 289
  modes, 106
  occurrence with little divergence at loci sampled by electrophoresis, 358
  patterns, 106
  populations in reproductive isolation, 105
  rates, 106
  reproductive isolation, 97, 105
  sibling species and, 49
  time scale of, 112
Species. *See also* Key species, Populations
  abundance and mutation effects, 130
  adaptability, 126, 127
  adaptation, 95, 187
  adaptation to ecological variables, 95, 365
  adaptive differentiation of, 93
  adaptive genetic differentiation, 92
  allopatric, 101
  amounts of genetic variation, 131
  antagonistic interactions among, 373
  area limitations to population size, 49
  area required for large, 6
  asexually propagated, 436
  assessment of success of, 415
  assessment of vulnerability of, 119, 415
  average fitness, 322
  biological constraints, 313

Subject Index 711

body size, 115
capacity for evolution, 443
change over generations, 38, 39
character variation, 133
characteristics based on complex polygenic balances, 197
coevolution, 373
commonness of, 115
of convergent niche type, 440
criteria for classification, 358
cross-fertilizing, 192, 193
decline of abundance, 114
decline of fitness, 114
decline by habitat patch, 113
decline in numbers, 112, 437
decline in range, 112, 114
deleterious alleles in, 195
demes of a (model), 53
densities, 422
diploid, 190, 193, 194
dispersal powers and isolation, 49
divergence potential, 439
    with large park and reserve size, 439
diversity, 137, 429
diversity in ecosystem, 318
diversity of, 415, 439
    maintenance of, 415
diversity within, 318, 351
    measurements, 351
dominant in ecosystem, 10
dominant, and extinction, 10
ecosystem-dependent, 127
in environmentally distinct patches, 320
evolution of, 319, 439
evolution curtailment with small population, 319
evolution with large population maintenance, 319
evolutionary capacity, 320
existence and survival, 126
extinct, 118
extinct in wild, 187
extinction, 103, 160, 313, 414, 437
extreme environmental change adaptation, 95
fast-evolving, 146
few, in adaptation to extreme radical environmental changes, 94, 95
forage available and, 442
fragmentation of, 205
frequency (plants), 6
gene effect variance, 329

gene pools, 351, 414
generation time, 442
genetic content variation, 330
genetic differences, levels of, 51
genetic differences in isolated populations of, 66, 67
with genetic divergence among populations, 319
genetic diversity of, at a locality, 439
genetic diversity maintenance, 49
genetic diversity preservation, 439
    with large park and reserve site, 439
genetic diversity in survival of, 429, 430
genetic flexibility, 93
genetic options, 217
genetic population structure of, 39
genetic resources of, 51
genetic systems, 193
genetic variability, 330
    by mating pattern, 330
    and reproduction biology, 330
genetic variability in, 196
genetic variation, 63
genetic variation maintenance, 77
genetic variation within, plants, 82
genetically integrated groups, 351
geographic dispersal of, and environmental studies, 142
geometric increase in individual numbers of, 38
groupings below species level, 114
habitat increase for a, 437
on heavy metal-contaminated sites, 83
heterozygosity of, 142
    geographic populations and temperature variation, 142
high community impact, 10
highly specialized, 439
human impact on, 437, 442
important in regulation of ecosystem, 10
inbred, 244
    deleterious alleles with, 244
inbreeding avoidance, 139
inbreeding in normally outcrossing, 178
initial contact, 308
interdependence, 127
introduction of one in ecosystem, 127
introgression, 271

isolation and, 49
key, 10
keystone, 442
   loss of, 442
large, 6
   in large areas, 115
large-bodied, 441, 442
   area requirement, 442, 443
with large continental distribution, 300
with large number of zygotes per breeding pair, 130
in large populations, 319
life history characteristics, 109
long generation times, 93
   and de novo mutations, 93
long-term evolution potential, 439
   with large park and reserve size, 439
longevity, 415, 444
   prediction of, 415
low density, 10
maintenance of genetic diversity in populations of, 414
management implications of, 319
   in preserving species natural diversity, 319
managerial intervention in preservation of, 6
mating system of, 434, 499
migratory behaviors, 121
   extinction and, 121
monogamous, 77
morphological distinctions, 97
multiple populations of, 439
mutational input, 129
mutation rate of, 93
natural stability of, 430
number for protection, 435
number in community, 350
number surviving extreme conditions, 94
numbers in areas of known size, 421
origin of, 97, 125, 192
outbred, 244
   deleterious alleles with, 244
outbred/inbred advantage, 138
outcrossing, 436, 441, 442
parapatric, 101
phenotypic plasticity and, 92
phenotypic variances, 148
polygynous, 77
polyploid, 269, 436
population size for survival, 6

population size and variability of, 319
preadapted genotypes of low frequency, 85
predisposition to extinction, 109
preservation, 241
preservation of evolutionary potential, 443
preservation of natural diversity and, 330
protection, 415
rapidly reproducing, 439
rare, 118, 119
rarity of, 115
rate of extinction, 109
rate of origin of new, 109
recessive alleles with inbreeding, 184
reduction in gene pool size and, 187
reestablishment of, in a habitat, 187
relative abundance of, 350
relative adaptation ability, 126
relative advantage outbred/inbred, 137, 138
removal of one in ecosystem, 127
resilience to humanity, 127
resilience to human pressures, 127
response to selection pressures, 151
secondary contact, 308
self-fertilizing, 414, 436, 440, 441
sexually reproducing, 178, 192
sexual reproduction, 441, 442
sibling, 96–106
of similar trophic level, 115
slow-evolving, 146
small-bodied, 440–422
with small demands for gene flow, 440
small effective population sizes, 93
   de novo mutations and, 93
in small management areas, zoos, 68
with distribution limitations, 67
with small resource demands, 440
with small spatial demands, 440
space requirements for, 415
specific genes, 319
specific traits, 319
structure in populations, 330
subdivision of, 330
subdivision level of diversity, 332
subspecies from, 318
summary indices of genetic variation, 365
   allozyme heterozygosity, 365

survival, 103, 414
survival mechanism, 319
sympatric, 98
target, 7
of a taxonomic group, 115
temperature effects on, 142
threat of extinction, 313
at top of trophic ladder, 441, 442
of a trophic level, 115
units of adaptation, 351
variability, 38
variation within, 40
viability, 387
vitality, 387
vulnerability, 119, 415
zonation and variation, 131
Species, animal, supply from the wild, 434
Species architecture, alteration of, 318
Species/area relationships, and island biogeography, 422
Species characteristics, for specific ecosystems, 439
Species conservation, subdivision of populations, 372
Species densities, histories of, 443
Species differences
  by allelic divergence, 319
  by cryptic mating barriers, 319
  by maintenance of migration limitation, 320
  by selective differences, 319
  transient, 320
Species diversity
  distribution, 319
  among divergent populations, 319
  in parks and reserves, 415
  scale of ecological processes and, 121
  selective factors, 314
  variations in biotic environment, 314
Species diversity preservation, single population subsample and, 319
Species dynamics
  in diversity preservation, 330
  in evolutionary capacity, 330
Species evolution, dynamics of, 314
Species, exacerbating qualities of, in extinction analysis
  big (low r), 120
  highly social, 120
  insular, 120
  nonvagile, 120

rarity (clumped), 120
rarity (dispersed), 120
Species expansion from small isolated to large interbreeding, 319
Species extinction
  animal dependence on plants, 119
  compared with population extinction, 113
  dispersal abilities, 120
  habitat patch extinction and, 113
  plant dependence on animals, 120
  seed dispersal mechanisms and, 119
Species extinction, intrinsic factors
  behavioral dysfunction, 116–118
  demographic stochasticity, 116, 117
  genetic deterioration, 116, 118, 119
Species genetic structure
  between component populations, 330
  within single polymorphic system, 330
Species homeostasis
  within genetic loci, 330
  within individual genotypes, 330
  within populations, 330
Species interactions, 325
Species, large-bodied
  future prospects, 439
  level of demographic protection, 441, 442
  longevity, 441, 442
  park and reserve size and, 438, 439
  size of protected area for, 441, 442
Species maintenance
  long-term, 260
  short-term, 260
Species management
  maintenance for education, 435
  restoration to natural habitat, 435
Species mutualists, vulnerability of, 120
Species, plant, supply from the wild, 434
Species range, variable selective pressures, in, 82
Species size
  area requirements for, 415, 416
  large, space for, 443
  management protection level and, 438
  small, space for, 443
Species, small-bodied. *See also* Small organisms

asexually reproducing, 439
park and reserve size and, 438, 439
rapidly reproducing, 439–441
sexual polyploid, 439, 440, 441
typically inbreeding, 439–441
Species specificity, in minimum viable populations, 118
Species survival
long-term, 443
park and reserve size and, 438
population size relationships, 415
reserve size relationships, 415
size of demographic unit, 444
short-term, 443
Species Survival Plan (SSP), controlled migration between zoos, 435
Species system, definition, 330
Species vulnerability, large groups and catastrophic mortality, 117
Species vulnerability analysis. *See also* Population extinction
factors of environmental change, 119–120
factors intrinsic to species, 119, 120,
factors that characterize species-to-environment relationships, 119, 120
Sperm, 41
Spoil sites, microhabitat differentiation and, 94
SSP program, Grevy's zebra project, Texas ranch, 383, 384
SSP programs, 376
Standard population model of inbreeding depression, 247
Stanford University's Jasper Ridge Biological Preserve, California, 153
State and county administration
cooperative agreements with, 438
on elk, 438
Statistical methods
allele frequencies from genotype frequencies, 40
genotype frequencies from allele frequencies, 40
allele frequencies, Hardy-Weinberg equilibrium, 41, 42
genotype frequencies, Hardy-Weinberg equilibrium, 41–43
allelic diversity among/within populations, 337, 338
calculation of effective population sizes with dispersal distances (neighborhood effective size equation), 70
calculation of genotypic ratios, 24
frequencies of expected genotypes, 27
frequencies of expected phenotypes, 27
gene diversity statistics (Nei), 336, 337, 347
inbreeding coefficient (accumulation of inbreeding), 70
in Mendelian inheritance, 37
neighborhood effective size, 70
number of alleles at each polymorphic locus, 337, 338
Punnet's square, cross with genotype Rr, 24
Shannon information index, 350
Spearman rank correlation coefficients, heterozygosity and polymorphism, decapods, 140
standard inbreeding regression analysis, 246–250
standard t-test, for juvenile mortality, 247
stepwise linear discriminant analysis, 233
Sterility, pocket gopher hybrids, 103
Sterilization, death equivalence, 139
Stochastic events, genetic structure and, 88
Stochasticity
demographic stochasticity, 115
genetic drift, 115
Stocks, comparison of inbred and outbred, 146
Strains
adaptation capability, 213
homozygous lines, 213
elimination of, 213
genotypic variance, 213
hybridization between, 213
heterozygous lines, 213
persistence, 213
of one, 213
reproductive variance, 213
homozygous lines, 213
type
allopatric, 213
sympatric, 213
Structure of plant populations, analysis for diversity, 314
Structures (barriers)
buildings, 78

cities, 85
dams, 78, 85
fences, 78
reservoirs, 85
roads, 78
Stud books, endangered species, 404
Studies, butterflies, 161
Study site area, birds in tropical forest habitats, 119
Subpopulations
  allele frequency change in, 201
  for analysis of variation in species, 40
  average heterozygosity reduction in, 201
  definition, 40
  ecological associations and, 372
  extinction rates, 205
  founder number, 205
  founder number, small, 201
  inbreeding in, 202
  random genetic drift in, 201
  repeated colonization, 205
  size, 205
  variance between, 202
  variance within, 202
Subspecies
  allelic diversity within a species, 360, 362,
    gene diversity analysis, 360
  changes of characters, 358, 359,
    in laboratory populations, 359
    in managed populations, 359
    in transplanted populations, 359
  classification, 360, 362
  color or size characteristics, 363
  conservation, 359, 362, 363
  definition, 358
  diversity and preservation of species, 362
  by electrophoretic data, 360, 363
  endangered, 199
  preservation of characters, 199
  environmental effects and, 358, 359, 363
  evolutionary change and, 358, 359, 363
  extinction, 114
  genetic complements, 360, 361
    within/between populations, 360, 361
  genetic components in change, 359, 363
  geographic variation, 359
  indicator of genetic diversity, 359
  maintenance and realization of, 200
  manager selection of, 359
  origin, 192
  plant varieties equivalent to, 358
  systematics, 359, 363
  taxonomy by geographic variation, 359
  validity of concept, 358, 359
  value, 360
  variation indicators, 359
    below species level, 359
    in species variation preservation, 359–363
  variations by species level preservation, 318
Subvariation
  reproductive biology, 314
  selective differences, 314
Sudan, weedy sorghum, 273
Surinam, sea turtle farming and ranching, 282
Survival, 38
  in areas of small size, 442
  to breeding stage, 150
  of demographic units, 439, 440
  in elimination of inbreeding depression, 255
  enhancement of chances for, 1
  environmental changes and, 49
  evolutionary, of species, 151
  and generations after inbreeding, 197
  genetic diversity and, 429, 430
  genetic variation and, 49, 127, 149
  heterozygosity and, 127
  human effort in target species, 332
  with inbreeding, 246
    for one year, 246
    for thirty days, 246
  large-bodied species, 441, 442
  loci critical for, 320
  long-term prospects, 439
  genetic diversity and, 414
  with introduction of unrelated animals, 174
    for large-bodied species, 439
  mutation and, 37
  parent to reproductive age (model), 52
  by park and reserve demographic protection level, 434–439
  in parks of given size, 443, 444
  phenotypic plasticity and, 219
  of polyploid species, 436
  of populations, 241

populations in short-term, 438
probability of, 247
   and level of depression, 247
   by propagation, 103
   by protection, 1
   and rare genotypes, 130
   of rats with warfarin, 88
   to reproductive age, 44, 384
      in zoos, 172
   of self-fertilized species, 436
   short-term and inbreeding, 172, 173
   of small mammal populations, in areas of small size, 442
   of small populations in small reserves, 440–442
   of species, 319, 331, 415, 439, 440–443
   subdivided species vs. single system, 330
   taxa, 375
   variation and, 3, 4
Survival probabilities, for wildebeest herd, 184
Survival probability (model)
   of deleterious homozygous genotype, 53
   of homozygotes (selection coefficients), 53
Survival rates, physiological limitations, 325
Survivors, species pool of, 320
Survivorship. See also Viability
   age class, 394
   demographic stability and, 398
   differential in captive populations, 180
   sex class, 394
Suture, hybridization and introgression along, 276
Sympatry
   among avian species, 292
   avian populations, 308
   warblers, 296, 300
Systematics
   conservation goals, 350, 351
   molecular procedures, 349
   statistical procedures, 349
Systematists
   population terminology and, use of term population, 40

Tamaulipas, Mexico, 104
Tamils, India, inbred population survival, 244

Target (focal) species, 2, 3, 13
   choice, 9, 10, 332
   competitor population size and genetic diversity, 332
   cost of management, 10
   destabilization, 326
   incipient speciation, 9
   living fossils, 9
   management intervention, 7
   management of natural ecosystems, 7
   plant control, success and reproduction mode, 226
   population size and genetic diversity, 332
   rarity, 9
   scientific value, 10
   selection options, 9, 10
Taxa. See also Species
   allopatric populations, 358
   alteration, 313
   as biological populations, 386
   captive reproduction, 386
   endangered, 333, 375
   evolution of, 317
      in response to environmental change, 317
   extinction, 313
   measurement of genetic diversity in, 349
   morphological differences, 349
   morphological similarities, 349
   original unrecoverable, 313
   preservation, 386
   propagation, 386
   in subspecies category, 358
   swamping of, 313
   union of grasses, 271
Taxa, criteria for zoo selection
   captivity feasibility, 376
   distinction, 380
   diversity representation (genetic, taxonomic, zoogeographic), 376
Taxa mergence, reticulate evolution characteristic, 271
Taxonomic group, special characters for sibling species identification, 99
Taxonomic levels, genetic diversity measurement for various, 356, 357
Taxonomy
   distance between samples, 351
   evolutionary relationships, 349
   genetic relationships, 349

nomenclatural type vs. biological "type," 196
stepwise linear discriminant analysis, 233
use of genetic markers, 33
Taxoscope, hypothetical, and general/local extinction continuum, 113
Techniques
  biochemical analysis, 99
  electrophoretic analysis, 51, 97, 99, 101, 103
  karyotypic analysis, 99, 405
  maternal lineages, 187
  mitochondrial DNA analysis, 187
  morphometric analysis, 99
  for subspecies classification
    electrophoresis analysis, 363
    karyotypic analysis, 363
    morphometric analysis, 363
Technology, multivariate analysis, morphometric data, 104
Technology, animal number limitation (zoos)
  artificial insemination, 386
  cryogenic techniques, 386
  embryo transplantation, 386
Technology, sibling species
  identification of, 99, 101, 103
  karyólogy, 100, 101, 103, 104,
  meiotic analysis of hybrids, 103
  parasite distributions for recognition and evaluation of sibling species relationships, 101, 103
Teleology, "inner drives" dictate evolution, 39
Temperate species
  genetic variation, 131
  heterozygosity, 132
Temperate waters, turtles, 277
Temperate zone, species diversity in, 422, 429
Temperature
  sex determination in sea turtles, 278
  tolerance
    in *Drosophila*, 99
    in sibling species, 99
Tennessee, warblers, 296
Terrestrial slugs, breeding systems in, 205–210
Territorial displays, in sibling species identification, 99
Territorial expansion
  ranchcs, 383
  zoos, 383

Texas, 100
  ducks, 295, 296
Thailand, grass species, 271
Tigers
  for exhibition purposes, 179, 180
  white, 179, 180
Tiger subspecies, maintenance, 380, 383
Time scale, 2, 13, 429, 440
  de novo mutation dependence on, 93
  extinction, 112
  geological scale, 113, 439
  habitat mutualists and, 120
  intermediate term, 420
  long-term, 443, 444
  millennia, 84
  pattern of genetic divergence, 91
  relaxed, 13
  short-term, 13, 443
  speciation, 112
  years, 84
Tissue banks, 408, 409
Toxins
  DDT, 86
  insecticides, 130
  rodenticides, 88
  warfarin, 86
Tortugero, Costa Rica, sea turtles, 282
Trait changes
  response to inbreeding, 253
  transiliences, 253
Trait selection, local adaptation to ecological variables, 365
Traits, 319
  allozyme, 344, 345
  association, 344
  behavioral, 198
  combinations of, 338
  developmental, 198
  in different habitats, 372
  in ensemble of populations, 372
  in evolutionary change, 253
  genetic variability of, 338
  life history traits and, 338
  under local selection, 372
  phenotypic level effects, 346
  physiological, 198
  polymorphism for, 372
  preservation of, 372
  quantitative, 344, 345
  response to selection pressures, 372
Translocation
  attachment of chromosome segments, 36
  duplication, rcpeat, 36

Transplantation
  butterflies, 156
  fish species and morphological characters, 359
Trees
  extinction among, 126
  keystone role, 11
  tropical
    availability of habitat for, 442
    large, in reserves, 6
Trials, multiplicity, 315
Tristan da Cunha, colonization of, 192
Tropical forest, 104
  butterfly population decline, 161
  human activities in, 161
Tropical species, genetic variation, 131
Tropical trees, large, with low density, 10
Tropical zone, species diversity in, 422, 429
Twins, monozygotic, 15, 16
Type, concept of species, by false extension from nomenclatural standard, 195

Uganda, weedy sorghum, 273
Ungulates, 415
  effects of inbreeding in, 184
  extinction among, 126
  juvenile mortality in, 171, 172, 184
  level of demographic protection, 438
  space and survival in, 442
United Nations List of National Parks and Equivalent Reserves (IUCN), 6
United States, 104, 273, 295, 359, 367
  ducks, 296
  heavy metal tolerance on mine sites, 84
  plant colonizing success and allozyme variation, 219
  rodenticide, 88
  sea turtles, 283
  Sumatran tigers in, 180
United States, northeastern, 98
  population interactions, 308
United States, western, 363
  weed sunflowers, 274
U.S., Endangered Species Act of 1973, 280, 403, 437

U.S., endangered species list, 295, 296
U.S., Dept. of the Interior,
  Fish and Wildlife Service, 289
    legal decision on hybrids, 306
  Solicitor's Office, 295
U.S., National Academy of Sciences, National Research Council, 403, 409
U.S., National Marine Fisheries Service, shrimp trawl turtle excluder device, 286
U.S., national parks
  area, 416
  intensive management needs, 416
  sample
    by name, 417–419
    by size, 417–419
U.S., National Zoological Park's Conservation and Research Center, Front Royal, Va., facility enlargement, 383
U.S., Regional Dairy Cattle Breeding Project, inbreeding and selection studies, 175
U.S.S.R., zoo exchanges, 389

Vagile species
  dynamics, 162
  genetics, 162
Vagility
  lack of in pollen over long geographic distances, 66
  lack of in seeds over long geographic distances, 66
Variability, 38
  genetic recombination and, 194
  by mutation, 37
  of species, 38
Variants (individuals), 38
Variation. See also Genetic variation
  electrophoretic assay and, 149, 150
  hierarchy of, 318
    among intraspecific population differences, 318
  under human control, 130
  intraindividual, 149
  at loci, 149
  nongenetic, 356
  for population survival at loci, 149
  in response to selection pressures, 46

and survival, 3, 4
  techniques for assessment at subspecies level, 363
  units of, by social or environmental structuring, 77
Variation, inbred and outbred lines
  egg weight, 149
  morphological asymmetry, 149
Vascular plants
  heterozygosity (H), 131
  polymorphism (P), 131
  self-mating, 243
Vectors in pollination
  biological, 6
  physical, 6
Vertebrates
  behavior change, 319
  environmental change and survival of, 115
  to epidemic populations, 319
  genetic variation, 133
  heterozygosity, 131, 132, 133
  heterozygosity and evolutionary change, 146
  loci survey, 146
  long-term survival of, 439
  polymorphism (P), 131
  space for, 443
Vertebrates, large
  decline in population, 416
  eventual demise, 12
  intensive management for, 416
  limited facilities for, 435
  prospects for long-term survival, 439
  space requirements for, 416
Viability
  30-day for selected animals, 254
  30-day for unselected animals, 254
  births, 246, 247
  Darwinian fitness, 133
  deaths, 246, 247
  depression with inbreeding, 243
  environmental death, 247
  inbreeding coefficients, 247
  inbreeding depression effects, 260
  inbreeding level and, 247
  index age, 246, 247
  number of lethal equivalents, 247
  one-year, 248
  of parents, 260
  sample size and, 247
  survival to given age, 246, 247
  survival to reproductive age, 44
  values, 246, 247

Vigor, 150
Viruses, extinction question, 126
Vitamin K
  dietary requirement for, 89
  in rodenticide resistance, 89
Volcanic eruptions
  fauna in, 126
  flora in, 126
Vulnerability
  to catastrophe, 440
  of habitat, 440
  of species, 415, 438, 439, 443

Wahlund effect, 321
Wales, rodenticide, 88
Wallace, Alfred Russell, 38
Washington County, Texas, 101
Water management plan, founding populations in, 331
Weather
  and extinction, 117
  patterns, population size, variation and, 157
Weed races
  crop species, 272–275
  of cultivars, 274
  in erosion control, 272
  forage, 272
  mimics, 274
  in soil protection, 272
  thrive under disturbance, 272
Weeds, 225
  adaptations to environments, 224
  aggressive, 275
  allozyme variation, 338
    of high level, 338
    at polymorphic loci, 338
  annual, inbreeding, 227
  colonization of new environments, 218
  control, 226, 366
  crop losses, 273
  disturbance by human activities, 224
  human activity and, 225
  intercontinental colonization, 225
  ploidy in, 224
  population expansion, 225
  range expansion, 225, 275
  release from
    environmental constraints, 225
    native parasites, 226
  rice, 274
  roadside populations, 218

sorghum, 274
  in species extinction, oceanic
    islands, 111
Weight, inbreeding depression effects,
    260
Whales, in aid of conspecifics, extinction, 117
Wicken Fen, England, butterflies, 160
Wild collection, lack of additional,
    259
Wild populations
  demographic management, 392
  distinction from captive
    diminishing, 401
  effective population size, 391
  genetics studies, 363
  inbreeding in, 165, 183
    depression in, 109
    loss of fitness in, 183
    small, compared with captive
      populations, 183
  plant, 272
  regulation, 398
  sea turtles, 287
  systematic studies, 363
Wild population management
  administrative decisions, 238
  demographic regulation, 392–398
  political decision, 238
Wild population programs
  captive population programs and,
    400
  preservation of biotic diversity, 400
Wild relatives
  of domesticates, 14
  important resources, 14
Wildlands, destruction, 400
Wildlife management. *See also*
    Genetic management, Population management, Management
  behavioral problems, 8
  capture and release of animals, 183
    game animal founding populations, 183
  conservation, 415
  of dependent populations, 436
  environment control, 8
  habitat diversity increase, 151
  inbreeding in management of wild
    populations, 165
  inbreeding rate recommended, 178
    undomesticated populations, 178
  introductions, 8
  lessons from breeders, 191

mtDNA studies in lineage analysis,
    238
operation difficulties/operation
    diversity, 8
plans, 415
population subdivision, 151
practices with sea turtles, 286, 287
programs, 242
reproductive capacity maximization, 151
of restrained populations, 436
of small populations, 436
social problems, 8
in space and time, 8
territorial problems, 8
Wildlife management goals
  breeding programs, 242
  evolutionary survival of species,
    151
Wildlife managers
  integrated strategy to conserve
    wildlife, 375
  observation of extinction, 109
  population terminology and, 40
  recognition of inbreeding dangers,
    183
  use of genetic markers, 33
Wind Cave National Park, South
    Dakota
  antelope in, 420
  bison in, 420
  elk in, 420
  surplus animal removal, 420
Woodside, Calif., butterflies, 158
Woodwalton Fen, England, butterflies, 159
Woody species, obligate outbreeding, 227
World Conference on Sea Turtle Conservation 1979, conservation
    strategy plan, 278
Wright, Sewall, 191
Wright's coefficient of gene differentiation (F_{ST}), 360
Wright's shifting balance model and
    sympatric speciation, 79

Yellowstone National Park,
    Wyoming, 6
  elk management in, 438
  elk migration in, 420
  grizzly bear management in, 438
  grizzly bear protection in, 420, 421,
    442

Subject Index 721

level of demographic protection, 438
park and reserve size, 438
size of, 420
species adaptation to changes in, 421

Zonation
  temperate, 422, 429
  tropical, 422, 429
  variation, 131
Zone of contact
  assortative mating in, 292
  speciation in, 292
  species elimination at, 270, 305
  sympatry in, 292
Zone of secondary integration
  genetic variability, 292
  phenetic variability, 292
Zoo animals
  lack of collection, 260
  vital statistics, 383
Zoo breeding programs
  adaptation to captive environment, 259, 260
  adaptation to genetic environment, 259, 260
  adaptation to inbreeding, 259
  inbreeding avoidance (heterozygosity maintenance), 259
  inbreeding depression management, 259
  inbreeding to no inbreeding, generations of survival, 172–174
  individuals selected for domestic traits, 259
  population size increase, 383
  populations in different zoos, 260, 261
  reintroduction capabilities, 260
  survival chances, 260
  survival of species, 260
Zoo facilities, 375, 376
  inadequacy of, 380, 383
Zoo geneticists (North American), strategy for genetic management, captive populations, 387–392
Zoo guidelines,
  criteria for taxa selection, 375
  space availability, 375
  taxon carrying capacity, 375
Zoo management

interzoo cooperation, Speke's gazelle, 250, 251
long-term endangered, 250
Zoo management goals
  inbreeding depression elimination, 259
Zoo managers, integrated strategy to conserve wildlife, 375
  genetics, 241
  inbreeding depression, 241, 242
  management, 241
Zoos. *See also* Zoological parks
  animal number limitation, 386
    for effective population size, 386
  animal transfers between, 250, 251
  breeding designs for, 63
  breeding exchanges, 261
  breeding loans, 261
  capacity limitations, 376
  captive habitat for viable populations, 376
  carrying capacity, 172
    for each taxon, 376
  chromosome studies, 404–408
  conservation, 402, 403
  cooperation between, 260, 261
    avoidance of disease, 260, 261
    herds subdivided between, 260, 261
  cooperative arrangements, 384
  demographic management, 399–401
  effective extinction in, 394
    animal removal, 394
  effective population size management, 389–392
  enlargement need, 383
    space requirement, 383
  European, 403
  fixation of alleles in, 68
  founder effect in, 192
  future potential of, 164, 165
  history, 165
  genetic management, 184, 387–392, 401
  global cooperation, 411
    in reproduction techniques, 410, 411
  history of breeding stock in, 165
  importation of animals, 165
  inbreeding in, 67, 165–174
  inbreeding problems, 409
    remedies, 242, 411
  intensive genetic management, 387–392

juvenile mortality survey, 178
maintenance of genetic diversity in, 183
maximum genetic variability in herds of, 261
numbers of each species in, 164
population management for, 63
populations in, 434
priorities, 401
rare mammal species in, 164
research, 402–412,
  microbiological studies, 412,
  parasitological studies, 412
  virological studies, 412
subpopulations
  "neighborhood", 69, 389
  "population", 389
Zoos, listing
  Bronx Zoo, N.Y., 403
  Brownsville, Texas, 242, 243
  National Zoological Park, Wash., D.C., 403
  Philadelphia Zoo, Phila., Pa., 403
  St. Louis, Missouri, 242, 243
  San Antonio, Texas, 242, 243
Zoos for wildlife
  germ plasm repositories, 374
  natural population reinforcement, 374, 375
  propagules for natural habitat repopulation, 374, 375

public education, 375
refugia for endangered taxa, 374, 375
research, 375
Zoological gardens. *See also* Zoos
  allelic and species components, 333
  management of, 333
  natural areas, 333
  relict populations, 333
  source of new genotypes, 7
  stability of management system, 333
Zoological parks. *See also* Zoos
  carrying capacity expansion, 401
  construction, 331
  guidelines, 401
  priorities, 401
  taxonomic diversity, 401
  technological expansion, 386, 401
  territorial programs, 401
  wild population coordination, 401
Zygotes, 40, 41, 43
  allele frequencies in simulated model, 52
  haploid egg and halploid sperm, 18, 19
  initial, 150
  production time interval, 151
Zygotic diversity, 193